World Survey of Climatology Volume 15

CLIMATES OF THE OCEANS

World Survey of Climatology

Editor in Chief:

H. E. LANDSBERG, College Park, Md. (U.S.A.)

Editors:

H. ARAKAWA, Tokyo (Japan)
R. A. BRYSON, Madison, Wisc. (U.S.A.)
O. ESSENWANGER, Huntsville, Al. (U.S.A.)
H. FLOHN, Bonn (Germany)
J. GENTILLI, Nedlands, W. A. (Australia)
J. F. GRIFFITHS, College Station, Texas (U.S.A.)
F. K. HARE, Ottawa, Ont. (Canada)
H. E. LANDSBERG, College Park, Md. (U.S.A.)
P. E. LYDOLPH, Milwaukee, Wisc. (U.S.A.)
S. ORVIG, Montreal, Que. (Canada)
D. F. REX, Boulder, Colo. (U.S.A.)
W. SCHWERDTFEGER, Madison, Wisc. (U.S.A.)
K. TAKAHASHI, Tokyo (Japan)
H. VAN LOON, Boulder, Colo. (U.S.A.)
C. C. WALLÉN, Geneva (Switzerland)

World Survey of Climatology Volume 15

Climates of the Oceans

edited by

H. VAN LOON

National Center for Atmospheric Research
P.O. Box 3000
Boulder, Colo. 80307 (U.S.A.)

ELSEVIER Amsterdam-Oxford-New York-Tokyo 1984

ELSEVIER SCIENCE PUBLISHERS B.V.
Molenwerf 1
P.O. Box 211, 1000 AE Amsterdam, The Netherlands

Distributors for the United States and Canada:

ELSEVIER SCIENCE PUBLISHING COMPANY INC.
52, Vanderbilt Avenue
New York, NY 10017

With 395 illustrations and 218 tables

ISBN 0-444-41337-5 (Vol. 15)
ISBN 0-444-40734-0 (Series)

Printed in The Netherlands

World Survey of Climatology

Editor in Chief: H. E. LANDSBERG

Volume 1   General Climatology, 1
Editor: O. Essenwanger

Volume 2   General Climatology, 2
Editor: H. Flohn

Volume 3   General Climatology, 3
Editor: H. E. Landsberg

Volume 4   Climate of the Free Atmosphere
Editor: D. F. Rex

Volume 5   Climates of Northern and Western Europe
Editor: C. C. Wallén

Volume 6   Climates of Central and Southern Europe
Editor: C. C. Wallén

Volume 7   Climates of the Soviet Union
by P. E. Lydolph

Volume 8   Climates of Northern and Eastern Asia
Editor: H. Arakawa

Volume 9   Climates of Southern and Western Asia
Editors: K. Takahashi and H. Arakawa

Volume 10 Climates of Africa
Editor: J. F. Griffiths

Volume 11 Climates of North America
Editors: R. A. Bryson and F. K. Hare

Volume 12 Climates of Central and South America
Editor: W. Schwerdtfeger

Volume 13 Climates of Australia and New Zealand
Editor: J. Gentilli

Volume 14 Climates of the Polar Regions
Editor: S. Orvig

Volume 15 Climates of the Oceans
Editor: H. van Loon

List of Contributors to this Volume

R. G. BARRY
Cooperative Institute for Research in Environmental Sciences (CIRES)
University of Colorado
Boulder, Colo. 80309 (U.S.A.)

M. Á. EINARSSON
The Icelandic Meteorological Office
Bústadavegur 9
Reykjavík (Iceland)

M. HANZAWA
Maritime Meteorology Division
Japan Meteorological Agency
1-3-4, Ootemachi, Chiyoka-ku, Tokyo (Japan)

O. HÖFLICH
Seewetteramt
2 Hamburg 4
Postfach 180
Berhard-Nocht-Strasse 76
(Federal Republic of Germany)

C. S. RAMAGE
Department of Meteorology
University of Hawaii at Manoa
Honolulu, Hawaii 96822 (U.S.A.)

N. A. STRETEN
Australian Numerical Meteorology Research Centre
Melbourne, Vic. (Australia)

J. J. TALJAARD
Weather Bureau
Private Bag 193
Pretoria (South Africa)

K. Terada
Sagamihara Research and Engineering Center
Nippon Steel Corporation
5-9-1, Nishi-Hashimoto
Sagamihara City, Kanagawa Pref. (Japan)

G. B. Tucker
Division of Atmospheric Physics, CSIRO
Mordialloc, Vic. (Australia)

H. van Loon
National Center for Atmospheric Research
P.O. Box 3000
Boulder, Colo. 80307 (U.S.A.)

J. W. Zillman
Bureau of Meteorology
P.O. Box 1289K
Melbourne, Vic. (Australia)

Preface

This volume suffered many upsets; the first editor died, the second withdrew, and when I took over in 1974 — 11 years after the work had begun — only one chapter was close to the shape in which it is printed. In addition, new authors had to be found for four of the seven chapters.

The geographical arrangements of the oceans for the chapters had been chosen by my predecessors. Because the Equator is not a natural boundary between circulation regimes, the apportionment should rather have been into chapters about the regions on the polar sides of the subtropical ridges and about the tropical regions between the ridges; and perhaps the entire southern west-wind belt should have been treated as a separate chapter.

The illustrations in individual chapters are not coordinated to make isopleths continuous from one chapter's maps to the next. This was not done because the authors used different sources for their maps and calculated radiation and heat balances by somewhat different methods.

Paradoxically, the longest texts herein are those written about the southern oceans where the data are most limited and the opportunities for observation are comparatively low, as the facts of global distribution of population and commerce have relegated these regions to the backwaters of observational activity. Recent findings have highlighted the importance of southern and cross-equatorial circulations for the understanding of the global circulation and its changes, so I decided to retain this difference in length, for there is little chance in the near future that the southern oceans will be the object of comprehensive programs of observation, whereas the experiments being carried out or planned for the northern oceans soon will make revision and expansion of the chapters on the North Pacific and Atlantic Oceans necessary.

When considering the size of this volume, one must remember that it deals with the climates for about 70% of the earth's surface. I therefore do not regard it as being too long but imagine rather that those who are familiar with the complexity of the circulations, whose changes in space and time are summarized here, will be just as impressed by the omissions as by the contents.

Finally, I should like to acknowledge Mrs. Ann Modahl's invaluable, untiring, and thorough work in the editing and preparation of the manuscripts for publication.

Harry van Loon

Contents

PREFACE IX

Chapter 1. CLIMATE OF THE SOUTH ATLANTIC OCEAN
by O. HÖFLICH

Basic principles 1
   Historical review 1
   Climatology, 2—Synoptic meteorology, 3
   Problems of the data 5
   Origin of data, 5—Problems inherent in the data, 6—Inhomogeneities, 6—Selected ocean areas, 8
   Fundamentals of climate 8
   Climate-producing processes, 9—Planetary circulation, 10—Characteristics of marine climate, 11—Climate classification of the South Atlantic, 12
Climatic elements 13
   Air pressure 14
   Mean pressure field, 14—Annual variation, 15—Variability, 16
   Wind 17
   Wind speed, 17—Wind direction, 19—Wind roses, 21—Strong winds and gales, 22
   Oceanic elements 22
   Sea and swell, 23— Ocean currents, 25—Subsurface conditions, 27
   Temperature 30
   Sea surface temperature, 30—Air temperature over the ocean, 34—Air–sea temperature difference, 36
   Humidity 37
   Vapour pressure, 38—Equivalent temperature, 39—Relative humidity, 40
   Visible weather phenomena 42
   Cloudiness, 42—Visibility, 46—Precipitation frequencies, 47—Thunderstorms, 48
Water and heat balance 48
   Water balance 49
   Precipitation amounts, 49—Evaporation, 51—Water balance, 53
   Vertical heat flux at the sea surface 56
   Latent heat flux, 57—Sensible heat flux, 57—Total heat flux, 59
   Radiation balance 60
   Insolation at sea level, 60—Terrestrial radiation, 62 Atmospheric counter radiation, 62—Net long-wave radiation at the sea surface, 63—Radiation balance at the sea surface, 64—Radiation balance in the atmosphere, 65

Contents

Heat balance 66

Heat balance at the sea surface, 66—Heat balance of the earth–atmosphere system, 68

Climatic stations 70

Sao Tomé 71

Fernando de Noronha 73

Ascension Island 75

St. Helena 77

Tristan da Cunha 78

Gough Island 80

Falkland Islands 81

South Georgia 83

Off-shore climates 86

South American off-shore areas 86

North coast of Brazil, 87—Northeast coast of Brazil, 88—Southeast coast of Brazil, 89—Rio de la Plata, 91—Patagonian Shelf, 93

African off-shore areas 94

Coast of Gabon, 94—Coast of Angola, 95—Coast of Southwest Africa, 96—Coast of the Cape Province, 98

Aerology 100

Geopotential heights 101

Absolute topographies, 101—Meridional profiles, 102—Zonal profiles, 103

Wind 105

Zonal winds, 105—Meridional winds, 107

Temperature 107

Relative topographies, 107—Meridional profiles, 107—Zonal profiles, 109

Humidity 110

Vapour pressure, 110—Relative humidity, 111—Several humidity elements, 112

Synoptic climatology 113

Tropics 113

Intertropical Convergence Zone, 114—Trade winds, 114—Disturbances, 114

Anticyclones 116

High-pressure centres, 116—High-pressure tracks, 116

Cyclones of the westwind zone 117

Frontal zone, 118—Low-pressure centres, 118—Low-pressure tracks, 119—Weather systems, 121

References 123

Appendix—Climatic tables and charts 132

Chapter 2. CLIMATE OF THE NORTH ATLANTIC OCEAN
by G. B. TUCKER AND R. G. BARRY

Introduction 193

Main climatic features 193

XII

Main oceanic features 194
History of observations 196
Climatic change 199
Representative data 201
Use of January and July maps 202
Mean circulation patterns 203
Surface pressure field 203
Winds 206
Synoptic climatology 210
Annual cycle of synoptic events 210
Winter, 210—Spring, 211—Summer, 212—Autumn, 213
Circulation patterns and synoptic events in the westerlies 213
Climatology of blocking action, 215—Coastal storms of eastern United States, 217—Thermal troughs over the northeastern Atlantic, 218
Trade-wind region 218
Intertropical Convergence Zone, 220—Tropical storms, 224
Air–sea interactions 225
Air temperature 227
Air–sea temperature difference 231
Interactions between the ocean and atmospheric circulation 233
Moisture conditions 234
Humidity 234
Cloudiness 237
Visibility 238
Precipitation 238
Evaporation 241
Heat balance 243
General remarks 243
Solar radiation 244
Terms in the heat balance equation 245
Climatic zones 248
Equatorial zone, 249—Tropical and subtropical zone, 252—Temperate zone, 252—Subpolar zone, 252
References 253
Appendix—Climatic tables 258

Chapter 3. CLIMATE OF THE SOUTH PACIFIC OCEAN
by N. A. STRETEN AND J. W. ZILLMAN

Introduction 263
History of climatic description 263
Data sources 266
Surface climatic data, 266—Upper-air data, 266—Satellite data, 267—Synoptic analyses, 267—Tabulated data, 267
Treatment 267

Contents

The ocean 268
    Circulation 269
    Temperature 273
    Salinity 274
    Ocean fronts 275
    Sea ice 277
    Icebergs 280
Synoptic climatology 281
    The data base 281
    Synoptic regimes of the South Pacific 284
    The tropical easterlies 285
    Tropical cyclones 288
    The tropical anticyclone belt 292
    The westerlies 296
    The Antarctic trough 307
    The Pacific cloud band 308
Main climatic elements 310
    Surface air temperatures 310
    Surface dew point 312
    Surface pressure 312
    Surface wind 314
    Cloudiness and fog 317
    Precipitation 319
Climate of the upper air 323
    Geopotential height and temperature 323
    Moisture distribution 329
    Wind 329
    Jet streams and tropopause 329
    Meridional cross-sections 331
The heat budget 332
    Surface heat budget 335
    Surface radiation balance, 335—Latent heat flux, 338—Sensible heat flux, 339—Heat available for storage and transport by the ocean, 341—Influence of ocean fronts, 342—Diurnal variation of the surface heat budget, 343
    Atmospheric heat budget 344
    Radiation budget, 345—Condensation heating, 346—Net diabatic heating, 348
Large-scale circulations 348
    The Southern Oscillation 349
    Walker Circulation 355
    El Niño 358
Conclusion 361
Acknowledgements 361
References 362
Appendix—Climatic tables 375

Chapter 4. CLIMATE OF THE NORTH PACIFIC OCEAN
by K. Terada and M. Hanzawa

Introduction 431
    Geographic features 431
    Early data summaries 431
    Ocean currents 432
Mean pressure field 434
    High pressure 434
    Low pressure 437
Wind and depression tracks 438
    Major wind systems 438
    Monsoons 441
    Gale statistics and cyclogenesis 444
    Extratropical depression tracks 445
    Tropical and subtropical cyclones 447
Temperature and humidity 453
    Temperature distribution 453
    Difference between air and sea surface temperature 457
    Humidity 458
Cloudiness, visibility, fog, and precipitation 459
    Cloudiness 459
    Visibility 463
    Fog 464
    Precipitation 465
    Snow and ice 467
Large-scale energy exchange between ocean and atmosphere 470
    Evaporation 470
Marine climate studies 471
Acknowledgements 475
References 475
Appendix—Climatic tables for ocean weather ships and land or island stations 478

Chapter 5. CLIMATE OF THE INDIAN OCEAN SOUTH OF 35°S
by J. J. Taljaard and H. van Loon

Introduction 505
    General 505
    Geographical setting and bottom topography 505
    Currents, convergences, and pack ice 507
    Historical survey and data used 511
Temperature 514
    Sea surface temperature 514
    March–September range of sea temperature, 517—Anomalies of sea temperature

Contents

in March and September, 518—The annual variation of sea temperature along 43°S, 519—Sea temperature extremes, 520

Surface air temperature 521
Surface air temperature in January and July, 521—The annual variation of air temperature, 523—Difference between air and sea temperature, 526—Summer to winter air temperature range, 527—January and July air temperature anomalies, 528—Air temperature extremes, 529—Temperature variability at Marion, New Amsterdam, and Kerguelen, 531

Upper-air temperature 531
General, 531—Mean monthly, seasonal, and annual upper-air temperatures and standard deviations at Marion, New Amsterdam, Kerguelen, and Heard islands, 532—The temperature distribution at 500 mbar in January and July, 533—Temperature change from January to July at 500 mbar, 533—Temperature deviations from the hemispheric zonal means at 500 mbar, 534—Annual variation of upper-air temperature, 535—Meridional temperature gradient at 500 mbar, 537—Zonal temperature anomaly profiles along 35° and 55°S in January and July, 539

Pressure 539
Sea-level pressure 539
Mean sea-level pressure in January and July, 539—The subtropical ridge, 539—The circumpolar trough, 541—Annual course of pressure at the islands and at Mawson, 542—Mid-seasonal fluctuations of pressure, 543—Variability of daily, monthly, seasonal, and annual mean sea-level pressures, 546

Upper-air pressure 547
General, 547—Mean monthly, seasonal, and annual heights of constant-pressure surfaces and standard deviations, 547—Contour charts for 500 mbar in January and July, 547—January-July height change and zonal anomalies at 200 mbar, 549—Variability of pressure heights, 549

Wind 551
Surface wind 551
Wind in the troposphere 556
Precipitation and clouds 560

Synoptic climatology 563
General 563
Anticyclones 564
Anticyclogenesis, 564—Anticyclone movement, 565—Anticyclone frequency, 566

Cyclones 570
Cyclogenesis, 570—Cyclone movement, 573—Cyclone frequency, 573—Speeds of cyclones, 575—Fronts, 575—Movement of deep cold and warm air masses, 577

Heat balance 578
Data 578
The equations used, 579—Results and discussion, 581

Acknowledgements 588
References 588
Appendix—Climatic tables 592

XVI

Chapter 6. CLIMATE OF THE INDIAN OCEAN NORTH OF 35°S
by C. S. RAMAGE

Introduction 603
    Sea surface currents 605
    Monsoon region, 606—Non-monsoon region, 606 . . . .
    Data 607
Typical synoptic situations 608
    Subtropical cyclone 608
    Subtropical cyclone of 12 and 13 August 1964, 609
    Near-equatorial troughs 610
    Near-equatorial troughs of April–May 1964, 612
    Air–sea interaction over the western Arabian Sea 617
    Weather off Somalia in August 1964, 618 . . . .
Conditions during January and July 627
    January 627
    Monsoon region, 627—Australian summer monsoon, 628—Non-monsoon region 628
    July 630
    Monsoon region, 630—Non-monsoon region, 638
    Winter–summer differences 640
    Heat balance, 640—Trough and ridge shifts, 641—Trans-equatorial flow, 641
Tropical cyclones 641
    Tropical cyclones over the northern Indian Ocean 644
    Tropical cyclones over the southern Indian Ocean 644
    Tropical cyclone movement 644
March of climate through the year 645
    Northern Hemisphere monsoon region 649
    November to April, 649—May to mid-September, 649—Mid-September to October, 650
    Equatorial region 651
    Southern Hemisphere monsoon region 652
    Western Indian Ocean, 652—Eastern Indian Ocean, 653—Central Indian Ocean, 653
    Southern Hemisphere non-monsoon region 653
    Trade winds, 653
Acknowledgements 654
References 654
Appendix—Climatic tables 659

Chapter 7. CLIMATE OF ICELAND
by MARKÚS Á. EINARSSON

Introduction 673
Climatic factors 674

Contents

Weather types 676
    Southeastern 676
    Southwestern or western 677
    Southern with warm air mass 677
    Warm air mass originating in Europe 677
    Eastern 677
    Northeastern 677
    Northern 678
    A high over Iceland 678
Climatic variations 678
Temperature 680
Precipitation 684
Snow 686
Wind 687
Humidity 689
Cloudiness 689
Visibility and fog 691
Sunshine and radiation 691
Potential evapotranspiration 692
Acknowledgements 693
References 693
Appendix—Climatic tables 695

REFERENCE INDEX 699

GEOGRAPHICAL INDEX 709

SUBJECT INDEX 713

Chapter 1

Climate of the South Atlantic Ocean

O. HÖFLICH

Basic principles

Climate is the sum of all meteorological phenomena in an area, represented by the mean values of a period and by the characteristics of their temporal behaviour. It reflects the statistical properties of atmospheric circulations, and may be defined as the mean statistical property of all components of the climatic system over a long period (GATES, 1977). The World Climate Conference of the World Meteorological Organization gave the following definition: climate is the synthesis of weather over the whole of a period essentially long enough to establish its statistical ensemble properties (mean values, variances, probabilities of extremes, etc.) and is largely independent of any instantaneous state (WMO, 1979, p. 752).

The description of climate thus requires the collection and statistical evaluation of weather data. Because the South Atlantic Ocean belongs to those sea areas where the data are insufficient and heterogeneous, only preliminary results can be presented and these are unsatisfactory for the southern part. Supplementary studies may be found in the references.

A detailed climatic description is needed only in coastal regions where orographic effects play a role. Over the open ocean marine influences dominate, thereby demonstrating clearly the characteristics of the global climate zones.

Historical review

The South Atlantic Ocean was unknown to Europeans before the Age of Discovery. It was first navigated at the end of the 15th century when Portuguese and Spaniards explored the coasts of Africa and America. In the 16th century, knowledge of the ocean extended only to about 50°S where the northern coast of Antarctica was supposed to be. Step by step this was shifted to the south and at last to 65°S by James Cook's journey around the world.

At first the ocean was called the "Ethiopic Sea", and in the first Prussian sea atlas (SCHMETTAU and BRUCKNER, 1749) it was named "Ocean Meridional". The term "South Atlantic Ocean" was suggested by VARENIUS (1671) and introduced in 1845 according to a resolution of a commission of the London Geographic Society.

Along with the exploration of the ocean, climatological knowledge grew, especially that needed for navigation. Additional oceanographical and meteorological information and experience were gained by geographical expeditions, and the development of marine meteorology also advanced interest in synoptic meteorology over the ocean.

Climatology

The oldest chart of currents in the Atlantic Ocean stems from KIRCHER (1678), and the first chart of winds over the ocean from HALLEY (1688) who in 1695 edited the first magnetic declination chart too. These charts are based upon sailors' observations during voyages on sailing ships. They watched wind and weather and used their experience to make a voyage as short as possible, which also served the interests of the shipping companies and commercial houses.

Frequently, barometer and air and water thermometers were on board (MEUSS, 1913). In 1805, Beaufort developed his scale of wind force based on the amount of sail carried. The observations were entered in logbooks, and the first evaluation of such entries by the cartographer BERGHAUS (1842) increased geographical, nautical, and oceanographical knowledge. Full benefit could not yet be derived from the meteorological information, although the general features of the pressure distribution and wind systems were already known (see BUCHAN, 1869).

The systematic performance, collection, and evaluation of meteorological observations on board ships were first arranged by Maury. For this purpose, an international conference took place in Brussels in 1853. From 1849 to 1860, MAURY edited the nautical atlas *Wind and Current Charts,* the sailing instructions *Explanations and Sailing Directions to Accompany the Wind and Current Charts,* and in 1858 the book *The Physical Geography of the Sea.* These were models for all further nautical and scientific descriptions of the oceans and for the progress in marine meteorology.

In the seafaring nations, institutes soon arose (for example, in 1876 the Deutsche Seewarte in Hamburg) which equipped the vessels with observer's instructions, logbooks, and instruments. They collected meteorological observations gained during the voyages, evaluated them, and issued sailing instructions based on climatology to improve the economy of navigation. The large number of observations collected over more than 100 years makes it possible nowadays to describe the climate over extensive ocean areas such as in this volume.

The international cooperation which started at the Brussels conference continued. Soon after the Congress of Vienna in 1873, where the foundation had been laid, the International Meteorological Organization (IMO) was established. In 1951, this was renamed the World Meteorological Organization (WMO). Increasingly, it has sponsored progress and development in meteorology (DANIEL, 1973), not least in marine climatology. This has also influenced research activity in the South Atlantic Ocean. In earlier years, systematic oceanographic profile measurements were chiefly undertaken by single vessels that, in addition, observed atmospheric conditions, especially the RV *Meteor* (STOCKS, 1962). More and more international expeditions were arranged whereby several vessels simultaneously provided a dense network of observations in a special sea area for a specific purpose. Examples of this were the International Polar Year 1882—1883 and the International Meteorological Cooperation 1901—1904, comprising South Polar expeditions of five nations. Then followed the International Geophysical Years and the present Global Atmospheric Research Programme (GARP) of the WMO (WMO, 1969). The study of the exchange between sea and air connects oceanography and meteorology through such efforts.

Many special publications and climate atlases (e.g., MCDONALD, 1938) contain research results and evaluations of ship observations. The following such atlases were used in this chapter to describe the climate of the South Atlantic Ocean.

Koninklijk Nederlands Meteorologisch Instituut, *Oceanographische en Meteorologische Waarnemingen, Atlantische Oceaan,* De Bilt, 1931.

U.S. Hydrographic Office, *Sea and Swell Charts, South Atlantic Ocean,* Washington, 1948.

U.S.S.R., *Morskoj Atlas,* Moscow, 1953.

U.S. Office of Naval Operations, *Marine Climatic Atlas of the World, IV. South Atlantic Ocean,* Washington, 1958.

Meteorological Office, Air Ministry, *Monthly Meteorological Charts of the Atlantic Ocean,* London, 1959.

U.S.S.R., *Atlas on Sea and Wind in the Atlantic Ocean* (in Russian), Leningrad, 1967.

Deutsches Hydrographisches Institut, *Monatskarten für den Südatlantischen Ozean,* Hamburg, 1971.

After completion of the present text, the following appeared.

C. G. GORSKOV, *Oceanic Atlas, Atlantic and Indian Oceans* (in Russian), 1977.

U.S. Navy Climatic Atlas of the World, IV. South Atlantic Ocean, Washington, 1978.

Because the climatologies are useful for navigation, climatic descriptions of the South Atlantic Ocean may be found in handbooks (e.g., HAMBURG, 1981a) and in pilots for various sea areas. An effort is also made in this respect for international cooperation in the exchange of data.

An evaluation of internationally collected ship observations has taken place since 1961 (see p. 6) with the publication of yearly and decadal *Marine Climatological Summaries* (WMO, 1977). For the South Atlantic Ocean, they are made by the Seewetteramt in Hamburg (Federal Republic of Germany). These yearbooks contain tables for every month in selected areas with frequency distributions and mean values of all climatic elements of interest. These tables will be summarized for the 30-year period 1961 – 1990 to become part of a world marine climatic atlas.

For ship observations before 1961, an international data exchange and statistical evaluation will be provided by the WMO and named the Historical Sea Surface Temperature Data (HSSTD) Project. It will contain temperature and wind data from selected areas in all oceans for all single months in more than 100 years and will describe temporal and spatial fluctuations of the elements as background information for GARP.

Synoptic meteorology

Ship observations are used not only in climatology but also in synoptic meteorology. For the latter purpose, simultaneous observations are plotted on weather maps and analyzed. The first daily weather maps for the South Atlantic Ocean were analyzed by MELDRUM (1861) for three months of 1861. South of 30°S, daily weather charts for the International Polar Year 1882 – 1883 were produced by the Deutsche Seewarte. They were based on German, English, Dutch, and American ship observations and on land stations, but they were not sufficiently detailed to show the complex cyclonic activity of temperate latitudes.

Better daily weather charts were drawn and published for the period October 1901 – March 1904 within the scope of the International Meteorological Cooperation during the German, English, Swedish, Scottish, and French South Polar Expeditions (MEINARDUS and MECKING, 1911). The shipping firms were asked that their ships produce at least one carefully made observation of the weather daily at 1200 GMT during this period.

In the Southern Hemisphere Map Analysis Project of the Massachusetts Institute of Technology and the U.S. Weather Bureau, daily surface weather charts of the Southern Hemisphere were drawn for July 1948 to June 1951 (RUBIN, 1952). The South African Weather Bureau started the publication of daily Southern Hemisphere weather maps in December 1949 as part of the Southern Hemisphere Atmospheric Project (PRETORIA, 1952); in addition, daily upper-level charts were analyzed for a short period after December 1951 (SCHMITT, 1952), although the few radiosonde ascents hardly justified such analyses. The tropospheric behaviour had to be inferred from the surface analysis by using knowledge gained of the Northern Hemisphere (SCHERHAG, 1948). This task did not become worthwhile until additional stations and ship observations were available during the 18 months of the International Geophysical Year (IGY) 1957 – 1958 (PRETORIA, 1962 – 1966; HAMBURG, 1963 – 1967 for the tropics).

These charts were not completed until several years after the IGY since an effort was made to collect as many observations as possible. Ship observations could not be plotted before the logbooks arrived which caused a long delay. The charts thus did not serve operational synoptic meteorology, but were used for marine meteorological research to improve operational diagnosis and prognosis.

For climatological research, it is not immaterial if a weather chart has been analyzed currently or historically. The careful and thorough analysis of historical weather charts, made nowadays only in special cases, is frequently an important advantage when the network of stations is scant.

Current weather maps for weather prediction existed in the beginning only for land areas and were based on telegraphic weather reports from fixed stations in a region. The development of wireless telegraphy was necessary before current weather charts could be drawn over ocean areas.

Fast collection and evaluation of world-wide synoptic observations have become urgently necessary to supply forecasts for shipping and to analyze large areas for numerical prognoses. To achieve this goal and to expand the observational network, especially over the oceans on the Southern Hemisphere, the WMO initiated the World Weather Watch (WMO, 1966b).

Automation of the collection, checking, and evaluation of data speeded up the analyses of world-wide weather maps. Since 1960, cloud photos and radiation measurements from weather satellites have provided an additional valuable means of gathering synoptic knowledge and employing it for analysis and prediction (VAN LOON and THOMPSON, 1966; MARTIN, 1968). Numerical models simulating the atmospheric behaviour have furthered the understanding of effects of physical and dynamical processes in weather and in climate in a hitherto undreamt of manner (GATES, 1977).

Problems of the data

For climatological evaluation of the observational data and for correct interpretation of the statistical results, a critical examination of data is necessary. This takes into account the observing method, the accuracy of measurement, historical changes, and national differences as well as heterogeneities in density and distribution of data.

Origin of data

Apart from the continental coastal stations, in the South Atlantic Ocean the only data from fixed locations come from a very few islands. Over the sea, one therefore depends almost entirely on weather observations from ships. Drifting buoys were used extensively for the first time during the First GARP Global Experiment in 1979 and will undoubtedly become an indispensable tool of observation in the future, and an evaluation of satellite data for climatological purposes is just beginning. The WMO sponsors marine climatology by international exchange of data, expansion of the network, and introduction of modern methods of data processing.

The climate data of the South Atlantic coastal stations are available in the descriptions of the climates of Africa and South America (see Volumes 10 and 12 of this Series) and are not considered in this volume. They are used on the charts as marginal values for the analysis of the climatic elements.

A description of the climatic data for island stations in the South Atlantic Ocean begins on p. 70. There are eight such climate stations: the airport of Sao Tomé, Fernando de Noronha, Georgetown on Ascension Island, Jamestown on St. Helena, Tristan da Cunha, Gough Island, Port Stanley on the Falkland Islands, and Grytviken on South Georgia. Climate stations farther south appear in Volume 14 (SCHWERDTFEGER, 1970).

The data from fixed stations should be used with caution for descriptions of the ocean climates. Frequently, the influence of the land is so great that the data are not representative of the marine conditions. This is true when the station elevation is high, such as Hutts Gate on St. Helena, or when mountains deflect the wind or cause foehn. The mountains also affect temperature, humidity, cloudiness, and precipitation, and the climate data therefore depend on the site of the station on the windward or lee side of the mountains according to the prevailing wind direction. The distribution of precipitation on Sao Tomé (p. 73) illustrates this effect.

The only weather observations available to describe truly maritime conditions in the South Atlantic Ocean are those made from merchant ships, and these data, of course, do not present a time series of values fixed in space and time.

In earlier times, observations were taken every four hours according to the duration of the watch. After the observations became used in the synoptic service, they were taken every six hours according to world time. Nevertheless, ship observations can be used climatologically because the elements at sea show hardly any daily cycle and the observation time is thus irrelevant.

Systematic, usable ship observations have existed since the middle of the last century (p. 2). They are hidden in thousands of ship logs from which they must be extracted and transferred onto punchcards or magnetic tapes, because the statistical calculations based on these data have to be made on electronic computers. This process is still going on and it is not yet possible to include all observations in a climatological treatment. After all seafaring nations had set up their national climatic archives with the observations made by their vessels in all oceans, the WMO arranged an international exchange of these data among its members to enhance the effectiveness of climatological research by international cooperation and evaluation of the available observations. All ship observations since 1961 are collected and evaluated by eight nations, each responsible for one ocean area (WMO, 1963, 1977). In the Federal Republic of Germany, the Seewetteramt in Hamburg handles the data for the South Atlantic Ocean. For this chapter, one million international observations were available during 1961 – 1970, and from the archives of national data another four million observations for a period of about 100 years were at our disposal.

Problems inherent in the data

It is customary to use all existing data in the evaluation of ocean climates, although this should not be done without a thorough check of their quality. Not only physical units but also observing methods have changed several times during the last 100 years, and the recognition of possible climate fluctuations over the sea is difficult when these fluctuations are the same size as the uncertainties of the data. A discussion of the reliability of the data starts on p. 13 as part of the description of the mean values of the various elements, but a systematic treatment is omitted here. A change of units is taken into account when the data are archived, punched on cards or placed on magnetic tapes. Changes of methods, although noted in the observation lists, can hardly be heeded during the processing and evaluation so that in the interpretation of the results one has to pay attention to the possible consequences (WMO, 1954, 1955).

With the support of the WMO, a worldwide standardization of measuring methods, observing principles, units, and recording methods is being attempted. The international weather code is the present result of many efforts toward standardization in synoptic meteorology. In marine climatology, the corresponding achievement is the International Maritime Meteorological Punch Card (IMMPC) whose format has been used to exchange, collect, and evaluate all ship observations since 1961 (WMO, 1977).

Despite all these efforts to ensure quality in the data, the accuracy of the values is limited by numerous sources of errors such as interference by the vessel or flawed measurements or observations. Quality control of the values is therefore indispensable when they are placed in the archives. According to Recommendation 31 (78-CMM) of the WMO, all observational data from logbooks, their coding, the consistence among the elements in a set of observations, the consistence in time along the ship routes, and the reliability of extreme values should be checked and the data compared with simultaneous neighbouring values if possible. The scarcer the data, the more important it is to make these checks.

Inhomogeneities

In contrast with observations at a fixed station, which are taken continuously and which record regularly all weather phenomena occurring at a given place, the distribution of ships is so irregular in space and time that all observational series over the oceans are inhomogeneous. In the short term, this is owing to random fluctuations in shipping, and in the long term to the variability of trading relations between nations; the economical situation also influences shipping and thus the frequency and distribution of observations. This is especially so in wartime when the amount of data decreases considerably.

Certain routes selected according to climatological principles have always been recommended to the shipping industry, and over large areas outside these routes data therefore remain sparse. This inhomogeneity is time-dependent because the routes have changed during the centuries, but they also vary within a year due to seasonal changes in the weather. Whereas the sailing-ship routes depended strongly on wind conditions and thus varied much, the shipping lanes became narrower and shorter after the introduction of steam vessels. There may moreover be a weather bias, either involuntary because of the effect on the ship's movement of cross winds or cross currents, or voluntary by the change of route according to weather forecasts to avoid gales, fog, or ice and to achieve an optimal speed (WMO, 1958).

A normal and regular recording of weather situations over the sea cannot be guaranteed and the international exchange of data cannot remove these inhomogeneities, although it does achieve the best possible data coverage and quality.

The spatial inhomogeneities of the observations over the South Atlantic Ocean are demonstrated by Chart 1 (Appendix, pp. 158ff) which shows the distribution of observations. The areas with more than 1000 observations per 1°-square are hatched, and a distinction has been made between historical German observations which stem mainly from sailing ships, and international observations made since 1961 and obtained by the international exchange of data. The broken curve outlines an area with less than 100 observations per 1°-square.

The two main shipping routes along the continents are clearly seen. The Cape Horn route to the west coast of South America dominates in the historical observations. The route for the ships sailing to the west coast differed from that followed on the return because the most favorable wind field was used on either route (SCHÜCK, 1875; SCHOTT, 1895). Since the opening of the Panama Canal in 1914, the west Atlantic route reaches only to Buenos Aires and has become narrower and lies nearer the coast.

The shipping lane along the African coast leading to the Indian Ocean is now more frequented, especially when the Suez Canal is unavailable to shipping. Traffic has also increased as new commercial relations have opened up areas along the African coast. Data are scarce over the central and southern parts of the ocean (compare SCHUMANN, 1952). There is an infrequently used sailing route to the Cape of Good Hope along 30° to 40°S that avoids the southeast trade winds (SCHUMACHER, 1936), and during summer a few vessels go to the Antarctic to supply the stations there or the whaling operations. The lack of observations in the highly variable westerlies of higher latitudes is unfortunate because the greater variability of the weather elements there requires a large quantity of data for the calculation of reliable climatological means in contrast with the tropics where fewer observations suffice.

The use of stationary and drifting buoys and satellite observations will eventually benefit climatology in regions where conventional observations are scarce, although patience must be exercised until sufficiently long series of observations have been accumulated by these new observing techniques. For this reason, it is urgent to expand these facilities as recommended by the World Weather Watch (WMO, 1966b) and in the World Climate Programme (WMO, 1979).

Selected ocean areas

The collection of ship observations in selected areas for evaluation is possible because these observations represent a wider area than those at fixed locations on land. The likely small-scale variations and structures are statistically distributed and are of no climatological importance.

The size of these selected areas depends on the desired degree of resolution of the climatological conditions, on how large the gradients of the climatological elements are, and on whether enough observations are available to make accurate statements about the climate of the area. This is a problem of optimization. The frequently small number of observations necessitates a minimum size of the selected area within which, however, the climatological differences should not be greater than those associated with random fluctuations of weather. This should be so for all elements and for every season. The areas may cover the whole ocean without gaps between them such as within the scope of the HSSTD project, but mostly only a certain number of areas are chosen depending on the money available for computer time, illustration, and publication. The areas are generally distributed along the main shipping lanes where there are usually sufficient observations. Ideally, they should represent the main climatic properties of the ocean. Such selected areas may, for example, be found in the marine atlases published by the U.S. Navy (WASHINGTON 1958, 1978) where the frequency distribution of climatic elements is also for selected areas.

The 18 selected areas in Chart 2 are those for which monthly climatological values have been calculated to be presented in the tables; the areas differ in size according to the density of observations. The chart also shows the location of the eight island stations for which climatic tables are given (Appendix, p. 150ff), and the chart itself encompasses those parts of the South Atlantic Ocean for which climatic charts of the various elements have been drawn.

Fundamentals of climate

The South Atlantic Ocean stretches from the Equator to the Antarctic Ocean at about 60°S and thus embraces all climate zones from the tropics to the subpolar regions. It is the smallest ocean in the Southern Hemisphere, and with an area of nearly $46 \cdot 10^6$ km^2 it is nearly the size of the North Atlantic Ocean proper. It is bordered by South America and Africa and in the south by Antarctica. Continental influences on the climate are not as marked as in the Northern Hemisphere. Africa reaches only to 35°S and South America is rather narrow beyond 40°S. Nevertheless, some orographic effects do modify the maritime influence, especially in the coastal areas (p. 86f). Some zonal differences between sea and continent are added to the meridional differences of the planetary climate, which are also transmitted into higher tropospheric levels although this effect is relatively weak (p. 100).

Ocean currents deflected by the continents and exhibiting a meridional component and upwelling along some coasts add to the zonal differences of climate over the ocean. The extremely cold conditions over the antarctic ice intensify the meridional contrasts of air pressure and temperature and help to create the vigorous cyclonic activity over the sea with the associated weather features (p. 115).

Climate-producing processes

Insolation, the earth's main source of energy, heats the surface and the atmosphere, and its amount determines the total energy. The internal energy of the temperature fields and the potential energy of the pressure fields are the source of the energy of motion which manifests itself as turbulence and circulation on many scales (RIEHL, 1969).

Because the insolation is absorbed mainly in lower latitudes and in the surface layers of the sea, horizontal and vertical temperature gradients are created which act as the motor in the atmospheric heat engine. They produce the circulations in the atmosphere and sea that are the links in the exchange processes (GATES, 1977). These processes, in turn, attempt to reduce the temperature gradients through the exchange of heat. This is also true for the exchange of matter, e.g., for the water vapour of the atmosphere (p. 51 et seq.) or the salt of the sea. The climatological gradients represent the balance between the continuous generation of temperature differences, for instance by regional differences of radiation balances, and their dissipation by exchange and friction (ROLL, 1965).

This exchange first takes place on small scales (extremely close to the sea surface) by heat conduction or diffusion. Turbulence in a moving medium acts more effectively (DEACON and WEBB, 1962) and its efficiency increases with the strength of the gradient and the basic current (p. 56). If this kind of exchange does not keep the gradients constant, instabilities arise causing circulations (wind currents, large eddies) that transport heat, or the matter with which heat is associated, and this organized, large-scale exchange is the most effective one. At the same time, these circulations act as basic currents for small-scale turbulence so that, altogether, processes distributed over a wide spectrum take part in the exchange (LUMLEY and PANOFSKY, 1964).

The proportion between the horizontal and vertical dimensions of the troposphere determines the prevalence of horizontal circulations and motions. Nevertheless, vertical motions are no less important than horizontal ones (NIEUWOLT, 1977), but they are either small-scale, like the vertical currents in thunderstorms, or large-scale but very slow such as the subsidence above the trade inversion.

A peculiarity of the vertical heat exchange is the fact that the vertical transport of water vapour from the sea surface (p. 51) is also an exchange of latent heat (p. 57) which, in turn, is released into the atmosphere when clouds form (convection). This kind of heat exchange is generally more effective than the exchange of sensible heat (p. 58). Extreme heat exchange takes place in thunderstorms, tornadoes, and tropical cyclones.

Whereas weather is characterized by short-term changes of the atmospheric circulation, climate develops from equilibrium conditions whose long-term variations result in climatic fluctuations. In this respect, variations within the oceans and their influence on the atmosphere are decisive (HASSELMANN, 1977).

Planetary circulation

The climatic zones are primarily characterized by different radiation balances (p. 60f) associated with different latitude-dependent warming and cooling of the atmosphere and the ocean. The warming and cooling are modified by the march of the seasons, by the different effects of insolation on land and sea, and by the influence of wind systems and ocean currents.

The latitude-dependent radiation balances cause meridional climatic differences with a meridional temperature contrast. The compensating motions first produce a circulation with an equatorward flow near the surface, ascent in low latitudes, poleward motion in the upper troposphere, and descent in higher latitudes. This meridional circulation, however, is modified by the earth's rotation such that zonal circulations are created which contribute nothing to the heat exchange but form the basic current for superposed turbulence and secondary circulations.

Therefore, the main cell for meridional exchange in a vertical plane is confined to lower latitudes as a meridional component of the tropical circulation. This cell is called the Hadley circulation. The descent within this circulation takes place in the horse latitudes at about 30°N and S. There the atmosphere produces the anticyclonic belt near the surface, and the equatorward pressure gradient is expressed in the trade winds, steady easterly winds with a frictionally caused equatorward component. The convergence of the trade winds from both hemispheres near the Equator is called the Intertropical Convergence Zone (ITCZ) where upward motion and convection are widespread.

The deflection by the earth's rotation of the poleward flow in the upper branch of the Hadley circulation leads to a concentration of westerly zonal flow that is well developed in the high troposphere at about 30°S (p. 105f). This belt of maximal westerlies, the so-called subtropical jet stream, supplies kinetic energy for the development of higher latitude cyclones that achieve the necessary meridional heat fluxes by means of large-scale turbulence.

Poleward of the horse latitudes, temperature and air pressure decrease on the average. Here the tropospheric westerlies dominate also at sea level. Much of the temperature gradient is concentrated in the polar front zone (p. 118) and the tropospheric westerlies of the polar jet stream. In this frontal zone, instabilities occur that produce variable circulations—cyclones and anticyclones. The associated meridional flow provides the heat exchange between subtropical and polar latitudes. The westerly basic current appears as a zone of prevailing westerlies in the climatic mean only.

The following marked climate zones thus result over the South Atlantic Ocean.

(*1*) The zone of inner tropics near the Equator with high temperatures, weak winds, relatively low air pressure, and high precipitation. The ITCZ, where the trade winds of both hemispheres converge, is decidedly north of the Equator (p. 114).

(*2*) The trade wind zone with steady easterlies deflected equatorward and with relatively dry but cloudy weather conditions.

(*3*) The anticyclonic belt of the horse latitudes in the subtropics at about 30°S with weak winds, dry and fair weather, warming by subsidence, and strong insolation.

(*4*) The zone of westerlies in temperate latitudes with strongly varying air pressure decreasing poleward in the mean; variable, poleward decreasing temperature; unsteady winds, sometimes gales, from predominantly westerly directions; and lively sequences of weather with occasional precipitation.

(5) The polar zone with mostly low air pressure, low temperature, unsteady winds, and variable weather with frequent snow.

The zone of easterly winds farther south, often katabatically strengthened near the Antarctic continent, is not a subject of this volume (see Vol. 14).

Characteristics of marine climate

The planetary climatic zones are different over the oceans and the continents because the insolation has different thermal effects over land and sea. The marine climate may be modified by continental influences.

The penetration into the ground of the insolation on land is so small that only little heat is stored in the soil. Instead, a shallow surface layer is strongly heated and as quickly cooled by nocturnal radiation, and this warming and cooling are transmitted to the atmosphere. In this way, the characteristics of the yearly cycle of insolation are transmitted to the temperature, stability of stratification, and other meteorological parameters.

At sea, the greater penetration of the insolation and strong mixing within the moving sea surface layer cause a heat storage that nearly suppresses the daily cycle of temperature and reduces the annual cycle. The vertical heat exchange in the ocean, however, is limited by the formation of a 20–200 m thick warm surface layer (pp. 28, 62).

Because of the difference in radiation balance between ocean and continent, gradients of temperature and air pressure develop in coastal regions, causing special circulations with daily and yearly cycles. The daily cycle results in a sea breeze from sea to land during the day and a land breeze from the coast to the sea at night. The yearly cycle is reflected in circulations between continent and ocean which are on a larger scale and which are superposed on the planetary basic currents and may even reverse them. These seasonally changing winds are called monsoons. The larger the continent and the intenser the radiation effects, the more pronounced and extensive the monsoon will be. The monsoon plays only a subordinate role in the South Atlantic Ocean.

In summer, these circulations are directed cyclonically into the continents and in winter anticyclonically out of them. In contrast, over the sea on the poleward edge of the subtropics, anticyclonic circulations dominate over the comparatively cold water in summer and cyclonic circulations are more widespread in winter when the surface water is comparatively warm (p. 15).

Over the continent, the larger amplitude of the yearly temperature cycle results in a greater seasonal shift of climatic zones than over the ocean. This is especially so for the ITCZ. Over the South Atlantic, the yearly oscillation of the climate zones amounts to 5–10° latitude with a more southerly position in the southern summer and northerly position during southern winter.

The yearly cycle is caused not only by the sun's position but also by the length of the day, resulting in a weakening of the over-all meridional temperature contrasts in summer and a strengthening in winter. The yearly cycles of pressure gradients and wind speeds are analogous. The yearly ranges are relatively small over the South Atlantic Ocean because the influence of the adjacent continents is limited to coastal areas (p. 86).

A special difference between the North and South Atlantic is caused by the snow-covered, extremely cold Antarctic continent. The Arctic Sea is much warmer, and thus the meridional mean gradients of pressure and temperature are much steeper over the South than over the North Atlantic Ocean during the summer of either hemisphere. For the same reason, the ITCZ, which forms the thermal equator, is situated north of the geographical Equator with a mean position of 5°N. Therefore, the South Atlantic is colder than the North Atlantic Ocean and without tropical cyclones. The seasonal differences in the South Atlantic are smaller too because the albedo of Antarctica reduces the radiational warming during the summer (FLETCHER, 1969).

Differences between sea and land are not only thermally caused. Friction, which influences the wind field near the surface by exchange of momentum, is also greater over land than over sea and convergences are therefore generated by on-shore winds, divergences by off-shore winds. These alter the stability of the atmospheric layers and cause special circulations that influence weather and contribute to the heat exchange. In the west wind region, an anticyclonic flow (high pressure ridge in the troposphere) develops over the eastern side of the ocean, a cyclonic flow (trough) over the western side.

Such a planetary Rossby wave in the tropospheric westerly belt is but weakly developed over the South Atlantic Ocean (p. 101). It is most pronounced in the western part because the Andes Cordilleras produce a lee effect. The tropospheric high-level trough associated with that effect introduces a weak northerly component to the mean wind which is also reflected in the mean path of the cyclones (p. 120).

Climate classification of the South Atlantic

The climatic factors mentioned in the previous section determine the climatic zones of the South Atlantic Ocean according to astronomical and physical principles and can be used in a classification of the climate (KNOCH and SCHULZE, 1952). Here KÖPPEN's (1931) classification is the basis for description of the climates of the South Atlantic Ocean. In a sense this classification was designed only for land areas because it is based on monthly and yearly values of air temperature and precipitation, and at sea only coarsely estimated precipitation amounts are available (p. 49). Even more doubtful is the use of the so-called "rain value of temperature" as an evaporation index to evaluate the aridity of a climate: When one compares rainfall amounts and temperature levels, the aridity will be rated the higher the more of the annual precipitation which falls within the summer season when the evaporation is high, which of course raises the value of the aridity index. Only when water balances (p. 53f) and radiation balances (p. 64f) can be determined more precisely will better climate classifications be possible (BARRETT, 1974). So far Köppen's method is still the most suitable one for a rough but lucid outline of climate regions.

The following elements are the bases of the classification: R = yearly precipitation in cm (Chart 38); T = yearly mean air temperature in °C; T_w = temperature of the warmest month of the year (Chart 23); T_k = temperature of the coldest month of the year (Chart 24); T' = rain value of temperature (evaporation index).

With these values the following climates can be distinguished: A = tropical rain climates ($T_k \geq 18°C$); BW = desert climate ($R < T'$); BS = steppe climate ($R < 2T'$); C = temperate rain climate; ET = polar climate (tundra) ($T_w < 10°C$); with the following modifiers: s = summer dry ($T' = T$); f = moist throughout the year ($T' = T + 7$); w = winter dry ($T' = T + 14$); h = hot ($T \geq 18°C$); a = summer hot ($T_w \geq 22°C$);

l = mild (10°C < T_k, $T_w \leq$ 22°C); k = temperate (18°C$\leq T_w$ < 22°C); k' = cool (T_w < 18°C); b = moderate ($T_w \geq$ 10°C for at least 4 months); i = isothermal ($T_w - T_k$ < 5°C).

Chart 3 presents the climate zones over the South Atlantic Ocean. The main climates A, B, C, and E are separated by thick solid lines and are subdivided according to different temperature and precipitation conditions.

The southern boundary of the tropical climates A and B lies at about 30°S. The larger part of the area is occupied by the arid climate B, especially the region east of 30°W, and to the north it locally reaches the Equator. In the centre of this arid climate, precipitation amounts are so small that for the most part it has to be classified as a desert climate BW. This is because subsidence above the trade inversion makes the atmosphere thermally stable and dry so that deep convection with precipitation is excluded.

Along the African coast, upwelling creates relatively low temperatures so that mild (l) to temperate (k) climates can be demarcated in the southern parts, whereas elsewhere the hot (h) type prevails.

Within the tropical rainy climates near the Equator and along the South American coast, a narrow zone on both sides along the Equator, where the yearly shift of the ITCZ causes a rainy and a dry season, can be defined as savannah climate Aw. In the remaining area, the precipitation is distributed equally all over the year (Af).

In the western part of the ocean, the zone of temperate rain climate C extends to 50°S, in the eastern part to 45°S. It can be divided into a zone with a precipitation peak in winter and mild weather (Cfl) north of 40°S and a southern zone with almost equal amounts of precipitation in all seasons and lower temperatures (Cfb). This is an effect of the seasonal shift of the subtropical anticyclonic belt and the polar front zone.

In the west, off the South American coast there is a humid region with hot summers (fa) at about 30°S and a cooler steppe climate (k,k') at 45°S whose dryness is caused by a lee effect of the Andes. The obstruction to the westerly winds by the American continent results not only in aridity off the coast but also in a relatively mild climate which stretches about 100 sea miles out over the sea to the edge of the cold Falkland Current. Farther south lies a polar climate ET with the mean temperature of the warmest month between 0 and 10°C and precipitation all through the year, often falling as snow.

The annual range over the South Atlantic Ocean is relatively small; for this reason, the climates off the coasts can be classified generally as isothermal (i) because the highest and lowest monthly mean values often differ by less than 5°C.

Climatic elements

The climatology of the weather elements over the South Atlantic Ocean is described in this section. Thirty-four charts and eighteen tables serve this purpose. In the charts, the geographical distribution of each element over the ocean between 7°N and 57°S is represented, mostly for January and July. These months were chosen to conform with the other volumes of the *World Survey of Climatology*, although they do not reflect the total annual variation over the ocean. For conditions over the Antarctic Ocean, the reader is referred to SCHWERDTFEGER (1970) in Volume 14.

Monthly means, annual means, and mean monthly standard deviations of the climatic elements are compiled in the tables for the 18 selected areas (p. 132ff) whose geographical situation is shown in Chart 2. The climatic values were computed from international ship observations between 1961 and 1970 collected by the Seewetteramt in Hamburg (Federal Republic of Germany) through the exchange of data arranged by the WMO (p. 6). With these observations, mean values and standard deviations of all elements of interest were calculated within 5°-squares for all months and for the year. The monthly means were used for a harmonic analysis of amplitudes and phases of the annual variations.

The analyses of the charts are also based on the 2°-square means of these data and on the evaluation of historical German ship observations for more than 100 years (p. 5) for January and July. Averages and standard deviations of the most important elements were calculated from these data for 1°-squares for the whole period and for decades. Already published atlases (p. 3) were used to complete the analyses.

The climatic elements are:

(1) *Air pressure*, the most important element for synoptic meteorology, reflecting the conversion of thermal energy into dynamic energy.

(2) *Wind*, to describe atmospheric currents over the ocean, important for shipping.

(3) *State of the sea* and *ocean currents*, also important for shipping and responsible for thermal influences on the atmosphere.

(4) Air and sea surface *temperature*, to describe thermal conditions and the heat exchange between ocean and atmosphere.

(5) *Vapour pressure* and *relative humidity*, computed from the temperatures of dry and wet bulb thermometers, to characterize the atmospheric moisture content over the ocean and to compute the evaporation and exchange of latent heat.

(6) *Cloudiness* and frequencies of *fog, precipitation*, and *thunderstorms* to describe visible weather phenomena.

Besides ship observations, climate data from eight climatic stations on islands in the South Atlantic Ocean (p. 5) were evaluated. Long-period monthly means of these elements are arranged in tables (p. 150f) and described on pp. 71–86.

Air pressure

Air pressure, although not seen or felt, belongs to the most important meteorological elements. In synoptic meteorology, weather charts are based on analyses of the air pressure field, their isobars characterize the weather situation, and the charts elucidate the spatial and temporal developments used in forecasting weather.

In climatology, the air pressure field describes a balance between fields of available potential energy, associated with the thermal structure of the atmosphere, and their conversion into kinetic energy expressed in the wind field which is associated with the pressure field. Therefore, air pressure is suitable for describing the large-scale average circulations which determine climate.

Mean pressure field

Charts 4 and 5 represent climatic averages of air pressure over the South Atlantic Ocean for January and July. In both months, the subtropical anticyclone dominates the map; in January (southern summer) its centre of more than 1021 mbar is positioned at 32°S, 5°W and in July with 1025 mbar at 27°S, 10°W (p. 116).

The polar belt of low pressure with a mean pressure of about 985 mbar lies near 65°S, to the south of the area represented in the chart. Across temperate latitudes, the pressure falls in the mean about 1 mbar per degree of latitude, which is reflected in the strong mean westerly winds for which this zone is named, the west wind belt (p. 118).

North of the ridge, the decrease of pressure amounts to about 0.3 mbar per degree of latitude. The associated easterly winds are the trade winds (p. 114).

The tropical low pressure belt covers the ocean from the north coast of South America to the African coast. It lies near the Equator in January with a central pressure of about 1011 mbar in the west, somewhat farther south than in the east. It is situated at about 10°N in July with a central pressure of about 1013 mbar. The low pressure belt separates the trade winds of both hemispheres; their convergence is called the Intertropical Convergence Zone (ITCZ, p. 114).

The centre of the subtropical high lies to the east of the central meridian of the ocean and, accordingly, strong trades blow in the eastern part of the South Atlantic while the mean winds are weaker in the western part.

Another asymmetry is observed in middle latitudes where the isobars converge toward the east, especially in winter. The mean pressure gradient in the west wind belt off South America thus amounts to only 0.9 mbar per degree of latitude, whereas south of Africa it is 1.3 mbar per degree (in summer 1.2, in winter nearly 1.5 mbar). The weak gradient in the western part of the ocean is a lee effect of the South American Andes, because the flow across the mountains reduces the air pressure near the east coast (p. 12). In the eastern part of the ocean, the polar front zone and thus the pressure gradient are intensified by the warm Agulhas Current (p. 27) in the north and by the cold Antarctic Ocean in the south.

Annual variation

The gradient of pressure in middle latitudes shows no essential difference between January and July, but the position and configuration of the subtropical anticyclone do. In July, the high pressure centre lies 5° farther north than in January with a 4 mbar higher central pressure. Its displacement during the year is also evident from the tables containing the monthly mean values of 18 selected areas (Chart 2). The highest mean air pressure appears in October—April in area P relatively far southeast, in May in area H, in June—August farther west in area M, and in September between areas H and M. All three areas have almost the same value of 1019 mbar in the annual mean. The mean intensity of the subtropical anticyclone also varies during the year; the lowest value of 1018 mbar is observed in April and in December (area P), the highest of 1023 mbar in July (area M).

The shape of the anticyclone differs distinctly between the seasons. The pressure difference between the centre and the coast amounts to about 10 mbar in January and the west wind belt extends north approximately to 40°S. In July, the pressure difference between the high centre and the coast is reduced to about 5 mbar and the westerlies extend to 35°S. The west—east component of the circulation is thus more pronounced in winter than in summer when the meridional wind components off the coasts are so strong that the extent of the anticyclone is clearly defined over the ocean. This difference in the seasonal shape and extent of the subtropical high is influenced by the continents

(p. 11) and is associated with the annual variation of air pressure over them (see VAN LOON, 1972, fig. 4.5). The pressure is lower over the continents in summer than in winter because of the different thermal conditions in the atmosphere over land. In South America, e.g., the annual range of pressure amounts to 10 mbar at 30°S (SCHWERDT-FEGER, 1976, fig. 3). The accompanying seasonal changes in the wind field are monsoonal (p. 21): The lower summer values of pressure over the continent cause an on-shore wind component, and the higher continental pressure in winter produces off-shore components.

The pressure differences between July and January are shown in Chart 6. The air pressure is almost everywhere higher in July than in January, particularly near the coasts: 9 mbar in the bight of São Paulo, 7 mbar off South Africa. This reflects not only the monsoonal effect of the continents but also a planetary effect. Worldwide, the surface air pressure is higher in July than in January in the tropics and this difference amounts to about 3 mbar (HÖFLICH, 1974). It appears that the unlike distribution of land and water in the two hemispheres produces a seasonally different partial pressure of water vapour with a maximum in July (PFLUGBEIL, 1967). The highest monthly mean values of pressure in the selected areas appear mostly in July, whereas the lowest ones are found from December to April. The differences between the extreme monthly means range from 3.4 mbar in area A off the northeast corner of South America to 8.4 mbar in areas I and K and 8.1 mbar in area Q near the coast of South America, with an average of about 6 mbar.

A harmonic analysis of the annual course of air pressure yields the largest amplitude of the first harmonic in off-shore regions (3—4 mbar) with the maximum occurring between the middle and the end of August; over the open ocean the amplitude of 2—3 mbar reaches its maximum at the middle of July. In higher latitudes, a marked half-yearly period with maximum in spring and autumn is indicated (VAN LOON, 1967; see Gough Island, p. 80f). Unfortunately, the number of ship observations is insufficient for a more exact calculation.

The atmospheric tides, caused by the gravitational force of the sun and the moon, result in a daily variation of air pressure, the size of which depends on local time (longitude) and latitude. The amplitude is greatest at the Equator where it amounts to nearly 2 mbar; near the Cape of Good Hope, it is almost 1 mbar. The maximum appears at 10h00 and 22h00, and the minimum at 04h00 and 16h00 local time.

Variability

Chart 7 shows the annual means of the monthly standard deviations of individual observations of pressure. This quantity is a measure of the variability of pressure caused by the daily weather within a month. It amounts to about 2 mbar north of 20°S and increases to 12 mbar poleward, without much zonal difference. In the selected areas, the standard deviations of pressure increase from 2 mbar at low latitudes to 8 mbar in the southernmost area R. Maps of quarterly standard deviations of air pressure during the IGY show similar values (TALJAARD, 1966; KAUFELD and RUDLOFF, 1972). The day-to-day circulation is thus very variable in subpolar latitudes of the South Atlantic Ocean, and the large variability there implies that stable climatic statistics can only be achieved from long series of observations.

Wind

The mean air pressure field in Charts 4 and 5 indicates the direction of the mean wind over the South Atlantic Ocean according to Buys Ballot's law. A geostrophic west wind of $6-8$ m s^{-1} can be computed from the zone of strong pressure gradient in middle latitudes at about 45°S and a trade wind of 7 m s^{-1} north of the subtropical anticyclone at about 15°S. These computed wind velocities indicate the flow of air along the isobars. At sea level, however, the actual wind is weaker because of the influence of friction, and it is thus deflected toward low pressure. On the average, the observed winds show a component toward lower pressure—in the trades a northward component, in the west winds of higher latitudes a weak component toward the south.

The wind field over the South Atlantic Ocean is thus divided into three main belts—the trade-wind region with mostly southeasterly winds between the Equator and about 25°S, the horse latitudes with predominantly weak variable winds, and the west wind belt south of 35°S (p. 10). The last region extends to the low pressure trough off Antarctica from where easterly winds prevail farther south. The anticyclonic curvature of the isobars around the subtropical high strengthens the meridional wind component.

Wind speed

Normally, it is not the vector mean wind velocity that is of interest but the scalar mean wind speed, which is computed without considering wind direction and therefore is the larger of the two.

The wind force over the sea was of primary interest in the days of the sailing ships, as sails were set according to it. In 1805, Beaufort therefore based his scale of wind force on the amount of canvas that a full-rigged frigate could carry. In 1927, PETERSEN related the numbers of the Beaufort scale to the effect of the wind on the sea surface, and the scale could then be used by steamers and motor ships. The subjective estimate of the wind force by the Beaufort scale supplies quantitative data, homogeneously observed for more than 100 years by all nations on and for all oceans, although these estimates are relatively rough and the scale not exactly linear.

Such estimates are being increasingly replaced by *measurements* of wind speed. A measurement allows a precise statement in m s^{-1} or knots, but it does not necessarily represent an exact value because a measurement on a ship may contain various errors. A ship disturbs the wind field while the surrounding sea surface remains unaffected. The pitching and rolling of a vessel falsify the measurement too. It is only the difference between the speed of the wind and the movement of the ship which is measured, and when the ship's speed is subtracted from the measurement, the drift caused by wind and current is ignored. As the measurement is performed at the mast-top at a height of $20-60$ m in order to avoid the field disturbed by the vessel, the change in wind between mast-top and sea surface must also be considered. This change depends on the stability of the atmospheric layer over the sea surface and on the height of measurement, i.e., on the class of ship and its depth of immersion. To eliminate the effect of gustiness, the measured wind values have to be averaged over 10 minutes, which can be properly done only with automatic instrumentation.

Because both methods are still in use, the weather code prescribes an equivalent scale combining the Beaufort numbers with speed in knots. However, this connection is very problematical because the equivalent values are approximations. An improved equivalent scale has been proposed for approval (WMO, 1970); otherwise, the scale can be used only as a rough compromise.

The distribution of the mean wind speed over the South Atlantic Ocean is shown in Charts 8 and 9 for January and July. The isotachs are scalar averages in m s^{-1} and both Beaufort equivalents of estimated wind and measured values have been used.

The zone of strong west winds is clearly seen in the charts with mean values of $10-13$ m s^{-1} (6 Beaufort) south of 40°S and east of 40°W, although the force at single points can be determined only inaccurately because the number of observations is too small. A detailed analysis of higher latitudes is published by ZILLMAN (1967), but the lower wind speeds at the climatic stations have apparently been overemphasized by him (cf. p. 85).

The weaker wind off Argentina is a lee effect of the Andes. The difference in air pressure gradient between the western and eastern parts of the South Atlantic (p. 15) is connected with an increase in mean wind speed eastward within the west wind belt to a maximum south of Africa.

The centre of the subtropical anticyclone is reflected in a minimum of wind speed over several parallels of latitudes. The trade winds east and north of the subtropical high pressure centre have mean wind speeds of $6-8$ m s^{-1} (4 Beaufort). The position of the anticyclonic centre over the eastern part of the ocean causes the highest mean wind speeds off the African coast in those latitudes. Near the Equator, the ITCZ is represented by lower mean wind speeds in January, below 4 m s^{-1} (2 Beaufort).

In the selected areas, the annual mean values range between 4.3 m s^{-1} (3 Beaufort) in area C at the Equator and 7.5 m s^{-1} (4 Beaufort) in area L in the trade wind region. Higher averages are not found in the tables because in the zone of westerly winds no area, especially in winter, had a sufficient number of observations for computing reliable mean values.

The most important planetary wind fields which have thus been outlined are in agreement with the distribution of mean pressure in Charts 4 and 5. The seasonal differences, already discussed by means of the pressure charts, reappear in Charts 8 and 9, both the more northern position of all wind fields and the more zonal alignment of the isotachs in July (p. 16). In general, the relatively weak annual pressure variation over the South Atlantic results in only small seasonal differences in the wind field.

Remarkable differences occur in the equatorial region. The doldrums, with weak winds characterizing the ITCZ, are pronounced only in January and only east of 30°W, whereas off the north coast of South America the trade winds apparently cross the Equator and merge with trades of the Northern Hemisphere, which appear at the northern border of the chart with a mean wind speed of 8 m s^{-1}.

In July, a maximum of mean wind speed of 6 m s^{-1} occurs north of the Equator where the trade winds of the Southern Hemisphere, deflected after passing the Equator, increase with a monsoonal component. The northward-displaced ITCZ is visible only west of 30°W at the border of the chart.

A harmonic analysis of the annual variation of wind speed yields amplitudes of $0.5-1.5$ m s^{-1} with the maximum occurring mostly in August (southern winter). The smallest annual variation is observed in the trade wind region, the biggest in the west wind belt and in the inner tropics due to the displacement of the ITCZ. Compared with the North Atlantic, the annual variation is weak.

The monthly means in the selected areas clearly show the small annual variation. The difference between the highest and the lowest monthly mean speed is only 1.7 m s^{-1} averaged over all areas. Accordingly, the extremes occur in different months, although maxima are more common in southern winter and minima in southern summer.

The annual mean of the monthly mean standard deviations of wind speed, which gives the mean range of wind speeds within a month, is presented in Chart 10. These standard deviations vary from 2 m s^{-1} in the equatorial zone to more than 5 m s^{-1} in the west wind belt, which corresponds to $1-2$ numbers on the Beaufort scale. In the selected areas, the same standard deviations lie between 2 and 4 m s^{-1} depending on the latitude and mean value of the wind speed. On the average, the scalar mean wind speed is about twice as large as the standard deviation. This relation varies between 40% in lower latitudes and 60% in higher latitudes and reflects not only the daily variability of the wind but also the coarseness of the Beaufort scale.

The statistical meaning of the standard deviation of wind speed depends on whether or not its frequency distribution is normal (Gaussian). For the trades between the Equator and 5°S (areas A + B) and for the areas N and O south of the subtropical zone (30−35°S), which lie along the ship routes near South America and Africa, the frequency distributions of the observed Beaufort numbers and the averages and standard deviations of the wind speeds were computed. These frequency distributions were compared with the normal distribution determined by the average and standard deviation of the wind speed which in Beaufort numbers have the following values:

Area A + B, average 3.3, st. dev. 1.1.

Area N, average 3.6, st. dev. 1.6.

Area O, average 3.9, st. dev. 1.7.

The results in Fig. 1 show good agreement between the observed frequencies and the computed normal distribution in all three areas. The wind speeds are therefore almost normally distributed.

Naturally, the normal distribution cannot have zero Beaufort as the limit of the frequencies, and it therefore contains small values of unreal negative Beaufort numbers. If the frequency of the negative numbers is great, the Raleigh distribution may give a better description. In the above instances, the mean values are substantially larger than the standard deviations, and the frequencies of fictive negative wind speed are very small so that the normal distribution presents a good approximation to the distribution of wind speeds.

Wind direction

Predominant wind directions are presented in Charts 11 and 12 for January and July. These direction values were computed for 2°-squares as vector means but with the Beaufort number 1 given to each observation. The regions with the same predominant direction within 45° are separated by dotted lines.

The wind directions fit well the pattern of isobars in Charts 4 and 5, apart from the angle between gradient wind and true wind and from the equatorial zone where the geostrophic relationship is not valid. The predominant wind direction as computed here

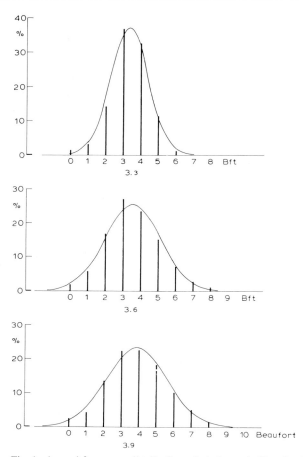

Fig. 1. Annual frequency distribution of wind speeds (Beaufort) and the Gaussian distribution according to their mean and standard deviation. In the tropics (area A and B, top), off South America (area N, middle), and off South Africa (area O, bottom).

may differ from the direction of the vector mean wind, which corresponds to the climatological pressure gradient, in areas where frequently changing synoptic situations introduce a large variability in the wind, for example, in the frontal zone at the northern border of the west wind belt.

The charts of prevailing wind directions indicate convergences and divergences in the wind field where air currents meet or part. In the middle of the South Atlantic, the subtropical anticyclone stands out as an area of divergence. The ITCZ can be seen in January as a wind convergence north of the Equator, whereas in July a divergence exists in-the same place in the trade winds which cross the Equator. The convergence associated with the frontal zone in the west wind belt is too variable in position to emerge from this rough analysis of the mean conditions. Detailed analyses put it at about 45°S, farther north in the west and farther south in the east.

The annual variation of wind direction is generally small. Only the south — north displacement of the subtropical ridge causes a seasonal change of wind direction, such as in areas Q and M where the winds are northeasterly in summer and westerly in winter. In area O, monthly wind directions vary from south in summer to south-southwest in winter. Small wind shifts are found off the coast of Brazil, e.g., in area A between east-southeast and southeast corresponding to the movement of the ITCZ, in area E between east and east-southeast, in area I between northeast and east for orographic reasons, and in area N between east and northeast. Within the west wind zone, the mean wind direction varies between west-northwest in summer and west in winter in areas P and R.

Seasonally changing wind directions near the coast are produced by the different thermal conditions over continent and ocean. They are part of a monsoon circulation characterized by a component toward land in summer and sea in winter. This circulation is similar to the small-scale daily variation which consists of on-shore winds at midday and off-shore winds at night; on steep coasts these winds are strengthened by anabatic winds at midday and katabatic winds at night. This daily circulation is the stronger the more intense the insolation, and thus its optimum conditions are in the subtropics and in summer (p. 11).

The ratio of resultant to scalar wind speed is called the *steadiness* of the wind. Contours of equal steadiness in percent are also plotted on Charts 11 and 12 for January and July. LAMB (1977) has published maps of steadiness over the tropical Atlantic for the four seasons.

The steadiness of the trade winds is high—more than 90%, whereas in the west wind belt 50% is rarely exceeded. The smallest steadiness is in the subtropical ridge. In January, the ITCZ is a zone of low steadiness wedged between the high steadiness of the trades in both hemispheres. In July, only the gradient of steadiness indicates the position of the ITCZ outside the map. In the zone of westerlies, the steadiness decreases from a peak in middle latitudes to a minimum in the low pressure zone over the Antarctic Ocean.

Corresponding values of steadiness appear in the monthly averages of the selected areas. While small values prevail off the east coast of South America—8% in area N in April, 10% in area Q in February—the steadiness of the trade winds exceeds 90%.

The lowest annual means are 21% in area Q, 23% in area N, and the highest ones are 96% in area F, 94% in area D, 93% in areas G and J, and 92% in area B.

Wind roses

A more detailed picture of the mean wind conditions is given by the wind roses in Charts 13 and 14 in 10°-squares for January and July. The arms of the roses indicate the mean relative frequencies of eight different wind directions divided into four different groups of wind force. From the centre outward they are Beaufort 1−3, 4−5, 6−7, and ≥ 8. The number within the circle is the relative frequency of calms and variable winds in percent of all wind observations in the square.

High steadiness is implied when one wind direction dominates the wind rose and this direction then represents the mean wind direction shown in Charts 11 and 12. On the other hand, low steadiness is indicated by similar frequencies in all directions. This difference between the two types of distribution is evident when the trade-wind region and middle latitudes are compared.

In parts of the west wind belt, the west direction, which appears in the resultant wind, is not always recognizable as the prevailing wind direction but emerges only when all west components are combined and offset the mean east wind component.

South of 40°S, in July already south of 30°S, the proportion of gales, plotted black at the end of the wind-rose arms, is very evident. The irregularities of the wind roses in the southern region are owing to the uncertainty that stems from the scarce observations (WASHINGTON, 1957).

Strong winds and gales

Frequencies of strong wind ($\geq$ 6 Beaufort, $\geq$ 11 m s^{-1}) and gales ($\geq$ 8 Beaufort, $\geq$ 17 m s^{-1}) are of special interest. They are marked in the wind roses but are also shown separately in Charts 15 and 16 for January and July with some details along the coasts that are important for shipping. The solid thin lines give the frequencies of strong wind, the broken lines are the frequencies of gale. The analyses show similarities to the isotachs of mean wind speed in Charts 8 and 9. In the southern winter, strong winds and gales are more frequent than in summer and the highest values are in the west wind belt in both seasons. The northern boundary of 5% gale frequency and 20% strong-wind frequency is situated at 40°S in summer, at 30 – 35°S in winter. In the trade wind zone, the 10% line of strong winds covers an extensive area in winter, whereas in summer it is limited to the region off southern Africa. Gales are rare in the trades with frequencies of 1% reached only south of 20°S. The coasts are mostly sheltered from strong winds, especially that of Argentina.

South of 50°S, the information is doubtful, and thus the value of the peak of gale frequency and the location of this peak (near 50°S, cf. WASHINGTON, 1978) cannot be determined with certainty. At the southern edge of the chart, the gale frequency probably decreases toward the south. However, weather bias contained in ship observations (p. 7) and the orographic reduction of the wind at land-based climate stations may give the impression of a stronger southward decrease of wind than is real.

Among the selected areas, the highest frequency of strong wind of 19% is in area L in the southern trade-wind region. Less than 10% appear in all tropical areas off the coasts of Brazil and Angola. In areas P and R at the border of the west wind belt, the annual mean of gale frequencies exceeds 2%, whereas north of 30°S 1% is not reached except in area I. In the annual march, the maximum is reached in August.

Oceanic elements

Because water can move, the exchange of momentum between atmosphere and ocean produces both periodic and translatory motions of water at the surface (DAVIS, 1977). The periodic motions are travelling waves caused by a balance between a forcing power and a reacting force. The wind stress effects the forcing and surface tension of water and gravity react. Capillary waves arise from the balance between wind stress and surface tension. They are rapidly running ripples that may be used in the estimate of wind speed in Beaufort numbers as they quickly develop and vanish. They contain only a small portion of the energy of the sea and are therefore neglected in the description of the state of the sea.

Waves generated by the local winds (sea) are in balance between the wind stress and gravity. Heavy sea hinders shipping; it reduces the speed of vessels and their dynamic stability in the sea. Wave predictions help ships to choose the most advantageous route for reaching their destination rapidly and safely (KRUHL, 1974).

The translatory movements develop into wind-forced currents. They are in balance between the Coriolis force and the horizontal density gradient of surface water which depends on temperature and salinity (p. 27f). Divergences and convergences in the current field cause vertical motions in the ocean which affect deeper levels so that currents also develop there. Because the ocean currents transport water of different temperature, they contribute to the global heat exchange and influence climate zones.

Depending on the wind direction, drift or stemming of water appears along the coasts. Drift causes upwelling of cold water and stemming raises the sea level.

Sea and swell

The state of the sea is gradually built up by the wind as wind waves till it is said to be mature, which requires from several hours to days according to wind force and fetch. Capillary waves of small height develop at first, then gravity waves of increasing height and length. Within this process, energy is transferred from shorter- to longer-period waves by nonlinear interactions. This mechanism of energy transfer is not yet explained in detail, and experiments are being performed to evaluate that mechanism in order to compute predictions of waves.

Mature conditions are possible only over the open ocean and with a steady wind. The optimum condition for the development of the sea is a wind field that moves in the direction of the waves and with their group velocity so that the sea is steadily incited. The sea is mature when the propagation speed of the waves coincides with the wind speed. Then the exchange of momentum is in balance with the dissipation of the waves. The sea becomes swell when the waves leave the wind field or the wind speed decreases. Like the process of maturing, the dissipation of the sea proceeds slowly. Swell may amount to a considerable portion of the sea state and may travel great distances across the ocean. It increases the mean wave height and moderates the differences between quiet and rough regions.

The sea state is reported as the *direction* from which the waves come and their height and period. *Wave height* is the difference between crest and trough, and *wave period* is the interval between the passage of two successive wave crests.

These elements have been routinely observed on board for only about the last 20 years. Grades of sea state were estimated previously on a qualitative scale of 10 degrees, and since these grades can be compared only roughly with modern wave data, relatively few observations are available for reliable wave statistics.

The processes of exchange and maturing cause a wide spectrum of different wave heights and periods, and therefore a "significant wave height" is estimated on board. This term relates to the one-third of highest waves of a given group and is defined by the average of their heights and periods (WALDEN, 1958). A "significant wave period" is analogously defined.

Single waves can exceed the significant wave height by 60% within a wave spectrum; such exceptionally high waves are called outsize waves. Several superposing wave trains cause interferences, and the wave heights of such a cross-sea may for a short time exceed the significant wave height by 100%.

In Charts 17 and 18, isopleths of mean significant wave heights are drawn for January and July. The analysis is based on modern international wave height observations. An evaluation of historical data was published by HOGBEN and LAMB (1967), among others. The pattern is similar to that of the mean wind speed for the same month (Charts 8 and 9). The trade winds meet the requirements of maturing best because of their great steadiness. Thus, mean wave heights of 2 m are found there and waves as high as 7 m have been observed. A detailed analysis of wave heights over the Benguela Current (QUAYLE and ELMS, 1979) demonstrated an anomaly of strong sea in the zone between 25°S 10°E and 30°S 15°E, where the pressure gradient and wind speed are strongest and the wind direction steadiest in the stably stratified air over the cold current.

The stronger winds of the westerlies and the wide extent of the wind fields around cyclones and anticyclones produce mean wave heights of more than 3 m, higher in winter than in summer, except near South America where the lee effect of the Andes and the short fetch result in only low wave heights. In the region of the subtropical anticyclone, swell gives rise to mean wave heights of more than 1 m. Only near the shore are the wave heights lower.

The monthly averages of wave heights in the tables of selected areas generally range below 2 m. Values of 2 m are exceeded only in areas J, L, and O off southern Africa, in winter also in areas M and P at the northern border of the west wind belt. The lowest mean of 1.0 m appears in area C off the equatorial coast of Africa, the highest one with 3.2 m in area O off southwest Africa. A comprehensive presentation of frequencies of wave heights, periods, and directions in selected areas of the South Atlantic Ocean appears in the U.S. Navy Atlas (WASHINGTON, 1978).

A harmonic analysis of the monthly means shows an annual variation of significant wave heights with maximum in July – August. Both months also have the highest values among the monthly means of the selected areas. The largest amplitudes of the annual variation (0.5 m) are found in the central South Atlantic (at 32°S 15°W), but in the west wind belt, where the number of observations in winter is too small for a trustworthy analysis, still higher values might be possible, at least in the west.

The monthly standard deviation of wave-height observations varies between 0.5 and 1.5 m. North of 30°S and off the coast of Argentina, values below 1 m dominate; elsewhere they are above 1 m. That agrees with the standard deviation values of the selected areas. Mean monthly standard deviations of 1 m are exceeded in the trade-wind region off the African coast and south of 30°S. Waves with longer periods move faster and may thus leave a wind field as swell which runs the more rapidly and far the longer its period is, until it dissipates. Therefore, swell on the average is observed with a greater period than a wind wave. Since long-period waves propagate more rapidly than the wind field of a cyclone, the swell runs ahead of the cyclone as a warning sign.

The *wavelength* is the horizontal distance between two crests. It may be computed as the product of speed of propagation and period and is thus proportional to the square of the wave period. Swell is characterized by large wavelengths. Unfortunately, estimated values of wave periods are unreliable and statistical values based on them are even more doubtful.

The ratio of wave height to wavelength is called *steepness* of the wave and is of special importance. When energy is transferred into the wave in the process of maturing, its height increases by constant period and the wave becomes steeper; however, if the limit of 1.7 is exceeded, the wave breaks. A wave generally steepens when its orbital motion is deformed. Strong wind accelerates the upper part of the orbital motion and the wave steepens and breaks. At a wind speed of 4 m s^{-1} whitecaps appear, becoming more numerous the stronger the wind. At 7 Beaufort, the foam of breaking waves determines the appearance of the sea. The foam arranges itself in bands along the direction of the wind as the convergent part of a cylindrical motion of the water (LANGMUIR, 1938). At gale force spray blows from the wave crests.

Steepening of a wave also occurs when the lower part of the orbital motion is retarded. This happens near coasts where the wave contacts the bottom of the sea and surf is then generated (WALDEN and SCHÄFER, 1969). Strong surf can be produced by swell which runs from distant storms such as occurs at Ascension Island (p. 77), at St. Helena (p. 78), on the coast of Angola (p. 96), and on the coast of southwest Africa (p. 97).

A current that runs opposite to the waves also deforms the orbital motion. Then steep and high "freak waves" arise which are dangerous to shipping (MALLORY, 1974). Such a sea is observed off the south coast of Africa when a southwesterly gale blows over the southwestward setting Agulhas Current (p. 99).

There are wave motions at all boundary surfaces and not only at the sea surface. Internal waves build up in the deep between different, stably stratified water bodies, particularly in the thermocline beneath the upper mixed layer (p. 28). These waves sometimes reach much greater amplitude and period than the surface waves. With decreasing stability, the turbulent exchange between the water bodies increases at the expense of the internal waves.

Ocean currents

Whereas over land horizontal exchange between regions of different temperature is possible only within the atmosphere, in oceanic regions it takes place within the water too. The fact that water can move means that a density gradient at the sea surface will start exchange processes that may lead to horizontal circulations. The large-scale ocean currents originate mainly by exchange of momentum because the wind drives the surface water. Apart from gradient currents between different ocean regions and at river mouths, it is the wind that mainly supplies the necessary energy to generate and maintain large-scale currents. Because they transport heat in their meridional branches, the currents contribute to the planetary heat exchange processes (p. 10).

The earth's rotation deflects the water drift in the sense of the Ekman spiral so that a horizontal pressure gradient is produced which balances the current. In the Southern Hemisphere, the current on the average is deflected counterclockwise by 30° from the wind direction, depending on the latitude, the variability and strength of the wind, and on the sea state.

The continuous interactions of wind, sea, and current result in corresponding variations of ocean currents which only in the climatological average exist as steady large circulations. In single instances, small-scale current patterns are observed, such as splits, meanders, and cut-offs.

Where currents meet coast, they are deflected. Therefore, closed circulations of surface water develop in an ocean framed by continents. Fig. 2 illustrates how nearly zonal wind-driven currents turn meridionally near the coasts, so that in ideal cases they combine to form closed circulations.

The large ocean currents of the South Atlantic are illustrated in Chart 19 which shows contours of equal velocity in nautical miles per hour, the prevailing directions, and the names of the currents.

Forced by the trades, the Benguela Current (*E* in Fig. 2) flows parallel to the west coast of Africa. It is a cold current setting northwestward. At lower latitudes, it merges with the South Equatorial Current (*B*) which sets westward across the South Atlantic Ocean. The configuration of the Brazilian coast causes the main part of this current to be deflected along the north coast of Brazil across the Equator into the North Atlantic where it continues as the North Equatorial Current.

The position of the ITCZ north of the Equator is associated with the fact that the South Equatorial Current driven by the southeast trades crosses the Equator, especially in the southern winter, and then turns eastward and flows into the Gulf of Guinea as the Guinea Current (*A*).

Only the weaker part of the South Equatorial Current remains in the South Atlantic Ocean. It flows southward as a warm current off the east coast of Brazil (*D*). This Brazil Current meets the cold Falkland Current (F) at about 40°S; the latter sets northward east of Cape Horn. Both currents turn eastward and form the northern boundary of the West Wind Drift or Antarctic Circumpolar Current (*C*). This current commonly sets east-northeastward as forced by the prevailing west winds in that zone and is part of the current system flowing around the globe at 15−20 cm s^{-1} between 40° and 60°S. It enters the Atlantic through the Drake Strait where the Falkland Current splits off toward the north.

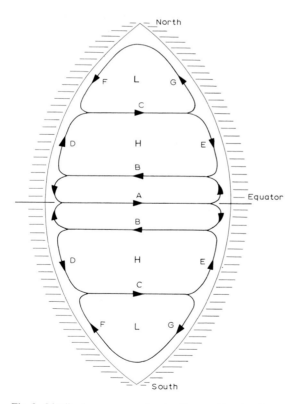

Fig. 2. Idealized ocean currents, *A − G*, caused by the planetary wind field in an ocean bordered by continents.

South of Africa the West Wind Drift meets the warm Agulhas Current which flows southwestward from the Indian Ocean along the African east coast. One part of the two currents turns northward into the Benguela Current, the other bends southeastward into the Indian Ocean and continues within the West Wind Drift.

Thus, an anticyclonic gyre flows around the subtropical high (*H*), but, in contrast, the cyclonic gyre (*L*) is only weakly developed in the South Atlantic Ocean. A coastal current sets westward along Antarctica; off the coast of the Antarctic Peninsula it turns northward from the Weddell Sea (*F*) to deflect the West Wind Drift northeastward in the region north of the South Orkneys (see WASHINGTON, 1957).

This picture of the currents in the South Atlantic Ocean is only valid as a climatological average since the variability of the wind field produces irregular current patterns on different scales.

The rivers from the surrounding continents also feed the currents, such as the Congo the Benguela Current and the Amazon the Equatorial Current. The Uruguay and the Paraná flow through the Rio de la Plata into the Atlantic near the place where the Brazil and Falkland currents meet and turn eastward.

Weaker tidal streams are superposed on the currents and move zonally in the open sea. They are deflected at the coasts where they are modified by rivers which intensify ebb currents and weaken flood currents.

The Benguela Current is stronger than the Canary Current (*E*), the Brazil Current is much weaker than the Gulf Stream (*D*), and the West Wind Drift (*C*) has a northward component in both the South and North Atlantic Ocean. All flow in the South Atlantic is thus enhanced toward the Equator.

While the West Wind Drift of the North Atlantic is principally supplied by the warm Gulf Stream and is relatively weak during the summer, cold water from antarctic regions prevails in the South Atlantic West Wind Drift. The surface water of the South Atlantic Ocean therefore is on the average colder than that of the North Atlantic (p. 33).

The circumpolar current participates only to a comparatively small extent in the circulation of water in the South Atlantic Ocean. Its peculiarity is to be the only zonal current that is continuous around the earth and that connects all oceans.

A comparison between the current and the wind charts shows that the current speeds amount to about 5% of the mean wind speeds (Charts 8 and 9) and the current directions are similar to the mean wind directions (Charts 11 and 12), considering that the currents are deflected counterclockwise from the wind direction by the earth's rotation.

In southern latitudes, the currents also carry icebergs. They are mostly tabular bergs with 30−50 m steep walls and nearly flat tops, they are formed by calving of the antarctic ice shelf, and they drift east-northeastward with the West Wind Drift. As they often are rather big (STRÜBING, 1978), they are able to undertake long voyages to latitudes as far north as 40°S, sometimes even to 35°S before they melt. The farthest northward extent is achieved in November−January and the least northward extent is reached in March−April (see WASHINGTON, 1967).

Subsurface conditions

The different water bodies are stratified according to their *density* which increases with falling temperature and growing salinity. Evaporation raises the salinity (p. 54) and precipitation lowers it (p. 49), but temperature is the most important factor in determining the density.

The heating of the sea surface (p. 11) forms a warm *upper mixed layer* which is 20 – 200 m thick depending on latitude and season. It is limited below by the thermocline where the temperature decreases strongly with depth, whereby a very stable stratification is set up.

Heating of the sea surface intensifies the stability of the upper mixed layer and thus diminishes the vertical water exchange in the ocean but strengthens the exchange with the atmosphere. In contrast, because cooling makes the surface water heavier and the stratification more unstable, it increases vertical exchange (GATES, 1977) such that cold water sinks and starts a circulation of the deeper water within the ocean, notably in the Antarctic Ocean.

In an ocean, *convergences* occur when the currents flow together and *divergences* when they flow apart. They produce vertical circulations that entrain water of deeper levels into the currents. In convergence zones, water sinks inducing compensation currents at deeper levels. In divergence zones, water wells up from deeper levels to the sea surface and participates in its currents.

A divergence is caused by the off-shore winds carrying water from the coast seaward. This water can be replaced only by water from below (DIETRICH, 1972). Pronounced zones of *upwelling* in the South Atlantic exist only in the trade-wind regions near Southwest Africa where the Benguela Current originates. The upwelling extends from Punta Albina (15°S) to the Cape of Good Hope. Another upwelling area is in a coastal region of South America (Cabo Frio) where the Brazil Current has an off-shore component. The Guinea Current produces weak upwelling near Accra and Lomé.

In summer, marked upwelling develops along the Equator where the South Equatorial Current crosses into the Northern Hemisphere and is anticyclonically deflected which causes a divergent flow. At the southern border of the West Wind Drift, in connection with the low-pressure trough over the Antarctic Ocean, upwelling is produced which feeds the circumpolar current.

Where the thermocline is marked, the upwelling water from below that discontinuity introduces a strong cooling of the surface. This is especially so in the upwelling areas of the trades. Along the African coast, water wells up from a depth of about 200 m and it is about 15°C colder at the surface than normal for the latitude. Upwelling water therefore presents an intense source of cold which strongly influences the climate of the region (HÖFLICH, 1972). The equatorial upwelling produces a cooling of the surface that sharpens the horizontal temperature gradient toward the ITCZ (p. 31f).

Upwelling is climatologically less important at higher latitudes because the weaker vertical temperature gradient of the surface layers does not lead to such extreme climate modification, but the less stable stratification of these water bodies facilitates vertical current components over greater depths.

Fig. 3 is a meridional cross-section illustrating the different *water bodies* of the South Atlantic Ocean with their vertical stratification and convergences and divergences at the surface (PEPPER, 1954; GORDON and GOLDBERG, 1970). The current arrows are only climatological averages of meridional current components.

While the surface currents are mostly wind-driven, the slower circulations of water bodies in the deeper ocean are caused by temperature and salinity differences through

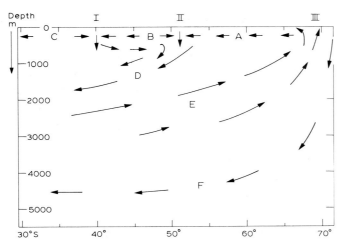

Fig. 3. Meridional section of ocean currents in the South Atlantic. *I* = Subtropical Convergence; *II* = Polar Front; *III* = Antarctic Divergence; *A* = Antarctic Surface Water; *B* = Subantarctic Surface Water; *C* = Subtropical Water; *D* = South Atlantic Intermediate Water; *E* = Circumpolar Deep Water; *F* = Antarctic Bottom Water.

which available potential and internal energy is changed into kinetic energy (MONIN et al., 1977). The heaviest water is the cold saline Antarctic Bottom Water (*F* in Fig. 3). It is generated at the border of the continental shelf, principally in the Weddell Sea (DEACON, 1934; GORDON, 1975), where it sinks and flows slowly northward at the bottom of the ocean. Above the bottom water, somewhat lighter North Atlantic Deep Water spreads from the north. It originates in the Arctic Ocean and flows southward until it mixes with the Circumpolar Deep Water (*E*). Rising in the Antarctic Divergence (*III*), this water mass is then separated into less saline Antarctic Surface Water (*A*) and into the more saline water that flows south and sinks along the continental slope to form Antarctic Bottom Water.

The lightest water is the Subantarctic Surface Water (*B*) in the west wind zone because it loses salinity by dilution from precipitation and melting icebergs. On its southern border at 50 — 55°S, but farther south near Cape Horn, the Antarctic Convergence (*II*) divides the Subantarctic from the colder Antarctic Surface Water. The latter flows north from its area of origin and sinks at the Antarctic Convergence (Oceanic Polar Front) where the West Wind Drift reaches its highest speed (KORT, 1968). On the average, the polar front zone of the atmosphere is situated there (p. 118f).

The Subtropical Convergence (*I*) lies at the northern boundary of the Subantarctic Surface Water at about 40°S. This is the northern border of the West Wind Drift and near the northern limit of icebergs. In the Subtropical Convergence, the warm saline Subtropical Water (*C*) slides under the Subantarctic Surface Water and forms the South Atlantic Intermediate Water (*D*), together with the descending Antarctic Surface Water. This intermediate water appears at the surface in the upwelling areas, and in the surface currents it is heated as in the Benguela Current on its path to a convergence. This happens, for example, when it is transported by the South Equatorial Current into the equatorial convergence where it sinks again as intermediate water. Opposing currents at the surface and in the intermediate water make up a closed circulation.

Exchange processes slowly lead to a balance between bottom, deep, and intermediate water. Convergences and divergences result in vertical circulations between the intermediate water and the surface layer, and small-scale variations of the currents contribute to the exchange between different water bodies.

As the South Equatorial Current carries warm water into the North Atlantic, there is a mean meridional current component from the Antarctic Ocean northward through the South Atlantic Ocean. The *water balance* of the surface currents (p. 55) shows, however, that even if the flow in this meridional component is fed by the West Wind Drift and, in particular by the Falkland Current, it is not compensated by a horizontal transport from the Pacific Ocean, but rather by a vertical circulation in which intermediate water flows southward (KORT, 1972). In the West Wind Drift, especially in the Antarctic Divergence, and within the upwelling water of the trade-wind region, as off the African coast, that cold intermediate water reaches the surface and enters into the circulation of the surface currents.

As indicated in Fig. 3 for the South Atlantic, there is a meridional circulation in the southern oceans in which surface water streams equatorward. It transports a mean water amount of $43 \cdot 10^6$ m^3 s^{-1}. The intermediate water flows poleward at the rate of $77 \cdot 10^6$ m^3 s^{-1} while the bottom water moves equatorward at $34 \cdot 10^6$ m^3 s^{-1}. The mean speed of these currents is 1 mm s^{-1} (DEACON, 1977).

Temperature

Air temperature at deck height and water temperature at the sea surface are measured routinely on board merchant ships. The difference between the air and water temperature is a measure of the stratification of the atmosphere above the sea surface and therefore a parameter for the heat exchange between the ocean and atmosphere. In synoptic meteorology, it is used to classify air masses.

Sea surface temperature

Since the beginning of systematic observations, the water temperature at the sea surface has been measured aboard ships with bucket thermometers at a depth of half a metre. Only recently have mounted recorders been used whose temperature sensors are placed somewhat deeper (WMO, 1954; JAMES and FOX, 1972). Radiation measurements of the temperature in the uppermost surface layer are not considered in this chapter (TAUBER, 1969).

Charts 20 and 21 show the distribution of sea surface temperature for January and July. The isotherms are labeled in °C (see TALJAARD, 1972a; ALEXANDER and MOBLEY, 1976; WASHINGTON, 1978; for the tropical Atlantic: MAZEIKA, 1968; LAMB, 1977). In the climatological mean, the temperature of the sea surface represents a balance between the net radiation at the sea surface (p. 64f), the vertical heat exchange with the atmosphere (p. 59f), and the advective heat exchange in the currents (p. 25).

South of 30°S, the average meridional temperature gradient amounts to 0.9°C/deg.lat. in summer and 0.7°C/deg.lat. in winter away from the coastlines where it is disturbed by continental influences and currents. The position of the ITCZ determines the temperature pattern in the tropics. The highest sea surface temperatures are measured there (HÖFLICH, 1974; LAMB, 1977) and the lowest ones at the ice edge (ALEXANDER and MOBLEY, 1976).

South of Africa, the temperature in the warm Agulhas Current (see Chart 19) is 21°C in summer and 18°C in winter. The presence of this current displaces the beginning of the steep meridional temperature drop to south of 35°S. On the other hand, cold antarctic water extends exceptionally far northward at this longitude (0°C at 55°S) because of the zonal asymmetry of Antarctica, and therefore the strongest mean temperature gradient of 1.1°C/deg.lat. is located south of Africa. An eastward convergence of the isobars of sea level pressure is associated with this steep temperature gradient, with the strongest pressure gradient (p. 15) and the strongest westerly winds (Charts 4, 5, 8, 9) over the eastern parts of the ocean.

North of the Agulhas Current, upwelling appears along the African south coast in summer with northeasterly winds. Off the west coast of Africa, the cold Benguela Current reduces the sea temperature by about 4°C below the latitude mean. Within the current, surface water drifts away from the coast and its place is taken by water from about a depth of 200 m (p. 28). This upwelling water further decreases the water temperature at the surface, in the climatological average by about 5°C in summer and 3°C in winter. It extends along the coast from the Cape of Good Hope to Punta Albina (15°S), where a strong temperature gradient (1°C/deg.lat.) separates the cold upwelling water from the warm tropical water, notably in summer when the ITCZ lies near the Equator.

Fig. 4 shows mean temperatures along the African coast between 10° and 35°S, based

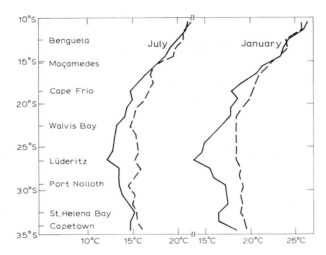

Fig. 4. Mean sea surface temperature (°C) in the upwelling area off southwest Africa: near the coast (solid line), 200 nautical miles away (broken line), January and July (HAMBURG, 1971).

on ship observations in January and July. The solid line refers to data directly at the coast, the broken line to data at a distance of 200 nm from the coast (HÖFLICH, 1972). The cooling by upwelling is most marked at the coast and the sea becomes warmer westward.

The effect of the cold Benguela Current reaches to the Equator where in January the relatively cool water meets the warm water of the ITCZ characterized by temperatures of 27−29°C from the Gulf of Guinea to the north coast of Brazil. In July, the ITCZ lies outside the range of the chart (HÖFLICH, 1974). On the Equator, a tongue of cold water with temperatures as low as 23°C marks the local upwelling in July. It is caused by a divergence of the South Equatorial Current as it passes the Equator and is clearly separated from the cold water zone of the Benguela Current (FLOHN, 1972).

The Brazil Current extends a tongue of warm water to about 45°S east of South America. It is more extensive in summer, but more evident in winter by the contrast with the colder water along the coast. From the south, along the eastern border of the Patagonian Continental Shelf, the cold water of the Falkland Current extends toward the Brazil Current. The two currents strengthen the temperature contrasts off the Rio de la Plata (farther north in winter than in summer) where in the mean the subtropical frontal zone is found in the atmosphere (p. 21f) and the subtropical convergence in the ocean (p. 29). Thus, the ocean currents show themselves in the sea surface temperatures which also means that the deviations of temperature from the zonal average are essentially determined by currents and vertical circulations. This is so not only for the large-scale climatological mean but also for small scales where the main currents disintegrate into smaller current patterns. These are recognizable in satellite pictures by the distribution of sea surface temperatures (LEGECKIS, 1978).

The zonal differences of sea temperature in the tropics (MAZEIKA, 1968) are associated with a vertical zonal circulation in the atmosphere first detected in the tropical Pacific (Walker circulation, p. 115; BJERKNES, 1969). It interacts with the Hadley circulation (BJERKNES, 1966; FLOHN, 1972). This circulation cell consists of ascending motion in the west (Amazon region) and subsidence in the east over the upwelling water, and of westward flow at lower levels and eastward flow at higher levels in the troposphere.

Knowledge of areas of strong gradients of sea surface temperature is of special importance for the climate of ship holds (MCDONALD, 1957) because damaging sweat may form in the cargo holds. As the position of the transition areas and the intensity of the gradients exhibit noticeable fluctuations, it is necessary to watch the temperatures carefully in order to start appropriate action to protect the cargo.

Knowledge of ocean temperatures is likewise essential for fishing because different species of fish prefer distinctly different sea temperatures (MONIN et al., 1977). The oxygen content, necessary for life in water, depends on the vertical temperature gradient. A stable stratification, especially the thermocline beneath the warm upper mixed layer (p. 28), hinders the exchange of oxygen and with that the sustaining of life in the water. In contrast, the upwelling regions are distinguished by an abundance of fish (STEINBACH, 1968).

The Antarctic Convergence outlines another temperature discontinuity (Fig. 3). In the mean it lies south of 50°S and represents the northern limit of cold antarctic water of low salinity produced by melting ice (p. 29). At the convergence, the temperature difference is 2−4°C. Owing to the small number of ship observations in the southern part of the South Atlantic outside the summer season, the pattern of mean sea surface temperature is not known with certainty and the climatological analyses are only approximations. For the summer months, VOWINCKEL (1957) has made a more accurate analysis of sea temperatures in the Antarctic Ocean.

The different mean sea temperatures of January and July indicate the annual range. A harmonic analysis of the monthly means gives the largest amplitudes of the first harmonic (5°C) near Argentina at Cabo Corrientes. Along the coast of Angola and in the central South Atlantic at 30°S 15°W, an amplitude of 3.5°C is reached with the extremes from the middle to the end of February and August. Otherwise, the maximum of the annual variations appears in March. At Cabo de Sao Roque, amplitudes of only 1°C are observed with maximum at the end of March. The charts of sea temperatures for January and July thus do not show the extremes of the year.

In the selected areas, the annual variations of monthly averages confirm this pattern—the highest means are in February and March, the lowest in August and September. On the average, their difference (double amplitude) amounts to 5°C and varies between low values of lower latitudes (2°C in area A, 3°C in areas D and E) and high values in middle latitudes (8°C in area R and 9°C in area Q over the Patagonian Shelf). In the westerly belt, the annual range reaches another low of less than 2°C (VAN LOON, 1972).

Compared with the North Atlantic, the sea temperatures of the South Atlantic Ocean are low. The annual averages are as much as 7°C lower at 60°S than at 60°N. Generally, the difference between the two oceans is about 4°C without a noteworthy seasonal course. The reason is the northward-directed component of the Atlantic currents that carry warm water across the Equator to the North Atlantic Ocean (p. 27).

Normally, the sea surface temperature has no essential daily variation because the strong exchange within the upper mixed layer and the large storage of heat equalize the daily variation of insolation (p. 11). Only in the upwelling area along the African coast does the sea temperature show an average daily variation of 2°C, since the alternation of sea and land breezes causes corresponding variations in the strength of the upwelling.

The annual means of the monthly standard deviations of sea temperature in the selected areas are given in the tables; they are computed from all variations within a month in all years. The highest monthly standard deviations are above 1.5°C and are found in middle latitudes and along both coasts. Otherwise, they amount to only about 1°C. The maximum is 2.5°C near San Antonio where on the average the frontal zone between subtropical and temperate latitudes is found. In a normal distribution, the mean monthly maximum is obtained by adding twice the standard deviation to the average, and the mean monthly minimum is found by subtracting twice the standard deviation from the average. The absolute extremes are computed in the same way using three times the standard deviation. These computed extremes may be better values than the observed extremes because of the limits set by quality control.

Sea surface temperatures have a spatial as well as a temporal scatter because the temperature field contains small-scale moving features, as revealed by aeroplane and satellite measurements. This variability is caused by currents, turbulence, wind, and waves as well as by convection within the ocean. Increasing wind speed, cooler air, precipitation, and upwelling lower the sea surface temperature. Cells of cooler water with a lower salt content generated by precipitation may be preserved in density balance with their environment (OSTAPOFF et al., 1973).

Because measuring methods have remained fairly homogeneous from the beginning of systematic observations until the advent of modern methods, and because water temperature values have comparatively small standard deviations, the ocean is suited for investigations of climate variations on a time scale of several years or decades. Long-term anomalies in the ocean are presumably reflected in the atmosphere (MONIN et al., 1977; HASSELMANN, 1977). Therefore, decadal averages of sea surface temperature were computed from which, unfortunately, the decade of 1940–1949 had to be omitted due to the scarcity of observations during the war. Seven such decadal averages are shown in Chart 22 for 20°-squares and for the whole South Atlantic Ocean as deviations from the mean value of each square for the total period.

The temperatures were generally lower before 1930 with a minimum in the decade of 1920 − 1929 at higher latitudes and 1900 − 1909 at lower latitudes. After 1930 they were higher with a maximum in the decade of 1950 − 1959 at higher latitudes and of 1930 − 1939 at lower latitudes. In the last decade, a weak decrease is apparent in the western part of the ocean, while in the east a further increase took place (see also DAMON and KUNEN, 1976). In some areas, the difference between decadal averages is more than 1°C (cf. MITCHELL, 1961).

For the whole of the South Atlantic Ocean, the sea surface temperatures were 0.25°C below the overall mean before 1930 and as much above it after 1930. The biggest change occurred between the decades of 1920 − 1929 and 1930 − 1939 with a mean rise of 0.6°C. The same computations made for 10°-squares show similar variations, which confirms the statistical validity of the results.

The penetration of insolation and turbulence in the upper layers generates a warm upper mixed layer which stores the absorbed heat. It reaches to the thermocline at a depth of 20 − 200 m, depending on location and season, beneath which the temperature decreases to about 2°C at the sea bottom (ISELIN, 1959). A detailed analysis of water temperature, salt content, and oxygen content to a depth of 3000 m was made by GORDON and GOLDBERG (1970) for the Antarctic Ocean south of 40°S.

Air temperature over the ocean

The air temperature over the ocean is measured on board ship at a height of approximately 10 m. While formerly the thermometer was placed in a screen aboard the ship, the sling-psychrometer has been used for about the last 40 years. This method is much better than the old one if it is applied on the windward side of the vessel (KUHLBRODT, 1936a). Therefore, historical and modern air temperature values are comparable only to a limited extent (KUHLBRODT, 1936b). In the tropics, a comparison showed that on the average the historical values were as much as 1°C higher than modern temperatures (HÖFLICH, 1974).

Over the ocean, the air temperature adjusts rapidly to the sea surface temperature. By the heat exchange between atmosphere and ocean, a 1°C change of air temperature affects the sea temperature by only 0.001°C (GATES, 1977). Accordingly, Charts 23 and 24, which present mean air temperatures over the South Atlantic Ocean, look like those of sea surface temperatures (Charts 20 and 21), which is also true for the monthly averages of the selected areas.

The highest monthly averages of 27°C occur in the ITCZ and in summer over the northern Brazil Current. The tables show values of 28°C in the tropical areas B and C in March and April, 27.5°C in area A in March and April, in area D in April, and in area E from February to April. The meridional temperature gradient in middle latitudes amounts to about 0.8°C/deg.lat. south of 30°S.

The relatively mild climate off the Argentine coast is obviously caused by the Andes, as shown also by the dryness of the air (p. 93). The warm Brazil Current, in contrast to the cold Benguela Current and the upwelling off Africa, causes higher air temperatures on the Brazil coast than those found along the African coast, and there is thus a mean zonal temperature difference of 3.5°C (in January 5°C, in July 2°C) between both coasts, largest with 7°C at 20°S. South America (see PROHASKA, 1976, fig. 7) south of 30−40°S is broad enough to produce a cool climate over the immediately adjacent ocean in winter, whereas south of Africa the warm Agulhas Current brings relatively high temperatures; in this region, the poleward-directed air temperature gradient is concentrated south of 40°S and reaches a maximum value of 1.0°C/deg.lat. on the average.

Fig. 5 shows the deviation of annual temperature averages along the South American coast (*W*) and along the African coast and the eastern border of the South Atlantic at

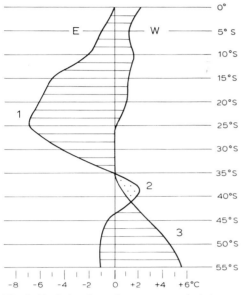

Fig. 5. Deviation from the mean annual air temperature (°C) at 20°W along South America (*W*) and along South Africa and 20°E (*E*). For further explanation see text.

20°E (*E*) from the average in the central South Atlantic at 20°W. Three regions stand out: (*1*) the cold upwelling water in the east (10−30°S), (*2*) the warm Agulhas Current in the east (40°S), and (*3*) the warm zone over the Patagonian Shelf in the west (45−55°S). The hatched area marks positive zonal temperature contrasts (*W* minus *E*) (zones *1* and *3*), the dotted area negative ones (zone *2*).

Chart 25 presents the annual variation of air temperature computed by harmonic analysis of monthly averages. The maxima of the amplitudes of the first harmonic are in the same positions as those of the sea temperatures. An amplitude above 4°C is found off South America between 30° and 50°S, and amplitudes above 3°C appear in the central South Atlantic at 30°S and off Angola. VAN LOON (1966) attributes the decreasing annual range between 30°S and 45−50°S to the effect of southward increasing cloudiness and mixing by winds and currents.

The highest air temperature within the year occurs generally in February, in lower latitudes in March, about half a month earlier than the maximum sea surface temperature. This is confirmed by the monthly averages of the selected areas. North of the Equator the change of seasons from one hemisphere to the other (0 amplitude in Chart 25) denotes the mean position of the ITCZ.

A noticeable mean daily range of air temperature over the ocean is only a product of radiation errors in the measurements, principally in the historical observations. Special measurements on research vessels with ventilated psychrometers reveal mean daily ranges of at most 0.5°C. They are caused by absorption of insolation in the humid atmosphere over the sea surface and they vary with region and season. At the coasts, the land may influence the daily range through advection but this effect is mitigated by on-shore winds.

The averages of monthly standard deviation of air temperature amount to less than 2°C north of 30°S and exceed 2°C in the south; the largest value of about 3°C is found near Cabo San Antonio, and the lowest one is 1.2°C in area A.

Averaged over all areas, the standard deviation of air temperature is about 0.3°C higher than that of the sea surface temperature. This difference is small off Africa, large near South America, in agreement with the difference of weather variability between the two coasts. There is a substantial difference between the size of the standard deviations over sea and over land. The storage of heat in the water causes a reduction of variability in the air temperature over the ocean, and not only the annual variations but also those on the time scale of weather change. The temperature of air, advected meridionally or from a continent, adapts rapidly to the sea surface temperature. In the U.S. Navy Atlas (WASHINGTON, 1978), cumulative relative frequencies of temperature values and their dependence on wind direction and speed are given for 40 selected areas in the South Atlantic.

Applying the temperature criteria of Köppen's climate classification (shown in Chart 3 and described on p. 12f) to the monthly averages of the selected areas, the areas can be assigned to the following climate regions: Areas A − F, H, I, and K have tropical climates (A,B); the remaining areas have a temperate climate (C). It is hot (h) in areas A − K, M, and N, mild (l) in L, O, P, and Q, and moderate (b) in R. The isothermal condition (i) is met in areas A, B, D, E, H, and I according to the monthly averages. However, the mean error (confidence interval) of monthly means stipulates a scatter of these means that is larger than their normal fluctuations and thus expresses too large a variability. Isothermal conditions may thus also be assumed in areas F, J, L, O, and P. On the other hand, Charts 23 and 24 give the impression of too small an annual range because the extremes do not appear in January and July. Chart 25 is a better indication of the range and according to that, areas with an amplitude of 3°C and more should not be termed isothermal.

The *air density* is computed from air pressure and virtual air temperature. Charts of mean air density over the South Atlantic Ocean in January and July have been published by BECKER (1944). The lowest values occur with 1.165 kg m^{-3} in January and 1.18 kg m^{-3} in July in the tropics. At 55°S, these values range about 0.1 kg m^{-3} higher. Off the African coast, they are about 0.02 kg m^{-3} higher than near South America, and the isopycnic lines therefore run east-northeastward across the ocean.

Air − sea temperature difference

The difference—air temperature minus sea surface temperature—is an important indicator of the stability of the air over an ocean and it determines the sensible heat flux into the atmosphere (p. 58). Positive values indicate a stable stratification which hinders the flux, and the atmosphere is then being cooled. Negative values indicate an unstable stratification that stimulates the heat flux from the ocean into the atmosphere by intensified turbulence.

Inasmuch as air—sea temperature differences are so important, climatological averages for January and July are presented in Charts 26 and 27. The isopleths are lines of equal difference between air and sea surface temperatures. Because the difference between two measurements is more susceptible to errors than the measurements themselves, the analyses are uncertain in some regions, especially at higher latitudes. Positive values dominate in January, negative ones in July. Averaged over the whole South Atlantic Ocean, the difference is $+0.1°C$ in January and $-1.0°C$ in July. This annual variation reflects the thermal inertia of water. The annual course of the difference can be seen in the monthly averages of the selected areas. The highest values appear in December and the lowest mostly in June. The mean differences are somewhat more than 1°C, reaching nearly 2°C in the southernmost areas Q and R. A harmonic analysis of the monthly mean values yields a mean amplitude of the annual variation of about 0.5°C, with maximum in December over most of the South Atlantic.

All through the year, positive air—sea temperature differences are found over the upwelling regions and over the cold currents, while the ITCZ and warm currents are characterized by negative values. In the annual mean, the air—sea temperature difference is $-0.5°C$ over the section of the South Atlantic covered in the chart. For comparison, a computation of this quantity using the North Atlantic weather ships and modern ship observations in the tropical North Atlantic gave a mean value of $-0.7°C$. This vertical temperature difference refers to a difference in height of measurement from about 10 m above to 0.5 m below the sea surface. A vertical temperature gradient is produced in the water between the surface and the depth of measurement by long-wave radiational cooling of the skin layer (p. 62) and by evaporation (p. 51f), whereas insolation (p. 60f) warms deeper layers as far as the optical depth of water. However, turbulence in the water, facilitated by instabilities and waves, tries to offset the radiational processes. The mean vertical temperature gradient in the surface water could be determined by a comparison with radiation measurements of the immediate surface (TAUBER, 1969) if these could achieve the necessary exactness.

The smaller air—sea temperature difference in the South Atlantic Ocean than in the North Atlantic is explained by the mean northward flow of relatively cold water through the South Atlantic which therefore is a weaker heat source than the North Atlantic.

The monthly standard deviation of air—sea temperature differences varies between 1°C at lower latitudes and 2.5°C at higher latitudes over the South Atlantic Ocean. Compared with the means, it is relatively large and random differences therefore tend to obscure the climatological ones which consequently can be estimated only with much uncertainty. This can be seen in both the mean distribution on the charts and in the annual variation of the monthly averages.

Humidity

The water vapour content of the atmosphere is measured on board with a sling-psychrometer containing dry and wet bulb thermometers. The humidity is determined from the difference between them by computation or by a table. This method has been in use for about 40 years and is successful provided that the measurement is made on the windward side, the thermometers are sheltered from spray, they are slung long enough, and the wick is kept wet. Previous methods of measuring the humidity, by hair hygrometer or in the screen aboard, are inferior (WÜST, 1950), and humidity measurements from that time are mostly absent. Historical observations are therefore only of limited use for climatological humidity data. The results presented here rely mainly on modern international ship observations. Over land, the humidity conditions were always better known (see world charts of vapour pressure and relative humidity by JOZSEF, 1938).

Humidity can be characterized by different measures. Vapour pressure is the partial pressure of water vapour in the atmosphere and is the most suited item for computing averages. The most illustrative measure is relative humidity, much used in climatology and defined as the ratio of the vapour pressure to saturation vapour pressure at a given air temperature.

The *dew point* can be determined unambiguously though not linearly from vapour pressure. It is the temperature at which the actual vapour pressure equals the saturation vapour pressure. Climatological charts of dew point temperature have been published by TALJAARD et al. (1969), among others. They look like those of vapour pressure and air temperature because the dew point depression is generally small over the ocean. In synoptic meteorology, the use of dew point is prescribed in the weather code.

The *specific humidity* remains constant in the vertical moisture flux and can therefore be used to describe the moisture flux in the atmosphere. It is thus a quantity suitable for computing evaporation (p. 51f) and latent heat flux (p. 57). It is the dimensionless ratio of the mass of water vapour to the total mass.

The *absolute humidity* gives the mass of water vapour in a unit volume of moist air. It is proportional to the vapour pressure, but also depends on the air temperature. This value is used in ship-holds meteorology to compute the moisture balance there.

The *equivalent temperature*, used to estimate human comfort, is computed from air temperature T (°C) and vapour pressure V (mbar) by the formula $E = T + 1.5\,V$ (°C).

Vapour pressure

The isopleths of vapour pressure in millibars for January and July over the South Atlantic Ocean are presented in Charts 28 and 29, and monthly averages for the selected areas and for the climate stations are found in the tables (p. 132ff). Like air temperature, vapour pressure is a function of latitude but, in contrast to the temperature, the gradient is stronger at lower latitudes and weaker at higher latitudes in agreement with the nonlinear relation between vapour pressure and dew point.

In January, the meridional vapour pressure gradient is about 0.6 mbar/deg.lat. south of 30°S and 0.3 mbar/deg.lat. in the tropics where, on the other hand, a zonal gradient of 0.2 mbar/deg.lat. exists, corresponding to a vapour pressure difference of 10 mbar between the two sides of the ocean at about 20°S. The highest values are observed near the Equator along the ITCZ off the coasts within the zone of high precipitation.

In July, the meridional gradient of vapour pressure, 0.4 mbar/deg.lat. hardly changes between the Equator and the southern border of the chart. The west−east difference in the tropics is weaker than in January, highest with 7 mbar (0.15 mbar/deg.lat.) at 20°S. A weak minimum at the Equator marks the upwelling there. Relatively high values of vapour pressure are observed off the north coast of Brazil, and on the Equator there is a zonal gradient of 0.15 mbar/deg.lat. west of 15°W.

According to the monthly averages of vapour pressure in the tables of selected areas, the annual maximum occurs in March north of 20°S and in February south of 20°S; the minima are generally in August so, like the sea temperature, the charts of January and July do not show the extreme conditions.

The highest monthly average of 30.7 mbar is observed in April in area C, the lowest one of 9.2 mbar in July in the southernmost area R. The annual variation between the highest and lowest monthly mean amounts to 6 mbar averaged over all areas. It is smaller in the northwestern and southeastern areas and bigger in the northeast and southwest.

The harmonic analysis of the annual variation of vapour pressure yields patterns similar to those of air temperature. The amplitude reaches 5 mbar at 30°S 25°W, and 4 mbar in the Bight of Angola as well as along the South American coast between Cabo Frio and Cabo Corrientes (areas K, N, and Q).

The annual mean of monthly standard deviations of vapour pressure ranges between 1 and 4 mbar as seen in Chart 30. It is small in the south where the vapour pressure is low and in the upwelling region off Africa; it is highest at 30°S off South America over the Brazil Current in the variable weather of the frontal zone.

In the selected areas, the annual means of monthly standard deviation vary between 2 and 4 mbar. It is lowest, 2.1 mbar, in area A at the Equator and in areas G, J, and L off Africa. The largest value, 4.0 mbar, appears in area K and 3.9 mbar is reached in area N.

Equivalent temperature

Human comfort is limited to a comparatively narrow zone of climatic conditions, bounded by critical temperatures and modified by humidity. The limits of comfort are often different from person to person and depend on acclimatization and stress. A sultriness limit of 19 mbar vapour pressure crosses the South Atlantic Ocean in the subtropics. In summer, it extends from Cabo Corrientes in the west to Cabo Frio in the east, in winter from São Paulo to Lobito in Angola. Using equivalent temperature as a measure of comfort and a value of 55°C as the upper tolerance limit, somewhat more northern borders are found over the ocean. In summer, this limit starts near Montevideo (35°S), turns a little southward off the coast over the Brazil Current, extends eastward across the central South Atlantic along 35°S, and then turns northward west of the cold upwelling water. It reaches the African coast near Moçamedes (15°S). In winter, it crosses the Brazilian coast north of Cabo Frio (20°S), but bulges southward over the warm Brazil Current to 30°S at 40°W. In the central South Atlantic, it extends north-eastward, reaching the African coast at Luanda (10°S).

Table I contains the highest and lowest monthly averages of equivalent temperature for the selected areas. The areas are arranged geographically, the western areas on the left, the areas of the central South Atlantic in the middle, and the eastern areas on the right. In winter, A, B, C, and E are in the zone of sultriness (>55°C) and D and I are at the boundary; F, H, K, M, and N are also in that zone in summer. A mean equivalent temperature of 70°C is exceeded in summer only in equatorial areas A to C, in winter nowhere.

TABLE I

MAXIMUM AND MINIMUM MONTHLY AVERAGES OF EQUIVALENT TEMPERATURES (°C)

Latitude (°S)	West			Centre			East		
	area	max.	min.	area	max.	min.	area	max.	min.
0									
	A	72	65	B	72	58	C	74	58
5									
10				D	71	56			
	E	70	60				F	62	46
15									
							G	53	38
20				H	60	44			
	I	68	54				J	52	39
25									
	K	66	48				L	49	36
30				M	60	39			
	N	58	38				O	47	34
35				P	46	32			
	Q	49	27						
40									
	R	40	23						
45									

The equivalent temperature drops below the comfort optimum of about 42°C in summer only in area R, but in winter this happens in G, J, and L to R. The annual mean equivalent temperature varies between 69°C in the tropical area A and 30°C in the southernmost area R. This value is regarded as the lower limit of comfort, and in summer it extends northeastward from the Magellan Strait and then eastward across the central ocean at 45°S. In winter, it lies about 5° farther north at 40°S in the centre of the ocean; it reaches the coast of Argentina at Buenos Aires and turns southward over the warm Agulhas Current to about 42°S.

Relative humidity

While the vapour pressure gives the absolute content of water vapour in the atmosphere, the relative humidity is a measure of the degree of saturation. Because there is always evaporation over the ocean, saturation can only be reduced by vertical exchange with dry air. Hence, the relative humidity over the ocean represents a balance between evaporation and vertical exchange in the atmosphere. This is a turbulent exchange proportional to the stability of the atmosphere. When the sea surface is warmer than the air above, the air is less stable, and conversely (p. 36f); relative humidity and air—sea temperature differences are therefore negatively correlated.

The correlation between relative humidity and air—sea temperature difference makes it possible to estimate humidity values from the temperature differences when accurate humidity measurements are not available, which is true of historical ship observations from the previous century. In this way, computations of evaporation and latent heat exchange are possible for climatological purposes (not for single observations) to estimate water and heat budgets over the ocean in former times.

Charts 31 and 32 demonstrate the distribution of relative humidity over the South Atlantic Ocean in January and July. The contours show equal relative humidity in percent. As noted (p. 38), not enough humidity data were available for a reliable analysis, especially in the central and southern parts of the ocean. The mean values of relative humidity vary between 70 and 90% and the annual means in the selected areas between 75 and 83%. Minima appear over the warm currents and in the subtropical anticyclone, and maxima are found in the upwelling regions and along the ITCZ as well as over the cold Falkland Current. Areas B and C at the Equator have the highest annual averages; D, F, and H in the trades have the lowest.

The inverse relationship between relative humidity and air—sea temperature difference is evident if Charts 31 and 32 are compared with Charts 26 and 27, except that the high averages of relative humidity in the ITCZ, about 83% over a warm sea, should be explained by convergence of the trades and moisture accumulation by evaporation of precipitation.

The harmonic analysis of the annual course of the relative humidity also agrees with that of the air—sea temperature difference, although with somewhat greater scatter of phase. The maximum of the annual component is generally found in December, except for the South American coast north of 10°S where it is in March to April, south of 30°S (August), and on the African coast north of 20°S (September). The biggest amplitude is 5% at 25°S 35°W, over the Brazil Current.

The monthly means of the selected areas show little difference over the year, on the average scarcely 5% and at the most 7% in areas F, I, M, and Q; the temporal position of the extremes is accordingly uncertain and variable.

The daily variation of relative humidity, considerable on land in agreement with the variation of temperature, is negligible over the sea. The decrease connected with a small warming of the air during the day is counteracted by an increase caused by a more stable air stratification since the sea surface temperature has no daily variation. Near the coast, the continental influence is seen in a daily variation of relative humidity when the wind blows off-shore.

The annual mean of monthly standard deviation of relative humidity is 10% south of 25°S, elsewhere it is less. That is confirmed in the selected areas where the annual means lie between 6 and 12%, increasing toward the pole.

The standard deviation is rather large, which reflects the uncertainty inherent in estimating the values from the psychrometer temperatures. Although a deviation of 10% results in a difference of only 1.2°C in the wet bulb temperature for an air temperature of 20°C and a relative humidity of 80%, and this corresponds to the standard deviation of the temperature values, the psychrometric temperature differences nevertheless should show a lower standard deviation than that of the temperature.

The *dew point depression* (VAN LOON, 1972, figs. 6.4 and 6.5) has a spatial distribution similar to that of relative humidity. It shows maxima off Southwest Africa at 30°S and off South America at 40°S in winter, at 30°S in summer. The lowest mean values are observed over the Antarctic Ocean.

Visible weather phenomena

Visible weather phenomena are such quantities as can be observed visually—cloudiness, visibility, fog, precipitation, and thunderstorms. They are described in the synoptic code and entered in ship logbooks. Before the code was introduced, they appeared in the logbooks under "remarks" or were denoted by certain letters as "Beaufort weather". Such entries were later converted into digits for data processing and numerical evaluation.

Apart from cloud amount, these weather phenomena are events whose distribution is given as mean frequencies; these frequencies are then the percentage of the ship observations in which the given phenomena are reported. The mean number of days in a month when a specified event occurred can only be given for land and island stations (p. 71), as no continuous observation series is available over the open ocean.

Cloudiness

Before the synoptic code was introduced, cloud cover was just described in the logbooks or indicated by a letter, or later given in tenths. Today the cover is coded in eighths and refers to the total cloud amount of all types as seen by the observer aboard. Clouds near the horizon are not considered because their bases are distorted by the perspective and thus only the cloudiness at the point of observing should be recorded.

On weather maps, the depiction of cloud cover in eighths is useful, but in climatology tenths of cloud amounts have the advantage that they can be interpreted as percentages and so can be immediately compared with the corresponding values of sunshine duration. Therefore, Charts 33 and 34 show the mean total cloud amount in percent of the sky covered for January and July (see FEAN, 1961; also VAN LOON, 1972, figs. 6.1 and 6.2). Frequencies of cloud in eighths and of low clouds alone are published in the U.S. Navy Atlas (WASHINGTON, 1978) for selected areas of the South Atlantic.

The highest amounts of cloud are observed in the west wind zone of middle latitudes, along the ITCZ, and over the cold Benguela Current. The lowest amounts appear over the warm currents, in the area of the subtropical anticyclone, and off some coastal districts. The annual means of cloud amount in the selected areas range from 47% in areas A and E off northeastern Brazil to 77% in area F in the trades.

An overall cloud amount of 62% was computed from all ship observations, averaged over the section covered by the map. Satellite data give lower values. Towering cumuli lead horizontally looking observers to report cloud amounts that are too high compared with satellites viewing the clouds vertically (BARRETT, 1974); and because of the angle most observers overestimate cloud amounts even though clouds near the horizon are not taken into account. Therefore, the value of the climatological cloudiness data depends on the observation method (SADLER et al., 1976).

The minimum of mean cloud amount in July at the Equator is noteworthy. It indicates the subsidence in the atmosphere over the upwelling water, caused by a divergence of the wind field when the southeasterly trades cross the Equator. In the other seasons, differences appear especially at the coasts.

Large cloud amounts are observed near Rio de Janeiro in January and in the trades over the Benguela Current at the latitude of Punta Albina. Minima occur off Brazil at 10°S and along the Argentinean coast, as well as along the coast of South Africa. Maxima are found off the coast of Angola in July and in the Gulf of Guinea, minima over the Brazil Current as far as Cabo Frio and along the African coast south of Swakop-

METEOSAT 1978 MONTH 4 DAY 13 TIME 1155 GMT (NORTH) CH. VIS 2
NOMINAL SCAN/PREPROCESSED SLOT 24 CATALOGUE 1001720100

Fig. 6. Cloud imagery of METEOSAT on 13 April 1978, 1155 GMT, visual range of 0.4 – 1.0 μm.

mund and near Luanda. A detailed analysis of cloudiness over the Antarctic Ocean in summer has been published by VOWINCKEL (1957) according to whom an area of heaviest cloud cover, about 90%, extends from 65°S 70°W toward 55°S 20°W and further. Synoptic cloud pictures are supplied daily by satellite. The impressive pictures from the stationary satellite METEOSAT in Fig. 6, the visible channel, and Fig. 7, the water vapour channel, show the distribution of clouds on 13 April 1978, 11h55 GMT. The cloud band of the ITCZ in the northern South Atlantic and the cloud field of the trades are immediately recognizable and the spiral cloud system of a cyclone is visible at higher latitudes. Scarcity of clouds marks subsidence in the atmosphere between these areas.

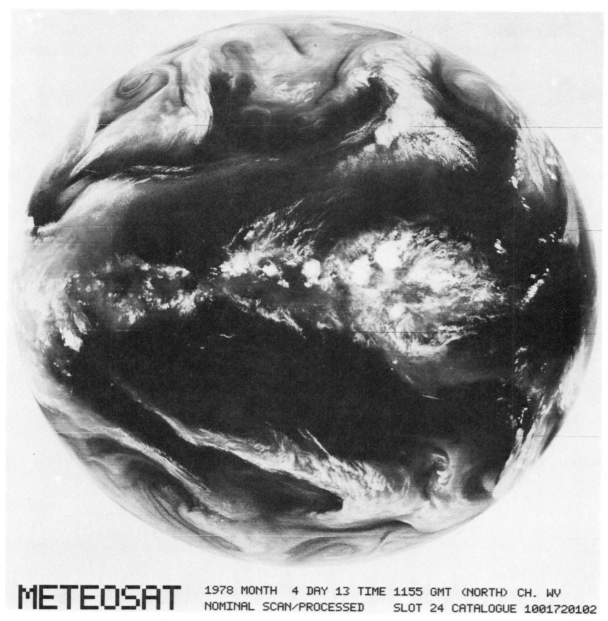

METEOSAT   1978 MONTH 4 DAY 13 TIME 1155 GMT (NORTH) CH. WV
            NOMINAL SCAN/PROCESSED   SLOT 24 CATALOGUE 1001720102

Fig. 7. Same as 6, but for the vapour channel at 0.5 μm.

The ITCZ and the frontal zone of middle latitudes are more distinct in the water vapour than in the visible channel because the high convective clouds of the ITCZ and the stratiform clouds of the frontal zone are accentuated by a high content of atmospheric water vapour (MORGAN, 1978). The flat clouds beneath the trade inversion are only recorded in the visible rànge (Fig. 6) and not in the vapour channel (Fig. 7) because of the low content of water vapour in the subtropical region of subsidence.

The monthly averages of selected areas show seasonal differences of cloudiness of more than 30% in area B at the Equator and F in the trades. The highest monthly averages in the annual course appear at the Equator in southern summer when the ITCZ is farthest south, in the subtropics mostly in September when the water temperature is near its minimum, and in the southern areas in winter when the frontal zone is farthest north and most pronounced.

This is confirmed by the harmonic analysis of monthly averages. An amplitude of 10% is exceeded only at 10°S between 20°W and 10°E with the maximum occurring in September—October, and at the Equator west of 30°W with maximum in March—April. The meridional differences in January and in July are shown by the latitudinal averages of cloudiness over the South Atlantic Ocean (in percent) in the following table:

	0°	5°	10°	15°	20°	25°	30°	35°	40°	45°	50°	55°S
January	57	51	50*	56	54	53*	53	60	68	75	82	*84*
July	42*	47	59	59	55	53*	58	67	74	80	*89*	81

Maxima are in italics, minima are marked by an asterisk.

In January, the different climate zones are well marked in the zonal mean cloud cover—strong cloudiness in the ITCZ near the Equator, minimum at 5—10°S, more clouds in the trades at 15°S, less cloudiness in the horse latitudes at 25—30°S, and southward increasing values in the west wind belt.

In July, the climate zones are shifted northward by about 5°; the low cloudiness at the Equator is pronounced.

The annual means of monthly standard deviations are more than 25% at the coasts, elsewhere mostly less. In the selected areas, the values lie between 26% in area F and 38% in areas Q and R; compared with the mean values, these standard deviation values are relatively large, as much as 60% of the mean.

Because the values of cloud cover are limited to between 0% and 100%, there can be no normal distribution of their frequencies. Rather, small and large cloud amounts are more frequent than the numbers between, which explains the large standard deviations in the coastal areas I, K, N, Q, and R along the South American coast as well as in L and O near Africa. The average is therefore not necessarily the most frequent value, as it would be if a normal distribution were assumed. It shows only a mean frequency from which no conclusion can be drawn about the distribution of the frequencies, and accordingly, the standard deviation does not have the importance it would have in a normal distribution and therefore is not shown on maps.

To illustrate the distribution of different amounts of cloudiness, three classes of cloud amounts were defined: (*1*) clear to fair, with at most 30% cloudiness (0—3 tenths or 0—2 eighths): (*2*) partly cloudy, with 35—65% (4—6 tenths or 3—5 eighths): and (*3*) cloudy to overcast with at least 70% (7—10 tenths or 6—8 eighths).

The most frequent clear to fair skies, occurring more than 40% of the time, are found off South Africa (area L), and in summer at the coast of northern Argentina (area Q), in winter at the Equator (area B) and near Cabo Frio. They are observed less than 20% of the time in the ITCZ, in the trades at 15°S (area F), and in the west wind belt.

Cloudy to overcast skies are found more than 60% of the time in the ITCZ and in the trades and more frequently than 80% in the west wind belt, in southern winter also in the Gulf of Guinea and in the trades between 10° and 15°S (area F). They are below 40% in the region of the subtropical anticyclone and off South Africa, below 20% in southern summer off northern Brazil, and in southern winter at the Equator (area B).

In comparison with those two categories, the frequency of partly cloudy sky is relatively small. If one averages over all 18 selected areas, the mean cloudiness is 57% with a standard deviation of 32%. However, 25% are in the first category, 28% are partly cloudy, and 43% are in the last category, so although the mean of 57% is in the partly cloudy category, this is obviously not the most frequent condition.

Cloudiness has a daily variation at the coasts where it is normally greater in the afternoon than at night. Cumuli predominate by day, stratiform clouds at night. Over the upwelling water, a cloud cover often forms at night and dissipates during the day outward from the land. In the tropical sea areas, convection is enhanced at night when the negative radiation balance at the upper cloud surface decreases the stability of the atmosphere (RUPRECHT and GRAY, 1976).

Visibility

The determination of visibility on the open sea is of necessity an estimate, as there are normally no fixed objects to define the visible range. Only the distinction between fog and good visibility is incontestable. The present-day definition of fog as visibility less than 1 km has not necessarily been adhered to in historical ship observations.

Outside the upwelling regions, fog over the South Atlantic Ocean is restricted to the west wind zone, where frequencies of 10−20% in summer and 5−10% in winter are found near the coast. Over the cold Falkland Current, fog occurs with frequencies of more than 5%. See also VOWINCKEL and OOSTHUIZEN (1953) on the occurrence of clouds, precipitation, and fog south of 50°S in summer.

Cold upwelling water easily causes fog (HÖFLICH, 1972). At the southwest African coast, fog frequencies of 3−5% are found. Here the fog forms mainly at night about 40−60 days of the year. In the morning, it lifts to become low stratus and during the daytime it is transformed into stratocumuli that migrate with the trade wind and account for the high cloud amount over the Benguela Current. The upwelling water on the Brazilian coast is also marked by some occurrence of fog. On the Argentinean coast in winter, continental fog moves with the wind toward the ocean.

Fog frequencies will not be shown on charts because the frequencies at middle latitudes are of doubtful accuracy, but fog frequencies of single months in the selected areas can be seen in the climatic tables. The regions in higher latitudes with frequent fog are not covered in the tables nor are those in the immediate vicinity of the coasts.

Apart from the regions mentioned, most areas are virtually free of fog, having annual frequencies below 1%. Fog frequencies of more than 1% are observed only at the South American coast south of 20°S, with as much as 3% in area Q. The month of greatest fog frequency here is mostly September when the water is coolest, with a peak of 9% in area N. At the African coast in areas G and L, the mean annual fog frequency is 0.8%; here May has the most fog with 2%.

Since the change between sea and land breeze causes a daily variation in the strength of the upwelling with a corresponding variation of sea surface temperature, fog frequency also acquires a daily variation with a maximum in the morning.

Fog indicates high relative humidity and frequent fog therefore points to a high average relative humidity. Charts 31 and 32 confirm this: the regions of high mean values of relative humidity are also favoured regions of fog, and the annual course with maximum in early summer is the same for humidity and fog.

The propensity for fog is obviously greater in higher than in lower latitudes (cf. Vo-WINCKEL and OOSTHUIZEN, 1953). In middle latitudes, fog occurs in the southward flow of warm air in front of cyclones where moist, warm air flows over cooler surface water and the dew point is higher than the water temperature. Fog caused by mixing of adjacent air masses and by evaporation into a cold air mass under stable conditions adds to the frequency. The wind force plays an important role in the frequency of fog over upwelling water because the intensity of the upwelling depends on it.

The climatic tables of selected areas also contain monthly frequencies of mist with visibility below 4 km, which occurs more often than fog—2% averaged over all areas. The distribution in space and time is similar to that of fog. Area N off Uruguay has the most frequent occurrence of mist with 20% in September.

In the Gulf of Guinea, the sand and dust carried by the Harmattan strongly reduce visibility in northern winter.

Spray and rain in the west wind zone and snow in the south reduce the visibility which is otherwise mostly good over the ocean, better at any rate than over continents where the wind raises dust and the atmosphere is polluted by human activities. In contrast with conditions over a continent, the frequencies of medium visibility are lower over the ocean.

Precipitation frequencies

Precipitation appears in the synoptic code under present weather. Before the code was introduced, the corresponding information consisted of remarks in the logbooks which were not always unambiguous or complete. Anyhow, light rain and spray often cannot be distinguished from each other under strong wind conditions. Precipitation amount (p. 49) is not measured aboard voluntary observing ships, so that only the frequency of precipitation can be compiled from ship observations.

The mean frequency of precipitation in whatever form is shown in Charts 35 and 36 for January and July. The averages south of 40°S are not reliable because of the scarce observations. All present weather with any kind of precipitation (in the synoptic code, ww $\geq$ 50) was counted. Somewhat higher frequencies would result if observations with precipitation within the last hour (ww = 20—27) were added (WASHINGTON, 1978).

Precipitation is most frequent in the ITCZ and in the west wind belt where the frequency exceeds 20%. The northern boundary of the precipitation in the westerlies extends in summer to 40°S, in winter to 35°S. North of that boundary, precipitation is infrequent, especially in the trades off the African coast where the minimum in July reaches to the Equator, but also along the Argentine coast in lee of the Andes (p. 93). Seasonally changing precipitation frequencies are marked farther north along the South American coast (p. 88f): there are autumn rains on the north coast of Brazil, summer rains near Cabo Frio, and winter rains at Cabo de Sao Roque and near Uruguay. Precipitation tends to be less intense in summer than in winter over the sea in these latitudes because in summer the atmosphere is more stable over the ocean and the frontal zone weaker.

Among the selected areas, P has most frequent rain with a 9% occurrence for the year. The highest monthly frequencies are in the west wind belt; area P has 12% in July and 11% in March and June, and at the American coast area N has 11% in September and area Q 10% in July and October. The lowest frequencies are observed in areas G, J, and L off Southwest Africa where the annual average is only 1%.

A harmonic analysis of the annual course of the frequency is shown in Chart 37. There are two regions with an amplitude of the first harmonic above 5%, namely, the west wind zone east of 50°W where the maximum is in July, and the equatorial zone west of 20°W with maximum in April. In area A, the range is between 12% in April and May and 1.5% in November and December; otherwise the annual range is weak.

A comparison of precipitation frequency and average cloud cover (Charts 33 and 34) shows great differences. If one defines the capability of producing precipitation as the fraction of mean cloudiness (in %) to precipitation frequency (in %), one obtains mean values of this fraction between 1% in areas G and J in the trades and 13% in area P at the northern boundary of the west wind belt. Values above 10% are also reached in area A at the Equator where the capability reaches a maximum in April and May, in area M, and off South America in the areas I, N, Q, and R. The low capability of the cloudiness in the trades is remarkable, but it is owing to the mostly flat stratocumulus clouds beneath the trade inversion that dominate the area (cf., satellite imagery, Fig. 6, p. 43).

The mean northern boundary of snowfall is at 40°S, and the extreme latitude is 35°S.

Thunderstorms

Thunderstorms are infrequent in the South Atlantic Ocean because the relatively cool surface water encourages a stable stratification. A frequency above 1% is observed in winter off South America and in the Gulf of Guinea where the frequency is about 2%, near Uruguay with about 5%, and in summer near Rio de Janeiro where it also reaches 5%. In the west wind belt, the number of ship observations is too small to yield reliable statistics.

The observed thunderstorm frequencies can be converted into the mean number of days with thunderstorms per month (WMO, 1956) as follows: less than 1 day per month in the trades and south of 40°S, 3−6 days over the Brazil Current, 5−10 days in January north of the Equator and on the tropical coasts. And the number of days per year: less than 1 south of 60°S, less than 5 between 50° and 60°S and in the trades, 20−70 over the Brazil Current, 10−100 in the ITCZ, increasing eastward, and a maximum of about 180 at Duala in the Gulf of Guinea.

Water and heat balance

To elucidate the connections between different climate elements and the different types of climates, the water and heat budget over the South Atlantic Ocean is discussed below. Estimates of the balances have been made by, among others, WÜST (1954), BUDYKO (1956, 1972), PRIVETT (1960), ALBRECHT (1961), LAEVASTU et al. (1969) as well as by BAUMGARTNER and REICHEL (1975). The balances presented here were computed by means of the values of climatological elements given in the previous section and are presented in Charts 38−53 and Figs. 8−10 and shown in the climatic tables for the selected areas.

Water balance

The water balance is the difference between evaporation and precipitation. The sign of the balance indicates if a climate is humid or arid. Regional deficits in the ocean are offset by runoff from the continents and by ocean currents. The water vapour transferred to the atmosphere from the ocean moves in the atmospheric circulation until it condenses to form clouds and to fall as precipitation. To compute the water balance, information is needed on precipitation amounts and the intensity of evaporation.

Precipitation amounts

Beginning on p. 47, the distribution of precipitation frequencies over the South Atlantic Ocean (Charts 35 and 36) based on ship observations is discussed. The amount of precipitation at sea is measured only on stationary vessels such as weather ships and light vessels. No useful method has been designed to measure precipitation routinely on moving ships (WMO, 1962). The movement of the vessel, the air circulation around it, wind, superstructure, and spray all influence the measurement. Hence, the precipitation amount over sea can only be estimated by methods that consider the measurements on the coasts and on islands (p. 70f) as well as the observed frequencies. In addition, attempts have been made to relate the type of precipitation as given in the synoptic code to the amount of precipitation (TUCKER, 1961). Because precipitation reduces the salt content of the surface water, measurement of the salt content makes it possible to deduce the precipitation amounts (OSTAPOFF et al., 1973).

Remote measurements with radar or from satellites open new possibilities which, however, still do not suffice for climatological statistics over the ocean (BARRETT, 1974). To estimate precipitation amounts from satellite observations, clouds are classified into categories of productiveness (MARTIN and SCHERER, 1973).

Estimates of precipitation amounts at sea were published by SUPAN·(1898), WÜST (1936), SCHOTT (1942), RIEHL (1954), ALBRECHT (1960), KNOCH (1961), DROZDOV (1964), JACOBS (1968), BAUMGARTNER and REICHEL (1975), and JAEGER (1976), among others. Jaeger even presented monthly charts, but such a detailed description is not warranted for the South Atlantic Ocean. The seasonal differences are relatively small and estimates of them are more problematic than estimates of the annual average.

Only annual amounts of precipitation are therefore presented in Chart 38 which shows the probable mean precipitation amount per year in centimetres. For precipitation bordering continents, see RATISBONA (1976, fig. 8) and PROHASKA (1976, fig. 12).

The largest amounts fall in the ITCZ north of the Equator, 300 cm in the eastern Gulf of Guinea and 200 cm off the equatorial coast of Brazil, and as much as 100−200 cm fall in a narrow strip along the coast of Brazil. More than 100 cm are estimated to fall in the central and eastern parts of the west wind zone.

The trade wind zone between 5° and 30°S east of 25°W is extremely dry with precipitation amounts below 25 cm per year. Another dry region with annual amounts of only 20−40 cm is situated over the Patagonian Shelf off the coast of Argentina. The dryness here is caused by the lee effect of the Andes.

The precipitation amounts vary not only regionally but also from one year to another, especially near the coasts where the precipitation is connected less with advective than with convective processes and often occurs in small-scale systems. A comparison of Chart 38 with the mean frequencies in Charts 35 and 36 shows large-scale similarities, as expected. The higher moisture content of the atmosphere in lower latitudes ensures more precipitation per event, whereas in the south the amount decreases relative to the frequency.

The ratio of amount (in cm) to frequency (in %) is a measure of the yield per event. Over the open sea, the yield is about 5 cm %$^{-1}$; it is lower in the trades, higher over the Brazil Current, and it reaches 20 cm %$^{-1}$ in the equatorial zone. Summing up all 10°-square averages in the South Atlantic Ocean between the Equator and 55°S weighted according to area, one gets a mean annual amount of about 75 cm, a mean frequency of 11%, and a yield of 7 cm %$^{-1}$.

Seasonal differences are small beyond the coastal regions. Near the Equator, the seasonal variations in position of the ITCZ determine the annual range, and most of the rain there thus falls at the end of the southern summer. Near the subtropical coasts, more precipitation is observed in summer than in winter. The seasonal migration of the frontal zone in middle latitudes is accompanied by a corresponding meridional expansion and contraction of the rainy zone with more precipitation in winter south of 30°S

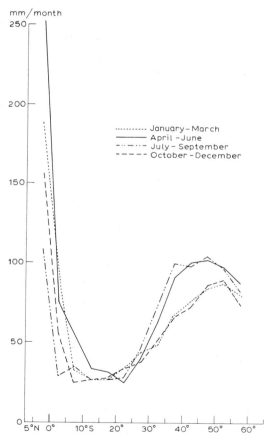

Fig. 8. Zonal averages over the South Atlantic of monthly precipitation amounts in mm per month during the four seasons.

at the northern border of the west wind belt and also on the coasts. South of 60°S and on the American coast south of 50°S, the precipitation reaches its peak in summer despite the stable surface conditions and fewer occurrences, because the higher moisture content of the air increases the amount which falls on single occasions.

In Fig. 8, zonal mean precipitation amounts for the different seasons are presented; the values are from JAEGER (1976) and clearly reflect the seasonal movement of the climate zones. South of 30°S, the amounts are markedly higher in autumn and winter than in spring and summer because the sea is warmer than the atmosphere in autumn and winter. A curve of the annual average is included in Fig. 9 which is similar to that produced by WÜST (1950) from SCHOTT's (1942) precipitation charts.

Evaporation

Evaporating water leaves the ocean as a gas, water vapour, which mixes with the atmosphere. As long as the vapour pressure in the atmosphere has not reached the saturation pressure, which is a function of temperature, evaporation is possible. The opposite process, condensation, seldom takes place at the sea surface. The effective evaporation, evaporation minus possible condensation, is used for balance calculations.

Measurements of evaporation at sea (WMO, 1966a) are even more difficult to make than those of precipitation amounts, and they can therefore not be made as routine observations on voluntary ships. In climatology, the evaporation over oceans is obtained by formulae derived through a parameterization of the process which makes use of those parameters measured routinely at sea.

Because evaporation increases the vapour pressure, the saturation pressure would soon be reached if the vapour pressure were not continuously reduced at the surface by turbulence and vertical exchange which mix dry air downward. The rate of evaporation depends on the vertical gradient of vapour pressure as well as on the intensity of the turbulent atmospheric exchange (NIEUWOLT, 1977). The difference of the specific humidity, or vapour pressure, between the sea surface and a height of about 10 m is the humidity gradient. The saturation pressure at the given water temperature is used at the surface. The wind velocity at 10 m is a measure of the turbulence.

An empirical formula has been developed to calculate the moisture exchange between ocean and atmosphere by means of the elements contained in the ordinary ship observations. These elements are sea surface temperature, measured with a bucket thermometer at a depth of about 1/2 m (p. 30), vapour pressure at deck height (p. 37f), and wind speed estimate in Beaufort numbers or measured at the mast of the vessel (p. 17).

The reader is referred to the works of LAEVASTU et al. (1969), WUCKNITZ et al. (1977), and KONDO (1977) for the details of these empirical relationships. Suffice it to say here that they greatly simplify the conditions so the evaporation values computed from them are not without problems. They describe, for instance, the exchange of water vapour only as a first approximation, independent of secondary influences such as temperature stratification, wind shear, and sea state; and they assume a neutral stratification of the atmosphere and a logarithmic vertical profile of wind speed and humidity.

Evaporation is lower in the case of stable stratification and higher when the atmosphere is unstable, and the vertical gradient of humidity is correspondingly small or large, the more so since the air—sea temperature difference, which is a measure of the stability, is used to compute the humidity at the sea surface (p. 40).

Breaking waves and spray, which depend on wave height, wave steepness, wind speed and current strength, strengthen the moisture exchange so that the linearity of its dependence on wind speed is not fulfilled in strong winds (WMO, 1976). The enormous heat exchange in hurricanes cannot otherwise be understood; on the other hand, such extremes are not important for climatological means. Other formulae try to take such influences into account, but they are very complicated and require parameters not available from routine observations. Only a rough survey is given in this text of the evaporation over the South Atlantic Ocean, as the quality and amount of observations do not justify a detailed treatment and the simple formulae are adequate for computing such approximate values. The question is whether the formulae should be used to compute the evaporation for each available observation after which a climatological mean should be computed from all the single values or whether evaporation should be computed from the mean of the observations.

As the product of wind speed and humidity difference enters the calculation, the evaporation intensity is not linearly dependent on these quantities. A comparison of the two approaches shows that the differences are small over the ocean because the variations of the parameters are small (NEUWIRTH, 1978). The computation of evaporation from single observations may be more reliable because it covers all variations. The computation from climatological averages yields somewhat lower values, but this method has the advantage that grid point values of mean maps can be used which are sufficiently smoothed to eliminate the defects and inhomogeneities of the single observations.

Wrong observational values can produce evaporation values that are more than 100% off owing to the progression of errors in the computations. This is so for historical observations that do not contain sufficient humidity measurements. An empirical relationship established by correlation between relative humidity and air—sea temperature difference could be used to estimate climatological averages of evaporation for earlier times. The main values of evaporation in this chapter were obtained by computations with the formula of Laevastu, using the international ship observations of 1961—1970. The monthly means were included in the climatic tables of the selected areas and were referred to in the analysis of the maps.

Contours of mean evaporation are presented in Charts 39 and 40 for January and July, labelled in mm month^{-1}. This unit has the advantage that one can compare directly with the amount of latent heat (p. 57) contained in the water vapour, as an evaporation of 1 mm month^{-1} corresponds to a latent heat of 0.95 W m^{-2}, using an average value for the heat of condensation.

The necessary smoothing of the contours in the evaporation charts was done with the aid of the climatological averages of wind speed (Charts 8 and 9), sea surface temperature (Charts 20 and 21), and vapour pressure (Charts 28 and 29), according to the formulae of Laevastu and Wucknitz.

There are several similarities between the two months. Areas with strong evaporation are found over warm currents in regions where the wind speed is sufficiently high too, as over the Agulhas Current, the Equatorial Current, and the Brazil Current. Weak evaporation is observed in upwelling regions on the African coast, in January also near Cabo Frio on the Brazilian coast, and in July on the Equator. In addition, the evaporation is weak over the cold Falkland Current and in the ITCZ. In the west wind belt, evaporation is weaker in January than in July.

The annual mean evaporation for the South Atlantic is plotted as a function of latitude in Fig. 9. There is a maximum of nearly 200 cm year^{-1} at 15°S, a weak minimum on the Equator (150 cm year^{-1}), and a continuous decrease into high latitudes (50 cm year^{-1} at 55°S). The bend in the curve at 25°S is associated with the upwelling areas at the African and South American coasts where the evaporation is reduced.

The average over all grid points is a total evaporation of 135 cm year^{-1}, but as the water temperature at the surface itself perhaps is about 0.5°C lower than the one measured, the evaporation could be about 15 cm less per year.

The charts of evaporation show some similarities to those of wind speed in the tropics and to those of sea surface temperature, especially near the shore. The charts of vapour pressure are less similar.

The patterns on Charts 26 and 27 of air — sea temperature difference and Charts 31 and 32 of relative humidity look very much like those of evaporation but with opposite sign, which confirms that the relative humidity is in balance with evaporation and turbulent exchange, depending on the stability of the lower atmosphere.

Monthly values of evaporation can be obtained from the tables of the selected areas. They are presented in mm day^{-1} and vary between 7.4 in January in area D, where the chart shows a distinct maximum, and 1.1 in November in the southernmost area R. The maxima of the annual range are mostly in June in lower latitudes and in May in higher latitudes. At the coasts, they appear earlier, sometimes already in March. A harmonic analysis of the annual course yields amplitudes of 1.2 mm day^{-1}; the largest ones are at 10°S (area D), over the Brazil Current (area M), and at the coast of Argentina (area Q) where the annual peaks are in June, May, and March, respectively.

In the selected areas, the monthly standard deviation of evaporation is 2.5 mm day^{-1} on the average. It is smaller in lower latitudes and in the eastern South Atlantic than in the regions of the subtropical anticyclone and along the South American coast where the maximum is reached in area N.

Water balance

The water balance is the difference between evaporation and precipitation. It is positive when precipitation is larger which leads to a higher sea level, and negative when evaporation is larger, in which case water is transferred from the sea to the atmosphere. The ocean currents are links in the budget such that when precipitation is larger than evaporation, divergence occurs in the current field, and when evaporation prevails, convergence takes place (p. 28).

Chart 41 shows the annual mean of the water balance in cm year^{-1}; it should be kept in mind that the values cannot be more exact than either of the components in the budget (WÜST, 1950). Positive values, caused by high precipitation amounts, are a feature of the ITCZ (100 cm year^{-1}) at the southwest coast of North Africa and in the adjacent Gulf of Guinea (150 cm year^{-1} at the coast near Duala), at South America north of 35°S (90 cm year^{-1} near Rio de Janeiro), and in the west wind belt (60 − 70 cm year^{-1} at 50°S 30°W). Negative values caused by strong evaporation prevail in the trades (−210 cm year^{-1} at 10°S 20°W) and over the warm currents (−200 cm year^{-1})

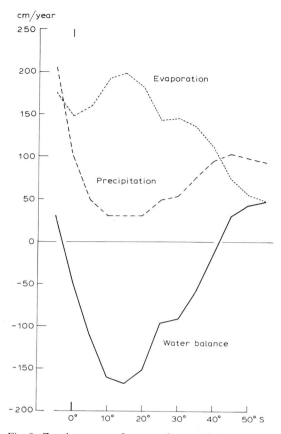

Fig. 9. Zonal averages of evaporation, precipitation amount, and water balance in cm per year in the South Atlantic.

south of Africa over the Agulhas Current. The lee effect of the Andes creates a negative water balance off the coast of Argentina as low as -60 cm year^{-1}, because of the low precipitation amounts there. The cold upwelling water along the coast of the Southwest Africa, where evaporation is suppressed, reduces the negative water budget to -20 cm year^{-1}.

Fig. 9 shows the annual zonal means of evaporation, precipitation, and water balance for the South Atlantic Ocean (for comparison with the whole hemisphere, see NEWTON, 1972). Owing to the high evaporation and low precipitation in the tropics and subtropics, the water balance there is negative. It becomes positive south of 40°S and also north of the Equator in the ITCZ (BAUMGARTNER and REICHEL, 1975).

Since evaporation increases the *salinity* of the surface water and precipitation decreases it, the water balance might be ascertained through the salinity. This relation, however, is valid only as a first approximation because several factors influence the salt content. Currents advect water of different salinity, a vertical gradient of salinity leads to an exchange and mixing with deeper layers, rivers supply fresh water, and the formation of ice increases the salt content whereas melting decreases it (WÜST, 1936).

In the mean, the salinity of surface water amounts to 35.5 °/$_{00}$ at the Equator and 37°/$_{00}$ in the trades at 15°S whence it diminishes southward, so that at 35°S 35°/$_{00}$ is observed and at 50°S only 34°/$_{00}$. This meridional structure is rather independent of season and is shaped like that of the zonal averages of the water balance (Fig. 9). There are also zonal deviations, e.g., in the South Atlantic north of 25°S the average salinity is $1-2^o/_{oo}$ higher in the west than in the east. Large variations exist along the coasts where the salinity is lowest in the mouths of great rivers, below $30^o/_{oo}$ in La Plata, for instance. It reaches about $35^o/_{oo}$ in upwelling regions, and the highest values are observed on the east coast of Brazil where $37.5^o/_{oo}$ is reached in summer.

The annual water balance is -60 cm year^{-1} for the whole of the South Atlantic Ocean. It may be no more than -45 cm year^{-1} if the sea surface temperature, from which evaporation is computed, were reduced by about 0.5°C. BAUMGARTNER and REICHEL (1975) suggest -41 cm year^{-1}. The negative water budget means that water is lost from the ocean; this is replaced by an inflow consisting partly of *runoff* from the continents and partly of advection at various levels from the neighbouring oceans.

The largest rivers feeding water into the South Atlantic are the Congo in Africa and the Paraná in South America. The Amazon does not contribute to the South Atlantic because it flows into the sea at the Equator and its water is carried into the North Atlantic Ocean with the Equatorial Current.

The Congo accounts for the largest influx from land with about 40,000 m^3 s^{-1} = 1,250 km^3 year^{-1} (BREITENGROSS, 1972), which for the South Atlantic area of $40\cdot10^6$ km^2 converts to a height of 3 cm year^{-1}. About 20,000 m^3 s^{-1} of fresh water flows through the mouth of the Paraná (La Plata) into the South Atlantic Ocean (BAUMGARTNER and REICHEL, 1975). These authors assume a total runoff from the bordering continents of 17 cm year^{-1} when spread over the area of the South Atlantic Ocean.

The transport in ocean *currents* is more important for the water balance, both the transport between different regions of the South Atlantic and that across its borders from the neighbouring oceans. If the water budget in one region causes a change in sea level, a gradient current develops in addition to the existing wind-driven and thermohaline currents; and thus a balance comes into being that seeks to keep a constant sea level superposed on the seasonal fluctuations. The net flow of all currents in the South Atlantic is toward the north and, in accordance with the negative water balance, the inflowing currents must thus be stronger than those flowing out.

Such inflowing currents are the Cape Horn Current from the Pacific Ocean, and the Agulhas Current and Antarctic Coastal Current from the Indian Ocean. Currents flowing out are the West Wind Drift into the Indian Ocean and the South Equatorial Current into the North Atlantic. The narrow Drake Passage restricts the inflow from the Pacific. According to an estimate of the transport by currents (KORT, 1962), it seems likely that surface currents contribute less to the water budget than vertical circulations in the ocean, so that currents at all depths take part in the water balance of the sea surface. Upwelling is an inflow from below. The Antarctic Divergence which brings water to the surface has a particularly large share in the water budget.

Because we are not certain about all these influences, their part in the water budget of the South Atlantic Ocean can only be determined as a residual.

A negative water balance at sea level means an input of *water vapour* into the atmosphere; this vapour must be carried away because in the mean water vapour is not stored in the atmosphere or at most only in short periods within the annual range. The water which stems from the negative water balance of the South Atlantic is transported in the atmosphere as water vapour or cloud droplets to the adjoining continents and oceans.

The vertical flux of water vapour in the atmosphere can be computed with aerological data from the horizontal divergence of the vertically integrated water vapour flow (RASMUSSEN, 1966; PALMÉN, 1967). This method is suitable for checking estimates of the water balance. It cannot be applied to climatological averages, but has to be used on daily aerological data (PALMÉN and SÖDERMANN, 1966). Over the South Atlantic Ocean, however, there are insufficient aerological stations, and satellite data have so far given too inexact information on wind and humidity in the atmosphere.

Only a part of the water evaporated over the oceans reaches the adjoining continents where it is precipitated and flows back to the ocean in the rivers. However, the cycle is not directly closed even over the ocean because the excess water in the atmosphere is not carried to those regions where the inflowing water originated. The winds and currents in a given area mostly come from the same direction, the only exception being the Agulhas Current. Otherwise the water flows to the adjoining oceans and is not returned directly to the South Atlantic Ocean by the atmosphere.

In the ITCZ, the water balance is positive because there is less evaporation than precipitation in spite of the high water temperature. This leads to a negative water balance in the atmosphere whose requirements then are supplied by the convergent flow of water vapour in the trades, as part of the circulation in the lower branch of the Hadley circulation (p. 10). The zonal Walker circulation handles the west — east differences in the tropical water balance. Evaporation is strong in the eastern South Atlantic and the surplus water vapour is brought west to Brazil by the lower branch of the Walker circulation where it condenses in the Amazon region. Such a circulation system exists on an even larger scale in the tropical Pacific (BJERKNES, 1969). In southern summer, the zonal differences are strengthened by the upwelling water on the Equator.

The largest part of the water from the deficit in the water balance over the South Atlantic Ocean is likely transported by the trades across the Equator to the ITCZ. This is particularly true for the southern summer. It is then carried to the North Atlantic Ocean or to the Amazon region where it precipitates. Another part flows southward into the cyclones of middle latitudes where the water balance is positive.

Knowledge of the water balance of all oceans and continents is needed to prove how the water exported from the South Atlantic returns to this ocean. BAUMGARTNER and REICHEL (1975) assume that the Pacific Ocean and the North Polar Sea have a positive balance, because the gain of water from precipitation combined with the runoff from the continents there is larger than the loss from evaporation. This surplus water flows into the South Atlantic as surface currents or as intermediate water. The inclusion of transport by currents at deeper levels should reveal which stationary circulations in the ocean balance the water budget.

Vertical heat flux at the sea surface

The vertical heat flux between ocean and atmosphere is composed of sensible and latent heat flux at the sea surface. Both are brought about by turbulent exchange (p. 9). Sensible heat is exchanged when a difference of potential temperature exists between sea surface and atmosphere, and the latent heat flux takes place in the exchange of water vapour when the specific humidity (or vapour pressure) shows a vertical gradient. Otherwise, the climatological averages of air — sea temperature difference (p. 36f) reflect a balance between heating of the sea surface by net radiation and sensible heat flux by turbulent exchange. Climatological averages of relative humidity (p. 40) reflect the mean vertical gradient of specific humidity and thus the balance between evaporation at sea surface and latent heat flux into the atmosphere.

Latent heat flux

In the evaporation process, the latent heat of vaporization is absorbed in the water vapour and it is, in turn, released to the atmosphere when the water vapour condenses so that the vertical flux of water vapour also leads to flux of latent heat. At the sea surface, this flux is proportional to the evaporation.

It is not the amount of water vapour but the amount of latent heat transferred to the atmosphere that is of interest here. This transfer W (in W m^{-2}) is computed as the product of the rate of evaporation E (in mm day^{-1}) and the latent heat of vaporization $L = 2.46 \cdot 10^6$ J kg^{-1} without regard to its dependence on temperature. With ρ_ω the density of water $= 10^3$ kg m^{-3}, 1 day $= 86,400$ s, and 1 mm $= 10^{-3}$ m, we get $W/E = 28.5$.

Evaporation is shown in mm month^{-1} on Charts 39 and 40; W/E is then 0.95 and the charts can also be read as approximate latent heat flux (in W m^{-2}).

It should be emphasized that high values of latent heat flux are observed in the tropics and over the warm ocean currents. The highest values appear over the warm Brazil Current and over the warm Agulhas Current, whereas low values occur over the cool upwelling water at the Equator and along the African coast and over the cold Falkland Current.

Evaporation values computed from single ship observations and monthly averages of latent heat flux are given in the climatic tables of monthly averages for selected areas. The latter were computed as follows: $W = \frac{1}{3} \times L \times (E + E_A + E_B)$, where E is evaporation averaged from computations by Laevastu's method for each observation, E_A is evaporation computed by Laevastu's method from monthly averages of wind speed, sea surface temperature, and water vapour pressure, and E_B is evaporation computed from monthly averages according to Wucknitz.

In spite of different computation methods for the tables and charts, the monthly averages of evaporation and latent heat flux are similar in both. The tables are more exact than the charts, although the period of 10 years is too short to yield representative numbers. The table values are based only on modern observations with psychrometers, whereas many of the historical observations used in the charts do not contain such reliable humidity values.

Sublimation and melting or evaporation of snow and ice, if taken into account, would have slightly increased the latent heat flux in high latitudes, but they are unimportant for the balance over the ocean.

Sensible heat flux

The vertical temperature gradient between sea surface and atmosphere causes a sensible heat flux. In the exchange formula, sensible heat flux is proportional to wind speed F (in m s^{-1}) as a measure of the turbulence, and to the difference between sea surface temperature T_w and air temperature T_a at about 10 m above sea level. This temperature difference (in °C) is a measure of the stability of the air.

Thus for the sensible heat flux W (in W m^{-2}), the following formula from WUCKNITZ (1974) and KRÜGERMEYER (1975) is valid: $W = aF(T_{w-}T_a)$ with $a = \rho_a c = 1.6 \cdot 10^{-3}$, the exchange coefficient, $\rho_a = 1.2$ kg m^{-3}, the density of air, and $c = 1.3 \cdot 10^{-3}$, the heat transport coefficient.

The general application of this formula is subject to restraints similar to those in computing evaporation. The formula is valid only for a logarithmic vertical profile of wind speed and air temperature (or more exactly, virtual potential temperature), for a height of 10 m above the sea level, and for an undisturbed sea state. The density of air decreases with decreasing air pressure, but increases with decreasing temperature; generally, it increases slightly poleward.

Monthly averages of the sensible heat flux at the sea surface were computed with the formula above, and the analyses for January and July are shown in Charts 42 and 43. Positive values are found over the warm currents and in the ITCZ and negative values are more common in January than in July, in agreement with the seasonal change in sign of the air−sea temperature differences (Charts 26 and 27). Nevertheless, for the South Atlantic and the year as a whole, positive values of sensible heat flux prevail because the water is mostly warmer than the air above (p. 37). The small flux values are subject to doubt, especially at higher latitudes. The formula yields only an approximation; errors in the temperatures due to insolation unfortunately cannot be excluded, and the measured water temperature does not represent the real sea surface temperature. Therefore, computation of the temperature difference and the values of sensible heat flux are fraught with uncertainty.

The sensible heat flux averaged over the whole South Atlantic is 6 W m^{-2}. That corresponds to a mean air−sea temperature difference of -0.5°C and a wind speed of 7.5 m s^{-1}. If one assumes that the observed water temperature is 0.5°C higher than that of the sea surface, which is really the proper water temperature for calculation of the exchange with the atmosphere, then there would be no net sensible heat flux over the South Atlantic Ocean. The adjustment between air and sea temperature is remarkably good. The radiative heating of the sea surface is offset by an advective flux of cooler water.

The highest values of sensible heat flux of 40 W m^{-2} are found in July on the coast of Argentina between Punta Rasa and Cabo San Juan and over the Agulhas Current with 35 W m^{-2}. The lowest values in January are -25 W m^{-2} on the coast of Argentina near Cabo Dos Bahias and -20 W m^{-2} over the African upwelling area.

The monthly averages for the selected areas were computed by the above formula from monthly means of wind speed and air and sea surface temperature (see the climatic tables). The annual means vary between 7 W m^{-2} in areas H and P in the subtropical anticyclone and in area N on the South American coast, and -1 W m^{-2} in areas L in the upwelling region and R in the west wind belt near the coast of Argentina. The annual range is biggest in area R where the monthly averages vary between 9 W m^{-2} in June and -11 W m^{-2} in November. The maps are similar to those of evaporation (Charts 39 and 40), but the sensible heat flux is substantially weaker than the latent heat flux which rarely becomes negative.

The ratio of sensible to latent heat flux is called the *Bowen ratio*. Averaged over the

South Atlantic Ocean, this ratio is 0.06, at lower latitudes about 0.03, at higher latitudes about 0.1. It is negative over regions of cold water, and as the evaporation furthermore is lower there, the Bowen ratio sinks to -0.3. In the west wind zone, a marked annual range occurs with values between -0.5 in summer and more than $+0.5$ in winter.

Total heat flux

Since the Bowen ratio is mostly small over the ocean, it is the latent heat flux that mainly determines the total heat flux, defined as the sum of both fluxes. The total flux is presented in Charts 44 and 45 for January and July in W m^{-2}. Naturally, the total flux cannot be more accurate than its components, and with this reservation the following may be stated. The strongest heat flux with values of about 250 W m^{-2} is observed over the warm Agulhas Current. Over the Brazil Current, the mean heat flux amounts to about 200 W m^{-2} in July and 150 W m^{-2} in January with high values of about 125 W m^{-2} extending to 35°S 45°W.

The lowest heat flux is over the upwelling water along Southwest Africa with values below 25 W m^{-2} near the coast; another minimum lies over the Falkland Current.

Otherwise the heat flux decreases poleward; in January the rate of decrease is greater than in July so that at 50°S the flux is about 0 in January and 100 W m^{-2} in July. The overall heat flux is larger in July because in the southern winter the water is relatively warm and the air—sea temperature difference negative, the relative humidity is low, and the vertical gradient of vapour pressure is large. The strength of the heat flux then indicates the intensity of the vertical exchange, which depends on the stability of the air over the sea but also on the turbulence, that is, wind speed.

In individual situations, the heat flux is subject to variations in space and time, according

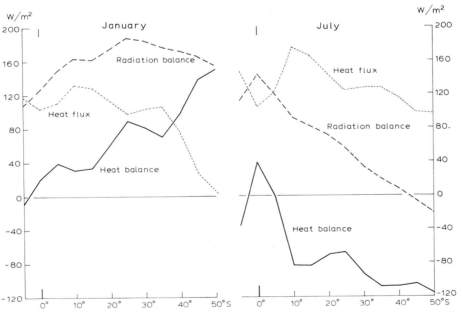

Fig. 10. Zonal averages of heat flux, radiation balance, and heat balance in W m^{-2} at the sea surface in the South Atlantic, January and July.

to weather conditions and water characteristics. Cool, dry air enhances the heat exchange into the atmosphere, and precipitation forms patches of cool water which is low in salt content and reduces the heat flux (OSTAPOFF et al., 1973).

In the selected areas, the annual heat flux reaches a maximum in area D of 174 W m^{-2} and a minimum in area R of 54 W m^{-2}. The largest annual variation of 110 W m^{-2} is in area Q and is caused by changes of the frontal zone off South America. The lowest annual range appears in area O off South Africa. The maximum in the annual range is mostly observed in autumn (May) and the minimum in spring (November).

The zonal averages of total heat flux over the South Atlantic Ocean are entered in Fig. 10 as dotted lines for January and July. The heat flux is strongest at $10-15°$S because the air over the Brazil Current is relatively unstable and the strong turbulence of the trade winds intensifies the exchange. In January, the heat flux declines sharply poleward, whereas in July it is still strong over the relatively warm water in middle latitudes. This results in an annual range with higher values in July and increasing southward from the lowest range at the Equator. The low annual range at the Equator is caused by the combination of cool upwelling water in the southern winter and weak winds in the southern summer.

Over the South Atlantic Ocean between the Equator and 55°S, the heat flux in May is about 80 W m^{-2} and in November 140 W m^{-2}, that is, almost twice the value in May.

Radiation balance

The heat balance at the sea surface is computed as the difference between the radiation balance and the heat flux, so in the following we consider the components of the radiation balance. These consist of the net short-wave insolation and the net outgoing long wave radiation; the latter is the infrared radiation from the sea surface reduced by the infrared back radiation from the atmosphere.

Some comments will also be made about the radiation balance of the earth—atmosphere system.

Insolation at sea level

The following equation was used to compute the net short-wave insolation R_1 (in W m^{-2}):

$$R_1 = (1 - r)(1 - aN)(bR^0 + (1 - b)R')$$

where r is the albedo of the sea surface, N is the total cloud amount in percent, R^0 is the extra-terrestrial solar radiation (W m^{-2}); R' is the short-wave diffuse sky radiation (W m^{-2}) reaching the surface; b is the ratio of the direct radiation to the loss of direct radiation under clear skies; and a is defined through c, the ratio of the global radiation under overcast skies to the global radiation under clear skies, such that $c = 1 - 100a$. BERNHARDT and PHILIPPS (1958) assume that $b = c = 0.36$; then $a = 0.0064$.

The mean albedo r was taken to be 0.06 over the ocean. An essentially higher albedo is found only when the sun is low. World maps of surface albedo, published by BAUM-GARTNER et al. (1976) give detailed values over the Atlantic Ocean and refer to BUDYKO (1974) and PAYNE (1972). Larger albedo values occur at higher latitudes in the southern winter such as 0.1 south of 40°S and 0.2 south of 50°S. Since radiation fluxes are small there, the exact value of the albedo is not important.

Values of extra-terrestrial solar radiation R^0 by month and latitude and those of the global short-wave radiation R' by month, latitude, and longitude were taken from tables published by BERNHARDT and PHILIPPS (1958). R^0 depends only on astronomical conditions (sun angle and length of day); R' is reduced by the turbidity factor, which takes into account the atmospheric content of dust and water vapour.

Using monthly averages of cloudiness (Charts 33 and 34) and the equation above, grid point values were computed for January and July to construct Charts 46 and 47 and to compile monthly and annual values in the climatic tables for the selected areas. The charts and tables present mean values of the insolation in W m^{-2}.

In January (southern summer), the values of insolation vary according to the cloudiness. High values of 280 W m^{-2} appear on the coasts of Southwest Africa and Argentina and more than 250 W m^{-2} in the central South Atlantic region of high pressure. The mean insolation is below 200 W m^{-2} in the west wind belt, in the ITCZ, and in the cloudy trade wind area over the Benguela Current. The relatively large insolation values in higher latitudes are caused by the greater length of day which compensates for the smaller sun angle.

In July (southern winter), the pattern is also to a large extent similar to the distribution of cloudiness, but there is a strong meridional gradient of insolation with the highest values of 200 W m^{-2} at the Equator and the lowest below 25 W m^{-2} south of 50°S. The latter should be somewhat lower because the actual albedo is higher than used in the computations; at 40°S, the values should be reduced by about 2, at 50°S by about 4 W m^{-2}. The steep gradient is a consequence of the decrease of sun angle and length of day from lower to higher latitudes in winter; these factors contribute toward making the annual range of insolation small in low latitudes and larger as one approaches the polar regions. This is confirmed by the monthly averages in the tables of the selected areas. South of 10°S all areas show a distinct annual variation of insolation with highest values in December and January, lowest values in June and July. In the southernmost area R, where the annual mean is 157 W m^{-2}, the difference between December and June amounts to 322 W m^{-2}.

The highest insolation is in areas A and E, the lowest in P. There is a striking west—east difference such that in the western areas A, E, and I, the insolation is considerably larger than in the eastern areas C, F, G, and J. That is evident in the charts of annual averages (BUDYKO, 1974), where the north—south contrast between 200 W m^{-2} at 5°S and 100 W m^{-2} at 50°S has superposed upon it a west—east difference between a maximum of 210 W m^{-2} over the Brazil Current and a minimum of 150 W m^{-2} over the Benguela Current. The continental influence is reflected in maxima of 175 W m^{-2} off South America in 30—40°S and of 200 W m^{-2} off Southwest Africa at 30°S.

The insolation averaged over the South Atlantic Ocean is 215 W m^{-2} in January and 105 W m^{-2} in July; in other words, the summer value is twice that of winter as in the instance of the heat flux, although the amounts are larger than those of the heat flux by about 40 W m^{-2} in the annual average.

The accuracy of the computation of insolation depends on the validity of the formula and on the given values of radiation, cloudiness, and the constants in the formula. The insolation decreases with higher albedo and increased cloudiness and turbidity. An albedo of $r = 0.1$ (KONDRATYEV, 1972) diminishes the insolation by about 5%, and a change of mean cloudiness from 60 to 65% results in a 5% reduction in the insolation.

Satellite observations suggest that the cloudiness is less than that derived from ship observations (p. 42).

The penetration depth of insolation in the ocean depends on its wavelength and on the transparency of water. Silt, foam, bubbles, and plankton obstruct the penetration. On the average, about 10% of the radiation reaches a depth of 10 m, and 3% a depth of 50 m (GROEN, 1967). Turbulence further extends the layer heated by insolation to about 100 m. This layer of water forms the warm, well-mixed layer above the thermocline in which the temperature is almost constant (p. 28).

Terrestrial radiation

Terrestrial radiation is the long-wave thermal radiation of the sea surface R_T (in W m^{-2}) proportional to its temperature T_w (in °C). The equation to compute this radiation is:

$$R_T = (1 - r) \, s(T_0 + T_w)^4$$

where r the albedo $= 0.06$, s the Stefan-Boltzmann constant $= 5.67 \cdot 10^{-8}$ W m^{-2} deg^{-4}, $T_0 = 273.15 =$ absolute temperature of 0°C, and T_w the ocean surface temperature in °C.

The mean values of the terrestrial radiation were computed with monthly averages of sea surface temperatures (Charts 20 and 21) at grid points for January and July and for the selected areas for all months and the year.

Averaged over the South Atlantic, the mean terrestrial radiation is 395 W m^{-2}. This corresponds to a mean sea surface temperature of 20°C, which is about 3°C higher than the real mean because the relation between radiation and temperature is not linear. If the temperature is lowered about 5°C, the long-wave radiation would diminish by about 7%. A warming causes an increase of the thermal radiation, that is, a loss of energy in the ocean with an accompanying cooling there, which thus would help to damp fluctuations of climate. Since the temperature of the radiating sea surface is about 0.5°C lower than the measured temperature, actual radiation values will be lower too but by less than 1%.

The mean thermal radiation over the ocean varies between 450 W m^{-2} (corresponding to a sea surface temperature of 28°C) in the ITCZ and 310 W m^{-2} (corresponding to 0°C) in the west wind belt. The seasonal differences amount to only 10 W m^{-2} with maximum in the southern summer.

Atmospheric counter radiation

The atmospheric counter, or back, radiation is the long-wave radiation R_G (in W m^{-2}) directed from the atmosphere to the surface. It is essentially a function of cloudiness as in:

$$R_B = (1 - r) \cdot ((100 - N) \cdot B^0 + N \cdot B')/100$$

where r the albedo $= 0.06$, N is total cloudiness in %, B^0 is the back radiation for cloudless sky in W m^{-2}, and B' is the back radiation for overcast sky in W m^{-2}.

The values of B^0 and B' by month, latitude, and longitude were taken from tables by BERNHARDT and PHILIPPS (1966). They take into account the radiation effects of water vapour and carbon dioxide in the atmosphere according to the air temperature. With these values and monthly averages of cloudiness, grid point values of counter radiation over the South Atlantic Ocean were calculated for January and July and monthly and annual averages for the selected areas. The accuracy of the mean values depends, of course, on the accuracy of B^0 and B', the cloudiness, and the albedo.

The counter radiation averaged over the South Atlantic Ocean is 340 W m^{-2}. The regional averages vary between 390 W m^{-2} in the ITCZ and 170 W m^{-2} at 55°S in the west wind belt with only little zonal difference. The annual variation shows a maximum in southern summer of about 200 W m^{-2} in the same way as the thermal radiation.

This computation can give only a rough outline of the back radiation. The use of total cloudiness without considering the type, height, and thickness of the clouds is in itself too coarse an approach. A comparison of the maps of counter radiation (not shown) with those of cloudiness shows that they are scarcely similar; it is those variables that enter into the computation of B^0 and B' whose influence prevails, and of these the distribution of water vapour in the atmosphere is particularly important.

Energy is gained at the sea surface by the atmospheric counter radiation which thus contributes to the heating of the surface water. It reflects the greenhouse effect of the atmosphere which depends on its content of water vapour and absorbing gases, especially CO_2. Since the latter increases steadily owing to industrialization, a long-term warming of the lower atmosphere is expected (MANABE and WETHERALD, 1975).

Net long-wave radiation at the sea surface

The net radiation R_N (in W m^{-2}) at the sea surface is the difference between the thermal radiation R_T and the atmospheric counter radiation R_B:

$$R_N = R_T - R_B$$

The difference between the two large values is relatively small and therefore uncertain. Averaged over the South Atlantic Ocean and weighted by area, it amounts to 55 W m^{-2}, only 15% of either of the two components.

Geographically and seasonally, the net radiation exhibits only small variations, as seen in Charts 48 and 49 for January and July and in the tables of the selected areas for single months and the year.

The highest values of the mean net long-wave radiation reach 75 W m^{-2} in July at the northeastern corner of South America over the Brazil Current and at the south coast of Africa over the Agulhas Current. Low values of about 20 W m^{-2} are found in January over the upwelling areas on the African coast and in the eastern part of the west wind zone at 55°S. Areas over warm currents generally have higher values than those over cooler water. The annual variation is small. Because the annual range of the counter radiation is twice that of the thermal radiation, the counter radiation determines the phase. The mean values are about 10 W m^{-2} lower in January than in July. The range becomes larger south of 40°S, although the opposite annual variation of the albedo tends to reduce it.

The monthly averages in the selected areas show maxima of the annual variation in May and June, minima in November and December, with a mean difference of 15 W m^{-2}. The highest monthly mean of 79 W m^{-2} occurs for June in area K over the Brazil Current, the lowest one of 30 W m^{-2} for November in area G in the northern part of the African upwelling region. The highest annual mean, 68 W m^{-2}, is in area O off South Africa, the lowest, 40 W m^{-2}, in area G off Cape Frio.

Radiation balance at the sea surface

The difference between insolation R_I and long-wave net radiation R_N at sea level is the radiation balance R (in W m^{-2}) between ocean and atmosphere (ZILLMAN, 1967; KESSLER, 1973):

$$R = R_I - R_N$$

Naturally, the same reservations apply to the values of the radiation balance as to its components. The distribution of the radiation balance is similar to that of the insolation because the mean net long-wave radiation is only half the size of the insolation. The balance averaged over the South Atlantic Ocean is 105 W m^{-2}, but there are strong regional and temporal variations.

The distribution of the radiation balance is illustrated in Charts 50 and 51 for January and July. Together the maps give an idea of the annual mean radiation balance which has a value of about 140 W m^{-2} near the Equator and decreases to 50 W m^{-2} at 55°S. The annual mean is everywhere positive, so that the sea surface in general can be considered a heat source for the atmosphere (PERRY and WALKER, 1977).

The annual means are high in the upwelling areas on the African coast (160 W m^{-2}) and near Cabo Corrientes at Argentina (120 W m^{-2}) where the cloudiness is light. The west—east differences are striking in the trades, from high averages (140 W m^{-2}) over the Benguela Current to low values (100 W m^{-2}) over the Brazil Current, which is similar to the distribution of insolation (p. 61, cf. Charts 46 and 47). A comparison of the radiation balance with Chart 3 of the climates shows that the boundary between tropical (climates A and B) and temperate climates (C) lies where the yearly radiation balance is 100 W m^{-2}. The combination of the radiation balance with aridity defined by the water balance (p. 53f) is a good determinant of marine climates.

When the sea surface warms, the thermal radiation (R_N) increases (p. 62), the radiation balance (R) decreases, and the sea surface temperature in turn drops again. Everything else being equal, there is thus a balance between a given sea surface temperature and the insolation (R_I).

In January (Chart 50), the radiation balance reaches the highest averages of 230 W m^{-2} at the southwest coast of Africa near Lüderitz and on the coast of Argentina near Punta Rasa. In the central South Atlantic, 190 W m^{-2} is the highest value at 20°S 25°W. In contrast to the subtropics, the balance is lower on the Equator, 130 W m^{-2}, and it reaches a minimum of 110 W m^{-2} in the Gulf of Guinea. It falls below 150 W m^{-2} in the west wind belt, but it is still higher at 55°S than on the Equator. The summer months have high balance values owing to the large sun angle and long days.

July (Chart 51) shows a strong meridional contrast between averages of 150 W m^{-2} on the Equator and negative values in the westerlies south of 40°S. This is in accordance with the low sun angle and the fact that the day becomes shorter as one approaches the polar region; together these reduce the insolation, whereas the long-wave radiation does not change much from summer to winter because of the small annual range of sea surface temperature. The radiation balance is high in the upwelling area on the African coast, almost 100 W m^{-2}; otherwise zonal differences are unimportant in comparison with the meridional ones.

The monthly averages in the tables for the selected areas show the highest values, 220 W m^{-2}, in December in area R and the lowest ones, -15 W m^{-2}, in July in area P, both areas on the northern border of the frontal zone. The biggest annual mean, 140 W m^{-2}, is in area A on the Equator, and the smallest, 80 W m^{-2}, in southern area P. The annual variation is pronounced south of 5°S with maxima in December and January and minima mostly in June. The southernmost area R experiences the largest range of 230 W m^{-2}. Averaged over areas D to R, the annual range is 130 W m^{-2}; it obviously depends on the elevation of the sun and the corresponding seasonal differences of insolation. It is insignificantly reduced by the opposite annual variation of the albedo south of 45°S.

The zonal averages of the radiation balance are shown in Fig. 10. Uniformly high values prevail in January with the maximum at 25−30°S where the sun is highest and cloudiness low. On the Equator, the cloudiness of the ITCZ (reduced insolation, but strong thermal radiation) lowers the balance. The large balance values in higher latitudes result from the long daily insolation and weaker thermal radiation.

The short winter days lead to the north−south difference of radiation balance in July. The highest averages are on the Equator; south of 45°S the balance is even negative. The cloudiness enhances this difference because it increases toward the south, and the annual range is thus smallest in low latitudes and increases poleward.

The values of the radiation balance are no more exact than the radiation components on which they are based. Substantial improvements may still be expected, notably from the use of satellite technology (BARRETT, 1974). A map of the global annual radiation balance for the surface based on satellite data was published by PERRY and WALKER (1977).

Radiation balance in the atmosphere

The satellites measure the different radiation components of the earth−atmosphere system at the top of the atmosphere to determine the radiation balance of the system. Annual means of that balance have been published by VONDER HAAR and SUOMI (1971) and by PERRY and WALKER (1977), monthly averages for January, April, July, and October by SASAMORI et al. (1972).

The atmospheric radiation balance in our area is characterized by a pronounced meridional gradient between positive values in the latitudes north of 40°S and negative values in higher latitudes, corresponding to the atmospheric horizontal temperature gradient. The highest balance values are in summer in the subtropics (BARRETT, 1974).

The difference between the radiation balance of the earth−atmosphere system and the radiation balance at sea surface gives the balance in the atmosphere over the ocean. It is everywhere negative in agreement with the vertical temperature gradient in the atmosphere (see Table II).

The effect of the distribution of both insolation and long-wave radiation is to set up a strong meridional gradient of the radiation balance for the earth−atmosphere system. Radiation energy is gained in the tropics, most at the Equator, and lost near the Pole; and the exchange in the meridional circulations produces an equilibrium given by the balance values and illustrated in the climate zones.

Globally, about half of this meridional heat exchange takes place as sensible heat transport consisting of warm air flowing into colder or cold air into warmer regions. The latent heat transport contributes about 35%, and the advection of heat in ocean currents

TABLE II

ANNUAL ZONAL AVERAGES OF THE RADIATION BALANCES (in $W\ m^{-2}$) AT THE SEA SURFACE, IN THE ATMOSPHERE, AND FOR THE EARTH–ATMOSPHERE SYSTEM OVER THE SOUTH ATLANTIC OCEAN BETWEEN THE EQUATOR AND 50°S

	0°	5°	10°	15°	20°	25°	30°	35°	40°	45°	50°S
Sea surface	135	135	130	120	120	120	110	100	90	80	65
Atmosphere	−45	−50	−50	−45	−55	−65	−70	−75	−85	−95	−95
Earth–atmos. system	90	85	80	75	65	55	40	25	5	−15	−30

15% (NIEUWOLT, 1977). Over the South Atlantic, the share of the sensible heat may be larger and that of the currents smaller than these numbers.

Heat balance

The heat balance, the difference between radiation balance and heat flux, is discussed first for the sea surface and then for the earth — atmosphere system over the ocean.

Heat balance at the sea surface

The radiation balance R at the sea surface (p. 64f) is a measure of the gain or loss of heat through radiation at the boundary between ocean and atmosphere. In contrast, the heat flux there (p. 59f) describes the loss of heat from the sea surface to the atmosphere by the exchange of sensible heat S and water vapour LE. The difference between the radiation and the heat flux is the heat balance (in $W\ m^{-2}$) at the sea surface (DEACON, 1969):

$$W = R - (S + LE)$$

It gives the net gain or loss of heat to the atmosphere at this boundary. It is positive when the gain by the radiation balance is greater than the loss by the heat flux. The heat gain is stored in the ocean, and as it is distributed horizontally and vertically by currents and turbulence, it indicates a divergence of heat transport in the ocean. HASTENRATH (1977) also computed the heat balance for the tropical Atlantic from ship observations (see HASTENRATH and LAMB, 1978) as the residual of the net radiation and the heat flux at the surface, and interpreted it as the mean sum of the oceanic heat flux divergence and heat storage.

Averaged over the South Atlantic Ocean, the radiation balance and heat fluxes counterbalance, yet the climatic averages have regional and seasonal differences. The latter are balanced by heat storage and the former by currents.

The averages of the heat balance over the South Atlantic Ocean are presented in Charts 52 and 53 for January and July. The heat balance tends to be negative over the warm currents and positive over the cold currents and in the areas of upwelling. Over warm currents there is a loss to the atmosphere because the positive vertical temperature and water vapour gradients enhance the heat flux, whereas heat is gained over cold water because the stable stratification reduces the heat flux and the thermal radiation.

The heat balance is positive nearly everywhere in January because the insolation produces a gain of heat while the still relatively cool water reduces the loss into the atmosphere. High values (more than 150 W m^{-2}) are observed in the west wind belt at 55°S. The highest values (above 200 W m^{-2}) appear over the upwelling water on the African coast. Over the Falkland Current, the mean heat balance reaches 180 W m^{-2}. The balance is negative only over the Agulhas Current and at the mouth of the Amazon which is reached by the negative heat balance of the North Atlantic.

Negative values predominate in July and reach an extreme over the Agulhas Current of −245 W m^{-2}. In the west wind zone east of 45°W, they vary between −100 and −140 W m^{-2}, and in the central South Atlantic at 10−15°S they are as low as −120 W m^{-2}.

The heat balance is less negative on the coast of South America. North of Cabo Corrientes it lies between −25 and −75 W m^{-2}. On the African coast near Swakopmund, it rises to about +80 W m^{-2}; otherwise a positive heat balance is found only on the Equator in 10−15°W and on the African coast near Pointe Noire (5°S).

The monthly averages in the climatic tables of the selected areas confirm these numbers. They vary between +172 W m^{-2} in December in area R and −158 W m^{-2} in June in area P at the northern border of the west wind belt. The largest annual variation is to be found here, 240 W m^{-2} between extreme monthly averages. South of 5°S the annual maximum falls in December and the minimum in June. From May to July, positive monthly averages appear only in the equatorial areas B and C. In December−January, the values are nowhere negative. Values of more than +100 W m^{-2} are found in areas L, O, Q, and R, and of less than −100 W m^{-2} in areas D, E, F, H, M, and P in the central South Atlantic and in K and N on the South American coast.

The zonal averages of the heat balance are mostly positive in January and negative in July when the amounts increase poleward. Fig. 10 shows the zonal averages of the heat flux, the radiation balance, and the heat balance for January and July. In both months, there are relative minima of the heat balance in 10−15°S and 35°S which is where the heat flux over the warm Brazil and Agulhas currents is strong.

In July, the difference is large between 5°N, the mean position of the ITCZ, and the Equator where upwelling occurs. In the ITCZ, the radiation flux is small, the heat flux large, and the heat balance thus negative. The ITCZ is then a place where energy is withdrawn from the ocean. The opposite is true at the Equator where the heat balance is positive in July when the water is cool through upwelling and the atmosphere cloudless through subsidence. South of 5°S, the heat balance is positive in summer and negative in winter, and the difference between the seasons increases poleward. In the west wind zone, the difference between January and July is more than 200 W m^{-2} at 40°S and more than 250 W m^{-2} at 50°S.

The positive balance values in January are mainly a result of the highly positive radiation balance in higher latitudes. The heat flux is comparatively small and decreases poleward. In July, on the other hand, the radiation balance decreases poleward while the heat flux is large over the relatively warm water so that the heat balance becomes negative. The annual range of temperature (Chart 25) increases poleward only to about 30−35°S after which it decreases to a minimum at 45−50°S owing to the combined effect of the mitigation of radiation influences by clouds and mixing of heat to greater depths in the west wind belt than in the subtropical ridge (VAN LOON, 1966).

The highest annual mean values are 130 W m^{-2} on the African coast in the centre of the upwelling (25°S), 80 W m^{-2} over the Falkland Current at 45°S, and 60 W m^{-2} on the Equator at 15°W. The lowest values are -150 W m^{-2} over the Agulhas Current, -50 W m^{-2} over the South Equatorial Current at 12°S 25°W, and -30 W m^{-2} over the Brazil Current at 30 $-$ 35°S 20°W. In the west wind belt, positive and negative values almost cancel each other over the year, but in the central part of the ocean the annual balance is negative with values between -10 and -30 W m^{-2} in 15 $-$ 40°S.

In the selected areas, the annual mean values of the heat balance vary between $+48$ W m^{-2} in area R over the cold Falkland Current and -35 W m^{-2} in area D over the Brazil Current. The annual heat balance is positive in areas A, B, C, L, O, and R, and negative in D, E, F, H, J, K, M, N, and P.

Overall, the heat balance values at the sea surface cancel. Averaged over the South Atlantic Ocean between the Equator and 55°S, the balance amounts to $+80$ W m^{-2} in January and -80 W m^{-2} in July with uncertainties as large as those of Charts 52 and 53.

A positive balance means that the sea surface absorbs more heat than it gives off to the atmosphere and conversely for a negative balance. The compensation in the ocean is carried out by currents, especially along the borders; in the atmosphere it is accompanied by the circulations and by radiation into space.

Areas with a positive heat balance at the sea surface are heat reservoirs in the ocean. Areas with a negative heat balance consume heat stored in the ocean but are a heat source for the atmosphere; the loss of heat in such areas is made up for by currents from areas with positive heat balance, and in this way the heat balance acquires a cycle not only between individual ocean areas but also between the different oceans and between seasons.

If the mean sea surface temperature were 0.5°C lower, which it would be if measured exactly at the surface, the annual latent flux over the South Atlantic would be reduced by about 12 W m^{-2}, the sensible heat flux by about 6 W m^{-2}, and the thermal radiation by about 3 W m^{-2}. That would result in a positive heat balance of about $+20$ W m^{-2} which could be explained by the relatively low water temperature of the South Atlantic. This surplus heat would flow to the North Atlantic in the South Equatorial Current, while the influx of cold water from the upwelling in the Antarctic Divergence would satisfy the water balance (p. 56).

Heat balance of the earth — atmosphere system

A heat balance for the atmosphere and for the whole earth including the atmosphere — ocean system can be similarly computed. From the time when satellite observations made the necessary radiation values available, our understanding of such estimates has grown. HOLOPAINEN (1977) has described the associated problems. Unfortunately, the oceans on the Southern Hemisphere are not yet explored sufficiently to allow us to draw final conclusions.

To calculate the heat balance of the earth — atmosphere system over the ocean, one must weigh the radiation balance (p. 65f) against the gain or loss of heat by condensation and evaporation. The sensible heat flux stays in the atmosphere and so does not contribute to the system. Further heat sources such as the heat of sublimation and chemical conversions play a secondary role. The balance is satisfied by heat storage and advection in the ocean and in the atmosphere (BARRETT, 1974). These processes are efficacious in annual and regional variations, but in the mean only on the borders of the area.

The effective heat gain by the earth — atmosphere system from condensation and evaporation depends on the difference between them in the sense that the latent heat tied up in water vapour is a loss, whereas the condensation heat released by the formation of clouds is a gain for the system. Dissipation of clouds, as it is evaporation within the atmosphere, counts as a loss. It is assumed that the effective formation of clouds happens at the location of precipitation, so that it can be described by the measured precipitation amount which, when multiplied by the condensation heat, can then be used as measure of the heat gain of the system (BUDYKO, 1974). In the same way, the evaporation multiplied by the condensation heat gives the heat loss (p. 57). The difference between precipitation P and evaporation E gives, as we have seen, the water balance $B = E − P$ (p. 53).

The heat balance W of the *earth — atmosphere system* is thus the sum of the radiation balance R of the system and the water balance B multiplied by the condensation heat L:

$$W = R + LB = R + L(E − P)$$

In Table III the annual zonal averages of these balances are shown for the South Atlantic Ocean between 5°N and 55°S.

The atmospheric balance of available heat energy has a distinct maximum value in the ITCZ near 5°N (PERRY and WALKER, 1977), where the convergence of water vapour brought by the trades plays a decisive role. In the subtropics, the energy of the intensive insolation is converted into the latent heat of evaporation, and in the ITCZ this latent heat is transformed into potential and kinetic energy by the formation of clouds (precipitation) to maintain the "heat engine" of the atmosphere. The latent heat energy released in the ITCZ is distributed by the Hadley circulation. It becomes available for all types of circulations (CORNEJO-GARRIDO and STONE, 1977), also in the polar front zone where the heat balance is positive too, although comparatively small.

The meridional distribution of the atmospheric heat balance is valid for all seasons, but the maximum moves with the ITCZ, and the north — south contrast is stronger in the southern winter than in summer.

The atmospheric heat balance W can be computed in the same way. It consists of the atmospheric net radiation R (first line of Table IV) and of the atmospheric heat gain from the sensible heat flux S and from the latent heat released by condensation of water vapour P (second line of Table IV). This heat is calculated from the precipitation amount

TABLE III

ANNUAL ZONAL AVERAGES OF RADIATION, WATER AND HEAT BALANCES FOR THE SOUTH ATLANTIC OCEAN (in $W\ m^{-2}$)

	5°N	0°	5°	10°	15°	20°	25°	30°	35°	40°	45°	50°	55°S
Radiation balance	85	90	85	80	75	65	55	40	25	5	−15	−30	−40
Water balance	25	−40	−90	−130	−140	−125	−80	−75	−50	−10	+25	35	40
Heat balance	*110*	50	−5	−50	−65*	−60	−25	−35	−25	−5	+*10*	5	0

The highest heat balance values are in italics and the lowest are marked by asterisk.

TABLE IV

ANNUAL HEAT BALANCE FOR DIFFERENT LATITUDES

	5°N	0°	5°	10°	15°	20°	25°	30°	35°	40°	45°	50°S
Net radiation	−25	−45	−50	−50	−45	−55	−65	−70	−75	−85	−95	−95
Heat gain	160	65	25	25	5	0	30	40	65	85	90	85
Heat balance	*135*	20	−25	−25	−40	−55*	−35	−30	−10	0	−5	−10

Maximum italicized, minimum with asterisk.

P, multiplied by the condensation heat L, as for the heat balance of the whole system:
$$W = R + S + LP$$
The annual heat balance of the atmosphere is positive only near the Equator (Table IV) with an annual mean of 135 W m^{-2} concentrated in the ITCZ at 5°N. It is everywhere negative over the South Atlantic Ocean. The lowest yearly average of −55 W m^{-2} is situated at 20°S where the subsidence above the trade inversion is strongest and where there is no heat gain in the atmosphere. South of this region, the heat gain and the net radiation of the atmosphere both increase so that the heat budget is balanced in higher latitudes.

The sum of the heat balance at the sea surface and in the atmosphere constitutes the heat balance of the earth−atmosphere system (Table V).

TABLE V

HEAT BALANCE OF THE EARTH–ATMOSPHERE SYSTEM AS THE SUM OF THE HEAT BALANCES AT THE SEA SURFACE AND IN THE ATMOSPHERE (in W m^{-2})

	5°N	0°	5°	10°	15°	20°	25°	30°	35°	40°	45°	50°S
Sea surface	−25	30	20	−25	−25	−5	10	−5	−15	−5	15	15
Atmosphere	135	20	−25	−25	−40	−55	−35	−30	−10	0	−5	−10
Earth–atmos. system	*110*	50	−5	−50	−65*	−60	−25	−35	−25	−5	10	5

Maximum in italics, minimum with asterisk.

Climatic stations

In this section, the climate at eight stations on South Atlantic islands will be described. There is one climatic table for each island (see Appendix, p. 150f) and the climatic means have been taken from publications listed in the text and references. The positions of the islands are shown in Chart 2.

Unfortunately, the climate data stem from different periods and could not be reduced to one uniform period for all stations, sometimes not even the different parameters at the same station since they had to be collected from several publications; and frequently, the series of observations was too short for representative statistics.

For each month and for the year, the following parameters are indicated in the climatic tables: mean air pressure (mbar) reduced to sea level; mean air temperature (°C); mean daily maximum and minimum of air temperature (°C); absolute maximum and minimum of air temperature (°C); mean dewpoint (°C); mean vapour pressure (mbar); mean relative humidity (%); mean evaporation (mm) computed according to Laevastu; mean wind speed (m s^{-1}); mean wind direction; mean precipitation (mm); maximum of daily precipitation (mm); mean number of days with precipitation (≥ 1 mm day^{-1}), with snowfall, with thunder, with fog; mean cloud amount (%); mean sunshine in h/day and in % of the astronomically possible.

The climate classification of Köppen (p. 12f), illustrated in Chart 3, was applied to each station's climate data. Special features of the stations are noted if they were available, especially those that affect the representativeness of the data. Most of the islands are small, but they are mountainous and their orographical influence often disturbs the ocean climate considerably.

If a climate station is situated in a selected area (Chart 2), its data are compared with the averages of the area, to test its representativeness of the marine conditions. Deviations from the area values are noted and related to orographical effects. If enough ship observations were available in an ocean area containing an island station, the climatological charts for January and July were used for a comparison with the station data. On the other hand, the island data were applied in the analyses of the charts if their marine character was ensured.

Sao Tomé

The island of Sao Tomé is situated in the Gulf of Guinea off the Bight of Biafra on the Equator at the northern boundary of area C. Its size is 836 km^2, and the island is dominated by the 2024 m high Pico de Tomé. The climate station is at the airport on the northeast coast (0°23'N 6°43'E).

The climate data are from the following publications: *Normals for the Period 1931–1960* (WMO, 1971); *O Clima de Portugal*, Fasciculo XIV (LISBOA, 1965); *Africa Pilot*, Vol. II (TAUNTON, 1977); and *Weather on the West Coast of Tropical Africa* (LONDON, 1949). The observations originate from the years 1939–1960.

The climate on the island is determined by the ITCZ which in southern summer and autumn lies in the latitudes of Sao Tomé whence it moves north at the beginning of the

TABLE VI

COMPARISON OF THE SUMMER MONTHS JANUARY–MARCH WITH THE WINTER MONTHS JULY–SEPTEMBER FOR SAO TOMÉ

	Summer	Winter	
Air pressure	1010	1014	(mbar)
Air temperature	26	24	(°C)
Vapour pressure	29.5	24	(mbar)
Relative humidity	85	80	(%)
Evaporation	75	110	(mm day^{-1})
Precipitation	100	6	(mm month^{-1})

southern winter. The South Equatorial Current approaches from the south and joins the eastward flowing Guinea Current north of the island (Chart 19).

These circumstances are reflected in the annual variation of the meteorological parameters. A comparison of the summer months with the winter months for the mean climate values is given in Table VI.

From the averages of air temperature and vapour pressure, mean equivalent temperatures of 70°C for summer and 60°C for winter were calculated (for definition, see p. 38). Normally, it is above the limit of sultriness of 55°C and below it only during a few nights in winter. Even the upper bearable limit of 80°C is not infrequently exceeded.

The climate of Sao Tomé is thus always sultry and only little more bearable in winter than in summer. On the average, the daily range of temperature is 7°C, that of the vapour pressure 2 mbar, hence that of the equivalent temperature 10°C. The extreme temperatures are 34° and 13°C.

The monthly precipitation amounts vary considerably more. The average in March is 130 mm on 9 days and only 0.1 mm in July. From October to May, a moist tropical climate prevails. When the island is surrounded by warm water and high humidity, convective conditions provide large amounts of rain (100 mm month^{-1}). From June to September, the cooler South Equatorial Current causes a drier tropical climate (less than 10 mm/month). The island thus lies in an isothermal, winter-dry tropical climate Awi.

While the wind speed scarcely varies through the year, the wind direction shifts slightly between south-southwest in January with a steadiness of 70% and south-southeast in August—October with a steadiness of about 85% in agreement with the trades of the Southern Hemisphere.

There is indeed hardly more sunshine during the dry than the rainy season. Neither cloudiness (with a mean of 71%) nor duration of sunshine (38% of possible) shows a noteworthy annual variation. It seems that the clouds of the ITCZ in the southern summer are replaced by the trade clouds in the southern winter, and although the latter cover as much sky they yield less rain. Days of clear sky (cloudiness $\leqslant$ 20%) are rare. On the average, 6 days per month have nearly overcast sky (cloudiness $\geqslant$ 80%).

Comparison of the data of Sao Tomé with the monthly averages of area C reveals that both monthly means and annual variations are similar but that the wind speed at sea is twice as high as that at the island station.

The same, not surprisingly, is true when the climate data for January and July are compared with those on the climate charts at the location of the island because the analyses took into account data of the station. Differences are found only in the vapour pressure which is about 1 mbar lower at the station and in a weaker evaporation. These differences show the influence of the mountainous island on the climate. The mountains shelter the station from the mostly southerly winds, and the air which crosses the mountains must descend whereby the vapour pressure decreases. The cloud base is obviously so high that the amount of cloudiness is not affected. The extreme situation of a foehn is not reached.

The descent of the air along the mountains is expressed clearly in the precipitation. The airport has the lowest amount of rain (88 cm/year) of all precipitation stations on the island.

Fig. 11 shows the dependence of the annual precipitation (abscissa) of 48 stations on

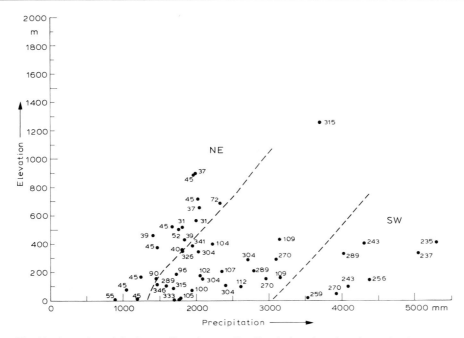

Fig. 11. Annual precipitation at 48 stations on Sao Tomé plotted against the station height (m), and annual precipitation amount (mm). The angle (degree) between the station and the Pico de Tomé is indicated next to the plot.

Sao Tomé (LISBOA, 1965) on their height (ordinate) and position. The number at each station indicates its direction from Pico de Tomé in degrees.

The northeastern stations on the lee side have little rain in contrast with the southwestern stations on the windward side. Most rain is observed on the steep slopes in the southwestern part of the island with the highest amount 530 cm/year. The amount does increase a little with height, but the location on the lee or windward side plays a bigger role.

Fernando de Noronha

The island of Fernando de Noronha lies east-northeastward of Cabo de Sao Roque off the northeast coast of Brazil (3°51'S 32°25'W) in the southern part of area A and in the South Equatorial Current. The main isle is 3 km at its broadest and 11 km at its longest, stretching from southwest to northeast, and its area is 18 km². Its small surface with steep but low mountains—the largest hill, O Pico, is 322 m high—causes only slight orographical disturbance. Therefore, this island is well suited for characterizing the marine climate around it.

The climatic data were taken from the following publications: *South America Pilot*, Vol. I (TAUNTON, 1975); *Climatological Normals for the Period 1931−1960* (WMO, 1971); *Tables of Temperature, Relative Humidity and Precipitation for the World*, Part II (LONDON, 1960a); *Normalis Climatologicas* (RIO DE JANEIRO, 1942); and *Climat do Brasil* (RIO DE JANEIRO, 1922). The observations stem from the years 1911−1960. Information on thunderstorms is not available.

The climate of Fernando de Noronha gets its character from the ITCZ and the trades.

TABLE VII

COMPARISON OF THE SUMMER MONTHS JANUARY–MARCH WITH THE WINTER MONTHS JULY–SEPTEMBER FOR FER-NANDO DE NORONHA

	Summer	Winter	
Air pressure	1011	1014	(mbar)
Wind speed	5.5	7.5	(m s^{-1})
Air temperature	26	25	(°C)
Vapour pressure	29	26	(mbar)
Relative humidity	85	80	(%)
Evaporation	140	190	(mm day^{-1})
Cloudiness	70	55	(%)
Precipitation	120	70	(mm month^{-1})

Winds blow with a steadiness of 90% from east or southeast. The annual variation at the station is similar to that of Sao Tomé, and according to the classification of Köppen, the island lies in a tropical climate Awi.

The mean values of the climate elements in the summer months of January—March and the winter months of July—September are shown in Table VII.

The mean equivalent temperature of about 67°C remains above the limit of sultriness, also at night. In the daytime, the upper limit of bearableness (80°C) is often exceeded and even in winter there is little thermal difference. The mean daily variation of air temperature is only 4°C and the extreme maximum and minimum are 31° and 18°C.

The annual course of precipitation is delayed with respect to the extreme seasons. The biggest monthly amounts of more than 260 mm on 18 days are in April and in May, the smallest ones of less than 10 mm on 3 days in October and November. In autumn, the ITCZ is nearest the island and the water is warmest so that convection is more

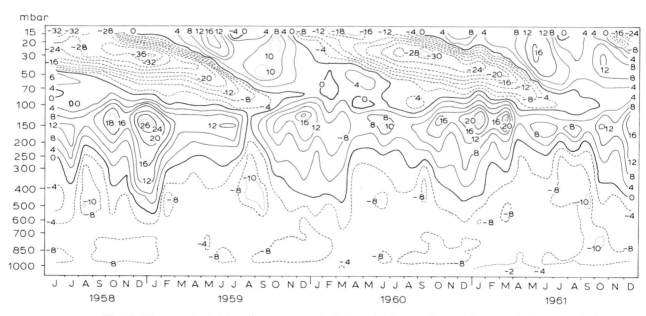

Fig. 12. Mean zonal wind (m s^{-1}) over Fernando de Noronha from surface to 15 mbar; solid line: west wind; dotted line: east wind (HENNING, 1969a).

pronounced than in spring when trade clouds prevail in stable air. The wind is weakest in autumn (March) and strongest in winter (July).

The upper wind data based on radiosonde measurements at the station from June 1958 to December 1961 are shown in Fig. 12 (HENNING, 1969a). The east winds reach a peak at 850 mbar where they have monthly means of 10 m s^{-1} in spring (September—December) and 5 m s^{-1} in autumn. Above 500 mbar, there is a distinct annual variation with easterly winds of 10 m s^{-1} in winter (August) and westerly winds in summer; the latter increase upward and dominate the year at 150 mbar where they vary between 20 m s^{-1} in summer and 10 m s^{-1} in winter.

Above 100 mbar, marked changes of the zonal wind occur between easterly and westerly with a period of about 2 years, as observed at other equatorial stations (e.g., Canton Island, WALLACE, 1966; BRIER, 1978). At 30 mbar, monthly averages of 30 m s^{-1} from the east alternate with averages of 15 m s^{-1} from the west. This stratospheric wind change propagates downward and disappears at 100 mbar (HENNING, 1969a).

The comparison of Fernando de Noronha with area A and the climatological charts for January and July shows many similarities. The temperature of the island is 1°C lower and the wind speed 1 m s^{-1} higher, which may be connected with the station height of 45 m. The evaporation values seem rather high; for the year they exceed those over the ocean by about 270 mm.

The ratio of the monthly precipitation at the station (in mm) to the precipitation frequency over the surrounding sea (in %) at Fernando de Noronha averages 20 mm %$^{-1}$ in April and 22 mm %$^{-1}$ in May which is unexpectedly high, but the precipitation amounts are probably somewhat increased by upslope motions. The cloud amount is also relatively high with 65%.

The precipitation at the station and that of Luanda (8°S) on the African coast during the rainy season (April—August at Fernando de Noronha and January—March at Luanda) are well correlated, suggesting a relationship between the precipitation in the ITCZ in southern summer over Africa and at Fernando four months later. Such a relationship does not seem to exist for the Brazilian stations, however. In the rainy months, the mean water temperature is a little higher and the wind blows from a more northerly direction which indicates that the ITCZ is then closer. In moist years, the Hadley circulation is weaker and displaced southward, whereas it is strengthened in dry years and extends farther north. The South Equatorial Current, coupled with the trades, may transport anomalies from Africa to the west; correlations over such distances with a time lag of four months are therefore understandable (EICKERMANN and FLOHN, 1962).

An evaluation of aerological observations at Fernando de Noronha (HENNING, 1969b) gives a net atmospheric heat flux across the Equator to the North Atlantic. It amounts to 30 · 10^{15}W integrated over the whole atmosphere. Its components are a flux of sensible heat (75%) and latent heat (25%). The annual variation has a peak in the southern summer, but in the lower troposphere the strongest flux is observed in August when the southeast trades cross the Equator in strength. Then a southward-directed flux appears in the upper troposphere as part of the Hadley circulation which is then well established.

Ascension Island

The area of Ascension Island is only 88 km^2. It lies within the South Equatorial Current and the southeasterly trades on the eastern border of area D. It is a volcanic island whose Green Mountain reaches a height of 860 m. The climate station is situated in Georgetown on the northwest coast (7°55'S 14°25'W).

The climate data were from the Climat collection of the Seewetteramt (Hamburg) and the following publications: *Africa Pilot*, Vol. II (TAUNTON, 1977); *Tables of Temperature, Relative Humidity and Precipitation for the World*, Part IV (LONDON, 1960b); *Weather on the West Coast of Tropical Africa* (LONDON, 1949); and *The Climate of Ascension Island* (BROOKS, 1931).

The data span the years 1899 — 1965, but unfortunately the period is variable for the different parameters. No data are available for visibility and thunderstorm frequencies. The climate of Georgetown does not belong to the inner tropics in spite of the short distance from the Equator. It is warm with an annual mean of 26°C, but dry during the whole year. The precipitation of 14 cm per year is lower than the "rain value of temperature" of Köppen's classification, and the climate is therefore a desert climate which in addition is hot (h) and isothermal (i) because the monthly averages of air temperature vary only between 25° and 27°C. The total description is BWhi. On the peak of the Green Mountain, the mean air temperature is 5°C lower, and 66 cm per year of rain fall there, quite evenly distributed with slight maxima in spring and autumn. The climate on the peak can thus be classified as Afi. The port suffers a chronic lack of drinking water, and cisterns on the peak of Green Mountain must supply the town.

The wind is weak. The mean speed amounts to 2 m s^{-1}. It blows with a steadiness of 95% from the southeast. The proportion of directions in 22.5° sectors are: 15% east-southeast, 52% southeast, and 14% south-southeast. Gales are unknown.

The annual variation is small. The summer (January—March) and winter (July—September) means are given in Table VIII.

The mean equivalent temperature is above the limit of sultriness, but frequently sinks below it at night, especially in winter. The climate is more comfortable than at Sao Tomé and Fernando de Noronha because of the drier air. The mean daily range of air temperature is 7°C, and the extremes reach 35° and 18°C.

The mean cloudiness amounts to 46%, the mean sunshine duration to 60% of the possible. A cloudless sky is rare (2%), however, but so is an overcast sky (11%) which occurs most frequently in September (30%) and least frequently in March (2%). A partly cloudy sky (4—6 tenths) is observed on 52% of the occasions. The monthly averages vary between a maximum of 55% in spring (September—November) and a

TABLE VIII

COMPARISON OF THE SUMMER MONTHS JANUARY–MARCH AND THE WINTER MONTHS JULY–SEPTEMBER FOR ASCENSION ISLAND

	Summer	Winter	
Air pressure	1013	1016	(mbar)
Wind speed	1.8	2.2	(m s^{-1})
Air temperature	27	25	(°C)
Vapour pressure	25	22	(mbar; 1 month later)
Equivalent temperature	64	58	(°C)
Precipitation	11	10	(mm month^{-1})

minimum of 40% in autumn (March—May). This is the opposite annual course of the precipitation which reaches a peak in April.

Morning fog reaches a frequency of 20% with an annual range similar to that of the cloud cover, and the peaks are often covered with clouds.

A comparison with area D and the charts for January and July shows some differences. The mean wind speed at Georgetown is only a third of that over the sea, and the vapour pressure is almost 2 mbar lower. The station is protected against the wind as it lies on the northwest coast of the island in lee of Green Mountain which also makes the precipitation amounts unrepresentative. Unfortunately, no precipitation station exists for comparison on the windward side of the island.

Georgetown is accordingly also protected from breakers; there are, however, dangerous "rollers"—a long swell with periods of more than 20 s which cause a heavy surf. In the annual mean, they appear on 22 days. In winter, when they approach from the southwest, they are born in the storms of the South Atlantic westerlies, but in summer they come from the northwest out of the North Atlantic. Sometimes they arrive from both directions and these cross seas may produce water pyramids as high as 5 m.

St. Helena

St. Helena is an island in the region of the trades and the Benguela Current on the border between areas F and H. Its area is 122 km² with several hills reaching 650—800 m; the highest peak, Mount Actaeon, is 818 m.

The permanent station on the island is Hutts Gate, but because its elevation is 604 m it cannot be used to describe the marine climate. Thus, the discontinued station in the port of Jamestown on the northwest coast of the island (15°55′S 5°43′W) is used here although only observations from 1853—1862 are available.

The climate data were taken from *Weather on the West Coast of Tropical Africa* (LONDON, 1949). Unfortunately, data on wind speed, sunshine duration, and the number of days with precipitation, thunderstorms, and fog are missing. The coast has a dry tropical climate with a mean annual temperature of 22°C and a precipitation of only 14 cm which, like the one on Ascension Island, can be classified as a desert climate BWhi.

The averages of Jamestown in summer (February—April) and in winter (July—September) are given in Table IX.

The mean daily range of temperature is 5—6°C, and the extremes are 34° and 14°C. The

TABLE IX

COMPARISON OF MEAN SUMMER AND WINTER VALUES FOR JAMESTOWN, ST. HELENA

	Summer	Winter	
Air pressure	1015	1019	(mbar)
Air temperature	24	20	(°C)
Vapour pressure	19.5	17	(mbar)
Relative humidity	65	75	(%)
Equivalent temperature	54	45	(°C)
Cloudiness	55	75	(%)
Precipitation	13	16	(mm month⁻¹)

annual vapour pressure of 18 mbar and a relative humidity of 70% give a mean equivalent temperature of 49°C. The limit of sultriness (55°C) is exceeded only in the summer months, but then frequently during the day.

In contrast, Hutts Gate in the mountains has a moist temperate climate (Cfli). It is about 6°C colder there with an annual mean of 16°C, and the precipitation is seven times higher (95 cm per year). The monthly amounts vary between 40 mm on 8 days in November and more than 100 mm on 16 days in March, and the monthly mean temperature has a range between 14°C in September and almost 19°C in March. The equivalent temperature stays on the average below the upper comfort limit even in March (49°C). The frequency of fog is about 10%.

The air pressure at Jamestown varies little both from year to year and from day to day. The standard deviation of pressure values remains below 2 mbar in each month (Chart 7), and the wind is correspondingly steady (90%, Charts 11 and 12). The cloud cover has a minimum of 50% in spring.

Jamestown is situated on the northwest coast of the island in lee of the mountains to shelter the port against the southeast trades, and the precipitation accordingly is low. Monthly values of 20 mm are observed in autumn (March—July) and only 4 mm in spring (September—January), and the variations from year to year are large. Annual amounts of up to 25 cm have been observed (LAMB, 1977). In moist months, the temperature is mostly above average. Thunderstorms are rare.

Like in Georgetown, rollers at St. Helena cause an enormous surf. The breakers approach in winter from the southwest, in summer also from the northwest out of the North Atlantic. This swell runs against the Benguela Current and is independent of wind and weather at the station.

Additional information can be gleaned from the charts for January and July and from the climatic tables of areas F and H. The monthly averages of air pressure and temperature in Jamestown are in this way corroborated, but the cloudiness is about 5 — 10% lower and the vapour pressure about 2 mbar lower at the station, which shows the lee effect of the mountains.

If the wind speed were undisturbed by the island, it would reach an average of 7 m s^{-1} in all months. The frequency of strong winds at sea is about 10%, somewhat more in winter and less in summer, but gales are rare. During the night and in the morning, the wind on the coast is strong and gusty owing to land and mountain breezes, but in the afternoon it is very weak.

Tristan da Cunha

Tristan da Cunha is a group of volcanic islands in the central South Atlantic Ocean. The area of the main island is 98 km^2 and it is situated on the northern border of the west wind zone in the southwest corner of area P. The climate station is near Edinbourgh at Herald Point on the northwest coast (37°3'S 12°19'W), and the 2062 m high volcanic mountain is 5.5 km from the station.

The climatic means are based on observations from the years 1942 — 1960, in part only from 1943 — 1947. Observations stopped in September 1962 after a volcanic eruption on the island and began again in 1972.

The data were taken from the following publications: *Africa Pilot*, Vol. II (TAUNTON,

1977); *Climatological Normals for the Period 1931 — 1960* (WMO, 1971); *On the Synoptic Climatology of the Tristan da Cunha Region* (VAN LOON, 1959); *Meteorological Observations at Tristan da Cunha* (PRETORIA, 1949).

Further discussions of the island's climate can be found in *Das Klima von Tristan da Cunha* (LOEWE, 1950) and the *Climate of Tristan da Cunha* (NEWNHAM, 1949). Besides that, Climat-reports of the Seewetteramt (Hamburg) are available.

The island has a moist temperate climate Cfl (Chart 3) with an annual mean air temperature of 14.5°C and an annual precipitation of 167 cm. The influence of the mountains undoubtedly enhances the amount of precipitation. Summer is somewhat drier (100 mm in January) than winter (173 mm in August). This difference may be a little bigger at sea. The classification l is valid because all the monthly temperature averages lie above 10°C.

The daily range of temperature is 4 — 5°C and the extremes are about 23° and 3°C. With a relative humidity of 80%, the mean vapour pressure is 13.5 mbar and the equivalent temperature 34.5°C. The climate of the island is mostly comfortable; sultriness happens only occasionally in the warm season, and more frequently conditions are below the lower limit of comfort with an equivalent temperature of 30°C.

The proximity of the frontal zone is felt in the annual variation of the meteorological elements (see Table X).

TABLE X

COMPARISON OF MEAN SUMMER (JAN–MAR.) AND WINTER (JULY–SEPT.) VALUES FOR TRISTAN DA CUNHA

	Summer	Winter	
Air pressure	1017	1019	(mbar)
Air temperature	17.5	12	(°C)
Vapour pressure	16	11	(mbar)
Equivalent temperature	42	28	(°C)
Precipitation amount	110	165	(mm month^{-1})
Precipitation frequency	12	20	(days month^{-1})
Wind speed	9	11	(m s^{-1})

The mean pressure shows the smallest variations in the course of a year, yet the monthly standard deviation of pressure values is 7.5 mbar (Chart 7). That reflects a strong variability of daily pressure, especially in winter, which in turn indicates that the weather at the station is rather changeable.

The winds blow mainly from northwest or southwest but with a steadiness of only 40%. At sea the frequency of strong winds ($\geq$ 6 Beaufort) is 30 — 35% in winter (Chart 16), but only 15% in summer (Chart 15); the frequency of gales ($\geq$ 8 Beaufort) is about 2% in summer and above 10% in winter.

The annual average cloud cover is 72% with a minimum in January of 60% and a maximum in June and September of 80%. In the annual mean, there are only 9 clear days (cloudiness $\leq$ 2 tenths), but 220 nearly overcast days ($\geq$ 8 tenths). The sunshine duration at Herald Point is correspondingly short with an annual average of 3.7 h per day = 31% of the possible time. The frequency of fog is 1 day per month on the island, 5% at sea and is quite evenly distributed over the year.

The annual amount of precipitation varies comparatively little from year to year. There were 196 cm in 1947, 148 cm in 1948. The monthly sums differ more. During the period of observation, the highest amount of 349 mm fell in March 1949 and the lowest of only 28 mm in February 1956. It rains on half of all days, in winter more frequently than in summer. Thunderstorms occur only in winter, about 1 per month and 4 per year, owing to the less stable stratification then in the air over the ocean.

The ratio between amount of precipitation in mm month^{-1} and frequency in % is on the average 7 mm %$^{-1}$ with the amount taken from Herald Point and frequencies from ship observations. Because the climate station is situated on the west coast which is the windward side, the precipitation amount is probably twice that over the sea, if the value on Chart 38 is to be believed. At Sandy Point on the east coast, however, a precipitation amount of 360 cm was measured in one year, twice that of Herald Point in the same year, although it is in the lee of the mountains.

A comparison with the climate data of area P yields distinct differences for all parameters, because the average position of the ship observations in P is farther north and the climate elements in the west wind belt have strong meridional gradients there.

These gradients make the weather highly variable as numerous depressions and fronts migrate through the area (p. 119f). Warm air from the north alternates with cold air from the region of the West Wind Drift; and the mean atmospheric frontal zone near Tristan da Cunha has its counterpart in the Subtropical Convergence in the ocean which divides subtropical from antarctic surface water (p. 29).

Gough Island

The area of Gough Island is only 73 km^2, but it has several mountain peaks as high as 900 m. It is only 3° south of Tristan da Cunha in the west wind belt south of area P. The climate station is situated on the east coast of the island (40°19'S 9°54'W), protected against wind and sea.

The climate data were taken from the annual reports of the Weather Bureau, Republic of South Africa, from 1956 to 1973. Wind data are not available, but they would not reflect the undisturbed marine conditions anyway.

The island has a moist temperate climate Cfb (Chart 3). The annual mean equivalent temperature of 28°C is derived from a mean temperature of 11.5°C and a relative humidity of 80%; it sometimes goes below the lower limit of comfort in summer and frequently in winter. The mean daily range of temperature is 5°C.

Pressure (1015 mbar), relative humidity (80%), cloud cover (80%), and precipitation (327 cm year^{-1}) are characterized by only small annual variations, although the pressure has a dominant half-yearly oscillation with maxima in spring and autumn and an amplitude of about 1.5 mbar (VAN LOON, 1967). In contrast, monthly values of air temperature and vapour pressure show a pronounced annual variation. Table XI shows mean values for the summer months January—March and the winter months July—September.

The wind blows mainly from westerly directions with a mean speed of 12 m s^{-1} at sea and a steadiness of 50%. The frequency of strong winds is about 35% and the frequency of gales 10%. In winter, there are more gales (15%) than in summer (5%).

The amount of cloudiness is high, and the monthly averages vary only slightly about

TABLE XI

COMPARISON OF MEAN SUMMER AND WINTER VALUES FOR GOUGH ISLAND

	Summer	Winter	
Air pressure	1015	1016	(mbar)
Air temperature	14	9	(°C)
Vapour pressure	13	9.5	(mbar)
Equivalent temperature	34	23	(°C)
Precipitation	225	300	(mm month^{-1})

80%. Accordingly, the sunshine duration is short with 5.5 h per day (about 35% of the possible) in summer and 1.5 h per day (15%) in winter. The fog frequency at the station is higher in spring and autumn than in summer and winter, whereas at sea it is low in winter and high in summer when frequencies of 5% are reached.

Precipitation falls on 226 days per year, that is, 62% of all days. Seven of those, or 2%, are days with snowfall. The mean amount per day of precipitation is 14.5 mm with little difference between months. The precipitation frequency at sea is 20% near the island which corresponds to a mean productiveness of 13 mm %$^{-1}$.

The annual precipitation of 327 cm at Gough Island is about twice that at sea (Chart 38), despite the fact that the station is on the lee side of the island, and it is reasonable to assume that the cloudiness at sea is lower too. Unfortunately, the number of ship observations in that area is too small and the scatter of the elements owing to weather variability is too large to allow a meaningful comparison.

Falkland Islands

The Falkland Islands are a group of more than 100 islands on the southeastern part of the Patagonian Shelf, northeast of the southernmost point of South America. The climate station is Port Stanley (51°42'S 57°52'W) at the southeast corner of East-Falkland in a protected bight; till 1945 directly on the shore at an elevation of 2 m and afterwards south of the town at an elevation of 53 m.

Regular observations are available since 1905, the first ones from 1842. The climate data are published in *Annual Meteorological Tables, Falkland Islands and Dependencies, Stanley*. Climate evaluations are presented in: *South America Pilot*, Vol. II, 15th Ed. (TAUNTON, 1971); *Climatological Normals for the Period 1931–1960* (WMO, 1971); *Tables of Temperature, Relative Humidity, and Precipitation for the World*, Part II (LONDON, 1960a); FABRICIUS, 1957; *The Meteorology of Falkland Islands and Dependencies 1944–1950* (PEPPER, 1954); *Sailing Directions for South America*, Vol. II, 5th Ed. (WASHINGTON, 1952); *Climatological Data for the Southern South America* (REED, 1929); and *Climate and Weather of Falkland Islands and South Georgia* (BROOKS, 1920).

The islands are located on the northern border of the snow climate ET (Chart 3), in the changeable weather of the west wind regime. The cold Falkland Current flows northward around the islands (Chart 19); it carries along icebergs, mostly in autumn. The cold current not only cools the overlying air in general but also intensifies the weather conditions in the rear of the passing cyclones, and the climate is therefore cooler and more humid than in the same latitude of South America (p. 93).

The means of the meteorological elements in the summer months January — March and the winter months July — September are shown in Table XII.

The mean daily range of temperature is 5 — 8°C, but the irregular variations are twice as large as the mean daily range. In summer, most of the scatter lies between 2° and 20°C, in winter between — 5° and + 10°C. Days with frost are most frequent in winter; they total 65 days per year, but frost may occur in any month.

The monthly averages of the equivalent temperature lie beneath the comfort limit of 30°C. The relative humidity has an annual mean of 83% between mean monthly values of 77% and 88% with average daily variations between 86% at night and 80% in the daytime.

The wind blows mainly from westerly directions with a steadiness of 40%. The mean wind speed amounts to 7 — 8 m s^{-1}. The frequency of gales reaches 9% in summer,

TABLE XII

COMPARISON OF MEAN SUMMER AND WINTER VALUES FOR THE FALKLAND ISLANDS

	Summer	Winter	
Air pressure	1001	1005	(mbar)
Air temperature	9	2.5	(°C)
Vapour pressure	9	6.5	(mbar)
Equivalent temperature	23	12	(°C)
Relative humidity	80	85	(%)
Precipitation	65	45	(mm month^{-1})

15% in winter, and the frequency of strong winds is about 20%; 35 m s^{-1} are observed about once a year.

The annual mean pressure of 1003 mbar reflects the proximity of the subantarctic trough and the position of the station near major storm tracks (p. 121). The variation of monthly mean values about the annual mean amounts to only ± 3 mbar with maximum in September, minimum in January. That annual difference is due to the variations in the subantarctic trough whose axis in January runs through the tip of the Antarctic Peninsula but in September is 6° closer to the Pole in that area. The mean variation of pressure from year to year is about 2 mbar, and the mean monthly standard deviation is 8.5 mbar (Chart 7).

The monthly average cloudiness is 70 — 78%, greater in summer than in winter. Low clouds with a base at 150 or 100 m are frequent in summer, spreading a dull and gloomy mood. Overcast or nearly overcast sky (cloud cover above 85%) appears on more than half of all days, clear sky (cloud cover less than 15%) on only 5% of all days.

The duration of sunshine is little more than 4 h per day in the annual mean (36% of the possible). Half of all days have less than 3 h of sunshine and about 30 days in summer more than 9 h.

The frequency of fog is about 5% at sea with a maximum in summer caused by cooling of moist, warm air over the Falkland Current. The visibility is also reduced by low clouds, especially in drizzle, rain or snow. The island has 43 days per year with fog that mostly originates from nocturnal radiation, so that in summer the tendency for fog is lower in the daytime. But even then snowfall and drifting snow occasionally cause bad visibility.

Precipitation falls on nearly half of all days, but the amounts are often small because of the low water vapour content in the atmosphere. Stable temperature stratification above the Falkland Current also decreases the possibility of convective precipitation. Monthly precipitation amounts vary between 70 mm in summer and 40 mm in spring, and the annual amounts between 50 and 100 cm.

The mean monthly amount of 54 mm and the mean frequency of about 17% result in a productiveness of only 3 mm $\%^{-1}$. Stanley does lie in the lee of the island, but the lee effect of the Andes may be more important as there is a persistent westerly flow in the troposphere and foehn may be traced as far as the Falkland Islands.

Thunderstorms occur on 4 days in summer on the average. Snow falls on almost half of all precipitation days, namely on 71 days out of 155 per year, and no month is without snow.

If one calculates wind roses of the different meteorological elements, that is, their dependence on the wind direction (BROOKS, 1920), the following may be noted.

Pressure is lowest with westerly or west-northwesterly winds, highest with easterly winds.

Air temperature is lowest with south wind, highest with northwest wind.

Relative humidity is lowest with northwest wind, highest with easterly wind.

Cloudiness shows little dependence on wind; it is somewhat less with easterly than with westerly winds.

The length of the record makes it possible to investigate climate fluctuations uninfluenced by civilization. Averages of air temperature of the following periods indicate no marked differences, only a slight decrease of temperature after 1940, some of which may be ascribed to the change of level of 51 m in 1945 (compare also with the decade averages of sea surface temperature on p. 33f, Chart 22): 1875−1877, 6.0°C; 1904−1910, 6.0°C; 1911−1920, 6.1°C; 1921−1930, 5.7°C; 1931−1940, 6.1°C; 1941−1950, 5.9°C; 1951−1960, 5.8°C; 1961−1964, 5.4°C. The differences of monthly and annual averages are also relatively small.

South Georgia

South Georgia lies north of the Weddell Sea in the arc of islands connecting South America with Antarctica, and is located 30°E of the southern point of South America, 40° west of Bouvet Island which has no climate station. With an area of 3755 km², South Georgia consists of a range of steeply rising mountains 2000 m in height and extending southeastward. Mount Paget is 2915 m.

The climate station is in Grytviken on the sheltered east coast (54°16′S 36°30′W). Observations have been made since 1905 plus a short series from 1882−1883. The climate data are published in the same annual reports as those for the Falkland Islands. Climate evaluations are published in *The Antarctic Pilot*, 4th Ed. (TAUNTON, 1974); *Climatological Normals for the Period 1931−1960* (WMO, 1971); *Tables of Temperature, Relative Humidity, and Precipitation for the World*, Part II (LONDON, 1960a); *Estadisticas Climatologicas 1901−1950* (BUENOS AIRES, 1958); *Climate of Subantarctic Islands* (FABRICIUS, 1957); *The Meteorology of the Falkland Islands and Dependencies 1944−1950* (PEP-

PER, 1954); *Sailing Directions for South America*, Vol. II (WASHINGTON, 1952); and *Climatological Data for the Southern South America* (REED, 1929).

The island has a snow climate ET in the region of variable weather within the west wind zone. It is washed by the waters of the Antarctic Circumpolar Current which sets east-northeastward in this region. Although only 2.5° farther south than the Falkland Islands, Grytviken is 4°C colder with an annual mean of 1.7°C and the precipitation amount (140 cm year^{-1}) is also twice that at Port Stanley.

The mountainous island, especially the western part, is mostly covered with snow and the valleys are filled with glaciers. Otherwise, the climates of South Georgia and the Falkland Islands are similar (STRETEN, 1977).

The mean pressure is lowest in November–January (993–994 mbar) and highest in August–October (1000–1001 mbar). The daily pressure wave is less than 0.1 mbar, with maximum at night and minimum in the afternoon, and is entirely negligible in comparison with the daily aperiodic variations.

The seasonal displacement of the west wind belt with its frontal zone causes an annual variation of all meteorological elements. The monthly mean values of summer (January–March) and of winter (July–September) are shown in Table XIII.

The daily range of air temperature is 6–7°C. The equivalent temperature is 10°C, far below the limit of comfort.

TABLE XIII

COMPARISON OF MEAN SUMMER AND WINTER VALUES FOR SOUTH GEORGIA

	Summer	Winter	
Air pressure	995	1000	(mbar)
Air temperature	5	−1	(°C)
Vapour pressure	6	4	(mbar)
Equivalent temperature	14	5	(°C)
Cloudiness	75	65	(%)
Precipitation	116	124	(mm month^{-1})

The strong meridional temperature gradient in the area is connected with the Antarctic Convergence in the sea. This is the northern border of the cold melting water of low salinity, which lies as a 100 m thick layer above heavier and saltier water. The Convergence, or the Polar Front in the ocean (p. 29, Fig. 3), is normally found between the Falkland Islands and South Georgia. The sea surface temperature difference of about 5°C across it is responsible for a similar air temperature difference between the two stations. On the average, the air temperature at South Georgia is about 2°C above normal with northerly winds and about 1°C below with southerly winds (FABRICIUS, 1957).

The extreme temperatures in a month differ by more than 20°C. The frequency distribution of temperature values shows that positive deviations from the mean are mostly larger (as much as +15°C) than the negative ones (−10°C), apart from rare extremes. Circulation patterns of long duration that may be associated with a meridional displacement of the Antarctic Convergence, give rise to a fairly large variability of monthly and annual averages. The mean variation of monthly averages from year to year amounts to about 2°C, that of annual averages to about 1°C.

For a short time a foehn can cause a sudden increase of temperature with a simultaneous decrease of vapour pressure on the downwind side of the mountain range, while it rains heavily on the upslope side. The foehn in Grytviken is enhanced by 623 m high Mount Hodges rising immediately west of the town. The occurrence of foehn is reflected in the extreme maxima which in many months are higher than those at Tristan da Cunha, 17° closer to the Equator. An example is given in Table XIV. The equivalent temperature increased by about 4°C and then decreased by 2°C. The foehn effect itself caused a warming of about 5°C, obtained by reduction to the same equivalent temperature, which corresponds to the raining out of a cloud mass about 1000 m thick.

In spite of higher equivalent temperature (the new air was somewhat moister and warmer), the foehn atmosphere became warm and extremely dry.

The wind at Grytviken blows chiefly parallel to the coast from northwest, in summer

TABLE XIV

AIR MASS APPROACHING FROM 290°, CAUSING A FOEHN (From PEPPER, 1954)

Date	GMT	Wind (°)	Speed (kts)	Temp. (°C)	Humid. (%)	Vap. pres. (mbar)	Equiv. temp. (°C)
22/9/50	18	20	3	0.6	83	5.3	8.5
23/9/50	0	290	22	8.9	22	2.5	12.7
23/9/50	6	290	24	8.2	14	1.5	10.5

with an on-shore and in winter with an off-shore component. The relatively northern position of the polar trough in summer results in an increased frequency of easterly and southeasterly winds, especially in January. The steadiness of the wind is only about 50%, indicating the variability of the weather.

The wind speed at the climate station is very disturbed by the topography. An annual average of only 4 m s^{-1} is far too low, compared with the ship observations of the adjacent sea which have an average of about 11 m s^{-1}. In winter, the cyclones are larger and more intense than in summer and, accordingly, the mean wind speed and the frequency of gales are higher in winter than in summer.

Over the year, 140 cm of precipitation is distributed on 154 days. With a frequency of 25% over the surrounding sea, productiveness is almost 5 mm %$^{-1}$. The higher precipitation amount than at the Falkland Islands indicates that South Georgia is not influenced by the lee effect of South America, but also that the island lies closer to a major storm track so that the cyclones will be more intense than in the west, especially in winter. The monthly precipitation amounts vary between 140 mm in autumn and winter and 90 mm in spring and summer. The difference between years but also between different coastal locations, even neighbouring ones, is considerable. The winter precipitation consists mostly of snow, 10–13 days with snow are observed per winter month. Even in January and February, there are still four days per month with snow. Drifting snow may distort the precipitation measurements so that the measured amounts in winter are less reliable. The snow limit on the south side of the island is at a mean elevation of 300 m. On the north side it varies more because the snow there frequently melts owing to the foehn, even in winter.

The annual mean cloudiness is 70% with more clouds observed in summer than in winter and spring. Clear days with a cloudiness below 15% are rare (4% of all days) and appear mostly with foehn. All the more frequent are overcast or nearly overcast days with a cloudiness above 85%—almost half of all days. It is possible that there are more clouds over the ocean because orographic influences, notably foehn, may reduce the cloud amount near the station.

The mean daily sunshine duration is 3.5 h (29% of the possible). Fog is more frequent in summer than in winter, on the average 3 and 1 days per month, and there are 25 days with fog per year.

The sea area around South Georgia lies south of the Antarctic Convergence and most often north of the sea ice limit (MACKINTOSH and HERDMAN, 1940; SCHWERDTFEGER, 1970) which approaches the island to within 100 nautical miles in late winter (September — October). It may reach 52°S in extreme years. The ice drifts nearly in the direction of the geostrophic wind and at 1.35% of the wind speed (SHULEIKIN, 1968). The ocean round South Georgia is frequented in summer by mostly tabular icebergs separated from the Antarctic shelf ice. The mean limit of the ice is associated with the atmospheric circulation and temperature, but the temperature is in turn influenced by the occurrence of ice. The mean temperature will thus vary from year to year according to the circulation and ice conditions between the antarctic cold source and the warmer surface water of the South Atlantic. This variability is characteristic of all seasons but is most pronounced in winter.

The month-to-month variability of all elements is likewise high; over the adjacent sea the standard deviation of pressure is about 12 mbar, of the wind 6 m s^{-1}, and of air temperature it is 2°C.

Off-shore climates

In the coastal areas of the South Atlantic Ocean, climate and weather are of special interest because these regions are economically important. Five regions along the South American coast will be described on p. 87f and four along the African Atlantic coast on p. 94f.

The climatological description is primarily based on the analyses in Charts 4 to 53, on the tabulated values of the selected areas, and on climate values of the applicable island stations. Further sources of information are climate descriptions for navigation and the volumes of the *World Survey of Climatology* for the adjacent continents. In addition, typical weather situations will be described on p. 113f.

For the antarctic sea areas, the reader is referred to the *Navy Pilots* (TAUNTON, 1974) and Volume 14 (SCHWERDTFEGER, 1970) of this series; for the Drake Strait, to SCHWERDTFEGER (1962) and VOWINCKEL (1957).

South American off-shore areas

Detailed descriptions for the sea areas along the South American coast can be found in *Sailing Directions* (WASHINGTON, 1952, 1981); *Naval Air Pilots* (WASHINGTON, 1945a, 1945b); *Navy Pilots* (TAUNTON, 1971, 1975); and *Seehandbücher* (HAMBURG, 1981b). The climate of the continent is described in the *World Survey of Climatology*, Volume 12 (SCHWERDTFEGER, 1976).

North coast of Brazil

The eastern part of the Brazilian north coast, east of the mouth of the Amazon, belongs to the South Atlantic Ocean if the Equator is the northern boundary. Climatological data for the region are contained in the tables of area A, the climate records of Fernando de Noronha, and the description by RATISBONA (1976) of the adjoining land.

In the sea area off this coastal segment, tropical warmth dominates throughout the year with mean water temperatures of 27 − 28°C and air temperatures of 26 − 27°C. The mean daily range amounts to about 10°C at coastal stations but to only 1°C over the sea.

The mean vapour pressure of 27 to 29 mbar and relative humidity of about 80% fit the tropical climate. The equivalent temperature calculated with these values is 70°C and indicates constant high sultriness.

The monthly mean pressure varies from 1010 − 1011 mbar in the southern summer to 1013 − 1014 mbar in winter with a standard deviation below 2 mbar. This is the lowest value along the American east coast and it reveals a very steady circulation, although pressure differences are always small at the Equator because Buys Ballot's law is not valid there.

The daily tidal variation dominates the pressure changes; it amounts to 3 − 4 mbar with maxima at 10h00 and 22h00 and minima at 04h00 and 16h00 local time. The yearly variation of air pressure depends not only on the movement of the ITCZ but also on a planetary seasonal course of vapour pressure which contributes to the total pressure. In the eastern part of the area, steady trade winds blow from southeast which back to east farther west, in summer partly to northeast. The steadiness of the wind is 80% during summer and 85 − 90% in winter. Gales are rare; a frequency of 1% is reached only in southern summer. The mean wind velocity over the sea is 3 − 4 Beaufort, and the daily alternation of land and sea breeze plays only a subordinate role. Backing winds are a sign of the ITCZ and are combined with more cloudiness and precipitation. The ITCZ crosses the coast at 3°S in March/April and north of the Equator from June to December (RATISBONA, 1976, fig. 2).

The average wave height is 1 − 1.5 m. The state of the sea contains a swell that frequently arrives from the direction of the trade wind, but in northern winter it also originates in the North Atlantic storm areas. On the coast it causes a corresponding surf.

In late summer (March − April), a cloudiness of 60% and a precipitation frequency of more than 10% show the proximity of the ITCZ. In winter to spring, when the ITCZ is situated well north of the Equator, cloudiness is reduced to 40% and the precipitation frequency to less than 5%. Then Cabo de Sao Roque has the highest amounts because of the strong on-shore trade winds which generate a coastal convergence.

The precipitation consists mostly of showers with daily amounts as high as 100 mm. On the average, 300 mm per month are measured during southern summer and less than 100 in winter. The showery character results in a strong variability of precipitation in space and time. The largest yearly sum of nearly 300 cm is observed in the region of the Amazon estuary where the highest frequency of thunderstorms is also found. Belem reports 63 days with thunder per year, mainly in the southern summer. There are fewer over sea than over land. Differences between years are linked to the intensity of the convergence. When the ITCZ stays north of the Equator, dry conditions prevail in northeastern Brazil, which may have catastrophic consequences for the economy of the country (RATISBONA, 1976).

Fog is rare off the north coast of Brazil, even during the rainy season. The trade wind and the mostly unstable air above the warm South Equatorial Current provide sufficient turbulent exchange. The visibility, however, is reduced in heavy showers.

Over the sea, precipitation and evaporation are about equal so that the water budget is nearly balanced. In the region of the Amazon mouth, the balance is weakly positive, because precipitation prevails over evaporation, but it is negative east of 45°W; near Cabo de Sao Roque, the yearly deficit amounts to 100 cm. Owing to evaporation, there is a heat flux from the ocean to the atmosphere with a yearly average of 150 W m^{-2} to which the sensible heat flux contributes only 5 W m^{-2}. In contrast, the radiation balance at the sea surface is only about 130 W m^{-2}, and its changes during the year are small. In southern winter, however, the highest values off the South American east coast are found here because the insolation is then strongest at the Equator and has a steep gradient toward the south.

The difference between radiation balance and heat flux results in a negative heat balance at the sea surface with a yearly mean of $10-20$ W m^{-2}. All other coastal areas of the South Atlantic show a positive annual heat balance. Along the north coast, the values are lower in the west than in the east during the southern summer, but in winter it is the opposite. The heat surplus, which is transferred to the atmosphere and carried to the ITCZ, in turn is made up for by advection within the warm Equatorial Current.

Northeast coast of Brazil

This coastal strip stretches from Cabo de Sao Roque at the northeast corner of Brazil to Cabo Frio at 23°S and is represented by area E. This region also has a tropical climate.

The air temperatures are determined by the trade winds which mostly have an on-shore component and by the sea surface temperature. The annual mean is about 26°C and the range of the monthly means is 3°C; the daily range is $6-8$°C on the coast. Within the area is a decrease of 3°C from north to south.

The temperature difference between the air and the warm Brazil Current below is -2°C along the coast during winter, and the current is also responsible for the small variation of the air temperature whose daily minimum hardly ever goes below 10°C and on the average stays about 20°C.

In the south, the Brazil Current leaves the coast where instead cold water wells up and water temperatures sometimes occur that are 10°C lower than their immediate surroundings.

The vapour pressure has mean values of 28 mbar during summer, whereas in winter it decreases to 26 mbar in the north and 21 mbar in the south. The corresponding equivalent temperatures are 69°C in summer near Cabo de Sao Roque and 53°C in winter at 20°S, which is below the upper limit of comfort (55°C). Everywhere else sultriness prevails. Over the Brazil Current, the vertical exchange is so intense that the average relative humidity lies below 80%.

The wind blows mostly from the east, in summer at 3 and in winter at 4 Beaufort, and corresponding to this the frequency of strong wind increases from 1% in summer to 10% in winter. In the north, strong winds are more frequent than in the south. Near Salvador, winter gales from southerly directions are called "Cambueiros"; gales are rare at other times. Water spouts sometimes occur in the vicinity of the cape. The daily land and sea breeze is mostly well developed.

The sea reaches mean wave heights of 1 to 1.5 m which at times are enhanced by swell from the trade wind region, during winter sometimes by swell from the storm regions of higher latitudes.

With a mean cloudiness of about 45%, the coastal region is sunny, especially in summer. The amount of sunshine at Cabo de Sao Roque is largest with 3000 h per year (RATIS-BONA, 1976, fig. 4). Fog over the sea is rare, but in the morning or after rain it occurs more frequently on the coast.

The precipitation falls mainly as showers that are intensified along the coast when the wind blows on-shore. The frequency of rain is about 5%. The highest frequencies and the largest amounts are observed near Cabo de Sao Roque in the spring and near Cabo Frio during summer as the dry season shifts from autumn in the north to winter in the south. During the rainy season, mean monthly amounts of 200—300 mm are measured on the coast. The yearly sum reaches 100—200 cm. The showery character of the precipitation is responsible for the variability between places, months, and even years. Thunderstorms are rare over the sea with a frequency of 1%; on the coast about 10 days of thunder occur per year, mainly in summer.

Evaporation over the sea is strongest at 10°S with a yearly average of about 200 cm and decreases southward to 120 cm near Cabo Frio. The water balance, negative near Cabo de Sao Roque, accordingly increases southward so that south of Salvador its yearly mean is zero. It is positive in summer and negative in winter. The heat flux from the ocean to the atmosphere along the South American coast reaches its maximum here. At 10°S, a yearly mean of 160 W m^{-2} is assumed, and farther seaward over the Brazil Current it increases to yearly mean values of more than 180 W m^{-2}. It decreases to about 100 W m^{-2} in the south with higher values in winter than in summer. An opposite seasonal course is shown by the radiation balance at the sea surface. It is about 170 W m^{-2} in summer and 70—100 W m^{-2} in winter when it decreases southward. The highest yearly values of 130 W m^{-2} occur at 10°S, but the decrease towards the south is slight. The heat balance is almost zero near the coast, but the values drop as one goes seaward and the yearly mean reaches a minimum of -50 W m^{-2} over the Brazil Current. Heat withdrawn from there is released as latent heat in the clouds over Brazil.

Southeast coast of Brazil

This is the region between Cabo Frio (23°S) and about 30°S. It is represented by area I in the north and area K in the south, and although it has a tropical climate (Af in Chart 3), it is noticeably cooler than those described above.

The air temperature is fairly uniform along the coast in summer, 24° to 25°C, but in winter it ranges between 15° and 20°C. The annual range on the coast goes as high as 7°C and the daily range is 5—10°C. The fluctuations are thus larger than north of Cabo Frio. Extremes in summer may reach nearly 40°C and in winter light frost can sometimes occur.

Over the Brazil Current, temperatures are 2—3°C higher than on the coast in winter and the annual range is only 5°C. Along the coast between Cabo Frio and Rio de Janeiro, the mean temperatures are 2°C lower because here the trade winds shear off the surface water and cause upwelling.

With mean values of vapour pressure between 23 and 27 mbar in summer and 15 to 21 mbar in winter, the monthly mean values of equivalent temperature range from 40° to 65°C. The upper comfort limit is everywhere exceeded in summer. During winter, it is reached in the north, but otherwise conditions are comfortable. The relative humidity is high with 80—85%, especially over the upwelling water. Farther seaward the relative humidity decreases.

The yearly cycle of the pressure reflects the continental influence; in summer the mean is 1011—1012 mbar and in winter 1020—1021 mbar.

The trade winds which mostly blow from the northeast attain a mean of 4 Beaufort. The steadiness is 40% in summer and in winter 30%. "Pampeiros" (strong-to-gale-force winds from southwesterly directions) frequently occur in winter, and sometimes the cold "Minuano" blows from the south. Gales are rare in the north; in the south they reach a frequency of 1% in winter. Close to the coast, the wind directions change under the influence of the sea breeze ("Brisa Maritima" or "Viracao") during daytime and the land breeze ("Terral") at night.

The wind which blows mostly parallel to the coast or even off-shore does not allow a significant sea to build up, and the mean wave heights reach only 1 m. There is a noticeable swell from the stormy higher latitudes, particularly in winter.

The mean cloud cover is 50—60%, in the spring more than 65% with a maximum at the coast near Rio de Janeiro. Fog occurs about 3% of the time in spring. During autumn, it is less frequent because the water then is relatively warm and the atmosphere above unstable. Upwelling of cold water increases the probability of fog. In winter, radiation fog may drift on to the sea.

The mean precipitation frequency is 5% with only little seasonal difference. In summer when the amount of rain is biggest, the coast receives about 200 mm per month, in winter only 50 mm per month. The yearly totals vary between 100 and more than 200 cm and are largest at Santos.

Thunderstorms occur mainly at the coast on summer afternoons or along cold fronts. Five days with thunder are observed on the coast in January, 30 in the year.

Cold fronts (called "Pampeiros" like the accompanying strong winds) approach from the south, occasionally with southwest gales as "Minuano". The cold air arrives with rain showers and vigorous squalls, and the cold front moves across the water as a towering squally wall of clouds which spans the sky in a wide arc. Such cold outbreaks occasionally reach even Recife. The air is not only cooler behind the cold front but drier and more stable (RATISBONA, 1976, fig. 3). The sudden cooling may amount to 4—8°C on the sea and 10°C at the coast.

Evaporation is down to 100 cm per year in the bay of Rio de Janeiro as a result of the upwelling. The water balance is thus positive there, but elsewhere on the coast it is zero in the yearly mean. Farther seaward the evaporation increases so that the water balance becomes negative, principally in winter.

The heat flux from the ocean to the atmosphere has a minimum in the bay of Rio de Janeiro with a yearly mean of 70 W m^{-2} due to evaporation. Farther out to sea, the heat flux increases to more than 100 W m^{-2}, especially in winter over the Brazil Current. The radiation balance at the sea surface has a pronounced yearly cycle with high values of 160—190 W m^{-2} in summer increasing southward, and 40—60 W m^{-2} in winter decreasing southward. In the bay of Rio de Janeiro, the yearly mean of the radiation balance reaches a minimum of 110 W m^{-2} over the upwelling water and increases north and south of there to 115 W m^{-2}.

The heat balance between the radiation balance and vertical heat flux at the sea surface is nearly zero if averaged over the whole sea area under consideration. It is positive at the coast with yearly mean values of 20—30 W m^{-2}. Seaward it decreases and becomes negative over the Brazil Current, where heat is extracted from the warm ocean surface and carried away in the atmosphere by the trade winds. Otherwise, positive values prevail in summer and negative ones in winter, which decrease seaward. This yearly cycle matches the seasonal course of the monsoon circulation in the coastal areas and the precipitation in Brazil.

Rio de la Plata

The region of Rio de la Plata, into which the Uruguay and the Paraná flow, is of special meteorological interest because here the subtropical frontal zone can be found in the mean (p. 118). In the ocean, this frontal zone is reflected in the Subtropical Convergence, which separates the subtropical surface water in the Brazil Current from the subantarctic water which arrives in the Falkland Current (p. 29, Fig. 3, Chart 19). Climatologically, this convergence outlines the border between the tropical climate A in the north and the temperate climate C in the south and separates the comparatively constant weather of the trades from the variable weather of the westerlies. The sea area is represented by the selected areas N in the north and Q in the south. The land climate is described by PROHASKA (1976).

The climate is temperate-warm with a marked seasonal variation. Sultriness sometimes occurs during the day in summer. A mean temperature of 21°C and a vapour pressure of 20 mbar yield an equivalent temperature of 51°C which is near the upper comfort limit of 55°C. On the other hand, in winter a temperature of 10°C and a vapour pressure of 12 mbar result in an equivalent temperature of only 28°C that is below the lower limit of comfort (30°C).

The yearly cycle is enhanced over land. Hot days with a maximum of 30°C or more are frequently observed in the summer, and in winter frost occasionally occurs at night, particularly in the south.

A mean pressure of 1011 mbar in summer and 1018 mbar in winter still points to subtropical conditions, but the region is situated on the northern boundary of the zone of strong meridional pressure gradients, and the mean standard deviation of pressure is relatively high, 6 mbar. This is a transition zone between the anticyclone of the horse latitudes in the northeast, the continental heat low, and the subpolar cyclones in the south.

Cyclonic disturbances develop in the trough which extends southeastward from the continental heat low situated north of 30°S over the Paraná lowlands. They cross the coast line (MARKGRAF, 1954) and occasionally deepen over the sea or move northward along the coast. The north winds on their front side are accompanied by heavy precipitation and the south winds on their rear side by cooling. The wind over this coastal area varies correspondingly. Northeast winds prevail during summer, as they also do off the southeast coast of Brazil, and extend to about Rio Negro. In contrast, this district is situated in the region of prevailing westerlies in winter. The frequency of strong winds is 5% in summer and 10% in winter. For gales, it is 1% in the summer and 3% in winter. The differences are large near the coast and increase seaward. On undisturbed days, a sea breeze develops during daytime ("Virazon") and a weaker land breeze ("Terral") in the night. Gale-force southeasterly winds are called a "Sudestado" (SCHWERDTFEGER, 1958), and they occur with low pressure to the north and high pressure to the south (WEGENER, 1927). They indicate the existence of a high-level, cut-off low north of Buenos Aires and are often connected with blocking over South America. Their frequency is about 10 per year.

Eighty to 100 cold fronts (Pampeiros) a year (HOEFER, 1925) move northward along the South American coast. Their arrival is announced by an extensive roll cloud (arcus) approaching from the southwest and the wind frequently reaches gale force, accompanied by heavy thundershowers. The southwesterly wind blows unrestrained by mountain ranges from the wide plains of the Pampas to the sea, and the air is cool and dry. The state of the sea is comparatively unimportant, apart from the individual weather events and the Sudestado, because the wind blows mostly parallel to the coast or even off the land. On the average, wave heights of 1 m or slightly more are observed, increasing seaward.

The cloud amount is relatively low over the sea in summer with a mean of 40%, owing to the influence of the subtropical anticyclone, whereas the westerlies in winter cause a cloud cover of 60%. Fog occurs in the spring when the water is relatively cold and frequencies of 8% are reached, but it is rare during summer.

The frequency of precipitation is 5% in summer and 10% in winter and spring, but the precipitation is heavier in summer and autumn. It amounts to about 100 cm per year. The water content of the atmosphere is decisive for the productiveness of precipitation. Moist and warm air from northerly to easterly directions brings rain when it is lifted by cooler air masses flowing in the opposite direction. The cold air approaching from the southwest is generally dry and during an outbreak of cold air no showers can be expected after the passage of the cold front with its thundershowers. Thunderstorms occur all through the year but are most frequent in summer; there are about 50 days a year with thunder at the coast.

The annual water balance is also equalized on this coast. Evaporation is stronger than precipitation in summer so that a negative balance results; in winter it is positive.

The vertical heat flux at the sea surface decreases toward the south of the area. It varies between 80 W m^{-2} at 30°S and 60 W m^{-2} near Cabo Corrientes and increases seaward. Values are higher in summer, lower in winter.

The radiation balance at the sea surface varies from 200 W m^{-2} in summer to only 30 W m^{-2} in winter. There is a secondary maximum off the coast of northern Argentina with a yearly mean value of about 120 W m^{-2}.

The heat balance at the sea surface, calculated from the radiation balance and the heat flux, is positive near the coast in this region with yearly values of about 50 W m^{-2}. It decreases seaward and becomes negative east of 50°W. The yearly range is larger: in summer the heat balance is a positive $100-150$ W m^{-2}, and in winter -50 W m^{-2}.

Patagonian Shelf

On the Argentine coast and the Patagonian Shelf, the climate is one of temperate latitudes. The westerly flow aloft is, however, disturbed by the Andes, and the foehn east of the mountains brings warmth and dryness extending across the coastline to the sea. The climate of this coastal segment acquires the character of a steppe and can be described by type B of Köppen's classification (Chart 3).

The selected area R supplies the data for the adjacent sea, and over land the climate has been described by PROHASKA (1976). This area is of special interest for fishing (BROMANN, 1968) as the wind coming off the land causes upwelling of water rich in nutrients and yet it is sheltered from strong winds.

The temperature has a marked annual range. The monthly mean values decrease between north and south from 20° to 10°C in summer and from 8° to 2°C in winter. The monthly mean values of vapour pressure vary between 16 and 8 mbar in summer and 10 and 5 mbar in winter, and therefore the equivalent temperature ranges between 44° and 10°C. The lower comfort limit of 30°C equivalent temperature is reached in summer at 50°S, in winter already near 35°S. The dryness of the air is demonstrated best by the low values of relative humidity, whose monthly means lie below 70% at the coast while they reach 90% over the distant Falkland Current.

The seasonal differences are weak in the pressure and wind field. The wind blows mainly from the west with mean velocities of $6-8$ m s^{-1} (5 Beaufort). On the coast, winds blowing off the land are strengthened in the morning and weakened during the day. The frequency of strong winds is about 20%. Gales occur with a 10% frequency in winter and 5% in summer, the frequency increasing seaward. The storm frequency increases to $15-20\%$ at the southern tip of South America and in the Drake Passage.

Cloudiness is low over land in summer even with low pressure, but it increases southward and over the sea. Between the northeast of Argentina and the Falkland Islands (p. 82), it increases from 50 to 70%. In the south, the yearly cycle is reversed since the greatest cloudiness is observed there during the summer. The frequency of fog is low at the coast and becomes greater over the sea to reach a maximum over the Falkland Current. The precipitation frequency increases in the same manner from 5% at the coast to 20% at the Falkland Islands, and the same holds true for precipitation amounts, which remain below 25 cm per year in the central area of the coast with slightly greater amounts in winter than in summer. The foehn effect weakens the fronts and warm fronts are not observed (p. 121).

South of 50°S, the yearly cycle of precipitation is reversed, because the efficiency of precipitation is increased there during the summer owing to the larger amount of moisture in the air. Warm fronts contribute to precipitation (WÖLCKEN, 1962). The frequency of thunderstorms decreases with increasing latitude. Snowfall is rare in the northern part of the ocean area but more frequent in the south. Snow falls on the average on $3-4$ days during the winter months.

The yearly evaporation of about 70 cm near the coast results in a negative water balance of about -50 cm which becomes positive farther out to sea. The water balance is $+20$ cm per year around the Falkland Islands and near Cabo San Juan.

The yearly heat flux from the ocean to the atmosphere varies between 50 and 70 W m^{-2} with a weak meridional gradient. It decreases over the sea to about 30 W m^{-2} over the

Falkland Current. The values are higher in winter than in summer.

During summer, the strongest insolation along the South American east coast is in 40°S where the radiation balance at the sea surface reaches the high value of 230 W m^{-2}. It decreases to the south and to the east and on the coast it falls to 160 W m^{-2} near Cabo San Juan. In winter, the radiation balance becomes negative south of Valdés Peninsula and reaches -40 W m^{-2} near Cabo San Juan. In the yearly mean, there is a meridional slope in the radiation balance from 120 W m^{-2} in the north to 60 W m^{-2} in the south. The heat balance at the sea surface in the coastal region has annual averages between 70 W m^{-2} in the north and 0 in the south. The yearly means reach 80 W m^{-2} over the Falkland Current. Seasonally, the heat balance varies between values of 120 to 180 W m^{-2} in summer and -50 to -130 W m^{-2} in winter.

The heat, which is fed into the Falkland Current, comes mostly from the continent and is partly the result of the warming by the foehn.

African off-shore areas

Thorough descriptions of the sea areas along the Atlantic African coast may be found in *Naval Air Pilots* (WASHINGTON, 1943), *Seehandbücher* (HAMBURG, 1981c), *Sailing Directions* (WASHINGTON, 1969), and *Navy Pilots* (TAUNTON, 1977). The climate of the adjacent continent is described in Volume 10 of this series by GRIFFITHS (1972) and SCHULZE (1972).

Coast of Gabon

The tropical humid climate Af extends southwestward along the African west coast from the Equator to the mouth of the Congo (6°S). The sea area off this coast is represented by area C (Chart 2) and the island station of Sao Tomé (p. 71f).

The tropical climate's most striking feature is its sultriness. The mean yearly temperature of 26°C and vapour pressure of 27 mbar give a mean equivalent temperature of 67°C, which is much above the comfort limit of 55°C. There is only little variation through the year, and with a relative humidity of 85% it is uncomfortable even in winter.

The fluctuations of temperature are small; the daily and yearly ranges amount to 7°C at the coast. The yearly range is the same for the sea area and is due to the upwelling that occurs during southern winter on the Equator. In the summer, the water is warm in the vicinity of the ITCZ.

The pressure has its minimum in summer with monthly mean values below 1010 mbar along the coast ascribable to the nearby ITCZ and the continental heat low. In contrast to the west—east pressure gradient in summer, the subtropical anticyclone causes a meridional gradient on the coast in winter when monthly means are about 1014 mbar. Whereas the pressure fluctuations caused by weather are small, the semi-diurnal oscillation is well established with maxima at 10h00 and 22h00 and 3—4 mbar lower minima at 04h00 and 16h00 local time.

The wind blows mainly from southerly directions. It is weaker and less steady in the coastal region than over the sea, particularly during southern summer. The variation of the wind speed is small, however, although occasionally so-called "tornados" occur in summer evenings. These are not narrow vortices but thundery squalls with strong winds or even gales. They cross the coast as a squall line at about 25 knots in the easterly flow with a strong dusty wind in front, enhanced by the sea breeze. Behind the roll cloud a squally and stormy easterly wind blows, which is accompanied by torrential rain lasting for about one hour (p. 115).

In summer, the "Harmattan" may blow from the continent as an extraordinarily dry wind carrying much fine dust and occasionally sand that reduces the visibility.

The cloud amount over the district in question is relatively high. The monthly means reach 60% in southern winter and as much as 70% in summer. This is the main rainy season when towering cumulus clouds produce heavy showers. The precipitation frequency is 5% and the monthly amounts more than 200 mm. It is dry during the winter with a precipitation frequency below 1%. Sometimes a month is completely dry. In this season, the dryness of the air above the trade inversion is reinforced by the divergence of the air near the Equator. Thunderstorms occur mainly at the beginning and end of the rainy season when they happen nearly every day, especially near the Equator.

Fog is rare at the Equator but increases southward in the southern winter. At the most, six days per month with fog are observed on the coast.

The evaporation over sea is relatively weak because of the high amount of cloud. The yearly means vary between 130 cm at the Equator and 70 cm at the mouth of the Congo and increase seaward. The annual mean water balance is thus positive because precipitation exceeds evaporation (NEUMANN, 1972). Like precipitation, the water balance decreases to the south and out to sea.

At the southern edge of the ITCZ, the heat flux from the ocean to the atmosphere decreases in winter from 100 W m^{-2} on the Equator to 40 W m^{-2} at the mouth of the Congo, whereas in summer it is uniformly distributed at $70-80$ W m^{-2}. The heat flux increases to 100 W m^{-2} over the sea near the Greenwich meridian.

The radiation balance at the sea surface is $110-120$ W m^{-2} with only few seasonal fluctuations; in winter these are the highest values along the African coast.

The heat balance is about 50 W m^{-2} all through the year. The surplus heat is stored within the ocean and transported to the ITCZ by the South Equatorial Current.

Coast of Angola

The coast between 6° and 17°S lies within the arid trade wind zone, and there is a steep meridional temperature gradient along it between the warm, moist tropics in the north and the cold upwelling area in the south. Because of the aridity, the climate of the northern part of this sea area is a steppe climate BS, and even a desert climate BW in the south (Chart 3).

The cool water of the upwelling area extends along the coast northward to Punta Albina (15°S). The air temperature difference along the coast between the upwelling area in the south and the region of warm water in the north is about 7°C with an annual range of about 5°C and a daily range of 6°C. Since the vapour pressure in this coastal sector varies in a similar way, meridionally by 9 mbar and seasonally by 6 mbar, there are also large differences in the equivalent temperature. Its monthly mean values in the north change seasonally between 55°C in winter and 72°C in summer, in the south between 35°C in winter and 50°C in summer. As the equivalent temperature values indicate, it is sultry in the north at any time during the summer, but in winter only during the daytime. Substantially more comfortable conditions prevail in the south. The relative humidity is rather high, however; it exceeds 80% in all months, even 85% in the summer.The trades blow along the eastern limb of the subtropical high at 3 Beaufort and a steadiness of 70%, which increases seaward to 4 Beaufort and 90%. A monsoonal influence of the continent is expressed in the annual range of 5−6 mbar and a wind shift on the coast to southwesterly directions as summer approaches. Variations in the wind field are caused by disturbances which (p. 115) pass from the continent to the sea and wander westward with the trades. They show as shifts in the wind which blows from the southeast in front of the disturbance and from northerly directions behind it. The "Bergwind" occurs especially in winter on the coast as a dry, gusty and hot wind from northeast to east, but gales are rare.

The wind which blows mostly parallel to the coast generates only a low sea with significant wave heights below 1 m, particularly in summer. Sometimes a high surf caused by the swell is observed. In winter, this swell rolls in from the southwest from the stormy areas of middle southern latitudes, but in summer it stems from the North Atlantic and runs against the trades (compare the rollers at St. Helena, p. 78).

The cloud amount is lower than that of the coastal areas of the inner tropics, and it increases seaward. In the region of the Benguela Current, these clouds are mainly the low stratus or shallow stratocumulus below the trade inversion out of which rain rarely falls, at most a little drizzle. This cloudiness is called "Cacimbo" off Angola. Otherwise, precipitation is caused by the trade disturbances, and its frequency is about 1% and even less in winter. The yearly amounts vary between 10 cm in the north and 1 cm in the south; and the mean number of thundery days decreases southward. There are still 34 days per year with thunder in Luanda. In contrast, the number of days with fog increases southward, because fog is more likely to form and endure over the cold water in the reigning stable conditions. Fog occurs frequently in the morning at the mouths of rivers, and occasionally the low stratus of the Cacimbo descends to sea level.

The evaporation of 80−100 cm per year at the coast, which increases seaward, causes a negative water balance reaching −100 cm per year in the central part of the coastal waters.

The heat flux from the sea to the atmosphere due to evaporation has seaward-increasing values of 50−100 W m^{-2} and is slightly higher in summer than in winter. The same seasonal course is found in the radiation balance at the surface with 140 W m^{-2} in summer increasing toward the south, and 100 W m^{-2} in winter decreasing toward the south. The heat balance of the sea surface resulting from these quantities is positive in summer and negative in winter with a yearly mean of about 20 W m^{-2}. It decreases over the sea and becomes negative west of 10°E in the yearly mean too.

Heat is extracted from the ocean in this area and transported by the trade winds to the ITCZ, but at the coast heat is fed into the ocean from the atmosphere over Africa and enters the ITCZ in the South Equatorial Current.

Coast of Southwest Africa

This coastal sector extending from Cabo Frio (18°S) to Paternoster Point (33°S) is characterized by a dry trade climate (type BW in Chart 3). The sea area off the coast is represented by the climatic tables of areas J and L, and the climate of the adjacent coastal region is described by SCHULZE (1972).

The water has a warm surface layer extending to the thermocline, where a strong vertical temperature gradient separates the upper mixed layer from the colder intermediate water. Near the coast, however, the trade winds and the Benguela Current cause upwelling in which this intermediate water ascends to the surface. The surface water in this region is therefore nearly 10°C colder than one would expect from the latitude (Fig. 4, p. 31).

This lowering of temperature owing to upwelling undergoes strong fluctuations, both in space due to the configuration of the coast and in time because of the fluctuations in the wind field. The small scale horizontal temperature gradients and their variations thus generated influence the climate of the ship's hold and may cause damage due to perspiration of water (HÖFLICH, 1972).

The low water temperatures are reflected in the low air temperature and they affect the humidity. The mean air temperature is about 18°C in summer and 14°C in winter but increases seaward. At the coast, the mean daily range exceeds 10°C.

The lowest values of air temperature are measured in the centre of the upwelling area between Walvis Bay (23°S) and Port Nolloth (29°S). The air – sea temperature difference here reaches positive mean values of 2°C in summer. The cold water stabilizes the atmosphere so that the relative humidity reaches values of more than 85%. In the seasonal course, the vapour pressure varies between 13 and 18 mbar and, accordingly, the equivalent temperature varies between 34° and 45°C, i.e., within the region of comfort. As the temperatures increase out to sea, the relative humidity decreases slightly. At 10°W, the air temperature is 20°C in summer and 16°C in winter, the relative humidity is about 80%, and the equivalent temperature varies between 49°C in summer and 38°C in winter.

The pressure gradient is fairly zonal in summer around a mean pressure of about 1012 mbar; in the winter it is more meridional with pressures between 1018 and 1021 mbar. The accompanying southerly trade winds have a yearly mean of 6 m s^{-1} (4 Beaufort) with a steadiness of 70–90% at the coast and 90–95% over the sea during summer. The steadiness decreases to 30–50% in winter near the coast in the southern part and to 70% over the sea. Sometimes the westerlies encroach in the southern parts of this area. The probability of strong winds is 10% and of gales 1%. The latter occur both in the core of the trades, occasionally as sand storms, and in the southern part of the area, in winter more frequently than in summer.

Land and sea breezes develop regularly along the coast unless anomalous winds occur, like the down-slope Bergwind or when occasionally the winds are deflected by the large-scale cyclones of the westerlies. The Bergwind approaches from east to northeast, loaded with dust from the deserts of Southwest Africa south of about 15°S. It occurs at all seasons but prefers winter. Sometimes it lasts only a few hours, but it can also blow for several days, although more weakly in daytime than during night. It is extraordinarily squally and may reach 7 Beaufort at Lüderitz. The visibility may be as low as 30 m in dust storms.

The persistent, strong trades give rise to mean wave heights of 2 m in this area (QUAYLE and ELMS, 1979), although they are lower near the coast because the winds mostly have a component off the land. The heavy swell runs from the southwest and originates in the storms of middle latitudes and may occasionally cause a strong surf during the winter.

The cloudiness is extremely variable. The yearly means are below 40% along the coast, but farther out they reach 60%. It generally consists of low stratus or stratocumulus below the trade inversion. The mean duration of sunshine at the coast is below 60% of the possible, but it increases quickly over the mainland. The probability of fog is high over the cold upwelling water, in the north predominantly in winter and in the south in summer. There are about 100 days per year with fog on the average. The frequency of fog decreases to about 1% over the sea. The highest frequency is observed near Swakopmund. Fog normally forms during the night and dissipates in the daytime, but it may also last for 2−3 days. Dust and sand particles occasionally give it a yellow appearance.

The precipitation frequency is low throughout the year, mostly below 2%; precipitation appears mainly as drizzle or damp low stratus (mist rain). Only in the southernmost region do the fronts of the westerly cyclones cause a slight increase in the rain frequency, and there the variability is fairly large as the standard deviation reaches nearly 80% of the mean value (SCHULZE, 1972, fig. 11).

The evaporation is strongly reduced over the cold upwelling water so that in places on the coast it remains below 50 cm per year. The water balance is nevertheless negative due to the lack of rain and over the sea it sinks to −100 cm per year at 10°W.

In agreement with the weak evaporation, the heat flux from the sea to the atmosphere is relatively low with a minimum of 20 W m^{-2} at the coast. The values rise to a yearly mean of 100 W m^{-2} away from the coast.

Owing to the strong insolation in summer, this coastal strip has the highest values of the radiation balance with 230 W m^{-2} at 25−30°S, whereas in winter the balance is only 50−100 W m^{-2}, with the values decreasing southward. The highest yearly mean values of the South Atlantic Ocean occur at the coast near 25°S while the trade clouds cause a decrease of the radiation balance values farther out to sea.

Accordingly, a positive heat balance exists at the sea surface and reaches its highest annual means of 150 W m^{-2} at the coast near 25°S. These are the highest values in the South Atlantic Ocean, as in the instance of the radiation balance. The heat balance varies between 215 W m^{-2} in summer and 80 W m^{-2} in winter and decreases from coast to open sea. It is zero in the annual mean at 5°E, being positive in summer and negative in winter. The positive heat balance in the coastal waters results in warming of the surface water which balances the cooling caused by the upwelling of cold intermediate water and the horizontal heat transfer in the northwestward setting Benguela Current.

Coast of the Cape Province

The southernmost part of the African coast has a temperate Mediterranean climate (type C)—dry and warm summers, rainy and cooler winters. The corresponding sea area is represented by selected area O.

The annual mean temperature is about 17°C with a yearly cycle of 5°C over the sea and 8°C at the coast. The daily range at the coast is 10°C. The monthly mean values of vapour pressure vary between 13 mbar in winter and 18 mbar in summer, yielding an equivalent temperature of 34°C in winter and 47°C in summer; the climate is therefore comfortable. The lower limit of human comfort (about 30°C) is reached in winter at night.

The subtropical ridge axis lies north of this area in winter and south of it in summer, and the storms of the westerly belt with their variable weather will thus frequently affect the coastal area in winter, but rarely in summer. In the higher troposphere, the anticyclonic axis is even farther north—near 20°S—and the coast of the Cape Province lies below a westerly flow which steers disturbances from the Atlantic to the continent (p. 105f).

The air pressure nevertheless reaches mean values in winter that are 6 mbar higher than in summer. This is partly a planetary effect associated with the seasonal change in position and strength of the subtropical anticyclone, and partly a monsoonal effect reflected in the yearly cycle of the continental pressure. The influence of the west wind zone in winter is seen in an increase of about 7 mbar in the monthly average standard deviation of pressure.

The wind over the sea varies correspondingly from summer to winter, evident by its steadiness of 50% in summer and 30% in winter off the coast. The prevailing wind direction is south during summer and southwest in winter, and the mean velocity is about 7 m s^{-1} (4 Beaufort). On the south coast, the trade winds are enhanced by the sea breeze and particularly by coastal pressure minima approaching from the north (Cape South Easter). The monthly mean standard deviation of wind speed increases to 5 m s^{-1} from summer to winter. The frequency of strong winds near the coast is nearly 20% in summer and that of gales about 5% in winter. The trades are more responsible for the strong winds in summer and the westerlies for the gales in winter.

The mean significant wave heights are above 2 m, which includes the swell from stormy areas of higher latitudes. On the coast, it occasionally gives rise to a heavy surf, and off the south coast it runs against the Agulhas Current so that the waves are steepened. This high steep sea, called freak waves, is dangerous to shipping (p. 25).

The mean cloudiness is about 45%; it increases both seaward and toward the south, but it has no noteworthy seasonal change. The duration of sunshine reaches about 70% of the possible on the west coast, but it is below 60% on the south coast.

Fog is rare as frequencies of 1% are exceeded only in autumn, when 5 days per month with fog are observed; otherwise 1 − 2 days are the average.

Cyclonic activity causes an increase of the precipitation frequency to more than 5% in winter, with monthly amounts of 80 mm on the coast, whereas it is relatively dry during summer with monthly amounts of about 10 mm. The annual amount increases southward and reaches 50 − 80 cm on the south coast. The standard deviation decreases as a percentage of the mean from the west coast, where it is 20% of the mean, to the south coast where it is 15%. There are only 2 − 4 thundery days on the coast. Rain occurs with the passage of a cold front or in a cut-off low. Examples of synoptic situations are shown by SCHULZE (1972).

Evaporation is enhanced over the warm Agulhas Current where the highest values exceed 250 cm per year. The water balance is thus strongly negative, notably during summer. At the Cape of Good Hope, the water balance is − 100 cm per year and it increases eastward and seaward. The balance becomes positive in the west wind zone, principally in winter.

The strong evaporation over the Agulhas Current is the major component in a high heat flux from the ocean to the atmosphere. In the yearly mean, it is nearly 250 W m^{-2} at 40°S, and there is a steep gradient of flux between this region and the upwelling north of 30°S, so that already at the Cape of Good Hope the heat flux is only 100 W m^{-2}. The radiation balance at the sea surface in the area off South Africa has a marked seasonal course; the monthly mean values are 180 W m^{-2} in summer but only 25 W m^{-2} in winter. Moreover, the latitude dependence of the insolation causes the radiation balance to decrease toward the south.

The meridional decline of both heat flux and radiation balance creates a correspondingly steep gradient of heat balance at the surface with positive values over the upwelling water in the north and negative values over the Agulhas Current. Superposed on this is a yearly cycle with predominant positive mean values in summer and negative values in winter.

Near the Cape of Good Hope, the monthly mean values of the heat balance are positive in summer, about 50 W m^{-2}, and negative in winter: -50 to -100 W m^{-2}. The lowest balance values are over the Agulhas Current where they attain annual means of -150 W m^{-2} which are also the lowest ones in the South Atlantic Ocean.

The heat from the vigorous exchange between ocean and atmosphere over the Agulhas Current does not benefit the South Atlantic to any great extent, as it is mainly carried by the atmospheric circulation of temperate latitudes to the Indian Ocean or used in cyclogenesis.

Aerology

The previous sections dealt almost exclusively with the climate at sea level, but climate is also influenced by the free atmosphere.

Already at sea level, observational data are inhomogeneous and frequently insufficient over the South Atlantic Ocean (p. 6f), but the situation in the free atmosphere is even less satisfactory (WMO, 1966b). Various research expeditions have indeed investigated the troposphere. Kite ascents revealed the aerological structure of the trade winds (REGER, 1927, 1963) and of the ITCZ (SCHNAPAUFF, 1937), and radiosonde ascents during Antarctic expeditions disclosed the extremely cold troposphere over Antarctica in comparison with the Arctic (FLOHN, 1950). But this knowledge, important though it was, was based only on sporadic, single observations. For climatic representations, longer homogeneous time series are needed, such as those gained by the daily radiosonde ascents of the meteorological services and, eventually, by satellite measurements.

A comparatively dense observational network became available for the first time during the IGY of 1957 – 1958. Daily upper-level charts of the 500-mbar level (HAMBURG, 1963 – 1967, for the tropics; PRETORIA, 1962 – 1966, for southern latitudes) and monthly charts (PRETORIA, 1958, 1959; RUDLOFF, 1970) were produced. Monthly mean values of temperature, geopotential, and wind were calculated for the tropical belt and presented on charts for various upper levels (RUDLOFF et al., 1965; RUDLOFF and HÖFLICH, 1970). Over the South Atlantic Ocean, there are only three island stations with radiosonde ascents of at least five years, and weather-ships do not exist. Because of this unsatisfactory situation, the climate of the free atmosphere can be sketched only in rough outline to be improved when additional observations by balloon, aeroplane, and satellite become available.

The present evaluation is based mainly on the publications *Climate of the Upper Air* by

Taljaard, van Loon, Crutcher, and Jenne. In Volume 1 (TALJAARD et al., 1969), maps and 5°-grid values are presented for seven levels of the Southern Hemisphere up to 100 mbar. Volume 2 (VAN LOON et al., 1971) and Volume 3 (JENNE et al., 1971) contain winds calculated from the gradients of geopotentials, and Volume 4 (CRUTCHER et al., 1971) presents zonal and meridional profiles for mean values of geopotential, temperature, and dew point. The climatology of the stratosphere is described by LABITZKE and VAN LOON (1972).

In the following sections, the values of the geopotential at different levels are examined, wind and temperature are described, and the moisture conditions of the lower troposphere are outlined.

Geopotential heights

In aerology, unlike at sea level, geopotential height of a constant pressure level is used, not the pressure at a constant height. Height H (in m) of the 1000-mbar level is determined with sufficient accuracy from pressure P (in mbar) at sea level according to the equation:

$$H = 8(P - 1000)$$

without considering the effect of varying temperature. Isobars spaced by 5 mbar may also be interpreted as contours with a spacing of 40 m. So the centre of the South Atlantic subtropical anticyclone of 1022 mbar in January (Chart 4) or 1025 mbar in July (Chart 5) may also be described as a maximum of geopotential of 176 m or 200 m at 1000 mbar. A meridional pressure gradient of 25 mbar between 30° and 55°S corresponds to a slope of 200 m of the 1000-mbar level.

Absolute topographies

The analysis of the geopotential of a constant pressure level is called its absolute topography, and climate charts of absolute topography are derived from mean values of such analyses (RUDLOFF, 1970). If the mean absolute topographies (TALJAARD et al., 1969) of the various levels over the South Atlantic Ocean are compared, similarities as well as distinct differences are found. At all levels and all seasons, a meridional gradient prevails which intensifies toward the Pole, and this gradient is steeper in the high than in the low troposphere. There is weak cyclonic curvature in a trough in the western South Atlantic.

The subtropical anticyclone, which dominates the eastern part of the South Atlantic at sea level, shifts equatorward and westward with increasing height (NIEUWOLT, 1977). Its core is situated at 20°W in the lower troposphere and it is thus displaced westward by more than 10° compared with sea level. Its vertical axis slopes for both thermal and dynamic reasons, as the water temperature is lower in the east than in the west and the trade inversion lies lower in the east than in the west.

The anticyclone expands with height to a zone of high geopotential values at 20°S in January. Its centre shifts westward in the lower troposphere and is situated over the continents in the upper troposphere and in the stratosphere. South of 30°S, the meridional gradient increases with height to a maximum in the upper troposphere. In July, the zone of highest geopotential values lies even closer to the Equator. It is situated at 10°S

at 500 mbar, at the Equator at 300 mbar, and over the North Atlantic in the upper troposphere.

In the stratosphere, a yearly cycle dominates with relatively high values of geopotential in summer and low ones in winter due to the large annual variation of the daily duration of insolation; the highest annual range is over Antarctica.

Like the standard deviation of pressure at sea level (p. 16), the standard deviation of geopotential increases poleward. On the average, the monthly mean values of standard deviation at 500 mbar vary between about 10 m in tropical latitudes and about 50 m in the polar region with only a weak seasonal change (VAN LOON and JENNE, 1974). The standard deviation of geopotential values is slightly higher over the South Atlantic than over the adjacent land; compared with the North Atlantic it is smaller in winter, larger in summer.

Meridional profiles

The meridional difference of absolute topography at various pressure levels is best shown as a meridional profile. Fig. 13 presents such profiles between the Equator and

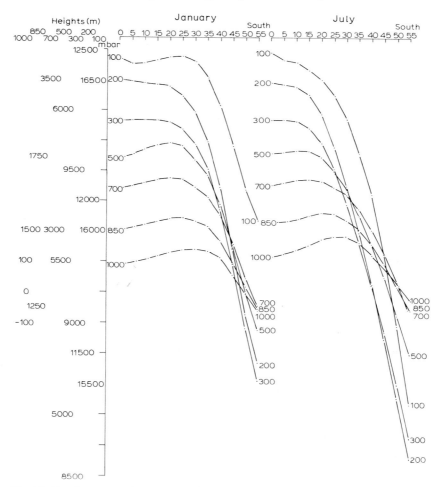

Fig. 13. Zonal averages of geopotential heights (m) at 100, 200, 300, 500, 700, 850, and 1000 mbar over the South Atlantic for January and July.

TABLE XV

COMPARISON OF GEOPOTENTIAL VALUES

Pres. level (mbar)	Latitude (°S)				Difference (m)			
	minimum		maximum		30–55°S		max–min	
	Jan.	Jul.	Jan.	Jul.	Jan.	Jul.	Jan.	Jul.
100	90	90	25	0	470	940	680	2080
200	90	90	0	0	825	890	1230	2330
300	90	90	15	5	780	770	1190	1560
500	70	90	20	10	560	535	830	1020
700	70	70	20	15	390	380	520	600
850	65	70	20	20	285	290	380	440
1000	65	65	30	30	205	210	290	320

55°S, averaged for the South Atlantic Ocean for January and July from the values in TALJAARD et al., (1969).

The most important data are compiled in Table XV. On the left is the latitude of the lowest and highest monthly mean value of geopotential for January and July in the South Atlantic. The right-hand side shows the difference of geopotential between 30° and 55°S and between the highest and lowest monthly mean. The maximum of geopotential, situated at 30°S at sea level, leans equatorward with increasing height. The minimum at 65°S at sea level leans poleward. The gradient of geopotential between 30° and 55°S, which amounts to 205 – 210 m at sea level, intensifies with increasing height. In January, it reaches a maximum at 200 mbar with 825 m, which is four times the value at sea level. In July, the maximum of 940 m at 100 mbar is five times as high. It is noteworthy that the geopotential height difference between 30° and 55°S is somewhat larger in summer than in winter in the middle and upper troposphere. This is the opposite of what is observed on the Northern Hemisphere and can be explained by the influence of the vast ocean areas on the circulation over the Southern Hemisphere (VAN LOON, 1966). The region of poleward geopotential gradient also expands toward the Equator and the Pole with increasing height so that the difference between the tropical maximum and the minimum over Antarctica at 200 mbar reaches a value of 1230 m in summer and 2330 m in winter, nearly twice as much. The above statements are illustrated by the graph in Fig. 13.

Zonal profiles

The zonal differences of the absolute topography at various pressure levels are shown in Fig. 14 for the longitudes between 55°W and 25°E for January and July. They are averaged over the tropical area between the Equator and 30°S. The South Atlantic subtropical anticyclone appears only in the lower troposphere. In July, it has nearly disappeared at 850 mbar and in January at 700 mbar; at higher levels it is replaced by a low. The opposite happens over the adjacent continents.

The zonal difference of geopotential between the central South Atlantic (20°W) and South America (60°W) or Africa (20°E) is given for various pressure levels in Table XVI. Positive values indicate higher geopotentials over the sea and negative values

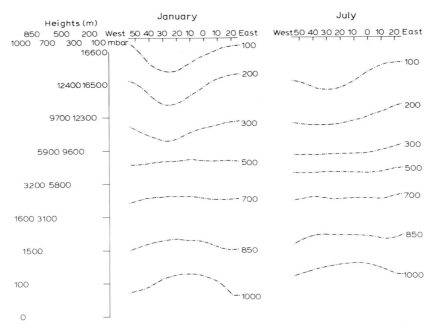

Fig. 14. Meridional averages of geopotential heights (m) at 100, 200, 300, 500, 700, 850, and 1000 mbar between the Equator and 30°S over the South Atlantic for January and July.

higher geopotentials over the continent. The single values are not quite reliable as the analysis is not based upon sufficient observations, but the sign should be real. The reversal of sign between the lower and upper troposphere can be explained by the different thermal conditions in the atmosphere over sea and land, particularly in summer when the differences are bigger. In the region considered, between the Equator and 30°S, the atmosphere heats more over the continents by the daily convection than over the ocean where no daily range produces such changes of stratification.

The observed eastward convergence of isobars at sea level in middle latitudes, which results in a maximum of meridional pressure gradient south of Africa (p. 15), is also found in the absolute topography of the whole troposphere. It is related to the asymmetry of Antarctica which displaces the cold pole in the direction of the Indian Ocean.

TABLE XVI

ZONAL DIFFERENCES OF GEOPOTENTIAL VALUES FOR VARIOUS PRESSURE LEVELS

Pressure (mbar)	Height differences (in m)			
	20°W–60°W		20°W–20°E	
	Jan.	Jul.	Jan.	Jul.
100	−100	−20	−90	−100
200	−100	0	−120	−60
300	−60	−20	−70	−30
500	20	0	20	−10
700	40	20	30	−10
850	60	60	60	10
1000	90	50	95	40

The lee effect of the Andes is reflected in the trough along the east coast of South America.

Wind

As mentioned above, the absolute topography over the South Atlantic Ocean has steep meridional gradients, especially in the upper troposphere in winter, but weak zonal gradients. Accordingly, the equilibrium condition between pressure gradient and wind demands strong zonal winds in the free atmosphere but only light mean meridional winds.

Zonal winds

Fig. 15 shows a meridional section of the zonal wind at 10°W between 10°N and 70°S from sea level to 100 mbar for January, and Fig. 16 shows the same for July (VAN LOON et al., 1971; NEWELL et al., 1972). Similar illustrations for South America are found in SCHWERDTFEGER (1976, figs. 4 and 5). Monthly meridional profiles along 10°E for the IGY have been published by HOFMEYR (1970).

In both months, maximum westerly winds at middle latitudes in the troposphere are connected with the polar front zone; in July there is also a distinct subtropical peak in the upper troposphere. The strength of the polar front westerlies is almost the same in both seasons. At 40° − 50°S, the westerly winds are even somewhat stronger in summer than in winter. The strong zonal wind during the whole year and the summer maximum in middle latitudes are a peculiarity of the Southern Hemisphere (VAN LOON, 1964, 1966). Above 200 mbar, the winter maximum westerlies appear south of 40°S and are

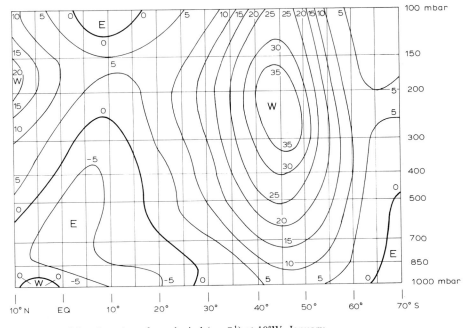

Fig. 15. Meridional section of zonal wind (m s^{-1}) at 10°W, January.

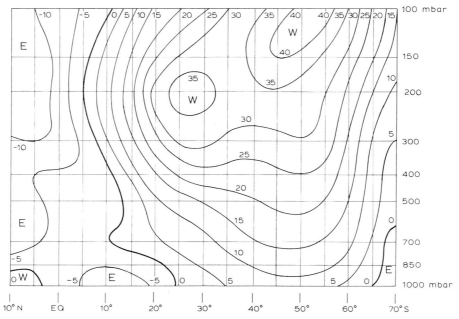

Fig. 16. Same as 15, but for July.

part of the polar vortex in the stratosphere. Like the gradient of geopotential, the domain of westerly winds expands horizontally in the upper troposphere.

The maximum of westerly winds, known as the *jet stream*, is more strongly marked on individual days than in the mean. Frequently, two jet streams occur, as indicated in the climatological mean of July. The wind maximum of lower latitudes is called the *subtropical jet* and that of temperate latitudes the *polar jet*. The polar jet stream marks the tropospheric frontal zone between subtropical and polar air masses where the circulations of the daily weather sequences develop. Since the position of the frontal zone varies over a wide range of latitude, directions, and strength, the wind maximum connected with it cannot be so well marked climatologically.

The trade winds, whose east—west component can be seen in Figs. 15 and 16, have a shallow vertical extent in winter; in summer easterly winds extend to 300 mbar in the region of the trades and are linked to the monsoon circulation over the continent in the region between the Equator and the ITCZ. They are best developed at the African coast. An analysis of wind observations during the IGY 1957—1958 confirms these conditions. The standard deviation of wind speed is about 10 m s^{-1} in the higher troposphere at tropical latitudes. In the region of the subtropical jet stream, it is about half the wind speed. The steadiness of the wind at low latitudes is 50% or less; in the zone of westerlies, it is near 90%; it decreases upward in the troposphere in lower latitudes but increases upward in temperate latitudes.

At the Equator, a quasibiennial alternation of westerly and easterly winds is observed in the stratosphere (Fig. 12). It is pronounced at 30 mbar with monthly mean values up to 35 m s^{-1} from the east and 15 m s^{-1} from the west. During the year, the phase descends from 10 mbar to the tropopause where it disappears (WALLACE, 1966; HENNING, 1969a).

106

Meridional winds

In the troposphere, the meridional component of the wind decreases upward in the mean and nearly disappears (cf. JENNE et al., 1971), so that in the higher troposphere weak meridional components occur only above the coastal regions. These components are southerly at the South American coast and northerly at the African coast of the South Atlantic, at least in summer (HENNING, 1969b).

Temperature

The analysis of the thickness of the layer between two pressure levels, that is, the difference of their absolute topography is called the relative topography. The relative topography, or thickness, is proportional to the mean temperature of the layer between the pressure levels.

Relative topographies

Maps of relative topography and maps of temperature (TALJAARD et al., 1969) are consequently very similar, in particular for the following levels:

Temperature:	200	300	700	850	(mbar)
Relat. topography:	100/300	200/500	500/1000	700/1000	(mbar)

The agreement between temperature at 500 mbar and relative topography 300/700 is less good.

The temperature charts of the various constant pressure levels reveal a strong temperature contrast between low and high latitudes with a few orographically caused exceptions.

The reader is referred to the atlas by TALJAARD et al. (1969) for the areal distribution of temperature over the South Atlantic. The meridional and zonal profiles described in the following have been constructed from the values in that atlas.

Meridional profiles

Fig. 17 shows the meridional profiles of relative topography for the layers 700/1000, 500/1000, 500/850, 300/700, 200/500, and 100/300 mbar, in addition to profiles of the temperature at sea level, 850, 700, 500, 300, 200, and 100 mbar, all averaged over the width of the South Atlantic Ocean between Equator and 55°S. The values of the relative topographies are given as layer-mean temperatures to make them comparable to the mean values of the pressure-level temperatures.

The two sets of curves fit each other well. From sea level to 300 mbar, all the meridional profiles have a rather smooth shape with the highest values at the Equator, a weak gradient in low latitudes, and a strong one in higher latitudes. The region of the strong gradient is narrower in January than in July. The slope between 30° and 55°S is 0.6° to 0.8°C per deg. lat.

The sign of the gradient reverses in January at 200 mbar, so that at 100 mbar a strong drop of temperature toward the Equator is observed with the highest temperature, about −40°C, at the Pole. In July, however, the maximum at 100 mbar occurs at 45°S, and the temperature then drops to the extreme value of −82°C at the Pole.

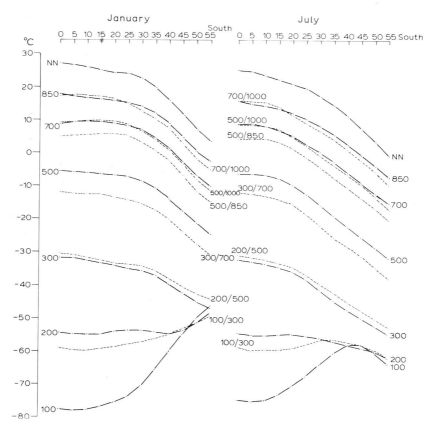

Fig. 17. Zonal averages of relative topography (°C, short dashed line) for the 100/300, 200/500, 300/700, 500/850, 500/1000, and 700/1000 mbar layers and of temperature (°C, long dashed line) at 100, 200, 300, 500, 700, 850, mbar and sea level (NN) over the South Atlantic in January and July.

Table XVII shows the mean temperature contrast between 30° and 55°S for various pressure levels in the South Atlantic Ocean and also between the Equator and the South Pole. The temperature contrast below 300 mbar across temperate latitudes does not vary much with height or season; if anything, it is slightly stronger in summer than in winter (VAN LOON, 1966). It weakens above 300 mbar and reverses in January.

TABLE XVII

TEMPERATURE DIFFERENCES BETWEEN VARIOUS ZONES AND FOR VARIOUS PRESSURE LEVELS

Pres. level (mbar)	Temperature difference (°C)			
	30°S–55°S		Equator–Pole	
	Jan.	Jul.	Jan.	Jul.
100	−24	0	−39	4
200	−5	6	−10	20
300	11	13	20	34
500	16	16	30	39
700	17	15	36	51
850	16	16	37	58
sea level	19	18	57	85

The temperature difference between the Equator and the Pole does show seasonal differences as it is larger in July than in January. It should be recognized that the value at the South Pole is based on data from one radiosonde station, while the Equator-value is a zonal mean for the ocean.

Zonal profiles

Zonal differences of temperature may be seen in Fig. 18 which contains the relative topographies and temperatures between 55°W and 25°E, averaged between the Equator and 30°S.

At sea level in January, the warm South American continent and Brazil Current are evident in the west, and the cold Benguela Current and upwelling as well as the warm African continent are clearly discernible in the east. The temperature level is lower in July, over the continents much lower.

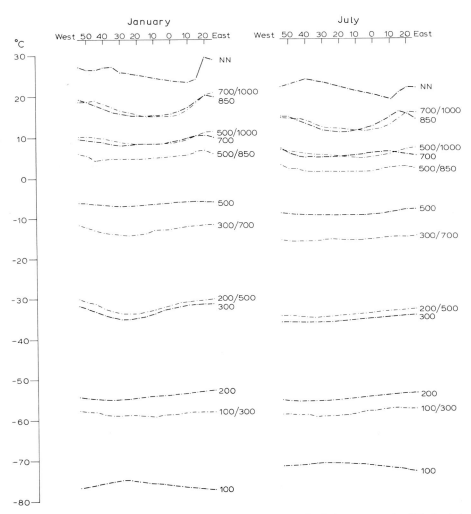

Fig. 18. Meridional averages between the Equator and 30°S of relative topography (°C, short dashed line) for the 100/300, 200/500, 300/700, 500/850, 500/1000, and 700/1000 mbar layers and of temperature (°C, long dashed line) at 100, 200, 300, 500, 700, 850 mbar and sea level over the South Atlantic in January and July.

In the troposphere, the meridian of relatively lowest temperature leans westward to about 20°W with increasing height. The zonal differences are nearly gone above 700 mbar, especially in July. In January at 300 mbar, a minimum lies at 25°W but at 100 mbar a maximum is indicated there (cf. VAN LOON, 1972, fig. 3.5).

It is warmer in the west than in the east in the lower troposphere, and conversely in the upper troposphere, which is an expression of the zonal Walker circulation (KRAUS, 1972). This circulation enhances the ascent of air over the Amazon Basin and reduces it over the northeastern part of the South Atlantic. In the west wind belt, the temperature is also higher in the west than in the east, but this difference decreases upward and disappears at 300 mbar (VAN LOON, 1972, fig. 3.17).

Humidity

Humidity may be expressed by different measures depending on the purpose. The various definitions are given on page 38 where the mean humidity at sea level is described. Monthly charts of dew point at sea level, 850, 700, and 500 mbar given in TALJAARD et al. (1969) are the basis for the climatological evaluations in this chapter.

Vapour pressure

Figs. 19 and 20 show profiles of zonal and meridional mean values of vapour pressure at sea level, 850, 700, and 500 mbar for January and July. It may be seen from both figures that the vapour pressure decreases with height and is lower in July than in January. Above 500 mbar, the vapour pressure is so low and the measurements of humidity by radiosondes so unreliable that an analysis is not worth while.

In the meridional profiles of zonal mean values (Fig. 19), the gradient is steep between the Equator and higher latitudes. The gradient in temperate latitudes is similar to that of the temperature; but at lower latitudes there is a difference between the moist tropics

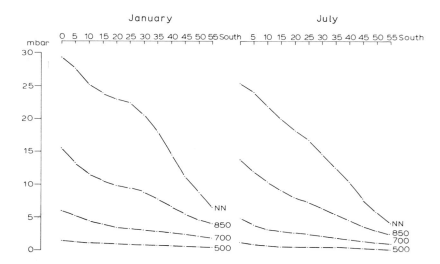

Fig. 19. Zonal averages of vapour pressure (mbar) at 500, 700, 850 mbar and at sea level over the South Atlantic, January and July.

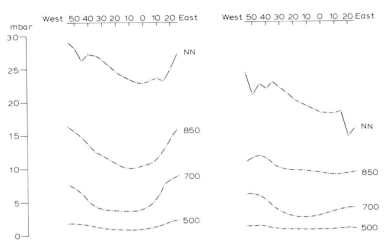

Fig. 20. Meridional averages between the Equator and 30°S of vapour pressure (mbar) at 500, 700, 850 mbar and at sea level over the South Atlantic, January (left) and July (right).

with high vapour pressure and the drier subtropics, corresponding to the rising and sinking air within the Hadley circulation.

In the zonal profiles (Fig. 20), the maritime troposphere has relatively low vapour pressure in January, in contrast to the high values over the neighbouring continents where the unstable layering over the heated surface promotes the vertical flux of moisture by convection. Convection is weaker over the sea, particularly over the cool Benguela Current and over the areas of upwelling.

The charts of vapour pressure at different levels have maxima in the ITCZ, which lean toward the continents with increasing height. Elsewhere the vapour pressure decreases toward the pole and from the west to the east side of the ocean. The Walker circulation in the tropics shows in the higher vapour pressure of the rising air in the west and the low vapour pressure in the descending air to the east.

Relative humidity

In Charts 54 and 55, the relative humidity at 700 mbar is presented for January and July. The dryness of the troposphere above the trade inversion, where the means are less than 25%, is evident over the eastern part of the ocean. The highest mean values in the ITCZ are 50—70% in January and 60—80% in July. In the westerlies, maxima of 60% are reached in January and 50% in July. The relative humidity in the troposphere obviously depends on the vertical flux of water vapour and therefore on the stability of the air.

At 850 mbar, the region of low relative humidity is limited to the zone along the African coast between 10° and 30°S; elsewhere the inversion lies above that level. Otherwise, the relative humidity in the low-level trade flow is 50—70% in January, at the Equator and south of 50°S the values range between 70 and 80%. The values in July vary between 60 and 70%, at the Equator they are about 75%, and south of 50°S about 70%.

At 500 mbar, the relative humidity decreases to mean values of less than 20% within the subtropics. Values of 50% are reached in the region off the north coast of Brazil and in January over the African coast near 10°S, where the vapour pressure attains a maximum of 2 mbar. Besides these areas, 50% relative humidity is measured in the westerlies where vapour pressures are only 0.5 mbar.

A comparison with sea level (p. 40f, Charts 31 and 32) reveals upward decreasing relative humidity everywhere. The difference between sea level and 700 mbar is 30−40% at the Equator over the central Atlantic, but only 10−20% along the coasts. Farther south the difference increases to 40−50% at about 20°S, 30% off the American coast, and a maximum of 60% along the African coast. In temperate latitudes, the relative humidity increases to the south at both sea level and 500 mbar, with a difference of 40% between the levels.

Several humidity elements

In Table XVIII mean values at the Equator, 25°S (subtropics), and 50°S (westerlies) are compiled for the following elements: vapour pressure *e* (in mbar), absolute humidity *a*

TABLE XVIII

MEAN HUMIDITY VALUES AT 0°, 25° AND 50°S FOR JANUARY AND JULY AT 500, 700, 850 AND SEA LEVEL*

	(mbar)	0°				25°S				50°S			
		e	*a*	*s*	*U*	*e*	*a*	*s*	*U*	*e*	*a*	*s*	*U*
Jan.	500	1.4	1.1	1.7	35	0.8	0.7	1.0	22	0.5	0.4	0.6	45
	700	5.9	4.5	5.3	53	3.3	2.5	2.9	31	2.1	1.7	1.9	58
	850	15.5	11.5	11.4	77	9.4	7.1	6.9	56	4.6	3.6	3.4	73
	SL	29.2	21.1	18.2	83	22.4	16.3	13.8	77	8.7	6.8	5.4	88
Jul.	500	1.2	1.0	1.5	32	0.5	0.4	0.6	22	0.3	0.3	0.4	50
	700	4.9	3.8	4.4	45	2.5	2.0	2.2	34	1.2	1.0	1.1	52
	850	13.8	10.3	10.2	78	7.4	5.7	5.4	66	3.0	2.4	2.2	67
	SL	25.4	18.5	15.4	82	16.7	12.3	10.2	74	5.8	4.5	3.6	74

* For explanation of symbols see text.

(g water vapour per m³ of air), specific humidity *s* (g water vapour per kg of moist air), and relative humidity *U* (in %).

The climatological means of absolute and specific humidity behave similarly to vapour pressure as they decrease with height and increasing latitude and are lower in winter than in summer. The relative humidity decreases with height too, but increases from the subtropics to higher latitudes.

The values are lower in July than in January, except for a few instances of the relative humidity. The seasonal differences are larger over land where the specific humidity is higher in summer, which is true also for the Northern Hemisphere (PEIXOTO, 1958).

The vertical gradient of the humidity elements is determined not only by the vertical exchange but also by advective fluxes, convergence, and divergence which all together establish a balanced distribution in the mean with moisture decreasing upward.

An integration of the absolute humidity with height yields the following mean water content of the atmosphere, in g cm⁻² for the following latitudes and months:

Latitude: 0°   25°S   50°S
January: 5   3   1.5
July: 4   2.5   1

As a mean, 2.5 g cm^{-2} is accepted (WMO, 1976).

If one compares the water content of the atmosphere with the evaporation at the sea surface (p. 51f) or with the precipitation amounts (p. 49f) over the South Atlantic Ocean, the mean residence time of water vapour in the atmosphere can be calculated; it can range from days to weeks and is about half a month on the average (LAEVASTU et al., 1969). Advective transports considerably prejudice this estimate, however.

Synoptic climatology

The climatic zones of the South Atlantic Ocean, as described on p. 10 and shown in Chart 3, can be described in terms of typical synoptic phenomena. In the tropical zone between the ITCZ and the subtropical anticyclone, trade winds dominate where only weakly developed and small-scale disturbances are embedded. Because the ITCZ itself remains outside the South Atlantic Ocean, the associated phenomena will not be dealt with here. Between the subtropical anticyclone and the low pressure trough off the Antarctic coast is the wide expanse of the westerlies whose weather is created by the continuous eastward drift of vortices in a frontal zone that separates subtropical from polar air masses.

Frequencies of cyclones and anticyclones and their tracks are presented in four charts, and the weather associated with them is described. The information is based on the analysis of daily synoptic weather maps and on ship logs describing the weather during the voyage. The lack of ship observations in the south renders difficult the analysis of weather maps, and often analysis models based on theory or on experience from other oceans have been applied. The analyses of weather maps over the South Atlantic Ocean are therefore not completely objective (TALJAARD, 1972b).

Tropics

The tropical zone, frequently defined as the region between the tropics of both hemispheres (NIEUWOLT, 1977), is regarded here as the region of the Hadley regime over the South Atlantic Ocean. It is bounded in the north by the ITCZ at or beyond the Equator and in the south by the subtropical anticyclone.

The tropics are characterized by horizontal variations of meteorological elements that are an order of magnitude smaller than the variation in temperate latitudes. Changes in the wind field are more important than those in pressure or geopotential. Where the stratification is unstable, convective heating plays a key role in development of perturbations (WMO, 1978), and the influence of water temperature anomalies is thus important over the ocean.

Intertropical Convergence Zone

In the Atlantic Ocean, the ITCZ remains north of the Equator throughout the year, touching the South Atlantic only in the southern summer (January—April) at the Brazilian north coast. The highest sea surface temperatures and vapour pressures are reached in the ITCZ (HÖFLICH, 1974), and the pressure there attains a minimum. The mean precipitation frequency is about 10% and 100 cm of rain fall annually over the open sea and about 200 cm near the coasts (Chart 38). The dominant feature of the ITCZ is the convergence of the trades of both hemispheres which leads to deep vertical overturning with corresponding convective clouds and rain. Within the ITCZ, cloud clusters with showers and thunderstorms propagate westward (MARTIN and SIKDAR, 1975) and form open or closed systems of cumulonimbus clouds, according to the strength of the vertical flux of heat. Their diameters range from 20 to 100 km (HUBERT, 1966).

The position and intensity of the ITCZ fluctuate with the variations of the trades both as regards daily weather and as regards seasonal and year-to-year changes. From the weather maps of the IGY (HAMBURG, 1963—1967), medium- and long-range variations have been studied, and analyses of monthly mean aerological observations for this period give an impression of the behaviour of the troposphere over the ITCZ (RUDLOFF et al., 1965a, b, 1970a, b). The most striking trait of the ITCZ is its seasonal movement from its southernmost position near the Equator to the northernmost position at about 10°N in July—October. This displacement over the ocean is small compared to that over the continents (HEISE and HÖHN, 1972).

Trade winds

The trade wind zone covers vast areas of the South Atlantic Ocean and is distinguished by a temperature inversion at a height of 1000—2000 m which separates sinking air above from rising air below. Radiational processes at the upper surface of the trade clouds further reinforce the inversion (AUGSTEIN, 1972; WMO, 1978).

The trades blow steadily in the boundary layer below the inversion with occasional weak perturbations showing as small pressure variations accompanied by a weak wind shift. The wind speed is more variable than the wind direction. The monthly mean standard deviation of the speed is 2.5 m s^{-1} (Chart 10) which is a fairly high percentage of the mean speed.

The trade wind region is characteristically dry, although the cloud amount is high. On the average it amounts to 80%, but it consists only of shallow stratocumuli below the trade inversion (REGER, 1927; AUGSTEIN, 1972) distributed as widespread cloud bands or cloud streets which can be 20—500 km long and 2—8 km broad (KUETTNER, 1971; LEMONE, 1973) and are oriented in the direction of the flow. The trade inversion may rise somewhat on occasion and isolated light showers may thus fall. In the mean, it slopes upward toward the Equator and toward the west and has also some seasonal variation (VON FICKER, 1936; VUORELA, 1950).

Disturbances

The position of the ITCZ north of the Equator and the low sea surface temperatures within the trade region imply that no tropical revolving storm of the hurricane or typhoon type forms over the South Atlantic. Even water spouts, common in the ITCZ, hardly occur south of the Equator apart from the area near Cabo Sao Roque.

There are indeed local squalls of gale force called tornados, but they have nothing to do with the funnel clouds which elsewhere are known by this name. They occur at the equatorial coast of Africa (p. 94f) as a squall line which propagates at 10−25 knots accompanied by thunder. They appear mostly at the beginning or end of the rainy season and more frequently during the night than the day. Thunderstorms are rare south of 5°S. The nightly maximum is owing to reduced stability because the negative radiation balance at the cloud top at night causes an additional cooling there; and as the sea surface temperature has no daily course, the nightly cooling in the troposphere suffices to destabilize the atmosphere so that showers and thunderstorms are enhanced or develop (NIEUWOLT, 1977).

Generally, tropical disturbances are of small scale in space and time. They gain their energy from the latent heat of vertical motions in an unstable air mass. Over sea, they develop mainly near the ITCZ induced by the convergence or over warm water triggered by turbulent convection. Other disturbances are caused by tropospheric cold cells which reduce the trade inversion from above to allow upward vertical motions. Since they are not embedded in bigger synoptic systems, they are difficult to predict (WMO, 1978).

With such systems the connection to large-scale triggering wave motions in the atmosphere is important. It is assumed that interaction between the triggering mechanism and the accompanying convective processes enhances the disturbances (KUO, 1975), and that thereby the ageostrophic effects in the vertical shear of the subtropical jet stream may generate the wave disturbances (RENNICK, 1976).

In the west, cold fronts move north along the coast of South America (VUORELA, 1950) and the troughs associated with them tend to develop into tropical disturbances as they enter low latitudes. These disturbances frequently give rise to a cloud band which extends from South America southeastward far into the South Atlantic (compare satellite photos, Figs. 6 and 7, p. 43f). This cloud band is also an extension of the cloud masses of the Amazon Basin, and it seems that such disturbances induce a southeastward flow of moist air aloft from this region while at sea level the trade winds blow against it, indicating a closed diagonal circulation over the western South Atlantic Ocean.

Circulation systems such as the frequently mentioned Hadley and Walker circulations and the one connected with the diagonal cloud band show short-term as well as yearly variations. These oscillations are similar to the Southern Oscillation of the Pacific and Indian Oceans and make long-range predictions possible for large areas, especially of the rain intensity over Brazil (WALKER, 1928; BJERKNES, 1966; ADEKODUN, 1978). These circulation systems, of course, do not function independently; fluctuations of one are transferred to the others. A strong temperature gradient between the tropics and temperate latitudes strengthens the meridional circulation and thus the trades and the Benguela Current. This transports cooler water and air toward the Equator, which lowers the temperatures in the tropics, encourages aridity, and enhances the Walker circulation. By this, the energy conversion in the atmosphere and the meridional temperature gradient are reduced so that, in turn, the meridional circulation weakens, the tropical water warms, and precipitation increases. Such pulsations can be influenced and steered by extreme circulations of temperate latitudes when outbreaks of cold air penetrate far northward and amplify the meridional temperature gradient in the subtropics. On the other hand, a stronger Hadley circulation may also stimulate the cyclonic activity of higher latitudes because the subtropical jet stream causes meanders in the higher-latitude westerlies.

Anticyclones

The mean position of the subtropical anticyclone is shown on Charts 4 and 5 for January and July. Its centre is near 30°S 10°W with a pressure of about 1023 mbar. This climatological mean does not inform us about the stability of the anticyclone. The monthly standard deviation of pressure (Chart 7) is about 5 mbar in this area; the monthly pressure must therefore in 95% of all cases vary between 1013 and 1033 mbar, so that the subtropical anticyclone cannot be as stable as the mean map makes it appear. This is especially true in winter. The statistical treatment below of the high-pressure centres and their tracks will show the variability of the anticyclone.

High-pressure centres

Charts 56 and 57 show the average number per unit area of high-pressure centres in January and July. The centres were counted in the analyses for the period 1960–1964, published in *Notos* by the South African Weather Bureau (PRETORIA, 1965–1970). The amounts indicate the number per month of high-pressure centres, summed up over the 5-year period per 5°-square, and reduced to the area of an equatorial 5°-square (308,642 km^2). Similar statistics have been published by TALJAARD (1967) for the IGY 1957–1958.

The highest numbers are in a zone centred on 35°S in January and 30°S in July. The zonally averaged numbers of anticyclonic centres per 10^6 km^2 over the South Atlantic are:

Latitude:	15°	20°	25°	30°	35°	40°	45°	50°	55°	60°S
January:	0.2	0.9	4.0	9.8	*12.6*	9.6	5.1	1.8	0.9	1.3
July:	0.4	2.9	7.1	*10.6*	8.7	4.1	2.8	2.2	1.8	1.9

The meridional profiles of the numbers are shown in Fig. 21 (p. 117).

In the region of the subtropical ridge, a maximum is found in the eastern part of the ocean near the position of the mean high-pressure centre. In contrast to the mean maps, a zone of high frequency extends west to a secondary maximum off South America which, on the average, lies about 5° farther south. Between both centres of maximum frequency, fewer anticyclonic centres are observed, particularly in January (TALJAARD and VAN LOON, 1962).

High-pressure tracks

Charts 58 and 59 from the monthly mean charts for the South Atlantic Ocean (HAMBURG, 1971) show the movement of highs derived from the daily 00 GMT maps for February and August 1963 published in *Notos* (PRETORIA, 1968–1969).

On these daily maps, nearly all anticyclones wander eastward at speeds as high as 20 m s^{-1}, but in the mean at about 10 m s^{-1} (cf., MEINARDUS, 1929). In summer, they

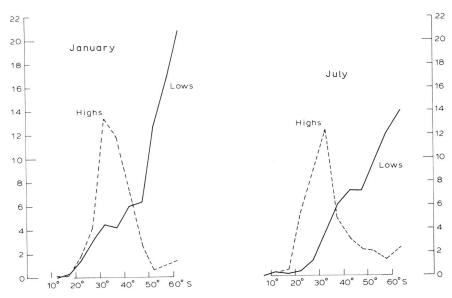

Fig. 21. Zonally averaged numbers of surface pressure highs (broken line) and lows (solid line) per 10^6 km^2 over the South Atlantic, January and July.

come from the Pacific or form over the sea, and in winter some form over South America. In summer, the tracks are concentrated between 35° and 40°S and in winter near 30°S. In the western part of the ocean, most tracks have a southerly, in the eastern part a northerly component.

The highs frequently slow down their eastward progression or even retrograde over the eastern part of the ocean. This stagnation along the tracks is the reason for the location of the mean high-pressure centre there. Those anticyclones that continue eastward from this position often accelerate over the sea area south of Africa in summer, and in winter they frequently move across the continent. The mean subtropical high thus has a complex origin. On the one hand, it develops in the descending region of the Hadley circulation and is statically supported by the cool water of the Benguela Current whose existence in turn is associated with the presence of the subtropical highs; on the other, the advection of negative vorticity contributes to its formation too, as shown by the equatorward component in the movement of the anticyclonic centres (NIEUWOLT, 1977). The tracks shown in Charts 58 and 59 are broadly representative of their respective seasons. A connection with the shallow antarctic anticyclone was not found in any of the months investigated; such a connection is most likely to exist in winter near 40 – 60°W in the region of the Drake Passage, where the westerlies could be interrupted by blocking episodes in which antarctic air penetrates to subtropical latitudes (VAN LOON, 1956; STRANZ and TALJAARD, 1965).

Cyclones of the westwind zone

In this chapter, cyclonic pressure disturbances and the associated weather phenomena are described which occur in the westerlies of the South Atlantic Ocean and play a major role in the analysis and prediction of daily weather wherever their paths cross shipping lanes or impinge on land areas.

Frontal zone

The planetary temperature contrast between the tropics and Antarctica is concentrated in temperate latitudes south of the anticyclonic belt and separates tropical and antarctic air masses. The daily meridional pressure gradient and wind field of the frontal zone are very variable. The zonal alignment of isobars and the strongly zonal wind exist only in the climatological mean maps. The climatological pressure differences indicate only the mean balance between the strengthening of the temperature gradient by the planetary heat balance and the weakening of the temperature gradient by turbulent exchange. At the same time, the mean-pressure gradient reflects the basic zonal flow used in the required exchange of heat (p. 9).

The daily wind field is shaped as large-scale meanders, and the pressure field is composed of the cyclones and anticyclones which are the turbulent elements of the large-scale heat exchange between the tropics and the antarctic and which develop in the frontal zone. In contrast to subtropical disturbances (p. 114f), depressions of the west wind zone gain their energy predominantly from the basic flow, that is, from the meridional temperature difference of the frontal zone. The vertical heat flow also plays an important role for the development of cyclones, but apparently not a crucial one (PALMÉN and NEWTON, 1969), although vertical heat fluxes and a strong temperature gradient at sea level strengthen the baroclinity in the atmosphere.

Unlike the North Atlantic, where the warm Gulf Stream is an intense source of heat and intensifies cyclonic development, there is no such concentrated heat source over the South Atlantic Ocean. The zonal basic current is pronounced all through the year due to the meridional temperature gradient and guarantees continuous development of large-scale moving cyclones.

The daily weather charts of the IGY from July 1957 to December 1958 (PRETORIA, 1962–1966) and the daily maps from the South African Weather Bureau published in *Notos* are suitable for studying synoptic weather situations. In *Monatskarten des Südatlantischen Ozeans* (HAMBURG, 1971), some typical and extraordinary weather situations are shown and described, and cloud pictures taken by satellites make it possible to recognize and follow fronts and vortices (Figs. 6 and 7).

Low-pressure centres

The number of low-pressure centres per unit area over the South Atlantic Ocean is shown in Charts 60 and 61 for January and July. As for Charts 56 and 57, the analyses of the years 1960–1964, which are published in *Notos* (PRETORIA, 1965–1970), have been used. A similar count has been made for the IGY by VAN LOON (1966) and TALJAARD (1967). The numbers are the frequency of cyclonic centres for the month averaged over 5 years in 5°-squares and reduced to an equatorial square.

The highest frequencies are at higher latitudes. The northern boundary of one low per month extends from the south tip of Africa northwestward to the coast of Brazil near 20°S. The frequency is comparatively high off the South American coast and in a zone stretching southeastward from there so that in the central and western parts of the ocean there are two zones of high frequency, one of which crosses the anticyclonic belt.

In the *polar zone* of high frequency, there is one peak south of Cape Horn and one south of Africa in summer; in winter, when the analysis is unreliable over the Antarctic Ocean, in contrast with summer when several whalers frequented the area, the frequencies are of doubtful value south of 45°S in the central and eastern parts. On the average, the cyclones which approach Antarctica from the north slow down and fill in the region of the climatological low-pressure trough along 60–65°S.

The *subtropical zone* of high frequency begins off South America at relatively low latitudes and extends southeastward to the circumpolar trough south of Africa. This zone reflects the tracks of wave disturbances that originate over South America or off the coast. The cold fronts of these disturbances mark the northern boundary of the frontal zone frequently seen on satellite photos as an extensive band of clouds (Figs. 6 and 7, p. 43f). This band is a more stable feature and is situated farther north in summer than in winter. Converging and rising air in this region is indicated by the comparatively high values of relative humidity and by high precipitation frequency.

Frequently, the cloud band, in situations when it is not connected with a front, lies north of the cyclonic centres, especially in summer (TALJAARD, 1972b) and it is then an indication of the upper branch of the Walker circulation between the rising air of the Amazon Basin and the sinking air over the central and eastern South Atlantic (p. 115). In the subtropical zone, cyclones alternate with travelling anticyclones, which are the components of the mean subtropical anticyclone. Occasionally, the wave pattern amplifies, the eastward progression slows down, and a situation akin to blocking develops. The following table contains the zonally averaged number of lows per 10^6 km² and month over the South Atlantic; as mentioned above, the numbers south of 45°S in winter are not reliable.

Latitude:	15°	20°	25°	30°	35°	40°	45°	50°	55°	60°S
January:	0.2	0.8	2.3	3.8	4.4	5.1	6.2	9.6	14.5	18.5
July:	0.2	0.3	0.9	2.6	4.9	6.6	7.2	8.5	10.9	13.0

In the zonal mean, the largest numbers appear in the polar low-pressure zone. In contrast, the subtropical maximum, which runs diagonally across the ocean, disappears in the average.

Fig. 21 shows a comparison with the frequencies of high pressure centres which clearly dominate in the subtropics. Their number exceeds that of the lows north of 45°S in January and north of 40°S in July.

Tropical disturbances (p. 115), hardly noticeable in the daily pressure field, do not appear in the frequencies as they are only marked by troughs or by cut-off lows in the high-level flow. The fact that the frequency of cyclones is shown lower in *winter* than in summer does not mean that cyclonic activity is lower in winter. It indicates rather that either the cyclones are larger and more intense in winter than in summer, which agrees with the higher frequency of gales in winter, or that the scarce observations in winter do not permit an analysis as detailed as that in summer. The area in which low-pressure centres exist is wider in winter than in summer, both toward the south and toward the north, in agreement with the steeper overall temperature gradient in winter (p. 107f).

Low-pressure tracks

The tracks and speed of movement of cyclones were derived from the daily 00 GMT position of low-pressure centres for February and August 1963, using the daily analyses published in *Notos* (PRETORIA, 1968 — 1969). These data are plotted in Charts 62 and 63 (HAMBURG, 1971) in the same manner as for the highs in Charts 58 and 59. The length of the line between two points expresses the 24-h displacement. The tracks run mainly west — east with a northerly component and gradually approach Antarctica. The two zones of a high frequency of low-pressure centres are clearly outlined by the tracks; from a planetary point of view, they are parts of the spiral arms round the South Pole (TALJAARD and VAN LOON, 1963).

In February 1963, the higher latitude cyclones were more numerous than those at lower latitudes. In the subtropics, the disturbances quickly leave South America, moving southeastward to the polar low-pressure belt south of Africa at speeds of more than 15 m s^{-1}, whereas the high-latitude cyclones in and near the trough move more slowly at 5 — 10 m s^{-1}.

In August 1963, there were more lows in the subtropics, many of which formed in the frontal zone over the Brazil Current and moved southeastward across the paths of the wandering anticyclones. Both cyclonic zones can still be distinguished south of Africa. The speed of movement was on the average about 10 m s^{-1} at temperate latitudes and 8 m s^{-1} at higher latitudes (10 — 20 degrees longitude per day).

The Andes are clearly a barrier to Pacific cyclones and the cause of the double frequency maximum and two major tracks, especially in winter when lows move unobstructed through the Drake Passage and others develop on the fronts in the trough at lower latitudes. Some of the latter are high-level lows whose low-level components have been blocked by the mountains. As the upper trough reaches the east coast, new development takes place at 30 — 40°S.

The Antarctic Peninsula also acts as a barrier to cyclones. In contrast to the Ross Sea, no stationary cyclone is found over the Weddell Sea which, in the mean, is covered by a weak extension of the subantarctic trough in the South Atlantic (VAN LOON, 1962). The interannual variation of the storm tracks is large in both zones. For instance, in the Drake Passage the axis of the polar cyclone tracks was at 60°S in July 1957 and at 50°S in July 1958 (TALJAARD, 1967).

Cyclogenesis in the period covered in Charts 62 and 63 took place predominantly over the continents, over the Brazil Current, and south of Cape Horn in summer. In winter, the frontal zone in the western South Atlantic off the coast of South America is a primary region of cyclogenesis, where even the mean values of wind direction, northeast in the north and west in the south (Charts 11 and 12), indicate the mean position of the frontal zone.

For the IGY 1957 — 1958, TALJAARD and VAN LOON (1962, 1963) have analyzed the geographical distribution of cyclones and anticyclones and of cyclogenesis in the Southern Hemisphere. They also show a marked centre of cyclogenesis off the South American coast over the southern part of the Brazil Current at 25 — 40°S, and another near Cape Horn at 55°S. The storm tracks south of Africa were limited to the zone between 45° and 70°S where cyclogenesis was rare. On the average, the axis of most frequent cyclogenesis is located 4° farther north in winter than in summer. STRETEN (1968) analyzed satellite photos of ESSA 3 for December 1966 — February 1967 and for areas of cyclonic formation and dissolution. According to him, there is a major area of formation of cyclonic eddies in summer over the sea off Argentina at 45°S 40°W and one of dissolution off the Antarctic coast at 65°S 10°W.

Weather systems

The frequency of fronts can also be analyzed (VAN LOON, 1965). The air masses which they separate over the southern oceans have been described by TALJAARD (1969). Maximum frequencies of fronts are found in a belt which extends from South America at about 30°S southeastward to the area south of Africa at 45°S, somewhat farther north in winter than in summer. A second zone of high frequency lies west—east at about 50°S. Coming from the Pacific Ocean, it crosses the southern tip of south America and joins the other maximum south of Africa. This distribution is similar to that of the storm tracks and cyclonic centres.

The shape of the frequency maxima of fronts also has a spiral structure (VAN LOON, 1965; TALJAARD, 1968). The spiral arms are channels of exchange between the tropics and temperate latitudes, and indicate that the kinetic energy of the horizontal exchange in temperate latitudes is supplied by the energy source of the tropics (p. 9).

During a 7-week expedition of the fishery research vessel *Anton Dohrn* in November—December 1977, 18 cold fronts were observed crossing the Andes from the Pacific to the Atlantic Ocean. Thunderstorms were often observed along with the fronts and eleven cyclones formed over the sea around 40°S 60°W in the frontal zone, three of which intensified in situ, whereas the others quickly moved away to the east.

Over the Patagonian shelf (p. 93), the foehn of the Andes reduces the intensity of eastward-moving fronts; warm fronts are never seen here and only weak cold fronts.

At the Hydrometeorological Centre in Moscow, weather maps from the four years of the IGY and the International Geophysical Cooperation were classified (DAVIDOVA, 1967) into six classes of meteorological processes. The classification showed that meridional weather situations were frequent in which cyclones intensified over the central South Atlantic at about 40°S 10°W. The mean frequency of meridional patterns was 70% with a maximum of 87% in July, but even with meridional types the zonal component remained strong. Both meridional and zonal circulations were markedly stronger than in the North Atlantic and seasonal changes were comparatively weak.

In his classification of synoptic situations in the region of Tristan da Cunha from daily maps for nine years analyzed in the South African Weather Bureau, VAN LOON (1959) found that anticyclonic situations prevail in the region especially in summer and winter, but that they are less frequent in autumn and least frequent in spring. The zonal situations reach their highest frequency during summer. In general, the air temperature is lower in anticyclonic situations and in meridional flow with southerly wind, i.e., on the rear sides of cyclones or front side of highs. The relative humidity is on the average higher in situations with relatively high temperature in air with stable layering, and lower in cold unstable air. Precipitation occurs more frequently in cyclonic and meridional situations than in anticyclonic and zonal ones, and precipitation in meridional situations tends to indicate a change of air masses.

Zonal indices are used to characterize the general weather situation (Grosswetterlage). They state the mean gradient between specified parallels of latitude of pressure at sea level or geopotential height at a tropospheric pressure level. A high index indicates strong meridional gradients and a lively zonal flow in which fronts tend to be orientated more zonally and wave disturbances to propagate eastward at high speed, sometimes without a complete circulation. In a low index, the flow meanders strongly and there are marked zonal pressure contrasts; highs tend to be farther south and lows farther north than normal, and in extreme instances the zonal flow collapses and slow-moving closed circulation cells predominate. This situation is called blocking, a name which refers to the obstruction, on a large scale, of the normal (west-to-east) progression of the migratory pressure systems. In a low-index situation, warm and cold air masses are displaced meridionally far beyond their region of origin and appear as anomalies of the relative topography, and large-scale meridional exchange of atmospheric properties occurs (RUBIN and VAN LOON, 1954).

During the IGY 1957–1958, high index was more frequent in winter than in summer, and the amplitude of the daily fluctuations of the zonal index was larger in winter than in summer (VAN LOON, 1965). Blocking is most frequent from July to September and has a secondary maximum from March to May (VAN LOON, 1956). Blocking in the South Atlantic occurs less often than in the North Atlantic and does not persist as long because of the strong basic zonal current.

The differences in the general atmospheric circulation between the North and South Atlantic (REITER, 1970) are altogether instructive as regards their dependence on different heat balances and monsoonal influences. The strong meridional differences in the heat balance over the South Atlantic Ocean (p. 66), associated with the vast expanse of the southern oceans and the antarctic cold source, result in a strong frontal zone and lively cyclonic activity in temperate latitudes and in a powerful Hadley circulation in the tropics. The weak continental influences, especially on the east side of the ocean, make the maritime and planetary character of the climates over the South Atlantic Ocean stand out clearly.

References

ADEKODUN, J. A., 1978. West African precipitation and dominant atmospheric mechanisms. *Arch. Meteorol. Geophys. Bioklimatol.*, Ser. A, 27: 289–310.

ALBRECHT, F., 1960. Jahreskarten des Wärme- und Wasserhaushaltes der Ozeane. *Ber. Dtsch. Wetterdienstes*, 9 (66): 19 pp.

ALBRECHT, F., 1961. Der jährliche Gang der Komponenten des Wärme- und Wasserhaushaltes der Ozeane. *Ber. Dtsch. Wetterdienstes*, 11 (79): 24 pp.

ALEXANDER, R. C. and MOBLEY, R. L., 1976. Monthly average sea-surface temperatures and ice-pack limits on a 1° global grid. *Mon. Weather Rev.*, 104: 143–148.

AUGSTEIN, E., 1972. Untersuchungen zur Struktur und zum Energiehaushalt der Passatgrundschicht. *Ber. Inst. Radiometeorol. Mar. Meteorol.*, 19: 71 pp.

BARRETT, E. C., 1974. *Climatology from Satellites*. Methuen, London, 418 pp.

BAUMGARTNER, A. and REICHEL, E., 1975. *The World Water Balance*. R. Oldenburg Verlag, München und Wien, 180 pp.

BAUMGARTNER, A., MAYER, H. and METZ, W., 1976. Globale Verteilung der Oberflächenalbedo. *Meteorol. Rundsch.*, 29: 38–43.

BECKER, R., 1944. Die Verteilung der Luftdichte auf dem Südatlantischen Ozean. *Ann. Hydr. Mar. Meteorol.*, 72: 346–347.

BERGHAUS, H., 1842. *Sechs Reisen um die Erde*. Grass, Barth and Cy, Breslau, 181 pp.

BERNHARDT, F. and PHILIPPS, H., 1958. Die räumliche und zeitliche Verteilung der Einstrahlung, der Ausstrahlung und der Strahlungsbilanz im Meeresniveau, Teil 1. *Abhandl. Meteorol. Hydrol. Dienstes DDR*, 45: 277 pp.

BERNHARDT, F. and PHILIPPS, H., 1966. Die räumliche und zeitliche Verteilung der Einstrahlung, der Ausstrahlung und der Strahlungsbilanz im Meeresniveau, Teil 2. *Abhandl. Meteorol. Hydrol. Dienstes DDR*, 77: 266 pp.

BJERKNES, J., 1966. A possible response of the atmospheric Hadley circulation to equatorial anomalies of ocean temperatures. *Tellus*, 18: 820—829.

BJERKNES, J., 1969. Atmospheric teleconnections from the equatorial Pacific. *Mon. Weather Rev.*, 97: 163—172.

BREITENGROSS, J. P., 1972. Saisonales Fliessverhalten in grossflächigen Fluss-Systemen. *Mitt. Geogr. Ges., Hamburg*, 60: 92 pp.

BRIER, G. W., 1978. The quasi-biennial oscillation and feedback processes in the atmosphere—ocean—earth system. *Mon. Weather Rev.*, 106: 938—946.

BROMANN, E., 1968. Witterungsverhältnisse auf den Fangplätzen Südostamerikas. *Seeverkehr*, 8: 298—299.

BROOKS, C. E. P., 1920. Climate and weather of Falkland Islands and South Georgia. *Geophys. Mem.*, 15: 146 pp.

BROOKS, C. E. P., 1931. The climate of Ascension Island. *Q. J. R. Meteorol. Soc.*, 57: 85—89.

BUCHAN, A., 1869. The mean pressure of the atmosphere and the prevailing winds over the globe. *Trans. R. Soc.*, 25: 575—673.

BUDYKO, M. I., 1956. *Teplovogo Balansa Zemnogo Shara (The Heat Balance of the Earth's Surface)*. Gidrometeor., Leningrad, Izdatel'stvo, 259 pp. (English translation by N. A. Stepanova, U.S. Dept. Commerce, Washington, D.C., 1958).

BUDYKO, M. I., 1963. *Atlas Teplovogo Balansa Zemnogo Shara (Atlas of the Heat Balance of the Earth's Surface)*. Gidrometeoizdat, Main Geophysical Observatory, Moscow, V plus 69 charts.

BUDYKO, M. I., 1972. The water balance of the oceans. In: *Proc. Reading Symp. 1970 World Water Balance*. Inst. Assoc. Sci. Hydrol., UNESCO-WMO, No. 92, Gentbrügge, pp. 24—33.

BUDYKO, M. I., 1974. *Climate and Life*, Academic Press, New York, N.Y., 508 pp.

BUENOS AIRES, 1958. *Estadisticas climatologicas 1901—1950*. Direccion de Meteorologia, Geofisica e Hidrologia, Buenos Aires, 44 pp.

CORNEJO-GARRIDO, A. G. and STONE, P. H., 1977. On the heat balance of the Walker circulation. *J. Atmos. Sci.*, 34: 1155—1162.

CRUTCHER, H. L., JENNE, R. L., TALJAARD, J. J. and VAN LOON, H., 1971. *Climate of the Upper Air: Southern Hemisphere, Vol. IV, Selected Meridional Cross Sections of Temperature, Dew Point and Height*. NAVAIR 50-1C-58, U.S. Navy, 62 pp.

DAMON, P. E. and KUNEN, S. M., 1976. Global cooling? *Science*, 193: 447—453.

DANIEL, H., 1973. *One Hundred Years of International Co-operation in Meteorology (1873—1973). A Historical Review*. WMO, Geneva, 345: 60 pp.

DAVIDOVA, N. G., 1967. Types of synoptic process and associated wind fields in oceanic regions of the southern hemisphere. In: *Polar Meteorology, WMO No. 211, Tech. Note 87*, Geneva, pp. 263—291.

DAVIS, R. A., Jr., 1977. *Principles of Oceanography*, 2nd ed. Addison-Wesley, Reading, 505 pp.

DEACON, E. L., 1969. Physical processes near the surface of the earth. In: H. FLOHN (Editor), General Climatology, 2. *World Survey of Climatology*, Vol. 2. Elsevier, Amsterdam, pp. 39—104.

DEACON, E. L. and WEBB, E. K., 1962. Interchange of properties between sea and air; small scale interactions. In: *The Sea*, New York and London, I: 43—87.

DEACON, G. E. R., 1934. Die Nordgrenze des antarktischen und subantarktischen Wassers im Weltmeer. *Ann. Hydrogr. Mar. Meteorol.*, 62: 129—136.

DEACON, G. E. R., 1977. Meridional transport in the Southern Oceaan. *Ocean Modelling*, 3: 3—5.

DE BILT, 1951. *Oceanographische en Meteorologische Waarnemingen, Atlantische Oceaan*. K. Ned. Meteorol. Inst., De Bilt, 4 Vols.

DIETRICH, G., 1972. Upwelling in the ocean and its consequences. *Geoforum*, 11: 3—4.

DROZDOV, O. A., 1964. *Atlas Mora*. Akad. Nauk, SSSR, Moscow, 298 pp.

EICKERMANN, W. and FLOHN, H., 1962. Witterungszusammenhänge über dem äquatorialen Südatlantik. *Bonner Meteorol. Abh.*, 1: 65 pp.

EREDIA, F., 1932. La meteorologia a l'aerologia degli oceani; l'Oceano Atlantico Sud. *Suppl. al fasciolo della Rivista Marittima*, 65: 171 pp.

FABRICIUS, A. F., 1957. Climate of the sub-antarctic islands. In: *Meteorology of the Antarctic*. Weather Bureau, Pretoria, pp. 111—135.

FEAN, C. R., 1961. *Seasonal Survey of Average Cloudiness Conditions over the Atlantic and Pacific Oceans*. Scripps Institution of Oceanogr., La Jolla, Calif., 34 pp.

FLETCHER, J. O., 1969. *Ice Extent on the Southern Ocean and its Relation to World Climate. The Rand Corporation Memorandum*, RM-5793-NSF, Santa Monica, Calif., 119 pp.

FLOHN, H., 1950. Grundzüge der allgemeinen atmosphärischen Zirkulation auf der Südhalbkugel. *Arch. Meteorol. Geophys. Bioklimatol.*, Ser. A, 2: 17–64.

FLOHN, H., 1972. Investigations of equatorial upwelling and its climatic role. A. L. GORDON (Editor), *Studies in Physical Oceanography*. Gordon and Breach Science, New York, N.Y., 1: 93–102.

FORTAK, H. G., 1979. Entropy and climate. *Devel. Atmos. Sci.*, 10: 1–14.

GATES, W. L., 1977. Modelling the ocean-atmosphere system and the role of the ocean in climate. *Rep. Mar. Sci. Affairs*, 11, WMO, No. 472, Geneva, pp. 1–29.

GORDON, A. L., 1975. General ocean circulation. *Proc. Symp. Numerical models of Ocean Circulation, Durham, 1972*. National Academy of Sciences, Washington, pp. 39–53.

GORDON, A. L. and GOLDBERG, R. D., 1970. Circumpolar characteristics of Antarctic waters. *Antarct. Folio Ser.*, 13: 5 pp., 19 plates.

GORSKOV, C. G., 1977. *Oceanic Atlas, Atlantic and Indian Oceans, 2*. Ministerstro oborony SSSR, Voennomorskoj flot, Moskva, 306 pp.

GRIFFITHS, J. F., 1972. General introduction. In: J. F. GRIFFITHS (Editor), *Climates of Africa. World Survey of Climatology*, Vol. 10. Elsevier, Amsterdam, pp. 1–35.

GROEN, P., 1967. *The Waters of the Sea*. Van Nostrand, New York, N.Y., 328 pp.

HALLEY, E., 1688. An historical account of the trade winds and monsoons observable in the seas between and near the tropics. *Philos. Trans. R. Soc. London*, 16: 153–168.

HAMBURG, 1963–1967. Tropical Zone. *International Geophysical Year 1957–1958. World Weather Maps, II*. Deutscher Wetterd., Seewetteramt, 18 Vols.

HAMBURG, 1971. *Monatskarten für den Südatlantischen Ozean*. Deutsches Hydrographisches Inst., Hamburg, 50 pp.

HAMBURG, 1981a. *Handbuch des Atlantischen Ozeans*. Seehandbuch 2057, 5th ed., Deutsches Hydrographisches Inst., Hamburg, pp. 1–184.

HAMBURG, 1981b. *Handbuch der Ostküste Südamerikas*. Seehandbuch 2051/52, Deutsches Hydrographisches Inst., Hamburg, pp. 35–123.

HAMBURG, 1981c. *Handbuch der Westküste Afrikas*. Seehandbuch 2062, Deutsches Hydrographisches Inst., pp. 31–111.

HASSELMANN, K., 1977. The dynamical coupling between the atmosphere and the ocean. *Rep. Mar. Sci. Affairs*, 11, WMO Geneva, 472, pp. 31–44.

HASTENRATH, S., 1977. On zonal asymmetries in the heat budget of the tropical oceans. *J. Meteorol. Soc. Jpn.*, II, 55: 168–173.

HASTENRATH, S. and LAMB, P. J., 1978. *Heat Budget Atlas of the Tropical Atlantic and Eastern Pacific Oceans*. University of Wisconsin Press, Madison, Wisc., 104 pp.

HEISE, G. and HÖHN, R., 1972. On the position of the ITCZ and its oscillations during the IGY from July 1957 to December 1958. *Einzelveröff. Seewetteramtes*, 79: 2–29.

HENNING, D., 1969a. Zonalkomponente des Windes über Fernando Noronha. *Meteorol. Rundsch.*, 22: 41–43.

HENNING, D., 1969b. Verfrachtung von Enthalpie, latenter Wärme und Wasserdampf. *Beitr. Phys. Atmos.*, 42: 94–131.

HOEFER, A., 1925. Im Pampero-Sturm, eine Strandung vor der La-Plata-Mündung. *Meereskunde*, 14, Heft 7: 32 pp.

HÖFLICH, O., 1972. Die meteorologischen Wirkungen kalter Auftriebswasser. *Geoforum*, 11: 35–46.

HÖFLICH, O., 1974. The seasonal and secular variation of the meteorological parameters on both sides of the ITCZ in the Atlantic Ocean. *GATE Rep.*, 2, Part VI, 36 pp.

HOFMEYR, W. L., 1970. Monthly mean aerological cross sections between the equator and the south pole along the 10E meridian for the period of the IGY. *Notos*, 19: 63–84.

HOGBEN, N. and LAMB, F. E., 1967. *Ocean Wave Statistics*. H.M. Stationery Office, London, 263 pp.

HOLOPAINEN, E. O., 1977. Energy balance of the earth. *Aust. Meteorol. Mag.*, 25: 89–103.

HUBERT, L. F., 1966. Mesoscale cellular convection. U.S. Dept. Commerce, Environmental Science Service Admin., *Satellite Lab. Rept.*, 37: 68 pp.

ISELIN, C. O., 1959. The basic thermal structure of the oceans. In: *Lectures I*. Woods Hole Oceanogr. Inst., pp. 9–22.

JACOBS, C. W., 1968. The seasonal apportionment of precipitation over the ocean. *Yearb. Assoc. Pacific Coast Geogr.*, 30: 63–78.

JAEGER, L., 1976. Monatskarten des Niederschlags für die ganze Erde. *Beitr. Dtsch. Wetterdienstes*, 18 (139): 38 pp.

JAMES, R. W. and FOX, P. T., 1972. Comparative sea-surface temperature measurements. *Rep. Mar. Sci. Affairs*, 5, WMO, No. 336, Geneva, 28 pp.

JENNE, R. L., CRUTCHER, H. L., VAN LOON, H. and TALJAARD, J. J., 1971. *Climate of the Upper Air: Southern Hemisphere, Vol. III. Vector Mean Geostrophic Winds.* NAVAIR 50-1C-57 and NCAR TN/STR-58, National Center for Atmospheric Research, Boulder, Colo., 68 pp.

JOZSEF, S. K., 1938. Verteilung der Luftfeuchtigkeit auf der Erde. *Ann. Dtsch. Hydrogr. Mar. Meteorol.*, 66: 373–378.

KAUFELD, L. and RUDLOFF, W., 1972. Quarterly charts of the standard deviation of the sea-level pressure for the IGY 1957/58. *Einzelveröff. Seewetteramtes*, 79: 31–49.

KESSLER, A., 1973. Zur Klimatologie der Strahlungsbilanz an der Erdoberfläche. *Erdkunde*, 27: 1–10.

KIRCHER, A., 1678. *Mundus Subterraneus.* 3rd Edition, Amsterdam, 2 Vol.

KNOCH, K., 1961. Rainfall and distribution of temperature throughout the world. In: *Welt-Seuchen-Atlas*, III: 116–118.

KNOCH, K. and SCHULZE, A., 1952. Methoden der Klimaklassifikation. *Ergänzungsheft 249, Petermanns Geogr. Mitt.*, 78 pp.

KÖPPEN, W., 1931. *Grundrisse der Klimakunde*, 2nd ed. De Gruyter, Berlin und Leipzig, 388 pp.

KONDO, J., 1977. Comparison of the Kondo's bulk transfer coefficient with the recently-made direct observations of fluxes on the sea surface. *J. Meteorol. Soc. Jpn.*, 55: 319–323.

KONDRATYEV, K. Y., 1972. *Radiation Processes in the Atmosphere.* WMO, No. 309, 220 pp.

KORT, V. G., 1962. The Antarctic Ocean. *Sci. Am.*, 207: 113–128.

KORT, V. G., 1968. Frontal zones of the Southern Ocean. *Symp. Antarct. Oceanogr., Cambridge, 1966*, pp. 3–7.

KORT, V. G., 1972. On the structure of deep currents. In: A. L. GORDON (Editor), *Studies in Physical Oceanography*. Gordon and Breach Science, New York, N.Y., 2: 115–121.

KRAUS, E. B., 1972. *Atmosphere–Ocean Interaction.* Clarendon Press, Oxford, 275 pp.

KRÜGERMEYER, L., 1975. Vertikale Transporte von Impuls, sensibler und latenter Wärme aus Profilmessungen über dem tropischen Atlantik während APEX. *Ber. Inst. Radiometeorol. Mar. Meteorol.*, 29: 84 pp.

KRUHL, H., 1974. Meteorologische Navigation auf See, Schiffsroutenempfehlungen. *Denkschr. 175. Jahrestag d. Gründung d. Hochschule f. Nautik*, Bremen, pp. 34–52.

KUETTNER, J. P., 1971. Cloud bands in the earth's atmosphere. *Tellus*, 23: 404–425.

KUHLBRODT, E., 1936a. Schleuderthermometer für Bordgebrauch. *Ann. Hydrogr. Mar. Meteorol.*, 64: 57–60.

KUHLBRODT, E., 1936b. Kritik der Lufttemperatur-Bestimmung auf See. *Ann. Hydrogr. Mar. Meteorol.*, 64: 259–264.

KUHLBRODT, E., 1962. Die Höhenwinde über dem tropischen und südlichen Atlantischen Ozean. *Wiss. Ergeb. Dtsch. Atl. Exped. "Meteor" 1925–1927*, Nr. 16, 1, Berlin, pp. 93–167.

KUHLBRODT, E. and REGER, J., 1933. Die aerologischen Methoden und das aerologische Beobachtungsmaterial. *Wiss. Ergeb. Dtsch. Atl. Exped. "Meteor" 1925–1927*, Nr. 15, Berlin und Leipzig, 305 pp.

KUHLBRODT, E. and REGER, J., 1938. Die meteorologischen Beobachtungen. *Wiss. Ergeb. Dtsch. Atl. Exped. "Meteor" 1925–1927*, Nr. 14, Berlin und Leipzig, 392 pp.

KUO, H. L., 1975. Instability theory of long-scale disturbances in the tropics. *J. Atmos. Sci.*, 32: 2229–2245.

LABITZKE, K. and VAN LOON, H., 1972. The stratosphere in the Southern Hemisphere. *Meteorol. Monogr.*, 13: 113–138.

LAEVASTU, T., CLARKE, L. and WOLFF, P. M., 1969. Oceanic part of the hydrological cycle. *Rep. WMO/IHD Projects*, 11, Geneva, 71 pp.

LAMB, H. H., 1957. Some special features of the climate of St. Helena and the trade-wind zone in the South Atlantic. *Meteorol. Mag.*, 86: 73–76.

LAMB, P. J., 1977. On the surface climatology of the tropical Atlantic. *Arch. Meteorol. Geophys. Bioklimatol.*, Ser. B, 25: 21–31.

LANGMUIR, I., 1938. Surface motion of water induced by wind. *Science*, 87: 119–123.

LEGECKIS, R., 1978. A survey of worldwide sea surface temperature fronts detected by environmental satellites. *J. Geophys. Res.*, 83: 4501–4522.

LEMONE, M. A., 1973. The structure and dynamics of horizontal roll vortices in the planetary boundary layer. *J. Atmos. Sci.*, 30: 1077–1091.

LENINGRAD, 1967. *Atlas of Sea Waves and Wind in the Atlantic Ocean.* Gidrometeorol. Inst., 79 pp. (in Russian).

Lisboa, 1965. O clima de Portugal. *Fasciculo*, 14: 243 pp.

Loewe, F., 1950. Das Klima von Tristan da Cunha. *Meteorol. Rundsch.*, 3: 147–149.

London, 1949. *Weather on the West Coast of Tropical Africa*. Meteorol. Office, London, 281 pp.

London, 1959. *Monthly Meteorological Charts of the Atlantic Ocean*. Meteorol. Office, Air Ministry, London, 120 pp.

London, 1960a. Central and South America. *Tables of Temperature, Relative Humidity and Precipitation for the World, II*. Meteorol. Office, London, 617b, 53 pp.

London, 1960b. Africa. *Tables of Temperature, Relative Humidity and Precipitation for the World, IV*. Meteorol. Office, London, 617d, 220 pp.

Lumley, J. L. and Panofsky, H. A., 1964. *The Structure of Atmospheric Turbulence*. New York, N.Y., 239 pp.

Mackintosh, N. A. and Herdman, H. F. P., 1940. Distribution of pack ice in the Southern Ocean. *Discovery Rep.*, 19: 285–296.

Mallory, O. I., 1974. Abnormal waves on the south-east coast of South Africa. *Int. Hydrogr. Rev.*, 51: 99–129.

Manabe, S. and Wetherald, R., 1975. The effects of doubling the CO_2 concentration of the climate of a general circulation model. *J. Atmos. Sci.*, 32: 3–15.

Markgraf, H., 1954. Das La-Plata-Tief. *Ann. Meteorol.*, 6: 244–250.

Martin, D. W., 1968. Satellite studies of the cyclonic developments over the Southern Ocean. *Int. Antarct. Meteorol. Res. Centre, Tech. Rep.*, 9: 64 pp.

Martin, D. W. and Scherer, W., 1973. Review of satellite rainfall estimation methods. *Bull. Am. Meteorol. Soc.*, 54: 661–674.

Martin, D. W. and Sikdar, D. N., 1975. A case study of Atlantic cloud clusters. *Mon. Weather Rev.*, 103: 691–708.

Maury, M. F., 1858. *The Physical Geography of the Sea*. Samson Lau, New York, N.Y., 348 pp.

Mazeika, P. A., 1968. Mean monthly sea surface temperatures and zonal anomalies of the tropical Atlantic. *Serial Atlas of the Marine Environment*. Am. Geograph. Soc., New York, N.Y., Folio 16, 6 plates.

McDonald, W. F., 1938. *Atlas of Climatic Charts of the Oceans*. U.S. Weather Bureau Pub., No. 1247. Government Printing Office, Washington, D.C., 130 pp.

McDonald, W. F., 1957. Notes on the problems of cargo ventilation. *Tech. Note 17*, WMO, No. 63, Geneva, 38 pp.

Meinardus, W., 1929. Die Luftdruckverhältnisse und ihre Wandlungen südlich von 30° südlicher Breite. *Meteorol. Z.*, 46: 41–49; 86–96.

Meinardus, W. and Mecking, L., 1911. Tägliche synoptische Wetterkarten der höheren südlichen Breiten, von Oktober 1901 bis März 1904. Deutsche Südpolar-Expedition 1901–1903. *Meteorol. Atlas*, Berlin, 119 tables.

Meldrum, C., 1861. *Synoptic Weather Charts for the Indian Ocean*. Edinburgh.

Meuss, J. F., 1913. Die Untersuchungen des Königlichen Seehandlungs-Instituts zur Emporbringung des preußischen Handels zur See. *Veröff. Inst. Meereskunde, N. F.*, B2: 329 pp.

Mitchell, J. M., 1961. Recent secular changes of global temperature. *Ann. N.Y. Acad. Sci.*, 95: 235–250.

Monin, A. S., Kamenkovich, V. M. and Kort, V. G., 1974. *Variability of the Oceans*. Hydrometeorisdat 1974, English translation, 241 pp.

Morgan, J., 1978. METEOSAT 1 in orbit. *WMO Bull. 27*, Geneva, pp. 250–253.

Morize, H., 1927. *Contribuicado ao Estudo do Clima do Brasil*. Observatorio Nacional, Rio de Janeiro, 114 pp.

Moscow, 1953. *Morskoj Atlas*, Physical-Geographical Part. Izdanic Glavnago Staba Voenno-Morskich SIL, Moscow, 76 charts.

Neumann, G., 1972. Precipitation, evaporation and monthly salinity variations in the inner Gulf of Guinea near the equator. In: A. L. Gordon (Editor), *Studies in Physical Oceanography*. Gordon and Breach Science, New York, N.Y., 1: 19–48.

Neuwirth, F., 1978. Die Bestimmung der Verdunstung einer freien Wasserfläche aus längerfristigen Mittelwerten. *Arch. Meteorol. Geophys. Bioklimatol.*, B, 25: 337–344.

Newell, R. E., Kidson, J. W., Dayton, G. V. and Boer, G. J., 1972. *The General Circulation of the Tropical Atmosphere and Interactions with Extratropical Latitudes*. M.I.T. Press, Cambridge, Mass., 258 pp.

Newnham, E. V., 1949. The climates of Addu Atoll, Agaleya Islands and Tristan da Cunha. *Professional Notes*, 101, Vol. VII, 20 pp.

NEWTON, C. W., 1972. Southern hemisphere general circulation in relation to global energy and momentum balance requirements. *Meteorol. Monogr.*, 13, (35): 215—246.

NIEUWOLT, S., 1977. *Tropical Climatology*. Wiley, Chichester, 207 pp.

NORQUIST, D. C., BECKER, E. E. and REED, R. J., 1977. The energetics of Africa wave disturbances as observed during phase III of GATE. *Mon. Weather Rev.*, 105: 334—342.

OSTAPOFF, F., TARBEYEV, Y. and WORTHEM, S., 1973. Heat flux and precipitation estimates from oceanographic observations. *Science*, 180: 960—962.

PALMÉN, E., 1967. Evaluation of atmospheric moisture transport for hydrological purposes. *WMO/IHD Rep.*, 1, Geneva, 63 pp.

PALMÉN, E. and NEWTON, C. W., 1969. *Atmospheric Circulation Systems, Their Structures and Physical Interpretation*. Academic Press, New York, N.Y., 603 pp.

PALMÉN, E. and SÖDERMANN, D., 1966. Computation of the evaporation from the Baltic Sea from the flux of water vapour in the atmosphere. *Geophysica*, 8: 261—279.

PARIS, AIR FRANCE, 1965. *Climatologie, Amérique du Sud*. Direction de l'Exploitation, Paris, 80 pp.

PAYNE, R. E., 1972. Albedo of the sea surface. *J. Atmos. Sci.*, 29: 959—970.

PEIXOTO, P., 1958. *Hemispheric Humidity Conditions During the Year 1950*. Dept. Meteorol., Massachusetts Institute of Technology, Rpt AFCRC-TN-58-609.

PEPPER, J., 1954. *The Meteorology of the Falkland Islands and Dependencies, 1944—1950*. Hodgson, London, 249 pp.

PERRY, A. H. and WALKER, J. M., 1977. *The Ocean—Atmosphere System*. Longman, London and New York, 160 pp.

PETERSEN, P., 1927. Zur Bestimmung der Windstärke auf See. Für Segler, Dampfer und Luftfahrzeuge. *Ann. Hydrogr.*, 55: 69—72, 394.

PFLUGBEIL, C., 1967. Hemisphärische und globale Luftdruckbilanzen. *Ber. Dtsch. Wetterdienstes*, 14 (104): 21 pp.

PRETORIA, 1949. *Meteorological Observations at Tristan da Cunha, 1943—47*. Weather Bureau, Pretoria, 71 pp.

PRETORIA, 1952. Tables and charts of Southern Hemisphere sea-level pressure. *Notos*, 1: 23—64.

PRETORIA, 1958, 1959. Monthly mean maps of sea-level-pressure and 500-mb height for the IGY. *Notos*, 8 and 9.

PRETORIA, 1962—1966. *Southern Hemisphere South of 20°S. International Geophysical Year 1957—1958, III*. Weather Bureau, Pretoria, 18 Vols.

PRETORIA, 1965. Historical weather charts for the southern hemisphere for the year 1960. *Notos*, 14: 93—474.

PRETORIA, 1966. Historical weather charts for the southern hemisphere for the year 1961. *Notos*, 15: 99—478.

PRETORIA, 1967. Historical weather charts for the southern hemisphere for the year 1962. *Notos*, 16: 71—450.

PRETORIA, 1968. Historical weather charts for the southern hemisphere for the year 1963. *Notos*, 17: 141—384; 18: 105—227.

PRETORIA, 1969. Historical weather charts for the southern hemisphere for the year 1964. *Notos*, 18: 228—348; 19: 107—352.

PRIVETT, D. W., 1960. The exchange of energy between the atmosphere and the oceans of the southern hemisphere. *Geophys. Mem.*, 13 (104): 61 pp.

PROHASKA, F., 1976. The climate of Argentina, Paraguay and Uruguay. In: W. SCHWERDTFEGER (Editor), *Climates of Central and South America. World Survey of Climatology*, Vol 12. Elsevier, Amsterdam, pp. 13—112.

QUAYLE, R. G. and ELMS, J. D., 1979. High waves in the Benguela Current. *J. Phys. Oceangr.*, 9: 858—865.

RASMUSSON, E. M., 1966. *Atmospheric Water Vapor Transport and the Hydrology of North America*. Dept. Meteorol., Massachusetts Institute of Technology, Report A1, 1970 pp., 118 charts.

RATISBONA, L. R., 1976. The climate of Brazil. In: W. SCHWERDTFEGER (Editor), *Climates of Central and South America. World Survey of Climatology*, Vol. 12. Elsevier, Amsterdam, pp. 219—293.

REED, W. W., 1929. Climatological data for southern South America. *Mon. Weather Rev.*, Suppl., 32: 23 pp.

REGER, J., 1927. Der Südostpassat. *Beitr. Physik Freien Atmos.*, 13: 59—63.

REGER, J., 1963. Der statische Aufbau der Luft über dem Südatlantischen Ozean. *Wiss. Ergeb. Dtsch. Atl. Exped. "Meteor" 1925—1927*, 16 (2): 65—96.

REITER, E. R., 1970. Meteorologische Unterschiede zwischen Nord- und Südhalbkugel. *Umsch. Wiss. Tech.*, 70: 219—220.

RENNICK, M. A., 1976. The generation of African waves. *J. Atmos. Sci.*, 33: 1955—1969.

RIEHL, H., 1954. *Tropical Meteorology*. McGraw-Hill, New York, N.Y., 392 pp.

RIEHL, H., 1969. Mechanisms of the general circulation of the troposphere. In: H. FLOHN (Editor), *General Climatology, 2. World Survey of Climatology,* Vol. 2. Elsevier, Amsterdam, pp. 1—37.

RIO DE JANEIRO, 1922. *Boletim de Normales.* Ministerio da Agricultura, Rio de Janeiro, 66 pp.

RIO DE JANEIRO, 1942. *Normais Climatologicas.* Ministerio da Agricultura, Rio de Janeiro, 167 pp.

ROLL, H. U., 1965. On the present state of knowledge in air — sea boundary layer problems. *U.S. Dept. Commerce, Environ. Sci. Serv. Admin., Tech. Note,* 9-SAIL-1: 31—63.

RUBIN, M. J., 1952. Southern hemisphere synoptic analysis. *Bull. Am. Meteorol. Soc.,* 33: 409—415.

RUBIN, M. J. and VAN LOON, H., 1954. Aspects of the circulation of the southern hemisphere. *J. Meteorol.,* 11: 68—76.

RUDLOFF, W., 1970. Monthly charts of the sea level pressure and the 500 mb geopotentials of the tropical zone and monthly world charts of the sea level pressure for the IGY 1957—1958. *Einzelveröff. Seewetteramtes,* 72: 75 pp.

RUDLOFF, W. and HÖFLICH, O., 1970. Aerologische Monatskarten der Tropen für das IGJ. Teil 3, Winde für das 200- und 300-mb-Niveau. *Einzelveröff. Seewetteramtes* (German and English), 70: 84 pp. Teil 4, 100 mb und 100/200 mb. *Einzelveröff. Seewetteramtes* (German and English), 71: 79 pp.

RUDLOFF, W., HÖFLICH, O. and HEISE, G., 1965. Aerologische Monatskarten der Tropen für das IGJ. Teil 1, 300 mb und 300/500 mb. *Einzelveröff. Seewetteramtes* (German and English), 50: 79 pp. Teil 2, 200 mb und 200/300 mb. *Einzelveröff. Seewetteramtes* (German and English), 51: 79 pp.

RUPRECHT, E. and GRAY, W. M., 1976. Analysis of satellite-observed tropical cloud clusters. *Tellus,* 28: 391—413.

SADLER, J. C., ODA, L. and KILONSKY, B. J., 1976. *Pacific Ocean Cloudiness from Satellite Observations.* Dept. Meteorology, University of Hawaii, 137 pp.

SASAMORI, T., LONDON, J. and HOYT, D. V., 1972. Radiation budget of the southern hemisphere. *Meteorol. Monogr.,* 13 (35): 9—23.

SCHERHAG, R., 1948. *Neue Methoden der Wetteranalyse und Wetterprognose.* Springer, Berlin, 424 pp.

SCHMETTAU, S. and BRUCKNER, I., 1749. *Nouvelle atlas de Marine.* (Der erste Preussische Seeatlas). L'Académie Royale des Sciences à Berlin, Berlin, 11 charts.

SCHMITT, W., 1952. Preliminary mean absolute topography 500 mb, southern hemisphere 70°W to 120°E, summer 1951—52. *Notos,* 1: 7—17.

SCHNAPAUFF, W., 1937. Untersuchungen über die Kalmenzone des Atlantischen Ozeans. *Veröff. Meteorol. Inst. Univ. Berlin,* II, 4: 35 pp.

SCHOTT, G., 1895. Die Verkehrswege der transozeanischen Segelschiffahrt in der Gegenwart. *Z. Ges. Erdkunde,* 30: 235—300.

SCHOTT, G., 1942. *Geographie des Atlantischen Ozeans.* 3rd ed., Hamburg, 438 pp.

SCHÜCK, A., 1875. Die Wege des Oceans für Segelschiffe. 2. *Jahresber. Geogr. Ges. Hamburg 1874—75,* pp. 110—128.

SCHULZE, B. R., 1972. South Africa. In: J. F. GRIFFITHS (Editor), *Climates of Africa. World Survey of Climatology,* Vol. 10. Elsevier, Amsterdam, pp. 501—586.

SCHUMACHER, A., 1936. Die Fahrten eines deutschen Seemannes aus der Segelschiffszeit. *Ann. Hydrogr. Mar. Meteorol.,* 2, Köppen-Heft, pp. 75—78.

SCHUMANN, T. E. W., 1952. The meteorological network of the southern quarter of the globe. *Notos,* 1: 66—69.

SCHWERDTFEGER, W., 1958. *Lectures on the Meteorology of Extratropical South America.* Univ. of Melbourne, Meteorol. Dept., Melbourne, 17 pp.

SCHWERDTFEGER, W., 1962. *Meteorologia del area del Pasaje Drake.* Publico H. 410, Buenos Aires, 76 pp.

SCHWERDTFEGER, W., 1970. The climate of the Antarctic. In: S. ORVIG (Editor), *Climates of the Polar Regions. World Survey of Climatology,* Vol. 14. Elsevier, Amsterdam, pp. 253—355.

SCHWERDTFEGER, W., 1976. Introduction. In: W. SCHWERDTFEGER (Editor), *Climates of Central and South America. World Survey of Climatology,* Vol. 12. Elsevier, Amsterdam, pp. 1—12.

SHULEIKIN, 1968. *Fizika Morya* (Physics of the Sea). Nauka, Moscow, 1083 pp. (in Russian).

STEINBACH, W., 1968. *Kernfischerei vor Südwestafrika.* DHZZ-Arbeitskreis Seefunk 8/9, Wustrow.

STEWART, R. W., 1969. The atmosphere and the ocean. *Sci. Am.,* 221: 76—86.

STOCKS, T., 1962. Zur Vollendung des "Meteor"-Werks. *Naturwiss. Rundsch.,* 15: 315—318.

STRANZ, D. and TALJAARD, J. J., 1965. Analysis of an abnormal winter situation in South Africa. *Notos,* 14: 17—32.

STRETEN, N. A., 1968. Some aspects of high latitude southern hemisphere summer circulation as viewed by ESSA 3. *J. Appl. Meteorol.*, 7: 324—332.

STRETEN, N. A., 1977. Seasonal climatic variability over the southern oceans. *Arch. Meteorol. Geophys. Bioklimatol.*, Ser. B, 25: 1—19.

STRÜBING, K., 1978. Trift eines bemerkenswerten Eisberges in den Südatlantik. *Der Seewart*, 39: 186—195.

SUPAN, A., 1898. Die jährlichen Niederschlagsmengen auf den Meeren. *Petermanns Geogr. Mitt.*, 44: 179—182.

TALJAARD, J. J., 1966. Standard deviation of daily sea-level pressure and 500 mb height over the southern hemisphere during the IGY. *Notos*, 15: 29—36.

TALJAARD, J. J., 1967. Development, distribution and movement of cyclones and anticyclones in the southern hemisphere during the IGY. *J. Appl. Meteorol.*, 6: 973—987.

TALJAARD, J. J., 1968. Climatic frontal zones of the southern hemisphere. *Notos*, 17: 23—34.

TALJAARD, J. J., 1969. Air masses of the southern hemisphere. *Notos*, 18: 79—104.

TALJAARD, J. J., 1972a. Physical features of the southern hemisphere. *Meteorol. Monogr.*, 13 (35): 1—8.

TALJAARD, J. J., 1972b. Synoptic meteorology of the southern hemisphere. *Meteorol. Monogr.*, 13 (35): 139—213.

TALJAARD, J. J. and VAN LOON, H., 1962. Cyclogenesis, cyclones and anticyclones in the southern hemisphere during the winter and spring of 1957. *Notos*, 11: 3—20.

TALJAARD, J. J. and VAN LOON, H., 1963. Cyclogenesis, cyclones and anticyclones in the southern hemisphere during summer 1957—1958. *Notos*, 12: 37—50.

TALJAARD, J. J., VAN LOON, H., CRUTCHER, H. L. and JENNE, R. L., 1969. *Climate of the Upper Air: Southern Hemisphere, Vol. I. Temperatures, Dew Points and Heights at Selected Pressure Levels.* NAVAIR 50-1C-55, 134 pp.

TAUBER, G. M., 1969. The comparative measurements of sea-surface temperature in the U.S.S.R. *Geneva, Tech. Note*, 103, WMO No. 247, pp. 141—151.

TAUNTON, 1971. *The South America Pilot II.* 15th ed., Hydrographer of the Navy, London, pp. 6—21.

TAUNTON, 1974. *The Antarctic Pilot.* 4th ed., Hydrographer of the Navy, London, pp. 54—100.

TAUNTON, 1975. *The South America Pilot I.* 11th ed., Hydrographer of the Navy, London, pp. 14—45.

TAUNTON, 1977. *The Africa Pilot II.* 12th ed., Hydrographer of the Navy, London, pp. 24—60.

TREWARTHA, G. T., 1968. *An Introduction to Climate*, 4th ed., McGraw-Hill, New York, N.Y., 408 pp.

TUCKER, G. B., 1961. Precipitation over the North Atlantic Ocean. *Q. J. R. Meteorol. Soc.*, 87: 147—158.

VAN LOON, H., 1956. Blocking action in the Southern Hemisphere. *Notos*, 5: 117—119; 171—177.

VAN LOON, H., 1959. On the synoptic climatology of the Tristan da Cunha region. *Arch. Meteorol. Geophys. Bioklimatol.*, Ser. B, 9: 313—322.

VAN LOON, H., 1962. On the movement of lows in the Ross and Weddell Sea sectors in summer. *Notos*, 11: 47—50.

VAN LOON, H., 1964. Mid-season average zonal winds at sea level and at 500 mb south of 15°S, and a brief comparison with the Northern Hemisphere. *J. Appl. Meteorol.*, 3: 554—563.

VAN LOON, H., 1965. A climatological study of the atmospheric circulation in the Southern Hemisphere during the IGY, Part I: 1 July 1957—31 March 1958. *J. Appl. Meteorol.*, 4: 479—491.

VAN LOON, H., 1966. On the annual temperature range over the southern oceans. *Geogr. Rev.*, 56: 497—515.

VAN LOON, H., 1967. The half yearly oscillations in middle and high southern latitudes and the coreless winter. *J. Atmos. Sci.*, 24: 472—486.

VAN LOON, H., 1972. Temperature, pressure, wind, cloudiness and precipitation in the Southern Hemisphere. *Meteorol. Monogr.*, 13 (35): 25—111.

VAN LOON, H. and JENNE, R. L., 1974. Standard deviations of monthly mean 500 and 100 mb heights in the Southern Hemisphere. *J. Geophys. Res.*, 79: 5561—5564.

VAN LOON, H. and THOMPSON, A. H., 1966. A note on southern hemisphere analysis incorporating satellite data. *Notos*, 15: 91—97.

VAN LOON, H., TALJAARD, J. J., JENNE, R. L. and CRUTCHER, H. L., 1971. *Climate of the Upper Air: Southern Hemisphere, Vol. II. Zonal Geostrophic Winds.* NAVAIR 50-1C-56 and NCAR TN/STR-57, National Center for Atmospheric Research, Boulder, Colo., 43 pp.

VARENIUS, B., 1671. *Geographia generalis.* Amsterdam, 784 pp.

VONDER HAAR, T. H. and SUOMI, V. E., 1971. Measurement of the earth's radiation budget from satellites during a five-year period, Part I. Extended time and space means. *J. Atmos. Sci.*, 28: 305—314.

VON FICKER, H., 1936. Die Passatinversion. *Veröff. Meteorol. Inst. Univ. Berlin*, 1 (4): 33 pp.

VOWINCKEL, E., 1957. Climate of the Antarctic Ocean. In: M. P. VAN ROOY (Editor), *Meteorology of the Antarctic.* Weather Bureau, Pretoria, pp. 91—110.

VOWINCKEL, E. and OOSTHUIZEN, C. M., 1953. Weather types and weather elements over the Antarctic Ocean during the whaling season. *Notos*, 2: 157—182.

VUORELA, L. A., 1950. Synoptic aspects of the tropical regions of the Atlantic Ocean, West Africa and South America. *Ann. Acad. Sci. Fennicae*, A79: 130 pp.

WALDEN, H., 1958. Die winderzeugten Meereswellen, Teil 1. Beobachtungen des Seegangs und Ermittlung der Windsee aus den Windverhältnissen. *Einzelveröff. Seewetteramtes*, 18: 40 pp.

WALDEN, H. and SCHÄFER, P. J., 1969. Die winderzeugten Meereswellen, Teil II. Flachwasserwellen. *Einzelveröff. Seewetteramtes*, 67: 66 pp.

WALKER, G. T., 1928. Ceará (Brazil) famines and the general air movement. *Beitr. Phys. Freien Atmos.*, 14: 88—93.

WALLACE, J. M., 1966. Long period wind fluctuations in the tropical stratosphere. *Planetary Circulation Project, Rept.*, 19, 167 pp.

WASHINGTON, 1943. Naval air pilot South Africa. *Weather Summary*, H.O. No. 264, U.S. Navy Department, Wash., 299 pp.

WASHINGTON, 1945a. Naval air pilot Brazil. *Weather Summary*, H.O. No. 527, U.S. Navy Department, Wash., 205 pp.

WASHINGTON, 1945b. Naval air pilot South America, southern part. *Weather Summary*, H.O. No. 529, U.S. Navy Department, Wash., 178 pp.

WASHINGTON, 1948. *Atlas of Sea and Swell Charts, South Atlantic Ocean*. U.S. Navy, Hydrographic Office, Wash., 12 pp.

WASHINGTON, 1952. *Sailing Direction for South America*, 2. Defense Mapping Agency Hydrographic/Topographic Center, Washington, Pub. 24, 5th ed., 350 pp.

WASHINGTON, 1957. *Oceanographical Atlas of the Polar Seas*, 1. *Antarctica*. U.S. Navy, H.O., No. 705, Wash., 70 pp.

WASHINGTON, 1958. *Marine Climatic Atlas of the World*, IV. *South Atlantic Ocean*. U.S. Office of Naval Operations, NAVAIR 50-1C-531, 284 pp.

WASHINGTON, 1969. *Sailing Direction for Southwest Coast of Africa*. Defense Mapping Agency Hydrographic Center, H.O., No. 50, Wash., 300 pp.

WASHINGTON, 1978. *Marine Climatic Atlas of the World*, IV. *South Atlantic Ocean*. U.S. Navy, NAVAIR 50-1C-531, 325 pp.

WASHINGTON, 1981. *Sailing Directions (enroute) for the East Coast of South America*, 1st. ed. Defense Mapping Agency Hydrographic/Topographic Center, Washington, 326 pp.

WEGENER, K., 1927. Die Plata-Stürme vom 10.—11. Juli 1923 und 10.—11. Januar 1925. *Ann. Hydrogr. Mar. Meteorol.*, 55: 33—36.

WÖLCKEN, K., 1962. Regenwetterlagen in Argentinien. *Südamerika*, 12: 83—140.

WORLD METEOROLOGICAL ORGANIZATION, 1954. Methods of observation at sea, Part 1. Sea surface temperature. *Tech. Note*, 2, WMO No. 26, 35 pp.

WORLD METEOROLOGICAL ORGANIZATION, 1955. Methods of observation at sea. Part. 2: Air temperature and humidity, atmospheric pressure, cloud height, wind, rainfall and visibility. *Tech. Note*, 2, *WMO* No. 40, 35 pp.

WORLD METEOROLOGICAL ORGANIZATION, 1956. World distribution of thunderstorm days, Part 2. WMO No. 21, 83 pp.

WORLD METEOROLOGICAL ORGANIZATION, 1958. Meteorology as applied to the navigation of ships. *Tech. Note*, 23, *WMO* No. 76, 26 pp.

WORLD METEOROLOGICAL ORGANIZATION, 1962. Precipitation measurement at sea. *Tech. Note*, 47, *WMO* No. 124, 18 pp.

WORLD METEOROLOGICAL ORGANIZATION, 1963. Fourth World Meteorological Congress, Resolution 35. *WMO* No. 142, pp. 79—80; 157—166.

WORLD METEOROLOGICAL ORGANIZATION, 1966a. Measurement and estimation of evaporation and evapotranspiration. *Tech. Note*, 83, WMO No. 201, 120 pp.

WORLD METEOROLOGICAL ORGANIZATION, 1966b. Meteorological observations from mobile and fixed ships. *WWW Planning Rep.*, 7, 100 pp.

WORLD METEOROLOGICAL ORGANIZATION, 1969. An introduction to GARP. *GARP Pub. Ser.*, 1, 22 pp.

WORLD METEOROLOGICAL ORGANIZATION, 1970. The Beaufort scale of wind force. *WMO Rep. Mar. Sci. Affairs*, 3, 22 pp.

WORLD METEOROLOGICAL ORGANIZATION, 1971. Climatological normals (Clino) for climate and climate ship stations for the period 1931—1960. WMO 117.

WORLD METEOROLOGICAL ORGANIZATION, 1976. Oceanic water balance. WMO 442, 112 pp.

WORLD METEOROLOGICAL ORGANIZATION, 1977. Guide to marine meteorological services. WMO 471, 155 pp.

WORLD METEOROLOGICAL ORGANIZATION, 1978. Numerical modelling of the tropical atmosphere. *GARP Pub. Ser.*, 20, 78 pp.

WORLD METEOROLOGICAL ORGANIZATION, 1979. Proceedings of the World Climate Conference, WMO 537, 789 pp.

WUCKNITZ, J., 1974. Bestimmung der turbulenten Flüsse von Impuls und sensibler Wärme aus Fluktuationsmessungen und Struktur des Windfeldes über den Wellen über dem tropischen Atlantik während APEX. *Ber. Inst. Radiometeorol. Marit. Meteorol.*, 25: 159 pp.

WUCKNITZ, J., HASSELMANN, D. and KÖNIG, M., 1977. Wassernahe turbulente Vertikalflüsse bei gestörten und ungestörten Bedingungen während GATE. *Ann. Meteorol., N. F.*, 12: 23—25.

WÜST, G., 1936. Oberflächensalzgehalt, Verdunstung und Niederschlag auf dem Weltmeer. *Länderkundl. Forsch., Festschr. N. Krebs, Stuttgart*, pp. 347—359.

WÜST, G., 1950. Wasserdampf und Niederschlag auf dem Meer als Glieder des Wasserkreislaufs. *Deut. Hydrogr. Z.*, 3: 111—127.

WÜST, G., 1954. Gesetzmässige Wechselbeziehungen zwischen Ozean und Atmosphäre in der zonalen Verteilung von Oberflächensalzgehalt, Verdunstung und Niederschlag. *Arch. Meteorol. Geophys. Bioklimatol.*, A7: 305—328.

ZILLMAN, J. W., 1967. The surface radiation balance in high southern latitudes. In: *Polar Meteorology,Tech. Note, 87, WMO* No. 211, pp. 142—171.

Appendix—Climatic tables* and charts

TABLE XIX

CLIMATIC TABLE FOR AREA A
Latitude 0–5°S, longitude 35–30°W

Month	Mean wind speed (m s⁻¹)	Wind steadi-ness (%)	Mean wave height (m)	Mean air press. (mbar)	Mean temperature (°C)			Mean vapour press. (mbar)	Mean relat. humid. (%)	Mean evapor. (mm day⁻¹)	Mean cloud-iness (%)
					sea	air	mean diff.				
Jan.	4.8	89	1.2	1011.1	27.1	26.8	−0.3	28.4	81	4.1	48
Feb.	4.4	83	1.1	1010.8	27.6	27.2	−0.4	29.3	81	4.0	50
Mar.	3.9	75	1.2	1011.0	28.1	27.5	−0.7	29.6	81	4.2	57
Apr.	3.6	66	1.1	1011.0	28.3	27.4	−0.9	29.6	81	4.3	63
May	4.7	73	1.2	1011.9	28.1	27.2	−0.9	29.4	82	4.8	60
June	5.6	86	1.5	1013.3	27.2	26.6	−0.6	28.1	81	4.9	46
July	6.2	90	1.6	1014.4	26.4	26.0	−0.4	26.9	80	5.0	44
Aug.	6.8	93	1.6	1014.2	26.1	25.8	−0.3	26.2	79	5.4	40
Sep.	6.4	93	1.6	1013.8	26.0	25.7	−0.2	26.1	79	5.1	37
Oct.	6.4	93	1.4	1012.8	26.2	26.1	−0.2	26.8	80	4.9	40
Nov.	6.3	91	1.4	1011.6	26.5	26.3	−0.2	27.4	80	4.9	41
Dec.	5.7	91	1.3	1011.3	26.8	26.6	−0.2	27.9	80	4.5	42
Annual	5.4	85	1.3	1012.3	27.0	26.6	−0.4	28.0	80	4.7	47
St. dev.	2.2		0.7	2.1	0.9	1.2	1.2	2.1	6	2.0	27

Month	Frequency (%) of					Mean heat flux (Wm⁻²)		Mean radiation (Wm⁻²)		Mean radiat. balance (Wm⁻²)	Mean heat balance (Wm⁻²)
	strong wind	storm	mist	fog	precip-itation	latent	sensible	solar	terrestr.		
Jan.	0.3		1.5	0.3	3.4	114	2	205	60	145	29
Feb.	0.6		1.5	0.3	5.5	110	3	208	65	143	30
Mar.	0.2		1.7	0.1	8.4	112	4	195	59	136	20
Apr.	0.6		2.3	0.7	12.1	113	5	175	56	119	1
May	1.3		2.6	0.4	11.7	130	7	167	63	104	−33
June	1.3		1.2	0.1	6.0	139	5	181	65	116	−28
July	3.3	0.1	1.4	0.1	4.8	139	4	190	61	129	−14
Aug.	4.1	0.1	2.4	0.9	3.1	152	3	209	64	145	−10
Sep.	2.2		0.8	0.3	1.8	142	2	226	63	163	19
Oct.	2.1		1.3	0.1	2.6	136	2	226	60	166	28
Nov.	1.9		1.2	0.3	1.5	136	2	218	56	162	24
Dec.	1.0		1.1		1.4	126	2	213	60	153	25
Annual	1.6		1.6	0.3	5.2	129	3	201	61	140	8

* For definitions see pp. 13ff.

TABLE XX

Latitude 0–5°S, longitude 10–5°W

Month	Mean wind speed (m s⁻¹)	Wind steadiness (%)	Mean wave height (m)	Mean air press. (mbar)	Mean temperature (°C)			Mean vapour press. (mbar)	Mean relat. humid. (%)	Mean evapor. (mm day⁻¹)	Mean cloudiness (%)
					sea	air	mean diff.				
Jan.	4.8	91	1.1	1011.4	26.4	26.2	−0.2	27.8	82	3.7	54
Feb.	4.7	91	1.1	1010.7	27.3	27.2	−0.1	29.0	81	4.1	47
Mar.	4.4	86	1.2	1010.6	28.1	27.8	−0.3	29.6	80	4.4	48
Apr.	5.1	89	1.3	1010.9	28.2	27.8	−0.4	29.9	80	4.9	57
May	6.0	93	1.5	1012.1	27.3	27.0	−0.3	28.6	81	4.9	53
June	5.6	94	1.5	1014.4	25.1	25.2	0.1	26.2	82	3.6	40
July	4.7	93	1.4	1015.7	23.4	23.6	0.2	24.0	83	2.7	32
Aug.	4.8	93	1.4	1015.3	23.0	23.1	0.2	23.5	83	2.6	45
Sep.	4.8	92	1.4	1014.6	23.3	23.4	0.1	23.8	83	2.7	56
Oct.	5.4	94	1.3	1013.5	24.1	24.1	0.0	24.3	81	3.5	64
Nov.	5.7	95	1.3	1012.5	24.8	24.7	−0.1	25.4	82	3.6	65
Dec.	5.6	94	1.2	1012.2	25.4	25.3	−0.1	26.4	82	3.7	61
Annual	5.1	92	1.3	1012.8	25.5	25.4	−0.1	26.5	82	3.7	52
St. dev.	2.1		0.7	2.1	1.3	1.5	1.4	2.3	7	2.0	31

Month	Frequency (%) of					Mean heat flux (Wm⁻²)		Mean radiation (Wm⁻²)		Mean radiat. balance (Wm⁻²)	Mean heat balance (Wm⁻²)
	strong wind	storm	mist	fog	precip- itation	latent	sensible	solar	terrestr.		
Jan.	0.3		0.3	0.2	2.0	102	2	196	55	141	37
Feb.	0.2		0.3		1.9	111	1	215	64	151	39
Mar.	0.3		0.4	0.1	3.5	120	2	213	68	145	23
Apr.	0.5		0.6	0.1	5.5	133	3	186	63	123	−13
May	1.0		0.4	0.2	3.3	139	3	178	59	119	−23
June	0.8		0.2		0.9	98	−1	192	60	132	35
July	0.5		0.8	0.6	0.3	73	−2	214	64	150	79
Aug.	0.4		0.3	0.1	0.7	71	−2	206	61	141	58
Sep.	0.5		0.1	0.1	1.4	84	−1	195	54	136	40
Oct.	0.7		0.3	0.1	1.4	96	0	185	49	130	28
Nov.	0.5		0.1		1.6	101	1	178	48	130	28
Dec.	0.8		0.3		1.3	102	1	180	51	129	26
Annual	0.5		0.3	0.1	2.0	103	1	195	58	137	34

TABLE XXI

CLIMATIC TABLE FOR AREA C
Latitude 0–5°S, longitude 0–10°E

Month	Mean wind speed (m s⁻¹)	Wind steadi-ness (%)	Mean wave height (m)	Mean air press. (mbar)	Mean temperature (°C)			Mean vapour press. (mbar)	Mean relat. humid. (%)	Mean evapor. (mm day⁻¹)	Mean cloud-iness (%)
					sea	air	mean diff.				
Jan.	4.0	82	0.8	1010.2	27.6	27.0	−0.6	29.8	84	3.7	69
Feb.	3.7	76	0.8	1009.4	28.2	27.6	−0.5	30.1	82	3.9	62
Mar.	3.7	77	0.8	1009.4	28.6	27.9	−0.8	30.4	81	4.3	63
Apr.	3.8	71	0.9	1009.5	28.6	27.9	−0.7	30.7	82	4.2	65
May	4.5	86	1.0	1011.1	27.3	26.9	−0.4	29.4	83	3.9	61
June	4.7	89	1.1	1013.5	24.8	24.7	−0.1	25.9	84	3.2	55
July	4.2	91	1.0	1014.3	23.2	23.0	−0.3	23.6	84	2.6	57
Aug.	4.1	90	1.1	1014.4	23.3	22.9	−0.5	23.4	84	2.7	68
Sep.	4.5	89	1.0	1013.2	24.1	23.5	−0.6	24.4	84	3.1	72
Oct.	4.9	91	1.1	1012.8	25.3	24.6	−0.7	25.8	84	3.7	75
Nov.	4.9	90	1.0	1010.9	26.2	25.5	−0.7	27.4	84	3.7	75
Dec.	4.6	87	0.9	1010.6	26.7	26.3	−0.5	28.7	84	3.6	71
Annual	4.3	85	1.0	1011.6	26.2	25.7	−0.5	27.5	83	3.5	66
St. dev.	2.1		0.6	2.8	1.3	1.4	1.4	2.4	7	2.1	30

Month	Frequency (%) of					Mean heat flux (Wm⁻²)		Mean radiation (Wm⁻²)		Mean radiat. balance (Wm⁻²)	Mean heat balance (Wm⁻²)
	strong wind	storm	mist	fog	precip-itation	latent	sensible	solar	terrestr.		
Jan.	0.2	0.1	0.8	0.2	4.8	106	4	166	54	112	2
Feb.	0.1		0.4		4.2	113	3	185	61	124	8
Mar.	0.4		0.8	0.1	3.9	112	5	183	61	122	5
Apr.	0.5		1.6	0.5	6.8	110	4	172	61	111	−3
May	0.8		0.1	0.1	3.0	103	3	165	57	108	2
June	0.4	0.1	0.1	0.1	0.5	80	1	170	57	113	32
July	1.3	0.1	0.4	0.3	0.5	68	2	172	54	118	48
Aug.	0.2	0.1	0.4		0.9	72	3	164	51	113	38
Sep.	0.4		0.1	0.1	1.5	83	4	164	52	112	25
Oct.	0.3		0.1	0.1	2.9	100	5	163	53	110	5
Nov.	0.6		0.4		5.1	102	5	158	43	115	8
Dec.	0.3		0.3	0.1	3.3	95	4	160	53	107	8
Annual	0.5		0.5	0.2	3.1	95	4	169	55	114	15

TABLE XXII

CLIMATIC TABLE FOR AREA D
Latitude 5–15°S, longitude 25–15°W

Month	Mean wind speed (m s⁻¹)	Wind steadi-ness (%)	Mean wave height (m)	Mean air press. (mbar)	Mean temperature (°C)			Mean vapour press. (mbar)	Mean relat. humid. (%)	Mean evapor. (mm day⁻¹)	Mean cloud-iness (%)
					sea	air	mean diff.				
Jan.	6.0	94	1.2	1013.4	25.7	25.8	0.0	25.5	77	5.0	47
Feb.	5.8	91	1.2	1013.0	26.7	26.5	−0.2	26.7	77	5.5	39
Mar.	6.3	95	1.5	1012.7	27.3	27.1	−0.2	27.8	78	5.8	41
Apr.	6.5	90	1.3	1013.2	27.5	26.9	−0.6	27.4	77	6.7	44
May	6.8	92	1.6	1014.0	27.0	26.4	−0.6	26.2	76	6.9	47
June	7.7	94	1.8	1016.1	26.1	25.5	−0.7	24.7	76	7.4	49
July	7.7	94	1.9	1017.2	25.3	24.5	−0.8	23.0	75	7.3	48
Aug.	7.8	95	1.8	1017.2	24.6	23.9	−0.7	22.5	75	6.7	53
Sep.	7.6	93	1.8	1016.5	24.5	24.0	−0.4	22.7	76	6.3	59
Oct.	7.4	94	1.6	1015.7	24.4	24.3	−0.1	23.4	77	5.5	63
Nov.	6.9	95	1.5	1014.4	24.7	24.6	−0.2	23.6	76	5.7	59
Dec.	6.3	95	1.3	1014.3	25.2	25.1	−0.1	24.2	76	5.4	53
Annual	6.9	94	1.5	1014.8	25.7	25.4	−0.4	24.8	76	6.2	50
St. dev.	2.4		0.7	2.1	1.1	1.7	1.7	2.7	8	2.8	29

Month	Frequency (%) of					Mean heat flux (Wm⁻²)		Mean radiation (Wm⁻²)		Mean radiat. balance (Wm⁻²)	Mean heat balance (Wm⁻²)
	strong wind	storm	mist	fog	precip-itation	latent	sensible	solar	terrestr.		
Jan.	0.9		1.4		2.1	137	0	223	53	170	33
Feb.	1.2		0.1		1.7	149	2	234	62	172	21
Mar.	1.7				2.6	160	2	235	64	171	9
Apr.	2.2	0.1	0.1		2.5	182	6	220	64	156	−32
May	2.6	0.3	0.4		2.7	190	7	174	66	108	−89
June	10.9		0.1		1.4	203	9	160	66	94	−118
July	9.3	0.1	0.1	0.1	4.0	203	10	173	72	101	−112
Aug.	7.2	1.0	0.4		3.1	187	9	180	65	115	−81
Sep.	10.6	0.2	0.5	0.1	3.0	176	5	185	62	123	−58
Oct.	4.6		0.3		3.2	141	1	192	54	138	−4
Nov.	3.7		0.2	0.1	2.0	150	3	202	53	149	−4
Dec.	0.4		0.1		1.2	149	1	213	53	160	10
Annual	4.6	0.15	0.3		2.4	169	5	199	61	138	−36

TABLE XXIII

Latitude 10–15°S, longitude 40–30°W

Month	Mean wind speed (m s⁻¹)	Wind steadiness (%)	Mean wave height (m)	Mean air press. (mbar)	Mean temperature (°C)			Mean vapour press. (mbar)	Mean relat. humid. (%)	Mean evapor. (mm day⁻¹)	Mean cloudiness (%)
					sea	air	mean diff.				
Jan.	5.2	43	1.1	1012.8	27.5	27.1	−0.4	27.6	77	5.4	45
Feb.	5.1	41	1.2	1012.2	28.0	27.5	−0.5	28.1	77	5.6	43
Mar.	5.0	43	1.1	1012.7	28.2	27.5	−0.8	28.4	78	5.7	49
Apr.	4.8	39	1.1	1012.6	28.1	27.4	−0.8	28.3	78	5.5	49
May	5.9	50	1.5	1014.3	27.6	26.7	−0.9	27.3	78	6.1	52
June	7.0	60	1.6	1016.1	26.8	25.9	−1.0	25.9	77	6.9	51
July	6.9	61	1.8	1017.3	26.0	25.2	−0.8	24.2	76	6.9	47
Aug.	6.5	58	1.6	1017.3	25.5	24.8	−0.8	23.5	75	6.5	45
Sep.	6.6	52	1.6	1016.2	25.7	25.0	−0.7	24.7	78	5.7	45
Oct.	5.7	46	1.4	1015.1	26.0	25.7	−0.3	25.9	78	4.9	42
Nov.	5.7	45	1.3	1013.1	26.5	26.3	−0.2	26.7	78	5.2	47
Dec.	5.4	41	1.2	1012.8	27.1	26.7	−0.4	27.1	78	5.3	47
Annual	5.8	48	1.4	1014.3	27.0	26.3	−0.6	26.6	77	5.8	47
St. dev.	2.5		0.7	2.5	1.0	1.4	1.4	2.6	8	2.6	28

Month	Frequency (%) of					Mean heat flux (Wm⁻²)		Mean radiation (Wm⁻²)		Mean radiat. balance (Wm⁻²)	Mean heat balance (Wm⁻²)
	strong wind	storm	mist	fog	precipitation	latent	sensible	solar	terrestr.		
Jan.	1.1		1.8	0.6	2.2	148	3	230	65	165	14
Feb.	1.0		1.9	0.1	2.6	155	4	230	72	158	−1
Mar.	1.1	0.1	1.9	0.6	4.2	155	6	207	68	139	−22
Apr.	1.8	0.2	2.0	0.2	5.5	151	6	185	69	116	−41
May	4.9	0.3	3.5	0.7	6.6	171	8	160	72	88	−91
June	10.1	0.2	2.7	0.4	5.3	191	11	149	71	78	−124
July	8.0	0.1	2.2	0.8	5.3	190	9	166	70	96	−103
Aug.	5.9	0.2	2.8	1.0	2.9	177	8	184	70	114	−71
Sep.	2.5	0.2	3.6	1.0	4.9	163	7	206	69	137	−33
Oct.	1.7	0.1	3.1	0.2	2.5	135	3	231	66	165	27
Nov.	2.4		1.7	0.4	2.4	140	2	227	63	164	22
Dec.	1.1	0.2	2.5	0.4	2.6	146	3	226	61	165	16
Annual	3.5	0.1	2.5	0.5	3.9	160	6	200	68	132	−34

TABLE XXIV

CLIMATIC TABLE FOR AREA F
Latitude 10–15°S, longitude 5°W–5°E

Month	Mean wind speed (m s⁻¹)	Wind steadiness (%)	Mean wave height (m)	Mean air press. (mbar)	Mean temperature (°C) sea	air	mean diff.	Mean vapour press. (mbar)	Mean relat. humid. (%)	Mean evapor. (mm day⁻¹)	Mean cloudiness (%)
Jan.	4.8	95	1.3	1013.0	23.7	23.7	0.0	22.4	77	3.9	66
Feb.	5.2	95	1.5	1012.3	25.1	24.8	−0.3	24.0	76	4.7	60
Mar.	5.8	95	1.6	1012.4	25.5	25.2	−0.4	24.7	78	5.1	66
Apr.	6.3	96	1.6	1013.2	25.1	24.5	−0.6	23.4	77	5.7	70
May	6.3	96	1.7	1014.9	24.2	23.4	−0.9	21.1	74	6.2	70
June	6.0	95	1.7	1017.2	23.0	22.0	−0.9	19.0	71	6.1	69
July	6.2	96	1.7	1018.1	21.9	21.0	−0.9	17.9	72	5.8	82
Aug.	6.7	97	1.8	1017.6	20.8	20.0	−0.8	17.4	74	5.3	93
Sep.	6.5	96	1.7	1016.7	20.5	19.8	−0.7	17.7	77	4.5	94
Oct.	6.4	96	1.6	1015.9	20.6	20.1	−0.5	17.9	76	4.8	94
Nov.	5.6	96	1.5	1014.8	21.3	21.0	−0.3	18.9	76	4.0	87
Dec.	5.2	96	1.3	1014.3	22.3	22.3	0.0	20.2	75	4.0	76
Annual	5.9	96	1.6	1015.1	22.9	22.3	−0.5	20.4	75	5.0	77
St. dev.	2.4		0.8	2.0	1.6	1.5	1.5	2.4	8	2.2	26

Month	Frequency (%) of strong wind	storm	mist	fog	precip-itation	Mean heat flux (Wm⁻²) latent	sensible	Mean radiation (Wm⁻²) solar	terrestr.	Mean radiat. balance (Wm⁻²)	Mean heat balance (Wm⁻²)
Jan.	0.4		0.4	0.2	1.9	106	0	192	46	146	40
Feb.	0.7		0.3		1.8	129	2	201	56	145	14
Mar.	1.2		0.3		3.0	140	4	180	57	123	−21
Apr.	2.7	1.0	0.3	0.1	1.6	159	6	152	53	99	−66
May	2.8		0.2	0.1	0.6	172	9	136	59	77	−104
June	3.0		0.1		0.7	167	9	127	59	68	−108
July	3.1	0.1	0.2	0.1	1.2	158	9	113	51	62	−105
Aug.	5.3	0.1	0.2	0.1	1.7	134	9	109	43	66	−77
Sep.	5.1	0.2	0.2		2.4	106	7	121	37	84	−29
Oct.	3.4	0.2	0.2	0.1	1.0	125	5	131	37	94	−36
Nov.	1.0		0.2	0.1	1.3	110	3	149	37	112	−1
Dec.	0.5		0.2		1.5	109	0	173	39	134	25
Annual	2.4	0.1	0.2	0.1	1.5	135	5	149	48	101	−39

TABLE XXV

CLIMATIC TABLE FOR AREA G
Latitude 15–20°S, longitude 5–15°E

Month	Mean wind speed (m s⁻¹)	Wind steadi-ness (%)	Mean wave height (m)	Mean air press. (mbar)	Mean temperature (°C)			Mean vapour press. (mbar)	Mean relat. humid. (%)	Mean evapor. (mm day⁻¹)	Mean cloud-iness (%)
					sea	air	mean diff.				
Jan.	5.3	93	1.4	1013.1	20.7	20.9	0.2	20.1	80	3.3	70
Feb.	5.7	93	1.5	1013.0	21.8	21.7	−0.2	20.7	79	3.9	72
Mar.	6.3	95	1.7	1013.2	22.0	21.2	−0.8	21.2	81	4.2	77
Apr.	6.8	94	1.8	1013.5	21.2	20.6	−0.7	19.9	81	4.3	65
May	6.3	92	1.7	1015.3	19.6	19.2	−0.5	18.1	81	3.6	58
June	6.0	90	1.7	1017.6	18.5	17.6	−0.9	16.2	79	5.4	62
July	6.2	89	1.7	1018.4	17.2	16.7	−0.5	15.4	81	3.2	73
Aug.	6.5	92	1.8	1017.8	16.5	15.9	−0.6	15.0	82	2.8	84
Sep.	6.2	93	1.6	1017.0	16.4	16.1	−0.3	15.3	83	2.5	81
Oct.	6.8	95	1.7	1016.1	16.6	16.9	0.3	15.8	83	2.5	75
Nov.	6.0	93	1.6	1014.9	17.8	18.1	0.3	17.1	82	2.7	74
Dec.	5.6	94	1.4	1014.3	19.7	19.7	0.1	18.5	81	2.9	68
Annual	6.1	93	1.6	1015.3	19.0	18.7	−0.3	17.8	81	3.3	71
St. dev.	3.3		0.9	2.8	1.7	2.0	2.0	2.1	8	2.1	33

Month	Frequency (%) of					Mean heat flux (Wm⁻²)		Mean radiation (Wm⁻²)		Mean radiat. balance (Wm⁻²)	Mean heat balance (Wm⁻²)
	strong wind	storm	mist	fog	precip-itation	latent	sensible	solar	terrestr.		
Jan.	2.2	0.2	5.1	0.6	1.8	78	−2	192	37	155	79
Feb.	3.0	0.1	1.5	0.1	0.8	99	2	181	42	139	38
Mar.	6.0	0.2	1.1	0.3	1.3	105	8	156	39	117	4
Apr.	7.7		1.3	0.4	0.6	111	8	155	47	108	−11
May	6.7	0.1	2.8	2.0	0.4	92	5	143	50	93	−4
June	8.9	0.6	1.7	1.2	0.5	112	9	125	51	74	−47
July	8.9	0.7	1.6	0.8	0.9	82	5	117	45	72	−15
Aug.	11.5	0.6	2.9	1.3	1.5	75	6	117	37	80	−1
Sep.	10.0	0.5	2.4	1.5	1.8	64	3	144	37	107	40
Oct.	10.4	0.8	1.6	0.5	1.1	64	−3	171	31	140	79
Nov.	6.4	0.2	1.4	0.3	0.6	65	−3	183	30	153	91
Dec.	3.7	0.2	2.0	0.5	0.9	77	−1	199	38	161	85
Annual	7.1	0.3	2.1	0.8	1.0	85	3	157	40	117	29

TABLE XXVI

CLIMATIC TABLE FOR AREA H
Latitude 15–25°S, longitude 15–5°W

Month	Mean wind speed (m s⁻¹)	Wind steadiness (%)	Mean wave height (m)	Mean air press. (mbar)	Mean temperature (°C) sea	air	mean diff.	Mean vapour press. (mbar)	Mean relat. humid. (%)	Mean evapor. (mm day⁻¹)	Mean cloudiness (%)
Jan.	5.3	88	1.3	1017.0	23.6	23.6	0.0	22.5	77	4.1	55
Feb.	5.9	90	1.5	1016.2	24.6	24.4	−0.2	23.6	77	4.9	54
Mar.	6.7	91	1.6	1016.6	24.7	24.4	−0.4	23.1	75	5.7	56
Apr.	6.0	85	1.7	1017.0	24.5	23.8	−0.8	22.0	75	6.0	54
May	6.0	80	1.7	1018.5	23.5	22.5	−1.0	20.4	74	6.0	54
June	5.9	79	1.7	1020.6	22.1	20.9	−1.2	18.2	73	5.9	58
July	6.5	78	1.8	1021.9	21.2	20.2	−1.1	17.5	74	5.7	59
Aug.	6.6	82	1.8	1022.0	20.8	19.8	−1.1	17.1	74	5.5	71
Sep.	7.1	85	1.9	1021.4	20.7	19.8	−1.0	17.2	75	5.3	76
Oct.	6.9	87	1.6	1020.4	20.7	20.3	−0.5	18.1	76	4.7	74
Nov.	6.0	89	1.5	1018.5	21.5	21.2	−0.3	18.8	75	4.5	71
Dec.	5.3	86	1.4	1017.5	22.7	22.6	−0.2	20.5	75	4.3	59
Annual	6.2	85	1.6	1019.0	22.6	21.9	−0.6	19.9	75	5.2	61
St. dev.	2.9		0.8	2.5	1.3	1.7	1.7	2.6	9	2.6	29

Month	Frequency (%) of strong wind	storm	mist	fog	precip- itation	Mean heat flux (Wm⁻²) latent	sensible	Mean radiation (Wm⁻²) solar	terrestr.	Mean radiat. balance (Wm⁻²)	Mean heat balance (Wm⁻²)
Jan.	0.3		1.3	0.6	1.8	110	0	227	59	168	58
Feb.	2.5	0.1			3.1	132	2	217	58	159	25
Mar.	4.2				3.1	158	4	192	60	132	−30
Apr.	3.8		0.2	0.1	2.1	161	8	168	66	102	−67
May	3.3		0.3		3.6	159	10	141	72	69	−100
June	6.4	0.1	0.6	0.5	4.1	155	11	121	72	49	−117
July	6.9	0.2	1.3	0.5	4.9	152	11	131	60	71	−92
Aug.	8.6	0.6	1.5	0.2	4.9	129	12	133	59	74	−67
Sep.	11.7	0.4	0.6		4.0	148	11	148	55	93	−66
Oct.	8.4	0.7	0.1		3.4	129	6	172	50	122	−13
Nov.	5.4	0.5	0.1	0.1	2.2	124	3	191	50	141	14
Dec.	1.2		0.4	0.1	2.3	117	2	223	58	165	46
Annual	5.3	0.2	0.5	0.2	3.3	140	7	172	60	112	−34

TABLE XXVII

CLIMATIC TABLE FOR AREA I
Latitude 20–25°S, longitude 45–35°W

Month	Mean wind speed (m s^{-1})	Wind steadiness (%)	Mean wave height (m)	Mean air press. (mbar)	Mean temperature (°C) sea	air	mean diff.	Mean vapour press. (mbar)	Mean relat. humid. (%)	Mean evapor. (mm day^{-1})	Mean cloudiness (%)
Jan.	5.9	61	1.2	1012.3	25.6	25.9	0.2	27.6	84	3.3	57
Feb.	5.6	60	1.3	1012.6	26.2	26.2	0.0	28.0	82	3.6	53
Mar.	5.1	51	1.2	1014.4	26.6	26.3	−0.3	27.7	81	4.1	51
Apr.	5.1	34	1.2	1015.3	25.9	25.4	−0.6	25.1	78	5.1	52
May	5.2	34	1.3	1017.6	25.0	24.0	−1.0	23.2	77	5.3	48
June	5.4	33	1.3	1019.8	23.9	23.2	−0.8	22.2	79	4.6	48
July	5.7	42	1.4	1020.7	23.0	22.3	−0.7	20.9	78	4.5	50
Aug.	6.1	44	1.4	1019.4	22.5	22.3	−0.3	21.3	80	3.9	47
Sep.	6.8	53	1.5	1018.0	22.6	22.6	−0.1	22.2	81	3.6	52
Oct.	6.4	51	1.4	1015.7	23.1	23.2	0.1	23.6	83	3.0	62
Nov.	6.3	35	1.4	1012.9	24.0	24.0	0.0	24.7	83	3.3	65
Dec.	6.3	48	1.4	1013.1	24.7	24.8	0.2	26.2	83	3.3	61
Annual	5.8	45	1.3	1015.9	24.4	24.2	−0.3	24.4	81	4.0	54
St. dev.	3.2		0.8	3.8	1.5	1.8	1.8	3.2	9	2.8	34

Month	Frequency (%) of strong wind	storm	mist	fog	precipitation	Mean heat flux (Wm^{-2}) latent	sensible	Mean radiation (Wm^{-2}) solar	terrestr.	Mean radiat. balance (Wm^{-2})	Mean heat balance (Wm^{-2})
Jan.	5.4	0.1	5.6	1.5	6.6	92	−2	219	57	162	72
Feb.	4.3	0.1	3.2	0.6	4.8	102	0	214	62	152	50
Mar.	2.6		2.4	0.4	4.4	113	2	193	66	127	12
Apr.	3.6	0.1	2.4	0.2	5.6	135	5	161	68	93	−47
May	4.6	0.2	5.1	1.1	4.3	141	8	139	77	62	−87
June	4.4	0.3	5.8	1.3	4.3	126	8	123	77	46	−88
July	4.8	0.3	4.5	0.8	3.2	125	6	134	73	61	−70
Aug.	6.4	0.3	7.8	1.4	2.9	110	4	156	65	91	−23
Sep.	10.4	0.6	8.1	1.6	4.5	103	1	182	63	119	15
Oct.	8.4	0.5	7.0	2.0	6.9	87	−1	191	56	135	49
Nov.	8.2	0.2	6.2	1.1	8.2	88	0	201	55	146	58
Dec.	7.0	0.2	7.0	1.8	6.5	92	−2	213	53	160	70
Annual	5.8	0.3	5.4	1.2	5.2	110	2	177	64	113	1

TABLE XXVIII

CLIMATIC TABLE FOR AREA J
Latitude 20–25°S, longitude 5–10°E

Month	Mean wind speed (m s⁻¹)	Wind steadiness (%)	Mean wave height (m)	Mean air press. (mbar)	Mean temperature (°C)			Mean vapour press. (mbar)	Mean relat. humid. (%)	Mean evapor. (mm day⁻¹)	Mean cloudiness (%)
					sea	air	mean diff.				
Jan.	7.0	95	2.0	1014.5	21.3	21.5	0.2	19.6	77	4.1	72
Feb.	7.3	96	2.1	1014.0	22.0	22.0	0.0	20.3	77	4.6	75
Mar.	7.8	96	2.2	1014.5	21.8	21.7	−0.1	20.2	78	4.8	72
Apr.	7.2	95	2.0	1015.6	21.2	20.8	−0.5	18.9	77	4.8	67
May	6.9	92	2.1	1017.1	20.1	19.5	−0.6	17.5	77	4.5	63
June	6.6	88	2.2	1019.3	18.8	18.1	−0.7	15.7	76	4.3	63
July	7.1	91	2.2	1020.3	17.8	17.1	−0.7	15.1	77	3.9	65
Aug.	7.7	90	2.3	1019.9	17.1	16.7	−0.4	14.6	77	3.9	74
Sep.	8.0	93	2.2	1018.8	16.9	16.7	−0.2	14.9	79	3.5	75
Oct.	7.5	94	2.1	1017.9	17.3	17.4	0.1	15.4	78	3.3	74
Nov.	7.2	93	2.1	1016.1	18.4	18.6	0.2	16.7	78	3.3	73
Dec.	6.8	95	1.9	1015.6	19.9	20.2	0.3	18.1	77	3.6	74
Annual	7.3	93	2.1	1017.0	19.4	19.2	−0.2	17.2	77	4.0	71
St. dev.	3.3		1.0	2.4	1.2	1.5	1.4	2.1	9	2.3	30

Month	Frequency (%) of					Mean heat flux (Wm⁻²)		Mean radiation (Wm⁻²)		Mean radiat. balance (Wm⁻²)	Mean heat balance (Wm⁻²)
	strong wind	storm	mist	fog	precip- itation	latent	sensible	solar	terrestr.		
Jan.	8.5		0.2		1.4	116	−2	194	45	149	35
Feb.	9.0	0.2	0.3	0.1	1.4	128	0	176	46	130	2
Mar.	12.2	0.2	0.2		0.7	133	1	161	47	114	−20
Apr.	11.0	0.5	0.2	0.1	0.6	132	6	142	54	88	−50
May	11.0	0.3	0.3	0.2	1.0	122	7	124	61	63	−66
June	13.0	0.5	0.1		1.6	118	7	109	61	48	−77
July	14.0	1.3	0.3	0.1	0.6	109	8	114	59	55	−62
Aug.	18.7	1.3	0.1	0.1	0.8	109	5	123	50	73	−41
Sep.	19.0	1.4	0.4	0.1	0.5	98	3	172	41	131	1
Oct.	13.8	1.0	0.3		0.7	93	−1	190	42	148	39
Nov.	10.6	0.4	0.4	0.1	0.6	94	−2	190	44	148	56
Dec.	6.0	0.2	0.3	0.1	0.8	103	−3	192	44	148	48
Annual	12.2	0.6	0.3	0.1	0.9	113	2	154	50	104	−11

TABLE XXIX

CLIMATIC TABLE FOR AREA K
Latitude 25–30°S, longitude 50–40°W

Month	Mean wind speed (m s⁻¹)	Wind steadiness (%)	Mean wave height (m)	Mean air press. (mbar)	Mean temperature (°C)			Mean vapour press. (mbar)	Mean relat. humid. (%)	Mean evapor. (mm day⁻¹)	Mean cloudiness (%)
					sea	air	mean diff.				
Jan.	5.9	38	1.3	1011.9	24.9	24.9	−0.1	25.6	81	4.0	59
Feb.	5.9	41	1.4	1012.1	25.8	25.5	−0.3	26.8	82	4.1	56
Mar.	5.7	41	1.4	1014.1	25.7	25.0	−0.6	25.7	80	4.8	59
Apr.	5.8	33	1.4	1016.0	24.5	23.6	−1.0	22.7	77	5.4	54
May	5.8	13	1.5	1018.0	23.1	22.0	−1.2	20.3	76	5.4	51
June	6.0	25	1.4	1019.7	21.9	20.8	−1.1	19.6	79	4.4	55
July	6.6	21	1.6	1020.3	20.8	19.9	−0.9	18.7	79	4.3	54
Aug.	6.5	33	1.6	1019.2	20.8	20.2	−0.6	19.0	80	3.9	59
Sep.	6.6	39	1.7	1017.4	20.8	20.4	−0.5	19.8	82	3.4	67
Oct.	6.6	39	1.6	1015.6	21.5	21.3	−0.2	20.5	81	3.4	62
Nov.	6.6	49	1.5	1013.6	22.6	22.5	−0.2	22.0	81	3.5	62
Dec.	6.3	45	1.4	1013.0	23.8	23.6	−0.2	23.7	80	4.0	61
Annual	6.2	35	1.5	1015.9	23.0	22.5	−0.6	22.0	80	4.2	58
St. dev.	3.1		0.9	4.7	1.6	2.1	2.0	4.0	10	3.5	34

Month	Frequency (%) of					Mean heat flux (Wm⁻²)		Mean radiation (Wm⁻²)		Mean radiat. balance (Wm⁻²)	Mean heat balance (Wm⁻²)
	strong wind	storm	mist	fog	precip- itation	latent	sensible	solar	terrestr.		
Jan.	6.0	0.5	2.2	0.8	5.3	107	2	222	56	166	57
Feb.	8.4	0.5	2.8	0.1	4.8	114	3	209	61	148	31
Mar.	5.3	0.1	1.7	0.4	6.4	129	5	175	62	113	−21
Apr.	5.9	0.2	2.7	0.5	5.7	143	10	147	72	75	−78
May	9.3	1.1	1.9	0.3	3.9	143	12	122	76	46	−109
June	8.8	0.8	3.8	1.5	5.4	121	11	101	79	22	−110
July	11.9	1.3	3.6	0.9	3.6	116	10	111	77	34	−92
Aug.	11.7	0.4	6.7	2.7	4.0	108	6	129	68	61	−53
Sep.	11.3	1.5	11.4	4.3	7.9	94	5	150	59	91	−8
Oct.	11.0	0.8	5.1	1.2	6.0	96	2	189	57	132	34
Nov.	10.5	0.4	4.7	0.2	5.4	102	2	213	58	155	51
Dec.	8.1	0.4	3.4	0.6	4.6	109	2	223	56	167	56
Annual	9.0	0.6	4.2	1.1	5.3	115	6	166	65	101	−20

TABLE XXX

CLIMATIC TABLE FOR AREA L
Latitude 25–30°S, longitude 10–15°E

Month	Mean wind speed (m s⁻¹)	Wind steadi- ness (%)	Mean wave height (m)	Mean air press. (mbar)	Mean temperature (°C)			Mean vapour press. (mbar)	Mean relat. humid. (%)	Mean evapor. (mm day⁻¹)	Mean cloud- iness (%)
					sea	air	mean diff.				
Jan.	8.1	93	2.3	1014.1	19.7	20.2	0.5	19.0	81	3.4	54
Feb.	8.5	95	2.4	1014.0	19.8	20.3	0.4	19.1	81	3.7	61
Mar.	8.4	94	2.4	1014.7	19.6	19.8	0.2	19.0	82	3.5	53
Apr.	7.4	87	2.2	1015.7	18.9	18.8	−0.1	17.7	81	3.4	53
May	6.5	81	2.1	1017.5	17.7	17.4	−0.3	16.4	82	3.1	46
June	6.2	72	2.2	1019.7	16.6	16.3	−0.4	14.7	79	3.1	42
July	6.7	76	2.2	1021.1	15.8	15.5	−0.3	14.0	80	2.9	43
Aug.	7.0	78	2.3	1020.2	15.5	15.3	−0.2	13.8	80	3.0	50
Sep.	7.9	87	2.2	1019.0	15.3	15.3	0.0	14.2	81	2.7	49
Oct.	7.5	87	2.1	1017.8	15.8	16.2	0.3	14.8	80	2.7	56
Nov.	8.3	91	2.3	1016.0	17.2	17.5	0.4	16.3	81	2.7	52
Dec.	7.8	92	2.2	1015.3	18.6	19.0	0.4	17.8	81	2.9	55
Annual	7.5	86	2.2	1017.1	17.5	17.6	0.1	16.4	81	3.1	51
St. dev.	4.1		1.2	3.1	1.8	1.7	1.6	2.1	9	2.6	36

Month	Frequency (%) of					Mean heat flux (Wm⁻²)		Mean radiation (Wm⁻²)		Mean radiat. balance (Wm⁻²)	Mean heat balance (Wm⁻²)
	strong wind	storm	mist	fog	precip- itation	latent	sensible	solar	terrestr.		
Jan.	21	1.7	0.9	0.5	1.2	91	−6	243	50	193	108
Feb.	24	1.7	1.4	0.9	1.3	98	−5	207	43	164	71
Mar.	23	1.9	1.3	0.7	0.9	92	−3	192	50	142	53
Apr.	20	1.5	1.9	1.2	1.2	90	1	153	60	93	2
May	14	1.3	2.0	1.6	0.9	78	3	130	64	66	−15
June	12	1.4	1.3	0.8	1.6	81	4	116	70	46	−39
July	15	1.8	0.8	0.5	1.1	78	3	124	67	57	−24
Aug.	18	2.2	1.4	0.7	1.2	80	2	144	61	83	1
Sep.	20	1.9	1.3	1.0	0.7	73	0	185	54	131	58
Oct.	19	2.0	1.1	0.7	0.8	67	−4	208	50	158	95
Nov.	24	2.4	1.1	0.6	1.1	77	−5	243	56	187	115
Dec.	18	1.2	0.8	0.4	0.9	81	−5	246	54	192	116
Annual	19	1.8	1.3	0.8	1.1	82	−1	183	57	126	45

TABLE XXXI

CLIMATIC TABLE FOR AREA M
Latitude 25–35°S, longitude 35–15°W

Month	Mean wind speed (m s⁻¹)	Wind steadi-ness (%)	Mean wave height (m)	Mean air press. (mbar)	Mean temperature (°C) sea	air	mean diff.	Mean vapour press. (mbar)	Mean relat. humid. (%)	Mean evapor. (mm day⁻¹)	Mean cloud-iness (%)
Jan.	5.5	37	1.5	1017.6	23.0	22.7	−0.2	22.4	80	3.4	60
Feb.	5.9	31	1.6	1016.2	24.2	23.8	−0.3	24.1	81	3.7	56
Mar.	5.8	33	1.5	1018.2	23.7	22.9	−0.8	22.4	79	4.5	57
Apr.	6.3	18	1.8	1016.5	22.5	21.6	−0.9	19.6	76	5.3	61
May	7.2	30	2.0	1017.7	20.9	19.9	−1.0	17.2	74	5.6	61
June	6.6	32	2.2	1021.4	19.5	18.2	−1.2	15.9	74	4.8	61
July	6.9	33	2.2	1023.1	18.4	17.4	−0.9	14.8	74	4.6	63
Aug.	7.2	26	2.2	1022.5	18.0	17.1	−0.8	15.0	75	4.2	64
Sep.	6.9	19	2.0	1021.4	18.1	17.3	−0.8	15.3	76	3.9	70
Oct.	6.2	29	1.8	1019.3	18.4	18.1	−0.3	16.2	77	3.2	68
Nov.	5.9	29	1.7	1016.9	19.5	19.5	0.0	18.4	81	2.6	68
Dec.	6.0	24	1.6	1017.8	21.1	21.0	−0.1	20.0	79	3.0	66
Annual	6.4	29	1.9	1019.0	20.6	20.0	−0.6	18.5	77	4.1	63
St. dev.	3.5		1.0	5.4	1.5	2.0	1.6	3.5	11	2.8	31

Month	Frequency (%) of strong wind	storm	mist	fog	precip-itation	Mean heat flux (Wm⁻²) latent	sensible	Mean radiation (Wm⁻²) solar	terrestr.	Mean radiat. balance (Wm⁻²)	Mean heat balance (Wm⁻²)
Jan.	6	0.4	0.8		8.9	96	1	225	59	166	69
Feb.	8	0.2	1.0	0.5	5.3	107	3	210	63	147	37
Mar.	7	0.2	0.9		4.6	123	7	178	63	115	−15
Apr.	8	0.8	0.9		7.2	145	9	133	65	68	−86
May	15	1.8	0.7		6.7	156	12	104	70	34	−134
June	8	1.3	0.8		4.3	133	13	88	69	19	−127
July	12	2.0	0.3		5.2	129	10	93	64	29	−110
Aug.	15	1.5	1.4	0.2	6.2	118	9	118	62	56	−71
Sep.	12	1.0	1.1		8.7	111	9	145	59	86	−34
Oct.	10	1.6	1.4	0.6	7.6	92	3	180	59	121	26
Nov.	8	0.8	2.3	0.5	9.1	75	0	206	58	148	73
Dec.	8	0.6	1.5	0.4	8.9	88	1	218	59	159	70
Annual	10	1.0	1.1	0.2	6.9	114	6	158	63	95	−25

TABLE XXXII

CLIMATIC TABLE FOR AREA N
Latitude 30–35°S, longitude 55–45°W

Month	Mean wind speed (m s⁻¹)	Wind steadiness (%)	Mean wave height (m)	Mean air press. (mbar)	Mean temperature (°C)			Mean vapour press. (mbar)	Mean relat. humid. (%)	Mean evapor. (mm day⁻¹)	Mean cloudiness (%)
					sea	air	mean diff.				
Jan.	6.3	34	1.3	1012.0	23.4	23.1	−0.3	22.6	79	4.3	49
Feb.	6.4	34	1.4	1012.1	24.0	23.6	−0.5	23.1	79	4.8	52
Mar.	6.7	20	1.6	1014.3	23.4	22.5	−0.9	21.9	79	5.1	56
Apr.	6.5	8	1.5	1015.8	21.9	20.9	−1.1	19.5	78	5.0	52
May	6.2	16	1.4	1017.7	19.8	18.6	−1.2	17.0	79	4.3	51
June	7.2	21	1.6	1017.9	18.1	16.7	−1.4	15.5	80	4.2	57
July	7.2	13	1.7	1019.2	16.4	15.7	−0.7	14.9	82	3.3	58
Aug.	7.0	13	1.8	1018.0	16.1	15.6	−0.5	14.6	81	2.9	59
Sep.	6.5	26	1.5	1017.6	16.6	16.4	−0.3	15.6	83	2.4	63
Oct.	6.9	25	1.6	1015.5	17.9	17.9	−0.1	16.7	81	2.8	54
Nov.	7.1	32	1.6	1013.7	19.7	19.5	−0.2	18.4	81	3.3	59
Dec.	7.0	33	1.5	1013.3	21.7	21.4	−0.3	20.6	80	3.9	54
Annual	6.8	23	1.6	1015.5	19.9	19.3	−0.6	18.4	80	3.9	55
St. dev.	3.8		1.0	5.5	2.0	2.4	2.2	3.9	11	3.7	36

Month	Frequency (%) of					Mean heat flux (Wm⁻²)		Mean radiation (Wm⁻²)		Mean radiat. balance (Wm⁻²)	Mean heat balance (Wm⁻²)
	strong wind	storm	mist	fog	precipitation	latent	sensible	solar	terrestr.		
Jan.	9	0.5	1.5	0.9	3.1	137	3	248	60	188	48
Feb.	10	0.3	1.8	0.2	5.4	129	5	216	65	151	17
Mar.	12	1.3	1.9	0.4	6.8	138	10	174	66	108	−40
Apr.	10	0.8	1.6	0.1	4.1	134	11	137	76	61	−84
May	12	1.7	2.2	0.2	3.6	116	12	106	76	30	−98
June	19	3.4	6.3	3.4	6.3	113	16	86	74	12	−117
July	17	2.2	5.7	1.5	6.0	83	8	92	70	22	−69
Aug.	14	2.2	7.8	3.9	8.3	75	6	116	62	54	−27
Sep.	11	1.1	20.0	8.6	10.8	65	3	148	58	90	22
Oct.	14	1.2	5.3	2.3	7.1	77	1	202	60	142	64
Nov.	14	0.7	7.4	1.7	6.1	92	2	220	56	164	70
Dec.	12	0.8	3.6	0.7	4.9	110	3	245	58	187	74
Annual	13	1.4	5.4	2.0	6.1	106	7	166	65	101	−12

TABLE XXXIII

CLIMATIC TABLE FOR AREA O
Latitude 30–35°S, longitude 10–20°E

Month	Mean wind speed (m s⁻¹)	Wind steadiness (%)	Mean wave height (m)	Mean air press. (mbar)	Mean temperature (°C)			Mean vapour press. (mbar)	Mean relat. humid. (%)	Mean evapor. (mm day⁻¹)	Mean cloudiness (%)
					sea	air	mean diff.				
Jan.	7.7	78	2.4	1014.8	19.7	19.8	0.1	18.3	79	3.6	50
Feb.	7.7	80	2.3	1015.0	19.7	19.8	0.1	18.3	80	3.6	54
Mar.	7.5	78	2.3	1016.1	19.1	19.4	0.3	18.2	81	3.2	47
Apr.	7.2	66	2.2	1017.0	18.2	18.0	−0.2	16.6	80	3.4	54
May	6.2	46	2.2	1018.4	17.5	17.1	−0.5	15.7	81	3.0	51
June	6.2	26	2.4	1020.0	16.6	15.8	−0.8	14.0	78	3.4	49
July	6.5	31	2.4	1021.7	15.7	15.1	−0.6	13.3	77	3.3	47
Aug.	7.0	38	2.5	1020.6	15.4	14.8	−0.6	13.1	78	3.3	52
Sep.	7.1	56	2.3	1020.1	15.5	15.1	−0.4	13.5	78	3.2	51
Oct.	7.4	64	2.3	1018.6	16.2	16.1	−0.1	14.4	79	3.0	57
Nov.	8.0	75	2.4	1017.0	17.3	17.4	0.1	15.8	79	3.2	48
Dec.	7.7	74	2.2	1015.9	18.6	18.8	0.2	17.3	80	3.2	49
Annual	7.2	59	2.3	1017.9	17.4	17.3	−0.2	15.7	79	3.3	51
St. dev.	4.0		1.3	4.4	1.5	1.7	1.8	2.5	10	2.8	35

Month	Frequency (%) of					Mean heat flux (Wm⁻²)		Mean radiation (Wm⁻²)		Mean radiat. balance (Wm⁻²)	Mean heat balance (Wm⁻²)
	strong wind	storm	mist	fog	precipitation	latent	sensible	solar	terrestr.		
Jan.	18	1.5	0.7	0.3	1.7	101	−1	257	62	195	95
Feb.	18	1.1	1.0	0.6	2.0	101	−1	221	56	165	65
Mar.	18	1.7	1.1	0.7	1.5	86	−4	195	61	134	52
Apr.	18	1.3	2.3	1.5	2.3	91	2	139	67	72	−21
May	11	1.7	1.4	1.0	3.1	81	5	108	75	33	−53
June	11	1.3	0.8	0.4	3.7	92	8	95	75	20	−80
July	12	1.8	0.8	0.3	2.8	88	6	103	74	29	−65
Aug.	14	2.3	0.7	0.3	3.8	90	7	126	69	57	−40
Sep.	15	1.3	0.5	0.3	2.5	86	5	171	66	105	14
Oct.	17	1.4	0.9	0.3	2.4	85	1	201	68	133	47
Nov.	22	1.7	2.3	0.3	1.6	89	−1	256	72	184	96
Dec.	18	1.6	1.5	0.5	1.4	90	−2	267	70	197	109
Annual	16	1.6	1.1	0.5	2.4	90	2	178	68	110	18

TABLE XXXIV

CLIMATIC TABLE FOR AREA P
Latitude 30–40°S, longitude 15°W–5°E

Month	Mean wind speed (m s⁻¹)	Wind steadiness (%)	Mean wave height (m)	Mean air press. (mbar)	Mean temperature (°C) sea	air	mean diff.	Mean vapour press. (mbar)	Mean relat. humid. (%)	Mean evapor. (mm day⁻¹)	Mean cloudiness (%)
Jan.	6.4	36	1.7	1020	19.6	19.1	−0.5	17.3	77	3.8	67
Feb.	6.1	30	1.7	1019	19.9	19.3	−0.6	18.0	78	3.6	69
Mar.	7.2	27	1.9	1019	19.8	19.3	−0.6	18.0	80	3.8	66
Apr.	7.2	36	1.9	1018	18.8	18.2	−0.6	16.7	80	3.6	69
May	7.7	35	2.2	1018	17.8	17.0	−0.7	15.5	79	3.8	68
June	8.3	53	2.2	1020	16.9	15.9	−0.9	13.7	77	4.3	66
July	7.9	54	2.6	1021	15.5	14.7	−0.8	13.1	78	3.9	74
Aug.	7.7	56	2.3	1017	15.0	14.1	−1.0	12.4	75	4.0	67
Sep.	7.0	38	2.3	1021	15.0	14.3	−0.7	12.6	77	3.1	76
Oct.	7.1	35	2.0	1021	14.8	14.3	−0.5	12.9	79	2.7	72
Nov.	7.0	35	1.7	1019	16.4	16.4	0.0	15.0	80	2.2	72
Dec.	7.1	44	2.1	1018	17.5	17.3	−0.2	16.0	79	2.9	64
Annual	7.2	40	2.0	1019	17.3	16.7	−0.6	15.2	78	3.5	69
St. dev.	3.9		1.1	7	1.8	2.4	1.7	3.2	12	2.8	30

Month	Frequency (%) of strong wind	storm	mist	fog	precipitation	Mean heat flux (Wm⁻²) latent	sensible	Mean radiation (Wm⁻²) solar	terrestr.	Mean radiat. balance (Wm⁻²)	Mean heat balance (Wm⁻²)
Jan.	10	1.7	2.1	1.0	7.1	105	5	215	60	155	45
Feb.	8	0.4	0.4	0.1	6.5	97	6	184	56	128	25
Mar.	14	3.3	2.1	0.6	11.4	107	7	155	62	93	−21
Apr.	15	2.9	1.5	0.3	9.1	103	7	112	66	46	−64
May	19	3.9	0.6	0.1	9.7	108	9	84	72	12	−105
June	27	1.6	1.2	0.5	10.9	131	12	70	85	−15	−158
July	20	4.4	1.1	0.1	12.5	104	10	69	60	9	−105
Aug.	23	2.9	1.5		8.5	105	12	100	61	39	−78
Sep.	12	2.0	2.6	0.4	10.2	89	8	124	55	69	−28
Oct.	14	3.0	2.5	1.2	9.0	78	6	168	51	117	33
Nov.	15	0.7	5.1	1.2	10.3	70	0	197	58	139	69
Dec.	17	1.2	3.4	1.4	4.9	88	2	229	59	170	80
Annual	16	2.3	2.0	0.6	9.2	99	7	142	62	80	−26

TABLE XXXV

CLIMATIC TABLE FOR AREA Q
Latitude 35–40°S, longitude 60–50°W

Month	Mean wind speed (m s⁻¹)	Wind steadiness (%)	Mean wave height (m)	Mean air press. (mbar)	Mean temperature (°C)			Mean vapour press. (mbar)	Mean relat. humid. (%)	Mean evapor. (mm day⁻¹)	Mean cloudiness (%)
					sea	air	mean diff.				
Jan.	5.3	18	1.1	1012	20.5	20.6	0.1	19.1	78	3.3	46
Feb.	6.4	10	1.3	1012	20.6	20.1	−0.6	18.5	78	4.3	41
Mar.	6.8	17	1.5	1016	19.9	18.7	−1.3	16.8	77	4.9	47
Apr.	6.4	14	1.2	1016	18.6	17.6	−1.1	16.1	79	4.2	46
May	5.6	29	1.2	1016	15.8	14.9	−1.0	14.0	82	2.8	61
June	5.9	32	1.1	1015	13.1	12.3	−0.7	12.2	83	2.0	66
July	6.4	24	1.7	1019	10.9	10.8	−0.1	10.6	80	1.9	65
Aug.	6.5	19	1.4	1018	11.7	11.3	−0.5	10.9	80	2.4	54
Sep.	5.9	16	1.2	1018	12.0	12.1	0.0	11.7	84	1.8	55
Oct.	5.4	23	1.0	1015	13.6	13.9	0.2	13.5	84	1.6	49
Nov.	5.9	26	1.3	1014	16.2	16.5	0.3	15.5	81	2.2	58
Dec.	6.0	27	1.2	1011	18.7	18.5	−0.2	16.9	78	3.0	51
Annual	6.0	21	1.3	1015	16.0	15.6	−0.4	14.6	80	2.8	53
St. dev.	3.8		0.9	6	2.5	2.8	2.4	3.4	12	3.0	38

Month	Frequency (%) of					Mean heat flux (Wm⁻²)		Mean radiation (Wm⁻²)		Mean radiat. balance (Wm⁻²)	Mean heat balance (Wm⁻²)
	strong wind	storm	mist	fog	precip- itation	latent	sensible	solar	terrestr.		
Jan.	5	0.6	2.7	2.3	3.0	78	−1	265	62	203	126
Feb.	12	1.2	0.5	0.1	3.4	114	6	240	63	177	57
Mar.	16	1.7	0.1		3.5	131	15	182	66	116	−30
Apr.	10	1.5	3.4	1.3	3.6	106	12	133	72	61	−57
May	8	0.2	2.5	1.2	4.2	71	10	82	66	16	−65
June	9	2.7	8.5	4.4	9.2	53	7	62	55	7	−53
July	13	1.4	2.6	2.4	10.0	48	1	69	51	18	−31
Aug.	14	1.1	4.5	3.2	3.8	60	5	107	54	53	−12
Sep.	8	1.1	9.4	6.6	5.5	44	0	150	49	101	57
Oct.	9	0.7	11.1	9.0	10.4	38	−2	207	53	154	118
Nov.	7	0.2	6.9	1.4	7.1	60	−3	226	51	175	118
Dec.	12	0.8	3.5	2.5	3.6	85	2	261	57	204	117
Annual	10	1.1	4.6	2.9	5.6	74	4	165	58	107	29

TABLE XXXVI

CLIMATIC TABLE FOR AREA R
Latitude 40–45°S, longitude 65–55°W

Month	Mean wind speed (m s⁻¹)	Wind steadiness (%)	Mean wave height (m)	Mean air press. (mbar)	Mean temperature (°C)			Mean vapour press. (mbar)	Mean relat. humid. (%)	Mean evapor. (mm day⁻¹)	Mean cloudiness (%)
					sea	air	mean diff.				
Jan.	6.1	22	1.4	1009	16.1	16.8	0.8	15.5	79	2.2	48
Feb.	6.7	29	1.4	1009	16.5	16.8	0.3	15.1	78	3.0	47
Mar.	5.9	26	1.4	1014	15.9	15.5	−0.4	14.1	79	2.8	46
Apr.	7.2	43	1.5	1013	14.3	13.9	−0.5	12.9	80	2.7	52
May	7.1	54	1.6	1012	12.6	11.9	−0.7	11.4	81	2.6	59
June	6.9	48	1.5	1012	10.6	9.9	−0.8	10.2	82	2.1	62
July	7.4	41	1.7	1014	9.2	8.6	−0.7	9.2	81	2.0	69
Aug.	7.1	46	1.4	1013	8.6	8.7	0.2	9.4	82	1.5	56
Sep.	6.4	37	1.4	1015	8.7	9.0	0.4	9.3	81	1.4	50
Oct.	6.4	22	1.2	1013	9.7	10.3	0.7	10.5	82	2.0	51
Nov.	6.0	26	1.5	1012	11.7	12.7	1.1	12.2	82	1.1	55
Dec.	6.3	32	1.3	1008	14.2	14.8	0.6	13.5	79	1.9	49
Annual	6.6	35	1.4	1012	12.4	12.4	0.1	11.9	81	2.0	54
St. dev.	4.2		1.0	8	2.0	2.5	2.2	2.5	11	2.4	38

Month	Frequency (%) of					Mean heat flux (Wm⁻²)		Mean radiation (Wm⁻²)		Mean radiat. balance (Wm⁻²)	Mean heat balance (Wm⁻²)
	strong wind	storm	mist	fog	precipitation	latent	sensible	solar	terrestr.		
Jan.	13	0.9	3.3	2.2	4.0	54	−8	266	54	212	166
Feb.	14	2.0	1.0	0.8	2.4	76	−3	224	51	173	100
Mar.	10	1.8	3.9	2.5	3.8	74	4	174	59	115	37
Apr.	15	2.4	1.6	1.1	3.5	73	6	111	66	45	−34
May	19	3.7	3.4	2.5	5.9	69	8	68	67	1	−76
June	18	3.1	3.7	2.4	5.5	55	9	51	62	−10	−74
July	22	3.5	4.0	2.5	5.8	53	8	52	54	−2	−63
Aug.	18	3.3	1.4	0.8	1.8	39	−2	90	54	36	−1
Sep.	13	1.9	3.2	2.7	3.9	37	−4	144	49	95	62
Oct.	11	2.2	2.5	0.8	4.2	39	−7	197	46	151	119
Nov.	9	0.7	5.7	3.2	3.1	29	−11	234	45	189	171
Dec.	11	1.1	3.7	2.5	2.4	56	−6	273	51	222	172
Annual	14	2.2	3.1	2.0	3.9	55	−1	157	55	102	48

TABLE XXXVII

CLIMATIC TABLE FOR SAO TOMÉ, AEROPORT
Latitude 0°23′ N, longitude 6°43′ E, elevation 8 m

Month	Mean press. (mbar)	Temperature (°C)					Mean dew-point (°C)	Mean vapour press. (mbar)	Mean relat. humid. (%)	Mean evap. (mm)
		daily mean	mean daily max.	mean daily min.	extremes max.	min.				
Jan.	1011	25.9	29.4	22.4	32.0	19.1	23.7	29.4	86	72
Feb.	1010	26.2	29.9	22.5	33.6	19.6	23.7	29.4	85	69
Mar.	1010	26.4	30.2	22.6	33.5	19.2	23.7	29.4	84	84
Apr.	1010	26.4	30.1	22.6	33.4	19.4	23.8	29.6	85	78
May	1012	26.0	29.3	22.6	33.9	18.5	23.0	28.2	84	89
June	1014	24.7	28.0	21.4	31.0	14.0	21.0	24.9	80	107
July	1015	23.8	27.3	20.4	30.7	14.0	19.8	23.2	78	119
Aug.	1014	24.1	27.7	20.5	31.0	13.4	20.3	23.9	78	112
Sep.	1013	25.0	28.6	21.3	31.7	16.0	21.6	25.8	81	99
Oct.	1012	25.2	28.7	21.8	31.5	18.3	22.4	27.2	84	84
Nov.	1011	25.5	29.0	22.0	31.6	18.6	22.9	28.0	86	73
Dec.	1011	25.6	29.1	22.1	32.0	19.6	23.4	28.9	86	75
Annual	1012	25.4	28.9	21.8	33.9	13.4	22.4	27.2	83	1061

Month	Wind		Precipitation (mm)		Number of days with				Mean cloud-iness	Mean sunshine	
	mean speed (m s⁻¹)	preval. direct.	mean	max. 24 h	precip.	snow-fall	thunder-storm	fog		(days)	(%)
Jan.	2.1	SSW	81	63	6		10	1	69	4.6	38
Feb.	1.9	S	84	82	6		9		68	4.8	39
Mar.	2.0	S	131	117	9		13		70	4.5	37
Apr.	2.1	S	121	84	8		21		70	4.2	35
May	2.6	S	113	97	8		6		71	4.7	39
June	3.1	S	19	61	3				69	5.5	46
July	2.6	S	0.1	2					69	5.2	44
Aug.	2.2	SSE	1	3				1	72	4.8	40
Sep.	2.1	SSE	17	51	5			1	75	4.0	33
Oct.	2.2	SSE	110	170	8		3	1	74	3.7	31
Nov.	2.4	S	99	72	9		8		70	4.4	36
Dec.	2.1	S	108	86	6		11	1	68	4.6	38
Annual	2.3	S	884	170	68		81	5	71	4.6	38

TABLE XXXVIII

CLIMATIC TABLE FOR FERNANDO DE NORONHA
Latitude 3°51′ S, longitude 32°25′ W, elevation 45 m

Month	Mean press. (mbar)	Temperature (°C)					Mean dew-point (°C)	Mean vapour press. (mbar)	Mean relat. humid. (%)	Mean evap. (mm)
		daily mean	mean daily max.	mean daily min.	extremes					
					max.	min.				
Jan.	1012	26.4	28.4	24.4	30.9	21.1	23.6	29.2	85	160
Feb.	1011	26.3	28.3	24.3	30.4	20.6	23.7	29.4	86	136
Mar.	1011	26.3	28.3	24.3	30.6	20.6	23.5	29.1	85	137
Apr.	1011	26.0	27.9	24.0	30.6	20.0	23.6	29.2	87	120
May	1012	25.8	27.8	23.9	29.8	20.0	23.2	28.6	86	136
June	1013	25.5	27.2	23.8	29.6	20.6	22.7	27.7	85	153
July	1014	25.0	26.7	23.3	29.6	20.4	21.9	26.3	83	183
Aug.	1014	24.8	26.7	22.9	29.6	20.4	21.3	25.4	81	186
Sep.	1014	25.2	27.2	23.3	30.0	20.4	21.9	26.3	82	195
Oct.	1013	25.8	27.8	23.8	30.0	21.0	22.0	26.6	80	206
Nov.	1012	25.9	27.9	23.9	30.7	18.3	22.3	27.2	81	191
Dec.	1012	26.2	28.1	24.3	30.1	21.0	22.6	27.6	81	181
Annual	1012	25.8	27.7	23.8	30.9	18.3	22.6	27.6	83	1985

Month	Wind		Precipitation (mm)		Number of days with				Mean cloud-iness	Mean sunshine	
	mean speed (m s⁻¹)	preval. direct.	mean	max. 24 h	precip.	snow-fall	thunder-storm	fog		(days)	(%)
Jan.	6.2	ESE	45	81	7				69	8.7	71
Feb.	5.8	ESE	123	125	10				74	8.7	72
Mar.	5.3	ESE	191	193	15				74	8.0	66
Apr.	5.5	ESE	266	105	18		1		86	7.3	61
May	6.2	ESE	261	130	18		1		74	8.0	67
June	7.2	ESE	177	128	13		1		64	8.8	75
July	7.8	ESE	143	164	11		1		57	9.4	80
Aug.	7.3	ESE	45	56	8		1		57	9.4	79
Sep.	7.0	ESE	17	43	3				55	10.3	86
Oct.	7.0	ESE	8	36	3				53	10.5	87
Nov.	6.5	ESE	9	33	3				57	10.5	86
Dec.	6.6	ESE	12	18	4				62	9.8	80
Annual	6.5	ESE	1297	193	113		5		65	9.1	76

TABLE XXXIX

CLIMATIC TABLE FOR ASCENSION ISLAND, GEORGETOWN
Latitude 7°55′ S, longitude 14°25′ W, elevation 17 m

Month	Mean press. (mbar)	Temperature (°C)					Mean dew-point (°C)	Mean vapour press. (mbar)	Mean relat. humid. (%)	Mean evap. (mm)
		daily mean	mean daily max.	mean daily min.	extremes max.	min.				
Jan.	1013	26.1	29.5	22.7	31.7	19.4	20.2	23.6	70	
Feb.	1013	26.9	30.5	23.3	32.8	19.4	20.7	24.5	69	
Mar.	1012	27.3	31.0	23.7	34.4	18.9	21.0	25.0	69	
Apr.	1012	27.4	31.0	23.8	35.0	19.4	21.1	25.2	69	
May	1014	27.0	30.5	23.4	33.3	19.4	20.5	24.2	68	
June	1015	26.2	29.6	22.9	32.2	18.3	19.8	23.2	68	
July	1016	25.5	28.8	22.2	31.7	19.4	18.9	21.9	67	
Aug.	1016	25.0	28.3	21.8	31.1	18.3	18.6	21.5	68	
Sep.	1016	24.8	27.9	21.7	31.1	18.9	18.7	21.6	69	
Oct.	1015	25.0	28.3	21.7	31.1	18.3	18.8	21.8	69	
Nov.	1014	25.0	28.3	21.7	31.1	18.3	18.9	21.9	69	
Dec.	1014	25.5	28.9	22.1	31.7	18.9	19.3	22.5	69	
Annual	1014	26.0	29.4	22.6	35.0	18.3	19.8	23.2	69	

Month	Wind		Precipitation (mm)		Number of days with				Mean cloud-iness	Mean sunshine	
	mean speed (m s⁻¹)	preval. direct.	mean	max. 24 h	precip.	snow-fall	thunder-storm	fog		(days)	(%)
Jan.	1.9	ESE	4	6	2				44	7.4	59
Feb.	1.7	ESE	10	41	2				43	8.0	65
Mar.	1.9	ESE	18	59	3				41	8.9	73
Apr.	1.9	ESE	33	210	4				40	8.9	74
May	2.3	ESE	16	38	3				41	8.5	72
June	2.6	ESE	15	41	3				42	8.4	72
July	2.3	ESE	12	41	3				44	7.7	66
Aug.	2.1	ESE	11	10	3				48	7.0	59
Sep.	2.1	ESE	8	8	2				54	5.5	46
Oct.	2.5	ESE	7	5	3				56	5.2	42
Nov.	2.5	ESE	4	3	1				55	5.3	42
Dec.	2.1	ESE	4	5	1				47	6.4	51
Annual	2.2	ESE	142	210	30				46	7.3	60

TABLE XL

CLIMATIC TABLE FOR ST HELENA, JAMESTOWN
Latitude 15°55′ S, longitude 5°43′ W, elevation 12 m

Month	Mean press. (mbar)	Temperature (°C)					Mean dew-point (°C)	Mean vapour press. (mbar)	Mean relat. humid. (%)	Mean evap. (mm)
		daily mean	mean daily max.	mean daily min.	extremes					
					max.	min.				
Jan.	1015	23.5	26.5	20.4	31.7	17.2	16.0	18.2	63	
Feb.	1015	24.0	27.1	21.0	32.2	18.9	17.0	19.4	65	
Mar.	1015	24.6	27.6	21.5	33.3	18.9	17.1	19.5	63	
Apr.	1015	24.3	27.3	21.2	33.9	17.2	17.1	19.5	64	
May	1016	22.0	24.5	19.5	33.9	16.1	16.5	18.8	71	
June	1018	20.7	23.2	18.2	28.3	16.1	15.2	17.3	71	
July	1019	19.8	22.3	17.3	27.2	14.4	14.9	16.9	73	
Aug.	1019	19.8	22.3	17.3	26.1	15.0	15.2	17.3	75	
Sep.	1018	19.8	22.3	17.3	25.6	14.4	14.7	16.7	72	
Oct.	1017	20.4	22.9	17.9	25.6	15.6	15.0	17.0	71	
Nov.	1016	20.9	23.4	18.4	25.6	16.7	15.5	17.6	71	
Dec.	1016	21.6	24.4	18.8	26.7	15.6	14.8	16.8	65	
Annual	1017	21.8	24.5	19.1	33.9	14.4	15.8	17.9	69	

Month	Wind		Precipitation (mm)		Number of days with				Mean cloud-iness	Mean sunshine	
	mean speed (m s⁻¹)	preval. direct.	mean	max. 24 h	precip.	snow-fall	thunder-storm	fog		(days)	(%)
Jan.		SE	8	13	3.8				55		
Feb.		SE	10	8	3.6				55		
Mar.		SE	20	23	5.5				52		
Apr.		SE	10	8	3.2				52		
May		SE	18	18	4.0				60		
June		SE	18	20	5.8				70		
July		SE	33	15	7.6				78		
Aug.		SE	10	8	3.3				72		
Sep.		SE	5	10	2.2				62		
Oct.		SE	3	5	0.7				56		
Nov.		SE	0	1	0.3				46		
Dec.		SE	3	5	1.3				52		
Annual		SE	138	23	41.3				59		

TABLE XLI

CLIMATIC TABLE FOR TRISTAN DA CUNHA
Latitude 37°3′ S, longitude 12°19′ W, elevation 23 m

Month	Mean press. (mbar)	Temperature (°C)					Mean dew-point (°C)	Mean vapour press. (mbar)	Mean relat. humid. (%)	Mean evap. (mm)
		daily mean	mean daily max.	mean daily min.	extremes					
					max.	min.				
Jan.	1017	17.4	19.7	15.1	23.9	8.3	14.1	16.1	81	
Feb.	1018	18.2	20.6	15.9	23.3	8.7	14.7	16.7	80	
Mar.	1019	17.2	19.6	14.9	22.8	9.4	13.3	15.3	78	
Apr.	1017	16.0	18.2	13.9	22.2	7.8	12.3	14.4	79	
May	1017	14.1	16.3	11.8	19.4	4.4	10.5	12.7	79	
June	1018	12.8	14.9	10.6	18.3	5.0	9.4	11.8	80	
July	1019	12.1	14.2	10.1	17.8	5.0	8.9	11.4	81	
Aug.	1018	11.5	13.7	9.3	16.7	4.2	8.2	10.9	80	
Sep.	1019	11.7	13.9	9.5	18.0	3.3	8.1	10.8	79	
Oct.	1018	12.7	15.0	10.4	18.3	5.0	9.4	11.8	80	
Nov.	1017	14.4	16.7	12.1	20.0	6.5	11.0	13.1	80	
Dec.	1017	16.3	18.6	14.0	22.2	7.2	13.0	15.0	81	
Annual	1018	14.5	16.8	12.3	23.9	3.3	11.1	13.3	80	

Month	Wind		Precipitation (mm)		Number of days with				Mean cloud-iness	Mean sunshine	
	mean speed (m s⁻¹)	preval. direct.	mean	max. 24 h	precip.	snow-fall	thunder-storm	fog		(days)	(%)
Jan.	9	W	100	33	12			2	60	4.4	31
Feb.	9	W	104	98	12			1	65	4.3	32
Mar.	9	W	119	91	11			1	65	4.6	37
Apr.	10	W	141	62	14			1	70	3.7	33
May	10	W	153	69	18	1		1	75	2.9	29
June	10	W	156	96	18	1		1	80	2.4	25
July	11	W	159	110	19	1		1	75	2.9	29
Aug.	11	W	173	77	21	1		2	75	3.4	32
Sep.	10	W	161	145	18			1	80	3.3	28
Oct.	10	W	143	59	16			1	75	4.1	32
Nov.	10	W	127	180	14			1	70	4.5	32
Dec.	10	W	135	50	12			2	70	4.4	30
Annual	10	W	1671	180	185	4		15	72	3.7	31

TABLE XLII

CLIMATIC TABLE FOR GOUGH ISLAND
Latitude 40°19' S, longitude 9°54' W, elevation 6 m

Month	Mean press. (mbar)	Temperature (°C)					Mean dew-point (°C)	Mean vapour press. (mbar)	Mean relat. humid. (%)	Mean evap. (mm)
		daily mean	mean daily max.	mean daily min.	extremes					
					max.	min.				
Jan.	1014	13.9	16.8	11.3	25.4	5.3	10.6	12.8	80	
Feb.	1015	14.6	17.4	12.1	25.7	6.2	11.4	13.5	81	
Mar.	1016	13.6	16.2	11.2	24.7	5.3	10.4	12.6	81	
Apr.	1015	12.5	14.9	10.3	22.0	4.3	9.6	11.9	82	
May	1014	11.3	13.5	9.1	20.5	1.4	8.6	11.1	83	
June	1013	9.9	12.3	7.6	19.0	0.1	7.1	10.1	84	
July	1015	9.0	11.4	6.7	17.3	−0.9	6.3	9.5	83	
Aug.	1016	8.9	11.1	6.6	18.6	−2.7	6.2	9.5	83	
Sep.	1017	9.1	11.4	7.0	19.3	0.5	6.3	9.5	82	
Oct.	1016	10.1	12.5	7.8	21.4	1.8	7.1	10.1	82	
Nov.	1014	11.7	14.3	9.4	22.9	2.4	8.6	11.1	81	
Dec.	1013	13.1	16.0	10.6	24.2	4.1	9.7	12.1	80	
Annual	1015	11.5	14.0	9.1	25.7	−2.7	8.6	11.1	82	

Month	Wind		Precipitation (mm)		Number of days with				Mean cloud-iness	Mean sunshine	
	mean speed (m s⁻¹)	preval. direct.	mean	max. 24 h	precip.	snow-fall	thunder-storm	fog		(days)	(%)
Jan.			216	98	16	0.1	0.1	0.3	80	5.4	37
Feb.			205	131	14		0.1	0.4	78	5.0	36
Mar.			251	189	16			1.0	80	3.9	31
Apr.			272	148	18		0.2	0.8	80	2.7	25
May			255	81	20	0.3	0.5	0.5	79	1.9	19
June			334	120	22	0.5	0.7	0.6	80	1.4	15
July			293	98	22	1.8	0.4	0.4	80	1.6	16
Aug.			337	118	22	2.1	0.1	0.7	81	2.3	22
Sep.			290	96	21	1.7	0.1	0.6	82	3.0	25
Oct.			286	151	19	0.3	0.2	0.9	81	4.1	31
Nov.			268	108	18	0.1	0.2	1.1	80	5.1	35
Dec.			262	127	17	0.1	0.1	0.9	79	5.5	37
Annual			3270	189	226	7.0	2.7	8.2	80	3.5	29

TABLE XLIII

CLIMATIC TABLE FOR FALKLAND ISLANDS, PORT STANLEY
Latitude 51°42′ S, longitude 57°52′ W, elevation 52 m

Month	Mean press. (mbar)	Temperature (°C)					Mean dew-point (°C)	Mean vapour press. (mbar)	Mean relat. humid. (%)	Mean evap. (mm)
		daily mean	mean daily max.	mean daily min.	extremes					
					max.	min.				
Jan.	999	9.1	13.3	5.6	24.4	−1.1	5.6	9.1	79	
Feb.	1002	8.9	12.8	5.0	23.3	−1.1	5.8	9.2	81	
Mar.	1002	8.0	11.7	4.4	21.1	−2.8	5.3	8.9	83	
Apr.	1003	6.1	9.4	2.8	17.2	−6.1	4.1	8.2	87	
May	1003	3.8	7.2	1.1	14.4	−6.7	2.0	7.0	88	
June	1002	2.4	5.0	−0.6	10.6	−11.1	0.4	6.3	87	
July	1004	2.0	4.4	−0.6	10.0	−8.9	0.2	6.2	88	
Aug.	1004	2.3	5.0	−0.6	11.1	−11.1	0.3	6.3	87	
Sep.	1006	3.6	7.2	0.6	15.0	−10.6	1.3	6.7	85	
Oct.	1005	5.2	8.9	1.7	17.8	−5.6	2.0	7.1	80	
Nov.	1002	6.9	11.1	2.8	21.7	−3.3	3.2	7.7	77	
Dec.	1000	7.9	12.2	3.9	21.7	−1.7	4.3	8.3	78	
Annual	1003	5.5	8.9	2.2	24.4	−11.1	2.9	7.5	83	

Month	Wind		Precipitation (mm)		Number of days with				Mean cloud-iness	Mean sunshine	
	mean speed (m s⁻¹)	preval. direct.	mean	max. 24 h	precip.	snow-fall	thunder-storm	fog		(days)	(%)
Jan.	7.5	W	71	30	16	2	1	3	78	5.8	36
Feb.	7.5	W	57	43	13	1	0.5	2	74	5.7	39
Mar.	7.5	WNW	62	28	14	3	0.5	3	71	4.8	39
Apr.	7.0	WNW	60	25	13	3	0.5	4	72	3.4	34
May	7.0	WNW	60	38	15	13		4	74	2.3	27
June	7.1	WNW	53	28	13	6		5	71	2.0	26
July	7.3	WNW	46	30	12	11		5	71	2.2	27
Aug.	8.0	WNW	49	25	13	10		4	74	2.8	29
Sep.	7.6	WNW	38	13	11	7		4	73	3.9	34
Oct.	8.0	WNW	38	20	10	5		4	70	5.8	42
Nov.	8.3	W	47	36	11	6	0.5	2	73	6.5	42
Dec.	7.4	W	70	41	14	4	1	3	78	5.8	36
Annual	7.5	WNW	651	43	155	71	4	43	73	4.3	36

TABLE XLIV

CLIMATIC TABLE FOR SOUTH GEORGIA, GRYTVIKEN
Latitude 54°16′ S, longitude 36°30′ W, elevation 3 m

Month	Mean press. (mbar)	Temperature (°C)					Mean dew-point (°C)	Mean vapour press. (mbar)	Mean relat. humid. (%)	Mean evap. (mm)
		daily mean	mean daily max.	mean daily min.	extremes					
					max.	min.				
Jan.	993	4.6	8.4	1.4	24.5	−4.1	0.7	6.4	76	
Feb.	995	5.1	9.1	1.7	26.5	−3.7	0.9	6.5	74	
Mar.	997	4.5	8.4	1.0	28.8	−6.3	0.5	6.3	75	
Apr.	996	2.4	5.6	−0.8	19.1	−9.8	−1.4	5.6	76	
May	996	0.1	2.9	−3.1	17.5	−11.4	−3.8	4.6	75	
June	998	−1.5	0.9	−4.6	14.0	−14.6	−4.8	4.2	78	
July	998	−1.5	1.2	−4.7	13.6	−15.2	−5.2	4.1	76	
Aug.	1000	−1.7	1.5	−4.9	13.2	−19.2	−5.3	4.1	76	
Sep.	1001	−0.1	3.5	−3.3	17.0	−18.4	−3.8	4.6	76	
Oct.	1000	1.7	5.4	−1.8	20.0	−11.0	−2.4	5.1	74	
Nov.	994	2.8	6.5	−0.5	22.5	−6.4	−1.7	5.4	72	
Dec.	994	3.7	7.5	0.4	21.5	−5.4	−0.1	6.1	76	
Annual	997	1.7	5.1	−1.5	28.8	−19.2	−2.2	5.2	75	

Month	Wind		Precipitation (mm)		Number of days with				Mean cloud-iness	Mean sunshine	
	mean speed (m s⁻¹)	preval. direct.	mean	max. 24 h	precip.	snow-fall	thunder-storm	fog		(days)	(%)
Jan.	3.9	N	97	97	12	4		3	78	4.9	30
Feb.	4.4	NNW	114	89	13	4		3	74	5.7	39
Mar.	4.4	NNW	137	102	14	6		3	71	4.1	33
Apr.	4.2	NW	137	112	14	9		3	70	2.2	22
May	3.9	NW	144	56	12	12		3	67	1.1	13
June	3.9	WNW	133	127	15	11		1	68	0.4	6
July	4.2	WNW	142	94	15	13		1	66	0.7	9
Aug.	4.2	WNW	135	97	14	14		1	67	2.4	25
Sep.	4.7	NW	94	81	11	10		2	65	4.1	36
Oct.	4.4	NNW	82	64	12	9		2	68	5.5	39
Nov.	4.4	NNW	93	64	11	10		1	74	5.8	37
Dec.	3.9	NNW	87	66	11	7		2	76	5.4	32
Annual	4.2	NW	1395	127	154	109	0.5	25	70	3.5	29

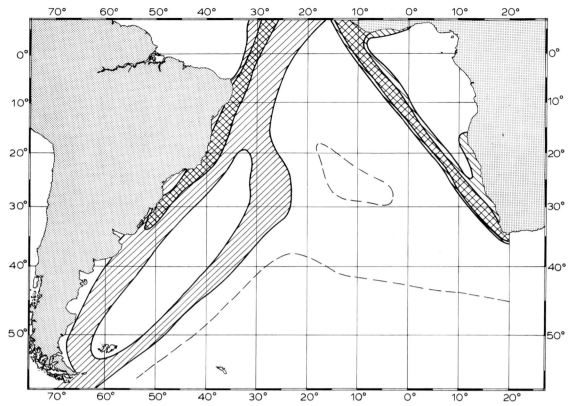

Chart 1. Frequency of observations by ships. Hatched: more than 1000 historical observations by German ships per 1° square; cross-hatched: more than 1000 modern international observations per 1° square; dashed: less than 100 observations per 1° square.

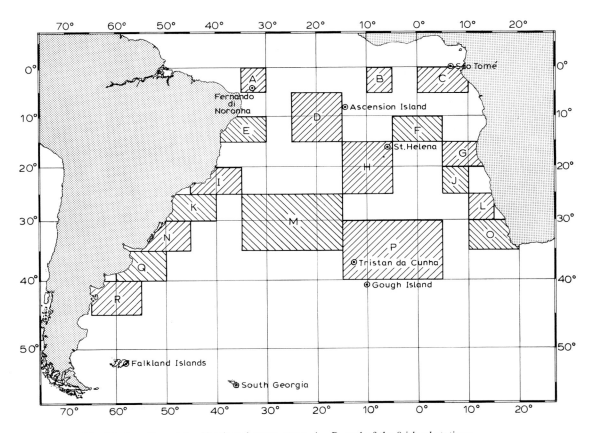

Chart 2. Location of the 18 selected ocean areas, A – R, and of the 8 island stations.

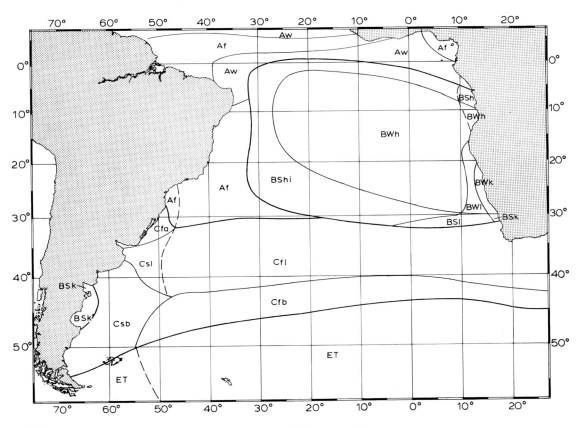

159   Chart 3. Climatic classification after KÖPPEN (1931); see p. 12f).

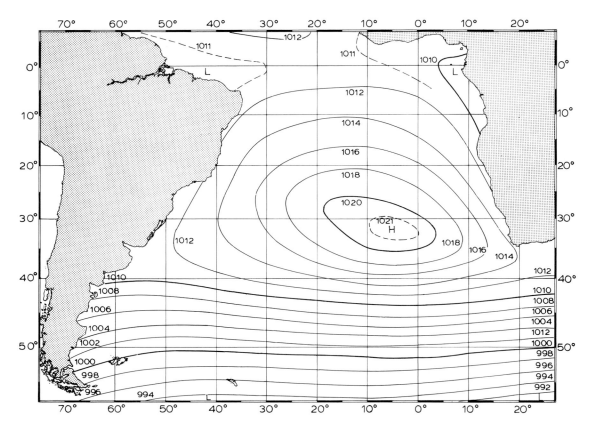

Chart 4. January sea level pressure (mbar).

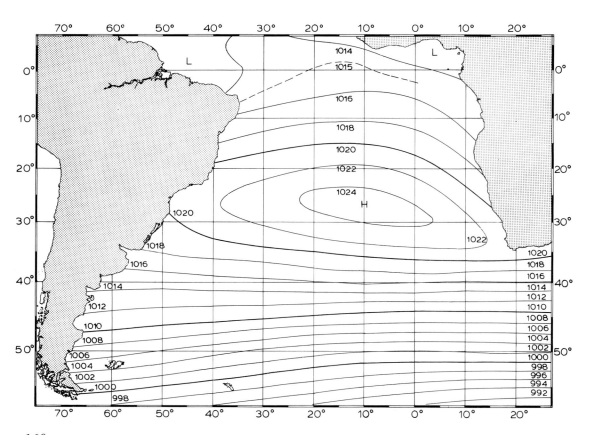

Chart 5. July sea level pressure (mbar).

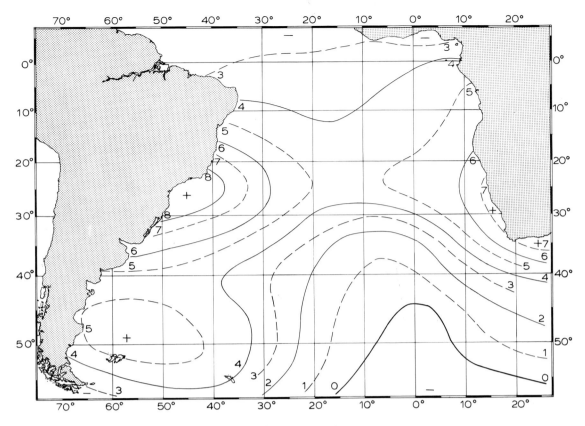

Chart 6. Sea level pressure difference July−January (mbar).

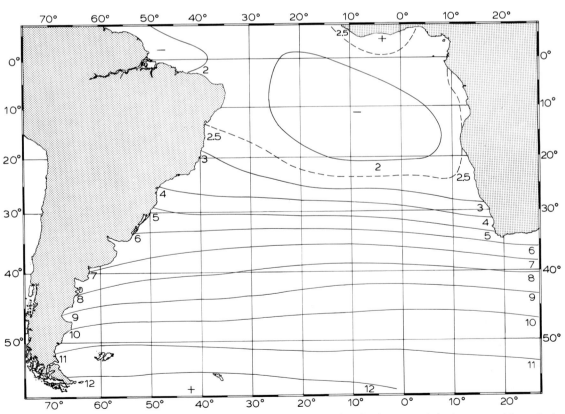

Chart 7. Annual averages of monthly standard deviations of sea level pressure (mbar), computed from single
observations.

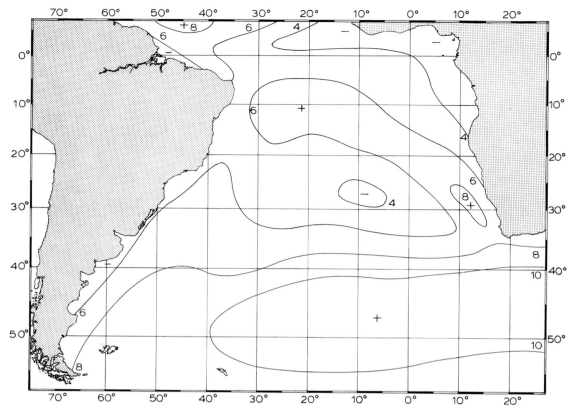

Chart 8. Scalar mean wind speed (m s^{-1}), January.

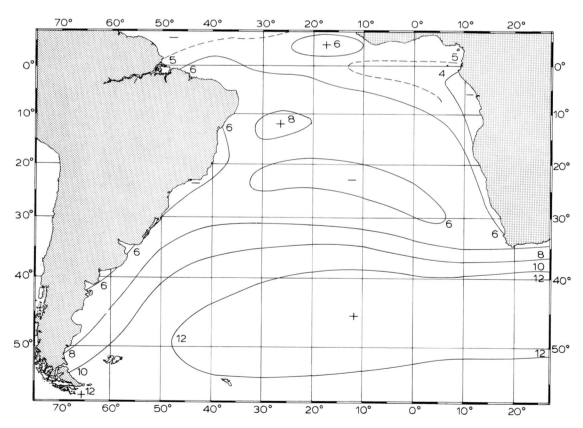

Chart 9. Same as 8, but for July.

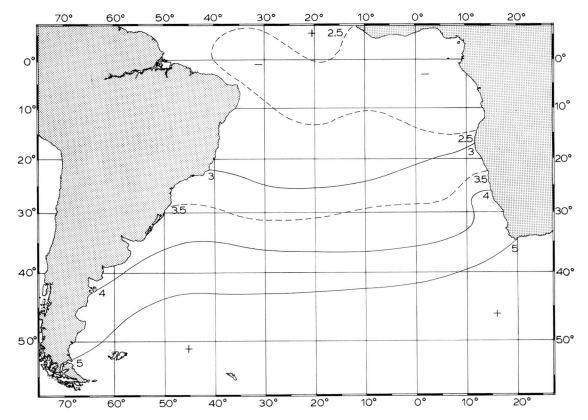

Chart 10. Annual averages of monthly standard deviations of scalar wind speed (m s^{-1}), computed from single observations.

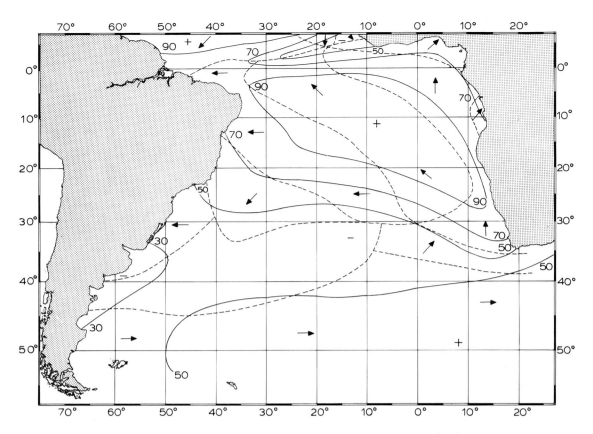

Chart 11. Mean steadiness of wind (%) and *predominant* wind (*scalar mean*) direction (January).

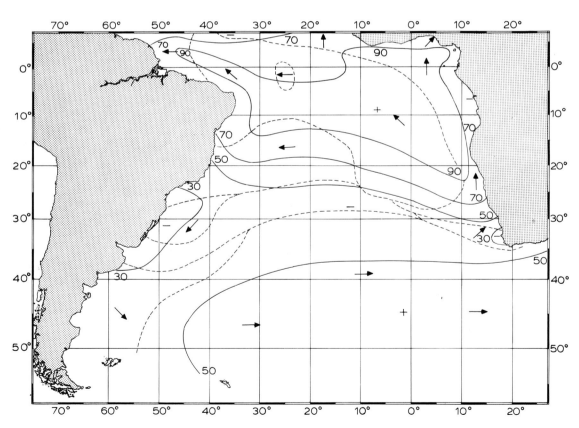

Chart 12. Same as 11, but for July.

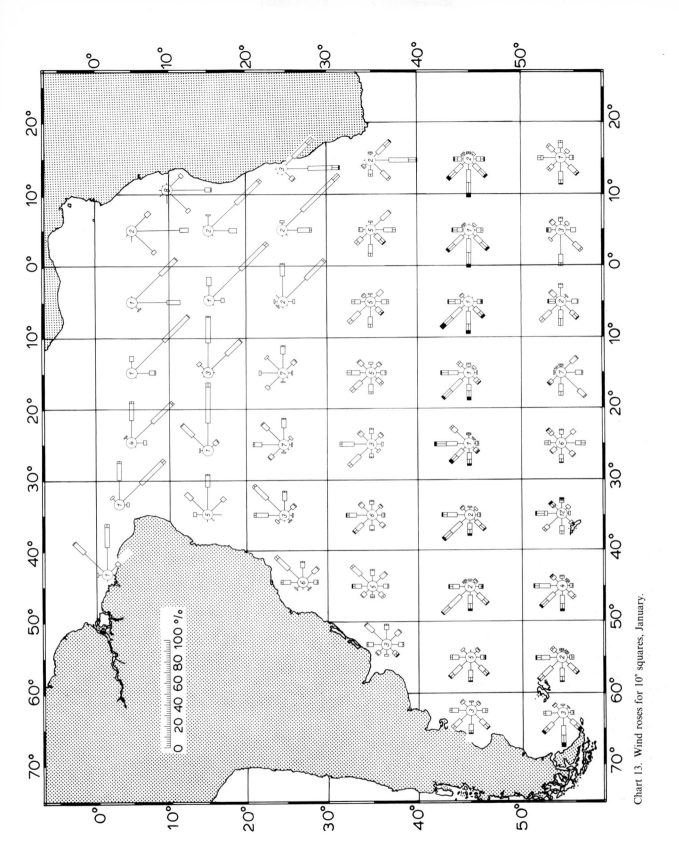

Chart 13. Wind roses for 10° squares, January.

0 20 40 60 80 100 %

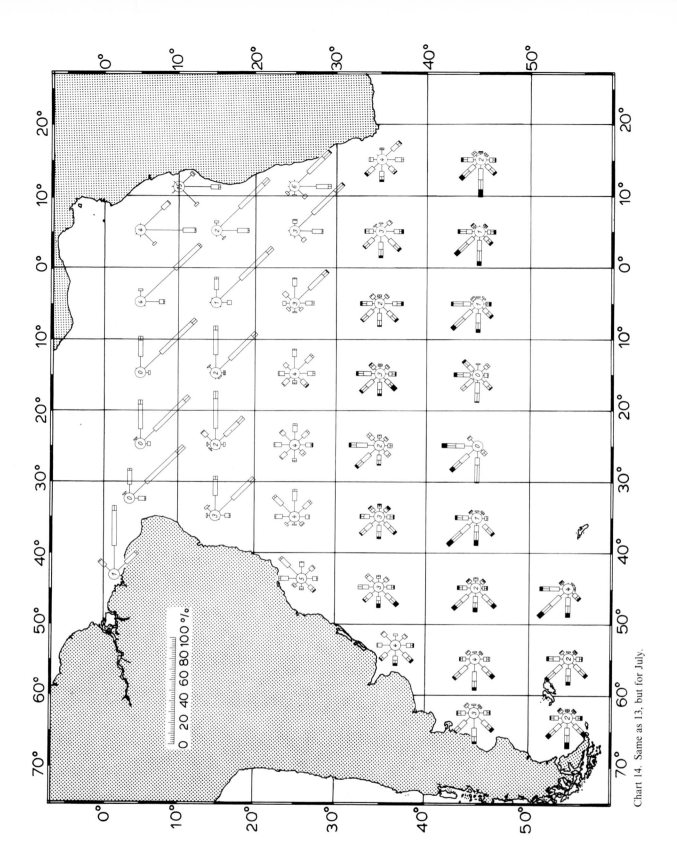

Chart 14. Same as 13, but for July.

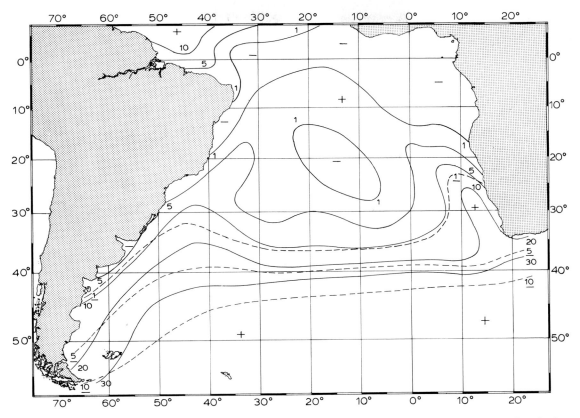

Chart 15. Percent frequency of strong winds (≥6 Beaufort, solid line) and of gales (≥8 Beaufort, broken line), January.

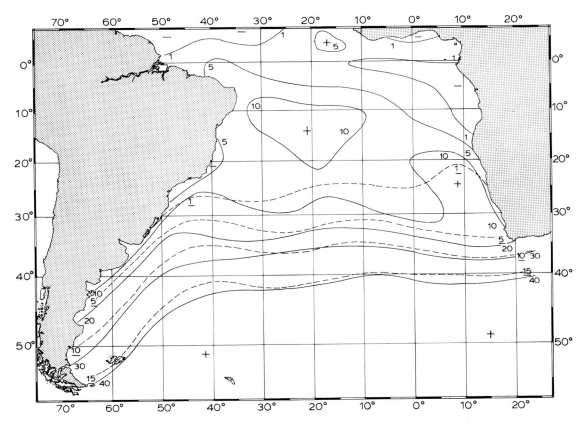

Chart 16. Same as 15, but for July.

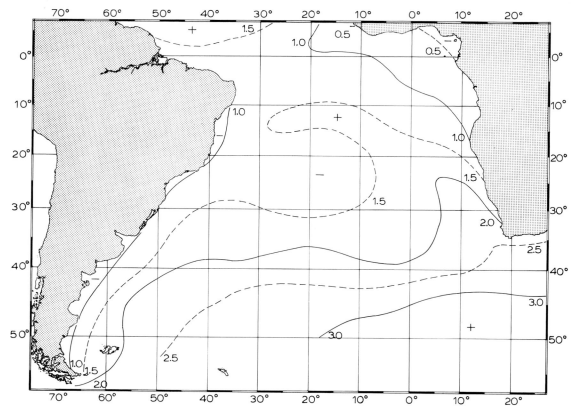

Chart 17. Wave height (m), January.

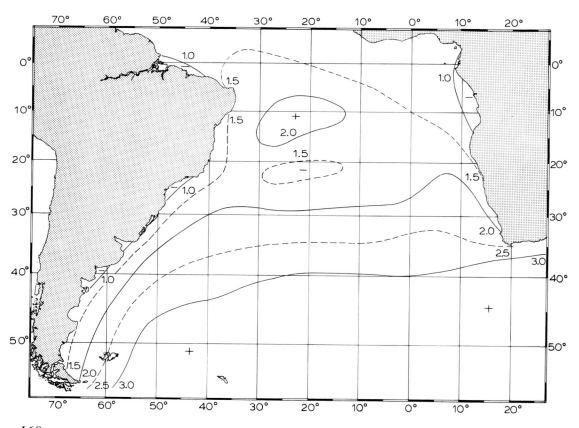

Chart 18. Same as 17, but for July.

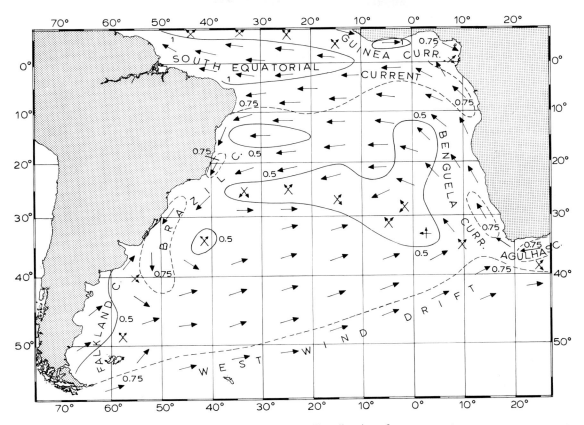

Chart 19. Annual mean speed (n.mile/h) and prevailing direction of ocean currents.

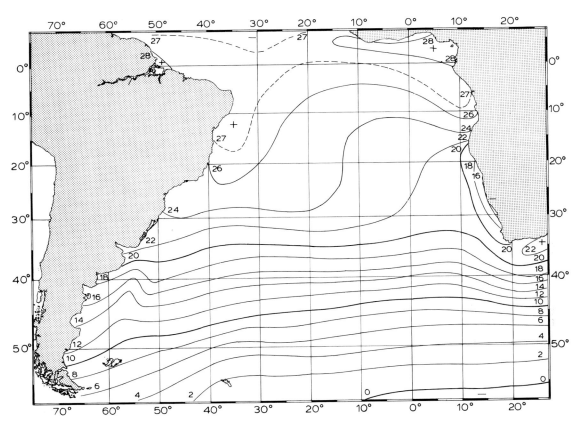

Chart 20. Sea surface temperature (°C), January.

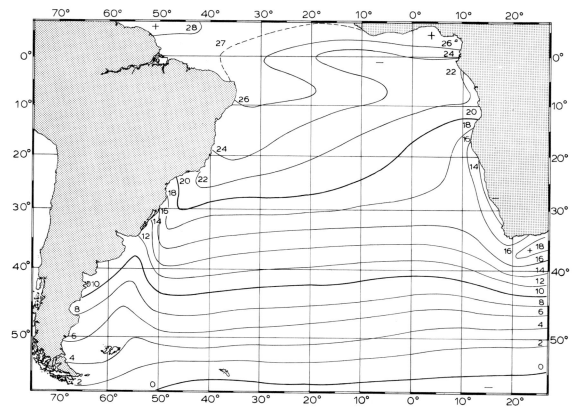

Chart 21. Same as 20, but for July.

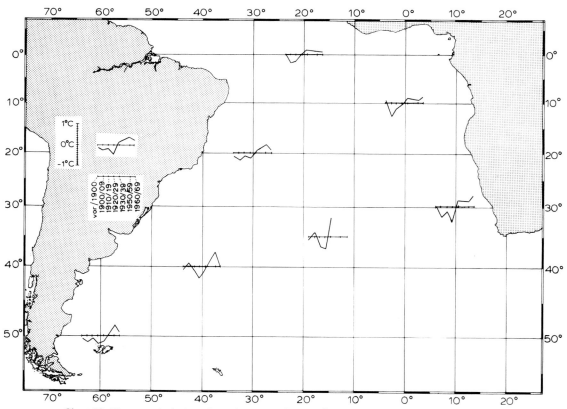

Chart 22. Ten-year deviations from the mean of sea surface temperature (°C) for seven 20° squares and for the whole South Atlantic.

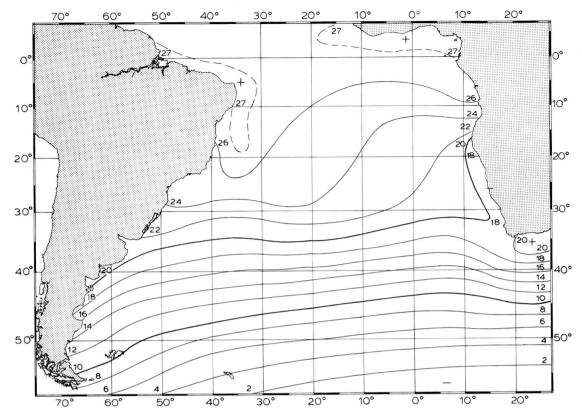

Chart 23. Surface air temperature (°C), January.

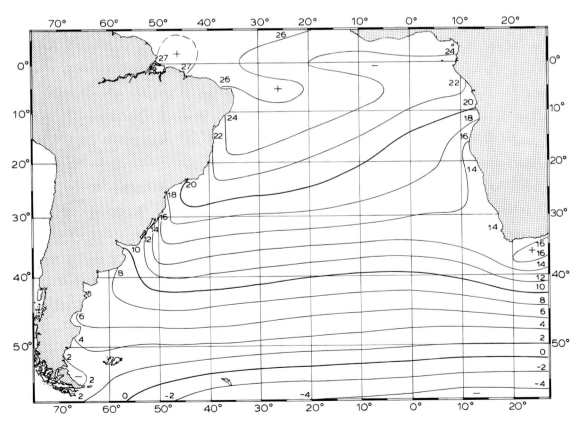

Chart 24. Same as 23, but for July.

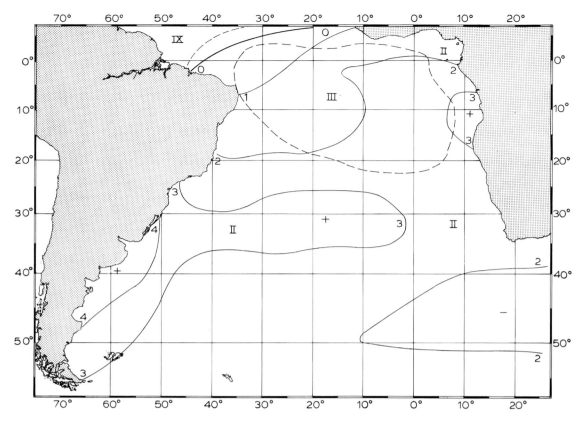

Chart 25. Amplitude (°C) and phase (month of maximum) of the first harmonic (annual component) of surface air temperature.

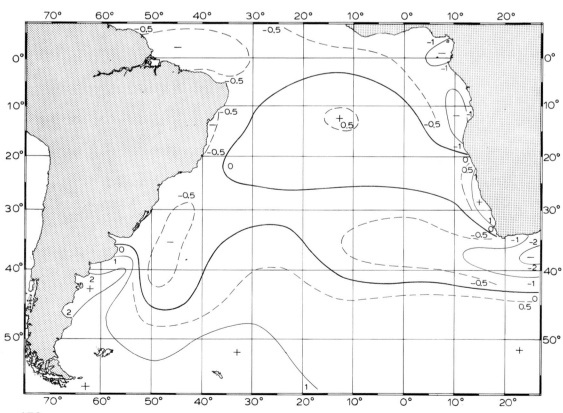

Chart 26. Air—sea temperature difference (°C), January.

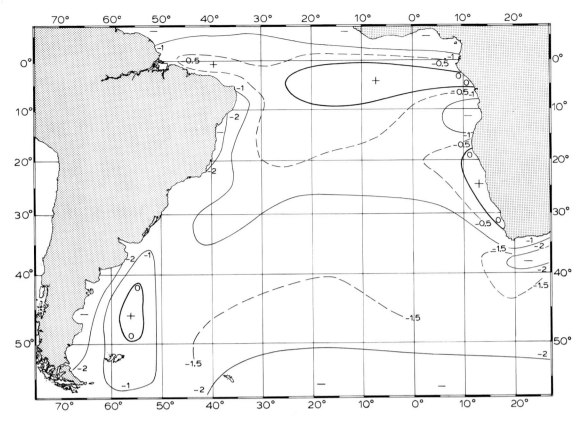

Chart 27. Same as 26, but for July.

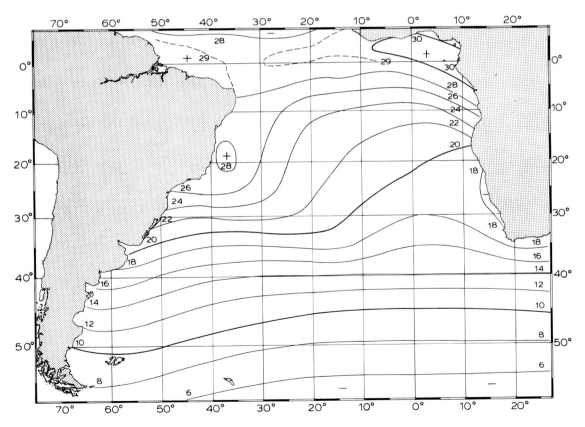

Chart 28. Vapour pressure (mbar), January.

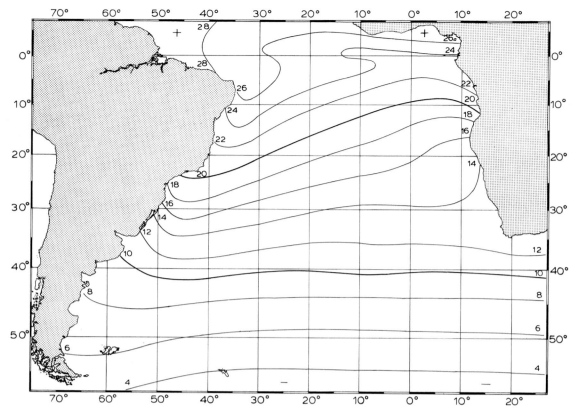

Chart 29. Same as 28, but for July.

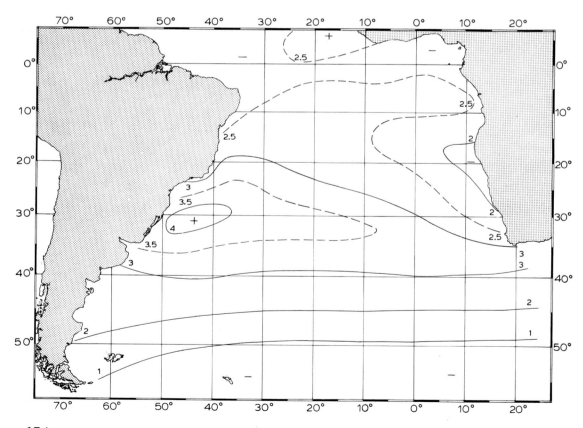

Chart 30. Annual averages of monthly standard deviations of vapour pressure (mbar).

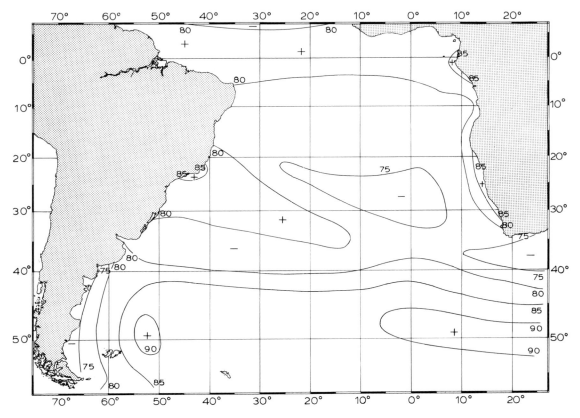

Chart 31. Relative humidity (%), January.

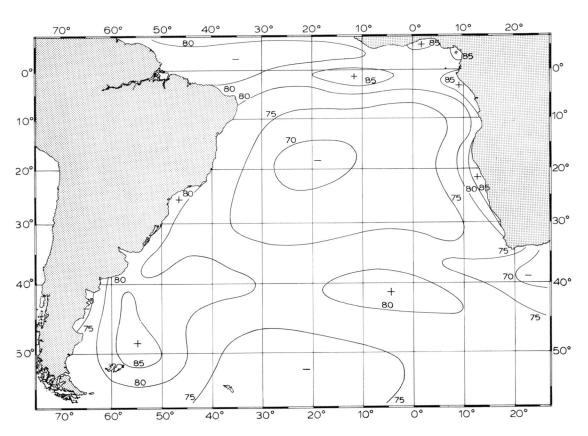

Chart 32. Same as 31, but for July.

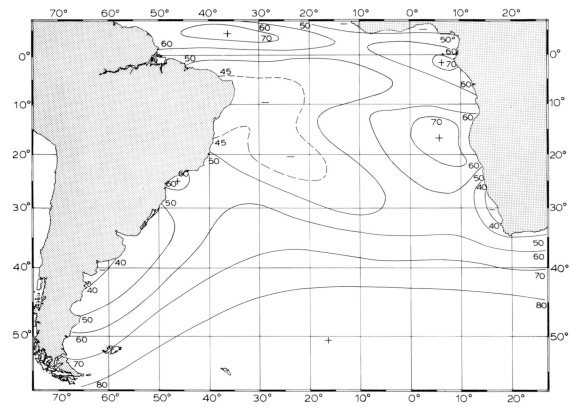

Chart 33. Cloud cover (%), January.

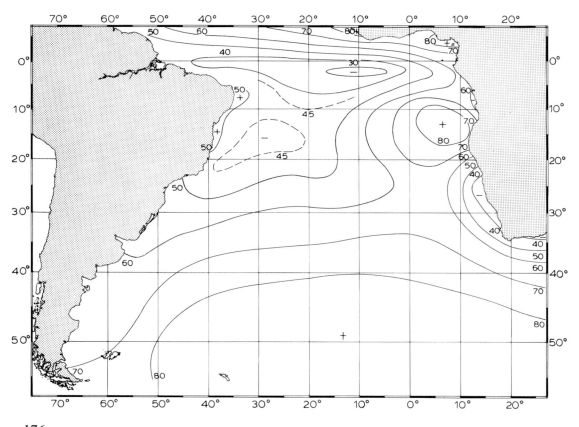

Chart 34. Same as 33, but for July.

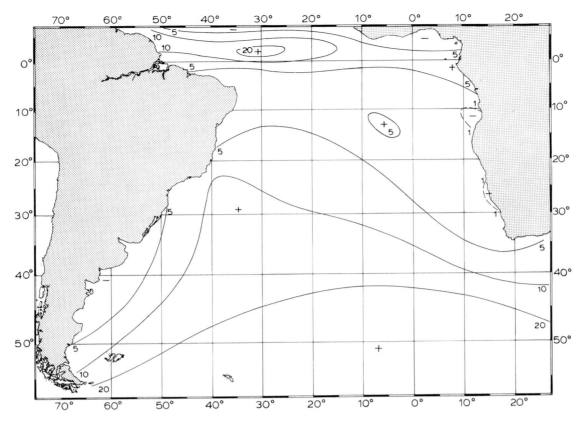

Chart 35. Frequency of precipitation (%), January.

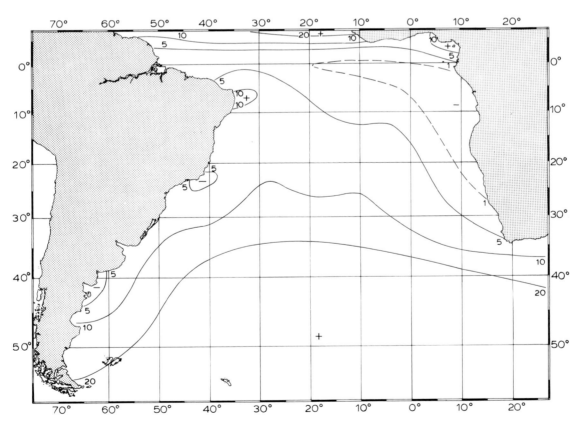

Chart 36. Same as 35, but for July.

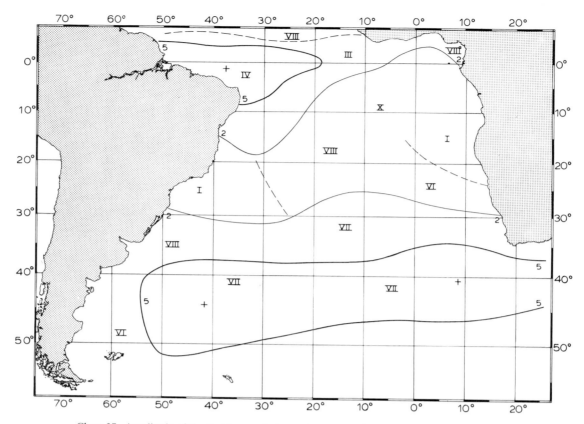

Chart 37. Amplitude of the first harmonic (annual component) and phase (month of maximum) of the precipitation frequency (%).

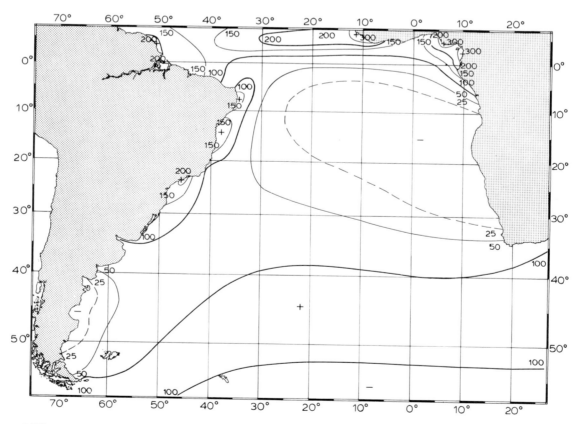

178   Chart 38. Estimated annual amount of precipitation (cm).

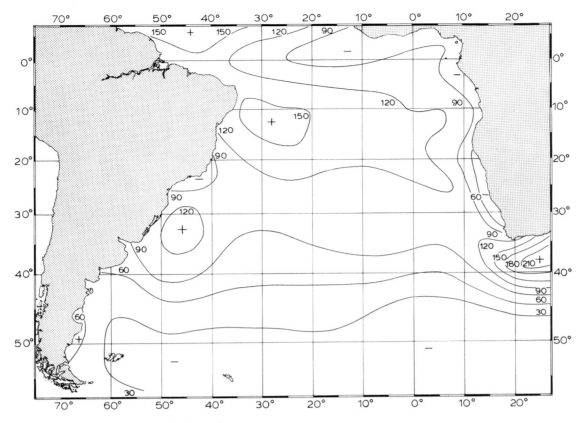

Chart 39. Evaporation (mm), January.

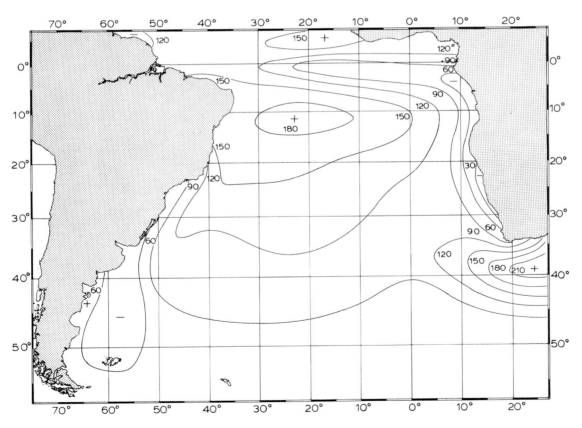

Chart 40. Same as 39, but for July.

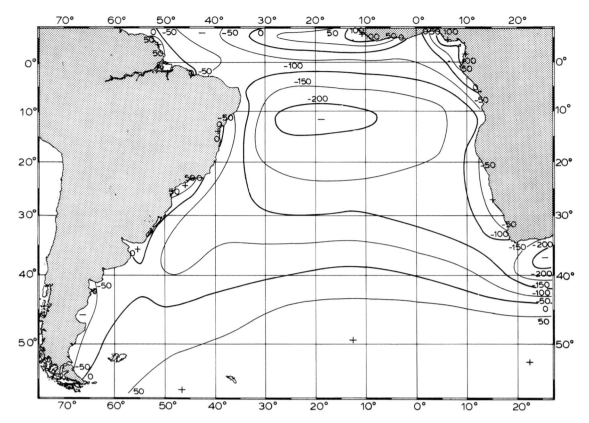

Chart 41. Annual water balance (cm).

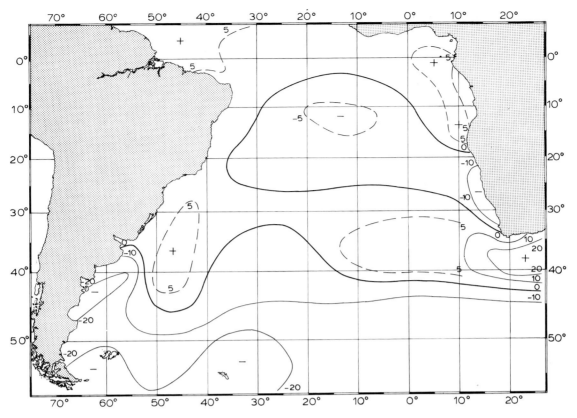

Chart 42. Sensible heat flux at the sea surface (W m^{-2}), January.

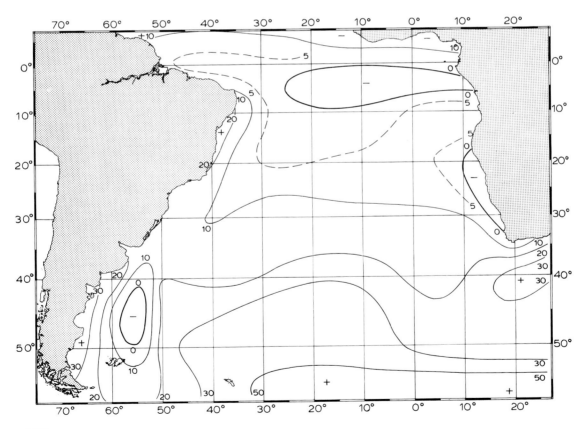

Chart 43. Same as 42, but for July.

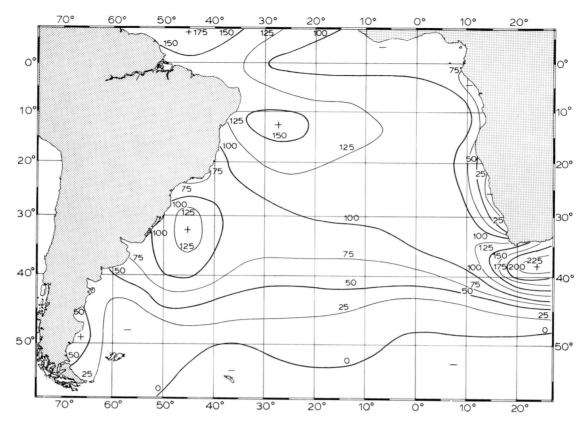

Chart 44. Total heat flux at the sea surface (W m^{-2}), January.

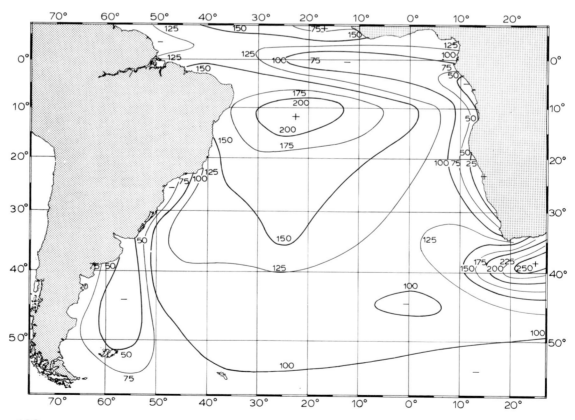

Chart 45. Same as 44, but for July.

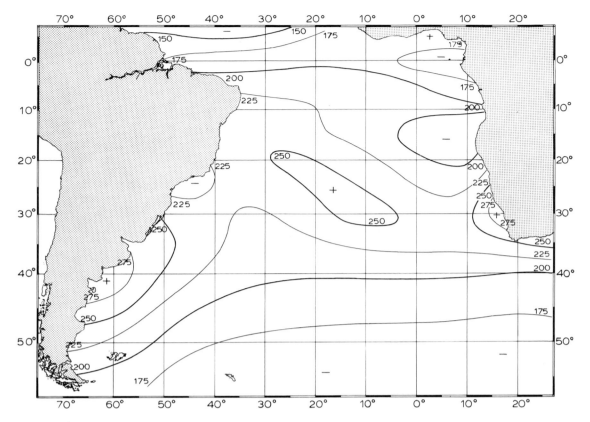

Chart 46. Insolation at the sea surface (W m^{-2}), January.

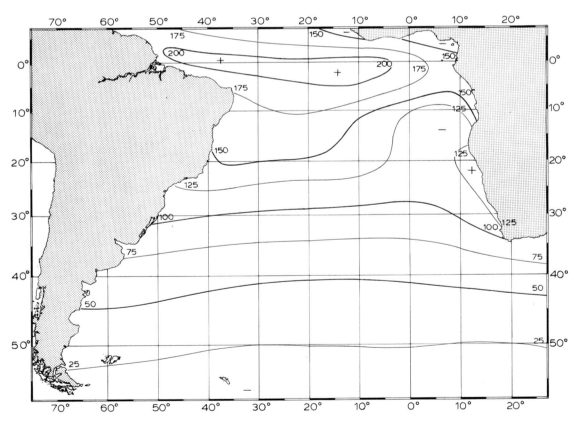

Chart 47. Same as 46, but for July.

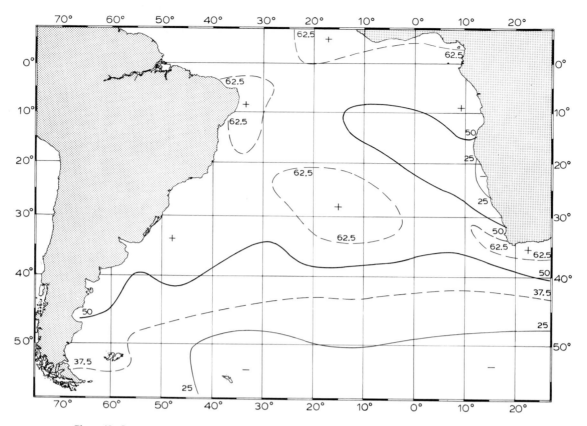

Chart 48. Long-wave net radiation at the sea surface (W m⁻²), January.

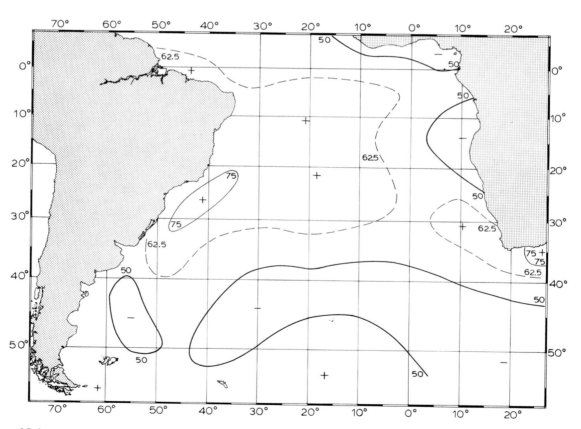

Chart 49. Same as 48, but for July.

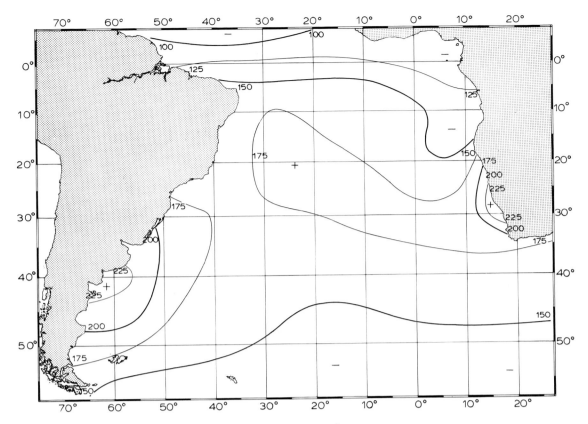

Chart 50. Radiation balance at the sea surface (W m^{-2}), January.

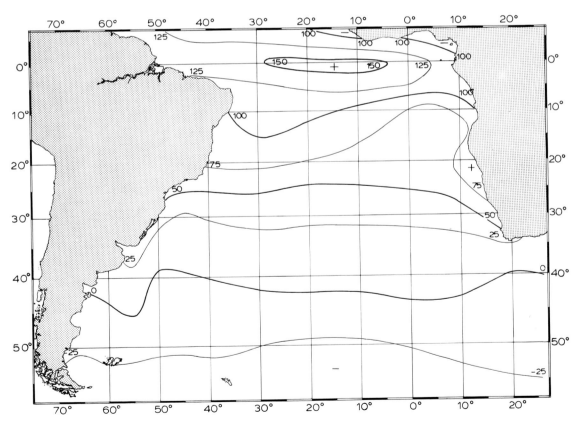

Chart 51. Same as 50, but for July.

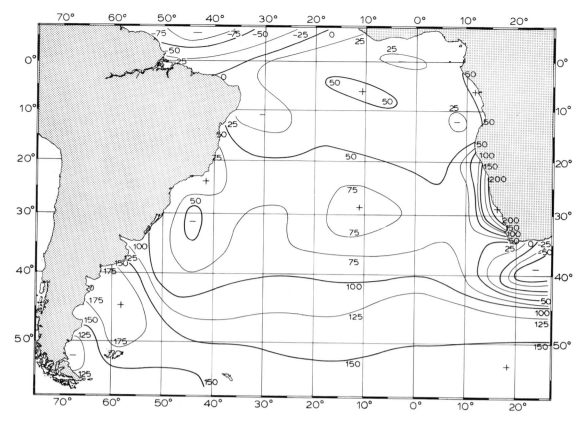

Chart 52. Heat balance at the sea surface (W m^{-2}), January.

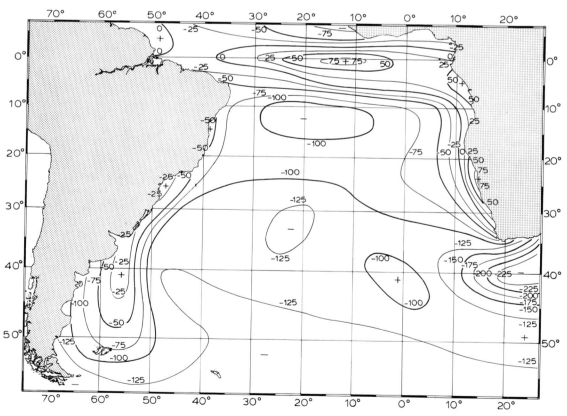

Chart 53. Same as 52, but for July.

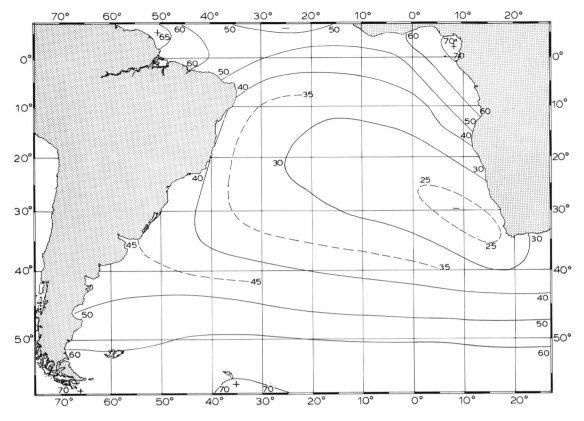

Chart 54. Relative humidity at 700 mbar (%), January.

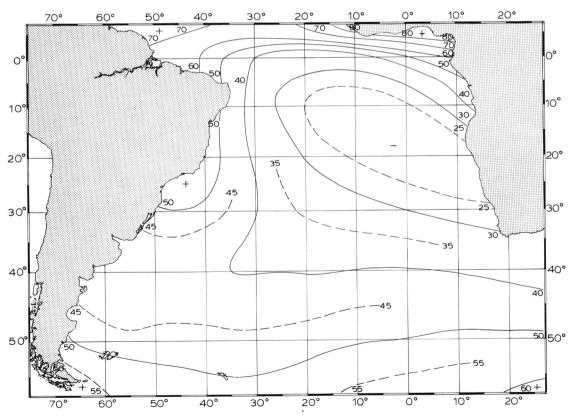

Chart 55. Same as 54, but for July.

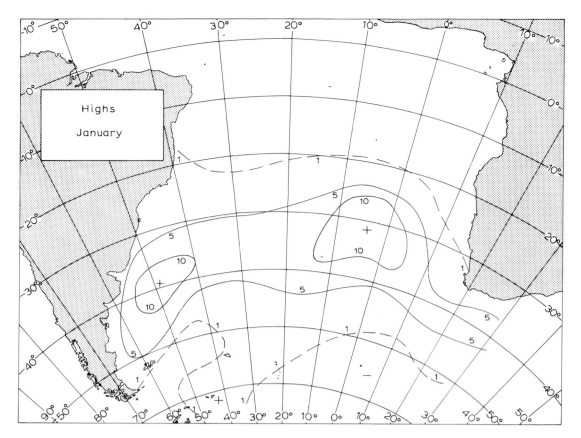

Chart 56. Average number of anticyclonic centres in January per 5° equatorial square, for 1960–1964.

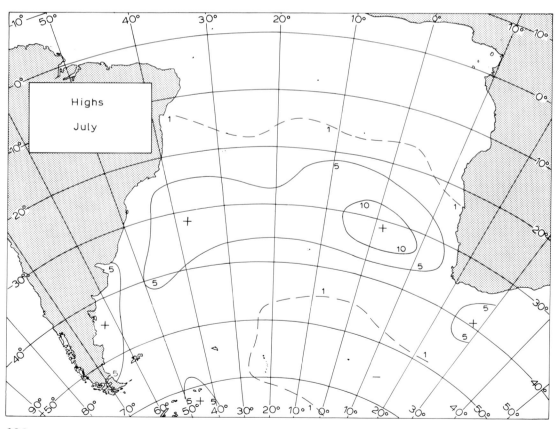

Chart 57. Same as 56, but for July.

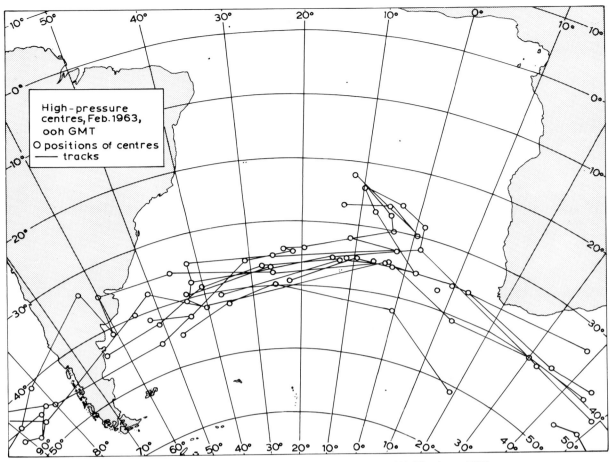

Chart 58. Daily positions and tracks of anticyclonic centres, February 1963 (Hamburg, 1971).

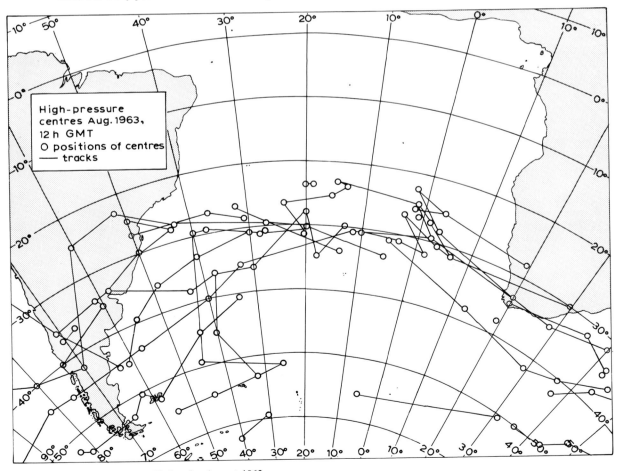

Chart 59. Same as 58, but for August 1963.

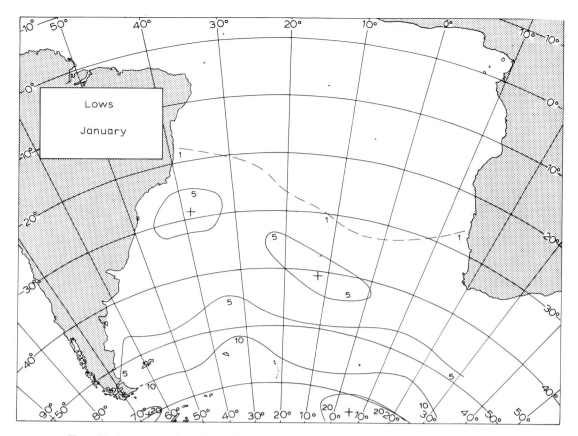

Chart 60. Average number of cyclonic centres in January per 5° equatorial square, for 1960—1964.

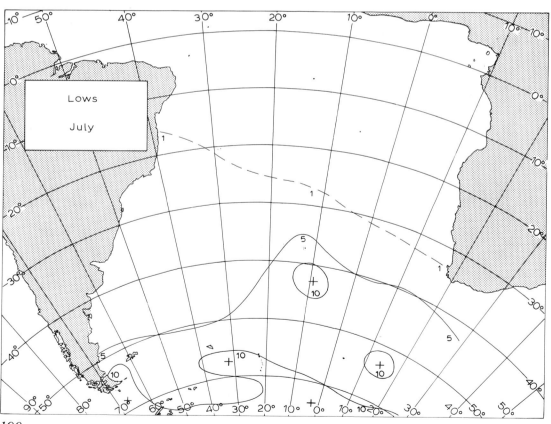

Chart 61. Same as 60, but for July.

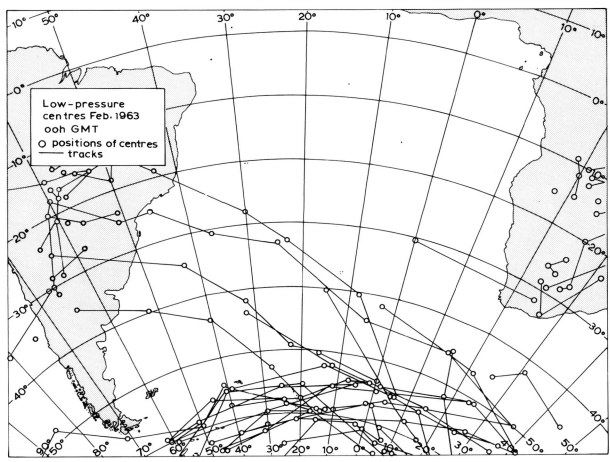

Chart 62. Daily positions and tracks of cyclone centres, February 1963 (Hamburg, 1971).

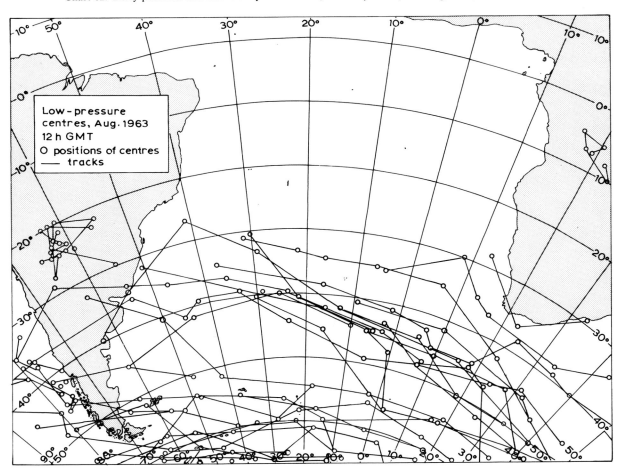

Chart 63. Same as 62, but for August 1963.

Chapter 2

Climate of the North Atlantic Ocean*1

G. B. TUCKER AND R. G. BARRY

Introduction

Main climatic features

Most elements used to describe tropospheric conditions in high and middle latitudes of the Northern Hemisphere show a marked longitudinal variation in addition to a gradient from Equator to Pole. This is due partly to the distribution of land and sea and partly to the occurrence of major orographic barriers in the path of the zonal circulation. The longitudinal variation is particularly noticeable in the 1000−500 mbar thickness pattern which is, effectively, a representation of the mean temperature of the lower half of the troposphere. In the Northern Hemisphere, thickness lines generally exhibit a pronounced sinusoidal pattern with a ridge over the North Atlantic Ocean in winter located to the east of a trough that is centred in the vicinity of the eastern seaboard of North America. The average monthly position and intensity of the trough vary more or less systematically throughout the year, being strongest and farthest west in winter and weakest and farthest east in summer.

Because the temperature gradient plays a major part in determining the tropospheric flow pattern, the general orientation of flow in the middle and upper troposphere over the North Atlantic north of about 30°N is from southwest to northeast. In these latitudes, thermal effects in the form of large-scale horizontal temperature gradients play an important part in the existence and behaviour of the major synoptic patterns that characterize the general circulation of the atmosphere. South of 30°N, horizontal temperature gradients are weak and global dynamical factors become more important in maintaining the surface subtropical high-pressure belt and the northeasterly surface trade winds to the south.

The climate of the North Atlantic Ocean is thus characterized by three main features. First is a highly turbulent regime north of about 40°N consisting mainly of migratory cyclones and anticyclones moving in an eastward or northeastward direction. A statistical result of these phenomena is the appearance on mean monthly surface pressure maps of a well-defined low-pressure area centred southwest of Iceland. Second is a relatively quiescent area between about 25°N and 40°N dominated by subsiding air. On mean monthly charts, this appears as a surface high-pressure region with a west-southwest east-northeast orientation. Third, south of the high-pressure belt is the trade wind region of northeasterly surface winds. This area is the source of tropical cyclones in late summer

---

*1 This chapter was originally written by G. B. Tucker and later updated by R. G. Barry.

and early autumn. In summer, the equatorial low-pressure (or trade wind) trough occurs well north of the Equator, and the northernmost parts of the southeasterly trade wind system of the Southern Hemisphere extend well into the North Atlantic.

The above features vary in position and intensity from month to month and their precise form is modified from day to day by transient, synoptic features. This complex, together with the associated variations in the meteorological elements, constitutes the "climate of the North Atlantic" described in this chapter.

Main oceanic features

The primary oceanographic feature of climatic significance is the spatial distribution of surface temperature and its seasonal variation. This distribution is determined, in part, by the pattern of surface currents which is, in turn, related to the surface wind stress. However, the horizontal gradients of sea surface temperature and their large-scale anomaly patterns may influence considerably the intensity and pattern of tropospheric air motion. Such feedback processes are discussed further on p. 233.

The patterns of sea surface temperature in January and July (Figs. 1 and 2) show lowest temperatures in the north and west with strong horizontal gradients over the western ocean at latitudes 35—45°N. The warm waters of the northeastern Atlantic in winter reflect the Gulf Stream — North Atlantic Drift system. The annual range of sea surface temperature is greatest off the east coast of North America in response to air—sea exchanges and ocean current systems (KIRK, 1953). It should be noted that these charts are not intended to be representative of coastal waters where, particularly in winter, strong gradients of sea surface temperature can occur (LUMB, 1961).

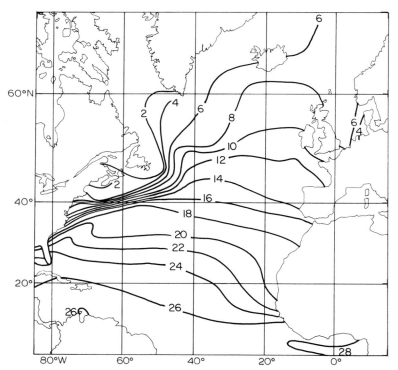

Fig. 1. Mean sea surface temperature (°C), January. Based on data for 1941—1972 (BUNKER, 1975).

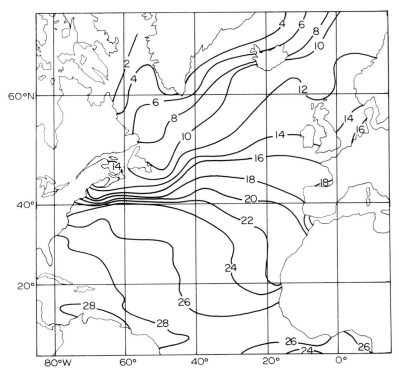

Fig. 2. Mean sea surface temperature (°C), July. Based on data for 1941–1972 (BUNKER, 1975).

The surface salinity averages about 35.5‰ in mid-latitudes, but 37.5‰ in the subtropics as a result of the high evaporation rates there. Exceptions occur near coasts where freshwater runoff reduces salinities to 32‰ or less. In higher latitudes, where the sea surface temperatures are below 5°C throughout the year, salinity differences have a significant effect on density. In lower and middle latitudes, surface heating intensifies the stable stratification by creating a warm, less dense layer at the surface; in high latitudes, freezing results in the addition of salt into the water immediately beneath the ice, making it sufficiently dense to sink.[1] It is this density factor that, in appropriate situations, permits the southward penetration of ice and cold, comparatively fresh meltwater over the warmer, more saline waters of the North Atlantic Drift and Irminger Current.

The vertical structure of temperature and density of the ocean's surface layer is important for seasonal changes of sea surface temperature. In summer in middle latitudes, there is typically a steep density and temperature gradient, referred to as the seasonal thermocline, at about 25–50 m (J. D. PERRY, 1968). Absorbed radiation can warm the shallow surface layer above the seasonal thermocline quite rapidly during fine, calm weather in spring and summer, but as the surface layer cools in winter the seasonal thermocline is removed and surface water can now sink to great depths; as it is replaced by warmer water from below, a limit is thus placed on the further cooling of the surface layer.

---

[1] At 0°C and 30‰ salinity, a salinity change of .0625‰ has the same effect on density as a temperature change of 1°C.

Sea ice, a further expression of the sea surface temperature conditions, is especially important because of its high surface reflectivity to solar radiation (albedo $\simeq 0.60$ compared with <0.10 for sea water; PAYNE, 1962). The maximum and minimum seasonal extent is illustrated in Fig. 3, but it is essential to note that these limits vary considerably between individual years and between decades. The ice extent in the western and northern North Atlantic, including the Greenland Sea and Barents Sea areas, decreased by 18% between 1901–1920 and 1921–1939, according to SCHELL (1956). In the 1960s, drift ice again became much more extensive, especially in the Denmark Strait and Davis Strait (MALMBERG, 1972; DUNBAR, 1972).

History of observations

The salient features of the surface wind pattern over much of the North Atlantic Ocean have been known to mariners for many centuries. The earliest attempt to synthesize these into a pictorial meteorological representation is Halley's map of the winds in intertropical regions, first published in 1688. The pattern of surface currents was also fairly well-known by this time; Benjamin Franklin noted the strong variation in sea surface temperature in the vicinity of the Gulf Stream, but proposed that it was basically a wind-driven current (BRYAN, 1963). Over a century later, MAURY (1857) suggested that the Gulf Stream was mainly density-driven. There is still some divergence of opinion on this point.

The first systematic collections of observations of winds and currents were begun in the 1830s by M. F. Maury of the United States Navy and by the British Admiralty. It was realized that the chance of getting more than one or two independent observations from the same small area of ocean surface in any one month was very low; hence, in 1831 William Marsden proposed dividing the ocean into a grid, thus affording a means of

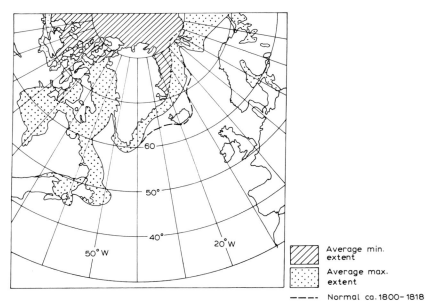

Fig. 3. Average spring maximum and autumn minimum extent of sea ice (SATER, 1969). Inferred normal spring extent ca. 1800–1818 is also shown (LAMB and JOHNSON, 1959).

grouping and averaging observations. These "Marsden squares" are bounded by the 10° meridians and parallels and are further subdivided into four 5°-quadrangles and, for detailed work, one hundred 1°-quadrangles. This system is still used in marine climatology.

By 1850, largely as a result of Maury's work, a number of American merchant and naval ships were making routine meteorological observations. The first international conference to discuss a system of observations at sea was held at Brussels in August 1853 when nine European countries agreed to collaborate with the United States. A register of observations was devised, and the quantities to be reported included position, time, pressure, currents, wind, precipitation type, state of sea, water (surface temperature, specific gravity, temperature at depth), state of weather, and miscellaneous remarks. It was also agreed that the table of wind force from 0 to 11 should accord with the specifications previously set out by Admiral Beaufort in 1805.

For nearly one hundred years after this first conference, meteorological observations at sea were made almost entirely by voluntary observers in merchant ships, apart from those made aboard the relatively small number of naval vessels. During 1925 – 1927, a German Atlantic expedition using the motor vessel *Meteor* made a series of invaluable measurements, including pilot balloon ascents, of meteorological conditions over tropical waters of the Atlantic Ocean (VON FICKER, 1936). A few years later, between 1936 and 1939, the possibilities of fitting up vessels as stationary meteorological ships were being investigated by several countries, and for a time both France and Germany operated a stationary vessel making surface and upper-air observations (by radiosonde) that were transmitted by radio to meteorological analysis centres. The 1939 – 1945 war caused a temporary break in these activities, although toward the end of the war several naval vessels were being used as stationary meteorological ships. During this period, meteorological reconnaissance flights by aircraft over the North Atlantic were most valuable. When the special meteorological ships were withdrawn soon after the end of the war, marine meteorological observations were available once again only from merchant vessels. It was soon obvious that routine upper-air observations would be required for trans-Atlantic air routes, and it was agreed by the United States, Canada, and European countries adjacent to the North Atlantic that stationary meteorological vessels should be established there. The signatories to the "Ocean Weather Ship" (O.W.S.) agreement in 1946 decided that thirteen ocean weather stations should be set up by July 1947.

The advent of O.W.S. enabled a standard framework of regular observations to be established over the North Atlantic Ocean. These ships proved to be very expensive items in national meteorological budgets and, for reasons of economy, their number was reduced to ten in 1949. It was further decreased to nine in 1954 (Fig. 4) and the position of one ship was changed. In 1972, the United States permanently withdrew ships from stations A, B, C, D, and E, but a revised network was established in July 1975 with stations C and M reinstated and new stations L (57°N 20°W) and R (47°N 17°W).

Although the observations provided by these quasi-stationary ships are of primary importance, the main difference between marine and land climatology is that on land data generally consist of a regular series of homogeneous long-term observations made at regular times and at fixed locations. Most oceanic observations are, of necessity, het-

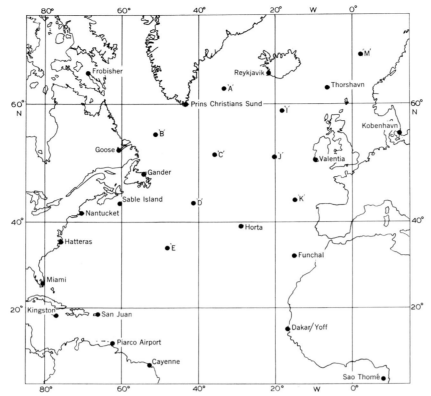

Fig. 4. Ocean Weather Ship stations and place names referred to in the text.

erogeneous. Marine climatology is based to a great extent on a large number of samples of weather isolated in space and time. These samples are not random because there is a strong correlation between consecutive observations from the same ship. To select the data according to the position of the ship would reduce considerably the number of observations and there are already too few over many parts of the ocean. Numbers of observations per month can range from several hundreds in a Marsden square on a main shipping route to nil in parts of the northernmost North Atlantic. Fortunately, the effect of this inhomogeneity is minimized by the absence of strong local differences over the sea (except near a coastline). This has been demonstrated quantitatively by BROCKS (1963) who found a very high correlation between simultaneous surface measurements of wind speed, humidity, and air temperature at two points over the North Atlantic at least 50 km apart.

Comparison of observations from merchant and weather ships shows that reports by merchant ships are 32% low for precipitation occurrence and 5% low for median wind speeds (QUAYLE, 1974). Using O.W.S. stations D and E, BUNKER (1975) finds that sea surface and dewpoint temperatures by merchant ships were measured slightly higher, and wind speeds and cloud cover slightly lower. Wind speed observations at sea based on the Beaufort scale (judged by state of the sea) are apparently no less reliable than actual shipboard measurements (VERPLOEGH, 1967). However, TAUBER (1969) reports a difference of 0.4° to 2°C between standard bucket or intake measurements of sea surface temperature in the uppermost metre and determinations by radiation thermometers. The latter demonstrates the occurrence of a 1−2 cm film of cold surface water.

Climatic change

Because "climate" is a set of long-term statistical parameters of the atmosphere, before describing climate it is desirable to consider to what extent the time series is stationary during the period of study. That is, is the region undergoing a "climatic" change? Attempts to deduce systematic trends in marine climatic data must be treated with reserve because of the difficulty in obtaining for a given area a sufficient number of observations that can be considered representative of a relatively short period.

One of the most detailed and painstaking analyses of climatic change over an area including the North Atlantic is that of LAMB and JOHNSON (1959, 1961, 1966). They argue that the surface pressure map is the best tool for reflecting the integrated effect of atmospheric motions, and they have produced a series of 10-year mean maps of sea level pressure for January and July extending back to 1750—1759. The earliest four decades extend over only the eastern North Atlantic Ocean, but from 1790—1799 onward most of the ocean is covered. The information derived from this series of charts has been combined with the results of ancillary studies of air temperature, sea surface temperature, ice, etc. to produce a comprehensive account of apparent climatic variation and observed changes in the general circulation. These interpretations must be regarded as somewhat tentative. The general features of the mean surface flow patterns are depicted by decadal mean maps, but monthly and interannual variations are also important in determining the regional climate.

In January over the region of prevailing westerlies, there is evidence for two main periods of relatively weak westerly circulation during the early part of the 19th century and again in the latter half. Two periods of strong westerly circulation occurred with maxima in the middle of the 19th and the beginning of the 20th centuries. In July, the decadal variations in the circulation index obtained from these maps are much smaller than in January and there is only slight correspondence between the two. Both show maximum circulation indices in the early 20th century.

Lamb and Johnson suggest that during the period of weaker circulation the Gulf Stream pursued a more southerly course with a much greater penetration eastward toward the Azores. They present maps of the difference in sea surface temperature between the periods 1780—1820, 1887—1899, and 1921—1938 for both January and July. Most of the North Atlantic north of 50°N was colder in the earlier period and there appears to have been a much greater temperature gradient across the Gulf Stream south of Newfoundland. The maximum difference in the Labrador Current (apparently broader than now) is −3°C while the maximum difference in the Gulf Stream reaches + 3°C (warmer in the earlier period). It is argued that the colder waters in the tropical North Atlantic (some 1°C colder in the earlier period) were probably due to a southward displacement of the North Equatorial Current, less warm water thus being diverted into the Caribbean and North Atlantic. The difficulty here appears to be to reconcile the lower temperatures of the water in the tropics with the much higher temperature of the Gulf Stream in middle latitudes.

Lamb and Johnson report that the annual mean sea surface temperature of the tropical North Atlantic fell about 1°C between 1880 and 1910. This appears to coincide with a marked decrease in tropical rainfall (KRAUS, 1955). The sea surface temperature minimum from 1910 to 1925 was followed by a rise of some 2°C in the next 20 years. This

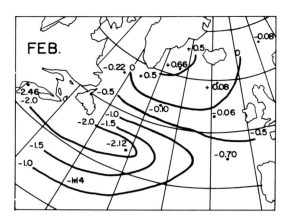

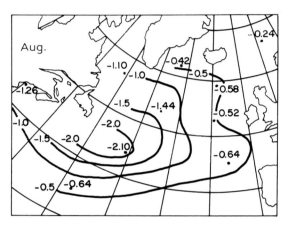

Fig. 5. Change of sea surface temperature between 1951–1955 and 1968–1972 for February and August. (RODEWALD, 1973).

relatively sharp rise in tropical sea surface temperatures from 1920–1940 is also a feature of the northernmost North Atlantic (BROWN, 1953, 1963). However, a significant cooling trend between 1951–1955 and 1968–1972 has also been documented (RODE-WALD, 1972, 1973; PERRY, 1974). This amounted to −0.85°C for all nine weather ships and −2.1°C at O.W.S. D in August (Fig. 5). There was a similar cooling in February at D but a slight warming in the Irminger Sea.

Since the 1930s, there has been a decreasing zonality of circulation over the North Atlantic as part of a global weakening of atmospheric circulation (DICKSON and LAMB, 1972). This has been accompanied by more frequent spells of northerly airflow over the European and eastern North American sectors, particularly in winter. The greater meridionality of atmospheric circulation in the 1960s brought about increased southward transport of polar water and ice in the Norwegian–Greenland Sea. In the late 1960s, this gave rise to record ice extent off northern Iceland in spring and early summer, with corresponding changes in the distribution and abundance of marine organisms (MALM-BERG, 1972; DICKSON and LAMB, 1972). However, the most recent data show that these conditions have undergone a reversal in the 1970s (DICKSON et al., 1975).

Representative data

Climatological statistics are determined over a finite interval of time which, because of the availability of observations, is usually less than 50 years. As discussed in the last section, there are apparently significant variations in climate on the scale of decades, so that maps based on observations extending over different periods may be expected to give different results. However, the variations in time are generally very much smaller than the variations in space and, consequently, the overall features of the patterns on climatic maps are not critically dependent upon the period chosen. This is important in marine climatology because it means that heterogeneous data, based on ship observations collected over extended and often irregular periods, nevertheless result in usable climatological maps.

Two major marine climatic atlases covering the North Atlantic are: (*1*) United States Navy's *Marine Climatic Atlas of the World,* Vol. 1, (U.S. WEATHER BUREAU, 1955), and (*2*) British Meteorological Office's *Monthly Meteorological Charts of the Atlantic Ocean* (METEOROLOGICAL OFFICE, 1948).

There are differences between data used in compiling these atlases, although there is considerable overlap. The earlier British atlas is based mainly on observations from British ships and these are limited to the period 1888–1936. For the United States Navy atlas, observations from ships of many nations were grouped together for a period from 1856 to 1953, and use was also made of the first few years of data from the relatively fixed O.W.S. locations.

Comparison of the maps in these two atlases shows them to be very similar. Table I gives the January and July mean monthly values of pressure, air temperature, and sea surface temperature taken from both atlases for the positions occupied by O.W.S. A and E. Also included are the 10-year mean values (1951–1960) for these two O.W.S. (PFLUGBEIL and STEINBORN, 1963). The pressures given by the two atlases are very similar and the temperatures do not differ by more than 1°C. The 10-year mean values subsequently derived for the O.W.S. are also similar to the atlas values with the exception of the January pressures. A decrease in the January surface pressure gradient appears to have occurred between the period represented by the atlases and the 1950s. The

TABLE I

COMPARISON OF MONTHLY VALUES TAKEN FROM THE TWO ATLASES AND THE 10-YEAR MEAN AT OCEAN WEATHER SHIPS A AND E

	Pressure (mbar)		Temperature (°C)		Sea surface temperature (°C)	
	A	E	A	E	A	E
January						
M.O. Atlas		1021		18		18
U.S. Atlas	994	1021	1	17	6	19
O.W.S. 1951–1960	999.0	1016.8	2.1	17.8	5.2	19.3
July						
M.O. Atlas	1008	1025	8	25	8	25
U.S. Atlas	1009	1025	9	25	9	25
O.W.S. 1951–1960	1010.6	1024.9	9.4	25.0	9.9	25.1

TABLE II

MEAN MONTHLY SURFACE PRESSURES FOR JANUARY FROM 1951 TO 1960

Year	1951	1952	1953	1954	1955	1956	1957	1958	1959	1960
O.W.S. A	989.8	996.5	999.0	1000.0	1004.0	1003.3	983.6	998.7	1012.5	1003.1
O.W.S. E	1021.9	1021.7	1022.5	1022.3	1004.9	1014.2	1023.2	1010.4	1015.3	1012.0
Pressure diff.	32.1	25.2	23.5	22.3	0.9	10.9	39.6	11.7	2.8	8.9

surface pressures at O.W.S. A and E are given in Table II for individual Januaries between 1951 and 1960. There is a decrease in pressure difference between these stations during this period, the average values for 1951—1953 being quite close to the values taken from the atlases.

Charts from both atlases are used in this chapter, preference being given to the United States Navy atlas where maps of the same element exist.

Arithmetic mean temporal values give only a partial and often misleading idea of the actual climate of an area. In addition to mean charts, some way of representing the variability of the atmosphere is required. The need to specify this variability was recognized in the preparation of the atlases; maps of the "range" (difference between maximum and minimum observed values) for some elements are given in the United Kingdom atlas, and the United States atlas includes cumulative frequencies. The frequency distribution of many of these quantities is, to a good approximation, close to normal. Thus, the standard deviations of pressure, temperature, and humidity are useful climatological parameters. A notable exception to this is rainfall where the distribution over land and, presumably, also over the sea is markedly skew. The standard deviations of individual observations of some elements about the monthly mean have been calculated from selected British ship data for 1951—1960 and are presented in this chapter.

Use of January and July maps

In temperate latitudes, it is customary to use January and July mean monthly maps when representing winter and summer patterns of various meteorological phenomena, but these should not be interpreted as representing the extremes in the annual cycle. In lower latitudes of the North Atlantic, the annual variation of mean monthly pressure takes the form of a double cycle, maxima in about January and July and minima in about April and October (see, for example, data for Horta and San Juan, Appendix, p. 258f). In high latitudes, the lowest pressures tend to occur in December and the highest in spring or early summer.

The annual temperature cycle also has maxima and minima that occur later than the "mid-season" months. For each O.W.S., the mean monthly surface air temperature maximum is in August (Table XIV). The minimum is in January in high latitudes and February in middle latitudes (March for O.W.S. D). There is a uniformity in the mean monthly sea surface maximum temperature that is also evident in August at each O.W.S.

However, the minimum sea surface temperature tends to occur later than the minimum air temperature—in February or March, or even in April for O.W.S. D (PRESCOTT and COLLINS, 1951). Maximum vapour pressures occur in August and minimum values in January (high latitudes) and February or March (middle and low latitudes).

Although it follows from the above paragraphs that extreme monthly values of surface elements do not generally occur in January and July, nevertheless these months are characteristic of the extreme seasons. This is particularly true of elements representing the activity of the atmosphere; for example, January is the most stormy month (having the highest frequency of gales) in middle and upper latitudes and July is the least stormy. Therefore, to conform with other chapters in this volume, maps for January and July will illustrate the spatial variation of climate.

Mean circulation patterns

Surface pressure field

The general features of the tropospheric circulation over the North Atlantic Ocean have already been outlined. There is a tripartite zonation of subpolar low pressure, subtropical high pressure, and equatorial low-pressure trough. The January map of mean sea level pressure (Fig. 6) is not quite representative of the winter extreme. The lowest value

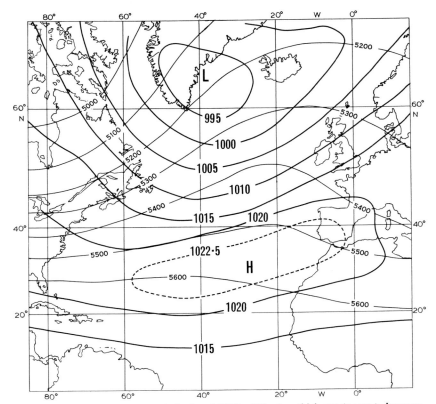

Fig. 6. Sea level mean pressure (mbar) and 1000 − 500 mbar thickness (g.p.m.), January.

of the Icelandic low-pressure area occurs in February; this increase in intensity toward the end of the winter is accompanied by a westward development. At the same time, there is a slight decrease in the intensity of the subtropical high-pressure belt from January to February. These changes are associated with a corresponding decrease in the strength of the southwesterly airflow across the central part of the ocean.

The high latitude low-pressure area decreases in intensity throughout spring and by April exists as a broad belt over the northern part of the ocean. There is little change in the high-pressure belt and the geostrophic flow pattern is weaker, becoming more westerly (especially over the western ocean) than southwesterly.

As summer approaches, the subtropical high-pressure area extends northward and increases in intensity until in July (Fig. 7) the highest pressure (in excess of 1027 mbar) and northernmost position are attained. In this month, the main low-pressure area, now centred in Davis Strait, has reached its extreme western position.

Throughout autumn, the subtropical high becomes weaker and the subpolar low deepens as its centre moves farther east. In October, two distinct low-pressure centres exist—one off the south coast of Iceland and another off the west coast of Greenland, reflecting the two main storm tracks and the positions of the deepest synoptic low-pressure systems. The weakest high-pressure belt (1023 mbar) occurs in November, although it is still a major feature of the mean monthly pressure pattern (CHASE, 1956). By mid-winter, only one low-pressure area appears on the mean monthly chart. This is a result of deep depressions at this time affecting the area east of the tip of Greenland

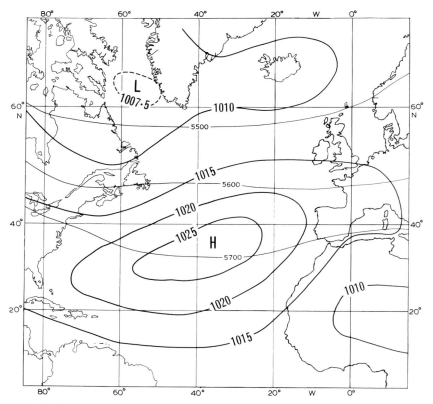

Fig. 7. Sea level pressure (mbar) and 1000–500 mbar thickness (g.p.m.), July.

and generally travelling into the northeastern ocean region; some move into the Davis Strait (Fig. 8).

The annual variation in latitude of the subtropical high-pressure area and the zone of interaction between the northeasterly and southeasterly trades is given in Table III.

To get a more complete climatological picture of the pressure distribution, it is necessary to know the extent to which day-to-day pressure variations differ from the monthly mean value and also the variability of monthly values from year to year. SCHUMANN and VAN ROOY (1951) have produced figures for both types of variation. Figs. 8 and 9 give the standard deviation of daily pressure (in mbar) about the monthly mean for January and July. These can be interpreted as representing the amount of synoptic activity throughout the area. For middle and high latitudes, much higher values exist in January than in July, reflecting the greater frequency and intensity of depressions in winter than in summer. The primary and secondary tracks of storms given in the United States Weather Bureau marine atlas for these months have been included on the maps. How these tracks pass through the areas of large pressure variability is evidence of the consistency between the two publications.

The standard deviations of monthly mean pressures about the "normal" for January and July (Figs. 10, 11) are a measure of the larger scale variations of climate. They provide a quantitative assessment of the year-to-year variability of monthly surface

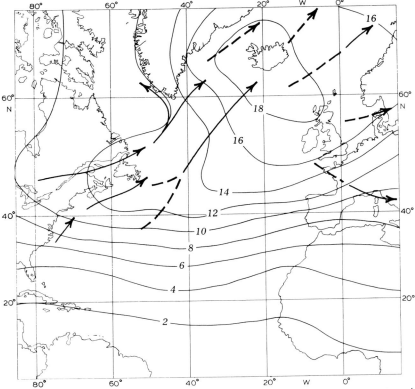

Fig. 8. Standard deviation of daily pressure (mbar) about the monthly mean for January (SCHUMANN and VAN ROOY, 1951). Storm tracks are indicated by lines terminating in arrowheads. Solid lines are primary tracks along which there has been maximum concentration of individual storm centre paths. Dotted lines are secondary tracks along which there has been moderate concentration of individual storm centre paths. (Data from U.S. Marine Atlas).

TABLE III

MEAN LATITUDES OF AXES OF DOLDRUMS AND SUBTROPICAL ANTICYCLONES IN THE NORTH ATLANTIC
(GORDON, 1951)

	Jan.	Feb.	Mar.	Apr.	May	June	July	Aug.	Sept.	Oct.	Nov.	Dec.
Northern anticyclone	32°	30°	28°	32°	33°	32°	33°	36°	36°	33°	36°	30°
Doldrums	2°N	1°S	1°N	4°N	8°N	12°N	12°N	12°N	12°N	12°N	12°N	4°N

pressures. Apart from the obvious seasonal and latitude differences, an interesting feature is the similarity between these patterns and those of the daily variations (Figs. 8, 9), the latter being larger in middle and high latitudes by factors of 2 to 2.5. In terms of surface pressure, the atmosphere over the extratropical North Atlantic at a given latitude is more variable on both a daily and seasonal scale than over the continents to the east or west.

Winds

The most obvious feature of the fields of monthly resultant winds (Figs. 12, 13) is the existence of the low-latitude belt of strong easterlies and northeasterlies throughout the

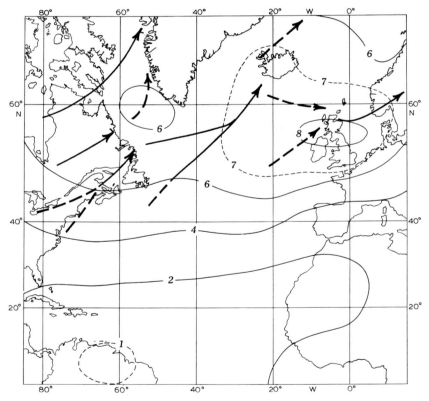

Fig. 9. Standard deviation of daily pressure (mbar) about the monthly mean for July (SCHUMANN and VAN ROOY, 1951). Storm tracks are indicated by lines terminating in arrowheads. Solid lines are primary tracks along which there has been maximum concentration of individual storm centre paths. Dotted lines are secondary tracks along which there has been moderate concentration of individual storm centre paths. (Data from U.S. Marine Atlas).

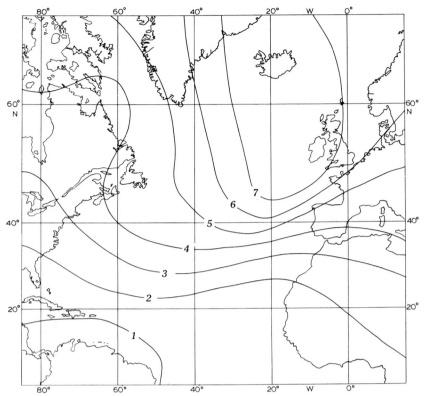

Fig. 10. Standard deviation of monthly mean surface pressure, January (mbar).

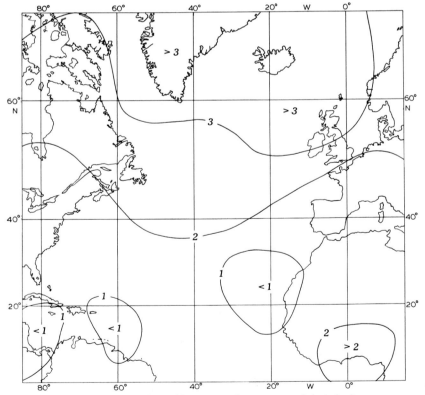

Fig. 11. Standard deviation of monthly mean surface pressure, July (mbar).

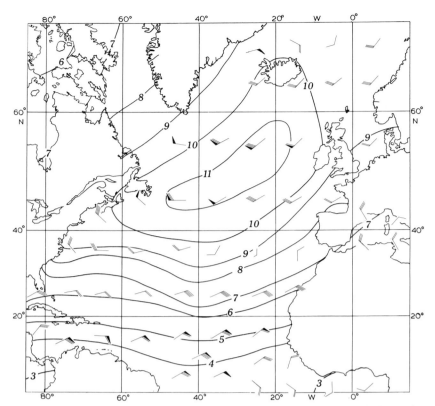

Fig. 12. Surface resultant wind vectors and isopleths of standard vector deviation at the top of the friction layer (m s^{-1}), January. A barb on the vector represents 1 m s^{-1}, a solid triangle 5 m s^{-1}. Isopleths after JENKINSON (1956).

year. North of the subtropical high-pressure zone, the strong westerlies in January are replaced by much lighter winds in July.

SHEPPARD (1954) has reported some work on the relation between actual and geostrophic mean winds at sea level which were compiled by A. H. Gordon. From an analysis of ship observations in the North Atlantic and South Atlantic over a long period, mean values of geostrophic and actual winds for all months and the angular departure α between them were obtained, α being positive for the geostrophic veered on the actual wind. Because there were no clearly defined effects of season or ocean, Sheppard combined all months at each latitude and included South Atlantic values for January and July (Table IV). He noted that in the zone of maximum baroclinicity in the North Atlantic Ocean, where analyses are most reliable, α = −1.3°; that is, the actual wind blows away from the low at 1.3°. This was interpreted as the effect of a strong downward transfer of momentum in the atmosphere. At low latitudes, the opposite was observed with values of α from 20° to 40° at 15° to 5° of latitude.

The scatter of instantaneous daily vector winds about the seasonal resultant wind is close to a normal bi-variate distribution. Hence, the standard vector deviation[1] is a good measure of the daily variability of wind. JENKINSON (1956) demonstrated the re-

[1] The standard vector deviation σ is given by σ^2 = V^2_R − $\overline{V^2_S}$, where V_R is the resultant magnitude, V_S is an individual scalar value, and the overbar denotes the time average over a period.

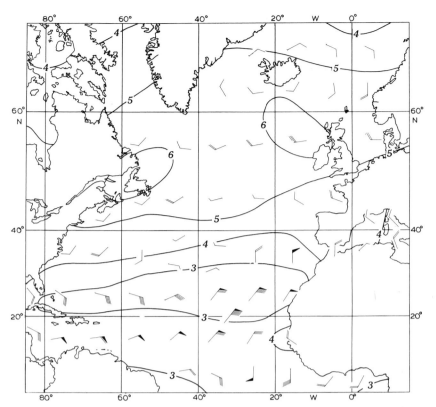

Fig. 13. Surface resultant wind vectors and isopleths of standard vector deviation at the top of the friction layer (m s^{-1}), July. A barb on the vector represents 1 m s^{-1}, a solid triangle 5 m s^{-1}. Isopleths after JENKINSON (1956).

lation between the standard vector deviation of wind and the standard deviation of daily values of 1000 mbar height contours. Using this, he produced January and July charts with isopleths of the standard vector deviation of wind at the top of the friction layer. These isopleths have been included in Figs. 12 and 13 on the assumption that surface wind variability is two-thirds that calculated by Jenkinson. Although there is an important latitude factor to be considered, one would expect that the correspondence between these patterns and those of standard deviation of daily surface pressure (cf. Figs. 8, 9) should have been greater; the differences may be in the sampling. It is obvious, however, that the wind variability is quite unrelated to wind speed; the southwesterlies

TABLE IV

ANGULAR DEPARTURE α OF SURFACE RESULTANT WIND V FROM THE SURFACE GEOSTROPHIC WIND J AS A FUNCTION OF LATITUDE OVER THE ATLANTIC OCEAN*
(SHEPPARD, 1954)

	Mean latitude (degrees):					
	5	15	25	35	45	55
α(degrees)	43	24	10	2	0	2
V(m s^{-1})	5.3	5.6	3.7	2.7	4.1	4.1
J(m s^{-1})	13.8	8.9	4.2	4.7	5.8	5.1

* $\alpha > 0$ for J veered on V.

are roughly twice as variable (using the standard vector deviation as an index) as the trades, even in summer when the trades are very much stronger.

An important factor in the dynamics of both atmosphere and ocean is the wind stress field over the ocean surface. Uncertainties in the climatological estimation of this field lie not in the description of the surface wind field and its variation but rather in the drag coefficient and the form of its variation with wind speed and stability. HELLERMAN (1965) has shown that, depending on which determinations of this quantity are used, the forcing function for wind-driven currents may vary by as much as a factor of two. The net result of this torque over the North Atlantic (and other extra-polar oceans) was computed by PRIESTLEY (1951). In general terms, the stress distribution over the North Atlantic shows negative (easterly) stress in high latitudes with greatest magnitudes, in excess of 0.3 N m^{-2} at 50°N in the western North Atlantic in January. Zero stress occurs at about 30°N in all seasons with positive (westerly) stress in lower latitudes approaching 0.1 N m^{-2} at 15−20°N in all seasons (BUNKER, 1975). This pattern is very similar to, if not somewhat stronger than, that over the Pacific. Also, the maximum easterly stress (westerly winds) is found farther north in the Atlantic, reflecting the more northerly position of the Icelandic as compared with the Aleutian low-pressure area. The zone of low stress separating northeasterly from southeasterly trades is well-marked.

Synoptic climatology

Annual cycle of synoptic events

Studies of climate are often based on mean monthly data, but this subdivision into calendar periods is merely for convenience; the limitations of such a representation of the annual weather cycle are obvious. The period over which a synoptic situation exists in a given area in middle and upper latitudes is highly variable, and shorter-period variations and mid-month changes of synoptic type will be blurred by monthly averaging. Much of the transport of various quantities important in the general circulation, and thus in determining climate, is accomplished by mechanisms with time scales of appreciably less than one month. Thus, the characteristics of most of these synoptic phenomena cannot be inferred from the mean monthly patterns. In an attempt to obviate this difficulty, the ensuing discussion of the normal seasonal trend of weather is based on a survey of day-to-day synoptic charts, a study by LAMB (1973) of average 5-day surface pressure variations for 1919−1938, and information in the bi-monthly *Mariners Weather Log* (published by the Environmental Data and Information Service, National Oceanic and Atmospheric Administration, Washington, D.C.).

Winter

The maximum intensity of the atmospheric circulation in the North Atlantic region occurs during mid-winter, although the westerlies do not attain their lowest latitudes until February−March. The most intense surface pressure gradients appear in early to mid-January, and the lowest surface air temperatures occur about three or four weeks later. At this time of year, the source regions of very cold air over the northern continents are well-established, whereas the warmer areas of the ocean surface do not achieve their annual minima of temperature until several weeks later. This is evident by comparing the mean monthly values of earth temperatures at Resolute Bay, Canada, and sea surface temperatures at O.W.S. E (Table V); the minimum at sea occurs one month later than on land.

The strongest thickness gradients also occur in late January and are situated close to the eastern coast of the United States. This position coincides with the zone of maximum sea surface temperature gradient over the western North Atlantic Ocean (Fig. 1) and with the area of maximum cyclonic development as shown by indices of cyclogenesis (KLEIN, 1957) and frontal frequency (WHITING, 1959; REED, 1960). This cyclogenetic area exists in all months and has an extension into the mid-Atlantic which is strongest in December when it links with another such zone off Greenland (KLEIN, 1957). The frequent occurrence of a strong baroclinic zone near the east coast of North America and the corresponding existence of cold air over the continent result in the Canadian cold trough being a well-marked feature of the January mean monthly thickness pattern (Fig. 6). This period of strong circulation is often associated with a long wavelength pattern in the upper westerlies. Depressions forming over the western ocean are frequently very large and deep and maintain vigorous development as they move northeastward, bringing milder air and storminess to the northeastern parts of the ocean. Occasionally, depressions that have moved across the southeastern United States are regenerated off Cape Hatteras. Although some depressions, particularly those that have moved across Canada, continue northward into Baffin Bay, most pass northeastward over the area southwest of Iceland (Fig. 8). This results in the "Icelandic low" in the January mean monthly surface pressure chart (Fig. 6). The speed of movement of depressions in winter is greater than in other seasons; over the central ocean, about 70% of surface depressions move at speeds exceeding 12 m s^{-1}. The incidence and mobility of deep depressions in the North Atlantic reach a maximum in February; this is reflected in the very high January values of the standard deviation of surface pressure about the monthly mean value (Fig. 8).

Stormy weather over the middle and northern area is associated with wave heights greater than 3.6 m occurring 30% of the time during December, January, and February over most of the ocean between 50 and 60°N. A maximum frequency of 60% occurs in a small area south of Iceland.

Gales are rare in the low latitudes, although strong northerly outbreaks of cold air sometimes occur over the Gulf of Mexico. South of about 30°N, the northeast trades are very constant, achieving 3 — 5 Beaufort more than 70% of the time. Between 1886 and 1965, less than ten tropical storms appear on the charts during the three winter months.

Spring

As spring approaches, the large cold anticyclone, which in winter often forms in situ over northern Canada, occurs more frequently in the vicinity of the Queen Elizabeth Islands and northern Greenland. Consequently, easterly winds are more frequent in extreme northerly latitudes; the mean pressure patterns show a high pressure area over

TABLE V

MEAN MONTHLY VALUES OF EARTH TEMPERATURE (AT 10 CM IN °C) AT RESOLUTE BAY, CANADA (74°43′N 94°59′W) AND SEA SURFACE TEMPERATURE AT OCEAN WEATHER SHIP E (35°00′N 48°00′W)

	Jan.	Feb.	Mar.	Apr.	May	June	July	Aug.	Sept.	Oct.	Nov.	Dec.
Resolute (1951–1955)	−26.8	−29.9	−28.6	−23.1	−14.6	−2.5	+5.6	+4.4	−0.8	−9.3	−19.1	−23.4
O.W.S. E (1951–1960)	19.3	18.6	17.6	18.4	19.6	21.8	25.1	26.6	26.1	24.2	22.5	21.0

the Canadian Arctic Archipelago in March and early April; this high-pressure area in late April and May moves closer to the Pole. In late spring (April – May), slow-moving depressions often exist close to Newfoundland and warm air is advected to very high latitudes. This corresponds to the time of most frequent blocking action over the eastern North Atlantic (see p. 216). Gales are notably less frequent and storm centres generally move at speeds less than 12 m s⁻¹. Depressions continue to develop off the east coast of the United States but are less intense. These quieter conditions are associated with a notable increase in frequency of low visibility between March and April, particularly over the Grand Banks south and east of Newfoundland.

In April, the trade winds extend to 30°N in the eastern North Atlantic Ocean. During March and April, the frequency of tropical cyclones is the lowest for the year. In the 77 years prior to 1965, only one tropical cyclone was reported in April anywhere in the North Atlantic and none was reported in March. In May, the number of tropical cyclones begins to increase, ten being reported between 1886 and 1965.

Summer

Over the central ocean, the minimum surface circulation intensity normally occurs in late spring or early summer, being somewhat later in more northerly latitudes. About mid-May, the maximum surface pressure gradient for westerly winds in the mid-Atlantic is only about one-quarter of its winter value. The frequency of fog approaches its peak, especially in the area from the Grand Banks to Greenland; in July visibilities of less than 8 km are reported here on more than 40% of occasions. This fog is usually observed in warm moist air brought northward into the region of low sea surface temperatures. June and July have the most intensive fog formation.

There is still some cyclogenesis over the western part of the northern ocean where slow-moving depressions develop near Newfoundland, again in association with the zone of maximum sea surface temperature gradient (Fig. 2). These quasi-stationary systems are often associated with relatively short wavelengths in the upper westerlies and with northerly outbreaks in the Norwegian Sea area, particularly in May and June. From June to July, there is a significant northward displacement in the mean depression track and in the belt of westerlies, which in late summer are at their highest latitude in the western North Atlantic. Over the eastern North Atlantic and Europe, however, the westerlies shift southward from June to July (to about 58°N) with the readjustment of the eccentricity of the circumpolar vortex. Conditions over the central ocean are relatively settled in the summer months with little cyclogenesis and few gales.

The Azores high-pressure area is a prominent feature of the synoptic charts on most days in summer and achieves its seasonal maximum intensity in July. It is the dominant feature on the mean surface pressure map for July (Fig. 7).

The frequency of tropical disturbances and cyclones increases markedly during these summer months, August being one of the principal months in the North Atlantic hurricane season. In the 78 years from 1886–1963, 136 tropical storms occurred in August, 102 reaching hurricane intensity. However, the year-to-year variability in the incidence of tropical storms is large, e.g., for 14 of those 78 years no tropical storms were reported in August, whereas a maximum of seven occurred in August 1933.

Autumn

Early in autumn a gradual increase occurs in the incidence of cyclogenesis; depressions become deeper and more extensive and the amplitude of the waves in the upper westerlies increases. A cyclogenetic zone becomes established between Baffin Island and Iceland in October and persists until about March (KLEIN, 1957). Depression tracks in the eastern ocean occur farther north and the frequency of anticyclones in the Biscay region increases. As autumn progresses, northwesterly outbreaks of cold air extend farther south and southeast of North America over the western and southern North Atlantic Ocean. These incursions are part of a southward movement and intensification of the zone of maximum thermal gradient. This is reflected in the sharpening and intensification of the Canadian cold trough on the mean monthly thickness charts. Over the Grand Banks–Greenland area in September, visibilities of less than 8 km occur on 20% of occasions; however, as the season progresses the frequency of low visibility rapidly decreases over the whole ocean. There is a corresponding increase in the frequency of gales and by November periods of severe weather occur. Apart from tropical storms, most rough weather is confined to north of about 45°N, except in November along the United States coast where frequent winter-type depressions develop (MATHER et al., 1964).

South of the Azores high pressure, which is at its weakest in mid-to-late October, the northeast trades are well-established between 10° and 30°N. In low latitudes in the western ocean, the prevailing direction of the trades is from the east or southeast. South of about 5°N the southeast trades prevail.

The hurricane season reaches its peak in September; two to four tropical cyclones may be expected during the month and about two-thirds of all tropical cyclones reach hurricane intensity. The hurricane season continues well into October, but the probability of occurrence of tropical cyclones diminishes rapidly toward the end of the month. A somewhat more detailed account of the climatology of hurricanes will be given on p. 224.

The main features of the annual cycle of synoptic events described above parallel the regular seasonal course of the distribution of net radiation. The features appear to a greater or lesser extent each year but are subject to variations in intensity, timing, and extreme positions.

Circulation patterns and synoptic events in the westerlies

The frequencies of different types of atmospheric circulation have been analyzed for the extratropical North Atlantic region by SORKINA (1965) and VAN DIJK et al. (1974). Sorkina classified the patterns of surface pressure and wind field over the North Atlantic between approximately 10°N and 75°N for 1899–1939 and 1949–1963 into six basic types, whereas Van Dijk et al. used a modification of F. Baur's 500 mbar *Grosswetter* classification with 28 patterns to examine the sector 60°E–100°W 30–70°N for 1949–1971. Interestingly, the results of both studies show an increase in the meridionality of circulation in winter months in recent decades, with a weakened subtropical anticyclone.

The relative seasonal frequency of the 500-mbar types determined by Van Dijk et al. is given in Table VI. There is a clear dominance of the three zonal patterns in all seasons except spring. A ridge in the vicinity of 10–30°W, often occurring as a blocking pattern (discussed below), is common at this time of year. Most other groups show remarkably little variation throughout the year.

A different view of circulation regimes in the North Atlantic is provided by the recent studies of teleconnections associated with the well-known "seesaw" of winter temperatures between Greenland and northern Europe (VAN LOON and ROGERS, 1978). Each mode of this temperature oscillation coincides with significant pressure anomalies over the North Atlantic, the Mediterranean, and much of the North Pacific Ocean. For example, when winter temperatures are below average in Greenland and above average in Europe (the GB mode), there is a deeper than normal Icelandic low and an enhanced Azores anticyclone. The strong Atlantic westerlies which occur in this GB mode are also associated with strong northeasterly trades. Moreover, GB winters are character-

TABLE VI

SEASONAL FREQUENCY % OF 500 MBAR CIRCULATION PATTERNS OVER THE NORTH ATLANTIC–EUROPE SECTOR, 1949–1971 (VAN DIJK et al., 1974)

Type[1]	Winter	Spring	Summer	Autumn
Ridge 10°–30°W	27.7	32.0	22.1	20.1
Strong zonal flow	26.0	17.3	29.6	22.7
Zonal flow, SW'ly in E. Atlantic	7.3	6.1	4.9	9.2
Zonal flow, ridge over Europe	3.1	3.8	0.5	3.6
Trough over British Isles or central Atlantic	6.9	11.2	11.4	14.2
Deep trough W. Atlantic, ridge E. Atlantic	9.1	10.9	7.7	6.8
SW'ly flow	18.8	17.0	21.8	21.1
Unclassified	1.3	2.0	1.7	2.0

[1] The description refers to the North Atlantic.

ized by heavy sea-ice conditions in Davis Strait in the following summer, as well as by more numerous icebergs off Newfoundland during the spring. The opposite "Greenland above average" (or GA) mode has light ice conditions in Davis Strait and fewer icebergs. ROGERS and VAN LOON (1979) also find that anomalies of sea surface temperature developing during the course of seesaw mode winters tend to persist through the subsequent spring and summer.

These circulation modes are a major feature of the hemispheric circulation and climate. Since the mid-19th century, 27% of winter months (December – February) have been of the GB type and 22% of GA type. The former predominated from the 1880s to the 1930s, while subsequently the latter has been more frequent although in the 1930s there was an unusually high frequency of patterns with temperatures above average in *both* Greenland and Europe. This switch can clearly be related to the observed temperature fluctuations in high northern latitudes.

In addition to these studies of general characteristics, several investigations of specific phenomena have been made and three are outlined below.

Climatology of blocking action

One of the most distinct and important features of the tropospheric flow pattern not represented on climatological monthly maps is 'blocking action' in the mid-latitude circumpolar westerlies. Blocking has become a technical synoptic term to describe the existence of a 'warm' anticyclone in high middle latitudes adjacent to a 'cold' depression in lower middle latitudes and a local decrease of zonal westerly flow in the upper levels above these features. There is a strong tendency in the literature to imply a cause and effect by describing the general synoptic effect as that of a persistent anticyclone blocking the normal eastward progress of cyclones and anticyclones. Whatever the cause of the phenomenon, it is characteristically a very stable dynamical arrangement generally lasting for several weeks and often slowly retrogressing (moving westward). The

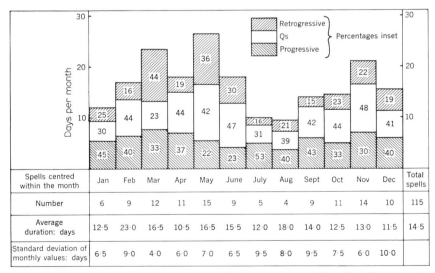

Spells centred within the month	Jan	Feb	Mar	Apr	May	June	July	Aug	Sept	Oct	Nov	Dec	Total spells
Number	6	9	12	11	15	9	5	4	9	11	14	10	115
Average duration: days	12·5	23·0	16·5	10·5	16·5	15·5	12·0	18·0	14·0	12·5	13·0	11·5	14·5
Standard deviation of monthly values: days	6·5	9·0	4·0	6·0	7·0	6·5	9·5	8·0	9·5	7·5	6·0	10·0	

Fig. 14. Average monthly distribution of blocking anticyclones, 1949 – 1956 (SUMNER, 1954). *Qs* = quasi-stationary.

frequency and distribution of blocking action are important features of the climatology of the North Atlantic Ocean.

Statistics of the occurrence of blocking anticyclones in the Atlantic – European sector of the Northern Hemisphere for 1949 – 1956 have been provided by Sumner (1954). The area considered was from 100°W eastward to 60°E and north of 50°N, and during the eight years studied there was an average of fourteen spells of blocking action each year, average duration 14.5 days. The variation from year to year was remarkably small; each year a blocking anticyclone existed somewhere in the area for about half the time. The average monthly distribution (Fig. 14) shows a spring (May) maximum, a late summer minimum, and a secondary maximum in autumn. Another interesting feature of this diagram is the long average duration of blocking action in February compared with the low value in April. Sumner's diagram of the frequency of blocking at various latitudes (Fig. 15) shows a pronounced maximum between 55 and 65°N. The only annual variation in latitudinal frequencies is that blocking appears to occur in lower latitudes in April with a tendency to higher latitudes in summer months. A particular feature of the Atlantic area (compared with the European area) was that the percentage of progressive cases decreases with latitude.

Well-developed blocking is rare over North America. The distribution of blocking with longitude shows a rapid increase eastward across the North Atlantic with a maximum frequency between 20°E and 0° longitude and falling off less rapidly over Europe (Fig. 16). Rex (1950), using a slightly different definition involving a split in the jet stream, found a similar sharp peak at the same longitude but a more rapid falloff to the east than to the west. Blocking highs over the extreme eastern North Atlantic were most frequent in late summer and autumn, whereas in spring they most frequently occurred in the central North Atlantic Ocean.

Coastal storms of eastern United States

Mather et al. (1964) have analyzed the seasonal and geographic distribution of storms that caused damage on the east coast of the United States from 1921 to 1962. In many

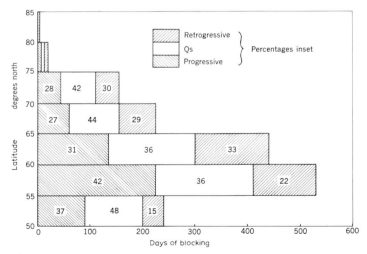

Fig. 15. Distribution of blocking with latitude, 1949 – 1956 (Sumner, 1954). *Qs* = quasi-stationary.

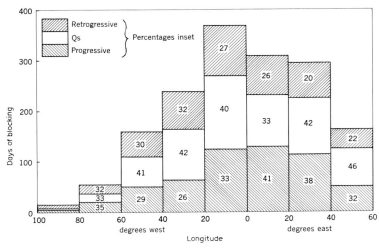

Fig. 16. Frequency of blocking in various longitude sectors, 1949–1956 (SUMNER, 1954). *Qs* = quasi-stationary.

ways, this is one of the most successful attempts at synoptic climatology because the classification used appears to be valuable in both describing and understanding the behaviour of the atmosphere in this region. Eight types of storms are distinguished.

Type 1: Hurricanes.

Type 2: Wave developments well to the east of the southeast coast or near Cuba.

Type 3: Wave developments over Florida or in the nearby Atlantic coastal waters.

Type 4: Wave developments in the Gulf of Mexico west of 85°W.

Type 5: Inland depressions that deepened on crossing the coast.

Type 6: Secondary disturbances in the Hatteras area.

Type 7: Intense cyclones moving northeastward, west of the coast.

Type 8: Strong cold fronts with associated line squalls.

The seasonal occurrence of these storms and the frequency and damage of each type between 1921 and 1962 are given in Tables VII and VIII. Coastal storms of moderate to severe intensity might be expected anywhere along the east coast of the United States

TABLE VII

SEASONAL OCCURRENCE OF STORMS BY CLASS OVER THE EAST COAST OF THE UNITED STATES, 1921–1962 (MATHER et al., 1964).

Class	Jan.	Feb.	Mar.	Apr.	May	June	July	Aug.	Sept	Oct.	Nov.	Dec.	Year
1	0	1	0	0	0	1	3	14	22	12	3	0	56
2	2	0	0	0	0	0	0	1	0	2	3	1	9
3	2	0	2	1	1	1	0	3	3	5	5	0	23
4	2	3	5	3	1	0	0	0	2	2	2	2	22
5	2	2	2	1	0	1	0	1	0	0	3	2	14
6	2	4	3	5	1	1	0	0	0	1	4	4	25
7	0	3	3	1	0	1	0	0	0	3	2	2	14
8	1	0	0	0	1	0	1	0	3	1	0	0	7
Total	11	13	15	11	4	4	4	19	30	26	22	11	170

TABLE VIII

FREQUENCY AND ESTIMATED DAMAGE BY STORM TYPE; ATLANTIC COAST 1921–1962
(MATHER et al., 1964)

Class	1	2	3	4	5	6	7	8
Total number of storms	56	9	23	22	14	25	14	7
Damage*	2.05	2.44	2.09	1.91	1.69	1.75	1.54	1.00

* Based on: 1 = "low" damage; 2 = "medium" damage; 3 = "high" damage

on an average from once every 1.4 years (New York and Jersey areas) to once every 4.2 years (area of Georgia).

Thermal troughs over the northeastern Atlantic

A study by MILES (1958) is a good example of the difficulty in analyzing the behaviour of transient synoptic features. Thermal (or "cold") troughs play an important part in synoptic development over the northern North Atlantic, but a large variety of structures within the same general classification makes their collective analysis difficult. Miles chose to analyze the surface pressure and the 1000 – 500 mbar thickness pattern over the northeastern Atlantic. He found that a surface trough usually containing a pronounced cold front occurs in advance of every thermal trough, and in some cases it was possible to interpret the characteristics of the front in terms of the movement and structure of the thermal trough. Four main types of trough were distinguished depending on isobaric structure, amplitude, and wavelength.

Troughs were most frequent in the winter half of the year and mainly mature, medium-to-long wave types; in July most troughs were small-scale perturbations moving in the periphery of slow-moving larger features.

Trade-wind region

A large surface anticyclone over the centre of the North Atlantic Ocean with an east – west axis in the vicinity of 30°N is a prominent feature on most days. The intensity, position, and form of this feature vary from day to day, but the variations are relatively small. Compared with the variability of the atmosphere farther north, the north-easterly trade wind regime to the south of this anticyclone is remarkably constant. Apart from the routine observations from selected ships, most of the information on conditions in the lower layers of the southern and southeastern part of the North Atlantic is provided by the results of the *Meteor* expedition (VON FICKER, 1936) and those of the Atlantic Trade-Wind Experiment (ATEX) (AUGSTEIN et al., 1973).

Three important climatological features of the North Atlantic trade-wind region are: (*1*) strength, constancy, and areal extent of the northeasterly and easterly wind regime; (*2*) vertical structure of wind and also of temperature, humidity, and cloudiness associated with the trades; and (*3*) equatorial zone of interaction between northeasterly and southeasterly flow (trade wind trough) and the extension of the southeasterly trades from the South Atlantic into the North Atlantic in all seasons. The axis of the surface subtropical high-pressure belt can be regarded as the northern boundary of the trade wind region. The seasonal variation of this is small and is generally exceeded by day-to-day variations. Over the central ocean, the latitude of the axis is about 35°N in summer and 30°N in winter; in the west, there is little seasonal movement with the axis lying between 25 and 30°N throughout the year. In the eastern ocean, the trades become more northerly with increasing latitude. This is particularly true in summer when strong "trade" northerlies (probably without the strong, overlying temperature inversion) can be recognized at 40°N and farther north; about 35°N appears to be the northern limit of the trades in winter.

In addition to tropical disturbances in summer and autumn, day-to-day synoptic variations occur in the trades and are most marked in winter. This is reflected by the difference between the low standard vector deviations of wind ($2.5-4$ m s^{-1}) in the tropical Atlantic in July (Fig. 13) and the significantly higher values in January (Fig. 12):3.5 m s^{-1} near the Equator increasing to nearly 8 m s^{-1} near the northern trade limit. The same maps contain monthly resultant winds; over wide areas these achieve $6-8$ m s^{-1} in both January and July. These resultant speeds are similar to those of the southwesterlies in the higher latitudes in winter which have much more synoptic activity.

Details of upper winds and temperatures over the centre of the tropical North Atlantic are still relatively unknown. Our knowledge of the vertical structure of the trades depends greatly on the *Meteor* data, inferences drawn from the Pacific studies of RIEHL et al. (1951, and later papers), the studies of CHARNOCK et al. (1956), and investigations during the ATEX in 1969 (AUGSTEIN et al., 1973).

The northeast trades are essentially a moist, relatively shallow air stream capped by a strong inversion of temperature due to subsidence in the mean flow. The classical map of the height of the base of this inversion (Fig. 17) shows it to increase from less than

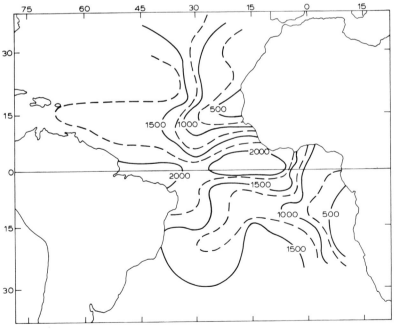

Fig. 17. Height of base of trade wind inversion (m) (VON FICKER, 1936).

500 m along the westernmost coast of North Africa to about 1500 m over the central ocean and 2000 m in the vicinity of the Equator. The height and structure depend on the balance of subsidence above and turbulent or convective mixing from below (WAGNER, 1975). Above the inversion, the air is much warmer and drier than in the lower layers. Where the inversion is low (and strong), the average increase from below to above the inversion exceeds 5°C, and the average relative humidity decrease exceeds 50%. Further details are given by RIEHL (1954) and AUGSTEIN et al. (1973).

The strong surface northeasterlies decrease with height; the level of change-over to westerlies does not coincide with the temperature inversion. The change-over level, zero at the axis of the high-pressure area in the north, increases with decreasing latitude (TUCKER, 1965), although in winter there are upper tropospheric westerlies continuous across the equatorial Atlantic (NEWELL et al., 1972). In summer, upper tropospheric easterlies occur over equatorial latitudes of the North Atlantic.

Intertropical Convergence Zone

Description of tropical weather systems has advanced rapidly since the advent of meteorological satellites, especially the geostationary Applications Technology Satellite (ATS) series. The latter has permitted detailed analyses of the evolution of tropical systems and the determination of low-level wind fields from cloud characteristics. Our information is again greatly increased by the 1974 GARP Atlantic Tropical Experiment (GATE) program in the eastern Atlantic (KUETTNER, 1974).

The term Intertropical Convergence Zone (ITCZ) originally implied the zone of convergence between the trade wind systems of the two hemispheres. Research has shown that this concept is oversimplified and that the fields of pressure, wind confluence, and cloudiness are complexly related in time and space. Analyses of daily cloud cover in the tropics carried out by the National Environment Satellite Service, Washington, D.C., since 1965 indicate that there is generally day-to-day continuity in the enhanced cloud distribution that can be associated with the zone of interaction between the northeasterly and southeasterly trades. However, at any one time the zone of enhanced cloudiness is often discontinuous in space. Modern studies distinguish three features: surface pressure trough (equatorial or trade wind trough), axis of maximum cloudiness, and asymptote of airflow confluence associated with the equatorial trough (SADLER, 1975). During winter, the axis of maximum cloudiness associated with convective activity in the equatorial trough occurs between about 0 – 5°N and there appears to be little organized synoptic development. In summer, this axis is located between 5 – 10°N (Fig. 18), orientated west-east and corresponding to the broad zone of highest sea surface temperatures (Fig. 2). Maximum cloudiness thus occurs south of the line of directional shear between the easterlies and westerlies within the pressure trough. The shear zone is apparent from data on wind steadiness. The directional steadiness ($|\bar{V}r|/\bar{V}$) in summer decreases sharply from 80 – 90% in the trades to below 40% along the WSW – ENE shear zone (HASTENRATH and LAMB, 1977). Sadler notes that the trough is a continuation of the low-level monsoon trough which extends from the heat low over West Africa southwestward to about 30°W. In August 1963, for example, soundings at 25°W showed the existence of a narrow zone of deep moist westerlies associated with a break in the trade wind inversion. The trough and the cloudiness embedded in the westerlies to the south of it

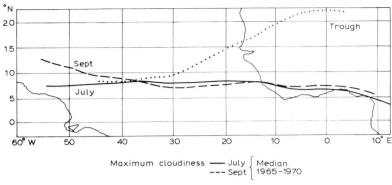

Fig. 18. Average location of lines of maximum cloudiness for July and September 1965–1970 from satellite observations. The mean August trough and shear line are also shown. The line of cloudiness for August is virtually identical with that for July (SADLER, 1975).

are semi-anchored with little month-to-month or interannual displacement in their monthly mean positions in summer. However, synoptic-scale disturbances form along the shear line in the trough, commonly over Africa, during June to September, and may intensify as they move west-northwestward into the easterly trade regime, with some developing into tropical storms. These disturbances travel at $5-10$ m s^{-1}, crossing the Atlantic in about four days.

A

00:00 238:74 01-A 0020-1801 4X4 IR IMAGE

B

00:00 241:74 01-A 0001-1821 4X4 IR IMAGE

C

Fig. 19. Infrared images from the Synchronous Meteorological Satellite 1 of the Atlantic Ocean tracing the evolution of hurricane Carmen from 22 to 31 August, 1974. Other systems are noted.
(A) 00GMT, 22 August. A disturbance is located over West Africa at 10°W. The southern cluster is a wave that drifted across Africa as part of extensive activity on the ITC prior to 18 August and then weakened on 21–21 (Thompson and Miller, 1976). A well-marked cold front cloud-band stretches southwestward from an occluding low at 55°N 25°W.
(B) 00GMT, 26 August. A vortex is now apparent north of the ITC cloud at 15°N 30°W. The tropical system which became hurricane Becky on 28 August is evident at 27°N 70°W. A developing low at 52°N 42°W shows a typical warm and cold front open wave form.
(C) 00GMT, 29 August. The system at 17°N 52°W is now recognized as a tropical depression and shows a pronounced spiral structure. Hurricane Becky is at 35°N 68°W. The central North Atlantic is dominated by an extensive subtropical anticyclone.
(D) 00GMT, 31 August. The system at 75°W is now named hurricane Carmen. It moved onto the Yucatan Peninsula on 2 September with 130-kt winds. It persisted in the Gulf of Mexico area, with regained hurricane strength from 5 to 8 September, and was the most intense cyclone since Camille of 1969. An extratropical cyclone is forming northwest of the remnant of Becky.

Tropical storms

Approximately 100 identifiable tropical systems per year affect the North Atlantic. About two-thirds of these are waves in the tropical easterlies; the remainder are waves in the Intertropical Confluence (ITC). Whereas these are not all persistent systems traversing the ocean from east to west, similar total numbers of waves have been observed over the 1968—1973 period at Dakar (56/yr), Barbados (54/yr), and San Andres (48/yr) (FRANK, 1975). In the Caribbean, the disturbances within the lower troposphere are known as easterly waves (RIEHL, 1954). They occur primarily in summer and autumn and bring clear weather ahead of the trough line, and maximum cloud buildup and precipitation after the passage of the trough line. Some 25 out of the 100 systems develop into tropical depressions and 8 into "named storms". The seedlings of these disturbances originate as both tropical and baroclinic systems. FRANK (1975) estimated that for 1967—1973 14 depressions per year were tropical, 11 of them of African origin and 3—4 forming on the ITC (Fig. 19), and 11 were baroclinic, nearly equally divided between upper and lower tropospheric systems. About 70% of the named storms developed from the tropical category. Frank also showed that 75% of the storms in the eastern Pacific originate over the Atlantic Ocean or Africa (50% derive from Africa).

The hurricane season in the North Atlantic Ocean lasts from July to October, commencing about one month earlier in the Caribbean Sea and the Gulf of Mexico than farther east. Hurricanes with winds in excess of 39 m s^{-1} are most prevalent south of 40°N and west of 50°W. The frequency of occurrence in the North Atlantic is given in Table IX. MILTON (1974) suggests that there has been a decline in frequency of North Atlantic hurricanes since 1945, following a general increase over the preceding 25 years. Comprehensive treatments of Atlantic hurricanes are given by DUNN and MILLER (1960), CRUTCHER and QUAYLE (1974), and ALAKA (1976).

The frequency of occurrence is similar in June and July, but in the latter month storms are usually greater in area and more likely to reach hurricane intensity. In August, both the occurrence and percentage of tropical storms reaching hurricane intensity are much greater than in July. Early September is the peak of the hurricane season; 35% of all tropical cyclones occur then and nearly two-thirds of these reach hurricane intensity. There is evidence of a decline in late September followed by a secondary maximum in early October. A fairly sharp decrease occurs after the middle of October with only 4% of all hurricane-intensity storms occurring after the end of October (ALAKA, 1976, table II).

The probability of at least one, two, or three tropical cyclones existing in any one month in the North Atlantic has been estimated by COLÓN (1953) based on observations between 1887 and 1950 (Table X). He has also prepared maps for each of the six months showing the total frequency of tropical cyclones with tracks starting at each 5°-square during 1887—1950 (Fig. 20). If correct, these represent the seasonal variation in the source regions of hurricanes. However, because of observational inadequacies during

TABLE IX

AVERAGE FREQUENCY OF TROPICAL CYCLONES IN THE NORTH ATLANTIC, 1899–1971
(CRUTCHER and QUAYLE, 1974)

	Jan.	Feb.	Mar	Apr.	May	June	July	Aug.	Sept.	Oct.	Nov.	Dec.	Year
Tropical storm	*	*	*	*	0.1	0.4	0.3	1.0	1.5	1.2	0.4	*	4.2
Hurricane	*	*	*	*	*	0.3	0.4	1.5	2.7	1.3	0.3	*	5.2

* Less than 0.05.

TABLE X

PROBABILITY OF TROPICAL CYCLONE OCCURRENCE PER MONTH
(COLÓN, 1953)

	May	June	July	Aug.	Sept	Oct.	Nov.
At least 1 storm	0.09	0.34	0.39	0.75	0.92	0.83	0.36
2 or more storms	0.02	0.06	0.11	0.52	0.72	0.59	0.03
3 or more storms	0.00	0.02	0.03	0.19	0.42	0.34	0.03

this period, particularly in the eastern area of the charts, it is doubtful that these give more than a rough guide to the frequency of formation of hurricanes. A different procedure was used by DUNN (1956) who plotted the points at which disturbances reached hurricane intensity from 1901—1957. His results are similar in some respects to those of Colón, but the distribution is somewhat more even over ocean areas and more storms are located east of 55°W. Current satellite studies mentioned above (p. 220) promise more accurate information on this topic.

The occurrence of tropical cyclones is essentially a seasonal phenomenon and the tracks of individual storms within the season are highly irregular. The general trend is for hurricanes to move westward or northwestward south of 15°N and then either to continue west-northwestward into the Gulf of Mexico, or to swing northward and then recurve northeastward, paralleling the Atlantic coast of North America (Fig. 21). The movements of hurricanes Carmen and Becky in 1974 (see Fig. 19) illustrate these two patterns.

Air—sea interactions

Interaction between the ocean surface and the marine atmosphere is highly complex, involving vertical and horizontal transfer in the atmospheric boundary layer, processes at the ocean interface, and the vertical and horizontal transport of ocean properties over a vast range of space and time scales. Before discussion of some of these interactions, the characteristics of basic temperature conditions are outlined.

Air temperature

Maps of mean air temperature in January and July are given in Figs. 22 and 23. In general, the coldest air over the North Atlantic is to the north and west and the warmest air to the south, similar to the patterns of sea surface temperature (Figs. 1 and 2). However, in relation to latitudinal average temperature, the air over the eastern Norwegian Sea is the warmest in the hemisphere in January with a deviation from latitude mean of +24°C near O.W.S. M. The largest gradient of air temperature (about 1°C per 25 km) occurs in the western ocean at about 40°N in January and February. The overall pattern in the western ocean is one of a relatively narrow zone separating cold air to the north and warm air to the south, but of a much more even south—north gradient of temperature in the east. This is characteristic of all months.

The main seasonal differences are weaker temperature gradients in the western ocean

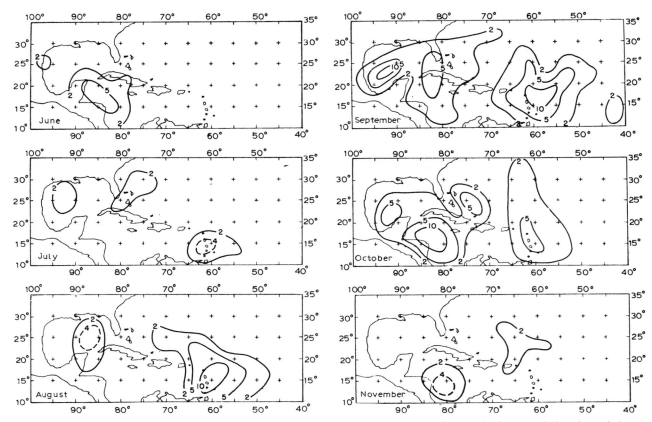

Fig. 20. Total frequency of tropical cyclones with a track starting from each 5° square during the period 1887–1950 (COLÓN, 1953).

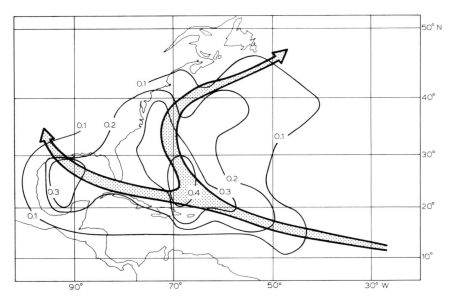

Fig. 21. Preferred tracks and mean frequency of tropical storms and hurricanes per 5° square, 11–20 September 1899–1971 (CRUTCHER and QUAYLE, 1974).

and higher temperatures in summer than winter. The interseasonal changes in both air and sea temperatures are greatest between early and late summer (May—June to July—August) and least in early autumn (PERRY, 1974). In low latitudes, the temperature varies little throughout the year; for example, at Sao Tomé, the average annual range is only 3.8°C, at Piarco 2.1°C, and at Cayenne 1.2°C. Over the northern oceans, O.W.S. B, A, I, and M have annual ranges of 10.7°C, 7.8°C, 6.6°C, and 7.9°C, respectively; on the continental coastline, Goose in Labrador has an annual range of 32.6°C while Emden in northwest Germany has 15.6°C. In this aspect, coastal stations in the east are more representative of oceanic climate than those in the west which are strongly affected by continental extremes.

To determine the variability of daily temperatures about the monthly mean, it was necessary to assume that daily temperatures are distributed normally. An approximation to the standard deviation of daily air temperature σ_T was then calculated from the range R between the upper and lower five percentile extremes given in the United Kingdom charts (METEOROLOGICAL OFFICE, 1948). The expression used was $\sigma_T = 0.61\ R$.

In January (Fig. 24), the zone of maximum variability corresponds to the zone of maximum temperature gradient. Maximum values are in excess of 5°C and lie slightly south of the prevailing storm tracks (Fig. 8). Values of less than 2°C characterize the southern and eastern half of the ocean. In July (Fig. 25), temperatures are less variable but the maximum still attains 3°C southeast of Newfoundland.

CRADDOCK (1964) notes that the distribution of monthly mean temperatures is effec-

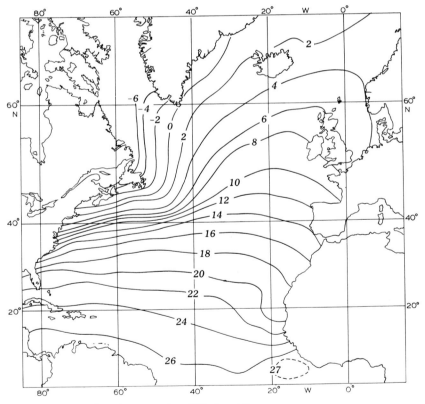

Fig. 22. Mean surface air temperature (°C), January.

227

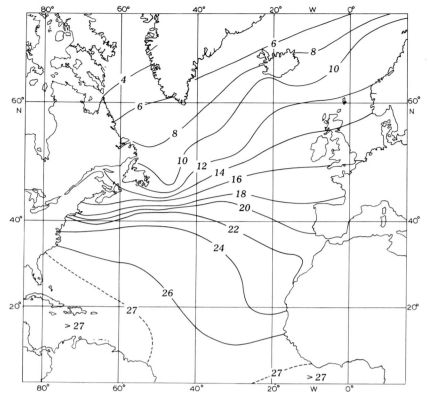

Fig. 23. Mean surface air temperature (°C), July.

tively Gaussian (though often with negative skewness in colder months and positive skewness in warmer months). He used the standard deviation as a measure of the interannual variability of monthly mean temperatures. These data have been used to construct maps of the standard deviation about the long-term mean of individual January (Fig. 26) and July (Fig. 27) temperatures. Sea areas adjacent to a continent have the highest interannual variability in both months, but particularly in winter. Values of less than 1°C occur over nearly all the ocean in July and over more than half the ocean in January.

Air — sea temperature difference

In January, the air is everywhere colder than the sea surface (Fig. 28) except for a small region off the West African coast, but just south of the zone of strongest temperature gradient in the western ocean the air is much colder than the sea—by more than 6°C in places. Temperature differences up to 8°C are shown in a second area of high contrast in the Davis Strait. In July, the temperature contrast between air and sea is very small except in a relatively narrow coastal zone off Canada where the air is some 2°C warmer than the sea.

For the central North Sea, HÖHN (1973) has shown how the sea — air temperature difference is dependent on season and wind direction (Fig. 29). Large positive differences arise in winter with cold north-to-east winds crossing the open sea. In spring when the

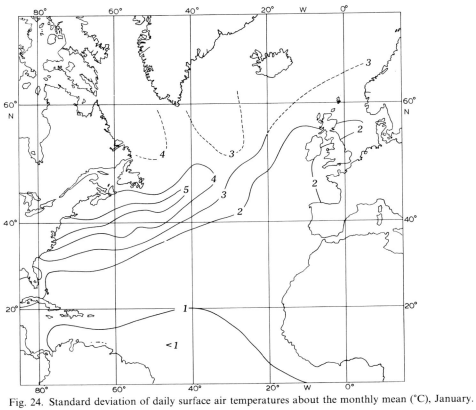

Fig. 24. Standard deviation of daily surface air temperatures about the monthly mean (°C), January.

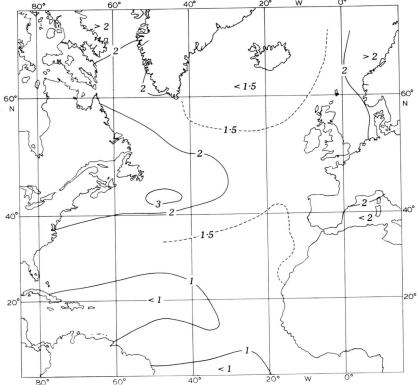

Fig. 25. Standard deviation of daily surface air temperatures about the monthly mean (°C), July.

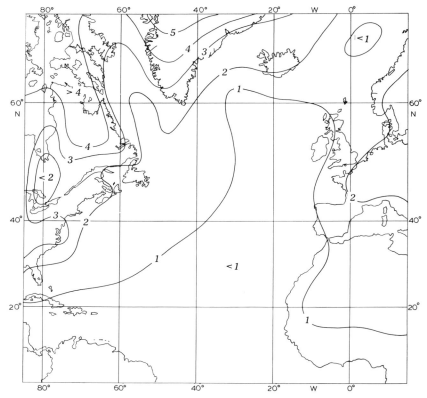

Fig. 26. Standard deviation of monthly mean surface air temperature (°C), January (CRADDOCK, 1964).

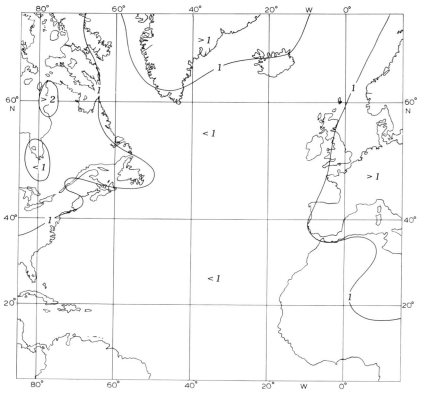

Fig. 27. Standard deviation of monthly mean surface air temperature (°C), July (CRADDOCK, 1964).

water is still cool, there are moderate negative differences with southwest winds and, likewise, in late summer with south to southeast winds from the continent.

Maps of the variability of daily sea surface temperature about the monthly mean (Figs. 30 and 31) have been obtained from the maps of five percentile extreme values (METEOROLOGICAL OFFICE, 1948) in the same way as for air temperature. The pattern of variability is very similar to that of air temperature in both months (Figs. 26 and 27), but the close similarity of patterns and magnitudes in the July maps is particularly noticeable. The variability of daily maximum air temperature is dominated by events on the synoptic time scale, whereas that of sea surface temperature is dominated by the seasonal change, being greatest during the early and late summer, according to analyses (without removal of seasonal trend) for O.W.S. C and D (JAKOBSSON, 1976).

Highly significant differences have been found between the same month in different years for both air and sea temperatures (KRAUS and MORRISON, 1966), but the between-year differences are larger for sea temperatures suggesting that long-term fluctuations (on a yearly scale) as compared with short-term fluctuations (on a monthly scale) are even more important in the sea than in the marine atmosphere.

Interactions between the ocean and atmospheric circulation

Research in the last decade has identified the association of ocean—atmosphere interaction with monthly and seasonal weather and climate anomalies. It is not yet clear

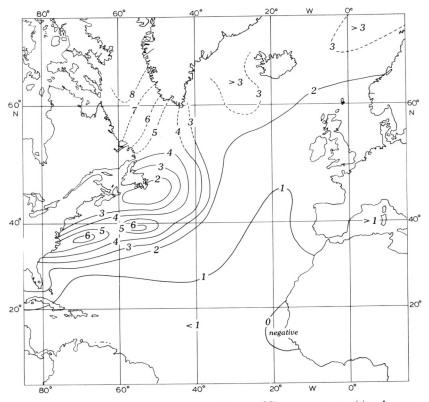

Fig. 28. Mean sea surface—air temperature difference (°C); warmer sea positive, January.

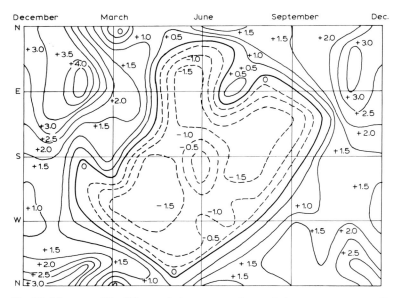

Fig. 29. Mean monthly difference between sea surface and air temperature according to wind direction in the central North Sea (Höhn, 1973).

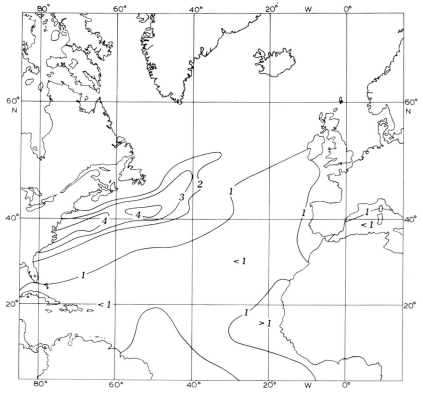

Fig. 30. Standard deviation of daily sea surface temperature about the monthly mean (°C), January.

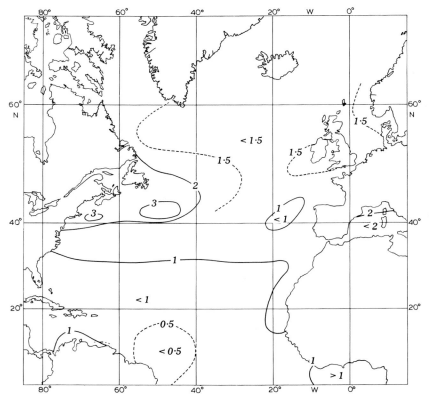

Fig. 31. Standard deviation of daily sea surface temperature about the monthly mean (°C), July.

what the exact mechanisms of interaction are, nor indeed how anomalies of ocean surface temperature are generated or maintained, but their apparent connection with weather and climate is well-documented. Namias (1964), for example, has demonstrated that recurrent blocking during 1958—1960 over northern Europe was associated with sea temperatures persistently below normal in the northwestern Atlantic and above normal in the northeastern Atlantic, which strengthened the southerly component of the thermal wind. Dickson and Lee (1969) suggest that, during this and other periods of anomalous southerly flow in the eastern Atlantic, the decrease in losses of turbulent heat from the ocean resulting from the reduced air—sea temperature difference maintains the positive anomaly of sea temperature in the northeast, thus creating positive feedback (or amplification). A key region of the North Atlantic for weather in northwestern Europe is located south of Newfoundland (Ratcliffe and Murray, 1970). During autumn and winter, positive departures of sea surface temperature by 1—2°C in this area are strongly correlated with pressures 3—4 mbar below normal in the following month over the northeastern Atlantic, and vice versa with negative departures of sea temperature. In the first case, the strengthened thermal gradient near 50°N intensifies the frontal cyclones and the upper westerlies in a northerly location; the cyclones tend to travel well eastward in response to an increase in the wavelength of the upper westerlies. The second case, with a negative anomaly off Newfoundland, enhances the likelihood of blocking in the eastern Atlantic as described above. Anomalies of monthly mean sea surface temperature in the critical area between 35—50°N, 40—60°W have been classified into six types, according to the distribution pattern and sign of the anomaly and for the period 1877–1970 by RATCLIFFE (1971), providing a valuable data base.

A persistent anomaly of sea surface temperature in the Newfoundland area must represent either a change in oceanic heat flux (from depth or by a progressive displacement of the current boundary) or a change in available energy and atmospheric heat fluxes at the surface. The mean duration of large-scale ocean anomalies in the North Atlantic during 1948–1959 is about two years, according to NAZAROV (1968). Some anomaly patterns do apparently spread eastward across the ocean with the winds.

It is recognized that a substantial part of the surface energy exchange takes place on the synoptic time and space scale, so that air–sea interactions need to be studied on this scale (PERRY, 1968; JAKOBSSON, 1976). Nevertheless, since it is the net effect of such transfers that is of climatic importance, oceanic temperature anomalies which persist for at least a month and extend across an area of at least 1000 km merit the prime attention. The nature of the atmospheric circulation response to changes in sea surface temperature gradient and energy fluxes on short and long time scales has been examined by BJERKNES (1963, 1964). He suggests that for anomalies of a few years' duration strong westerly airflow and cold air advection are associated with enhanced turbulent heat fluxes and since oceanic transport responds only slowly, the result is surface cooling. A similar negative correlation between wind speed and anomalies of sea surface temperature is apparent on the time scale of 50 years. However, in this case the primary control is exerted by upwelling of cold water that increases in response to the intensification of the Icelandic low.

Moisture conditions

Humidity

The representation of the amount of water vapour in the air in climatological studies is not satisfactory. Relative humidity depends on both the actual vapour pressure and the saturation vapour pressure; therefore, relative humidity alone is an unsatisfactory parameter. Vapour pressure as a parameter is preferable to dewpoint because it has a linear scale. However, charts of vapour pressure in the surface layers of the marine atmosphere do not seem to exist.

January and July maps of dewpoint and standard deviation of dewpoint were drawn from selected British ship data for 1951–1960 and agree well with the data published by KRAUS and MORRISON (1966) for Atlantic O.W.S.

In January (Fig. 32), dewpoints in excess of 20°C occur over the ocean south of 15°N corresponding to a vapour pressure of about 24 mbar; but some coastal stations report considerably higher values, e.g., Cayenne, 28.0 mbar, and Sao Tomé, 31.0 mbar. The zone of strong temperature gradient south of Newfoundland is also a zone of steep humidity gradient: the mean January vapour pressure at O.W.S. E is 15.6 mbar, while at Sable Island it is only 5.1 mbar. The lowest O.W.S. value is B which has 4.4 mbar; by contrast, in the Norwegian Sea, M has 5.9 mbar. The daily variation of humidity, as measured by the standard deviation of dewpoint values for January (Fig. 33), has a

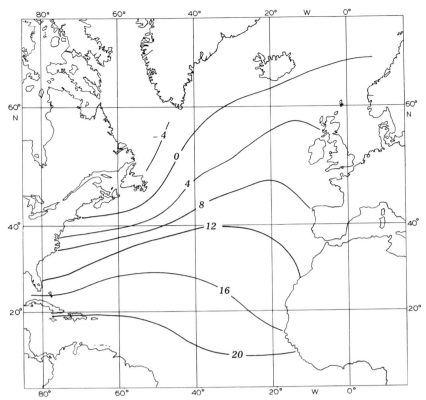

Fig. 32. Mean surface dewpoint (°C), January.

pattern similar to the variation of sea and air temperature, but with slightly higher values. The maximum is about 6°C, some 500 km northeast of Cape Hatteras; this corresponds to vapour pressure of about 4 mbar. The lowest variations, less than 2°C, occur south of 15°N and correspond to about 3 mbar. The difference between the western and southern areas is thus less than dewpoint variations imply. Over the northern areas, the standard deviation of dewpoint is about 4°C (1.5 mbar) and therefore less than in the tropics.

In July (Fig. 34), the highest dewpoints occur in the Caribbean, exceeding 24°C (vapour pressure, 30 mbar), but there is little horizontal gradient in the central region of the subtropical high. The gradient south of Newfoundland still exists, but is weaker than in January (E, 20.9°C, 25.0 mbar; Sable Island, 14.8°C, 16.8 mbar). B again has the lowest O.W.S. value of 10.1 mbar although Frobisher's value is 8.2 mbar. By contrast, M has 10.7 mbar. The standard deviation of dewpoint in July (Fig. 35) has a pattern similar to that of air and sea temperatures but, as in January, values are again a little higher. The maximum value occurs about 400 miles southeast of Newfoundland and exceeds 4°C (about 5 mbar). The northwest ocean regions have values of about 1.7°C, slightly higher than over tropical waters: 1.5°C. However, these correspond to vapour pressures of about 1 mbar in the north compared with 2.4 mbar in the tropics.

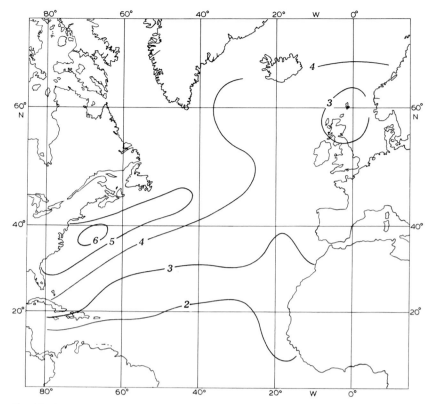

Fig. 33. Standard deviation of daily values of dewpoint about the monthly mean (°C), January.

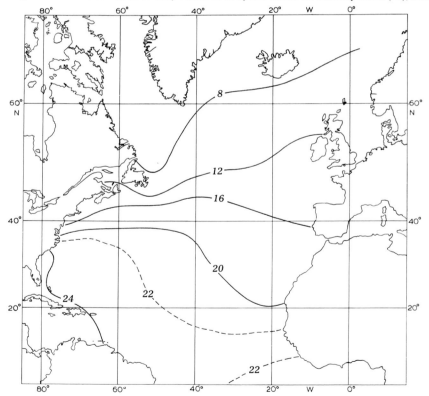

Fig. 34. Mean surface dewpoint (°C), July.

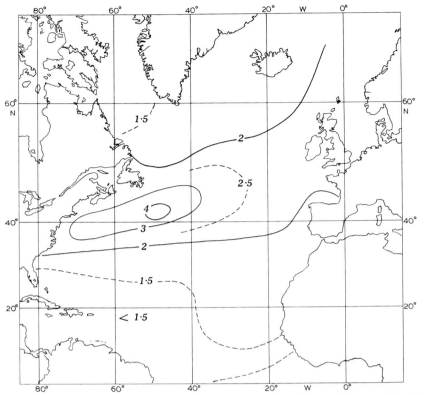

Fig. 35. Standard deviation of daily values of dewpoint about the monthly mean (°C), July.

Cloudiness

Monthly charts of cloudiness are published in the *Climatological and Oceanographic Atlas for Mariners* (U.S. WEATHER BUREAU, 1959). Isopleths are drawn for percentage frequency of total cloud amount equal to a cover greater than eight tenths and also equal to or less than two tenths; in the ensuing description "frequency of cloudiness" will refer to the former.

Cloud maps are not reproduced here. As might be expected, the most cloud-free area is that associated with the subtropical surface high-pressure area, particularly along its southern half. In January, a belt with less than 20% frequency of cloudiness of more than eight tenths is centred on 20°N and extends right across the width of the ocean. In July, this is limited to the central and western areas and in the centre it is slightly farther north.

The area with the greater cloud cover is associated with the high-latitude surface low pressure, and in January and July an area with more than 80% frequency of cloudiness exists over the ocean around weather ship B. The zone of greatest north—south gradient in frequency of ≥8/10 cloud cover exists at about 30—35°N in January and 40—45°N in July. In the equatorial area, a belt of cloudiness occurs in association with the ITCZ, located at 3—5°N in the winter months, 7—8°N in later summer. Satellite-derived maps by Sadler (ATKINSON and SADLER, 1970) show that in February more than 50% frequency of cloudiness exists west of about 25°W in the ITCZ. In summer, this belt is more extensive in the east due to the West African monsoon; values in excess of 60% occur over the West African coast in August.

Visibility

Monthly charts of the percentage frequency of visibility less than 8 km (referred to as poor visibility) are given in the *Climatological and Oceanographic Atlas for Mariners* (U.S. WEATHER BUREAU, 1959) and are not reproduced here.

A large part of the southern section of the ocean has very good visibility (less than 5% of occasions below 8 km visibility for every month), whereas in the northern parts of the ocean the frequency of poor visibility is in excess of 30% for most of the year.

In winter and spring, part of the North Sea, the ocean north of Iceland, and the Davis Strait down to the Grand Banks experience visibility less than 10 km on more than 30% of occasions. From May until August, areas with frequencies in excess of 30% occur along the coast of North America north of Nantucket Island, over the Grand Banks, in the Davis Strait, and in the northeast ocean. Advection fog is common, especially in summer, over the cold waters of the East Greenland Current and near the sea ice edge.

Precipitation

Because of the difficulties in measuring precipitation at sea, estimates of precipitation over the ocean are based on indirect methods (SCHOTT, 1926, 1935; MEINARDUS, 1934; MÖLLER, 1951). JACOBS (1951, 1968) has carefully produced seasonal precipitation charts over the oceans, but his method is open to three criticisms.

(*1*) He relies heavily on the areal distribution of mean annual precipitation in the charts of MEINARDUS (1934) which, in turn, are based on extrapolation from coastal and island stations with little reference to oceanic observations. The *Morskoi Atlas* (NAVY MINISTRY, MOSCOW, 1953) mean annual precipitation charts also rely on such data.

(*2*) WÜST's (1936) latitude profiles of annual precipitation are also used; these depend on estimates of mean zonal evaporation and salinity which are themselves uncertain.

(*3*) The seasonal distribution is calculated in relation to the seasonal frequency of a rather coarse classification of precipitation types.

The method assumes no significant seasonal variation in diurnal occurrence or mean intensity of the precipitation. JACOBS (1968) advances some evidence in support of the latter assumption, while the former has been shown valid by KRAUS (1963) who has analyzed the convective and non-convective "present weather" reports of the Atlantic weather ships and shown that in all seasons precipitation is significantly more frequent at night. This is particularly marked in summer months and over the western ocean. Kraus suggests that the absorption of solar radiation in the upper layers of cloud during the day may reduce their liquid water content through evaporation. This may be important in suppressing the precipitation-forming mechanisms, especially in stratiform cloud. At night, increased cooling of cloud tops below an inversion causes the level of the top to rise and therefore cloud depth to increase.

TUCKER (1961) developed a method of obtaining mean monthly values of precipitation at each O.W.S. by analyzing the frequency of various types of "present weather" observations. This method is likely to be more reliable than that using an extrapolation

TABLE XI

MEAN MONTHLY AND MEAN ANNUAL PRECIPITATION (MM) OVER ATLANTIC OCEAN WEATHER SHIPS DURING THE FIVE-YEAR PERIOD JANUARY 1953 TO DECEMBER 1957

(TUCKER, 1961)

O.W.S.	Jan.	Feb.	Mar.	Apr.	May	June	July	Aug.	Sept.	Oct.	Nov.	Dec.	Annual
A	120	148	77	69	51	71	95	127	107	138	101	141	1,245
B	146	169	123	93	75	68	59	93	97	105	105	170	1,303
C	108	78	61	54	65	76	72	95	85	80	76	102	952
D	75	77	76	67	66	65	38	49	65	73	57	113	821
E	61	52	51	20	30	24	15	15	33	39	24	38	402
I	97	86	50	42	55	65	61	72	92	92	78	103	893
J	78	55	49	43	55	64	56	77	69	68	64	92	770
K	35	47	52	24	35	26	20	29	28	27	28	34	385
M	142	146	88	92	78	73	59	52	102	121	116	125	1,194
H*	97	99	59	36	37	37	25	32	23	52	26	46	569

* 2 years

into ocean areas of precipitation at island and coastal stations. For the five-year period 1953 – 1957, a root-mean-square percentage error of less than 13% is claimed. Derived monthly and annual precipitation amounts for each weather ship are given in Table XI. Compared with JACOBS' (1951) chart of mean annual rainfall (Fig. 36), higher values are indicated by TUCKER (Fig. 37) in the northwest Atlantic (B) and lower values in the east (I, J, and K). This latter difference is particularly true in spring when Jacobs allocates a value in the 200 – 300 mm range for J, whereas Tucker obtains 147 mm. The most noticeable feature of Tucker's results are the very low values for K, generally less than half of previous estimates.

The high values of precipitation at the northerly ships (Table XI and Fig. 37) are a cause of some concern, and ANDERSSEN (1962) has suggested that an overestimation of the water content of snow may be the reason. However, a close investigation (TUCKER, 1962) shows this to be unlikely, and the problem remains unresolved. REED and ELLIOTT (1979) have recently prepared new seasonal precipitation maps which appear to eliminate the bias evident in Tucker's original analysis in the northern and northwestern North Atlantic.

Evaporation

The direct measurement of evaporation on board ship is particularly difficult; both the unstable platform and the measuring apparatus itself exert a considerable influence on observed values. In addition, these values then have to be corrected to apply to sea level; DEFANT (1961) estimates that on the average a factor of two is involved, evaporation on board ship being roughly double that at the sea surface. Defant has applied slight corrections to the direct measurements of WÜST (1920) to obtain the mean evaporation in 10°-latitude zones over the oceans. Values for the North Atlantic are given in Table XII; the limits of error are quoted as ±12%.

The evaporation occurring at any instant is obviously a complicated function of various meteorological parameters. Estimates based on the theory of small-scale turbulence near

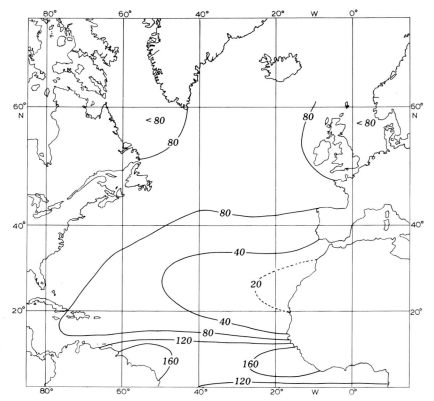

Fig. 36. Mean annual precipitation (cm) (JACOBS, 1951).

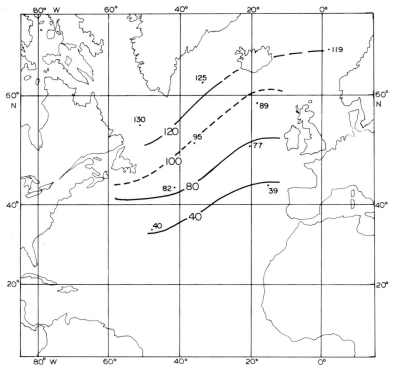

Fig. 37. Mean annual precipitation estimated from ship observations (cm) (TUCKER, 1962).

the surface are probably more reliable than direct measurements. However, the application of "instantaneous" formulae to evaporation on the climatological scale has been severely criticized by MONIN (1963). Of particular importance is the possible effect of using individual climatological parameters with no allowance for the correlation in time of the various elements concerned. Nevertheless, the climatic distribution of evaporation over the oceans is usually derived from a formula of the type:

$$E = Au_a(e_s - e_a)$$

where E is evaporation, e_s saturation vapour pressure at the ocean surface temperature, e_a vapour pressure of the air observed on board ship, and u_a scalar wind speed on board ship. Mean monthly climatological values are normally used. Quantity A is derived by applying the formula over all seasons and all oceans, or over an effectively closed area in which a total value of E can be derived from energy considerations.

One of the most detailed estimates has been carried out by ZUBENOK and STROKINA (1963). January and July maps are given in Figs 38 and 39. A more recent calculation for the annual total has been presented by BUNKER and WORTHINGTON (1976).

Evaporation is particularly large in the western North Atlantic Ocean in January and in the coastal waters of Cape Hatteras where the Gulf Stream carries warm water northward resulting in high vertical gradients of vapour pressure. In the western areas, conditions are quite complex—cold and warm sea areas are adjacent and winter cold air is advected southeastward from North America. This results in quite strong horizontal variations in the evaporation pattern. In general, evaporation in middle and high latitudes is characterized by summer values many times smaller than winter values.

In the trade wind zone, detailed measurements during the Barbados Oceanographic and Meteorological Experiment (BOMEX) in June 1969 indicate evaporation rates of 5–6 mm day^{-1}, 20% greater than climatological estimates, during undisturbed trade wind conditions (HOLLAND and RASMUSSEN, 1973). More data of this kind are needed as a check on the available climatological estimates.

Heat balance

General remarks

The oceans can store a vast amount of heat because of their high heat capacity. The resulting thermal inertia keeps down the amplitude of surface temperature variations, in contrast with continental climates. The heat input to the atmosphere depends to a

TABLE XII

ZONAL DISTRIBUTION OF EVAPORATION OVER THE NORTH ATLANTIC
(DEFANT, 1961)

Zone (N)	80–70°	70–60°	60–50°	50–40°	40–30°	30–20°	20–10°	10–0°
Evaporation (cm yr^{-1})	8	12	44	78	107	138	146	107

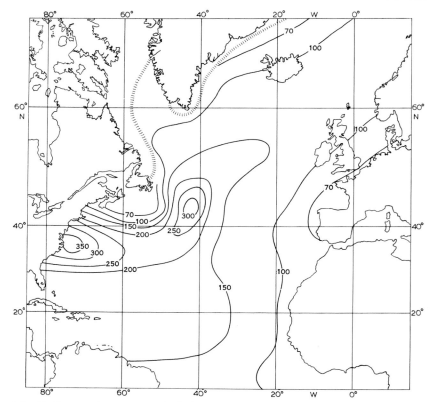

Fig. 38. Evaporation (mm month⁻¹), January (ZUBENOK and STROKINA, 1963).

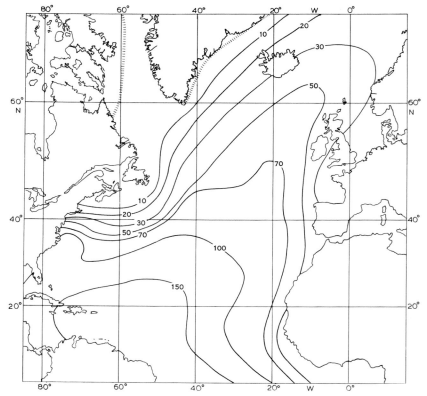

Fig. 39. Evaporation (mm month⁻¹), July (ZUBENOK and STROKINA, 1963).

large extent on the sensible and latent energy transferred from ocean to atmosphere; this transfer is of fundamental importance in the development of large-scale atmospheric dynamical systems. An adequate account of the heat balance at the ocean surface is thus essential for a description of ocean climate.

The heat balance equation for the ocean surface is:

$$R = LE + H + A$$

where R is net radiation, L latent heat of condensation, and E evaporation, the product representing the loss of heat at the ocean surface due to evaporation. H is the loss of sensible heat (or enthalpy) due to turbulent exchange across the ocean/atmosphere interface and A is the heat exchange with the deeper layers of the ocean and horizontal transport. On land the annual value of A is zero, but in the ocean there is an important annual net gain or loss of heat from the ocean surface as a result of water movement on all scales. This is usually estimated as a residual in the equation.

The net radiation or radiation balance is usually obtained from the equation:

$$R = Q(1-\alpha)-I$$

where Q is total solar radiation on a horizontal surface after depletion, α short-wave surface reflectivity (albedo), and I the effective terrestrial radiation (upward long-wave radiation minus the back radiation from the atmosphere).

SIMPSON (1929) was among the first to produce climatological maps of total solar radiation while JACOBS (1942 and later papers) published climatological maps for the oceans of elements in the heat balance equations. In recent years, much work has been done by Russian scientists and in 1956 the *Heat Balance Atlas* was produced by M. I. BUDYKO. STROKINA (1959, 1964) has continued this work paying particular attention to the North Atlantic, and her results are incorporated in the revised atlas (BUDYKO, 1963).

The methods employed to obtain climatological maps of elements in the heat balance equation were developed to use the large amount of meteorological data obtained by merchant vessels. A general outline of the method is given in numerous publications (for example, ROLL, 1964). As with evaporation, the methods involve the application of simple procedures for computing the vertical and eddy fluxes of heat and moisture based on measurements at a certain height above the sea and on sea surface observations. Since these methods utilize formulae that are valid only for instantaneous values of the elements concerned, substantial errors are involved if long-term averages are substituted in the equations without regard to the auto-correlations involved. Other limitations involve the assumed equality of the vertical exchange coefficient for heat, mass, and moisture and also the application of a standard factor representing the vertical exchange coefficient for all conditions of wind speed and stability. Although these objections are valid, the charts published by Budyko and Strokina are a useful basis for a discussion of the heat balance over the North Atlantic Ocean.

BUNKER (1975) has subsequently recalculated the energy exchanges over the North Atlantic using all available marine data between 1942 and 1972, grouped by 10 subdivisions of each Marsden square and by month. Computations were made by several methods—Budyko's formulae using a constant exchange coefficient; variable drag coefficients according to wind speed and stability (as indicated by air — sea temperature difference); and variable transfer coefficients for vapour flux and sensible heat flux. Radiation fluxes were calculated by Budyko's methods and by empirical formulae based on cloud amounts and precipitation. Bunker's results are incorporated in the following discussion where significant differences from the earlier studies have been determined.

Solar radiation

The charts of this quantity (Figs. 40 and 41) show a markedly zonal distribution in winter primarily due to latitudinal variations in solar radiation and length of day. Deviations from this—the decrease in values in the vicinity of the equator—are due to cloudiness. In summer, cloudiness associated with low-pressure systems causes a minimum at about 52°N, while the high values found in high latitudes show that the increased length of day more than compensates for the low sun angle. The only detailed study of the effects of cloud type and amount on solar radiation over the North Atlantic is that of LUMB (1964) for O.W.S. J and A. The influence of both the strong trade wind inversion and the stability associated with the cold Canary Current inhibits cloudiness near the coast; thus higher values of total radiation occur in the eastern part of the subtropical North Atlantic in summer. The range from high values in the tropics to low values in high latitudes is greatest in February; this is similar to the maximum thickness gradients (Fig. 6).

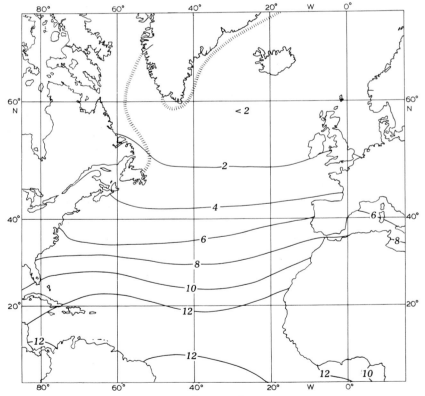

Fig. 40. Total solar radiation (kcal. cm^{-2} month^{-1}), January (BUDYKO, 1963).

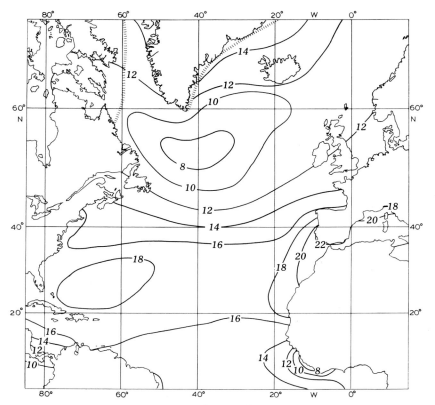

Fig. 41. Total solar radiation (kcal. cm^{-2} month^{-1}), July (BUDYKO, 1963).

Terms in the heat balance equation

Annual and seasonal patterns of net radiation also show a mainly zonal distribution, values decreasing with increasing latitude. Annual patterns are positive over the entire ocean, but winter values are negative north of about 45°N. The maximum latitude gradient occurs in the western region in January and February. Charts for January and July are given in Figs. 42 and 43.

The heat loss by evaporation attains maximum values in the middle of the Gulf Stream amounting to 170 kcal. cm^{-2} per year. Most evaporation occurs in autumn and winter, the effect of the warm currents being less marked in spring and summer. Charts of this quantity are not presented because the patterns are identically those of evaporation (1 mm = 59 cal. cm^{-2}).

Annual values of the sensible heat exchange are positive over the North Atlantic. Highest values occur in January (Fig. 44) when air—sea temperature differences are greatest, particularly in the northwest and in the Gulf Stream region where cold continental air moves over the warm ocean surface. In July (Fig. 45), very small air—sea temperature differences and generally lower wind speed values result in a very uniform distribution with values close to zero. Slightly negative values occur in the northwestern parts of the ocean and off the European and northwestern African coast.

The calculations by Bunker using variable transfer coefficients (according to wind speed

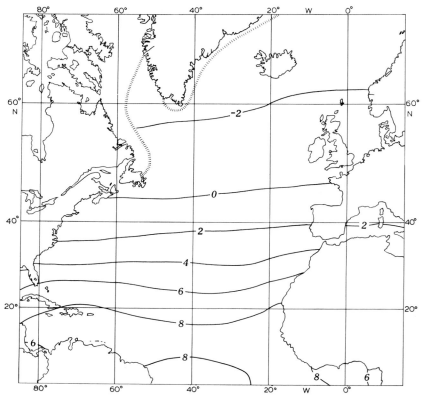

Fig. 42. Net radiation at the ocean surface (kcal. cm^{-2} month^{-1}), January (Budyko, 1963).

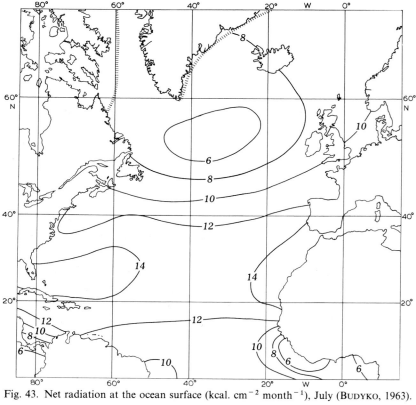

Fig. 43. Net radiation at the ocean surface (kcal. cm^{-2} month^{-1}), July (Budyko, 1963).

and stability) show major differences from heat fluxes computed by Budyko and Strokina over the western Atlantic. Fig. 46 shows sensible heat fluxes exceeding 8 kcal. cm^{-2} in January in a narrow zone following that of maximum air—sea temperature difference, with a maximum of 10 kcal. cm^{-2} about 38°N 62°W. The maxima on Strokina's maps are only 6 kcal. cm^{-2}. In the same location, the latent heat flux in January exceeds 25 kcal. cm^{-2} (Fig. 47), compared with 22 kcal. cm^{-2} according to Strokina.

Finally, it is instructive to examine the annual loss or gain of heat by the ocean. Fig. 48 shows Bunker's calculation based on the Budyko method for the radiational fluxes combined with variable transfer coefficient for the turbulent fluxes. This shows the influence of current systems and air flow off the continent on the net energy exchange. The substantial heat loss by the Gulf Stream system is sharply delineated on the northern side west of about 50°W and a secondary maximum of heat loss occurs between 40° and 50°W, north of 40°N, associated with the second current gyre (WORTHINGTON, 1976). The maximum loss of 200 kcal. cm^{-2} yr^{-1} compares with a figure of only 120 kcal. cm^{-2} yr^{-1} in the Budyko atlas. Heat gain by the Labrador Current as it flows southward is clearly shown; the Budyko maps show heat loss in the same area. In the subtropics, the ocean also gains heat as it flows slowly southwestward in the southern gyre (BUNKER, 1975).

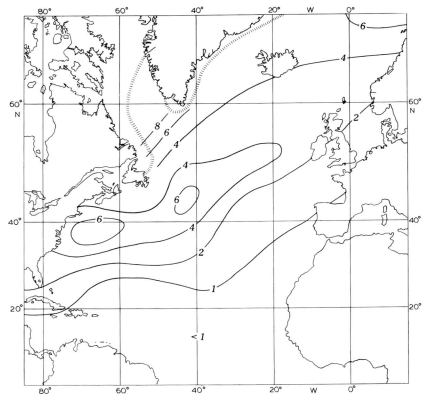

Fig. 44. Sensible heat exchange from the ocean to the atmosphere (kcal. cm^{-2} month^{-1}), January (BUDYKO, 1963).

247

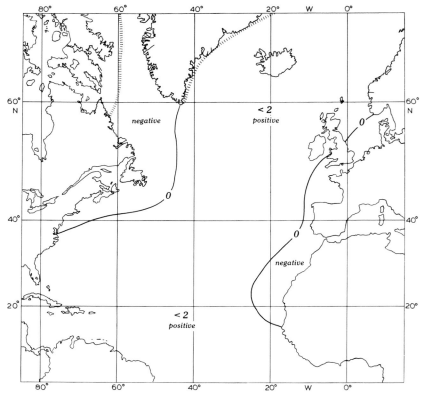

Fig. 45. Sensible heat exchange from the ocean to the atmosphere (kcal. cm^{-2} month^{-1}), July (BUDYKO, 1963).

Climatic zones

In order to describe the patterns and annual variations of the heat balance components, STROKINA (1959) divided the North Atlantic into climatic zones (Fig. 49). Major divisions are based on net radiation values. In the equatorial zone, net radiation throughout the year is positive with two maxima; in the tropical zone it is positive but with one maximum; in the temperate and sub-polar zones during one or more winter months, the net radiation is negative and there is a clearly defined summer maximum. A regional breakdown of these zones is made using differences in the evaporation and sensible heat flux characteristics. The seasonal variations of terms in the heat balance equation for each region are given graphically in Fig. 50.

Equatorial zone

In this region, the annual cycle of net radiation follows that of total radiation with equinoctial maxima and a primary minimum in December. Variations in evaporation depend on atmospheric activity associated with the ITCZ and also on seasonal air mass changes. There is a clearly defined summer minimum in evaporation during the incursion of cross-equatorial southerly winds and an equally well-defined maximum in winter associated with the northeasterly trade winds. Values of the sensible heat exchange are small in winter because of small air−sea temperature differences.

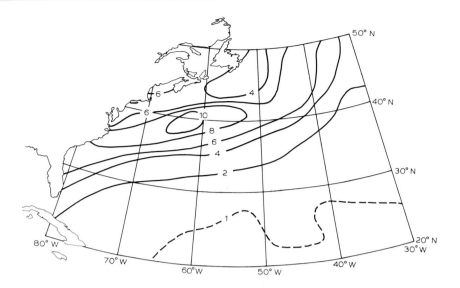

Fig. 46. Sensible heat exchange over the western Atlantic (kcal. cm^{-2} per 30 days), January (BUNKER, 1975).

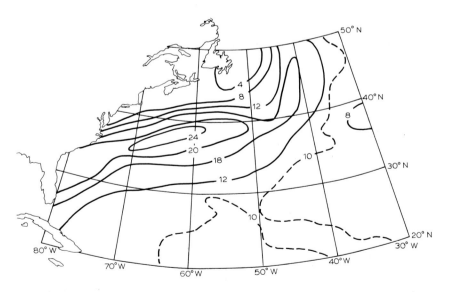

Fig. 47. Latent heat exchange over the western Atlantic (kcal. cm^{-2} per 30 days), January (BUNKER, 1975).

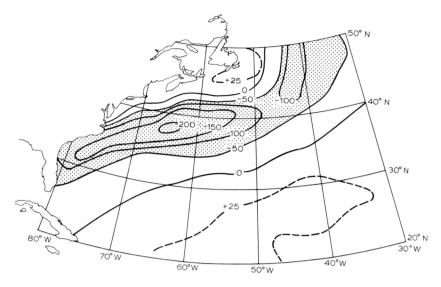

Fig. 48. Annual net heat gain by the ocean over the western Atlantic (kcal. cm^{-2} year^{-1}) (BUNKER, 1975).

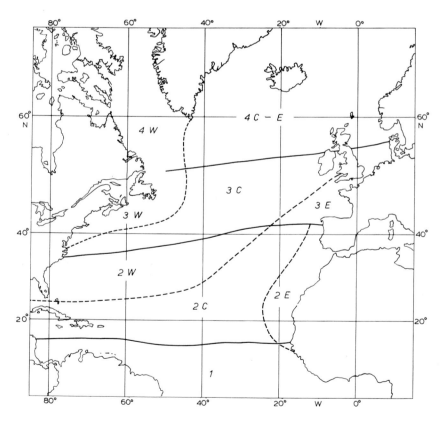

Fig. 49. Climatic zones based on the energy fluxes (see text) (STROKINA, 1959).

250

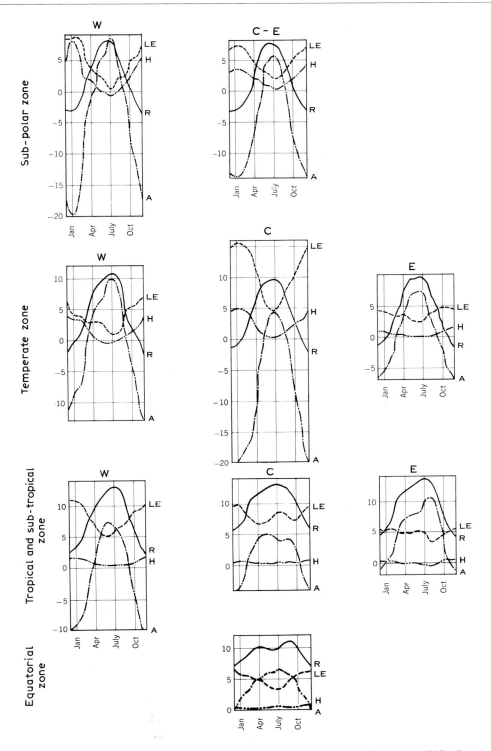

Fig. 50. Seasonal variation of energy budget components in climatic zones (STROKINA, 1959). R = net radiation; LE = loss of heat due to evaporation; H = loss of heat due to turbulent exchange across the ocean/atmosphere interface; A = heat exchange with the deeper layers of the ocean. Vertical scale is kcal. cm^{-2} month^{-1}).

Tropical and subtropical zone

Northwest region (W). The amplitude of both net radiation and heat exchange with deeper layers is greater than in the central region owing to the low winter values associated with the outbreaks of cold air from the American continent. The frontogenesis prevalent in the western ocean occurs to the north of this region, but results in more cloud cover than in other parts of the tropical zone. High values of evaporation occur in winter due to the cold continental air overlying the warm ocean surface and high values of wind speed associated with synoptic activity. Sensible heat exchange is positive throughout the year and winter values are higher than in any other part of the tropical zone.

Central region (C). Both net radiation and heat exchange with the deeper layers vary little during the summer months. The annual cycle of evaporation is similar to that in the equatorial region, and the sensible heat exchange is small but positive throughout the year.

Eastern region (E). The main physical factors affecting the heat budget are the cold Canary Current and the low and strong trade wind inversion. In the warmest part of the year, the net radiation is very high (about 13 kcal. cm^{-2} $mon.^{-1}$). Evaporation is relatively constant and limited because the high moisture content beneath the inversion leads to a small hydro-lapse at the surface. During the summer months, the sensible heat exchange is negative, that is, from the atmosphere to the sea.

Temperate zone

Western region (W). The net radiation is negative in winter and the annual range of all four elements in the heat balance equation is large. Both evaporation and sensible heat exchange have maximum values in winter and minimum values in summer; the sensible heat exchange becomes negative in one or two summer months—the effect of the northwest extension of warm air over a cold sea surface. The heat loss due to evaporation and sensible heat exchange in winter occurs at the expense of the heat reservoir in deeper waters, but these losses and gains throughout the area approximately balance.

Central region (C). The net radiation cycle is similar to that in the western region, but the evaporation is greater, especially in winter. The Gulf Stream and central Atlantic current evaporate more than any other ocean surface, but the warm water advection associated with these features plays an important part in the overall heat balance. There is an upward flux of sensible heat across the surface in winter.

Eastern region (E). The net radiation cycle is similar to that in other temperate zones, but the amplitude of the evaporation cycle is much less because of the lower surface water temperature and the greater water vapour content of the air which generally has had a long sea track. The sensible heat flux is also the smallest in the temperate zone and is effectively zero throughout the summer. This is a reflection of the very small air – sea temperature differences.

Subpolar zone

Western region (W). Net radiation has a well-defined summer maximum in this zone.

Both evaporation and sensible heat transfer are high in winter months and low in summer months, sensible heat transfer becoming negative in mid-summer. These low values are owing to the cold water associated with the Labrador Current and to reduced wind speeds. The negative values in sensible heat fluxes are associated with an air temperature slightly above the sea temperature. The strongly negative values of heat exchange with the deeper layers suggest that the ocean provides the energy for evaporation and heat transfer in winter.

Central and eastern regions (C−E). The main differences between this region and the west is that the annual range of both evaporation and heat transfer is smaller here.

This climatic classification suggested by Strokina provides an excellent descriptive summary of the physical climatology of the North Atlantic Ocean.

References

ALAKA, M. A., 1976. Climatology of Atlantic tropical storms and hurricanes. In: W. SCHWERDTFEGER (Editor), *Climates of Central and South America. World Survey of Climatology, 12.* Elsevier, Amsterdam, pp. 479−509.

ANDERSSEN, T., 1962. Precipitation over the North Atlantic Ocean: Correspondence. *Q. J. R. Meteorol. Soc.,* 88: 187.

ATKINSON, G. D. and SADLER, J. C., 1970. Mean cloudiness and gradient-level-wind charts over the tropics, Vol. 2. *Air Weather Service Technical Report 215.* U.S. Air Force, 48 charts.

AUGSTEIN, E., RIEHL, H., OSTAPOFF, F. and WAGNER, V., 1973. Mass and energy transports in an undisturbed trade-wind flow. *Mon. Weather Rev.,* 101: 101−111.

BJERKNES, J., 1963. Climatic change as an ocean−atmosphere problem. In: *Changes of Climate. Arid Zone Research Report No. 20.* UNESCO, Paris, pp. 297−321.

BJERKNES, J., 1964. Atlantic air−sea interaction. *Adv. Geophys.,* 10: 1−82.

BÖHNECKE, G., 1936. Temperatur, Salzgehalt und Dichte an der Oberfläche des Atlantischen Ozeans. Atlas. *Wiss. Erg. Dtsch. Atlantischen Expedition, "Meteor" 1925−1927,* Vol. 5. W. de Gruyter, Berlin, 76 maps.

BROCKS, K., 1963. Probleme der maritimen Grenzschicht der Atmosphäre. *Ber. Dtsch. Wetterdienstes,* 91: 340−346.

BROWN, P. R., 1953. Climatic fluctuations in the Greenland and Norwegian Seas. *Q. J. R. Meteorol. Soc.,* 79: 272−281.

BROWN, P. R., 1963. Climatic fluctuations over the oceans and in the tropical Atlantic. In: *Changes of Climate. Arid Zone Research Report No. 20.* UNESCO, Paris, pp. 109−123.

BRUMMER, B., 1976. The kinematics, dynamics and kinetic energy budget of the trade wind flow over the Atlantic Ocean. *"Meteor" Forschungsergebnisse,* B, 11: 1−26.

BRYAN, K., 1963. Ideas about ocean currents from Benjamin Franklin's day to the present. *Mariners Weather Log,* 8: 199−201.

BUDYKO, M. I., 1963. *Atlas teplovogo balansa zemnogo shara.* (Atlas of the Global Heat Balance.) Joint Geophys. Comm., Presidium of the Academy of Sciences, USSR, Moscow, 75 pp.

BUNKER, A. F., 1975. Energy exchange at the surface of the western North Atlantic Ocean. *Woods Hole Oceanogr. Inst. Mass. Tech. Rep.,* 75−3: 107 pp.

BUNKER, A. F., 1976. Computations of surface energy flux and annual air−sea interaction cycles of the North Atlantic Ocean. *Mon. Weather Rev.,* 104: 1122−1140.

BUNKER, A. F. and WORTHINGTON, L. V., 1976. Energy exchange charts of the North Atlantic Ocean. *Bull. Am. Meteorol. Soc.,* 57: 670−678.

CHARNOCK, H., FRANCIS, J. R. D. and SHEPPARD, P. A., 1956. An investigation of wind structure in the trades: Anegada 1953. *Phil. Trans. R. Soc. London, Ser. A,* 249: 179−234.

CHASE, J., 1956. The Bermuda-Azores high pressure cell, its surface wind circulation. *Misc. Geofis. Publ., Meteorol. Serv. Angola, Luanda,* (Comem. X Anivers. Serv. Meteorol. Nacional), pp. 29−54.

COHEN, T. J. and SWEETSER, E. I., 1975. The "spectra" of the solar cycle and of data for Atlantic tropical cyclones. *Nature,* 256: 295−296.

COLEBROOK, J. M., 1976. Trends in the climate of the North Atlantic Ocean over the past century. *Nature,* 263: 576−577.

COLÓN, J. A., 1953. A study of hurricane tracks for forecasting purposes. *Mon. Weather Rev.*, 81: 53−66.

CRADDOCK, J. M., 1964. The inter-annual variability of monthly mean air temperatures over the Northern Hemisphere. *Meteorol. Office, London, Sci. Pap.*, 20: 10 pp.

CRUTCHER, H. L. and QUAYLE, R. G., 1974. *Mariners Worldwide Climatic Guide to Tropical Storms at Sea.* Naval Weather Service Command, NAVAIR 50-IC-61, Washington, D. C., 114 pp. and 312 charts.

CUSHING, D. H., 1976. The impact of climatic change on fish stocks in the North Atlantic. *Geogr. J.*, 142: 216−227.

DANSKE METEOROL. INST., 1975. *Monthly Surface Temperatures of the Northern Atlantic.* Publ. danske meteorol. inst., Årbøger, København, 28 pp.

DEFANT, A., 1961. *Physical Oceanography.* Pergamon, London, 1, 729 pp.

DICKSON, R. and LAMB, H. H., 1972. A review of recent hydrometeorological trends in the North Atlantic sector. *Int. Comm. NW. Atl. Fisheries Spec. Pub.*, 8: 35−62.

DICKSON, R. and LEE, A., 1969. Atmospheric and marine climate fluctuations in the North Atlantic region. *Progr. Oceanogr.*, 5: 55−65.

DICKSON, R., LAMB, H. H., MALMBERG, S. A. and COLEBROOK, J. H., 1975. Climatic reversal in the northern North Atlantic. *Nature*, 256: 479−482.

DUNBAR, M., 1972. Increasing severity of ice conditions in Baffin Bay and Davis Strait. In: T. KARLSSON (Editor), *Sea Ice. Proceedings of an International Conference.* Nat. Res. Council, Reykjavik, pp. 87−93.

DUNN, G. E., 1956. Areas of hurricane development. *Mon. Weather Rev.*, 84: 47−51.

DUNN, G. E. and MILLER, B. I., 1960. *Atlantic Hurricanes.* State University Press, Louisiana, 326 pp.

FRANK, N. L., 1975. Atlantic tropical systems. *Mon. Weather Rev.*, 103: 294−300.

GORDON, H. A., 1951. Seasonal variation of the axes of low latitude pressure and divergence patterns over the oceans. *Q. J. R. Meteorol. Soc.*, 77: 302−306.

HASTENRATH, S. and LAMB, P., 1977. *Climatic Atlas of the Tropical Atlantic and Eastern Pacific Oceans.* University of Wisconsin Press, 112 pp.

HELLERMAN, S., 1965. Computation of wind stress fields over the Atlantic Ocean. *Mon. Weather Rev.*, 93: 239−244.

HÖHN, R., 1973. On the climatology of the North Sea. In: E. D. GOLDBERG . (Editor), *North Sea Science.* M.I.T. Press, Cambridge, Mass., pp. 183−236.

HOLLAND, J. Z. and RASMUSSEN, E. M., 1973. Measurements of the atmospheric mass, energy and momentum budgets over a 500-kilometer square of tropical ocean. *Mon. Weather Rev.*, 101: 44−55.

JACOBS, W. C., 1942. On the energy exchange between sea and atmosphere. *J. Marine Res.*, 5: 37−66.

JACOBS, W. C., 1951. The energy exchange between sea and atmosphere, and some of its consequences. *Bull. Scripps Inst. Oceanogr., Univ. Calif.*, 6 (2): 122 pp.

JACOBS, W. C., 1968. The seasonal apportionment of precipitation over the ocean. In: A. COURT (Editor), *Eclectic Climatology-Yearbook, Assoc. Pac. Coast Geogr.*, 30: 63−78.

JAKOBSSON, T. E., 1973. *Spectral Analysis of Marine Atmospheric Time Series.* Unpubl. Ph.D. thesis, McGill Univ., Montreal, 267 pp.

JAKOBSSON, T., 1976. Time sequence of energy exchange spectra at Ocean Stations Charlie and Delta. *Arch. Meteorol. Geophys. Bioklimatol.*, A25: 187−206.

JENKINSON, A. F., 1956. The relation between standard deviation of contour height and standard vector deviation of wind. *Q. J. R. Meteorol. Soc.*, 82: 198−208.

KIRK, T. H., 1953. Seasonal change of surface temperature of the North Atlantic Ocean. *Geophys. Mem. (London)*, 11 (5), 90: 34 pp.

KLEIN, W. J., 1957. Principal tracks and mean frequencies of cyclones and anticyclones in the Northern Hemisphere. *U.S. Weather Bur., Res. Pap.*, 40: 123 pp.

KRAUS, E. B., 1955. Secular changes of tropical rainfall regimes. *Q. J. R. Meteorol. Soc.*, 81: 198−210.

KRAUS, E. B., 1963. The diurnal precipitation change over the sea. *J. Atmos. Sci.*, 20: 551−556.

KRAUS, E. B. and MORRISON, R. E., 1966. Local interactions between the sea and the air at monthly and annual time scales. *Q. J. R. Meteorol. Soc.*, 92: 114−127.

KUETTNER, J. P., 1974. General description and central program of GATE. *Bull. Am. Meteorol. Soc.*, 55: 712−719.

LAMB, H. H., 1973. The seasonal progression of the general atmospheric circulation affecting the North Atlantic and Europe. *Climatic Res. Unit. Univ. East Anglia, CRU RP1, Norwich*, 83 pp.

LAMB, H. H. and JOHNSON, A. I., 1959. Climatic variation and observed changes in the general circulation. *Geograf. Ann.*, 49 (2/3): 94−134 (parts I, II).

LAMB, H. H. and JOHNSON, A. I., 1961. Climatic variation and the observed changes in the general circulation. *Geograf. Ann.*, 43 (3/4): 363−400 (part III).

LAMB, H. H. and JOHNSON, A. I., 1966. Secular variations of the atmospheric circulation since 1750. *Geophys. Mem., London*, 14 (5), 110: 125 pp.

LUMB, F. E., 1961. Seasonal variations of the surface temperature in coastal waters of the British Isles. *Meteorol. Office, London, Sci. Pap.*, 6: 21 pp.

LUMB, F. E., 1964. The influence of cloud on hourly amounts of total solar radiation at the sea surface. *Q. J. R. Meteorol. Soc.*, 90: 43−56.

MALMBERG, S. A., 1972. Annual and seasonal hydrographic variations in the East Icelandic Current between Iceland and Jan Mayen. In: T. KARLSSON (Editor), *Sea Ice. Proceedings of an International Conference.* Nat. Res. Council, Reykjavik, pp. 42−53.

MATHER, J. R., ADAMS, H. III and YOSHIOKA, G. A., 1964. Coastal storms of the eastern United States. *J. Appl. Meteorol.*, 3: 693−706.

MAURY, M. F., 1857. *The Physical Geography of the Sea.* Harper, New York, N.Y., 360 pp.

MAZEIKA, P. A., 1968. Mean monthly sea surface temperatures and zonal anomalies of the tropical Atlantic. *Serial Atlas of the Marine Environment.* Am. Geogr. Soc., New York, Folio 16, 6 plates.

MEINARDUS, W., 1934. Eine neue Niederschlagskarte der Erde. *Petermanns Mitt., Gotha*, 80: 1−4.

MESERVE, J. M., 1974. *U.S. Navy Marine Climatic Atlas of the World, Vol. 1. North Atlantic Ocean.* Naval Air Systems Command, Washington, D.C. NAVAIR 50-IC-528, 385 pp.

METEOROLOGICAL OFFICE, 1948. *Monthly Meteorological Charts of the Atlantic Ocean.* H.M. Stationery Office, London, M.O. 483.

MILES, M. K., 1958. Synoptic study of thermal troughs over the Atlantic and the British Isles. *Meteorol. Mag.*, 87: 1−12.

MILTON, D., 1974. Some observations of global trends in tropical cyclone frequencies. *Weather*, 29: 267−270.

MÖLLER, F., 1951. Vierteljahreskarten des Niederschlages für die ganze Erde. *Petermanns Geograph. Mitt.*, 95: 1−7.

MONIN, A. S., 1963. On the climatology of heat balance. *Izv. Akad. Nauk. U.S.S.R., Ser. Geograph.*, 5: 89−110.

NAMIAS, J., 1964. Seasonal persistence and recurrence of European blocking during 1958−60. *Tellus*, 16: 394−407.

NAVY MINISTRY, Moscow, 1953. *Morskoi Atlas* (Marine Atlas). Glavnyi Shtab Voenna-Morshikh Sil., Moscow, II: map 48(b).

NAZAROV, V. S., 1968. The duration and distribution of anomalies in temperature of the ocean surface (in Russian). *Akad. Nauk. Okeanograf. (Moscow)*, 8: 23−25.

NEWELL, R. E., KIDSON, J. W., VINCENT, D. G. and BOER, G. J., 1972. *The General Circulation of the Tropical Atmosphere, Vol. 1.* MIT Press, Cambridge, 258 pp.

PAYNE, R. E., 1972. Albedo of the sea surface. *J. Atmos. Sci.*, 29: 959−970.

PERRY, A. H., 1968. Turbulent heat flux patterns over the North Atlantic during recent winter months. *Meteorol. Mag.*, 97: 246−254.

PERRY, A. H., 1969. *A Synoptic Climatology of the Thermal Surface and Heat Balance of the North Atlantic.* Unpub. Ph.D. Thesis, Univ. of Southampton, 388 pp.

PERRY, A. H., 1974. The downward trend of air and sea surface temperatures over the North Atlantic. *Weather*, 29: 451−455.

PERRY, J. D., 1968. Sea temperatures at OWS "I". *Meteor. Mag.*, 97: 33−42.

PFLUGBEIL, C. and STEINBORN, E., 1963. *Zur Klimatologie des Nordatlantischen Ozeans.* Deutscher Wetterdienst, Hamburg, Einzelveröffentl., 38/40.

PRESCOTT, J. A. and COLLINS, J. A., 1951. The lag of temperature behind solar radiation. *Q. J. R. Meteorol. Soc.*, 77: 598−626.

PRIESTLEY, C. H. B., 1951. A survey of the stress between the ocean and the atmosphere. *Aust. J. Sci. Res.*, A4: 315−328.

QUAYLE, R. G., 1974. A climatic comparison of ocean weather stations and transient ship records. *Mariners Weather Log*, 18: 307−311.

RATCLIFFE, R. A. S., 1971. North Atlantic sea temperature classification 1877−1970. *Meteorol. Mag.*, 100: 225−231.

RATCLIFFE, R. A. S. and MURRAY, R., 1970. New lag-associations between North Atlantic sea temperatures and European pressure applied to long-range weather forecasting. *Q. J. R. Meteorol. Soc.*, 96: 226−246.

REED, R. J., 1960. Principal frontal zones of the northern hemisphere in winter and summer. *Bull. Am. Meteorol. Soc.*, 41: 591−598.

REED, R. K. and ELLIOTT, W. P., 1977. New precipitation maps for the North Atlantic and North Pacific Oceans. *J. Geophys. Res.*, 84: 7839−7846.

RESIO, D. T. and HAYDEN, B. P., 1975. Recent secular variations in mid-Atlantic winter extra-tropical storm climate. *J. Appl. Meteorol.*, 14: 1223−1234.

REX, D. F., 1950. Blocking action in the middle troposphere and its effects upon regional climate, II. The climatology of blocking action. *Tellus*, 2: 275−301.

RIEHL, H., 1954. *Tropical Meteorology.* McGraw-Hill, New York, N.Y., 392 pp.

RIEHL, H., MALKUS, J. S., YEH, T.-C. and LaSEUR, N. E., 1951. The north-east trade of the Pacific Ocean. *Q. J. R. Meteorol. Soc.*, 77: 598−626.

RODEWALD, M., 1972. Einige hydroklimatische Besonderheiten des Jahrzents 1961−1970 im Nordatlantik und im Nordpolarmeer. *Dtsch. Hydrogr. Z.*, 25: 97−117.

RODEWALD, M., 1973. Der Trend der Meerestemperatur im Nordatlantik. *Beil. Berliner Wetterkarte* SO29/73, 6 pp.

ROGERS, J. C. and VAN LOON, H., 1979. The seesaw in winter temperatures between Greenland and northern Europe, Part II. Sea ice, sea surface temperatures and winds. *Mon. Weather Rev.*, 107: 509−519.

ROLL, H. U., 1964. *Physics of the Marine Atmosphere.* Academic Press, New York, International Geophysical Series, 7: 426 pp.

SADLER, J. C., 1975. The monsoon circulation and cloudiness over the GATE area. *Mon. Weather Rev.*, 103: 369−387.

SATER, J. E., 1969. *The Arctic Basin.* Revised ed., Arct. Inst., North America, Washington, D. C., 337 pp.

SCHELL, I. I., 1956. Interrelations of arctic ice with the atmosphere and the ocean in the North Atlantic, Arctic and adjacent areas. *J. Meteorol.*, 13: 46−58.

SCHOTT, P. G., 1926. *Geographie des Atlantischen Ozeans.* Boysen, Hamburg, 368 pp.

SCHOTT, P. G., 1935. *Geographie des Indischen und Stillen Ozeans.* Boysen, Hamburg, 413 pp.

SCHUMANN, T. E. W. and VAN ROOY, M. P., 1951. Analysis of the standard deviation of atmospheric pressure over the Northern Hemisphere. *Union S. Africa, Weather Bur., Dept. Transport, W.B.*, 16: 21 pp.

SHEPPARD, P. A., 1954. The vertical transfer of momentum in the general circulation. *Arch. Meteorol. Geophys. Bioklimatol., Ser. A.*, 7: 114−124.

SIMPSON, G. C., 1929. The distribution of terrestrial radiation. *Mem. R. Meteorol. Soc.*, 3: 53−80.

SORKINA, A. I., 1965. Types of atmospheric circulation and wind field over the North Atlantic (in Russian). *Tr. Gos. Okean. Inst. (Moscow)*, 84: 1−133.

STROKINA, L. A., 1959. Teplovoi balans Severnoi Atlantiki (Heat balance of the North Atlantic, transl.). *Tr. Gl. Geofiz. Obs.*, 92: 27−49.

STROKINA, L. A., 1964. The heat balance of the North Atlantic by comparison with the heat balance of other regions of the world (in Russian). *Proc. Interaction between the Atmosphere and Hydrosphere in the North Atlantic, 2nd., Leningrad, Gidromet. Inst.*, 198 pp.

SUMNER, E. J., 1954. A study of blocking in the Atlantic-European sector of the Northern Hemisphere. *Q. J. R. Meteorol. Soc.*, 80: 402−416.

TAUBER, G. M., 1969. The comparative measurements of sea-surface temperature in the USSR. In: *Sea-Surface Temperature, W.M.O. Tech. Note 103*, (WMO No. 247T, TP. 135), World Meteorological Organization, Geneva, pp. 141−151.

THOMPSON, O. E. and MILLER, J., 1976. Hurricane Carmen: August−September 1974—development of a wave in the ITCZ. *Mon. Weather Rev.*, 104: 1196−1199.

TUCKER, G. B., 1961. Precipitation over the North Atlantic Ocean. *Q. J. R. Meteorol. Soc.*, 87: 147−158.

TUCKER, G. B., 1962. Precipitation over the North Atlantic Ocean: Correspondence. *Q. J. R. Meteorol. Soc.*, 88: 188.

TUCKER, G. B., 1965. The equatorial tropospheric wind regime. *Q. J. R. Meteorol. Soc.*, 91: 140−150.

U.S. WEATHER BUREAU, 1955. *U.S. Navy Marine Climatic Atlas of the World, Vol. 1. North Atlantic Ocean.* NAVAER 50-lC-528, Chief of Naval Operations, Washington, D. C.

U.S. WEATHER BUREAU and U.S. HYDROGRAPHIC OFFICE, 1959. *Climatological and Oceanographic Atlas for Mariners, I. North Atlantic Ocean.* U.S. Weather Bureau and U.S. Hydrographic Office, Washington, D. C.

VAN DIJK, W., SCHMIDT, F. H. and SCHUURMANS, C. J. E., 1974. Beschrijving en toepassingsmogelijkheden van gemiddelde topografieën van het 500 mbar-vlak in afhankelijkheid van circulatietypen. *Kon. Ned. Meteorol. Inst., Wetenschapp. Rapp.*, 74-3: 31 pp.

VAN LOON, H. and ROGERS, J. C., 1978. The seesaw in winter temperatures between Greenland and northern Europe, Part I. General description. *Mon. Weather Rev.*, 106: 296−310.

VERPLOEGH, G., 1967. Observation and analysis of the surface wind over the ocean. *Meded. Verh. Kon. Ned. Meteorol. Inst.*, 89: 67 pp.

VON FICKER, H., 1936. Die Passatinversion. *Veröffentl. Meteorol. Inst. Univ. Berlin*, 1(4): 33 pp.

WAGNER, V., 1975. Zusammenhänge zwischen tropophärischer Zirkulation und den energetischen Prozessen im Bereich der Hadleyzirkulation über dem Atlantik. *Ber. Inst. Radiometeorol. Marit. Meteorol., Univ. Hamburg*, 6: 83 pp.

WEARE, B. C., 1977. Empirical orthogonal analysis of Atlantic Ocean surface temperatures. *Q. J. R. Meteorol. Soc.*, 103: 467−478.

WHITING, G. C., 1959. Frontal passages over the North Atlantic Ocean. *Mon. Weather Rev.*, 87: 409–415.

WORTHINGTON, L. V., 1976. *On the North Atlantic Circulation.* Johns Hopkins University Press, Baltimore, Oceanographic Studies, 6, 110 pp.

WÜST, G., 1920. Die Verdunstung auf dem Meere. *Veröffentl. Inst. Meeresk. Univ. Berlin*, A6: 1–95.

WÜST, G., 1936. Oberflächensalzgehalt, Verdunstung und Niederschlag auf dem Weltmeere nebst Bemerkungen zum Wasserhaushalt der Erde. *Festschr. Norbert Krebs, Landerkundl. Forsch., Stuttgart*, pp. 347–359.

ZUBENOK, L. I. and STROKINA, L. A., 1963. Isparienie s poverkhnosti zemnogo shara (Evaporation from the earth's surface). *Tr. Gl. Geofiz. Obs.*, 139: 93–107.

Appendix – Climatic tables

TABLE XIII*

CLIMATIC TABLE FOR THE LAND STATIONS

Station		Jan.	Feb.	Mar.	Apr.	May	June	July	Aug.	Sept.	Oct.	Nov.	Dec.	Year
Frobisher	P	1011.9	1012.2	1018.2	1017.0	1015.1	1011.2	1007.7	1007.8	1007.6	1006.9	1009.3	1007.6	1011.0
64°N 69°W	T	−26.9	−24.8	−21.2	−13.2	−2.2	3.9	8.1	7.0	2.7	−5.0	−13.8	−21.7	−8.8
	VP					4.0	6.4	8.2	8.1	6.0				
	R	32	35	24	26	21	42	70	58	38	43	40	28	457
Reykjavik	P	999.6	1005.0	1006.7	1008.7	1014.3	1011.5	1008.7	1007.9	1005.5	1002.5	1001.8	998.5	1005.9
64°N 22°W	T	−0.4	−0.1	1.5	3.1	6.9	9.5	11.2	10.8	8.6	4.9	2.6	0.9	5.0
	VP	5.1	5.2	5.5	6.2	7.6	9.1	10.6	10.5	9.3	7.3	6.2	5.5	7.1
	R	90	65	65	53	42	41	48	66	72	97	85	81	805
Thorshavn	P	1003.6	1006.8	1009.9	1009.7	1015.2	1012.4	1009.9	1009.6	1008.3	1006.5	1004.9	1002.4	1008.3
62°N 07°W	T	3.9	3.7	4.6	5.4	7.3	9.2	11.0	11.1	10.0	7.9	6.1	5.0	7.1
	VP	5.4	5.2	5.8	7.4	9.4	12.0	14.5	14.6	12.4	9.7	7.8	6.4	9.2
	R	49	39	32	38	42	47	71	66	62	59	48	49	602
Goose	P	1011.6	1011.7	1013.0	1012.5	1013.1	1010.7	1008.9	1009.5	1010.2	1011.4	1010.5	1009.2	1011.0
53°N 60°W	T	−16.3	−14.1	−8.4	−1.7	4.9	11.3	16.3	14.6	10.0	3.3	−3.9	−12.9	0.2
	VP	1.3	1.5	2.3	3.9	5.8	9.0	12.6	11.6	8.8	5.7	3.5	1.7	4.5
	R	72	63	68	62	56	72	84	91	76	63	67	63	837
Valentia	P	1012.3	1014.7	1012.5	1015.5	1015.3	1016.4	1015.4	1014.7	1014.8	1014.1	1011.7	1011.8	1014.1
52°N 10°W	T	6.9	6.8	8.3	9.4	11.4	13.8	15.0	15.4	14.0	11.6	9.1	7.8	10.8
	VP	8.4	8.3	9.0	9.2	10.2	12.6	14.3	14.7	13.4	11.6	9.8	8.9	10.7
	R	101	62	124	30	75	62	93	78	134	88	80	136	1063
	R*	165	123	104	89	82	85	102	120	114	144	144	164	1436

TABLE XIII (*continued*)

Station		Jan.	Feb.	Mar.	Apr.	May	June	July	Aug.	Sept.	Oct.	Nov.	Dec.	Year
Sable Island	P	1012.7	1011.6	1011.6	1014.2	1015.5	1014.3	1016.4	1015.6	1018.0	1016.9	1015.3	1013.6	1014.6
44°N 60°W	T	-0.2	-1.1	0.4	3.4	6.8	10.9	15.7	17.8	16.1	11.6	7.3	2.4	7.6
	VP	5.1	4.6	5.2	6.7	9.0	12.1	16.8	18.7	15.9	11.5	8.7	6.0	9.1
	R	125	114	106	94	87	81	74	101	97	110	126	133	1248
Horta	P	1019.7	1018.4	1016.2	1020.5	1021.3	1023.5	1025.6	1023.6	1021.6	1020.1	1019.5	1021.5	1021.0
39°N 29°W	T	14.4	14.3	14.3	15.2	16.8	19.2	21.6	22.8	21.6	19.2	16.9	15.4	17.6
	VP	13.3	13.2	13.2	13.6	15.3	18.2	20.1	21.6	20.1	17.6	15.2	14.2	16.1
	R	125	106	122	68	70	48	32	44	81	110	109	113	1028
Funchal	P	1024.5	1022.7	1020.3	1020.4	1021.3	1022.4	1022.1	1021.0	1020.7	1019.8	1021.2	1023.5	1021.7
33°N 17°W	T	15.8	15.6	16.0	16.7	17.7	19.6	21.0	21.9	21.8	20.7	18.6	16.7	18.5
	VP	12.9	12.6	12.9	13.3	14.4	16.9	18.1	19.2	18.8	17.3	15.0	13.5	15.3
	R	84	85	71	44	21	5	2	2	29	82	97	94	616
Hatteras	P	1019.6	1018.3	1016.7	1016.7	1016.2	1015.4	1016.6	1016.3	1017.2	1017.8	1019.2	1020.0	1017.5
35°N 76°W	T	8.1	8.1	10.6	15.2	20.0	24.0	25.6	25.3	23.4	18.6	13.4	9.0	16.8
	VP	8.7	8.6	10.1	13.5	18.2	23.9	26.6	26.1	22.7	17.4	12.3	9.3	15.3
	R	99	100	106	58	101	105	156	163	150	108	104	116	1384
Miami	P	1019.8	1018.8	1017.5	1016.9	1015.8	1016.1	1017.4	1016.0	1014.3	1014.6	1017.2	1019.1	1017.0
26°N 80°W	T	19.4	19.9	21.4	23.4	25.3	27.1	27.7	27.9	27.4	25.4	22.4	20.1	23.9
	VP	16.4	16.7	18.1	20.1	22.9	26.5	27.5	27.4	27.7	24.3	19.8	17.4	21.6
	R	52	47	58	99	164	187	171	177	241	209	72	42	1518
San Juan	P	1016.8	1016.8	1016.4	1015.8	1015.0	1016.3	1016.8	1015.1	1013.9	1012.9	1013.1	1015.4	1015.3
15°N 75°W	T	23.6	23.6	24.1	24.7	25.9	26.7	26.9	27.2	26.9	26.7	25.7	24.6	25.6
	VP	21.8	21.8	21.9	23.3	25.7	27.3	27.6	28.1	27.6	27.3	25.8	23.2	25.3
	R	119	74	56	94	181	144	159	181	172	148	165	138	1631
Piarco	P	1013.2	1013.3	1012.7	1012.7	1012.8	1013.6	1013.7	1012.8	1012.0	1011.4	1011.1	1012.0	1012.6
11°N 61°W	T	24.5	24.7	25.4	26.3	26.6	26.1	25.9	26.1	26.2	25.9	25.4	24.8	25.7
	VP	25.2	24.9	25.0	26.3	27.5	28.7	28.4	28.7	28.9	28.4	28.2	26.9	27.4
	R	77	61	27	71	129	269	243	213	144	151	212	153	1750

TABLE XIII (continued)

Station		Jan.	Feb.	Mar.	Apr.	May	June	July	Aug.	Sept.	Oct.	Nov.	Dec.	Year
Cayenne	P	1011.7	1012.1	1012.1	1012.1	1012.3	1013.2	1013.7	1013.1	1012.6	1011.7	1010.7	1010.9	1012.2
5°N 52°W	T	25.1	25.2	25.5	25.7	25.3	25.2	25.1	25.6	26.1	26.3	26.0	25.4	25.5
	VP	28.0	27.9	28.1	28.7	29.0	28.5	28.0	28.2	28.1	28.1	28.2	28.5	28.1
	R	431	423	432	480	590	457	274	144	32	42	122	317	3744
Dakar	P	1012.7	1012.3	1011.6	1011.4	1011.8	1012.9	1012.6	1011.8	1011.6	1011.9	1011.6	1012.6	1012.1
15°N 17°W	T	21.2	20.4	20.7	21.6	23.1	26.0	27.2	27.2	27.5	27.4	26.0	23.1	24.3
	VP	17.4	18.0	18.5	19.9	22.0	25.5	27.4	28.1	29.0	28.8	24.5	18.1	22.8
	R	0	2	0	0	1	15	88	249	163	49	5	6	578
Sao Tomé	P	1010.5	1010.2	1010.0	1010.3	1011.5	1013.7	1014.4	1014.0	1013.0	1012.1	1011.2	1010.7	1011.8
00°N 07°E	T	27.6	26.2	26.4	26.4	26.0	24.7	23.8	24.2	24.9	25.2	25.4	25.6	25.5
	VP	31.0	28.2	28.2	28.6	27.9	24.3	22.4	22.9	24.6	26.3	27.2	27.6	26.4
	R	82	84	132	120	113	16	0	1	18	108	100	98	872

* Climatological Tables XIII and XIV: Data for the land stations were compiled from U.S. Weather Bureau World Weather Records, 1951–1960. Values of pressure and air and sea temperatures for the Ocean Weather Ships were taken from Pflugbeil and Steinborn (1963); vapour pressure values were calculated from dewpoint data compiled at the U.K. Meteorological Office, Bracknell. P is mean monthly sea level pressure in mbar; T is mean monthly surface air temperature in °C; T_s is mean monthly sea surface temperature in °C; VP is mean monthly surface vapour pressure in mbar. For land stations, VP is calculated from T and the mean monthly relative humidity; for Ocean Weather Ships, VP is calculated from mean monthly surface dewpoints; R is mean monthly rainfall in mm; R^* is 1871–1950

TABLE XIV*

CLIMATIC TABLE FOR OCEAN WEATHER SHIPS

O.W.S.		Jan.	Feb.	Mar.	Apr.	May	June	July	Aug.	Sept.	Oct.	Nov.	Dec.	Year
A	P	999.0	1003.3	1003.1	1007.5	1012.2	1011.2	1010.6	1009.8	1004.3	998.2	999.3	992.9	1004.3
62°N 33°W	T	2.1	2.7	3.2	4.0	5.8	7.7	9.4	9.9	8.7	6.3	4.8	2.9	5.6
	T_s	5.2	5.1	5.1	5.4	6.4	8.2	9.9	10.6	9.8	7.7	6.6	5.6	7.1
	VP	5.4	5.9	6.2	6.5	7.7	9.0	10.2	10.4	9.3	7.7	7.0	5.8	7.6
B	P	1004.2	1005.9	1007.1	1009.1	1012.6	1011.8	1010.0	1009.1	1007.4	1005.2	1004.8	1001.7	1007.4
57°N 51°W	T	−1.5	−0.9	0.2	1.7	3.5	5.6	8.3	9.2	7.8	4.8	2.5	0.1	3.4
	T_s	3.2	2.6	2.8	3.2	4.2	5.7	7.7	9.2	8.7	6.6	4.9	3.9	5.2
	VP	4.4	4.7	4.9	5.7	6.8	8.1	10.1	11.8	8.8	6.9	5.8	4.7	6.5
C	P	1005.7	1007.8	1003.8	1011.4	1012.8	1012.9	1015.6	1013.5	1011.1	1008.7	1009.1	1004.6	1009.8
53°N 35°W	T	5.4	5.3	5.5	6.8	8.3	10.0	12.1	12.9	12.1	9.6	7.7	6.0	8.5
	T_s	6.7	6.4	7.0	7.1	8.0	9.6	11.5	12.8	12.4	10.2	9.0	7.6	9.0
	VP	7.2	7.0	7.3	8.3	9.1	10.7	12.4	12.9	11.9	9.5	8.3	7.5	9.1
D	P	1011.9	1011.0	1007.0	1014.9	1015.8	1018.4	1021.8	1019.1	1017.7	1015.8	1017.2	1015.1	1015.5
44°N 41°W	T	12.1	11.1	11.2	12.6	14.1	16.3	19.9	21.4	19.1	16.2	14.6	12.7	15.1
	T_s	15.8	15.2	15.0	14.5	15.4	17.1	20.1	22.2	20.5	18.3	17.7	16.1	17.3
	VP	11.1	10.3	10.2	11.7	13.0	15.3	19.3	21.1	17.5	14.7	12.3	11.5	13.7
E	P	1016.8	1016.8	1014.9	1020.1	1020.0	1022.3	1024.9	1022.7	1020.5	1019.1	1020.3	1019.8	1019.9
35°N 48°W	T	17.8	16.7	17.2	17.7	19.2	22.1	25.0	26.2	25.1	23.1	21.1	19.5	20.9
	T_s	19.3	18.6	17.6	18.4	19.6	21.8	25.1	26.6	26.1	24.2	22.5	21.0	21.7
	VP	15.6	14.4	14.4	15.8	18.2	22.5	25.0	25.9	24.1	22.3	18.7	17.6	19.2
I	P	1003.2	1005.5	1004.3	1009.8	1013.7	1012.2	1010.7	1009.0	1006.6	1003.1	1002.5	996.1	1006.4
59°N 19°W	T	5.9	6.6	7.2	7.4	8.9	10.4	12.1	12.5	11.8	9.8	8.8	7.2	9.0
	T_s	9.2	9.1	9.0	9.1	9.9	11.3	12.7	13.2	12.7	11.4	10.5	9.9	10.6
	VP	7.5	7.8	8.0	8.1	9.5	10.5	12.0	11.9	11.3	9.8	9.1	8.2	9.4
J	P	1010.0	1010.1	1005.7	1015.1	1014.5	1014.9	1016.4	1012.6	1011.1	1010.8	1009.2	1005.3	1011.3
52°N 20°W	T	9.0	8.9	9.4	10.0	11.3	12.7	14.2	14.7	14.2	12.5	10.7	9.5	11.4
	T_s	10.8	10.6	10.7	10.9	11.9	13.1	14.7	15.3	14.8	13.5	12.3	11.4	12.5
	VP	9.1	8.7	9.4	9.8	10.6	12.3	13.8	14.1	13.3	11.6	10.3	9.6	10.9

TABLE XIV (*continued*)

O.W.S.		Jan.	Feb.	Mar.	Apr.	May	June	July	Aug.	Sept.	Oct.	Nov.	Dec.	Year
K 45°N 16°W	P	1017.0	1016.4	1010.4	1018.3	1018.2	1020.3	1022.1	1018.9	1018.2	1017.7	1016.6	1017.2	1017.7
	T	12.3	11.5	12.0	12.6	13.9	16.3	18.0	18.9	18.4	16.6	14.4	13.1	14.8
	T_s	13.0	12.6	12.6	13.0	14.0	16.3	18.2	19.0	18.8	17.2	15.4	13.9	15.3
	VP	11.4	10.4	11.4	11.5	13.0	15.1	17.0	17.5	16.7	14.9	12.4	12.1	13.5
M 66°N 02°E	P	1002.7	1006.1	1011.3	1009.7	1015.5	1013.2	1010.4	1009.7	1007.9	1005.9	1005.6	998.8	1008.1
	T	3.2	3.1	3.8	4.3	6.0	7.7	10.3	11.0	9.9	7.8	6.1	4.4	6.5
	T_s	6.6	6.3	6.3	6.4	7.5	8.9	10.9	11.9	11.0	9.3	7.9	7.2	8.4
	VP	5.9	6.0	6.3	6.7	7.6	8.8	10.7	11.2	10.4	8.5	7.7	6.5	7.8

* See footnote to Table XIII

Chapter 3

Climate of the South Pacific Ocean

N. A. STRETEN AND J. W. ZILLMAN

"I now gave up all hopes of finding any more land in this ocean."

JAMES COOK
Log of H.M.S. *Resolution*
November 27, 1774
Lat. 55°6′S 138°56′W

Introduction

The South Pacific Ocean (Fig. 1) including its marginal seas covers an area of approximately 105 million square kilometres, only slightly less than a quarter of the total surface area of the earth. On its western side lie the Bismarck ($0.338 \cdot 10^6$ km²), Solomon ($0.775 \cdot 10^6$ km²), Coral ($4.068 \cdot 10^6$ km²), Fiji ($3.177 \cdot 10^6$ km²), and Tasman ($3.336 \cdot 10^6$ km²) seas. Some 10,000–15,000 km to the east it is bounded by South America and the Antarctic Peninsula. To the south are Antarctica and the Ross ($0.440 \cdot 10^6$ km²), Amundsen ($0.098 \cdot 10^6$ km²), and Bellingshausen ($0.487 \cdot 10^6$ km²) seas. The northern boundary is the Equator, but by convention the islands of Kiribati and the Galápagos group which lie north of the Equator are included in the South Pacific.

The various small seas of the Indonesian Archipelago (e.g., the Java and Banda seas) are properly regarded as marginal seas of the South Pacific, but the Arafura and Timor seas are normally included with the Indian Ocean. The boundary between the Pacific and Indian oceans lies from South East Cape, the southern point of Tasmania, down the meridian 146°53′E to the Antarctic continent, but the climate of the southeastern portion of the Indian Ocean, including the Great Australian Bight, is treated in this chapter as a matter of convenience. Although not generally used by hydrographic authorities, the term "Southern Ocean" or "Antarctic Ocean" is often applied to the oceans around Antarctica south of the so-called Antarctic Convergence (Fig. 4). The boundary of the "Southwest Pacific Ocean" is usually regarded as extending south from the Equator through the Cook Islands to the region of the Subtropical Convergence (Fig. 4).

Fig. 1 identifies the main geographic features, including the marginal seas and principal island groups of the South Pacific, and shows the smoothed bathymetry of the ocean floor. General physical and geographical data on the South Pacific region may be found in most atlases and a comprehensive treatment of geographical, historical, administrative, ethnic, trade, and other information can be found in recent editions of the *Pacific Islands Year Book* (e.g., INDER, 1979).

History of climatic description

There have been few attempts to provide a comprehensive treatment of the climatology

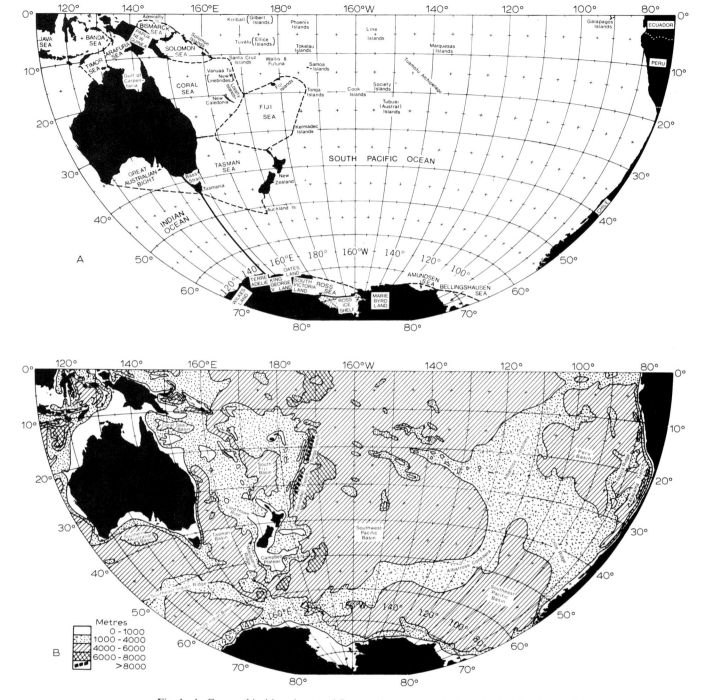

Fig. 1. A. Geographical locations, and B. smoothed deep-sea bathymetry (m) for the South Pacific.

of the entire South Pacific Ocean. This has been partly a data problem. No other ocean, indeed no other region of the world, contains so vast an area completely devoid of conventional meteorological observing stations as the ninety-degree quadrant of the southeast Pacific south of the subtropical ridge. On the other hand, although the tropical southwest Pacific is well endowed with island climatic records of relatively high quality extending back into the last century, systematic studies of the general climatology of the region have been few. Ship data have played a key role in the climatic analyses that do exist, but the number of ship reports in the southeast Pacific has decreased over the years, especially since the opening of the Panama Canal in 1914.

The earliest and most comprehensive studies by SCHOTT (1935, 1938) are still a key source of information for the region. In his definitive *Geographie des Indischen und Stillen Ozeans,* SCHOTT (1935) identified the major features of the climate of the South Pacific and mapped such basic climatic elements as air and sea temperature, ocean currents, surface wind and atmospheric pressure, cyclone tracks, and frequency of gales, cloudiness, and precipitation. His 114-page *Klimakunde der Südsee-Inseln* (SCHOTT, 1938) contains a large number of climatic tables and a detailed description of the climate of the major island groups of the South Pacific. Other important contributions to the general climatology of the region include:

Marine climatic atlases based on ship data

(*1*) The U.S. Weather Bureau's *Atlas of Climatic Charts of the Oceans* (McDONALD, 1938).

(*2*) The U.K. Meteorological Office's *Monthly Meteorological Charts of the Western Pacific Ocean* (METEOROLOGICAL OFFICE, 1947) and a companion volume for the Eastern Pacific (METEOROLOGICAL OFFICE, 1950).

(*3*) The Dutch atlas *Sea Areas Around Australia* (KONINKLIJK NEDERLANDS METEORO-LOGISCH INSTITUUT, 1949).

(*4*) The U.S. Navy's *Marine Climatic Atlas of the World,* volumes on the South Pacific (U.S. NAVY, 1965) and the world (U.S. NAVY, 1969). A revised edition of the South Pacific volume (U.S. NAVY, 1979) was issued in late 1979, but these newer data have not been used in this study.

(*5*) The U.S.S.R.'s *Ocean Atlas: Pacific Ocean* (U.S.S.R. MINISTRY OF DEFENCE, 1974).

Climatic surveys or summaries for all or part of the region

(*1*) Various publications by W. Meinardus including the chapter on the Antarctic in *Handbuch der Klimatologie* (MEINARDUS, 1938).

(*2*) A series of climatological notes for the South Pacific region (SEELYE, 1943–1944) issued by the New Zealand Meteorological Service. (See also ROYAL NEW ZEALAND AIR FORCE, 1943).

(*3*) The Australian wartime publication *Weather on the Australia Station* (ROYAL AUS-TRALIAN AIR FORCE, 1942) which provides detailed information on the climatology of the New Guinea and Solomon Islands regions and the Coral and Tasman seas.

(*4*) Detailed climatic summaries from ship data for restricted ocean regions such as the Tasman Sea (TRELOAR and NEWMAN, 1938).

(*5*) A series of papers by E. Vowinckel on the climate of the Antarctic Ocean based on ship data assembled by the South African Weather Bureau (e.g., VOWINCKEL, 1957).

(*6*) A short, informative review of the meteorology in tropical and middle latitudes of the South Pacific by C. S. Ramage (RAMAGE, 1970).

(*7*) The four-volume atlas on the *Climate of the Upper Air in the Southern Hemisphere* (TALJAARD et al., 1969; VAN LOON et al., 1971; JENNE et al., 1971; CRUTCHER et al., 1971) and the monograph on the *Meteorology of the Southern Hemisphere* (VAN LOON et al., 1972) which interprets and adds to it.

265

(*8*) A two-volume survey of the general circulation of the tropical atmosphere and interactions with extratropical latitudes (NEWELL et al., 1972, 1974).

(*9*) An atlas of upper troposphere circulation over the global tropics (SADLER, 1975).

Data sources

In addition to the various atlas presentations of climatic data and the general surveys listed above, the principal data sources used in this treatment of the climate of the South Pacific were the following.

Surface climatic data

(*1*) REED (1927), a comprehensive tabulation of climatological data for the tropical islands of the Pacific.

(*2*) SCHOTT (1938), which includes data tables for a large number of tropical island stations.

(*3*) BUREAU OF METEOROLOGY (1940), a tabulation of the results of rainfall observations made in Papua-New Guinea, the Solomon Islands, and the New Hebrides (Vanuatu).

(*4*) METEOROLOGICAL OFFICE (1958), a tabulation of temperature, relative humidity, and precipitation for the world (Part VI deals with the Pacific).

(*5*) BROOKFIELD and HART (1966), a compendium of rainfall data for the tropical western Pacific.

(*6*) PHILLPOT (1967), a compilation of selected surface climatic data for Antarctic stations.

(*7*) U.S. DEPARTMENT OF COMMERCE (1968), World Weather Records 1951 – 1960 (Volume 6 deals with the Pacific).

(*8*) SCHWERDTFEGER (1970), a compilation of climatic data for the Antarctic.

(*9*) NEW ZEALAND METEOROLOGICAL SERVICE (undated), a set of climatological summaries for stations in Fiji, Tonga, and Western Pacific High Commission Territories to the end of 1970.

(*10*) NEW ZEALAND METEOROLOGICAL SERVICE (1973), a publication of summaries of climatological observations for New Zealand to 1970 (supplemented by subsequent annual summaries).

(*11*) BUREAU OF METEOROLOGY, unpublished archived records for Australian climatic stations to the end of 1978.

(*12*) U.S. DEPARTMENT OF COMMERCE: ten annual volumes of marine climatic summaries of ship data to the end of 1970.

(*13*) U.S. DEPARTMENT OF COMMERCE, a regular publication of monthly climatic data for the world.

Upper-air data

(*1*) JENNE et al. (1974), a magnetic tape archive of upper-air data for the Southern Hemisphere.

(*2*) MAHER and LEE (1977), a publication of upper-air climatic data for Australian stations.

(*3*) U.S. DEPARTMENT OF COMMERCE, a regular publication of monthly climatic data for the world including most upper-air stations in the South Pacific.

Satellite data

(*1*) Daily mosaics of satellite imagery for U.S. orbiting satellites since 1966.

(*2*) Imagery from geostationary satellites over the Central Pacific (U.S.) and on 140°E (Japan).

(*3*) Processed imagery and satellite sounding data in various forms.

Synoptic analyses

(*1*) The South African Weather Bureau Chart series published in *Notos* (SOUTH AFRICAN WEATHER BUREAU, 1952).

(*2*) The IGY chart series for the Southern Hemisphere published by the South African Weather Bureau (SOUTH AFRICAN WEATHER BUREAU, 1962 — 1966).

(*3*) Microfilmed daily charts produced by the International Antarctic Analysis Centre in Melbourne 1959 — 1965.

(*4*) Microfilmed daily charts from the World Meteorological Centre, Melbourne, since 1966.

(*5*) The GARP Basic Data Set charts for November 1969 and June 1970 (PHILLPOT et al., 1971).

Various additional sources for special purposes are quoted at the appropriate places in the text.

Tabulated data

The set of climatic tables in the Appendix following this chapter was compiled by W. Haggard (personal communication, 1974) from the records of the U.S. National Climatic Center for the ocean stations and by the authors for the coastal and island stations. A location map of the stations for which climatic tables are included is given in Fig. 2.

Treatment

The treatment of the climate of the South Pacific in this volume is, of necessity, based on a selective synthesis and interpretation of a wide variety of earlier studies and climatic analyses, with a liberal sprinkling of recent and some unpublished work. The coverage is by no means comprehensive, but an attempt has been made to include representative references to original work on particular aspects.

In view of the strong coupling that exists between the circulation of the atmosphere and ocean, any interpretive treatment of marine climatology must take account of the characteristic features of the circulation and thermal structure of the underlying ocean. A brief description of the physical oceanography of the South Pacific thus provides a convenient starting point for a discussion of its climate. Features of special relevance include the main current systems, the various oceanic frontal zones, and the ice distribution in the South Pacific.

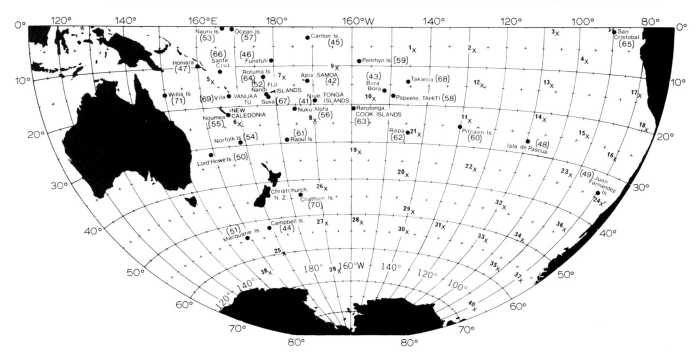

Fig. 2. Island stations (full circles) and marine area climate stations (×). Numbers refer to data tables in the Appendix, p. 375ff.

The climate of any region may be thought of as a long-term synthesis of its weather. It is thus appropriate to examine the main synoptic regimes and characteristic circulation features of the South Pacific including the tropical easterlies, the tropical cyclones of the southwest Pacific, the midlatitude cyclones and anticyclones and the various disturbances of the westerlies, and the Antarctic trough.

The next section of this chapter then examines the climate of the South Pacific in terms of the traditional map representation of climatic elements such as surface air temperature, pressure, wind, cloudiness, and precipitation. This is followed by a discussion of the climate of the upper air, presented in terms of circulation and temperature fields at 500 and 200 mbar and a selection of vertical meridional sections.

An examination of the heat budget of the South Pacific region based on a variety of early studies and more recent satellite and shipboard data follows and leads into a discussion of the large-scale circulation of the region and its relation to such well-known features as the Southern Oscillation and the El Niño phenomenon of the eastern Pacific.

The ocean

There are relatively few comprehensive accounts of the physical oceanography of the entire South Pacific, but such standard texts as SCHOTT (1938), SVERDRUP et al. (1942), and DEFANT (1961) contain a great deal of relevant information. The higher latitudes were the subject of intensive study during the long series of *Discovery* cruises (see, e.g., DEACON, 1937, 1963) and again over the decade 1962−1972 during some 600,000 km of research cruises by the U.S.N.S. *Eltanin* (GORDON, 1970; CAPURRO, 1973; GORDON and MOLINELLI, 1975). Reviews by KNAUSS (1963), WYRTKI (1965, 1966a), TSUCHIYA (1968), WOOSTER (1970), HAMON (1970), and PICKARD et al. (1977) deal with aspects of the oceanography of the tropical South Pacific. These, together with a review of the general ocean circulation of the entire South Pacific by WARREN (1970) and a series of papers presented at the 1972 New Zealand Symposium on the Oceanography of the South Pacific (FRASER, 1973), provide a good introduction to those aspects of the physical oceanography of the South Pacific relevant to the study of its climate.

Apart from the various atlas and other presentations of the *Eltanin* data (e.g., GORDON et al., 1978) for the high latitudes, probably the most useful and authoritative presentation of physical oceanographic data is that contained in the U.S.S.R. *Atlas of the Pacific Ocean* (U.S.S.R. MINISTRY OF DEFENCE, 1974). This beautifully produced atlas, based on all data accumulated in the U.S.S.R. up to the late 1960s, includes monthly, seasonal, or annual charts of surface and subsurface circulation, temperature, and salinity structure, wave and tidal data, and a large amount of other oceanographic information.

Circulation

The main features of the long-term mean surface water circulation of the South Pacific are shown in Fig. 3 and the schematic Fig. 4. Fig. 3 shows the mean annual dynamic topography of the sea surface relative to the 1000 dbar level expressed in dynamic metres. It is based on GORDON et al. (1978) for the region poleward of about 45°S and on WYRTKI (1974) for the remainder of the South Pacific. Other presentations of dynamic topography relative to the 1000-dbar surface and deeper levels have been given

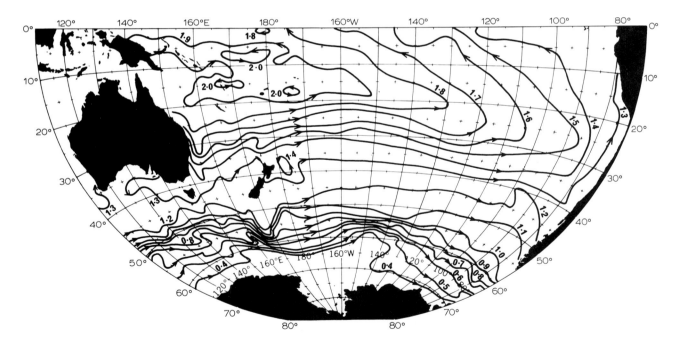

Fig. 3. Mean annual geopotential height anomaly of the sea surface relative to the 1000-dbar level expressed in dynamic metres.

by DEACON (1937), REID (1961), KORT (1962), GORDON (1967), and the U.S.S.R. atlases of the Antarctic (TOLSTIKOV, 1966) and the Pacific (U.S.S.R. MINISTRY OF DEFENCE, 1974). The isopleths of dynamic topography represent streamlines of geostrophic flow relative to the 1000-dbar level such that, on looking downstream, the dynamic height is higher to the left. It is important to realize that the geostrophic current flows that can be deduced from Fig. 3 represent flow relative to 1000 dbar. Whereas in the tropical regions the 1000-dbar surface is essentially flat and the relative topography accurately portrays the surface geostrophic flow, in the high latitudes the Antarctic Circumpolar Current extends to great depths and the 1000-dbar surface itself inclines steeply downward to the south.

Fig. 4 shows, albeit somewhat schematically, the surface currents as deduced mainly from ship drift. The current vectors in this diagram are based mainly on data from the U.S.S.R. atlas for February and August, but some account has been taken of other more detailed sources such as SCHOTT (1935) and the *U.S. Atlas of Pilot Charts for the South Pacific and Indian Oceans* (DEFENSE MAPPING AGENCY HYDROGRAPHIC CENTER, 1974).

Figs. 3 and 4 represent the long-term mean. Just as in the atmosphere, there is a significant seasonal variation in the pattern of ocean surface circulation and, in some parts of the ocean, a very large variability associated with an eddy structure corresponding, though acting on longer time scales, to the cyclones and anticyclones of the atmosphere. The eddies show up as regions of large standard deviation of dynamic height. The part of the South Pacific most subject to such eddies is in the western Tasman Sea in the region of the East Australian Current. Fig. 5, which indicates the displacements of a number of satellite-tracked drifting buoys during April—June 1979, gives some idea of typical current patterns on shorter time scales. The daily drift patterns show still finer structure, with numerous loops and oscillations observed in all the current systems.

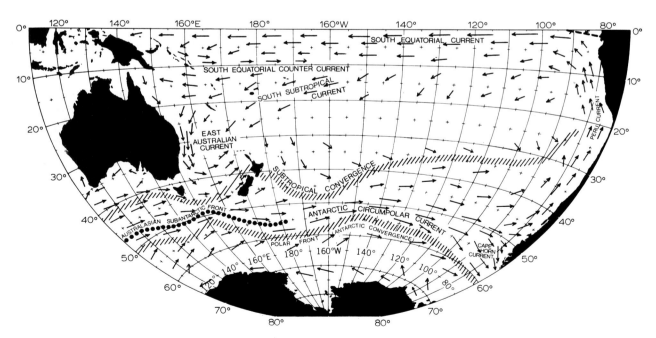

Fig. 4. Schematic diagram of observed ocean currents and convergence regions.

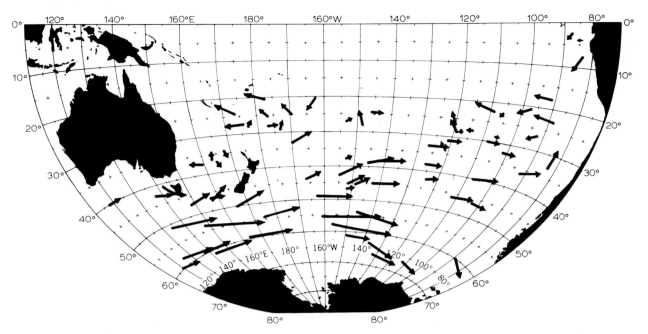

Fig. 5. Drift of satellite-tracked ocean buoys for the period 20 April to 22 June 1979. The arrows indicate buoy displacements during the period.

The surface currents of the South Pacific are driven basically by the southeast trade winds and the high-latitude westerlies. The main circulation is thus anticlockwise with the surface flow predominantly westward in low latitudes and eastward in high latitudes. At the eastern and western boundaries of the ocean, the currents are diverted meridionally to produce the poleward-flowing East Australian Current in the western Pacific and the northward-flowing Peru or Humboldt Current off the coast of South America. The South Equatorial Current extends from between 10° and 20°S to north of the Equator; its maximum velocities, exceeding one 1 kt., are close to the Equator in the eastern Pacific (Tsuchiya, 1970). There is evidence of considerable seasonal variation, the current being strongest in the southern winter. Embedded in the South Equatorial Current and partly separating it into two branches, the southernmost sometimes known as the South Subtropical Current, is the South Equatorial Counter Current. It was first identified by Reid (1959) who found evidence of eastward flow right across the Pacific near 10 – 12°S. It is most evident in the western Pacific from about December to April but is believed to be highly variable in location and chiefly subsurface, and it does not appear at all in the eastern Pacific in Fig. 3 (Wyrtki, 1975a). Another feature of the equatorial current system in the Pacific is the Equatorial Undercurrent or Cromwell Current (Cromwell et al., 1954; Knauss, 1966; Tsuchiya, 1970; Taft and Jones, 1973), a strong eastward-flowing subsurface current with its core at a depth of about 200 m (western Pacific), rising to about 50 m near the Galápagos Islands.

The South Pacific lacks a major western boundary current such as found in the North Pacific (Kuro Shio) or in the North Atlantic (Gulf Stream). The East Australian Current, although formerly thought of as a strong, narrow, south-flowing current near the edge of the Australian continental shelf between about 25° and 35°S is, in fact, a rather complex and variable system characterized by a series of large southward-moving anticyclonic eddies of diameter 300 – 500 km with lifetimes of at least several months and with surface currents up to 4 kt. (HAMON, 1970; BOLAND and HAMON, 1970; HAMON and TRANTER, 1971, NILSSON et al., 1977; HAMON and GODFREY, 1978).

On the eastern side of the Pacific, the equatorward-flowing Peru (or Humboldt) Current is notable for its latitudinal extent and its coldness which near the coast is mainly the result of upwelling. It has been discussed in some detail by DEFANT (1961), WOOSTER and REID (1963), SCHELL (1965), and WOOSTER (1966, 1970), among others. The Peru Current washes the shores of one of the most arid coastal regions on earth. It is fed by Subantarctic Water from the Antarctic Circumpolar Current which splits some distance from the Chilean coast at about 50°S, the southward-flowing branch becoming the relatively warm Cape Horn Current. The northward-flowing branch divides into the Peru Coastal Current and the Peru Oceanic Current, these being separated by a narrow southward-flowing counter current which reaches the surface only occasionally (WYRTKI, 1963).

The Antarctic Circumpolar Current is the major current system of the South Pacific. It is a strong, deep-reaching current with its axis between about 50° and 65°S, the fluctuations in space believed to be related to bottom topography. Much of our earlier knowledge of the Antarctic Circumpolar Current in the South Pacific derived from data collected during the *Discovery* cruises (DEACON, 1937), but the *Eltanin* program which began in 1962 provided an additional wealth of information on its structure, transport, and variability (e.g., GORDON, 1967; CALLAHAN, 1971; GORDON, 1972; GORDON and MOLINELLI, 1975; NOWLIN et al., 1977). South of Australia, the axis of the current lies near 50°S and the net flux of water through a section at 132°E is around 233 · 10^6 m^3 s^{-1} (CALLAHAN, 1971). On approaching the north – south trending Macquarie Ridge (Fig. 1), the current turns toward the south, passing within about 500 km of the Ant-

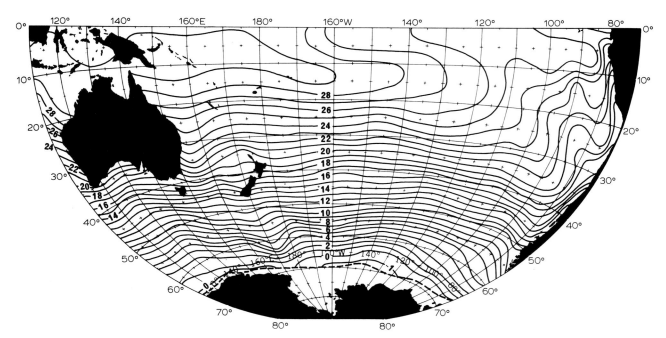

Fig. 6. Sea surface temperature in January (U.S.S.R. MINISTRY OF DEFENCE, 1974).

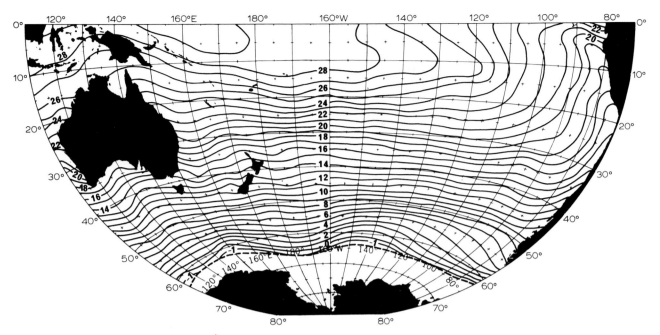

Fig. 7. Sea surface temperature in July (U.S.S.R. MINISTRY OF DEFENCE, 1974).

arctic coast. The main flow of the current then swings north into the central longitudes of the South Pacific before diverting again to the south to pass through Drake Passage.

Temperature

Figs. 6 and 7 show the distribution of sea surface temperature in January and July as given by the U.S.S.R. Atlas (U.S.S.R. MINISTRY OF DEFENCE, 1974). Various other comparable presentations based on slightly different data sets have been prepared by the U.K. Meteorological Office (METEOROLOGICAL OFFICE, 1947, 1950), the U.S. NAVY (1959, 1965, 1969, 1979) and others, in addition to the earlier charts of SCHOTT (1935) on a seasonal basis and MACKINTOSH (1946) on a monthly basis for the high latitudes. More detailed charts for certain parts of the South Pacific are also available. The Dutch atlas *Sea Areas Around Australia* (KONINKLIJK NEDERLANDS METEOROLOGISCH INSTITUUT, 1949) treats the southwest Pacific and southeast Indian Ocean regions. WYRTKI (1964) gives detailed monthly maps for the eastern tropical Pacific. So also does the atlas by HASTENRATH and LAMB (1977).

The outstanding feature of the sea surface temperature distribution in both January and July, and in fact through the entire year, is the extensive region of relatively cold water extending northward along the South American coast and then westward as a narrow tongue along the Equator. The cold water is partly the result of northward advection by the Peru Current, but the coldest water, off the shore of Peru, stems mainly from wind-induced upwelling. The warmest water, which changes little in temperature throughout the year, lies approximately east—west to the east of New Guinea. At high latitudes, the warmest water is to the south-southeast of New Zealand where the Antarctic Circumpolar Current turns southward around the Macquarie Ridge and the Campbell Plateau. The east-northeasterly trend of sea water isotherms across the Tas-

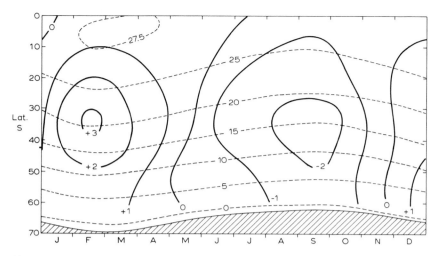

Fig. 8. Annual variation as a function of latitude of sea surface temperature zonally averaged across the South Pacific. Thin broken lines are isotherms (°C). Solid lines are the departures from the annual average. The hatched area indicates the mean northward extent of the pack ice based on U.S.S.R. MINISTRY OF DEFENCE (1974).

man Sea, with warmest water just off the East Australian coast, reflects the influence of the large warm anticyclonic eddies bringing warm water southward in the western Tasman Sea.

Fig. 8 summarizes the annual variation of the sea surface temperature as a function of latitude for the South Pacific as a whole (including the part of the southeast Indian Ocean south of Australia treated in this chapter), in terms of the seasonal migration of 5°C-interval isotherms and isopleths of departure from the zonal average for the South Pacific based on U.S.S.R. MINISTRY OF DEFENCE (1974). Examination of individual monthly maps indicates that the annual temperature range is largest (6 – 7°C) in a band lying from the Australian coast at about 30°S, north of New Zealand to about 45°S on the coast of South America. Another band of large annual temperature range is centred about 500 km off the South American coast between the Equator and 40°S.

Salinity

Variations in salinity have important implications for the density structure and hence the circulation patterns in the oceans. The distribution of surface salinity also has direct climatic implications in that it is related to the patterns of precipitation minus evaporation. Fig. 9 shows an approximation to the annual mean surface salinity for the Pacific based on the February and August patterns included in the U.S.S.R. atlas. The pattern does not differ much from that given by SCHOTT (1935).

The region of maximum salinity lies under the northern flank of the South Pacific subtropical anticyclone (Fig. 41) where high evaporation (Fig. 63) and limited input of fresh water as rain (Fig. 48) raise the salt content in the water above the thermocline. Similarly, as will be seen later, zones of low salinity lie beneath a band of high rainfall stretching southeastward across the Pacific from the Solomon Sea and just south of the zone of maximum rainfall in the westerlies.

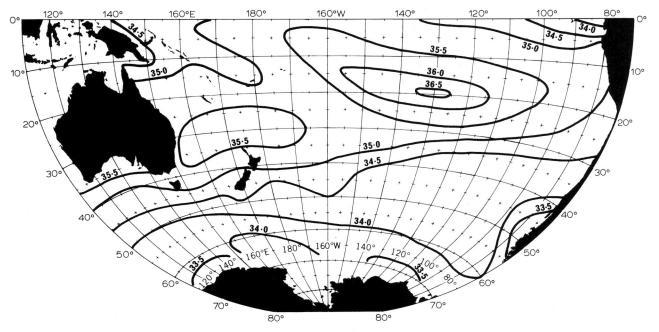

Fig. 9. Annual mean surface salinity (parts per thousand).

Ocean fronts

Fig. 4 depicts the main oceanic fronts of the South Pacific region. The Polar Front is undoubtedly the most intensively studied feature of the southern ocean. First identified as a circumpolar feature by MEINARDUS (1923) and subsequently mapped by DEACON (1937) and MACKINTOSH (1946), it became variously known as the Meinardus Line, the Polar Front, and, following DEACON (1937), the Antarctic Convergence. A definitive study of the Polar Front zone in the South Pacific was undertaken by GORDON (1967, 1971) on the basis of *Eltanin* data and its detailed structure has been described by EMERY (1977). The position shown in Fig. 4 follows TAYLOR et al. (1978) and is based on the location of the salinity minimum at 200 m as given by GORDON et al. (1978).

The Polar Front is of particular relevance to the study of Southern Ocean climate through its characteristic temperature structure and effects on ocean — atmosphere energy exchange and through its relation to the surface wind stress. There has been a fair amount of confusion over the definition and location of the Polar Front. Ideally, it might be expected to appear always as a sharp poleward drop of surface temperature but, in fact, the sea surface to the immediate north and south frequently exhibits transient temperature fluctuations which, from the data of a single crossing, may easily be mistaken for the quasi-permanent feature. The use of less wind-sensitive subsurface characteristics would be appropriate, but a problem of definition then arises. DEACON (1937) was the first to conclude that the recognizable surface phenomenon could be associated with features of the subsurface temperature, salinity, and circulation patterns; specifically, he found that the "Antarctic Convergence" could be placed at the latitude where the so-called Circumpolar Deep Water climbs sharply poleward over the colder Antarctic Bottom Water. Hence, he used the northern limit of the bottom water to locate the mean position of the "Convergence". GORDON (1967) found this an unsatisfactory criterion for the Pacific sector. MACKINTOSH (1946) sought to identify "the line at the surface along which the Antarctic Surface Water sinks below the less dense Subantarctic Water... distinguished by a more or less sharp change of temperature at the surface." He concluded that, when the surface temperature discontinuity was indistinct, a positive identification could be based on the location of the latitude where the minimum temperature layer sinks below 200 m. WEXLER's (1959) criterion was the location of a narrow subsurface minimum temperature zone south of the strong temperature gradient. Its relevance and Wexler's interpretation of its origin were rejected by HOUTMAN (1967). OSTAPOFF (1962) placed the Polar Front along the poleward edge of the 34.0 to 34.2°/oo salinity minimum at 200 m, and this approach has been broadly adopted by GORDON (1972) and others. Basically, the Polar Front is the boundary region between relatively homogeneous, hydrologically identifiable, Antarctic Surface Water and Subantarctic Water masses. Its main characteristics as they relate to the surface temperature field may be summarized as follows.

(*1*) The width of the zone of steep temperature gradient may vary from a few kilometres to several tens of kilometres. The temperature change across it (the "range") averages slightly less than 2°C (HOUTMAN, 1964).

(*2*) On occasion, the gradient may not be detectable or a steplike pattern may be found where it is not possible to identify positively the Polar Front gradient.

(*3*) The Polar Front gradient frequently terminates in a temperature minimum with patches or channels of warm water to the south. These cold cores usually extend to depths of several hundred metres. Similar reversals may be found immediately to the *north* of the Polar Front gradient.

(*4*) The Polar Front moves within a zone of some 2 to 4 degrees of latitude, and its location within this zone, although variable and unpredictable, appears to be related to the low-level winds. The displacement is greater than that of the associated subsurface features and in many cases appears to result from wind-induced southward extension of a very shallow over-riding layer of Subantarctic Water.

(*5*) There is some evidence that the mean position shifts seasonally (MACKINTOSH, 1946; IVANOV, 1959; MAKSIMOV, 1961; BOTNIKOV, 1964) and over longer periods (BOTNIKOV, 1964), but the magnitude and phase of the shift remain unresolved. It seems improbable that systematic seasonal shifts of more than 2 to 3 degrees of latitude occur.

(*6*) The location of the zone within which the Polar Front is found appears to be directly related to bottom topography (GORDON, 1967).

(*7*) The annual range of middle temperature is about 2°C, and the most rapid change (an average rise of 1.2°C through the month) occurs in November (HOUTMAN, 1964).

The so-called Subtropical Convergence is considerably less well documented than the Polar Front and more variable in structure and location (CRESSWELL et al., 1978; HEATH, 1981). Its mean position in the Pacific was delineated by DEACON (1937) and, in the absence of any more definitive treatment, the Deacon position is shown in Fig. 4. The position in the eastern Pacific is as given by WYRTKI (1966a). The Subtropical Convergence is variously taken to refer both to the sharp poleward surface temperature drop characteristically located around the 12–16°C isotherms and to the actual zone of convergence of surface currents. Unlike the Polar Front zone, it does not appear to be directly related to submarine topography or deep water circulation.

Another oceanic front of importance from the climatic viewpoint is the Australasian

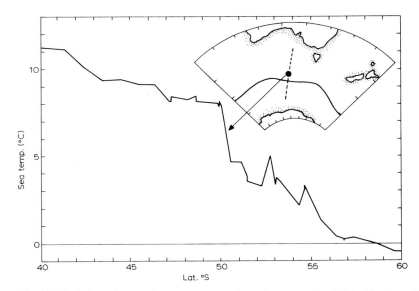

Fig. 10. Variation of sea surface temperature along the route of U.S.N.S. *Eltanin* at 130°E in August 1968 (after ZILLMAN, 1974a).

Subantarctic Front located on average a few degrees of latitude north of the Polar Front in the Australian – New Zealand sector. It was first identified by BURLING (1961) and HOUTMAN (1967) on the basis of hydrological data, and its sea surface temperature expression was mapped by ZILLMAN (1970a,b; 1974a). The location given in Fig. 4 is from GORDON (1973). Fig. 10 from hourly measurements of sea surface temperature aboard the U.S.N.S. *Eltanin* during Cruise 35 along 130°E in August 1968 illustrates the nature of the poleward temperature decrease. Typically, the sea surface temperature across the front drops between 3° and 5° over a degree of latitude.

Sea ice

The earliest substantial account of the ice extent of the South Pacific was published by MACKINTOSH and HERDMAN (1940) and a further analysis was made by TRESHNIKOV (1967). Because of the almost complete absence of shipping, aircraft, and coastal observations, a year-to-year surveillance of the ice extent has awaited the development of suitable satellite techniques. A review of visible, infrared, and microwave sensors for ice studies has been given by CAMPBELL et al. (1975). The development of microwave sensors such as ESMR (electronically scanned microwave radiometer) on NIMBUS 5 in 1972 promises a growing observational base unimpeded by cloud cover and which, with suitable "ground truth", will provide information on the extent, temperature, and physiochemical nature of the ice.

The analyses of longer-term data have so far been based largely on the visible and infrared data and have referred only to the broad characteristics of the ice extent. PREDOEHL (1966) utilized the data of NIMBUS 1 to establish ice – water boundaries; STRETEN (1973b) used 5-day averaged minimum brightness visual mosaics to observe the decay of the ice in three summer seasons, and to delineate the location and progressive changes of some of the large polynyas within the pack ice of the Pacific sector; and DE RYCKE (1973) used VHRR (very high resolution radiometer) data from NOAA 2 to

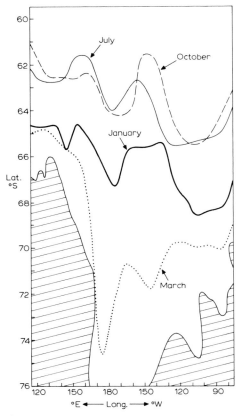

Fig. 11. Mean limit (south lat.) of the Antarctic sea ice in the Pacific sector for the first 4-day period of each indicated month based on 5 years (1973–1977) of U.S. Navy Weather Research Facility data. Shaded area represents continental Antarctica.

study the detailed mesoscale ice drift patterns of the eastern Ross Sea. Currently, a systematic mapping of the hemispheric ice extent is maintained by the U.S. Navy Fleet Weather Facility in Washington, D.C., utilizing all available satellite and conventional observations, thus providing an outstanding service to world climatology.

In Fig. 11 is shown the extent of ice at the beginning of selected months averaged over 5 years (1973–1977) from the U.S. Navy data and averaged over 10-degree longitude sectors.

The ice extent is least in February–March when the ice front retreats to the highest southern latitudes of the Ross Sea. At the same time, however, substantial ice usually remains along the coast to the east of and within the Amundsen and Bellingshausen seas. Similarly, the regions immediately to the west of the Ross Sea (the coasts of Oates Land, George V Land, and Terre Adélie) experience extensive summer ice and have always proved difficult of access to exploration ships. It seems probable that much of the ice along this coast is transported to the north and west by the cyclonic ocean circulation within the Ross Sea embayment. The northward growth of the ice proceeds rapidly in autumn, much more slowly in winter, and reaches a maximum in the early spring (September–October). The ice edge at this time follows the general trend of the Antarctic continental coast and lies from 61°S south of Western Australia to south of 65°S north of the Amundsen Sea, the latter being the region in the hemisphere of least extent of ice in spring.

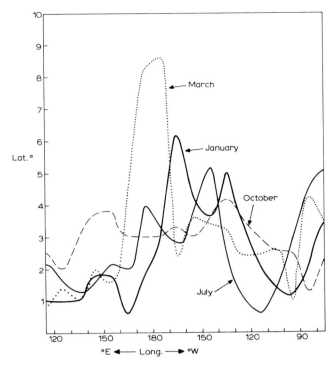

Fig. 12. Extreme range of the ice edge (deg. lat.) in the Pacific sector for the first 4-day period of indicated months (5 years 1973—1977).

The range of the ice edge throughout the 5-year period (Fig. 12) averages about 3 degrees of latitude in each month over all longitudes. It is greatest in the region to the north of the Ross Sea in late summer when the pattern is frequently complicated by the development of a large polynya within the embayment. This feature displays irregularities of growth which often result in open water close to the front of the Ross ice shelf but leave a temporary irregular band of ice across the head of the Ross Sea at around 69°S; this band was frequently the first ice encountered and described by the early Antarctic expeditions sailing southward to the McMurdo Sound region. Another region of substantial variation appears to exist in the extreme east of the region at the approaches to Drake Passage where, according to LAMB (1967), considerable variability in ice extent was observed by shipping during the late 19th century with particularly excessive ice between 1888 and 1907.

The importance of interactions between sea ice and atmosphere in the heat balance of high latitudes, and thus the broad-scale climate of the earth, has been discussed in some detail by FLETCHER (1969). As the heat transfer from ocean to atmosphere in winter is two orders of magnitude greater over open water than over unbroken ice, and as the seasonal and year-to-year variation of the ice is considerable, the ice may be an important factor in broad-scale climate. BUDD (1975) has studied records of annual mean temperature at coastal stations around Antarctica and suggests that a change of 1°C in annual mean temperature corresponds to approximately 2.5° of latitude variation in the maximum ice extent. Despite some limited investigation, no clear associations between ice extent and broad-scale atmospheric features have yet been firmly established and the question of cause and effect is unresolved.

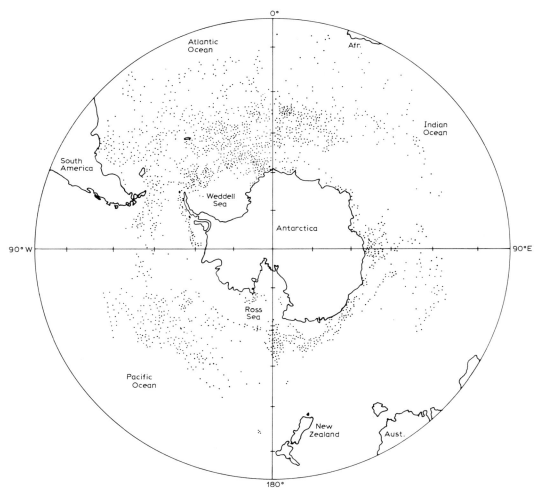

Fig. 13. Iceberg sightings 1773—1960 (after NAZAROV, 1962).

Icebergs

Present knowledge of the iceberg distribution of the South Pacific, as for all the southern oceans, is extremely limited. NAZAROV (1962) has accumulated records of sightings from the earliest observations of Cook in 1773 to those of 1960 (Fig. 13). Such records are obviously dependent on the frequency of shipping in various sectors, but may indicate a less dense pattern in the South Pacific by comparison with the South Atlantic Ocean. The frequency is much lower south of Australia than northeastward of the Ross Sea whose great ice shelf must be the source of many of the bergs. Large breakouts of ice have occurred along the ice shelf; e.g., SULLIVAN (1957) recorded a piece some 190 km by 16 km breaking off the shelf between 1948 and 1955 at the site of a former Little America station. Similarly, MAWSON (1915) records a berg near the coast of George V Land some 60 km long. The characteristic tabular icebergs of the Antarctic are typically of much more modest dimensions, perhaps about a kilometre in cross-section and 50 m above the water.

Information on the physical environment, melting rate, and life history of icebergs is almost totally lacking. Clearly, an observational program is required based on the satellite tracking of bergs using very high resolution imagery and the collection of data via satellite from automatic weather stations established on selected bergs. Such a program would have many difficulties, but with the increasing interest in Antarctic icebergs as a source of fresh water for the dry coastal regions of Australia and South America, it may eventually be undertaken.

Synoptic climatology

The data base

The absence of island stations and the low number of ship reports over most of the vast expanse of the South Pacific have meant a slow evolution of synoptic analysis. Progress has depended on the development of techniques for using fragmentary information to build up chart series on which a synoptic climatology can be based. The earliest significant attempt at regional weather chart analysis dates from 1887 when H. C. Russell in Sydney began to produce isobaric maps for the Australia and New Zealand area (RUSSELL, 1893). Using data from the early Antarctic expeditions to supplement information from ships and the lower latitude continents, MEINARDUS and MECKING (1911), HEPWORTH (1913), SIMPSON (1919), and KIDSON (1947) attempted a limited extension of the area of isobaric analysis over the middle and higher latitude oceans. In New Zealand, KIDSON and HOLMBOE (1935) first advocated the introduction of frontal techniques to the Southern Hemisphere, and this was extended by PALMER (1942) who produced the first detailed discussion of synoptic sequences in the Australia/New Zealand region and provided an interesting historical survey of the development of ideas on the Southern Hemisphere circulation up to that time. The establishment of additional meteorological stations in the tropical southwest Pacific during the Second World War and on the Subantarctic islands in the late 1940s provided a stimulus to Australian meteorologists to explore a wider area of synoptic analysis and to develop methods for the better utilization of point observations (see e.g., GIBBS et al., 1952; LANGFORD, 1957). Following an experimental period of Southern Hemisphere weather analysis conducted at the Massachusetts Institute of Technology in 1950, the first continuing analyses for the South Pacific as a whole came with the farsighted South African Weather Bureau (SAWB) project begun in 1950 and aimed at producing daily weather maps of the entire hemisphere; data in the form of grid point pressure values and monthly and seasonal charts were published in the SAWB journal *Notos*.

In 1957 as part of the International Geophysical Year (IGY), a "Weather Central" was established at Little America station on the Ross ice shelf where an international team of meteorologists attempted to produce surface and upper-level synoptic analyses for much of the hemisphere. The IGY data also provided the basis for the first substantial study of the Southern Hemisphere atmosphere since the work of Meinardus between 1911 and 1938. These studies were published in a series of papers by J. J. Taljaard and H. van Loon and later consolidated with other work as a monograph (VAN LOON et al., 1972), and as the first detailed atlas of Southern Hemisphere climatology (TALJAARD et al., 1969). Further analysis of the IGY data was also made by ASTAPENKO (1960). At the end of the IGY, the work of hemispheric synoptic analysis was continued in Melbourne, Australia, at the International Antarctic Analysis Centre (IAAC) (GIBBS, 1960) from 1959 to 1965 and, on its closure, from 1965 to 1972 at the Australian Bureau of Meteorology's Southern Hemisphere Analysis Centre (SHAC). In 1972, a fully numerical analysis system for the hemisphere was introduced for the first time in the newly reconstituted National Meteorological Analysis Centre (NMAC) in Melbourne (GAUNTLETT et al., 1972).

In all these hemispheric synoptic analyses, the South Pacific, especially the area south of 30°S between New Zealand and the coast of Chile, remained so devoid of information outside the summer months of the IGY that frequently no attempt could be made to

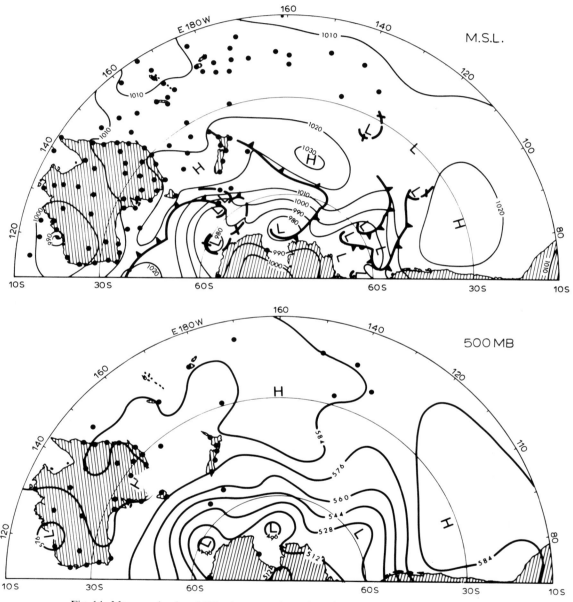

Fig. 14. Mean sea level and 500-mbar synoptic analyses for the South Pacific at 12h00 GMT, February 17, 1975. Full circles indicate locations from which observations were available. (National Meteorological Analysis Centre, Melbourne.)

produce an analysis over the entire region. As examples, typical synoptic charts over the South Pacific for recent years are shown in Fig. 14 for the surface and 500 mbar (L. B. Guymer, personal communication, 1979). The number of conventional observations is very low, and it is due entirely to the use of satellite data that an isobaric and frontal analysis is possible.

From the time of the earliest meteorological satellite, in 1960, attempts had been made to incorporate inferences drawn from the imagery into the conventional analysis scheme. In particular, after 1966 the invaluable daily mosaics based on the polar-orbiting ESSA satellite series became available, though not always in time to be incorporated in the day-to-day analysis. The reliability of the charts before the satellite era is probably best shown in Fig. 15 which still expresses our ability to map the South Pacific systems; satellite imagery now precludes the former possibility of failure to detect even major synoptic features. Development has continued in qualitative and semi-quantitative methods of using the satellite imagery over the Southern Hemisphere (ALVAREZ and THOMPSON, 1965; RUTHERFORD, 1966, 1969; MARTIN, 1968; GUYMER, 1969; ZILLMAN, 1969; TROUP and STRETEN, 1972; ZILLMAN and PRICE, 1972; STRETEN and KELLAS, 1973a; KELLY, 1978). A valuable summary of this work has been given by GUYMER (1978). A description of the use of satellite and other information in the basic data sets during the research-analysis periods of the so-called GARP (Global Atmospheric Research Programme) Basic Data Set Project in 1969—1970 is given by PHILLPOT et al. (1971) and LAMOND et al. (1972). Vertical temperature profiles from the NIMBUS radiometer data were first incorporated into the analysis system experimentally in 1976, and with further development of techniques for deriving accurate atmospheric temperature soundings from the radiance measurements, this data source may be of increasing importance in the future. The problems of analysis still remain immense (see e.g., TREN-

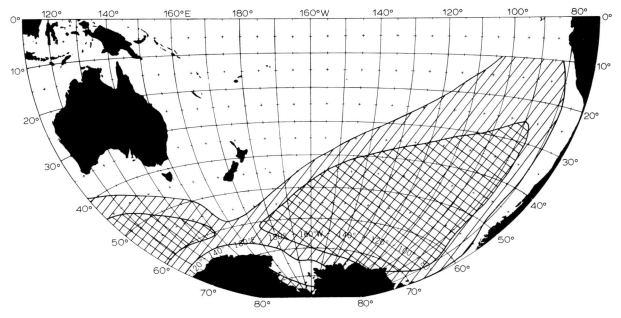

Fig. 15. Reliability of sea level analyses in winter for the International Geophysical Year (IGY). Blank areas indicate good reliability; hatched areas medium reliability; and cross-hatched poor reliability. (After TALJAARD and VAN LOON, 1964.)

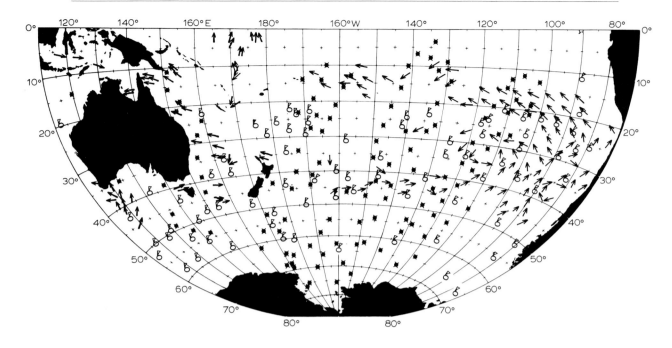

↖ Satellite derived winds from cloud movement

ᕃ Ocean buoys (temperature and pressure)

✶ Polar orbiting satellite (temperature profile measurement)

Fig. 16. Non-conventional meteorological data available at the National Meteorological Analysis Centre, Melbourne, for synoptic analysis at 00h00 GMT, February 2, 1979. Regional cloud patterns north of about 50°S could be viewed by the two geostationary satellites above the Equator at 135°W and 140°E.

BERTH and NEALE, 1977), but newer numerical systems now introduce a considerable measure of consistency into the process.

In December 1978, the First GARP Global Experiment (FGGE) began, with the added observational network of some 300 drifting buoys around the hemisphere measuring surface pressure and sea surface temperature and reporting via satellite. In addition, two geostationary satellites, some 36,000 km above the earth, viz., the United States' Synchronous Meteorological Satellite (SMS II) and Japan's Geostationary Meteorological Satellite (GMS) located at 135°W and 140°E, respectively, began a continuous surveillance of the weather of the Pacific with the capability of tracking cloud elements to measure low- and high-level winds. Fig. 16 shows some features of this new observational base. The forthcoming volume of data from these sources is likely to eclipse rapidly the sum of our previous knowledge of the South Pacific region and within a short time to provide detailed material for new concepts in the study of its synoptic climatology.

Synoptic regimes of the South Pacific

The mean sea level pressure distribution over the South Pacific for January and July, reproduced in Fig. 41 from TALJAARD et al. (1969), provides the basis for the discussion of the mean circulation and wind patterns over the ocean. For a better understanding of the circulation, it is important to consider: (*1*) variations in the mean monthly or seasonal patterns from month to month throughout the year and for the same months or seasons of different years: and (2) the geographical frequency, characteristics, and movement of the synoptic-scale pressure systems from day to day.

A very detailed description of pressure variations and, in particular, the yearly and half-yearly waves and latitudinal pressure gradients has been given by VAN LOON (1972), and the details will not be repeated here. Reference will be made to variations in the principal features of the Pacific region pressure distribution during a five-year (1972–1977) period of numerical weather analysis of the hemisphere (STRETEN, 1980). The overall representation of variations in the monthly mean pressure over the Southern Hemisphere has been considered by KIDSON (1975a) using eigenvector or principal component analysis of monthly mean station data for the ten years 1951–1960. Similar techniques have been employed by TRENBERTH (1975, 1976a) in examining the pressure variability in the Australia – New Zealand region where long and complete surface pressure records are available over a considerable span of latitude and longitude and where, as will be seen later, substantial pressure fluctuations take place on time scales of weeks or months.

Before the IGY, few investigations had been made of detailed day-to-day characteristics of synoptic-scale pressure systems over the South Pacific Ocean. Using data from the summers of 1955–1956 and 1956–1957, VAN LOON (1960) examined the data of the whaling fleets and the then recently established United States bases in the Ross Sea area of Antarctica to study the synoptic evolutions at high latitudes of the South Pacific; similar investigations were also carried out by LANGFORD (1960). KARELSKY (1954, 1956, 1960, 1961) employed definitions of the time spent by individual synoptic systems within specified geographical areas and published charts of the frequency and characteristics of cyclones and anticyclones within the Australia – New Zealand region from 1946 to 1960. This work was severely limited by the lack of reliability of the chart series over the oceans for this period before satellite observations, and by the restricted area covered by the maps. The IGY analysis and discussion of synoptic systems over the hemisphere (TALJAARD, 1972) incorporate much of the earlier work and will be referred to extensively here.

It is convenient to consider the mean monthly pressure distributions and the climatology of the synoptic systems in terms of the principal climatic regimes over the ocean, viz.: (1) the tropical easterlies: (2) the anticyclonic belt; (3) the westerlies of mid-latitudes; and (4) the Antarctic trough and the subpolar easterlies.

The tropical easterlies

The true northern limit of the Southern Hemisphere circulation is marked by the Intertropical Convergence Zone (ITCZ) or Intertropical Discontinuity (ITD). This is a zonally oriented band of active cumulonimbus clouds largely grouped in clusters and separated by frequent cloud-free areas. Climatically, the ITCZ lies within the envelope enclosing all such active clusters. In time-averaged satellite imagery, it appears as a bright band of cloudiness extending east – west across the Pacific primarily between 5°N and 10°N (e.g., MILLER and FEDDES, 1971). Over the central Pacific, little seasonal migration of the band is apparent, but in the western Pacific near 160°E the mean zone migrates southward in the Northern Hemisphere autumn to be just north of the Equator in January (GRUBER, 1972), although its place and strength fluctuate considerably. Thus, the ITCZ proper does not enter the South Pacific, but a very prominent convergence zone south of the Equator extends southeast from Indonesia where it may constitute a second and, often in summer, a more active band of the ITCZ. The cloud band which in the mean extends from the eastern tip of New Guinea to the vicinity of 30°S 130°W is sometimes known as the South Pacific Convergence Zone (SPCZ) and is important to the climate of the South Pacific; it will appear frequently in the following. Satellite observations have shown a two-band ITCZ structure in the eastern Pacific with one band south of the Equator between 90°W and 135°W and mainly during the southern autumn; further data suggest that this is an infrequent and not prominent feature of the circulation (HUBERT et al., 1969).

Theoretical studies of the role of the ITCZ in the general circulation and the reasons for its displacement into the Northern Hemisphere (e.g., PIKE, 1968; CHARNEY, 1969; MANABE et al., 1970; BATES, 1972) will not be discussed here. The numerical general circulation models indicate that the latitude and character of the ITCZ are sensitive to sea surface temperature and to the boundary-layer dynamics (HOLTON and WALLACE, 1971).

Charts of the mean distribution of pressure (Fig. 41) and wind (Fig. 43) show that in winter the southeast trades extend across the whole of the South Pacific north of the subtropical ridge. In summer, the effect of the low over the western part of the ocean results in a northwesterly circulation from New Guinea to Fiji (180°) and from the Equator southward to 15°S. Throughout the year, this area lies under the influence of the northwestern part of the Pacific cloud band (see p. 308). In summer, distinct surface low-pressure areas are evident on day-to-day charts, and the southern fringe of the zone is then frequently the breeding ground of tropical cyclones. In winter, periods of disturbed weather are still frequent (HILL, 1964), but apparently these disturbances originate at higher levels and are associated with the persistence of upper-level troughs; such troughs are only weakly shown by the limited upper-wind network. The persistence of the cloudy zone throughout the seasonally changing pattern of surface wind is a feature of considerable climatic importance. RAMAGE (1970) has suggested that the upper-level troughs in winter may be a result of the downstream influence of the upper-level northeast winds over India associated with the southwest monsoon. Invasion by these northeasterlies across the Equator into the Australian region may produce intensification of the subtropical jetstream (MUFFATTI, 1964) and instabilities leading to upper-level wave disturbances over the Coral Sea. The degree of persistence of the cloud feature is surprising, and further study of the winter weather of the region is necessary, particularly in relation to such possible Northern Hemisphere influences. This western Pacific region of disturbed weather is in stark contrast to the tropical zone east of the date-line, where throughout the year surface pressure in general falls off toward the trough north of the Equator. The southeast winds of the Southern Hemisphere are divergent as are the easterlies blowing over the equatorial waters (GORDON and TAYLOR, 1968). Thus, convection is only weakly developed, and it is further retarded by the upwelling of cold water not only along the coast of Peru but also westward along the Equator often as far as 165°W (see e.g., AUSTIN, 1960). This combination of cold water and divergent surface wind flow results in an area of persisting weak convective cloud development and the absence of substantial cloud cluster organization which is otherwise characteristic of the tropical oceans and of the ITCZ and the disturbed region of the western Pacific. In abnormal seasons in the eastern Pacific, this pattern of weak cloud development and low rainfall may be dramatically altered; these situations will be discussed later.

As viewed by satellite, the clouds of tropical weather systems display a bewildering variety of dimension, life cycle, and weather activity. Over most of the tropical Pacific, the air near the surface is slightly cooler than the sea, and the atmosphere continually receives heat and water vapour from the ocean (Figs. 63, 65). Convective cloud is formed chiefly as "trade cumuli." These clouds are usually of small horizontal extent and from 1 to 3 km thick. The effect of strong entrainment and vertical shear is to reduce their lifetime to less than half an hour (PALMÉN and NEWTON, 1969). The upper level of the cloud is capped by the trade inversion, but the characteristics of the distribution of this inversion are not well known for the South Pacific. North Pacific observations (NEIBURGER et al., 1961) suggest a decrease in frequency and an increase in height of the inversion over the western part of the ocean; it is probable that conditions are similar over the South Pacific.

The nature of tropical synoptic-scale weather systems below the intensity of tropical cyclones has been the subject of much controversy in meteorology. "Easterly waves," as shown by streamline flow patterns in the trade wind region such as those of the North Pacific and Caribbean (RIEHL, 1954; MALKUS and RIEHL, 1964; YANAI and NITTA, 1967), are observed over the South Pacific; published studies of such systems have been confined to the other tropical oceans of greater data availability. A typical easterly wave of this type is 3000 km long with a streamline amplitude of 2° of latitude moving westward with a speed of some 5 m s^{-1} (PALMÉN and NEWTON, 1969). Lack of understanding of the energy processes at work and the variability of observed weather in individual situations from that of the easterly wave "model" have led to much disagreement in the identification of the wave systems, particularly on a day-to-day basis of few observations.

PALMER (1952) described easterly waves over the equatorial Pacific with a wavelength of around 15 degrees of longitude and with westward movement at phase speeds of $5-7.5$ m s^{-1} which are thus similar to the waves in the trade wind region; ROSENTHAL (1960) has given a dynamical explanation of this type of wave. Another equatorial perturbation has been described by FREEMAN (1948) based on observations in the New Guinea area; this is a discontinuity in windspeed along with a distinct change in the depth of the trade wind inversion. It is apparently a gravity wave phenomenon and may be related to the wind patterns and orography of the specific region as it has not been described elsewhere (PALMER, 1951).

Over much of the tropical Pacific and west of the South Pacific cloud band in summer, active weather systems are in the form of "cloud clusters." Such systems have only been investigated in recent years and such studies are highly dependent on detailed satellite observations. Viewed this way, they appear as cold canopies of cirrus produced by the outflow from the lower-level cumulonimbi. Most of the observations refer to the western tropical North Pacific where the clusters (apparently similar to those in active regions south of the Equator) are from 3 to 6 degrees of latitude wide and may exist in identifiable form for various periods up to several days. Compositing of surface and upper-air soundings in the vicinity of such clusters (e.g., WILLIAMS and GRAY, 1973) has revealed some structural features and tends to support the view that low-level frictionally forced convergence is important in producing them (GRAY, 1968).

CHANG (1970) has used satellite photographic sequences in the western North Pacific to show clusters propagating westward with a phase speed of around 9 m s^{-1} and with a predominance of 4-day periods. Wind field fluctuations of similar period have been found in several investigations, e.g., by WALLACE and CHANG (1969). SIKDAR et al. (1972) and YOUNG and SIKDAR (1973) used time spectral analysis methods to study satellite cloud patterns from 140°E to 120°W for a region extending from 20°N to 20°S from April to July of 1967; they found occasional regimes of apparent eastward movement between 10°S and 20°S as well as a preponderance of westward motion. For systems with periods of 4 days, they found that westward propagation is prominent near the Equator, whereas eastward propagation is observed mainly beyond 10°S and 10°N. Such work is able to decompose the complicated space and time variations of tropical cloud patterns and to lead to identification of irregular and wavelike regimes. The study of tropical cloud clusters is only beginning, however, and the advent of geostationary satellite imagery with coincident wind measurements at high temporal frequency should do much to reveal the detailed nature of these weather systems.

Tropical cyclones

Tropical cyclones are a feature of the climate of the northwestern South Pacific where accounts of their past visitations are part of the local folklore of the island peoples. They have been the subject of a number of studies particularly in the Australian region, from the earliest substantial summary of VISHER and HODGE (1925). Since 1960, how-

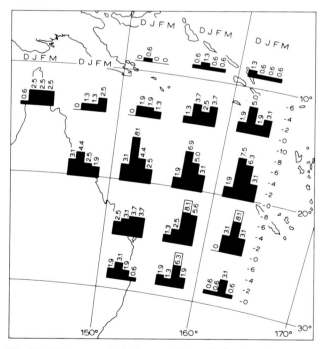

Fig. 17. Histograms of relative monthly tropical cyclone frequency per 5-degree block from July 1959 to July 1975. (Adapted from LOURENSZ, 1977.)

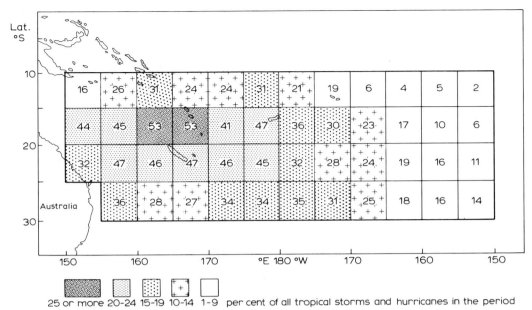

Fig. 18. Number of tropical cyclones that crossed each 5-degree square in the 30 "seasons" November 1939 to April 1969. Shading shows percentage of total storms. (After KERR, 1976.)

ever, observational data are much more reliable with the advent of satellite surveillance. The intensity to be reached by a tropical storm before it is classified as a tropical cyclone (or hurricane) has been defined in different ways by individual investigators (see e.g., the discussion by KERR, 1976); the World Meteorological Organization defines such storms as those having winds at least of 64 kt. (or Beaufort force 12), but such definitions cannot be rigorously applied.

In a worldwide study of the origins of tropical cyclones over a period of twenty years (1952–1971), GRAY (1975) identified the formation of some 213 tropical cyclones in the region extending from the Queensland coast to 155°W representing 11% of the global total for that period (see also CRUTCHER and QUAYLE, 1974). Detailed statistical information on frequency and movement for tropical cyclones in the Australian region from 1909 to 1975 has been given by LOURENSZ (1977). For the regions farther east, an account of the frequency, movement, and historical record of cyclones from 1939 to 1969 and their economic effects on particular island groups has been published by KERR (1976), following earlier work by GABITES (1956) and HUTCHINGS (1953). Figs. 17 and 18 summarize the frequency data from the more recent sources. In the Australian region, the systems are most frequent from December to March with peak frequencies occurring over the Coral Sea between the New Hebrides and New Caledonia in January. Occasional cyclones occur in April, but they are infrequent in other months and confined to lower latitudes. A rather similar pattern is revealed in the data for the region farther east (KERR, 1976). GIOVANNELLI and ROBERT (1964) and A. D'HAUTESERRE (1975, personal communication) have given information on some of the principal cyclones of French Polynesia. Their data suggest that in this region the systems originate in two areas, one to the west along the South Pacific cloud band and the other, in some years, to the north or northwest of the Tuamotu Archipelago. As noted earlier, the eastern South Pacific region of cold surface water and divergent surface winds is one of little

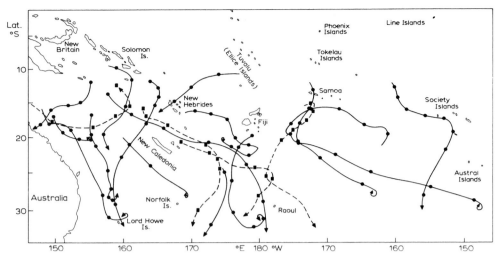

Fig. 19. Tracks of tropical storms and cyclones in March (1960–1969, inclusive); full circles show daily position at 00h00 GMT. (Adapted from KERR, 1976.)

cloud development, and tropical cyclones are not observed there.

GRAY (1975) indicates that seasonal cyclone genesis frequency is related to the product of (*1*) low-level vorticity, (*2*) Coriolis parameter, (*3*) the inverse of the vertical shear of the horizontal wind from the lower to the upper troposphere, (*4*) the ocean's thermal energy to 60 m depth, (*5*) moist stability from the surface to 500 mbar, and (*6*) middle troposphere relative humidity. Suitable conditions of surface water temperature and surface and upper-air circulation characteristics frequently apply over most of the tropical southwest Pacific in summer. The fact that the large-scale circulation of the tropics has a much longer time scale than that of tropical cyclogenesis is important in producing periods of weeks or months when tropical cyclogenesis tends to recur or, alternatively, to be suppressed. GABITES (1956) found that tropical cyclogenesis occurs between 8° and 18°S in the region with no significant seasonal changes. Simultaneous formation of cyclone pairs (one in each hemisphere) has been observed by satellite (LEIGH, 1969; STRETEN and KELLAS, 1973b).

Once formed, individual tropical cyclones tend to follow often erratic paths (see e.g., Fig. 19). Strike probabilities for certain regions have been published by CRUTCHER and HOXIT (1974). Overall, the tendency in the Coral Sea is for a high frequency of southwest movement at latitudes north of around 15°S and recurvature to a southeast movement south of this latitude (Fig. 20). This latitude of recurvature is considerably lower than that observed in the North Pacific (RAMAGE, 1970). KERR's (1976) data for the whole region indicate that 39% of cyclones initially moved westward and about 30% followed the "typical" recurving path. It therefore seems that a rather high proportion of cyclones, particularly in the east of the region, move eastward throughout their life cycle. The intense pressure gradient of tropical cyclones results in devastating winds and high seas at island and coastal stations. Examples of notable destructive storms are the following.

(*1*) That of January 1903 which devastated the islands of Hikuera, Marokau, Hao, and Napuka in the Tuamotu Archipelago with the loss of some 515 lives.

(*2*) That of January 1952 on Viti Levu (Fiji). Twenty-three persons were killed and 800

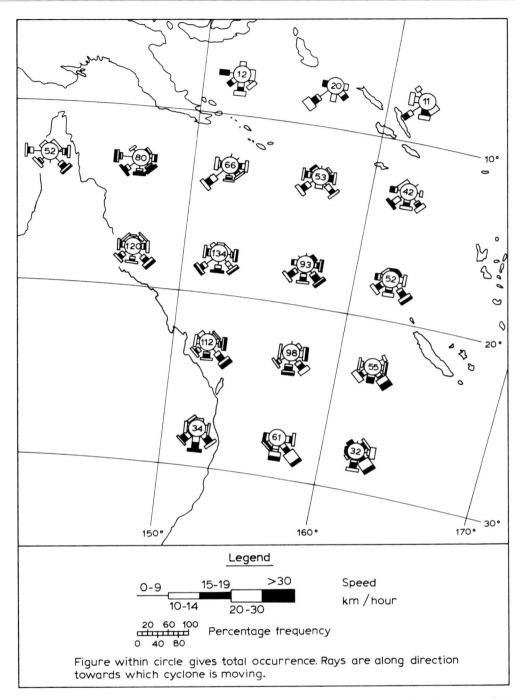

Fig. 20. Movement characteristics of tropical cyclones—Coral Sea region. (Adapted from Lourensz, 1977.)

treated for injuries in Suva, and estimates of the damage to crops and buildings on the island reached two million dollars.

(3) The so-called "Wahine storm" of April 1968 (cyclone "Gisele") which caused extreme damage on the east coast of North Island of New Zealand and resulted in the sinking of the 9000-ton ship *Wahine* at the entrance to Wellington harbour with the loss of 57 lives.

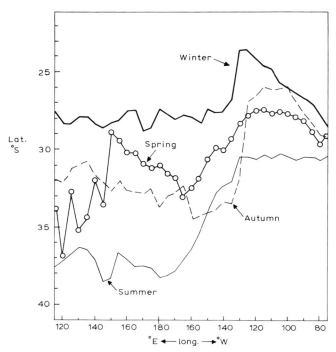

Fig. 21. Latitude of the mean axes of maximum pressure—South Pacific longitudes (1972–1977). (Adapted from STRETEN, 1980.)

(4) Cyclone "Bebe" of October 1972 which resulted in wide devastation on Funafuti and the deposition of a huge rampart of coral rock debris around the island lagoon foreshores (MARAGOS et al., 1973).

The subtropical anticyclone belt

Some of the principal circulation features of this region may be identified by examination of the monthly mean pressure fields derived from the summation of the daily 00h00 GMT mean sea level charts during the first five years of hemispheric numerical analysis (1972–1977) by the Australian Bureau of Meteorology (STRETEN, 1980). The data are grouped under the seasonal definitions originally used by TALJAARD (1967) in his analysis of the day-to-day IGY data, viz., summer (December–March), autumn (April and May), winter (June–September), and spring (October and November). Fig. 21 shows the location of the mean axis of highest pressure across the Pacific during this period. The contrasting range of latitudinal movement of the axis from winter to summer is evident between the western and eastern parts of the ocean. This results from the development of a heat low over northwest Australia in spring and the general monsoonal low over northern Australia and the Coral Sea throughout the summer. At this time, the subtropical ridge retreats to around 37°S over a span of longitudes from the Great Australian Bight to well to the east of New Zealand. By contrast, the axis of the East Pacific High shows much smaller southward retreat in summer. By winter, the ridge has advanced northward to lie across the Pacific at a fairly uniform latitude of around 28°S. The extreme range of the mean monthly axis of high pressure during the 5-year period is shown in Fig. 22. The most important feature is the large winter variation in the axis

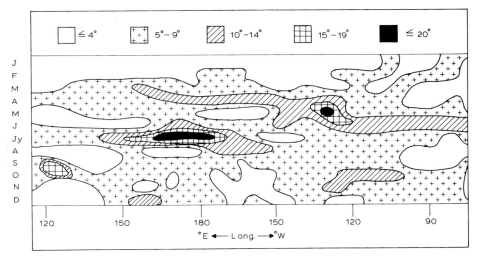

Fig. 22. Extreme range (1972–1977) in the monthly latitude of maximum pressure at South Pacific longitudes. (Adapted from STRETEN, 1980.)

in the region of the Tasman Sea and as far east as 150°W. This is associated with persistence in the tracks of anticyclones in different winters and the notable tendency of these systems to follow tracks either to the north or to the south of New Zealand (TALJAARD, 1967). This will be discussed further below. At all seasons a marked discontinuity in the axis occurs around 130°W and the variability of the axis location is also significantly greater in this area; this mid-ocean discontinuity is discussed on pp. 308–310.

An investigation into the characteristics of anticyclones in the Australian region—their mean latitude, rate of motion, and time intervals between the passage of their centres— was originally carried out by KIDSON (1925).

The axis of highest pressure for specific regions of comparatively high data density has been examined in more detail over longer time spans by LAMB and JOHNSON (1966) for Australia, New Zealand, and Chile, DAS (1956) for Australia, and RUBIN (1955) for South America. Later, PITTOCK (1971) investigated the latitudinal profile of mean monthly pressures along the east coast of Australia and the coast of Chile. His so-called "L index" (the latitude of maximum pressure) is important as its variations have been significantly correlated with the seasonal distribution of rainfall over eastern Australia and with some suitably exposed high rainfall stations in Chile.

An indication of the relative seasonal width of the subtropical anticyclone belt is given in Fig. 23 for 1972–1977. The diagram shows the mean latitudinal separation between the 1015-mbar isobars poleward and equatorward of the axis of highest pressure. A strongly marked maximum width is located over the eastern Pacific throughout the year, but the winter maximum over Australia decays rapidly in spring and summer with an increase in continental heating and the onset of the monsoon.

The longitudes of mean high pressure centres appearing on monthly mean pressure maps for 1972–1977 show a seasonal pattern for the South Pacific indicated in Fig. 24. The Southeast Pacific High persists on the mean charts throughout the year. Examination of the monthly data suggests that the centre exhibits a semiannual variation with the centre being farther west near 100°W in December and June and farther east

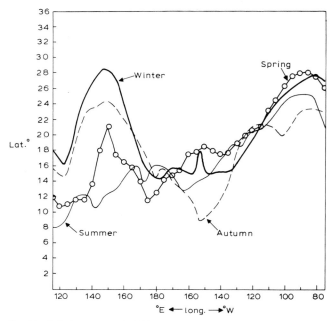

Fig. 23. Relative seasonal width (deg. lat.) of subtropical anticyclone belt at Pacific longitudes (mean distance between 1015-mbar isobars equatorward and poleward of the axis of highest pressure) (adapted from STRETEN, 1980); data for 1972—1977.

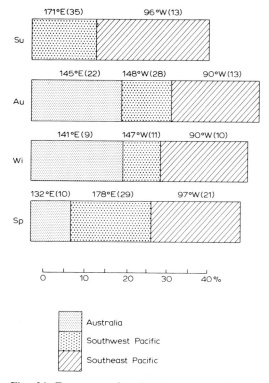

Fig. 24. Frequency of anticyclone centres in monthly mean charts occurring in individual regions (1972—1977). The frequency in particular regions is shown by the bar graphs and the related scale, and it is expressed as the percentage of the total number of anticyclonic centres shown in the monthly hemispheric maps for each season. For each season and region the median longitude and its range (bracketed figure) are also shown. (After STRETEN, 1980.)

near 90°W in April and August. Knowledge of the characteristics of the high pressure region is based on very limited data, and indications of its intensity used in determining indexes of the "Southern Oscillation" (see p. 349) are often based on single-point observations at Easter Island or Tahiti (see e.g., TRENBERTH, 1976a,b). The western boundary of the high near 130°W is marked by a discontinuity related to the zones of cyclogenesis and the Pacific cloud band to be discussed later.

The Southwest Pacific High is, in general, observed at a much lower frequency than that over the eastern part of the ocean. It is most evident in spring when it is in the mean to the north of the North Island of New Zealand and in the summer when it lies west of the island. In winter, it moves eastward to the central Pacific and becomes a less frequent feature of the mean charts. The variability in location from month to month is more marked than that of the other oceanic highs of the Southern Hemisphere. A closed anticyclonic centre appears in the monthly mean pressure chart over inland eastern Australia in early autumn and persists through the winter declining in frequency by spring and exhibiting a westward shift of some 13° in median longitude during this period; in summer, anticyclone centres in the monthly mean pattern are rare in the Australian region, the systems being located principally over the waters to the east of the continent but with occasional high pressure maxima over the waters of the Great Australian Bight.

The winter persistence of high pressure over Australia combined with a seasonal movement of the Indian Ocean anticyclone (Chapter 5, Fig. 32, p. 541) results in an annual pattern of changing longitudinal separation between the pressure maxima in the subtropical belt over the region. This pattern is shown in Fig. 25 for 1972—1977. From October to March, the two pressure maxima are separated by some 100° of longitude. In autumn and winter, high pressure over Australia and the eastward movement of the western Pacific anticyclone introduce a shortening of the separation between the eastern and western centres with corresponding increases in distance between the Australian and West Pacific maxima.

A census of anticyclones appearing on daily charts for winter and summer during the IGY has been published by TALJAARD (1967). This shows a band of maximum frequency encircling the hemisphere with a core of highest frequency located close to or a few degrees poleward of the axis of highest pressure. A marked departure from this pattern

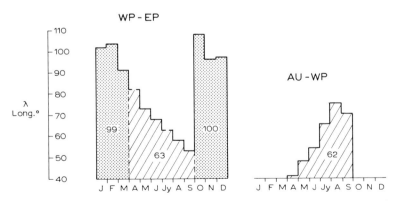

Fig. 25. Histograms showing mean monthly separation (deg. long.) between the subtropical anticyclone centres of the Pacific region (see Fig. 24); *WP* = western Pacific; *EP* = eastern Pacific; *AU* = Australia. Annual or shorter term means are shown by the figures on the face of the histograms. (After STRETEN, 1980.)

exists over the Tasman Sea and to the east of New Zealand in winter. In this area, there is a distinct double pattern of anticyclone frequency, one lying from the coast of southern New South Wales southeastward across New Zealand, and the other from the central coast of Queensland eastward across Norfolk Island to the Kermadec Islands. The detached maxima tend to be reunited in the circumpolar band in the vicinity of 150°W. This pattern is reflected in Fig. 22 showing the variability of the axis of maximum pressure. Very few anticyclones are observed south of 50°S, although Taljaard notes apparent extensions of high pressure from Antarctica resulting in closed anticyclonic patterns in the region to the north of the Bellingshausen Sea. However, anticyclones with the highest central pressure tend to form well south of the subtropical ridge, and winter systems with pressures reaching 1040 mbar are not infrequently observed over the southern Tasman Sea and to the east of New Zealand. KARELSKY (1965) has published maps of the mean and extreme pressures observed over the period 1952–1963 for the Australian and New Zealand region.

An analysis of anticyclones on daily charts for 1960–1969 has been made by BROWNE (1975) for the same area. His data show a high frequency of centres to the west of the North Island of New Zealand in spring, summer, and autumn, and in winter a high concentration to the east of the South Island.

The longitudinal variation in the frequency of anticyclones is related to the location of regions of formation, decay, and slow movement where high frequencies are recorded, and of regions of rapid translation where the frequency is low. In the Pacific region, anticyclones tend to form or to intensify in the Great Australian Bight frequently by "budding," i.e., extending eastward as a narrow ridge from a high pressure area in the Indian Ocean and thence forming new detached circulations which move steadily across Australia and New Zealand. This fairly rapid movement of cyclones through the Australian region was first recognized by RUSSELL (1893) as an important key to understanding the day-to-day weather evolution over the continent. Anticyclones also form over the northern Tasman Sea and to the northeast of New Zealand. The most significant decay region is apparently in the southeast Pacific near 90°W.

VAN LOON (1960) found that in summer anticyclones tended to move northeastward in the western part of the South Pacific but eastward to southeastward in the east. TALJAARD (1972) indicates that the average movement of South Pacific anticyclones is eastward with a small northward component, although individual systems display considerable variations in speed and direction of travel. Tracks published by TALJAARD (1967) for two months of the IGY show considerable variations in motion but display clearly the winter pattern in the southwest Pacific with two distinct types of tracks, one well to the north of New Zealand and the other across or to the south of the South Island. No further detailed long-term data have been published of day-to-day anticyclone tracks in the Pacific, and our knowledge in this area, particularly of the eastern region, is greatly deficient.

The westerlies

The broad stream of the Southern Hemisphere surface westerlies crosses the South Pacific at all seasons between 40°S and the latitude of the Subantarctic trough. Their strength measured by the pressure difference between 40°S and 60°S varies considerably

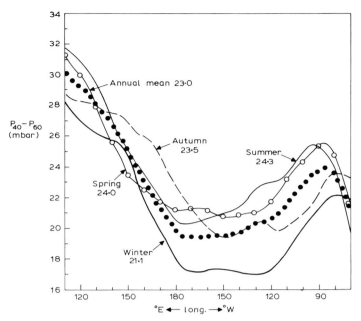

Fig. 26. Pressure gradient (mbar) between 40°S and 60°S at Pacific longitudes (1972 – 1977; after STRETEN, 1980).

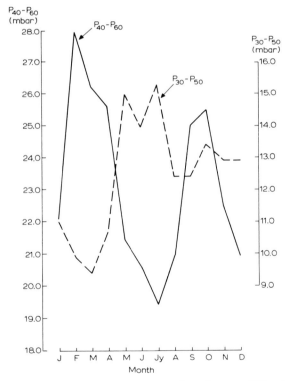

Fig. 27. Annual cycle of pressure gradients between 40°S and 60°S, and between 30°S and 50°S (1972 – 1977), averaged over Pacific longitudes 110°E to 70°W.

297

with longitude as shown in Fig. 26 based on the means from five years of numerical analyses 1972–1977 (STRETEN, 1980). The pressure gradients are strongest throughout the year to the south of Western Australia and weakest in the region between about 120° and 180°W. The pressure difference between 40°S and 60°S averaged over the wide span of Pacific longitudes (Fig. 27) varies on a semiannual cycle as shown by VAN LOON (1967a). Maxima occur in spring and autumn and minima in summer and winter to the south of about 50°S. Van Loon showed that this oscillation changed phase north of 50°S (see e.g., the corresponding pressure differences between 30°S and 50°S for Pacific longitudes also shown in Fig. 27). The pattern of variation of westerly strength at higher latitudes is related to the semiannual variation in the latitude and pressure in the Antarctic trough. These semiannual cycles are associated with the different seasonal temperature trends in middle and high latitudes resulting in the final analysis from differing responses of the earth's surface to the heat budget (VAN LOON, 1967a, 1971). In the latitudes near 50°S, the strongest westerlies tend on the average to occur in summer, as reflected in the 40–60°S pressure contrast in Fig. 26 (VAN LOON, 1966).

Examination of the variability of the mean monthly zonal current over the five years showed that the region south of eastern Australia and New Zealand (where the pressure gradients may be measured most reliably) exhibited the largest range of pressure gradient in the Pacific region and that this range was largest in spring. These observations confirm earlier longer-term data of pressure gradients between selected pairs of stations (STRETEN, 1977). By contrast, the region of strongest zonal westerlies south of Western Australia is also that of least variability.

This high variability in the mean monthly pressure gradient in the westerlies of the New Zealand sector corresponds to that in the axis of maximum pressure (Fig. 22) at similar longitudes. Both are manifestations of the blocking phenomena that are observed in this area with higher frequency than elsewhere in the Southern Hemisphere (VAN LOON, 1956; STRETEN, 1969a). WRIGHT (1974) carried out a detailed analysis of blocking over the Australian region and defined the phenomenon as having the following characteristics.

(*1*) The basic westerly current splits into two branches.

(*2*) The 5-day mean 500-mbar ridge at 45°S (defining the longitude of the block) has a rate of progression of less than 20° of longitude per week and progresses no more than 30° of longitude during the entire blocking.

(*3*) The ridge of high pressure at the longitude of blocking is at least 7° south of the normal position of the subtropical high pressure belt (as derived by TALJAARD et al., 1969) and is maintained with recognizable continuity.

(*4*) The occurrence lasts for at least six days.

Considering the data for the period 1950–1965, Wright compiled an annual distribution of the number of days with blocking in 10°-longitude sectors (Fig. 28). A maximum exists at the longitudes of the Tasman Sea and is most frequent there in June and July. On the average, some 21% of both summer and winter and 13% of spring and autumn are affected by blocking at these longitudes.

Wright's study did not extend to the longitudes of the eastern Pacific; the IGY and earlier studies (VAN LOON, 1956) and synoptic experience suggest that blocking is not as common in this region.

The characteristic synoptic features of the region of the mean westerlies are the great cyclones or depressions and their attendant fronts.

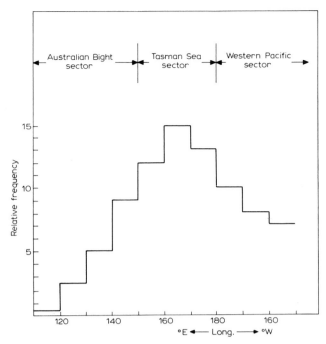

Fig. 28. Relative frequency of blocking pattern per 10°-longitude sector, southwestern Pacific. (Adapted from WRIGHT, 1974.)

These circulation systems form, travel, and decay with lifetimes around a week and their passage through a given location dominates the weather of the midlatitudes of the Pacific from the fringe of the anticyclonic belt to the coast of Antarctica. As we have seen, the frequency of anticyclones between 40°S and the Antarctic is low so that the depressions are usually separated by ridges rather than closed centres of high pressure. Thus, at any point on its northern fringe the regime of the westerlies is largely one of alternations of NW winds changing to W and SW with the passage of cyclones to the south, interspersed with periods of easterlies when individual systems move past at lower latitudes than normal.

The distribution of cyclones and cyclogenesis has been studied more than that of the anticyclones, but in many of the complete hemispheric analyses the South Pacific, particularly the eastern part, has received the least attention. SCHMITT (1957) used the charts of MEINARDUS and MECKING (1911) and KIDSON (1947) to study cyclone frequencies in the higher latitudes of the Pacific; and VAN LOON (1960, 1962, 1966) examined the movement and frequency of lows; a detailed analysis was produced for the U.S. Navy's operational requirements by PRANTNER (1962), based on the SAWB analyses. The IGY data yielded the most complete synoptic analysis series for its limited duration, and the cyclones and anticyclones occurring during this period were described in considerable detail by, e.g., ASTAPENKO (1960), TALJAARD and VAN LOON (1962), VAN LOON (1966), and TALJAARD (1967). Following the introduction of complete hemispheric satellite imagery, sequences of pictures were analyzed by STRETEN (1968a,b, 1969b) and STRETEN and TROUP (1973) for summer when the better illumination at higher latitudes enabled the television pictures to be most readily interpreted. These data tended to be very similar in broad features to those of the IGY. Satellite data have now been available

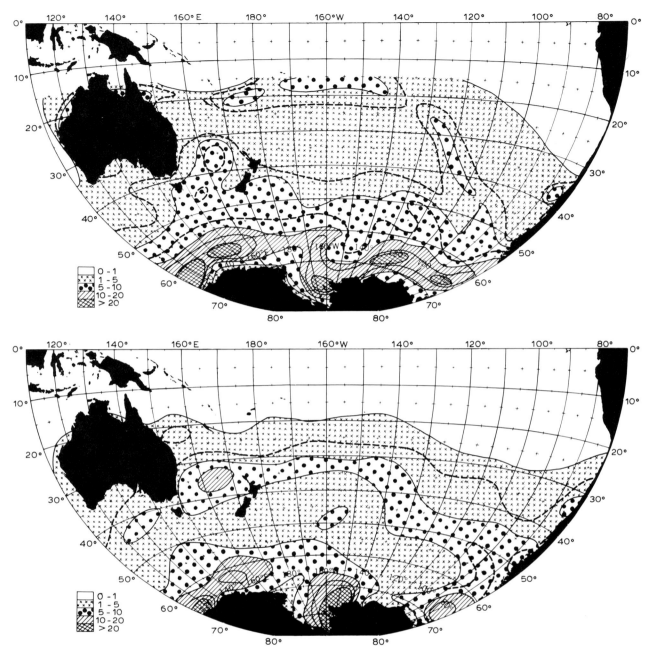

Fig. 29. Distribution of cyclones over the South Pacific per unit area (438,000 km²). Upper: per season in summer (December–March); lower: in winter during the IGY. (Adapted from TALJAARD, 1967.)

over the whole region for over a decade and, since late 1972, twice-daily thermal infrared imagery has made possible year-round surveillance of the development of cyclones and frontal systems. These data have been used qualitatively and semi-quantitatively in numerical weather chart series, and a census of the cyclone population over particular time spans may now be made by numerical processing of the chart data; no such complete analysis has yet been published. Thus, the analyses of VAN LOON (1966) and TALJAARD (1967) based on the IGY period still provide the best overall estimate of the cyclone distribution over the Pacific. Charts based on TALJAARD (1967) for summer (December—March) and winter (June—September) are reproduced in Fig. 29. Notable features are: (*1*) the steady increase in the number of cyclones from middle to higher latitudes over the ocean; (*2*) the extension of higher cyclonic frequency in a band reaching to lower latitudes in the central and western Pacific; and (*3*) the pronounced belt of high cyclonic frequency around Antarctica exhibiting maxima at particular longitudes.

An analysis of Southern Hemisphere cyclone vortex frequencies in relation to the location of the pack ice edge (SCHWERDTFEGER and KACHELHOFFER, 1973) has shown that there may be some relationship between the position of the ice—water border and the latitudinal band of maximum frequency of cyclones. North of 50°S, the vortex frequencies are consistently higher in the spring (maximum ice extent) than in the autumn (minimum ice extent) while the reverse is true south of 50°S. The Subantarctic trough is, however, nearer Antarctica and deeper in spring and autumn than in summer and winter (VAN LOON, 1967a, 1971).

The IGY data investigated by TALJAARD (1967) showed that over the Southern Hemisphere cyclogenesis occurs between 35° and 45°S. In the Central Pacific between 18° and 30°S and between 180° and 120°W, there is a strong tendency for depressions to form at much lower latitudes. This is related to the high frontal and cloud band frequency to be discussed later. Cyclogenesis frequency was also observed to fall off from midlatitudes toward Antarctica, and this was confirmed by the early satellite observations for the 1966—1967 summer (STRETEN, 1968a). Fig. 30 shows the frequency of developing cloud vortices, representative of the relative frequency of summer cyclogenesis, for three summers (STRETEN and TROUP, 1973). The northern extension in the Central Pacific is clearly separate from the zone of highest frequency around 45—50°S. Fig. 31 shows the axes of the zones of highest frequency of early vortex development in summer, spring,

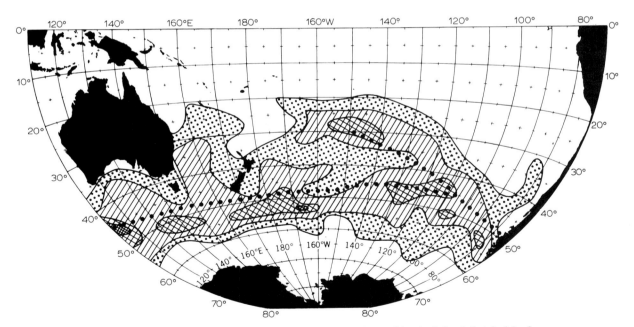

Fig. 30. Frequency of developing cloud vortices in summer (cyclogenesis): stippled < 1, hatched 1—2, cross-hatched > 2 per unit area and per season. (Adapted from STRETEN and TROUP, 1973). Dotted line is the mean position of the Polar Front in summer. (After TALJAARD, 1968.)

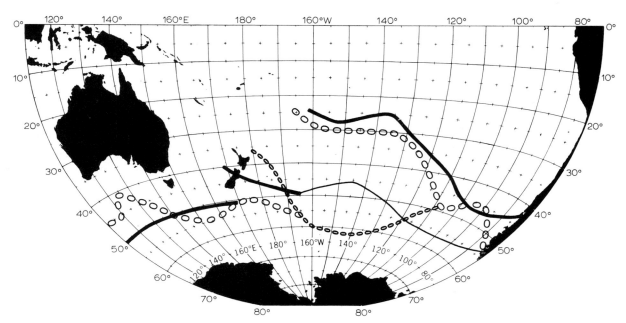

Fig. 31. Axis of the zone of highest frequency of early cloud vortex development (cyclogenesis) for summer (dotted) and an "intermediate season" of autumn plus spring (full line). Where a secondary maximum occurs, it is shown as a finer line. (After STRETEN and TROUP, 1973.)

and autumn over a three-year period. In this diagram, a median latitude of highest frequency was plotted for 10° longitude zones, and the major axes of these points are shown, together with such minor axes as occur at different latitudes. The central South Pacific at lower latitudes appears in this period as a principal region for cyclogenesis, in contrast to the western part of the ocean where the maximum of cyclogenesis appears

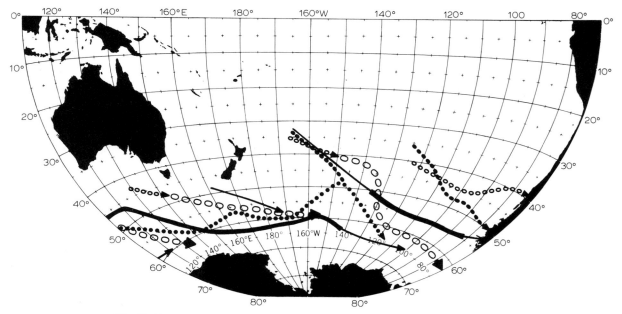

Fig. 32. Vortex (depression) track diagram. Heavy lines: major tracks; finer lines: "minor" tracks. November (open dotted); January (full dotted); March (full line). (After STRETEN and TROUP, 1973.)

302

at 45 – 50°S, the latitude commonly observed around most of the hemisphere.

By following sequences of satellite imagery displaying the cloud vortex evolution and movement, it has been possible to produce maps for some months of the tracks of depressions over a period of three years. A distinction is made between major and minor tracks depending on the frequency of individual tracks at particular longitudes. Such a diagram is shown in Fig. 32. The considerable variation in individual tracks should be noted; diagrams such as this can give only a very broad representation of the pattern of movement. As an example of winter tracks, Fig. 33 indicates the movement of individual systems for July 1957. Most track data for the Pacific show a general pattern with cyclones originating in the Indian Ocean or the western Pacific moving southeast toward the coast of Antarctica south of Australia or farther east toward the Ross Sea. Those originating eastward of New Zealand or in the Central Pacific and pursuing southeastward tracks arrive in high latitudes near the Bellingshausen Sea, Drake Passage, and southern Chile. In winter, depressions forming southwest of Western Australia or in the Great Australian Bight frequently pass close to, or cross inland over southeastern Australia and Tasmania, and into the Tasman Sea. It is common for upper-level cold pools to be cut off, to pass across southeastern Australia, and to lead to the formation or regeneration of surface depressions in the Tasman Sea. Such cyclones may become very intense and substantially affect the weather of New Zealand and the Tasman shipping routes. Although such systems may occur at any time, it is more common in summer for the depressions of the South Pacific to pass to the south of Australia and New Zealand. The region farther north is then close to the fringe of the westerlies and is more likely to experience only the passage of weaker troughs extending northward from the principal low pressure centres.

The movement of cyclones has been studied by MEINARDUS (1929) who found that the average zonal displacement in the Atlantic and Pacific sector was 10 m s^{-1}. The total motion (both zonal and meridional) was investigated by VAN LOON (1967b) for the sector from 140°E to 60°W using the IGY data. These data suggest a mean speed of movement varying little between summer and winter and ranging from about 9 m s^{-1} at 30° to 40°S and 60° to 70°S, to 13 – 14 m s^{-1} between 40° and 60°S. Cloud vortex

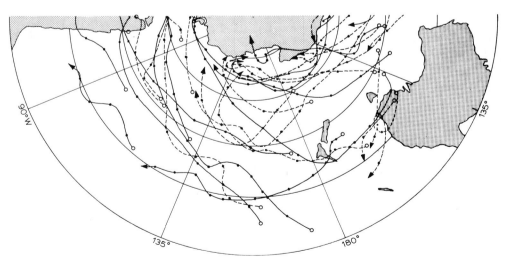

Fig. 33. Tracks of individual cyclones in July 1957. (After TALJAARD and VAN LOON, 1962.)

movement distributions for a later period published by STRETEN and TROUP (1973) have slightly smaller values, from 8 m s^{-1} at 30° to 40°S to 11–13 m s^{-1} south of 40°S. Cyclone decay or "graveyard" regions, investigated for the IGY and cyclone vortex data, show highest frequency of location in the Pacific region: (*1*) close to the coast of East Antarctica between 110° and 150°E with a maximum east of 140°E (Terre Adélie); (*2*) north of the central and eastern Ross Sea; and (*3*) north of the Bellingshausen Sea. Such high frequency of cyclonic decay in these regions is associated with stagnation in the antarctic trough and the general pattern of southeasterly track from the cyclone source regions of lower latitudes. The presence of antarctic coastal embayments is also probably of some importance (STRETEN, 1969a; TALJAARD, 1972). The complex interactions between ice cover, open water, and topography on one hand and cyclonic evolution, decay, and regeneration on the other for high southern latitudes remain to be investigated in the light of new information.

Much of the technique of synoptic analysis over the oceans of the Southern Hemisphere has been dependent on the identification of frontal systems and the subsequent tracking of their evolution and movement from one isolated observation point to the next. Having regard for the data base, objectivity of frontal definitions under these circumstances is very difficult. TALJAARD et al. (1961) provide a detailed discussion of the problem and define it for the IGY analysis, viz., "a narrow sloping layer with a vertical extent of at least 3 km, across which the temperature changes sharply in the horizontal direction by an average of at least 3°C in subtropical regions and 4–5°C in middle latitude and polar regions."

Papers from a conference (ROYAL METEOROLOGICAL SOCIETY, AUSTRALIAN BRANCH, 1977) suggest that a variety of opinion exists among practicing meteorologists on the problem of frontal analysis for sparse data regions. It is certainly true that over the past decade frontal analysis over the Pacific has been dependent not on rigid definitions which cannot be put into practice where there are no upper-air stations, but rather on the largely subjective utilization of satellite imagery, the general principles of which have been described by GUYMER (1978). ZILLMAN and MARTIN (1968) have published a description of a typical frontal passage across a point in the westerlies of the South Pacific. From the climatic point of view, a useful concept is the "climatic front," which may be regarded as a statistical entity based on the cumulative effect of individual circulation patterns and frontal zones at specific locations and times. TALJAARD (1968) defines such a front as a zone along which fronts (as identified in day-to-day analyses) are most numerous and along which the meridional temperature gradient in the lower and middle troposphere shows a maximum. Principal frontal zones were delineated from the daily analyses of the IGY data which did not include satellite data (VAN LOON, 1965). They and meridional gradients of thickness were used to define Taljaard's climatic Polar Front in summer in the South Pacific, which is reproduced in Fig. 34 together with the cloud extent averaged over nine summer months between 1967 and 1970, derived from computer-cumulated data of the ESSA satellites by MILLER and FEDDES (1971). The latter data are of limited value at higher latitudes in winter as they are based on the photographic brightness data derived from the television cameras then used on operational satellites. The satellite data indicate an extension of greater cloudiness into the subtropics of the central South Pacific and a substantial gradient of cloudiness near 45°S. The data tend to support the conclusions of Van Loon and of Taljaard, based

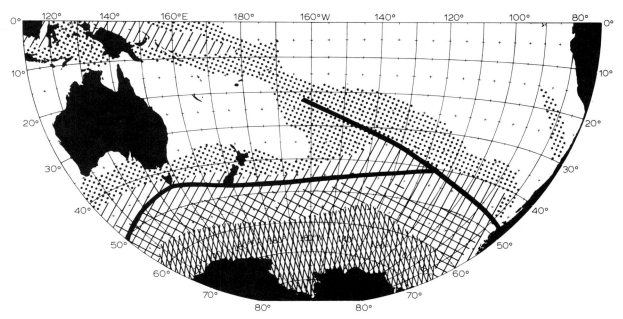

Fig. 34. Axes of the climatic Polar Front in summer: full line (after TALJAARD, 1972). Shaded areas are from MILLER and FEDDES (1971) and show mean summer cloud amount over the ocean: stippled >4 octas; hatched >5 octas; cross-hatched >6 octas; crossed >7 octas.

partly on synoptic chart sequences for a different time period and partly on long-term gradients of temperature.

An analysis of the frequency of cloud bands appearing on five-day averaged cloud brightness imagery extending over three years has been made by STRETEN (1973a). In

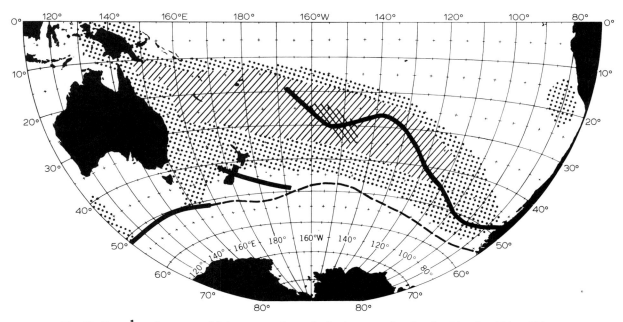

Fig. 35. Percentage frequency of 5-day averaged mosaics having axes of major cloud bands within a 5°-lat. by 10°-long. area for spring and autumn (stippled >10%; hatched >20%; cross-hatched >40%), after STRETEN (1973a). Axes of zones of highest frequency of early cyclonic development for the same seasons is shown by the full line (after STRETEN and TROUP, 1973).

305

this work, the definition of what constitutes a cloud band was designed to include only persistent, substantial, and quasi-meridionally oriented cloud features. The frequency of such bands in spring and autumn is reproduced in Fig. 35 as well as the axis of the zone of highest frequency of early cloud vortex development (STRETEN and TROUP, 1973). The zones of high frequency of cyclogenesis tend to be oriented in a manner similar to that of high band-frequency, and both extend to lower latitudes in the central Pacific. In the cloud vortex study, it was found that for the hemisphere as a whole less than half the cases of cyclogenesis occurred in the presence of a previously existing cloud band (in this instance, one observed 24 h previously). At higher latitudes, development from an inverted comma cloud alone was most frequent. In this latter case, a new cloud band exhibiting frontal characteristics comes into existence with the cyclogenesis and generally develops and decays with the depression. The decay of the parent vortex may leave a residual cloud band on which wave formation with new cyclogenesis may occur. Alternatively, a new comma cloud may develop in the rear zone of the cloud band, merge with it, and apparently initiate cyclogenesis.

In summary, TALJAARD's (1972) picture of the climatic fronts of the South Pacific is shown in Fig. 36. A frontal zone extends around the Pacific sector in summer along approximately 45°S with a major branch extending from around 30°S 150°W southeastward toward South America. In winter, the northern branch appears to be more active and displaced farther north and east (VAN LOON, 1965, cf. figs. 6 and 7). In the western region south of Australia, the frontal zone appears to be displaced somewhat farther south than in summer, and a winter "Antarctic front" extending across the ocean near 70°S is suggested.

The available evidence, in particular the pattern of cyclogenesis, does not on the whole suggest an "Antarctic front" as postulated by some investigators. Because the high latitudes of the South Pacific are the least known meteorologically of the oceans sur-

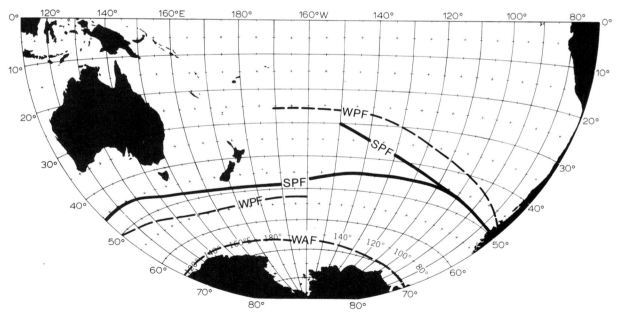

Fig. 36. Climatic frontal zones of the South Pacific (after TALJAARD, 1972). WPF: winter Polar Front; SPF: summer Polar Front; WAF: winter Antarctic Front.

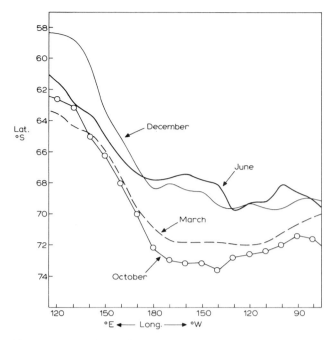

Fig. 37. Mean latitude of the axis of the Antarctic trough in indicated months (1972–1977). (Data from STRETEN, 1980.)

rounding Antarctica, specific synoptic analyses of the region utilizing the new satellite and buoy data will have to be studied to examine the characteristics of the regional atmospheric structure in more detail.

The Antarctic trough

One characteristic of the Southern Hemisphere mean circulation that distinguishes it from its northern counterpart is the circumpolar belt of low pressure at high latitudes between Antarctica and the broad stream of the westerlies. This zone is well marked in almost all mean charts of individual months with few situations when it is substantially disrupted over a broad longitudinal span by linkages of the higher pressure region at midlatitudes with that over Antarctica. The mean latitude of the axis of the trough in the Pacific Ocean for 1972–1977 is shown in Fig. 37. The pattern throughout the year reveals lower latitude locations of the trough over the western part of the Pacific rather than over the eastern, and thus it follows the general trend of the latitude of the Antarctic coast. The farthest equatorward extension (58°S) occurs in December near the longitude of Western Australia and the highest latitude location is at 74°S in October in the eastern Ross Sea. The movement of the trough has a clear half-yearly cycle such as it has for the hemisphere as a whole, being farthest north during the extreme seasons and farthest south during the transition seasons (VAN LOON, 1967a, 1971). In both March and October when the sea ice is either nearest to the continent or farthest away, the trough is nearest to the continent (VAN LOON, 1967a). The mean monthly location of the trough appears to be most variable at the longitudes of the Pacific in late winter (STRETEN, 1980). Centres of minimum pressure within the trough over the Pacific are clearly located between 130°W and 170°W north of the eastern Ross Sea. At some longitudes, pressure ridges extending south from the westerlies or northward from Antarctica impinge upon and weaken the trough. The most prominent of such zones in the Pacific lies between 130°E and 160°E. A less pronounced region of weakening of the trough lies at 70—90°W. Eastward of these favoured ridge positions are the locations of the most frequent incursion of cold air from high latitudes into the westerlies at midlatitudes. In the Pacific, the most notable of such regions is in the area south of Eastern Australia, the Tasman Sea, and New Zealand.

The Antarctic trough is essentially a region of cyclonic stagnation and decay. There is little evidence of frequent cyclogenesis occurring there, although regeneration and deepening of cyclones entering the region from lower latitudes are not uncommon. Detailed investigation of these synoptic developments should soon be possible with the advent of the new buoy and satellite observations.

The Pacific cloud band

In much of the previous discussion, reference has been made to the zone of discontinuity extending from northwest to southeast across the Pacific. In lower midlatitudes, this zone lies in the col between two regions of prevailing high pressure. Moist northeast flow to the west of the persistent southeast Pacific high is brought into the proximity of cooler and predominantly southeasterly winds associated with anticyclones moving eastward from the Australian and New Zealand region. This discontinuity was originally described by BERGERON (1930) and appears on the ocean climate maps of McDONALD (1938) as a region of maximum cloudiness and relatively low windspeed with its axis extending from the eastern tip of New Guinea to around 30°S 120°W. This zone has also been termed the South Pacific Convergence Zone (SPCZ) (TRENBERTH, 1976b).

It was not until the advent of meteorological satellites that the rather surprising persistence and meridional extent of a major band of cloud (e.g., Fig. 38) associated with the region became evident. HUBERT (1961) observed an extended cloud band on pictures of the region from the earliest meteorological satellite TIROS I, and subsequent development of techniques for the production of multi-day averaged imagery (KORNFIELD et al., 1967; LEESE et al., 1970) enabled the persistence of the band to be recognized as a major climatic feature (STRETEN, 1968b, 1970; KORNFIELD and HASLER, 1969).

The northern part of the time-averaged band becomes more zonally orientated in the region of Samoa and thence extends westward to merge with the general ITCZ cloudiness over Indonesia. The southeastward extension of the cloud band through the subtropics is the area of the South Pacific Convergence Zone (SPCZ) proper. The band of cloudiness may often extend to higher latitudes to the south of the anticyclonic belt; in these regions, it is apparently associated with a preferred occurrence of cyclogenesis, cyclone tracks, and possibly with jet stream cloud. Satellite observation of cyclonic development (STRETEN and TROUP, 1973) showed the whole band to be an area of substantial weather activity with wave disturbances and cyclogenesis frequently occurring along its length and with depressions then moving generally southeastward toward Cape Horn.

In a more general investigation into prominent cloud bands over the Southern Hemisphere and their relation to the upper flow (STRETEN, 1973a), it was found that the

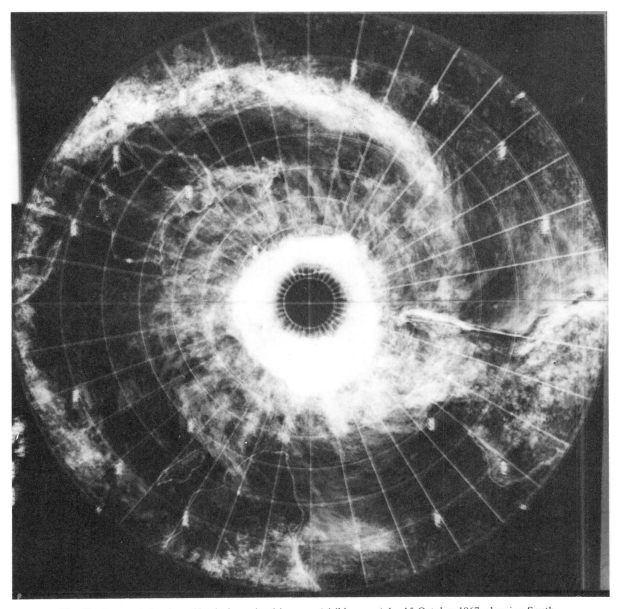

Fig. 38. Averaged Southern Hemisphere cloud imagery (visible range) 1–15 October 1967, showing South Pacific cloud band. (After KORNFIELD and HASLER, 1969.)

time-averaged bands tended to be located eastward of substantial troughs in the upper westerlies as shown on similarly time-averaged upper-level charts. The observational network in the past has been generally insufficient to define small variations in the rather low-amplitude waves in the time-averaged Southern Hemisphere upper flow, so the bands themselves provide "markers" (albeit crude) of changes in the circulation pattern (YASUNARI, 1977). Variation in the location of the band axis on monthly and seasonal time scales has given some overall indications of the broad-scale flow from one season to the next (STRETEN, 1975). Short-term variations (STRETEN, 1978) probably reflect oscillations in the behaviour of the atmosphere, such as have been observed in the

Northern Hemisphere (McGUIRK and REITER, 1976), and in the upper-level Southern Hemisphere flow at circa 200 mbar as revealed by the EOLE balloon data (WEBSTER and KELLER, 1975). Such oscillations are a probable manifestation of what has been described in ROSSBY and WILLETT (1948) and NAMIAS (1950) as the "zonal index cycle." The cloud band represents a substantial anomaly over a wide range of latitude and longitude across the South Pacific, and its intensity and location may be associated with changes in the intensity of the southeast Pacific High and with the jet stream structure; it may therefore be important in relation to the Southern Oscillation (TRENBERTH, 1976b, p. 351). The amount of the synoptic activity along the band and its orientation point to the zone as being one of considerable momentum and energy exchange between lower and higher latitudes. It may represent a climatological analogy of the short-term cloud band linkages between the tropics and midlatitudes that occur in specific synoptic situations and act as channels of poleward energy transfer (ERICKSON and WINSTON, 1972).

In the EOLE analysis, WEBSTER and CURTIN (1975) found that the poleward extension of the Southern Hemisphere cloud bands lies slightly upstream of the transient-wave momentum flux anomalies; these authors suggest that the band represents a region of periodic and intense wave energy flux toward the Equator. More detailed surface and upper observations are necessary to examine the nature of the cloudy zone and its relation to the general circulation.

Main climatic elements

The various characteristic circulation features just described represent the main synoptic influences on the overall climate of the South Pacific. We now describe briefly the resulting broad climatic characteristics of the region in terms of the long-term averages of the main climatic elements. Data will be presented primarily in chart form with only very brief discussion. Extensive published studies of the climate of island groups or stations are few although some contain more detailed data analysis, e.g., New Caledonia (GIRARD and RIGNOT, 1971), French Polynesia (D'HAUTESERRE, 1960), and Macquarie and Campbell Islands (GIBBS et al., 1952; FABRICIUS, 1957). Most data for the Pacific islands are contained in the various sources listed earlier, and a summary of these is presented in the climatic tables for particular selected stations following this chapter. Also included are tables for the selected ocean station regions shown in Fig. 2.

Surface air temperature

Over the greater part of the open ocean, the isotherms of mean monthly temperature closely follow the parallels of latitude (Fig. 39). The most notable exception to this occurs at lower and midlatitudes near the eastern boundary of the ocean where the isotherms in both seasons reveal the cool anomaly resulting from the southerly air trajectories on the eastern flank of the southeast Pacific High and the cooling effects of the coastal upwelling and the Peru (Humboldt) Current (Figs. 4, 6, and 7). This relatively cool region extends westward along the Equator to the mid-Pacific. The highest temperatures throughout the year are over the western oceanic tropics.

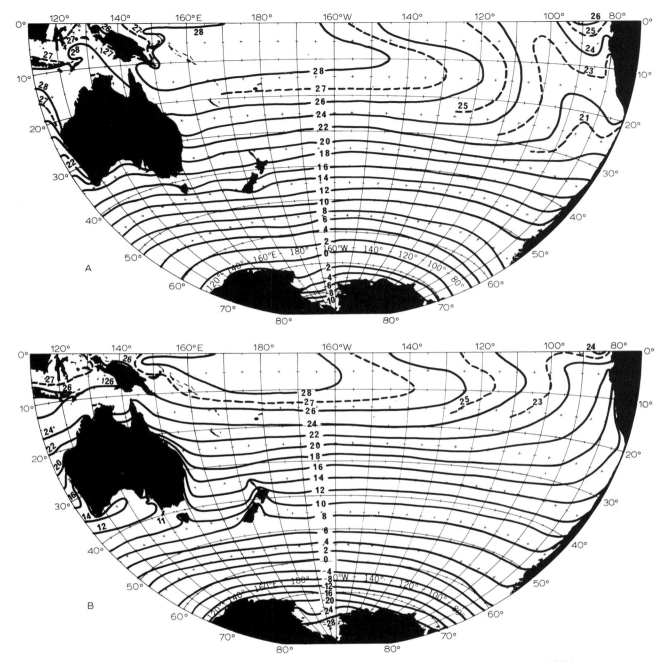

Fig. 39. Mean surface air temperature (°C) in January (A) and July (B). (After TALJAARD et al., 1969.)

Temperature gradients at midlatitudes are stronger in summer than in winter, but in winter the most substantial thermal gradients exist over the region of the winter pack ice south of latitude 60°S. In this season too, the contrast of a cool continental Australia and the warm southward-flowing East Australian Current produces a marked air temperature discontinuity on the western margin of the Tasman Sea. New Zealand also perturbs the otherwise zonal symmetry of the temperature field. Comparison with Figs. 6 and 7 indicates that over much of the tropical ocean the air temperature is slightly lower than that of the sea. Tables for the marine areas following this chapter indicate the variation of monthly mean temperature throughout the year, the corresponding interannual range, and the extremes of monthly mean. Analysis by VAN LOON (1955, 1966) indicates a threefold pattern of annual range of both sea surface and air temperature over the Pacific (*1*) steadily increasing southward from 0° to 5°C from the Equator to around 30°S, (*2*) a region of maximum range of 5° to 6°C between 30° and 40°S, and (*3*) a region with the range decreasing from 4°C to around 3°C southward toward the farthest northward boundary of the pack ice (see Fig. 8). An exception to this pattern exists off the coasts of Ecuador and Peru where upwelling cold water results in a region with 5 − 6°C range.

The first harmonic (yearly wave, VAN LOON, 1972) accounts for more than 90% of the total variance of temperature over the greater part of the Pacific, excluding the northwestern sector of the tropics equatorward of 15°S. VAN LOON (1966) estimated the heat storage in the mixed layer of the ocean at 35° and 50°S, and the resulting seasonal temperature change in the layer. The estimate suggests that the combined effect of the reduction of incoming radiation by cloud and the mixing of heat to greater depths in the ocean in the region of the westerlies than in the subtropics may account for the decreasing annual range of temperature with increasing latitude poleward of 40°S.

Surface dew point

The most readily mapped, although not always the most useful measure of atmospheric moisture, is the temperature of the dew point. This quantity as mapped for January and July (by TALJAARD et al., 1969) is shown in Fig. 40. In January, values in excess of 20°C lie over the whole ocean equatorward of 25°S with the exception of the cooler and drier region in the extreme east where conditions in the atmosphere reflect the colder ocean surface. In lower latitudes, a region of maximum dew point that is wider in summer than in winter extends from the northwest toward the central Pacific. In summer also, dew points are high on the subtropical coast of Australia, reflecting the inflow of lower-latitude air eastward of the monsoon low over the northern part of the continent. In winter, the core of high dew points retreats somewhat toward the Equator, although it retains the same shape. Around the margins of northern and eastern Australia, strong gradients of dew point are apparent, and significant periods of high precipitation over the continent are associated with synoptic patterns favouring the prolonged inflow of this moister air from the tropics to midlatitudes and its juxtaposition with colder, drier air from higher latitudes along frontal surfaces, in depressions or in cut-off upper lows; similar synoptic situations occur over the ocean region of the whole southwest Pacific in winter.

The most substantial changes in dew point from winter to summer occur at the margins of Antarctica where the decay of the sea ice effectively removes the wintertime extension of the continent's temperature and moisture regime over the ocean to the north.

Surface pressure

Mean sea-level pressure fields as derived by TALJAARD et al. (1969) for January and July are shown in Fig. 41. Seasonal and interannual variations in pressure distribution have

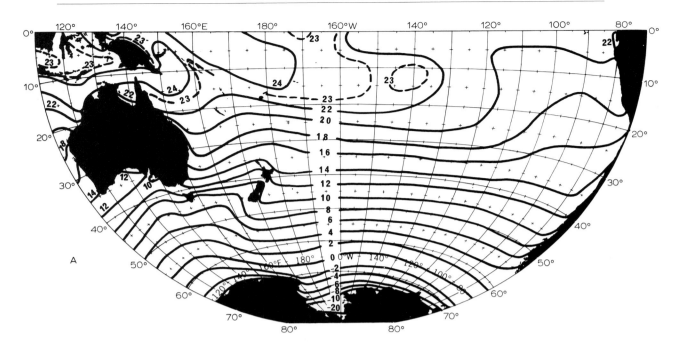

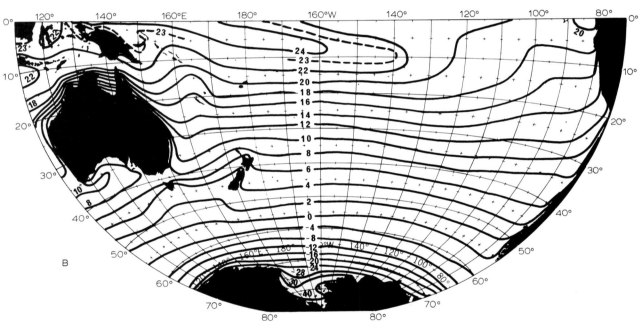

Fig. 40. Mean surface dew point temperature (°C) in January (A) and July (B). (After TALJAARD et al., 1969.)

been described as a basis for the discussion of the synoptic climatology of the Pacific (see pp. 284ff) and no further discussion will be given here. Pressure data for specific locations are given in the climatic tables for marine areas and stations following this chapter.

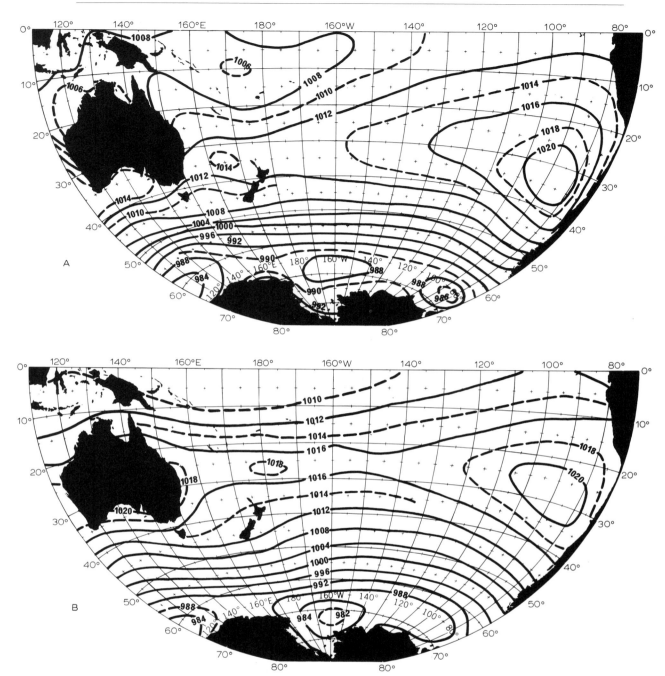

Fig. 41. Mean sea level pressure (mbar) in January (A) and July (B). (After TALJAARD et al., 1969.)

Surface wind

The pressure data of Fig. 41 (TALJAARD et al., 1969) show the general surface circulation pattern and thus the character of the wind flow over the Pacific for January and July. The mean geostrophic windspeeds derived from these data (JENNE et al., 1971) are shown in Fig. 42. The subtropical anticyclones are reflected in a region of light mean windspeed which moves in winter to lower latitudes. The trades on the northern flanks of the

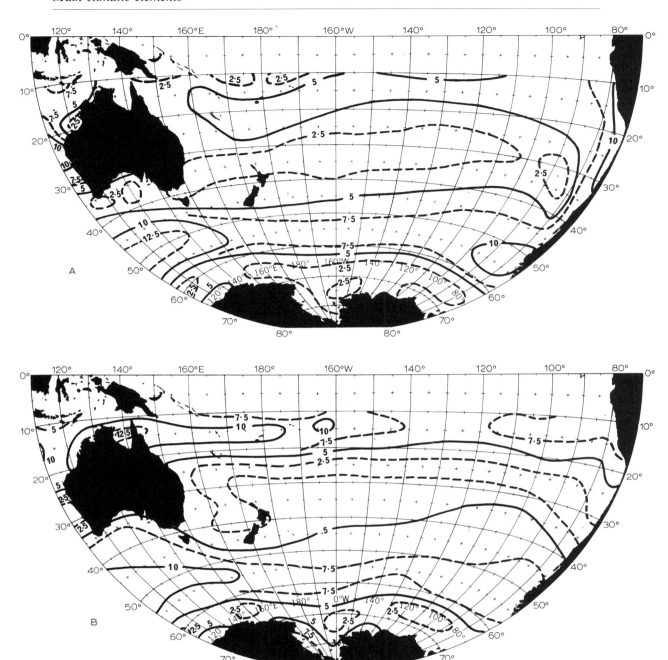

Fig. 42. Mean surface geostrophic windspeed (m s^{-1}) in January (A) and July (B). (After JENNE et al., 1971.)

subtropical highs are maintained with substantial strength throughout the year with stronger flow over the northern Coral Sea in winter than in summer, when this region is affected by monsoonal low-pressure regions and variable though generally weak pressure gradients are common. The general picture of the mean flow over the lower latitudes of the Pacific is probably best shown by the mean surface wind vectors of WYRTKI and MEYERS (1975) reproduced in Fig. 43.

Southward from the light winds of the subtropics, the core of the westerlies in midlati-

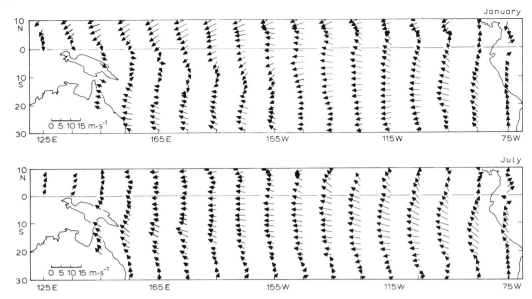

Fig. 43. Mean surface vector winds at lower latitudes of the South Pacific in January and July. (Adapted from Wyrtki and Meyers, 1975.)

tudes, as noted earlier, is stronger south of Australia than farther east, and somewhat stronger in January than in July. Southward of the westerlies, lower windspeeds mark the region of the Antarctic trough, with relatively light mean easterlies extending to the Antarctic coast, although the movement of depressions from midlatitudes into these regions produces periods of strong winds at frequent intervals.

Strong katabatic winds generated primarily by thermal differences between the atmosphere and icecap are common at many locations on the Antarctic coast, particularly at the foot of steep descents from the high plateau of East Antarctica (SCHWERDTFEGER, 1970); these winds generally dissipate within a few miles of the coast. Such winds are intensified by the movement of a depression close to the coast, but the most intense winds are limited to the near coastal zone. Some indication of the characteristics of these weather situations is given by STRETEN (1968c).

Information on the frequency of gales at sea (generally defined as observations of mean windspeeds in excess of Beaufort force 8 or 34 kt.) is limited to analyses of ship observations, and conclusions from such data must of necessity be very general. Fig. 44 adapted from U.S. Navy marine atlases displays a general pattern of a substantial frequency of gales throughout the year in the zone of the westerlies, with an increasing frequency in winter over a broader area. Such gales occur mainly in the wind directions from northeast through west to southwest and correspond to the strong pressure gradients associated with the near approach and passage of frontal disturbances associated with depressions. A description of such an event is given by ZILLMAN and MARTIN (1968). Occurrences of gale-force winds associated with tropical depressions and tropical cyclones are irregular at any given locality and are observed at a frequency lower than can be displayed in the charts. However, because of the extreme strength of such winds, they constitute a significant factor for consideration in the climate of the affected regions and can, or should, necessitate careful planning in the building construction methods and town planning of island and continental coastal communities.

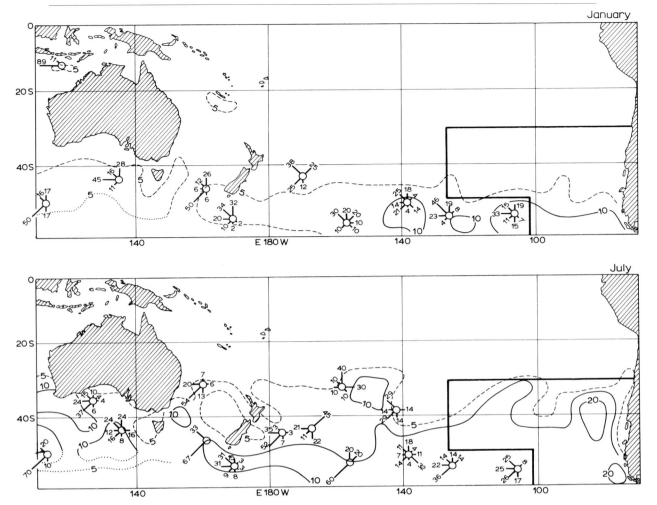

Fig. 44. Isopleths showing the frequency of gales in January and July, i.e., the percentage of observations having windspeeds greater than Beaufort force 8 (>34 kt.). Gale roses show the percentage proportion of gale force winds (Beaufort force 8—12) from each direction. The area of the southeast Pacific within the heavy frame has insufficient observations to show directional frequencies. (Adapted from U.S. Navy Marine Atlases of the South Pacific and Indian Ocean.)

Cloudiness and fog

An assessment of total cloudiness over the Southern Hemisphere for January and July based on earlier analyses of conventional surface observations has been made by VAN LOON (1972). A new climatology derived from satellite observations of daytime cloudiness has been published by MILLER and FEDDES (1971). In this work, computer analysis of television cloud imagery for 1967—1970 produced mean charts of monthly and seasonal cloudiness for the hemisphere and tropical regions (e.g., see the schematic representation in Fig. 34). This followed earlier studies of changes in cloudiness from month to month by KORNFIELD and HASLER (1969).

Charts of total cloudiness derived from conventional ship observations have also been published by a number of organizations, among them the U.S. Weather Bureau (MCDONALD, 1938), the U.S. NAVY (1959), and the U.S.S.R. MINISTRY OF DEFENCE

(1974). Although these charts differ in some important points owing to the varying observation samples used, the basic features of the analyses are similar. Maps for summer (December, January, and February) and winter (June, July, and August) derived from McDONALD (1938) are shown in Fig. 45.

The most prominent features are the relatively low cloudiness in both seasons over the

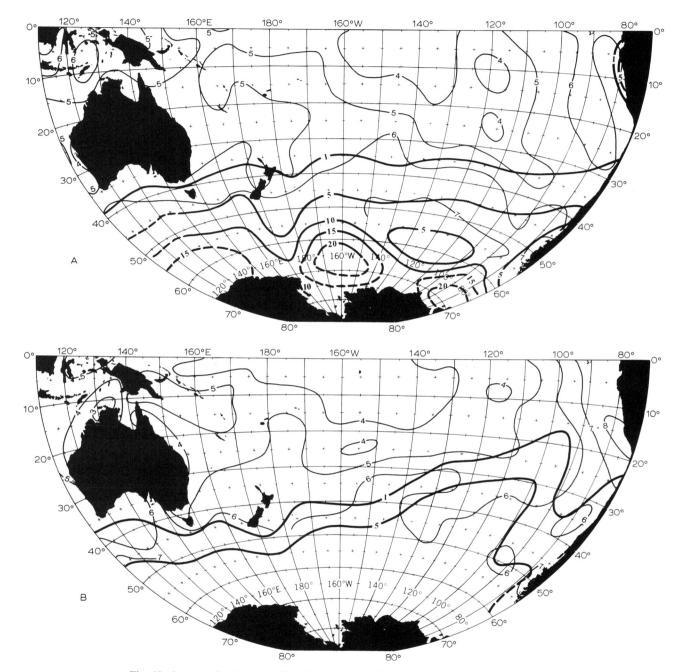

Fig. 45. Average cloud amount (fine lines; in tenths) in: A. summer (December—January—February); and B. winter (June—July—August) (based on McDONALD, 1938). Percentage frequency of observations of fog in A. January, B. July (heavy lines; adapted from Ocean Atlas—Pacific Ocean; U.S.S.R. MINISTRY OF DEFENCE, 1974). High latitude data are insufficient to show reliable cloud amount in both seasons or fog frequency in winter.

region of the southeast Pacific High, and the increasing cloudiness south of the anti-cyclonic belt. Although many analyses do not attempt estimates of cloudiness south of about 45°, it is clear from the charts of ZILLMAN (1967) and VAN LOON (1972) and from the satellite data that an average cloudiness of from 7 to over 8-tenths persists throughout the year in latitudes between 50°S and the Antarctic coastal waters of the South Pacific. In the immediate offshore zone of Antarctica, mean cloudiness is often lower due to the suppression of low cloud by off-shore downslope winds from the continent. Most of the oceanic cloud at higher latitudes is stratiform with the deeper cloud structure resulting from cyclonic and frontal activity. Further, advection of comparatively warmer and moister air masses over regions of much lower ocean temperatures leads to the frequent formation of extensive sheets of low stratocumulus and stratus cloud. By contrast, outbreaks of colder air into lower midlatitudes produce large fields of convective cloud to the rear of frontal zones.

At lower latitudes, the cloud is chiefly convective with active areas surrounded by cloudless regions of much larger extent. Such regions when viewed by satellite frequently appear to be less cloudy than ground observation suggests. This results from the limited resolution of the satellite cameras and radiometers (particularly the earlier instruments). It is generally accepted that satellite observation tends to underestimate cloudiness in comparison to surface-based observation—a value of 14% underestimation was found by BARNES (1966) for the United States; the quantitative value of the variation is doubtful (e.g., see the discussion by LEESE et al., 1970) and may be dependent on total cloudiness, type of cloud, and latitude.

Fig. 45 indicates that the principal variations with longitude of Pacific cloudiness are the mid-ocean cloud band, discussed previously (p. 308) as a significant feature of the circulation, and the high frequency of substantial cloudiness west of the coast of Chile and Peru, particularly in winter. The latter cloudiness is apparently associated with the cold Peru (Humboldt) Current which stabilizes the lower atmospheric layers and results in low-level inversions beneath which extensive though thin stratiform low cloud may frequently form.

The frequency with which fogs are observed by shipping is also shown in Fig. 45 based on analysis by the U.S.S.R. MINISTRY OF DEFENCE (1974). The incidence is small in the tropics and subtropics throughout the year. In summer (February), the frequency increases rapidly southward of 45° where fogs often occur with quasi-stationary pressure systems causing prolonged advection of warm moist air from subtropical latitudes over the cold waters. Data from the Subantarctic islands (e.g., Macquarie and Campbell Islands) show frequent occurrences of these weather types, which are probably higher at these longitudes because of frequent blocking patterns over and south of the Tasman Sea.

In winter, data at high latitudes are not sufficient to record reliable frequencies; it is probable that fog frequency, as at midlatitudes, is somewhat lower than in summer.

Precipitation

Considering the comparatively few observing stations and the abundant variability of rainfall in space and time, any representation of this element over the whole of the South Pacific must be most general. Earlier analyses of rainfall have been made by

MEINARDUS (1934), MÖLLER (1951), ALBRECHT (1951, 1960), SEKIGUCHI (1952), DROZ-
DOV (1953), VAN LOON (1972), KORZUM (1974), BAUMGARTNER and REICHEL (1975), and
DORMAN and BOURKE (1979), often as part of global or hemispheric studies of precipitation and evaporation. An analysis of the percentage frequency of ship observations
reporting precipitation was originally made by McDONALD (1938) and in a similar study
by the U.S.S.R. MINISTRY OF DEFENCE (1974). Maps for January and July from the
latter source are reproduced in Fig. 46.

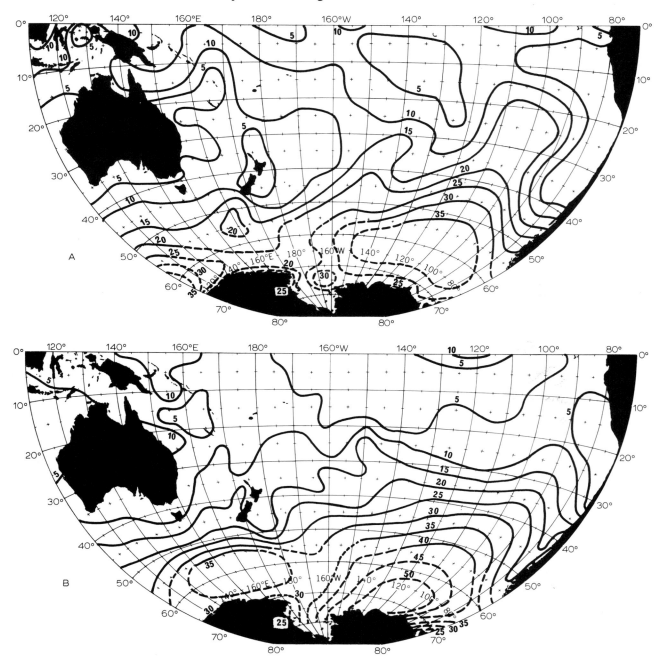

Fig. 46. Percentage frequency of observations recording precipitation in January (A) and July (B). (Adapted
from the Ocean Atlas—Pacific Ocean; U.S.S.R. MINISTRY OF DEFENCE, 1974.)

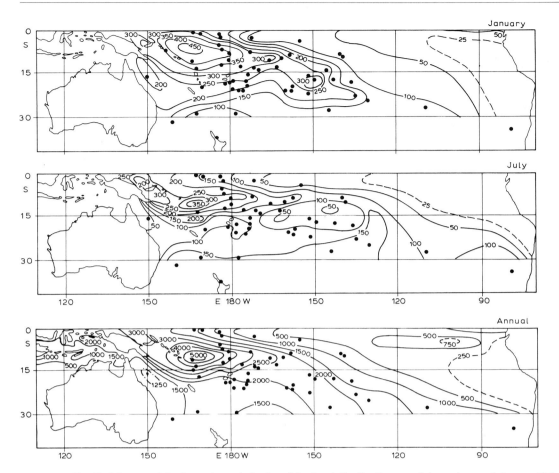

Fig. 47. Mean precipitation at lower latitudes of the South Pacific: January, July, and annual, in mm. (After
TAYLOR, 1973.)

4

Only over the island chains of the tropical South Pacific has it been possible to produce
a reasonably reliable indication of precipitation amount. This follows the very detailed
study of TAYLOR (1973) who collected all available records from the islands in conjunction with satellite cloud observations to enable mean monthly and annual estimates
to be produced as far south as 30°S. These maps for January, July, and the annual total
are reproduced in Fig. 47. Representation of annual totals over higher latitudes must
be based on many assumptions; a very broad indication is reproduced in Fig. 48
(U.S.S.R. MINISTRY OF DEFENCE, 1974) which differs in detail from the representation
of Taylor at lower latitudes, although the distribution patterns are similar in both maps.
Figs. 46, 47, and 48 thus indicate the principal features of Pacific precipitation.

(*1*) Minimum frequency of observation of precipitation and very low totals occur
throughout the year off the west coast of South America and extend westward over the
region of the southeast Pacific High corresponding to the region of low ocean temperature (cf. Figs. 6 and 7). This equatorial dry zone has its axis located slightly south of
the Equator westward of the dateline. TAYLOR (1973) indicates an anomalous minor
maximum near 5°S 90°W in March and April which appeared as a cloud band in six
years of cloud frequency observations of SADLER (1969). This may be associated with

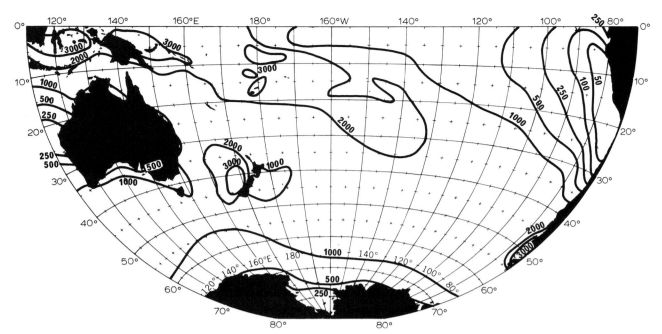

Fig. 48. Mean annual rainfall (in mm; adapted from the Ocean Atlas—Pacific Ocean; U.S.S.R. MINISTRY OF DEFENCE, 1974.)

the double intertropical convergence sometimes observed in this region (HUBERT et al.,1969). For the most part, the whole eastern Pacific is a zone of low precipitation throughout the year with the averaged cloud patterns clearly showing the extent of the dry area. In the extreme northeast, this pattern may be disturbed in an abnormal El Niño year (described later, see p. 358).

(*2*) A zone of high frequency of precipitation observations corresponding to the mid-Pacific cloud band (cf. Figs. 34, 35, and 45) lies with its axis from the Solomons to 30°S 140°W. Station data for the region of the cloud band indicate high rainfall reaching 5000 mm annually in the region of the New Hebrides but rapidly declining east of the dateline. At the fringes of the zone, considerable variation in rainfall occurs from year to year and month to month, reflecting slight changes in its orientation and movement.

(*3*) Low frequency of observations of precipitation is apparent in the western South Pacific at the latitudes of the subtropical anticyclones. The area of low precipitation changes in general in accordance with the changes in latitude of the high pressure belt, viz., expanding southward in summer and contracting equatorward in winter.

(*4*) Increasing frequency of precipitation occurs toward higher latitudes. In the Subantarctic, cyclonic and frontal activity is frequent, although the decreasing absolute humidity of the air results in smaller amounts of precipitation. Thus, frequent periods of drizzle and snow may result only in small precipitation totals. VAN LOON (1972) has produced some evidence to suggest that in the higher latitudes the half-yearly waves of pressure gradient, cyclonic frequency, and latitude of the Antarctic trough (cf. Figs. 27 and 37) may result in corresponding oscillations of precipitation at particular latitudes, but the data from Pacific longitudes do not appear to be sufficient for this to be directly observed.

Climate of the upper air

Geopotential height and temperature

The principal features of the upper-air circulation will be shown in map form for two representative levels (500 and 200 mbar). The data are based on those of TALJAARD et al. (1969) and JENNE et al. (1971). The mean geopotential height patterns of the 500-

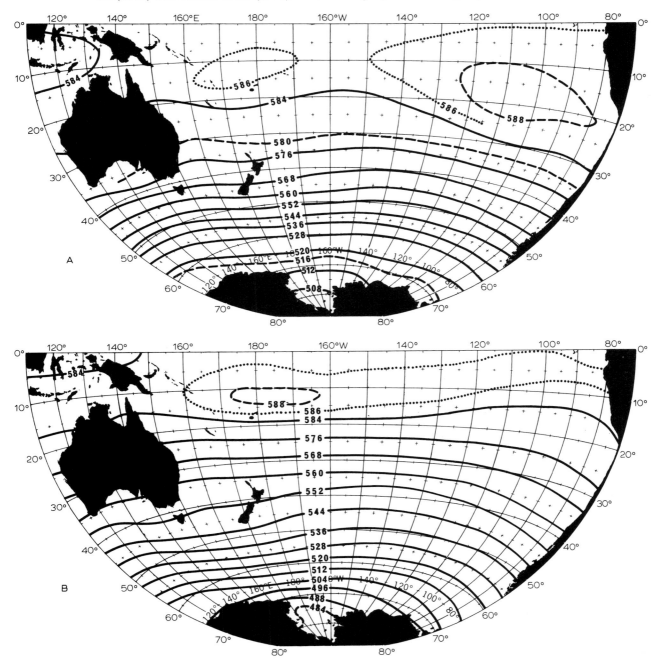

Fig. 49. Mean height of the 500-mbar surface (geopotential decametres) for: A. January, and B. July. (After TALJAARD et al., 1969.)

mbar surface in January and July are shown in Fig. 49 and the corresponding temperature fields in Fig. 50. Similar data for the 200-mbar level are reproduced in Figs. 51 and 52. Data on the standard deviation of the monthly mean 500-mbar heights are poor over the South Pacific; over the western part of the ocean, VAN LOON and JENNE (1974) indicate a general increase poleward from lowest values (circa 10 – 15 m) at the Equator reaching a maximum (circa 80 m) at the Antarctic coast in winter, and at around 50°S in summer (circa 60 m). Notable in the mean fields is the northward extension of the

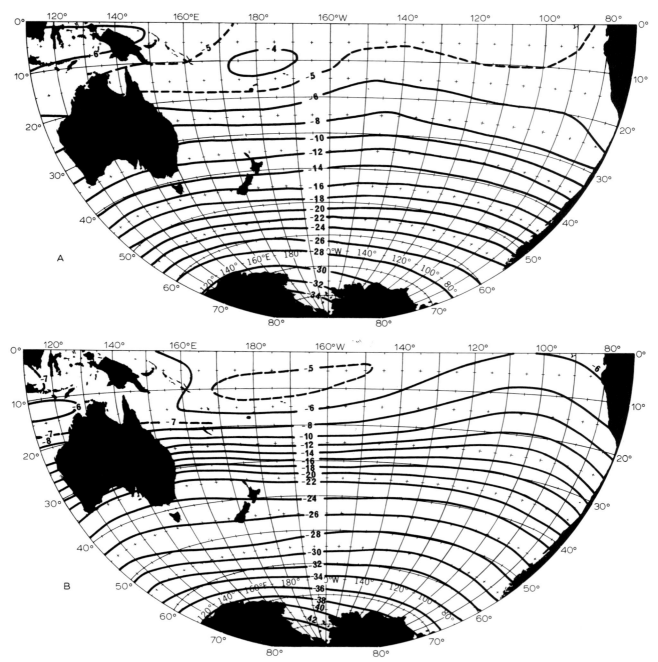

Fig. 50. Mean temperature (°C) at 500 mbar for: A. January, and B. July. (After TALJAARD et al., 1969.)

circumpolar vortex in July by some 10 degrees of latitude compared with the January pattern. This is accompanied by a cooling in middle and high latitudes by some 8° to 10°C (TALJAARD et al., 1968). Over the eastern tropical Pacific, the mid-tropospheric anticyclone of January disappears by July, but a weak detached upper anticyclonic circulation persists throughout the year over the region near the dateline at 10°S. Thus, for most of the year, mean easterly flow at 500 mbar is confined to the lowest latitudes

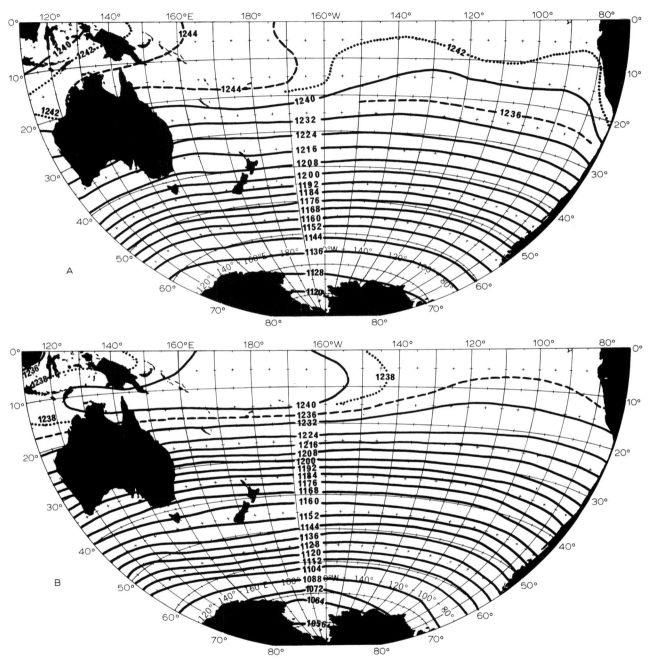

Fig. 51. Mean height of the 200-mbar surface (geopotential decametres) for: A. January, and B. July. (After TALJAARD et al., 1969.)

though with a wider latitudinal span in summer immediately westward of the coasts of Peru and Ecuador. South of 20°S in January and 10°S in July the Pacific is covered at 500 mbar by the broad sweep of the westerlies of the circumpolar vortex which remains centred over or southward of the Ross Sea. Little apparent disturbance of the mean flow by prominent troughs or ridges is evident. However, individual daily charts reveal considerable pattern variability. The zonal harmonic waves of the hemisphere have been examined by VAN LOON and JENNE (1972) following earlier studies (ANDERSSEN, 1965;

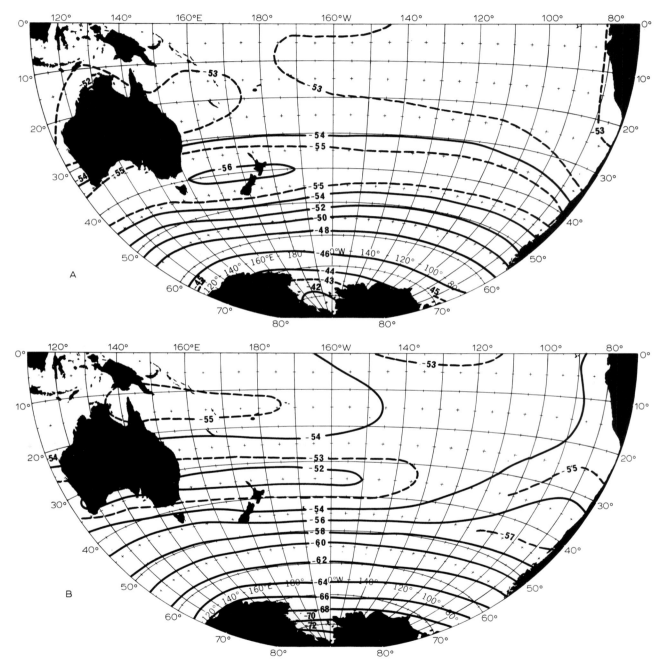

Fig. 52. Mean temperature (°C) at 200 mbar for: A. January and B. July. (After TALJAARD et al., 1969.)

VAN HAMME, 1970). Their analyses used long-term average heights of the constant pressure surface from sea level to 100 mbar and daily 500 mbar heights. The results indicate that waves 1, 2, and 3 have significant stationary components. Wave 1 has one peak in the subpolar regions with its ridge in the southeastern Pacific and another north of 45°S with its trough in the Pacific. Wave 2 has a standing component over Antarctica and wave 3 is well defined between 25° and 60°S throughout the year with ridges near the

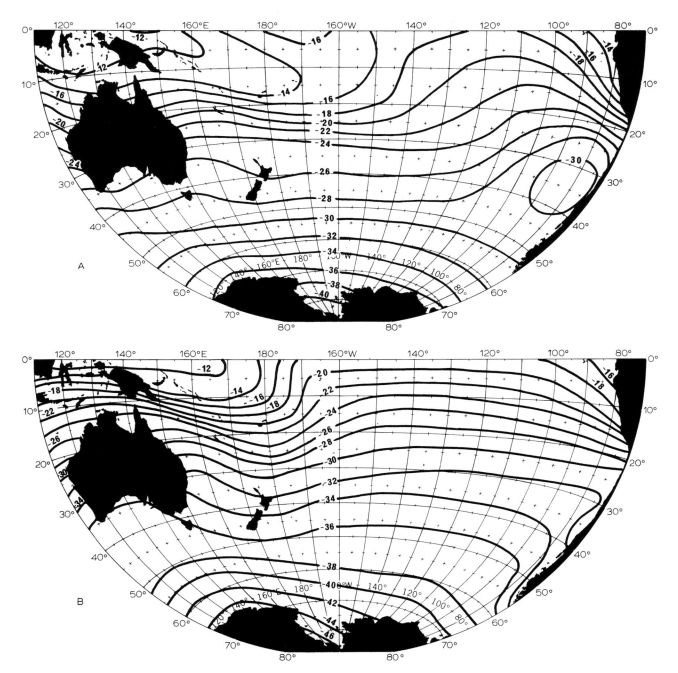

Fig. 53. Mean dew poiñt temperature (°C) at 500 mbar for January (A) and July (B). (After TALJAARD et al., 1969.)

three lower latitude continents. Over the Pacific at 500 mbar in the annual mean, the ridges of wave 3 lie, respectively, over and to the south of the Tasman Sea and along the west coast of South America. In the higher tropical troposphere (200 mbar) in January, wave number 3 also exhibits peaks in similar longitudes to those at higher latitudes.

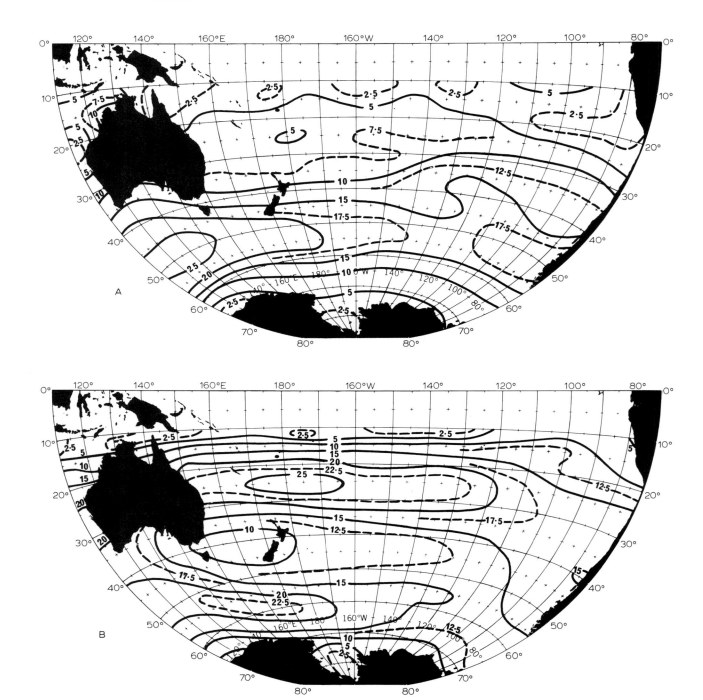

Fig. 54. Mean geostrophic windspeed (m s^{-1}) at 500 mb for January (A) and July (B). (After JENNE et al., 1971.)

Moisture distribution

Mean 500 mbar dew point temperatures (Fig. 53) indicate in both January and July a broad tongue of moister air extending into the Pacific to 20° or 30°S southeastward from the Indonesia – New Guinea region. This is reflected in the disturbed weather in the region between New Guinea and Fiji throughout the year (RAMAGE, 1970) and in the extension of the South Pacific cloud band from this region toward the southeast. By contrast, a region of relatively drier air appears in the mean charts westward of South America in both months between 30° and 50°S.

Wind

The geopotential height gradients of Fig. 49 result in a pattern of mean 500-mbar geostrophic windspeed as shown in Fig. 54. The strengthened gradient at lower midlatitudes in July results in a pronounced subtropical wind maximum. In summer, at higher midlatitudes the windspeed is greatest near 50 – 55°S and, as noted in the surface pressure gradients (Fig. 26) and windspeed (Fig. 42), the flow is considerably stronger south of Australia than over the eastern part of the ocean. In winter, the windspeed at these latitudes is somewhat weaker than in summer with maxima possibly slightly farther south and separated from the wind maximum at subtropical latitudes by a region of light winds with minimum speed over the Tasman Sea.

At 200 mbar (Fig. 55), the basic westerly stream south of 15°S persists throughout the year. SADLER's (1972) streamline isotach analysis (Fig. 56) shows more detail in the tropical regions and indicates a small amplitude trough-ridge pattern extending southeastward across the ocean in summer with the flow from the Southern Hemisphere merging into the equatorial westerlies; in July, with the southward advance of Northern Hemisphere northeast flow, the intertropical discontinuity at 200 mbar moves south of the Equator and the mid-Pacific upper trough is less pronounced. A discontinuity in the 200 mbar height field is apparent in July over the region around the South Island of New Zealand where the lighter winds observed at lower levels are also apparent in the high troposphere separating the subtropical jet stream from the polar front jet to the south at the longitudes of the Tasman Sea.

Jet streams and tropopause

The location and variation in the strength of the polar front jet have not been observed in detail over the higher latitudes of the Pacific. However, more information is available on the subtropical jet stream in the Australia – New Zealand region where surveys by several investigators, e.g., MUFFATTI (1964) and WEINERT (1968), have been reviewed by RADOK (1971) in Volume 13 of this series. NEAL (1972) using the more detailed observations for the preliminary GARP observing period of June 1970, carried out an analysis of the jet stream over the Pacific. For this period, there were somewhat weaker speeds than normal and a close association between the axis of high frequency of surface anticyclones and the latitude of the jet maximum for the region east of 160°W; by contrast, in the Australia – New Zealand region the jet was stronger and located 8° to 12° of latitude farther north than the axis of the surface anticyclone.

The EOLE constant-density balloon experiment from August 1971 to July 1972 (MOREL and DESBOIS, 1974) has provided a valuable set of wind data at 200 mbar for the hemisphere. Analysis by WEBSTER and CURTIN (1975) has indicated the seasonal changes of the EOLE winds. A long-term analysis of seasonal changes in the structure of jet streams over the Pacific must await analyses based on the new observing methods now becoming available.

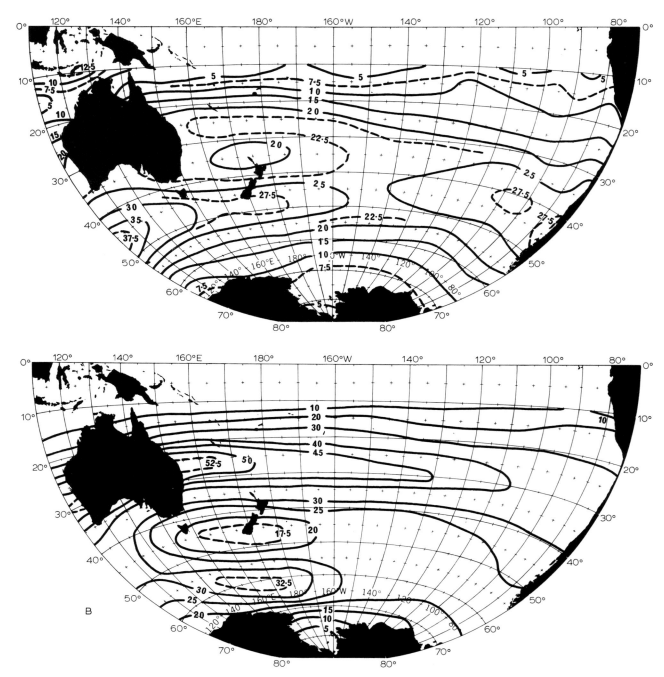

Fig. 55. Mean geostrophic windspeed (m s^{-1}) at 200 mbar for January (A) and July (B). (After JENNE et al., 1971.)

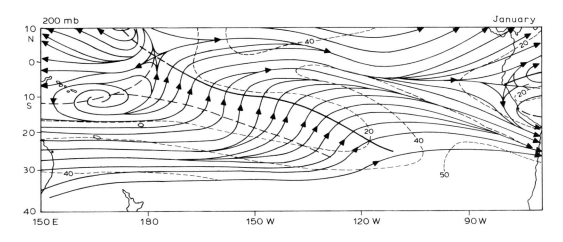

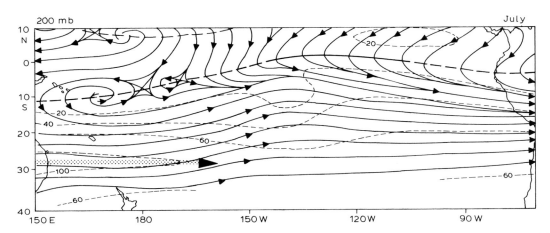

Fig. 56. Mean 200-mbar streamlines and isotachs (kt.) for the tropical Pacific. (Adapted from SADLER, 1972.)

A general picture of the variation in tropopause height is indicated in Fig. 57 for February and August as published in the U.S.S.R. Atlas of the Oceans (U.S.S.R. MINISTRY OF DEFENCE, 1974). In these months in lower middle latitudes, both polar and tropical tropopauses are observed. The tropical tropopause at around 110–120 mbar overlies the polar tropopause generally in the region of 200 mbar. In summer in the polar air, the tropopause height decreases toward the pole and lies below 300 mbar at the Antarctic coast. In winter, the polar air extends northward to around 35°S with the tropopause height between 240 and 260 mbar over most of the ocean to the south of that latitude; northward, over a narrow zone in the vicinity of the subtropical jet stream, the tropopause lies at greater heights and is overlain by the tropical tropopause.

Meridional cross-sections

As an indication of the mean temperature and dew point structure over the ocean, cross-sectional diagrams for longitudes 170°E and 100°W are reproduced from CRUTCHER et al. (1971) in Figs. 58 and 59.

The heat budget

The ultimate driving mechanism for the global atmospheric circulation and hence a major determinant of the worldwide pattern of climate is the distribution of solar radiation incident on the earth at the outer limits of its atmosphere. Some of the incident solar energy is absorbed in the atmosphere and some is absorbed by the underlying land

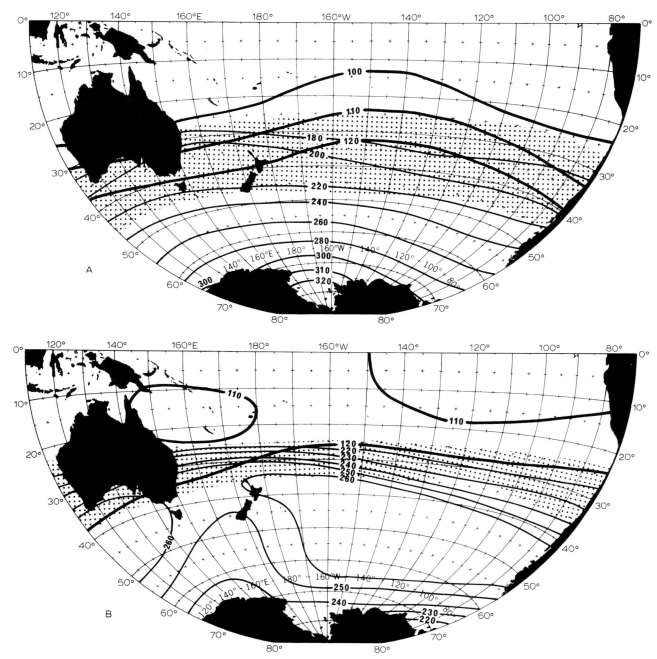

Fig. 57. Mean height (mbar) of the tropopause in February (A) and August (B). Tropical tropopause: heavy lines; polar tropopause: lighter lines. The region where a double tropopause may be observed is stippled. (Adapted from the Ocean Atlas—Pacific Ocean, U.S.S.R. MINISTRY OF DEFENCE, 1974.)

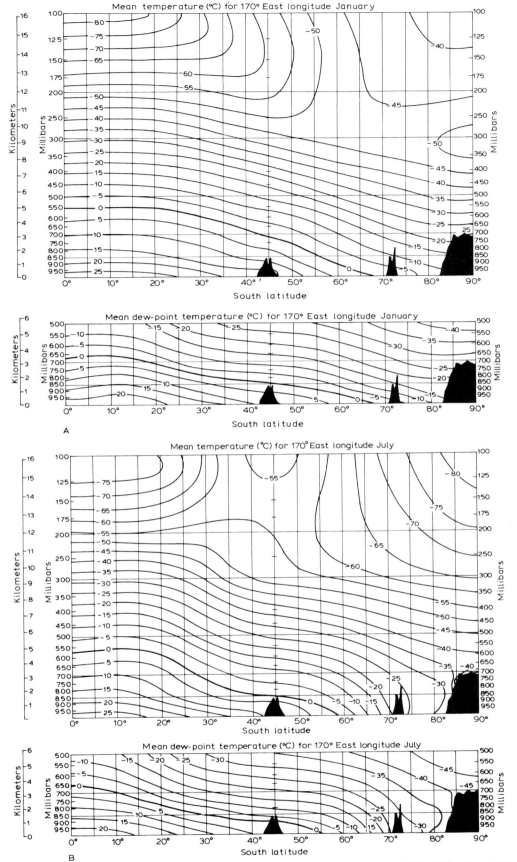

Fig. 58. Mean temperature (°C) and dew point (°C) section along 170°E in January and July. (After Crutcher et al., 1971.)

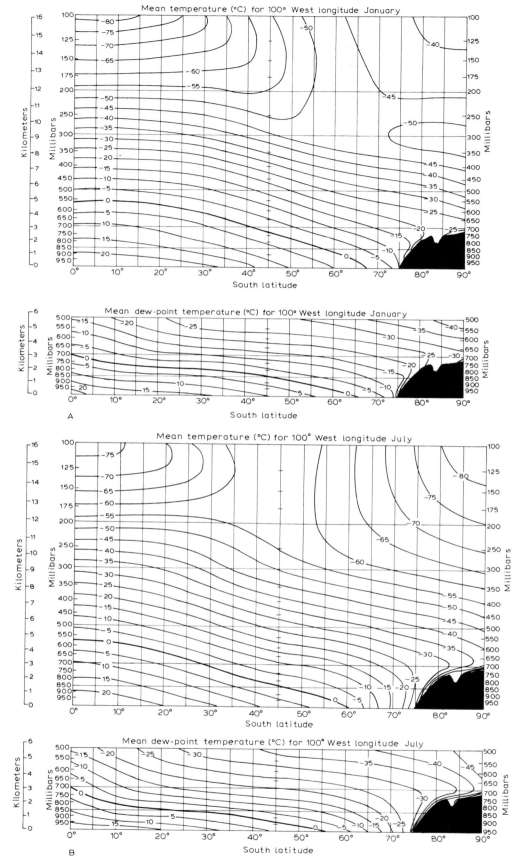

Fig. 59. Mean temperature (°C) and dew point (°C) section along 100°W in January and July. (After CRUTCH-ER et al., 1971.)

and ocean surfaces. The absorbed energy is redistributed in various forms and by various exchange processes between and within both the atmosphere and the ocean and eventually reradiated to space as long wave infrared radiation. To gain some understanding of the relationships between these energy flows and the patterns of atmospheric circulation and overall climate of the Pacific region, it is appropriate to examine the heat budget of the South Pacific Ocean and its overlying atmosphere.

Surface heat budget

The heat balance equation for the ocean can be written:

$$S_O - T_O = R - LE - H$$

where S_O is the rate of heat storage in a vertical column of ocean of unit cross-sectional area and T_O is the rate of transfer of heat out of the column through its walls. R is the net radiation balance at the surface (positive downward), LE is the latent (evaporative) heat flux from ocean to atmosphere (positive upward), and H the sensible heat flux from ocean to atmosphere (positive upward).

When considering annual average conditions, the storage term S_O is normally approximated to zero so that determination of the heat available for horizontal transport within the oceans depends on a knowledge of the three surface terms R, LE, and H.

The classical studies of the surface heat budget of the oceans including all of the South Pacific are those of ALBRECHT (1960, 1961) and BUDYKO (1963). Other more restricted studies for the high latitudes were undertaken by PRIVETT (1960), KANGOS (1960), VIEBROCK (1962), ZILLMAN (1967), and ZILLMAN and DINGLE (1969). DELISLE (1970) published a study of the sea—air heat transfer terms for the Tasman Sea, and ZILLMAN (1972a) examined the region south of Australia on the basis of data from research cruises by the U.S.N.S. *Eltanin* (ZILLMAN and DINGLE, 1973). The sea—air exchange terms for the tropical South Pacific north of 20°S were evaluated on a monthly basis by WYRTKI (1966b), whereas HASTENRATH and LAMB (1978) have produced a detailed climatology of all terms in the surface heat budget on a one-degree grid for the eastern tropical Pacific using an internally consistent data set for the period 1911—1970. ZILLMAN (1972b) included the South Pacific in a general broad-scale survey of the surface heat budget of the Southern Hemisphere oceans.

Surface radiation balance

The net radiation entering the ocean surface (R) may be evaluated as:

$$R = S_G - \alpha_G S_G + \varepsilon L_d - \varepsilon \sigma T_s^4$$

where the first term on the right (S_G) is the short-wave global radiation incident on the sea surface, the second term ($\alpha_G S_G$) is the short-wave radiation reflected upward from the surface, the third term εL_d is the absorbed component of the long-wave radiation downward from the atmosphere, and the fourth term ($\varepsilon \sigma T_s^4$) is the long-wave radiation emitted upward from the sea surface. Here α_G is the short-wave albedo of the sea surface, ε its long-wave emissivity, and σ the Stefan Boltzmann constant. T_s is the sea surface temperature.

The distribution of the net radiation input to the ocean is determined primarily by the variation of the global radiation (S_G). The global radiation, in turn, is a function mainly of latitude, time of year, and the distribution of cloud. A variety of empirical and other formulae has been developed for its evaluation in the absence of any suitable network for direct measurement of global radiation over the oceans (e.g., BLACK, 1956; QUINN et al., 1969). These have been compared and their applicability to the southern hemisphere oceans examined by ZILLMAN (1972b). Fig. 60 shows the mean annual distribution of global radiation interpolated mainly from the maps of BUDYKO (1963) and TOLSTIKOV (1966). BUDYKO's evaluation employed a cloud reduction of the global radiation of the form:

$$S_G = S_{Go}(1 - bC - 0.38C^2)$$

where S_{Go} is the equivalent clear-sky radiation, C is fractional cloud cover (0 − 1)), and b is a slowly varying function of latitude with average value about 0.38. Other more detailed annual and monthly maps exist for the South Pacific, e.g., ZILLMAN (1967) for the higher latitudes, but the Budyko charts, though heavily smoothed, adequately portray the broad features, and the Budyko data have been widely used in global energy budget studies.

The annual mean distribution of the net radiation (R) is shown in Fig. 61, and its annual variation as a function of latitude for the South Pacific region as a whole (including the part of the Indian Ocean south of Australia treated in this chapter) is summarized in Fig. 62, also on the basis of Budyko's results. Budyko's selection of short-wave albedo and his evaluation of the net long-wave radiation component of R, viz., $ε(L_d - σT_s^4)$,

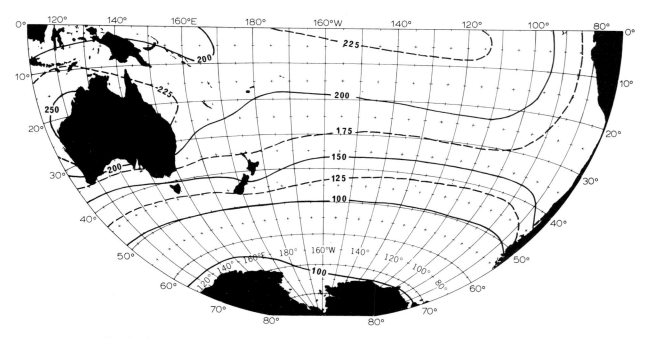

Fig. 60. Annual mean global radiation (W m^{-2}) reaching the ocean surface.

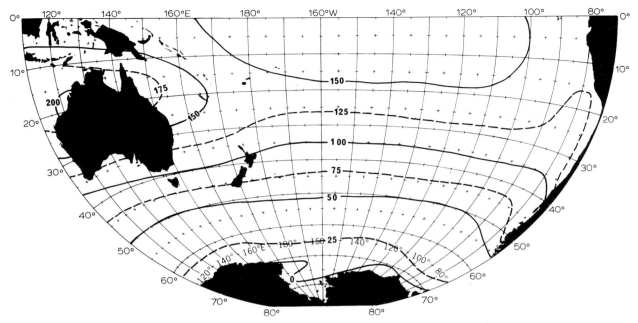

Fig. 61. Annual mean net downward radiation (W m^{-2}) at the ocean surface.

have also been fairly widely adopted by subsequent workers. It is worth noting that because the effect of cloud on the long-wave radiation works in the opposite direction to its effect on the solar radiation, the overall radiation balance is relatively insensitive to the cloud cover, except in the case of high intensities of short-wave radiation when the short-wave reduction dominates over long-wave back radiation and increased cloud cover thus reduces net radiation input to the ocean. The features of particular interest in this respect in Fig. 61 are the region of maximum net radiation input in the relatively cloud-free central tropical Pacific (Fig. 45) and the band of reduced net radiation input

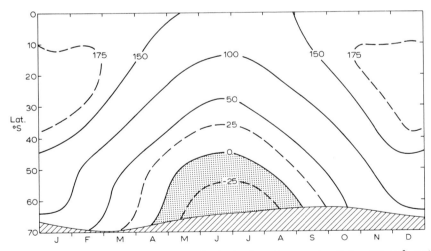

Fig. 62. Annual variation as a function of latitude of net downward radiation (W m^{-2}) at the ocean surface, zonally averaged across the South Pacific. The hatched area indicates the mean northward extent of the pack ice based on U.S.S.R. MINISTRY OF DEFENCE (1974).

lying southeastward through the Solomon Islands and Fiji in the region of the South Pacific cloud band (Fig. 38).

Latent heat flux

The latent heat flux LE represents the energy lost by the ocean in evaporating water from its surface into the atmosphere. Here L is the latent heat of evaporation $[(2.500 - 0.0024\ T_s) \cdot 10^6\ \text{J kg}^{-1}]$ and E the rate of evaporation. If E is expressed in the customary units of centimetres of water evaporated per year, the latent heat flux in Watts per square metre may be found approximately by multiplying by 0.77.

Various methods have been employed for evaluating evaporation over the oceans, none very satisfactory. The use of the so-called bulk aerodynamic formulae of the form:

$$LE = \rho L C_E (q_s - q_a)\ u_a$$

with climatological means of sea—air absolute humidity differences $(q_s - q_a)$ and wind speeds u_a to obtain long-term mean patterns of evaporative heat loss from the ocean has serious limitations (ZILLMAN, 1972b; POND, 1975; FRIEHE and SCHMITT, 1976), but it represents about the only practicable approach at present. PRIVETT (1960), KANGOS (1960), VIEBROCK (1962), BUDYKO (1963), WYRTKI (1966b), and HASTENRATH and LAMB (1978) have all applied this approach for the South Pacific, but different data sets have been used in each instance and the choice of bulk transfer coefficient C_E has ranged from about 0.0012 to 0.0022. The dependence of the transfer coefficient on windspeed and stability has been invoked in some studies.

For present purposes, the mean annual pattern of evaporative heat loss by the South Pacific has been derived by using a constant value of 0.0015 for the bulk transfer coef-

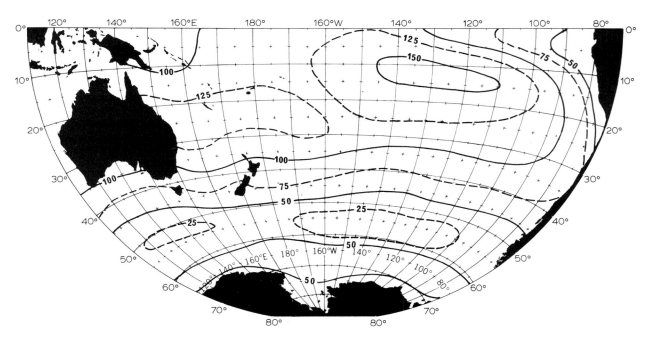

Fig. 63. Annual mean evaporative heat loss from the ocean (W m^{-2}).

ficient along with sea — air humidity differences and mean windspeeds from the U.S.S.R. atlas (U.S.S.R. MINISTRY OF DEFENCE, 1974). Using typical values for the air density (ρ) and latent heat of evaporation (*L*), the approximate working formula for use with mean sea — air vapour pressure differences ($e_s - e_a$) in millibars and windspeeds in metres per second is:

$$LE(\text{W m}^{-2}) = 2.75(e_s - e_a)\, u_a$$

The most notable feature of the resulting mean annual evaporation pattern (Fig. 63) is the region of very high evaporative heat loss by the ocean in the eastern South Pacific between about latitudes 10 and 20°S on the northern side of the subtropical anticyclone (Fig. 41).

The annual variation of the evaporative heat loss as a function of latitude for the region as a whole is shown in Fig. 64 based on the same data sources as Fig. 63. It is noteworthy that, at most latitudes for the South Pacific as a whole, evaporation is highest through the winter months.

Sensible heat flux

Fig. 65 shows the mean annual sensible heat flux from ocean to atmosphere over the South Pacific and Fig. 66 its annual variation as a function of latitude based on use of the bulk aerodynamic formula:

$$H = \rho c_p C_H (T_s - T_a)\, u_a$$

with a bulk transfer coefficient C_H of 0.0015 and monthly mean values of the sea — air temperature difference ($T_s - T_a$) and windspeed (u_a) from the U.S.S.R. atlas. Using typical values of the air density (ρ) and specific heat at constant pressure (c_p), the sensible heat flux in Watts per square metre with mean sea — air temperature differences in degrees Celsius and windspeeds in metres per second is given by:

$$H = 1.84(T_s - T_a)\, u_a$$

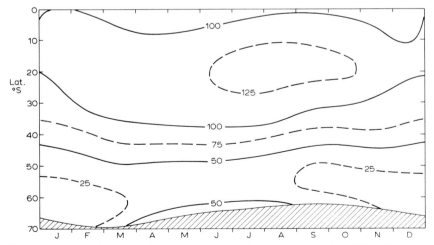

Fig. 64. Annual variation as a function of latitude of evaporative heat loss (W m^{-2}) from the ocean surface, zonally averaged across the South Pacific. The hatched area indicates the mean northward extent of the pack ice based on U.S.S.R. MINISTRY OF DEFENCE (1974).

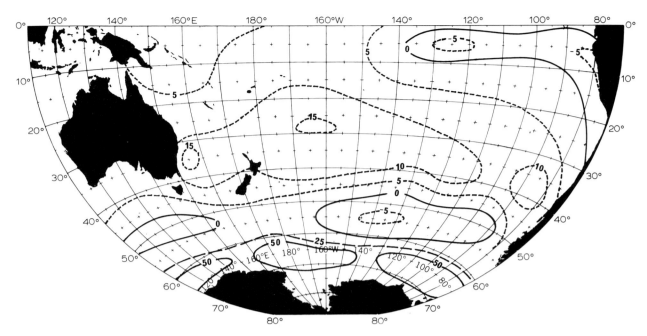

Fig. 65. Annual mean sensible heat flux (W m⁻²) from ocean to atmosphere.

Although the net heat flux is upward from the ocean to the atmosphere over most of the South Pacific during most of the year, there is a significant net sensible heat flux downward into the ocean in the high latitudes in summer and a narrow band of small downward heat flux into the tongue of cold water which extends along the Equator in the eastern Pacific (Figs. 6 and 7). The upward heat flux is greatest in high latitudes during autumn and winter as a result of the strong outflow of cold air over still relatively warm water northward from the edge of the pack ice.

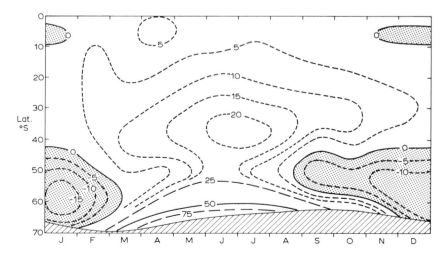

Fig. 66. Annual variation as a function of latitude of upward sensible heat flux (W m⁻²) from the ocean surface, zonally averaged across the South Pacific. The hatched area indicates the mean northward extent of the pack ice based on U.S.S.R. MINISTRY OF DEFENCE (1974).

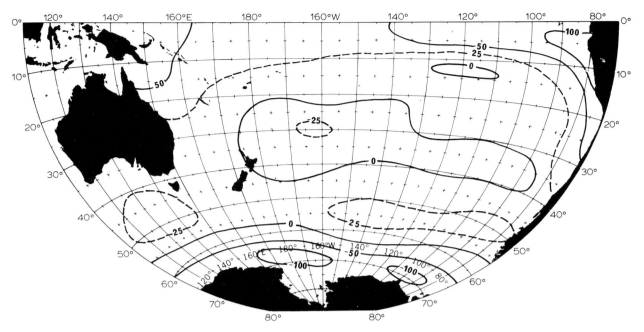

Fig. 67. Annual mean surface heat balance of the ocean (W m^{-2}; downward heat flux positive).

Heat available for storage and transport by the ocean

Figs. 67 and 68 based on Figs. 61—66 summarize the surface heat balance $R - LE - H$ for the South Pacific. It may be seen from Fig. 67 that, in the annual average, there is a considerable net downward flux of heat into the ocean in the low latitudes where the radiative input considerably exceeds the oceanic heat loss by evaporation and sensible heat transfer. This excess heat is available for transport within the ocean to make up the heat deficit at higher latitudes.

The annual variation of the surface heat balance $R - LE - H$ shown in Fig. 68 represents

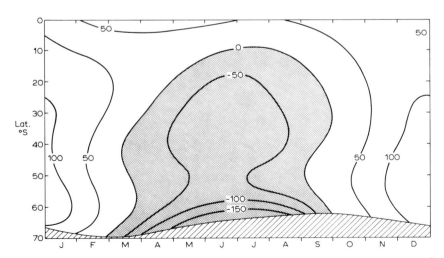

Fig. 68. Annual variation as a function of latitude of the surface heat balance (W m^{-2}; downward flux positive) of the ocean, zonally averaged across the South Pacific. The hatched area indicates the mean northward extent of the pack ice based on U.S.S.R. MINISTRY OF DEFENCE (1974).

heat available for both storage and transport within the ocean. The ocean continues to warm (Fig. 8) and the pack ice continues to recede through the summer as long as there is a net downward heat flux into the ocean. As the sign of the surface heat balance reverses, the ocean begins to cool and the pack ice spreads northward.

Influence of ocean fronts

One interesting aspect of the surface heat budget of the South Pacific is the role of the Antarctic Convergence and other oceanic fronts in producing elongated bands of enhanced sea – air heat fluxes in the high latitudes. Although these are below the resolution of the broad-scale presentations of Figs. 63 and 65, they represent well-defined climatic features of the region. Fig. 69, based on 14 crossings of the Australasian Subantarctic front south of Australia (Fig. 4), shows the north – south variation, with respect to the middle of the frontal zone, of sea surface temperature, sea – air temperature difference, sea – air vapour pressure difference, and windspeed averaged in 1°-latitude bands. Fig. 70 shows the resulting variation of the latent and sensible heat fluxes. It is evident that the region just north of the frontal zone constitutes an area of greatly enhanced energy flow to the atmosphere. It appears to be closely related to the zone of maximum frequency of cyclogenesis identified by STRETEN and TROUP (1973) and depicted in Fig. 30.

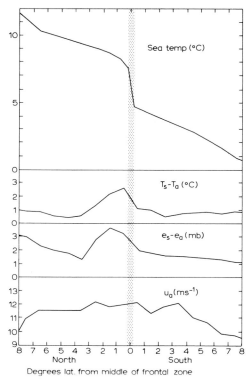

Fig. 69. Meridional variation of sea surface temperature, sea – air temperature difference, sea – air vapour pressure difference, and windspeed averaged with respect to the middle of the Australasian Subantarctic front, based on 14 crossings (1968 – 1972) by the U.S.N.S. *Eltanin*.

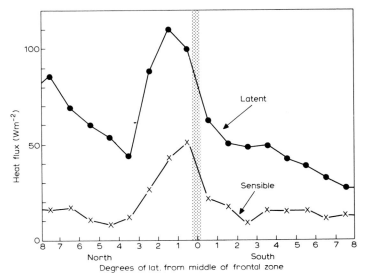

Fig. 70. Latent and sensible heat fluxes averaged with respect to the middle of the Australasian Subantarctic front based on the data of Fig. 69.

Diurnal variation of the surface heat budget

The main diurnal influence on the surface heat budget arises from the diurnal cycle of the solar radiation. This is illustrated in Fig. 71 where the individual components of the radiation balance are shown as a function of time of day based on measurements on six days in January 1970 aboard the U.S.N.S. *Eltanin*, south of Australia (ZILLMAN, 1972a). Fig. 72 includes the remaining terms of the surface heat budget for the *Eltanin* data (average noon latitude 59.8°S) and similar data averaged over 18 days of a cruise by the

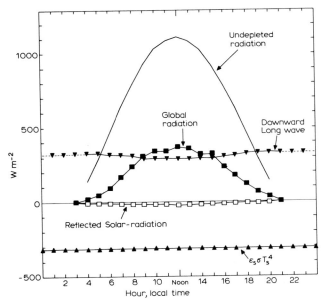

Fig. 71. Averaged diurnal variation of the extraterrestrial solar radiation and surface radiation balance components as measured during 7 January and 10—14 January 1970 during a cruise by the U.S.N.S. *Eltanin* south of Australia.

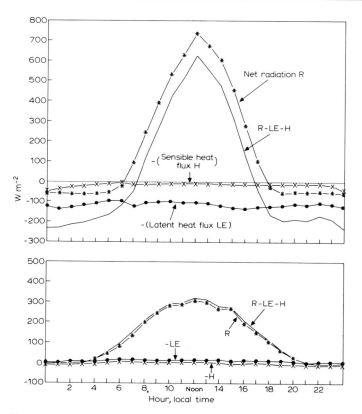

Fig. 72. Averaged diurnal variation of the surface heat balance components for the period 7 January and 10–14 January 1970 during a cruise by the U.S.N.S. *Eltanin* south of Australia (lower) and similar data averaged over 18 days in February–March 1961 during a cruise by the H.M.A.S. *Gascoyne* (upper). Average noon latitudes were 59.8°S (*Eltanin*) and 37.4°S (*Gascoyne*).

H.M.A.S. *Gascoyne* in the Great Australian Bight and Bass Strait in February–March 1961 (average noon latitude 37.4°S).

Atmospheric heat budget

The heat balance equation for the atmosphere can be written:

$$S_A - T_A = (R_T - R) + LP + H$$

where S_A is the rate of heat storage in a vertical column of the atmosphere of unit cross-sectional area and T_A is the rate of transfer of heat out of the column through its walls. The three terms on the right are: (*1*) the net radiative heating of the atmospheric column which is the difference between the net downward radiation at the top of the atmosphere (R_T) and at the surface (R); (*2*) the release of latent heat by condensation of water vapour (LP) where P is the rate of precipitation; and (*3*) the sensible heat flux into the atmosphere from below (H).

Their sum is frequently referred to as the net diabatic heating rate of the atmospheric column, and its spatial pattern indicates the distribution of the major atmospheric heat sources and sinks which drive the large-scale circulation.

There has been a number of studies for the Southern Hemisphere of the atmospheric heat budget and its various components (e.g., GABITES, 1960; BRIDGMAN, 1969; DOPPLICK, 1970; SASAMORI et al., 1972: HAURWITZ, 1972; NEWTON, 1972; NEWELL et al., 1974), but these have concentrated on consideration of zonal means and do not provide detailed information on the heat budget of the atmosphere of the South Pacific.

Radiation budget

An estimate of the spatial distribution of the annual mean atmospheric radiation balance for the South Pacific ($R_T - R$) can be derived by subtracting Fig. 61 from a corresponding representation of the annual radiation balance of the earth—atmosphere system derived from satellite measurements (e.g., VONDER HAAR and SUOMI, 1971; VONDER HAAR and ELLIS, 1974). For the present purpose, we have used as an approximation to the climatological distribution of R_T the 45-month (June 1974—February 1978) annual mean given by WINSTON et al. (1979). This is shown in Fig. 73.

Fig. 74 summarizes the annual variation of the net radiation at the top of the atmosphere for the South Pacific as a whole based on the same data as Fig. 73. The data have been interpolated from the monthly maps given by WINSTON et al. (1979) and slightly smoothed.

Comparison of Figs. 61 and 73 and Figs. 62 and 74 shows that there is a net radiative cooling ($R_T - R$ negative) of the atmosphere at all latitudes throughout the year.

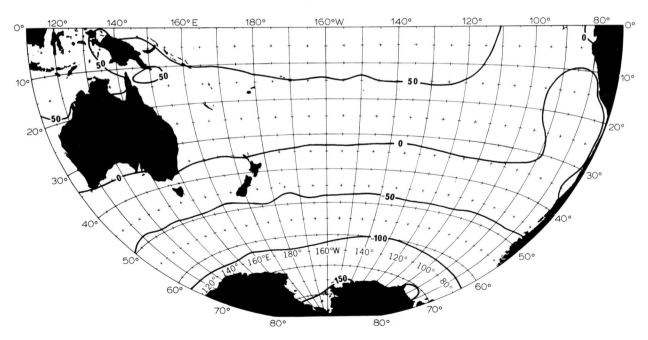

Fig. 73. Annual mean net radiation (W m^{-2}) at the top of the atmosphere for the period June 1974 to February 1978 as given by WINSTON et al. (1979).

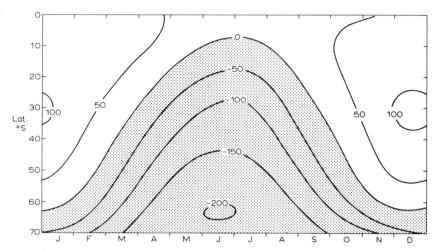

Fig. 74. Annual variation as a function of latitude of net downward radiation (W m^{-2}) at the top of the atmosphere, zonally averaged across the South Pacific (interpolated from the maps of WINSTON et al., 1979).

Condensation heating

The climatological distribution of the latent heat release by condensation within the atmosphere may be evaluated approximately in W m^{-2} as:

$$LP = 0.0793 \, P$$

where P is the annual precipitation in millimetres.

The pattern of annual precipitation over the South Pacific is poorly known and the chart in Fig. 48 gives only a broad indication of the major features. The smoothed distribution of annual mean rate of latent heating of the atmosphere shown in Fig. 75

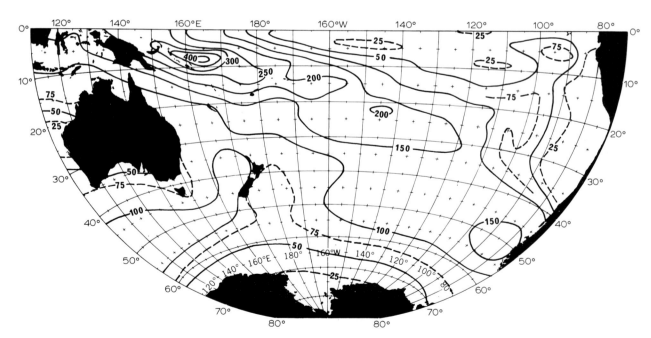

Fig. 75. Annual mean distribution of condensation heating of the atmosphere (W m^{-2}).

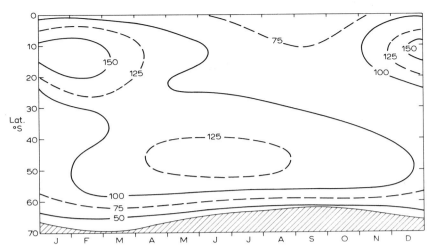

Fig. 76. Annual variation as a function of latitude of the condensation heating of the atmosphere (W m^{-2}), zonally averaged across the South Pacific. The hatched area indicates the mean northward extent of the pack ice based on U.S.S.R. MINISTRY OF DEFENCE (1974).

is based mainly on an evaluation by DORMAN and BOURKE (1979) north of 30°S and ZILLMAN (1972c) for the higher latitudes. Dorman and Bourke employed the same approach to obtain the amount of rainfall from present weather observations as did Tucker (1961) for the North Atlantic. Their results are in good agreement with those of TAYLOR (1973) shown in Fig. 47. The outstanding feature of Fig. 75 is the band of strong condensation heating lying northwest to southeast across the South Pacific Ocean.
The annual variation of zonally averaged latent heating over the Pacific shown in Fig. 76 is also rather uncertain. It is based on a selective combination of the results of SEELYE

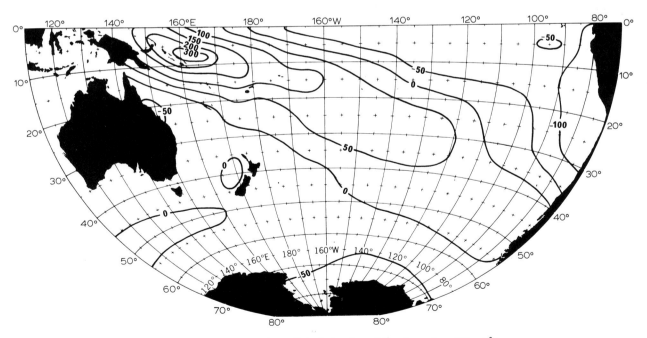

Fig. 77. Annual mean distribution of net diabatic heating of the atmosphere (W m^{-2}).

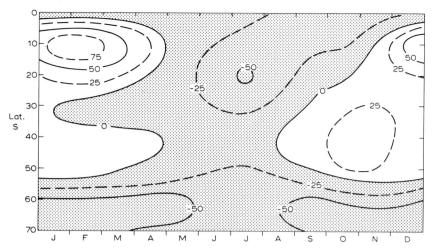

Fig. 78. Annual variation as a function of latitude of the net diabatic heating and cooling (W m⁻²; heating positive) of the atmosphere, averaged across the South Pacific.

(1950), ALBRECHT (1951), MÖLLER (1951), TAYLOR (1973), and DORMAN and BOURKE (1979) in conjunction with monthly precipitation frequency charts and island records. In general, the seasonal distribution as a function of latitude is similar to that given by ALBRECHT (1961) though the implied heating rates are some 10−20% higher.

Net diabatic heating

Figs. 77 and 78 show the annual distribution and variation through the year of the net diabatic heating of the atmosphere based on Figs. 61, 65, 73, and 75 (annual maps) and Figs. 62, 66, 74, and 76 (annual variation), respectively.

The outstanding features of Fig. 77 are the band of net diabatic heating associated with the precipitation maximum lying southeastward from the Solomon Islands (Fig. 73) and the region of dominant radiative cooling over the central and eastern tropical Pacific. As will be seen in the following section, this distribution of heat sources and sinks is closely coupled to the large-scale east−west circulation of the Pacific region.

Large-scale circulations

The subtropical and tropical parts of the South Pacific are remarkable for the strength of the east−west contrasts in temperature, rainfall, and various other atmospheric and oceanographic parameters and for the existence of a major well-defined pattern of interannual variability (e.g., SEELYE, 1950; RODEN, 1963) centred in the Pacific Basin but extending to and affecting the greater part of the globe. These unique features are associated with the large-scale atmosphere−ocean circulation system first described by Sir Gilbert Walker (WALKER, 1923, 1924, 1928a,b,c) and subjected to many detailed studies over the years. It is frequently referred to as the "Southern Oscillation," a large-scale exchange of atmospheric mass between the tropical and subtropical parts of the South Pacific Ocean and the Indian Ocean with a time scale of several years, or the

"Walker Circulation," the slow east — west vertical circulation cell associated with it. Major changes associated with the Southern Oscillation also take place in the ocean, one remarkable manifestation being an occasional dramatic warming of the ocean off the coast of Peru—the so-called El Niño.

Although the broad features of the Southern Oscillation, Walker Circulation, and El Niño are fairly well documented, much of the detail of this complex planetary-scale oscillation has yet to be revealed and no satisfactory explanation of all aspects of the phenomenon yet exists. It is clearly related to the distribution of the continents and oceans, and its dominance in the Pacific region would seem to be attributable to the peculiar geography and broad longitudinal extent of the Pacific Basin. Whereas there is little doubt that the various phenomena are part of a single coherent large-scale atmosphere — ocean circulation system, it is convenient to consider their separate features in turn.

The Southern Oscillation

The Southern Oscillation (SO) can be simply thought of as "a swaying of pressure on a big scale backwards and forwards between the Pacific and Indian Oceans..." (WALKER, 1928a). In his extensive investigations of correlations between climatic elements at various widespread locations over the earth, Sir Gilbert Walker discovered that a particularly strong negative correlation existed in his data between pressure anomalies in the southeast Asian — Indonesian region and those in the eastern Pacific (WALKER, 1924; WALKER and BLISS, 1932, 1937), such that when pressures were high in the eastern Pacific they tended to be low over the Indonesian region and vice versa. He found that in general rainfall varied inversely with pressure. The pattern of pressure anomalies was

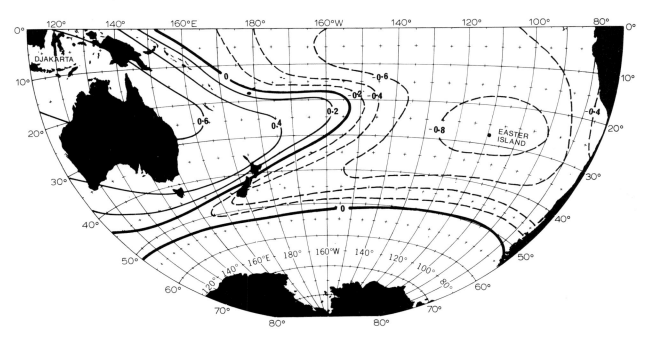

Fig. 79. Correlation of monthly surface pressure anomalies with those at Djakarta (6°11′S 106°50′E) for the period April — August. (Adapted from BERLAGE, 1966.)

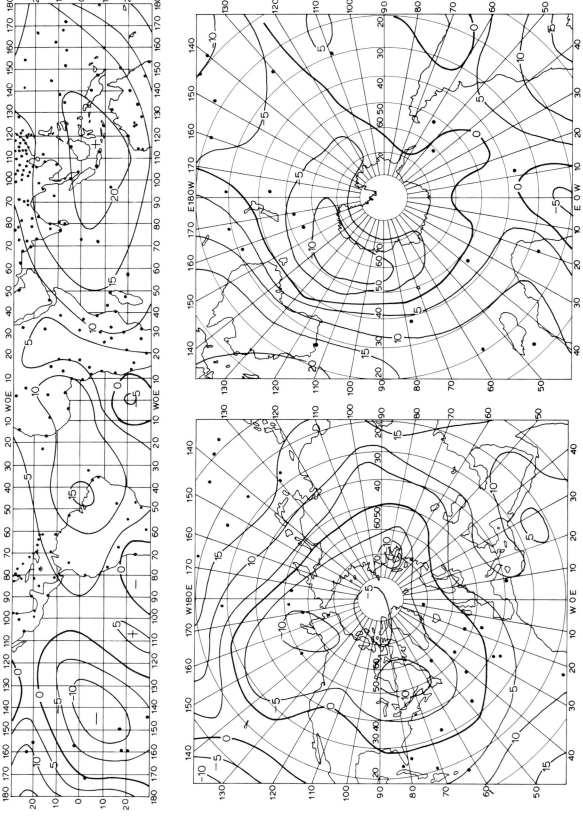

Fig. 80. First component of normalized monthly surface pressure departures defined over a global grid. (After KIDSON, 1975a.)

observed to oscillate on a scale of several years and, because of its dominance in the Southern Hemisphere, Walker referred to it as the "Southern Oscillation."

In subsequent studies, the Southern Oscillation has usually been described and mapped in terms of a variously defined Southern Oscillation Index (SOI) consisting of mean monthly, seasonal, or annual pressure differences between particular stations or groups of stations, frequently Djakarta (Batavia) or Darwin in the west, and Tahiti, Easter Island, or Apia in the east (BERLAGE, 1957, 1961, 1966; TROUP, 1965; BJERKNES, 1969, 1972; FLOHN, 1971; QUINN and BURT, 1972; WRIGHT, 1975; QUINN et al., 1978). Fig. 79 illustrates the general pattern of the Southern Oscillation in terms of the correlation of monthly mean station pressure with the pressure at Djakarta. KIDSON (1975a,b) found that the Southern Oscillation appeared as the most significant component in an empirical orthogonal function analysis of normalized monthly pressure, temperature, and rainfall in the tropics for the period 1951—1960. The patterns that he obtained (Fig. 80) were similar to those of earlier investigations, but they showed another centre of action over northeast Brazil in phase with the Indian Ocean region.

A more detailed examination of the Southern Oscillation in the South Pacific has been carried out by TRENBERTH (1975, 1976a,b) following evidence that changes in the North Pacific and southeast Pacific highs lead changes at Darwin by several months (BERLAGE, 1957; ALLISON et al., 1972; KIDSON, 1975b). TRENBERTH (1976b) has drawn attention to some of the difficulties with previous Southern Oscillation indices and the complications related to apparent phase differences between different stations. He used weighted values of seasonal surface pressure at Darwin and Tahiti as a reference. Fig. 81 shows correlations of yearly pressures at Darwin with other stations in the South Pacific. Correlations for the annual and seasonal (3-monthly) periods are higher than those for individual months which include "noise" unrelated to the Southern Oscillation. TRENBERTH

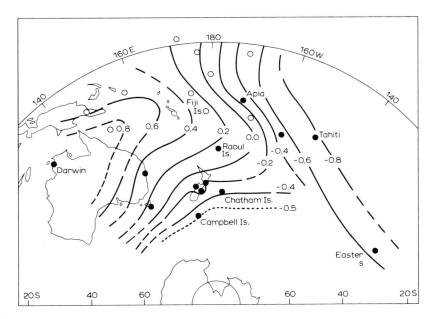

Fig. 81. Correlations of yearly pressures at Darwin with stations in the South Pacific. Solid circles indicate stations where 30 years (1944—1973) were used. Open circles indicate incomplete records. Dashed isopleths show less confidence. (After TRENBERTH 1976b.)

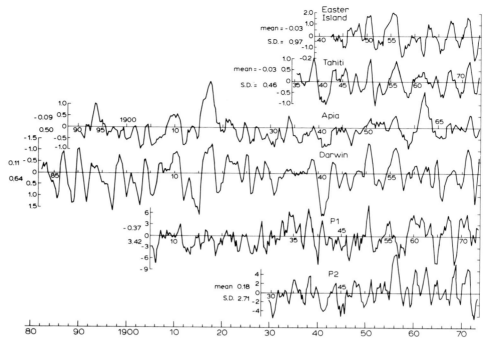

Fig. 82. Time series of Easter Island, Tahiti, Apia, Darwin, *P1*, and *P2* anomalies, 12-month running means. The mean and standard deviations are given at left in mbar. The scale (in mbar) differs for each series, and for Darwin is the inverse of other series. (After TRENBERTH, 1976b.)

(1976b) examined the coherence of observed periodicities among different stations and found that most of the variance is concentrated around 28 months (corresponding to a quasi-biennial oscillation) and 3- to 5-year periodicities. Similarly, BERLAGE (1957) found an average period of 2.33 years and TROUP (1965) reported 3- and 5-year periods in Darwin pressure. Time series of pressure anomalies show temporal variations in the Southern Oscillation (Fig. 82) and reveal a number of periods when the Southern Oscillation was clearly evident and others when it appeared to be inactive. Notable also are the periods of association between the Southern Oscillation and the time series of *P1* and *P2,* which are the coefficients of the first two patterns of an empirical orthogonal function analysis of departures from normal of monthly mean surface pressure in the

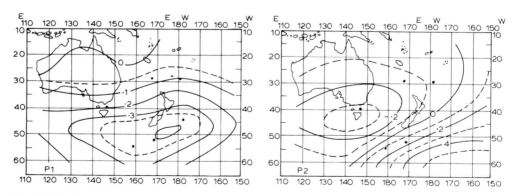

Fig. 83. The first two patterns, *P1* and *P2*, of the empirical orthogonal function analysis of departures from normal monthly mean sea level pressure in the Australasian region. (After TRENBERTH, 1975.)

352

Australasian region (TRENBERTH, 1975, 1976a); these patterns are reproduced in Fig. 83. To investigate further the historical record of the Southern Oscillation and the lag correlations between the data, TRENBERTH (1976b) examined short-period correlation coefficients calculated each year for 5-year periods and 20-year correlations calculated every five years. The time series of these for the Apia—Darwin relationship is shown in Fig. 84. The strong inverse relationship is notable in the period 1900—1920 which marked the period of Walker's investigation; in the periods 1937—1950, 1956—1959, and after 1970 the relationship was also clearly negative but in the early 1960s the influence was quite weak. Similar time series of correlation coefficients for different station pairs are shown in Fig. 85, and, e.g., that for Darwin—Easter Island with Easter Island leading shows the negative correlation to become a maximum at lags of 1 or 2 seasons. It is apparent from Figs. 82, 84, and 85 that without a physical understanding of the reversals in the circulation, the Southern Oscillation indices are of little value for forecasting beyond the use of persistence in some regions.

BJERKNES (1966a,b, 1969) has presented evidence of connections between the midlatitude circulation of the Northern Hemisphere and events in the equatorial Pacific, and Figs. 81—83 indicate an association between the pressure over northern Australia and that at higher latitudes, particularly near New Zealand. TRENBERTH (1976b) has drawn attention to the strong gradients in the isopleths of correlation south of New Zealand in Fig. 81 and in the region of the SPCZ (see pp. 308—310). He suggests that the Southern Oscillation may be related to changes in the characteristics of the SPCZ which is itself further related to the strength and position of the southeast Pacific High (see also KRISH-NAMURTI, 1971a,b; KRISHNAMURTI et al., 1973) and also to the positioning of the long-wave trough in the Pacific (STRETEN, 1973a). KIDSON's (1975a) empirical orthogonal function analysis also showed that when pressures were low in the east Pacific, the midlatitude westerlies in each hemisphere tended to be stronger. Thus, it is clear that although regarded generally as a tropical and subtropical feature, the Southern Oscillation is coupled with the westerly circulation at higher latitudes in both hemispheres. This is further demonstrated by the widespread correlations described by WRIGHT (1975), by the study of tropical interaction with higher latitudes in the Australian region (NICHOLLS, 1977), and by the numerical simulations of ROWNTREE (1972) and JULIAN and CHERVIN (1978).

TRENBERTH (1976b) has further examined the temporal changes in the relationship of

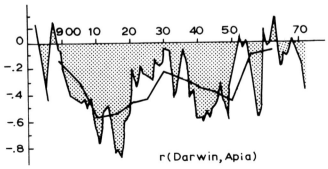

Fig. 84. Time series of correlation coefficients between Darwin and Apia. The stippled area shows 5-year correlations calculated every year and the heavy line shows 20-year correlations calculated every five years. (After TRENBERTH, 1976b.)

353

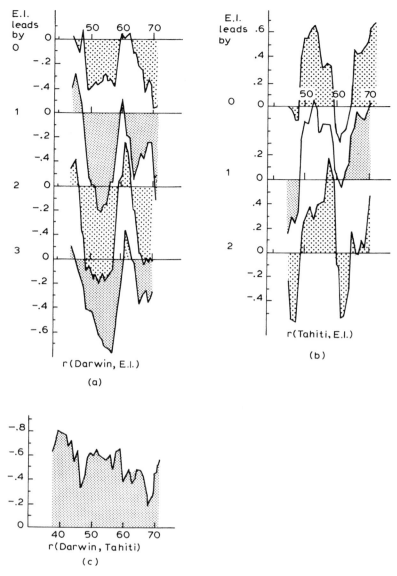

Fig. 85. Time series of 5-year correlation coefficients calculated every year for (a) Darwin and Easter Island with Easter Island leading by 0, 1, 2, and 3 seasons; (b) Tahiti and Easter Island with Easter Island leading by 0, 1, and 2 seasons; and (c) Tahiti and Darwin at zero lag. (After TRENBERTH, 1976b.)

the Southern Oscillation to his *P1* and *P2* indices for the Australasian region. The sequence occurring with most variance in a 3- to 6-year periodicity during the 20 years before 1976 is such that, if T is the period, then *P1* (Fig. 83) leads Easter Island by about $T/10$, Darwin and Tahiti by $T/5$, and *P2* by $T/4$; the shape of the Southern Oscillation pattern depends on the phase chosen. Some $8\frac{1}{2}\%$ of the monthly MSL pressure variance is explained by the *P1* part of the Southern Oscillation and $4\frac{1}{2}\%$ by the *P2* part. Trenberth's description of the sequence of pressure change across the South Pacific during a Southern Oscillation cycle is depicted in Fig. 86. Features of the global patterns associated with the Southern Oscillation have been given by KRUEGER and WINSTON (1974) and by VAN LOON and MADDEN (1981) and VAN LOON and ROGERS (1981).

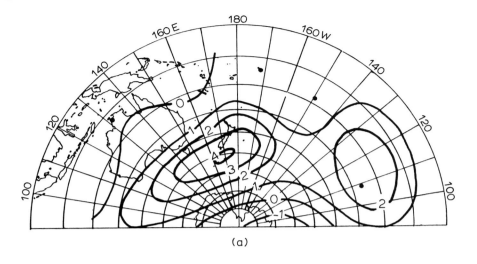

(a)

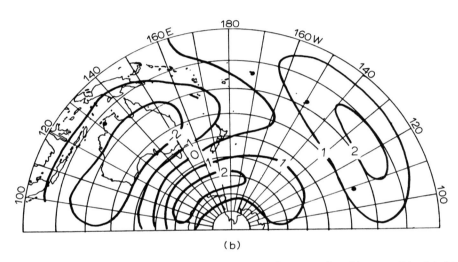

(b)

Fig. 86. Schematic representation of actual departures of pressure (in arbitrary units) related to different phases of the Southern Oscillation. (a) Pattern occurring as the pressure at Darwin changes from positive to negative and the Southern Oscillation index is increasing. (b) Pattern corresponding to maximum of the Southern Oscillation, occurring about 12 months after (a). (After TRENBERTH, 1976b.)

Walker Circulation

Although the concept of a large-scale vertical zonal circulation cell was implicit in the original work of Sir Gilbert Walker, it was TROUP (1965) who first explicitly interpreted the Southern Oscillation as being due to fluctuations in the strength of a solenoidally driven mass circulation between the warmer eastern and cooler western hemispheres. BJERKNES (1966a, 1969) further developed the concept of a thermally driven zonal circulation embedded within the tropical meridional Hadley Circulation, its main cell consisting of a region of ascent over the warm western Pacific linked by low-level easterlies and upper westerlies to a region of descent over the cold waters off the west coast of South America. BJERKNES (1969) named this Pacific cell "the Walker Circulation" since it can be shown to be an important part of the mechanism of Walker's "Southern

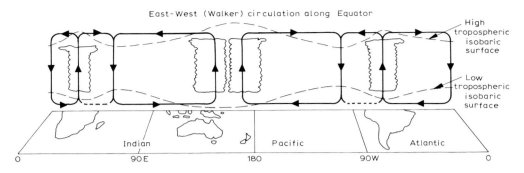

Fig. 87. Schematic illustration of the Walker Circulation along the Equator.

Oscillation," but the term is often applied to the east – west circulation right around the equatorial belt.

The structure of the Walker Circulation along the Equator is shown schematically in Fig. 87. This extended view of the tropical east – west circulation with ascent over Brazil, Central Africa, and the western Pacific and subsidence over the eastern Pacific, the eastern South Atlantic, and the central Indian Ocean is generally supported by the work of FLOHN (1971) and KIDSON (1975a,b) and by studies of atmospheric heating rates (KRUEGER, 1970) and seasonal mean vertical motions (BOER and KYLE, 1974; NEWELL et al., 1974). KRISHNAMURTI (1971a,b) and KRISHNAMURTI et al. (1973) have identified some of the detailed structure of the tropical east – west circulations from actual wind

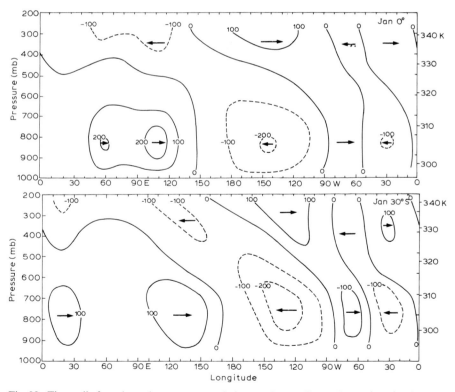

Fig. 88. Thermally forced zonal mass transport in isentropic coordinates ($kg\,m^{-1}\,K^{-1}\,s^{-1}$) along the Equator and 30°S in January. (ZILLMAN and JOHNSON, 1983.)

356

data, whereas WEBSTER (1973) and MURAKAMI (1974) have examined their relation to prescribed fields of thermal forcing using a two-layer linear and a nine-layer nonlinear model, respectively. ZILLMAN (1972c, 1974b) and ZILLMAN and JOHNSON (1983) have isolated the tropical east—west Walker Circulation as part of a planetary-scale thermally forced mean mass circulation in isentropic coordinates extending even to the high latitudes of the South Pacific. According to their view, the Pacific Walker Circulation is to be regarded as an ocean-wide thermally driven vertical circulation cell carrying excess heat from the strong west Pacific heat source centred near the Solomon Islands to offset the net atmospheric cooling in the eastern Pacific (Fig. 77). Fig. 88 shows the thermally forced east—west mass transport in isentropic coordinates (units of kg m^{-1} K^{-1} s^{-1}) in January along the Equator and 30°S based on a three-dimensional solution of the time-averaged continuity equation in isentropic coordinates and modeled diabatic heating fields compatible with those summarized in Figs. 77 and 78. The mass fluxes involved suggest mean zonal windspeeds of the order of 1 – 3 m s^{-1} as characteristic of

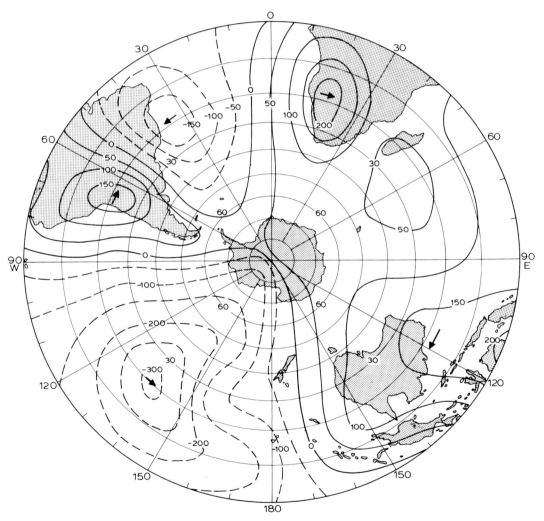

Fig. 89. Thermally forced zonal mass transport in isentropic coordinates (kg m^{-1} K^{-1} s^{-1}) on the 310°K isentropic surface in January. (ZILLMAN and JOHNSON, 1983.)

357

the upper (westerly) and lower (easterly) branches of the Walker cell in the Pacific. Fig. 89, which gives the thermally forced horizontal mass transport (kg m^{-1} K^{-1} s^{-1}) on the 310°K isentropic surface in January, shows how an east−west circulation associated with the Walker cell may be isolated even in the high latitudes of the South Pacific.

It is evident that the Pacific Walker Circulation is, as BJERKNES (1969) suggests, an integral part of the mechanism of the Southern Oscillation. Variation in its position and strength would be expected to be reflected in the pattern and phase of the Southern Oscillation and in such associated characteristics as the strength of the southeast trades and the westward extent of the equatorial dry zone (MUSK, 1976). Clearly also, it is part of a much more extensive mass circulation associated with the wider-scale correlations between the Pacific region and eastern South America (WRIGHT, 1975; HASTENRATH, 1976; HASTENRATH and HELLER, 1977; COVEY and HASTENRATH, 1978) as well as other parts of the globe (VAN LOON and MADDEN, 1981).

El Niño

Under the influence of the northward running Peru Current and intense coastal up-welling, very cold water is found over an extensive region off the coast of Peru for most of the year (Figs. 6 and 7). Through the southern summer, however, a pool of water above 25°C develops southeast of the Galápagos Islands with a tongue of warm offshore water extending southeastward as far as 20°S. A strip of cool upwelled water normally persists close inshore along most of the Peruvian coast, but around the end of December a weak surface current of warm water extends southward along the coast of Ecuador and northern Peru to about 6°S. Known locally as "El Niño" (the Child) because of its association with the Christmas season, it is an annual event of relatively minor significance. In exceptional years, this pattern is subject to catastrophic dislocation. The up-welling along the coast of Peru apparently ceases completely, the warm water extends much farther south and large positive temperature anomalies develop over a large part of the tropical eastern Pacific and persist for a year or more (Fig. 90). The warming of the coastal waters causes the death or migration of a large fraction of the fish stock while the bird population, dependent on fish for food, is decimated. At the same time, frequent heavy rains and electrical storms develop over the coastal desert. The failure of the fish and guano industries, together with the effects of flood damage, places the region's economy under great duress. It is to this infrequent dramatic warming of the entire eastern tropical portion of the South Pacific gyre and the spectacular accompanying biological and climatic events that the term El Niño is now generally applied. An extensive account of the El Niño phenomenon has been given by SCHÜTTE (1968), and useful bibliographies are included in papers by WYRTKI (1966a,b, 1975b), BJERKNES (1966b), QUINN (1974), WOOSTER and GUILLEN (1974), RAMAGE (1975), and QUINN et al., (1978). Mention of some interesting historical sidelights on El Niño and an account of its climatic effects in Peru are included in SCHWERDTFEGER (1976). The following discussion is concerned with the El Niño phenomenon as a major climatic event affecting a large part of the tropical south Pacific.

Major El Niños of the past century have occurred in 1878, 1884, 1891 (SCHOTT, 1931), 1917, 1925−1926 (MURPHY, 1926), 1932, 1939−1941 (SCHWEIGGER, 1942), 1943, 1951, 1953 (WOOSTER and JENNINGS, 1955; POSNER, 1957), 1957−1958 (WOOSTER, 1960;

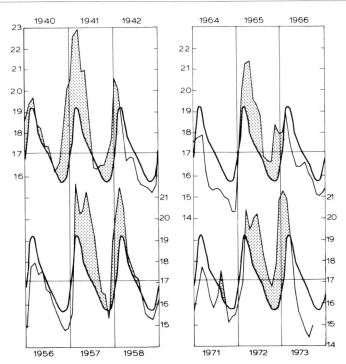

Fig. 90. Sea surface temperature (°C) at Puerto Chicama (7°S) during four recent El Niño events. The thin line gives observed monthly mean temperature, the heavy line the mean annual variation (1925–1973). Temperatures above normal are shaded. (After WYRTKI, 1975b.)

BJERKNES, 1966b), 1965, and 1972–1973 (WOOSTER and GUILLEN, 1974; CAVIEDES, 1975; RAMAGE, 1975). Although attributed to an approximate 7-year periodicity by weather lore on the coast of northern Peru, it is neither a simple periodic phenomenon nor does each major El Niño invasion occur in the same way. Stratigraphic evidence suggests that these quasi-periodic wet seasons may have characterized the northern coast of Peru far back into Tertiary times.

A full understanding of the mechanisms and sequence of events involved in the generation of El Niño does not yet exist, although the 1972–1973 occurrence has been subject to detailed study with use being made of the extensive satellite data now available (WOOSTER and GUILLEN, 1974; RAMAGE, 1975; STRETEN, 1975). According to SCHOTT (1931, 1932), El Niño can be traced to a weakening of the South Pacific High. The associated slackening of the southeast trades allows the Intertropical Convergence (ITC) and associated rains to be displaced southward over northern Peru. The North Equatorial Counter Current shifts southward and, with winds from a northerly or northwesterly direction displacing the normal southeasterlies over the waters off Peru, a southward flow of warm equatorial water displaces the cooler water normally present there. The El Niño rains are not seen as a direct consequence of the ocean warming; rather both are viewed as a result of the anomalous atmospheric circulation pattern associated with the weakening of the South Pacific High. FLOHN and HINKELMANN (1952), on the other hand, suggest that the El Niño rains may be associated with the development of a second ITC south of the Equator separated from the north by a belt of equatorial westerlies. WOOSTER (1960) attributed the ocean warming to a weakening of the circulations which cause the upwelling, coupled with the formation of a thin warm surface layer by local heating, with an associated shallow and sharp thermocline as well as shoreward movement of warmer water. More recently, papers by BJERKNES (1961, 1966b), SCHELL (1965), RAMAGE (1970, 1975), QUINN (1974), WHITE and McCREARY (1974), WYRTKI (1975a,b, 1979), BARNETT (1977), and QUINN et al. (1978) picture El Niño as but part of a complex chain of events involving anomalies of the general circulation over the entire Pacific and perhaps the globe. LETTAU (1976), however, has attempted to interpret the El Niño phenomenon in terms of a temporary breakdown of a thermally forced local atmospheric circulation associated with the maintenance of the Peruvian coastal desert. WYRTKI (1975b) explains El Niño as a result of the reaction of the equatorial Pacific Ocean to the relaxation of the southeast trades after a prolonged period of excessively strong winds and accumulation of water in the western Pacific. During a period of strong trades lasting more than one year, the water circulation in the subtropical gyre of the South Pacific is intensified, in particular the South Equatorial Current. There is an increase of east—west slope of sea level and an accumulation of water in the western Pacific, most likely in the area between Samoa and the Solomon Islands. Fig. 91 shows significant variations in sea level in the western Pacific preceding the El Niño events of 1957—1958 and 1972—1973. Wyrtki reasons that as soon as the wind stress in the central Pacific relaxes, the accumulated water will flow eastward, probably in the form of an internal seiche or an internal equatorial Kelvin wave. He draws on a theoretical study of ocean spindown by GODFREY (1975) for an estimate of about two months' travel time for the Kelvin waves across the equatorial Pacific. Wyrtki also assumes that eastward flow in those currents that normally transport water to the east (South Equatorial Counter Current, North Equatorial Counter Current, Equatorial Undercurrent) will intensify, leading to an accumulation of warm water off equatorial Peru and a depression of the usually shallow thermocline. He attributes the abnormally high temperatures observed offshore during El Niño to the greater solar radiation absorption by

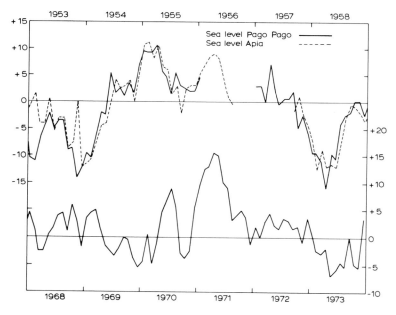

Fig. 91. Monthly mean sea level (cm) at Pago Pago and at Apia on Samoa during the periods 1953—1958 and 1968—1973, showing abnormally high sea level before El Niño. (After WYRTKI, 1975b.)

the water in its longer residence time in the offshore region. Wyrtki's theory is in line with the findings of QUINN (1974) that the periods of unusual equatorial Pacific activity during which El Niños occurred followed unusually high peaks of a Southern Oscillation index given by the MSL pressure difference Easter Island minus Darwin.

Observational studies by HICKEY (1975) and WYRTKI et al. (1976) and empirical orthogonal function analysis of Pacific sea surface temperature by WEARE et al. (1976) have tended to support Wyrtki's theory, as have further statistical investigations by BARNETT (1977). Further, numerical ocean models (McCREARY, 1976; CANE and SARACHEK, 1976, 1977) also appear to confirm Wyrtki's theory. However, the atmospheric effects with changing rainfall patterns in the equatorial Pacific must occur in a more complex manner; RAMAGE (1977) has pointed out that there is little clear relationship between rainfall and changing sea surface temperatures in the region.

El Niño and the Southern Oscillation are clearly related phenomena, the former being an ocean response to changes of atmospheric circulation pattern described by the latter, and primarily to changes in the southeast Pacific High and in the location and characteristics of the mid-ocean trough over the South Pacific. An explanation of the observed phenomena does not yet exist, but as we have seen, such an explanation must be related to the broad-scale circulation of the atmosphere in both hemispheres and at extratropical as well as tropical latitudes.

Conclusion

This survey of the climate of the South Pacific has sought mainly to provide a broad overview of the climatic features of the region on the basis of the rather fragmentary data assembled by various workers of the past century. There is still a good deal of uncertainty about many aspects of the Pacific climate particularly south of the subtropical ridge. Although the increasing attention now being paid to the role of the oceans in the context of the World Climate Programme, together with the continued development of satellite technology, promises to throw early light on some of the unknown or uncertain characteristics of the climate of the South Pacific, it may be several decades before a fully definitive treatment of the entire region is possible.

We have included only a minimum of discussion of the conventional climatic elements summarized in data tables and no detailed treatment at all of the climate of the various island groups. Both may readily be found in the list of references.

Acknowledgements

The authors acknowledge the provision of climatological data and information by various National Meteorological Services in the Pacific region and, in particular, data provided for several Pacific Islands by the Director of the New Zealand Meteorological Service. The data tables for the ocean areas were kindly provided by Mr. W. Haggard, formerly of the U.S. National Environmental Data Service. Valuable information and advice on various aspects of the climate of the South Pacific were provided by Messrs. A. B. Neal, W. Schwerdtfeger, K. Wyrtki, J. Sadler, and A. J. Troup. Mr. D. Pike, Mr.

P. Yew, Miss G. Burt, and Miss S. Kraemers provided a great deal of assistance with various aspects of this review.

References

ALBRECHT, F., 1951. Monatskarten des Niederschlages im Indischen und Stillen Ozean. *Ber. Dtsch. Wetterdienstes*, 29: 21 pp.

ALBRECHT, F., 1960. Jahreskarten des Wärme- und Wasserhaushaltes der Ozeane. *Ber. Dtsch. Wetterdienstes*, 9: 19 pp.

ALBRECHT, F., 1961. Der jährliche Gang der Komponenten des Wärme- und Wasserhaushaltes der Ozeane. *Ber. Dtsch. Wetterdienstes*, 11: 24 pp.

ALLISON, L. J., STERANKA, J., HOLUB, R. J., HANSEN, J., GODSHALL, F. A. and PRABHAKARA, C., 1972. Air — sea interaction in the tropical Pacific. *NASA Tech. Note D-6684*. Goddard Space Flight Center, Greenbelt, Maryland, 66 pp.

ALVAREZ, J. A. and THOMPSON, A. H., 1965. Improvement of weather analyses in isolated areas of the Southern Hemisphere by use of meteorological satellite information. *Notos*, 14: 33 — 42.

ANDERSSEN, E. C., 1965. A study of atmospheric long waves in the Southern Hemisphere. *Notos*, 14: 57 — 65.

ASTAPENKO, P. D., 1960. Atmosfernye protsessy v vysokikh shriotakh iuzhnogo polushariia. (Atmospheric processes in the high latitudes of the Southern Hemisphere.) *Akad. Nauk SSSR*, Moscow, 286 pp. (Israel program for scientific translations, Jerusalem, 1964.)

AUSTIN, T. S., 1960. Oceanography of the east central equatorial Pacific as observed during expedition EASTROPAC. *U.S. Fish Wildlife Serv., Fishery Bull.*, 60: 257 — 282.

BARNES, J. C., 1966. Note on the use of satellite observations to determine average cloudiness over a region. *J. Geophys. Res.*, 75: 6137 — 6140.

BARNETT, T. P., 1977. An attempt to verify some theories of El Niño. *J. Phys. Oceanogr.*, 7: 633 — 647.

BATES, J. R., 1972. Tropical disturbances and the general circulation. *Q. J. R. Meteorol. Soc.*, 98: 1 — 16.

BAUMGARTNER, A. and REICHEL, E., 1975. *The World Water Balance*. Elsevier, Amsterdam, 179 pp.

BERGERON, T., 1930. Richtlinien einer dynamischen Klimatologie. *Meteorol. Z.*, 47: 246 — 262.

BERLAGE, H. P., 1957. Fluctuations of the general atmospheric circulation of more than one year, their nature and prognostic value. *Mededel. Verhandel. K.N.M.I.*, 69, 152 pp.

BERLAGE, H. P., 1961. Variations in the general atmosphere and hydrospheric circulation of periods of a few years duration affected by variations of solar activity. *Ann. N.Y. Acad. Sci.*, 95: 354 — 367.

BERLAGE, H. P., 1966. The Southern Oscillation and world weather. *Mededel. Verhandel. K.N.M.I.*, 88, 152 pp.

BJERKNES, J., 1961. El Niño study based on analysis of ocean surface temperatures, 1935 — 1957. *Inter-Amer. Tropical Tuna Commiss. Bull.*, 5: 219 — 303.

BJERKNES, J., 1966a. A possible response of the atmospheric Hadley circulation to equatorial anomalies of ocean temperatures. *Tellus*, 18: 820 — 829.

BJERKNES, J., 1966b. Survey of El Niño 1957 — 58 in its relation to tropical Pacific meteorology. *Inter-Am. Tropical Tuna Commiss. Bull.*, 12: 25 — 86.

BJERKNES, J., 1969. Atmospheric teleconnections from the equatorial Pacific. *Mon. Weather Rev.*, 97: 165 — 172.

BJERKNES, J., 1972. Large scale atmospheric response to the 1964 — 65 Pacific equatorial warming. *J. Phys. Oceanogr.*, 2: 212 — 217.

BLACK, J. N., 1956. The distribution of solar radiation over the earth's surface. *Arch. Meteorol. Geophys. Bioklimatol.*, Ser. B, 7: 165 — 189.

BOER, G. J. and KYLE, A. C., 1974. Cloudiness, precipitation and vertical motion. In: R. E. NEWELL, J. W. KIDSON, D. G. VINCENT and G. J. BOER, *The General Circulation of the Tropical Atmosphere and Interactions with Extratropical Latitudes*. The M.I.T. Press, Cambridge, Mass., Vol. 2, pp. 143 — 158.

BOLAND, F. M. and HAMON, B. V., 1970. The East Australian Current, 1965 — 1968. *Deep-Sea Res.*, 17: 777 — 794.

BOTNIKOV, V. N., 1964. Seasonal and long-term fluctuations of the Antarctic Convergence Zone. *Soviet Antarct. Exped. Inform. Bull.* (English translation by American Geophysical Union), 45: 92 — 95.

BRIDGMAN, H. A., 1969. The radiation balance of the southern hemisphere. *Arch. Meteorol. Geophys. Bioklimatol.*, Ser. B, 17: 325 — 344.

BROOKFIELD, H. C. and HART, D., 1966. *Rainfall in the Tropical Southwest Pacific*. Australian Nat. Univ., Dept. Geography, Publ. G, 3, 25 pp. and maps.

BROWNE, M. L., 1975. Anticyclones in the Australian New Zealand region. *N. Z. Meteorol. Serv., Wellington, Tech. Note*, 228, 9 pp. and 7 figs.

BUDD, W. F., 1975. Antarctic sea ice variations from satellite sensing in relation to climate. *J. Glaciol.*, 15: 417–426.

BUDYKO, M. I., 1963. *Atlas Teplovogo Balansa Zemnogo Shara*. Glavnaia Geofizicheskaia Observatoria A. I. Voeikova, Moscow, 69 pp.

BUREAU OF METEOROLOGY, 1940. *Results of Rainfall Observations Made in Papua, Mandated Territory of New Guinea, Solomon Islands, New Hebrides etc.* Commonwealth of Australia, Melbourne, 74 pp.

BURLING, R. W., 1961. Hydrology of circumpolar waters south of New Zealand. *N. Z. Dept. Sci. Ind. Res., Bull.*, 149, 66 pp.

CALLAHAN, J. E., 1971. Velocity structure and flux of the Antarctic Circumpolar Current south of Australia. *J. Geophys. Res.*, 76: 5859–5864.

CAMPBELL, W. J., WEEKS, W. F., RAMSEIER, R. O. and GLOERSEN, P., 1975. Geophysical studies of floating ice by remote sensing. *J. Glaciol.*, 15: 305–328.

CANE, M. A. and SARACHEK, E. S., 1976. Forced baroclinic ocean motions, I. The linear equatorial unbounded case. *J. Marine Res.*, 34: 629–665.

CANE, M. A. and SARACHEK, E. S., 1977. Forced baroclinic ocean motions, II. The linear equatorial bounded case. *J. Marine Res.*, 35: 395–432.

CAPURRO, L. R. A., 1973. USNS Eltanin's 55 cruises—scientific accomplishment. *Antarct. J. U. S.*, 8: 57–61.

CAUCHARD, G. and INCHAUSPÉ, 1977. Le Temps dans L'Archipel des Marquises. *Aperçu Climatologique, Monographie de la Météorologie Nationale, Paris*, 100, 31 pp.

CAVIEDES, C. N., 1975. El Niño 1972: Its climatic, ecological, human and economic implications. *Geograph. Rev.*, 65: 493–509.

CHANG, C. P., 1970. Westward propagating cloud patterns in the tropical Pacific as seen from time composite satellite photographs. *J. Atmos. Sci.*, 27: 133–138.

CHARNEY, J., 1969. The intertropical convergence zone and the Hadley circulation of the atmosphere. In: *Proc. WMO/IUGG Symp. Numerical Weather Prediction, Japan Meteorol. Agency, Tokyo*, 3: 13–79.

COVEY, D. L. and HASTENRATH, S., 1978. The Pacific El Niño phenomenon and the Atlantic circulation. *Mon. Weather Rev.*, 106: 1280–1287.

CRESSWELL, G. E., GOLDING, T. J. and BOLAND, F. M., 1978. A buoy and ship examination of the subtropical convergence south of Western Australia. *J. Phys. Oceanogr.*, 8: 315–320.

CROMWELL, T., MONTGOMERY, R. B. and STROUP, E. D., 1954. Equatorial undercurrent in Pacific Ocean revealed by new methods. *Science*, 119: 648–649.

CRUTCHER, H. L. and HOXIT, L. R., 1974. *Southwest Pacific and Australian Area Tropical Cyclone Strike Probabilities*. Environmental Data Service, NOAA, Asheville, N. C., 208 pp.

CRUTCHER, H. L. and QUAYLE, R. G., 1974. *Mariners Worldwide Guide to Tropical Storms at Sea*. NAVAIR, 50-1C-61, U.S. Navy, Washington, D.C., 114 pp. and 312 charts.

CRUTCHER, H. L., JENNE, R. L., TALJAARD, J. J. and VAN LOON, H., 1971. *Climate of the Upper Air: Southern Hemisphere*. NAVAIR, 50-1C-58, U.S. Navy, Washington, D.C., 4, 2 pp. and 60 figs.

DAS, S. C., 1956. Statistical analysis of Australian pressure data. *Aust. J. Phys.*, 9: 394–399.

DEACON, G. E. R., 1937. The hydrology of the Southern Ocean. *Discovery Rep.*, 15: 1–24.

DEACON, G. E. R., 1963. The Southern Ocean. In: M. N. HILL (Editor), *The Sea. Ideas and Observations on Progress in the Study of the Seas*. Interscience Publishers, New York, N.Y., Vol. 2, pp. 281–296.

DEFANT, A., 1961. *Physical Oceanography*. Pergamon, New York, N.Y., Vol. 1, 729 pp.

DEFENSE MAPPING AGENCY HYDROGRAPHIC CENTER, 1974. *Atlas of Pilot Charts: South Pacific and Indian Oceans. Publication No. 107* (3rd ed., 1966, reprinted 1974). United States Dept. Defense, Washington, D.C.

DELISLE, J. F., 1970. Heat transfer from ocean to atmosphere in the Tasman Sea area. *N. Z. J. Sci.*, 13: 166–176.

DERYCKE, R. J., 1973. Sea ice motions off Antarctica in the vicinity of the eastern Ross Sea as observed by satellite. *J. Geophys. Res.*, 78: 8873–8879.

D'HAUTESERRE, A., 1960. Les perturbations atmosphériques. Climat de la Polynésie française. *Monographie de la Météorologie Nationale, Paris*, 18, 62 pp.

D'HAUTESERRE, A., 1972. Le vent en altitude au-dessus de la Polynésie française. *Monographie de la Météorologie Nationale, Paris*, 85, 53 pp.

DOPPLICK, T. G., 1970. Global radiative heating of the earth's atmosphere. *M.I.T., Dept. Meteorol., Cambridge, Planetary Circulations Project Rep.*, 24, 128 pp.

DORMAN, C. E. and BOURKE, R. H., 1979. Precipitation over the Pacific Ocean, 30°S to 60°N. *Mon. Weather Rev.*, 107: 896–910.

DROZDOV, O. A., 1953. Annual amounts of precipitation. *Morskoi Atlas*. Moscow, II, Chart 48b.

EMERY, W. J., 1977. Antarctic polar frontal zone from Australia to the Drake Passage. *J. Phys. Oceanogr.*, 7: 811–822.

ERICKSON, C. O. and WINSTON, J. S., 1972. Tropical storm, mid latitude cloud band connections and the autumnal buildup of the planetary circulation. *J. Appl. Meteorol.*, 11: 23–36.

FABRICIUS, A. F., 1957. Climate of the sub-antarctic islands. In: M. P. VAN ROOY (Editor), *Meteorology of the Antarctic*. Weather Bureau, Pretoria, pp. 111–135.

FLETCHER, J. O., 1969. *Ice Extent on the Southern Ocean and its Relation to World Climate*. The Rand Corporation Memorandum, RM-5793-NSF, Santa Monica, Calif., 107 pp.

FLOHN, H., 1971. Tropical circulation pattern. *Bonner Meteorol. Abhandl.*, 15, 55 pp.

FLOHN, H. and HINKELMANN, K., 1952. Äquatoriale Zirkulationsanomalien und ihre klimatische Auswirkung. *Ber. Dtsch. Wetterdienstes*, 42: 114–121.

FRASER, R., 1973. *Oceanography of the South Pacific 1972*. N. Z. National Commission for UNESCO, Wellington, 524 pp.

FREEMAN, J. C., 1948. An analogy between the equatorial easterlies and supersonic gas flows. *J. Meteorol.*, 5: 138–146.

FRIEHE, C. A. and SCHMITT, K. F., 1976. Parameterization of air–sea interface fluxes of sensible heat and moisture by the bulk aerodynamic formulas. *J. Phys. Oceanogr.*, 6: 801–809.

GABITES, J. F., 1956. A survey of tropical cyclones in the South Pacific. *Proc. Tropical Cyclone Symp., December 1956, Brisbane*, pp. 19–24.

GABITES, J. F., 1960. The heat balance of the Antarctic through the year. In: *Antarctic Meteorology*. Pergamon Press, Oxford, pp. 370–377.

GAUNTLETT, D. J., SEAMAN, R. S., KININMONTH, W. R. and LANGFORD, J. C., 1972. An operational evaluation of a numerical analysis-prognosis system for the southern hemisphere. *Aust. Meteorol. Mag.*, 20: 61–81.

GIBBS, W. J., 1960. International Antarctic Analysis Centre. *Polar Record*, 10: 86–89.

GIBBS, W. J., GOTLEY, A. V. and MARTIN, A. R., 1952. *Meteorology of Heard and Macquarie Islands, 1948*. Part 1(c) Discussion ANARE Reports, Series D, Vol. 1, Antarctic Division, Department of External Affairs (Australia), Melbourne, 67 pp.

GIOVANNELLI, J. and ROBERT, J., 1964. Quelques aspects des depressions et cyclones tropicaux dans le Pacifique Sud-ouest. *Monographie de la Météorologie Nationale, Paris*, 33, 24 pp.

GIRARD, J. and RIGNOT, D., 1971. Climatologie de la Nouvelle Caledonie. *Monographie de la Météorologie Nationale, Paris*, 82, 212 pp.

GODFREY, J. S., 1975. On ocean spindown I: A linear experiment. *J. Phys. Oceanogr.*, 5: 399–409.

GORDON, A. H. and TAYLOR, R. C., 1968. Numerical (steady state friction layer trajectories over the oceanic tropics as related to weather. International Indian Ocean Expedition. *Meteorol. Monogr.* 7. East West Center Press, Honolulu, 112 pp.

GORDON, A. L., 1967. Structure of Antarctic waters between 20°W and 170°W. *Antarctic Map Folio Series*. Folio 6, American Geographical Society, New York, N.Y.

GORDON, A. L., 1970. Physical oceanography on *Eltanin* cruises 32–37. *Antarctic Journal of the United States*, 4: 183–184.

GORDON, A. L., 1971. Antarctic polar front zone. In: J. L. REID (Editor), *Antarctic Oceanology, I. Antarctic Research Series*. American Geophysical Union, Washington, D.C., Vol. 15, pp. 205–221.

GORDON, A. L., 1972. Physical oceanography of the Southeast Indian Ocean. In: D. E. HAYES (Editor), *Antarctic Oceanology, II. The Australian–New Zealand Sector, Antarctic Research Series*. American Geophysical Union, Washington, D.C., Vol. 19, pp. 3–9.

GORDON, A. L., 1973. USNS *Eltanin* cruise 50: Physical oceanography in the southeastern Indian Ocean. *Antarctic Journal of the United States*, 7: 38–40.

GORDON, A. L., 1975. Antarctic oceanographic section along 170°E. *Deep-Sea Res.*, 22: 357–377.

GORDON, A. L. and MOLINELLI, E., 1975. *USNS Eltanin Southern Ocean Oceanographic Atlas*. Contribution No. 2256, Lamont-Doherty Geological Observatory, Columbia Univ., Palisades, N.Y.

GORDON, A. L., MOLINELLI, E. and BAKER, T., 1978. Large-scale relative dynamic topography of the Southern Ocean. *J. Geophys. Res.*, 83: 3023–3032.

GRAY, W. M., 1968. Global view of the origin of tropical disturbances and storms. *Mon. Weather Rev.*, 96: 669–700.

GRAY, W. M., 1975. *Tropical Cyclone Genesis*. *Atmos. Sci. Pap.*, 234. Colorado State University, Fort Collins, Colo., 121 pp.

GRUBER, A., 1972. Fluctuations in the position of the ITCZ in the Atlantic and Pacific Oceans. *J. Atmos. Sci.*, 29: 193–197.

GUYMER, L. B., 1969. Estimation of 1000-500 mb thickness patterns from satellite pictures of convective areas. In: *Satellite Meteorology—Proc. Inter-regional Seminar on the Interpretation of Meteorological Satellite Data, Bur. Meteorol., Melbourne*, pp. 51 — 55.

GUYMER, L. B., 1978. Operational application of satellite imagery to synoptic analyses in the southern hemisphere. *Bur. Meteorol., Dept. Sci., Melbourne, Tech. Rep.*, 29, 83 pp.

HAMON, B. V., 1970. Western boundary currents in the South Pacific. In: W. S. WOOSTER (Editor), *Scientific Exploration of the South Pacific—Proc. of a Symposium held during the Ninth General Meeting of the Scientific Committee on Oceanic Research, 18 — 20 June 1968.* Scripps Institution of Oceanography, La Jolla, Calif. National Academy of Sciences, Washington, D.C., pp. 50 — 59.

HAMON, B. V. and GODFREY, J. S., 1978. Role of the oceans. In: A. B. PITTOCK, L. A. FRAKES, D. JENSSEN, J. A. PETERSON and J. W. ZILLMAN (Editors), *Climatic Change and Variability: A Southern Perspective.* Cambridge Univ. Press, Cambridge, pp. 31 — 52.

HAMON, B. V. and TRANTER, D. J., 1971. The East Australian Current. *Aust. Nat. Hist.*, 17: 129 — 133.

HASTENRATH, S., 1976. Variations in low latitude circulation and extreme climatic events in the tropical Americas. *J. Atmos. Sci.*, 33: 202 — 215.

HASTENRATH, S. and HELLER, L., 1977. Dynamics of climatic hazards in northeast Brazil. *Q. J. R. Meteorol. Soc.*, 103: 77 — 92.

HASTENRATH, S. and LAMB, P. J., 1977. *Climatic Atlas of the Tropical Atlantic and Eastern Pacific Oceans.* Univ. Wisc. Press, Madison, Wisc., 97 pp.

HASTENRATH, S. and LAMB, P. J., 1978. *Heat Budget Atlas of the Tropical Atlantic and Eastern Pacific Oceans.* Univ. Wisc. Press, Madison, Wisc., 104 pp.

HAURWITZ, F. D., 1972. *The Distribution of Tropospheric Infrared Radiative Fluxes and Associated Heating and Cooling Rates in the Southern Hemisphere.* Tech. Rep., College of Engineering ORA project 031640, Univ. Mich., East Lansing, Mich., 168 pp.

HEATH, R. A., 1981. Oceanic fronts around southern New Zealand. *Deep-Sea Res., Oceanogr. Res. Pap.*, 28, 6A, pp. 547 — 560.

HEPWORTH, M. W., 1913. *Meteorology Part II, National Antarctic Expedition, 1901 — 1904.* The Royal Society, London, 26 pp. and charts.

HICKEY, B., 1975. The relationship between fluctuations in sea level, wind stress and sea surface temperature in the equatorial Pacific. *J. Phys. Oceanogr.*, 5: 460 — 475.

HILL, H. W., 1964. The weather in low latitudes of the southwest Pacific associated with the passage of disturbances in the middle latitude westerlies. In: J. W. HUTCHINGS (Editor), *Proc. Symp. on Tropical Meteorology, Rotorua.* New Zealand Meteorol. Service, Wellington, pp. 265 — 352.

HOLTON, J. R. and WALLACE, J. M., 1971. On boundary layer dynamics and the ITCZ. *J. Atmos. Sci.*, 28: 275 — 280.

HOUTMAN, T. J., 1964. Surface temperature gradients in the Antarctic convergence. *N.Z. J. Geol. Geophys.*, 7: 245 — 270.

HOUTMAN, T. J., 1967. Water masses and fronts in the Southern Ocean south of New Zealand. *N..Z. Dept. Sci. Ind. Res., Wellington, Bull.*, 174, 40 pp.

HUBERT, L. F., 1961. A subtropical convergence line of the South Pacific. A case study using meteorological satellite data. *J. Geophys. Res.*, 66: 797 — 812.

HUBERT, L. F., KRUEGER, A. F. and WINSTON, J. S., 1969. The double intertropical convergence zone—fact or fiction? *J. Atmos. Sci.*, 26: 771 — 773.

HUTCHINGS, J. W., 1953. *Tropical Cyclones in the Southwest Pacific.* N. Z. Meteorol. Service, Wellington, Note 37, 114 pp.

INDER, S. (Editor), 1979. *Pacific Islands Year Book.* Pacific Publications, Sydney, 13th ed., 512 pp.

IVANOV, J. A., 1959. Position and seasonal variation of frontal zones in Antarctica. *Dokl. Akad. Nauk SSSR.*, 120: 777 — 780.

JENNE, R. L., CRUTCHER, H. L., VAN LOON, H. and TALJAARD, J. J., 1971. *Climate of the Upper Air, Southern Hemisphere, Vol. IV.* NAVAIR 50-1C-57, U.S. Navy, Washington, D.C., 3 pp. and 60 figs.

JENNE, R. L., CRUTCHER, H. L., VAN LOON, H. and TALJAARD, J. J., 1974. *A Selected Climatology of the Southern Hemisphere: Computer Methods and Data Availability.* NCAR-TN/STR-92, National Center for Atmospheric Research, Boulder, Colo., 91 pp.

JULIAN, P. R. and CHERVIN, R. M., 1978. A study of the Southern Oscillation and Walker Circulation phenomenon. *Mon. Weather Rev.*, 106: 1433 — 1451.

KANGOS, J., 1960. A preliminary investigation of the heat flux from the ocean to the atmosphere in the Antarctic regions. *J. Geophys. Res.*, 65: 4007 — 4012.

KARELSKY, S., 1954. Surface circulation in the Australasian region. *Bur. Meteorol. Melbourne, Meteorol. Stud.*, 3, 45 pp. and diagrams.

KARELSKY, S., 1956. Classification of the surface circulation in the Australasian region. *Bur. Meteorol. Melbourne, Meteorol. Stud.*, 8, 36 pp. and diagrams.

KARELSKY, S., 1960. The surface circulation over Southern Ocean, Southern Indian Ocean, Australasia and Southern Pacific Ocean regions. In: *Antarctic Meteorology*. Pergamon, Oxford, pp. 293–313.

KARELSKY, S., 1961. Monthly and seasonal anticyclonicity and cyclonicity in the Australian region—15 year (1946–1960) averages. *Bur. Meteorol. Melbourne, Meteorol. Stud.*, 13, 11 pp. and diagrams.

KARELSKY, S., 1965. Monthly geographical distribution of central pressures in surface highs and lows in the Australian region, 1952–1963. *Bur. Meteorol. Melbourne, Meteorol. Summary*, 39 pp.

KELLY, G. A. M., 1978. Interpretation of satellite cloud mosaics for southern hemisphere analyses and reference level specification. *Mon. Weather Rev.*, 106: 870–889.

KERR, I. S., 1976. Tropical storms and hurricanes in the southwest Pacific November 1939 to April 1969. *N.Z. Meteorol. Serv., Wellington, Misc. Publ.*, 148, 113 pp.

KIDSON, E., 1925. Some periods in Australian weather. *Bur. Meteorol. Melbourne, Bull.*, 17, 33 pp.

KIDSON, E., 1947. Daily weather charts extending from Australia and New Zealand to the Antarctic continent. Australian Antarctic Expedition 1911–1914. *Sci. Rep., Ser. B, Vol. VII*. Government Printing Office, Sydney, 31 pp. and 365 charts.

KIDSON, E. and HOLMBOE, J., 1935. *Frontal Methods of Weather Analysis Applied to the Australia–New Zealand Area. Part I, Discussion; Part II, Weather Charts*. New Zealand Dept. Sci. and Ind. Res., Meteorol. Branch, Wellington, 20 pp. and charts.

KIDSON, J. W., 1975a. Eigenvector analyses of monthly mean surface data. *Mon. Weather Rev.*, 103: 177–186.

KIDSON, J. W., 1975b. Tropical eigenvector analysis and the Southern Oscillation. *Mon. Weather Rev.*, 103: 187–196.

KNAUSS, J. A., 1963. Equatorial current systems. In: M. N. HILL (General Editor), *The Sea: Ideas and Observations on Progress in the Study of the Seas*. Interscience, New York, N.Y., Vol. 2, pp. 235–252.

KNAUSS, J. A., 1966. Further measurements and observations on the Cromwell Current. *J. Marine Res.*, 24: 205–240.

KONINKLIJK NEDERLANDS METEOROLOGISCH INSTITUUT, 1949. *Sea Areas Round Australia, No. 124*. Staatsdrukkerijen Uitgeverijbedrijf, 's-Gravenhage, 79 pp.

KORNFIELD, J. and HASLER, A. F., 1969. A photographic summary of the earth's cloud cover for the year 1967. *J. Appl. Meteorol.*, 8: 687–699.

KORNFIELD, J. A., HASLER, A. F., HANSEN, K. J. AND SUOMI, V. E., 1967. Photographic cloud climatology from ESSA III and V computer produced mosaics. *Bull. Am. Meteorol. Soc.*, 48: 878–883.

KORT, V. G., 1962. The Antarctic Ocean. *Sci. Am.*, 207: 113–128.

KORZUM V. I., 1974. *Atlas of World Water Balance*. Hydrometeor. Publ. House, Moscow, 65 pp.

KRISHNAMURTI, T. N., 1971a. Observational study of the tropical upper tropospheric motion field during the northern hemisphere summer. *J. Appl. Meteorol.*, 10: 1066–1096.

KRISHNAMURTI, T. N., 1971b. Tropical east–west circulations during the northern summer. *J. Atmos. Sci.*, 28: 1342–1347.

KRISHNAMURTI, T. N., KANAMITSU, M., KISS, W. J. and LEE, J. D., 1973. Tropical east–west circulations during the northern winter. *J. Atmos. Sci.*, 30: 780–787.

KRUEGER, A. F., 1970. The zonal variation of cloudiness and convection over the tropics. *Proc. Symp. Tropical Meteorol., Honolulu*. Am. Meteorol. Soc., Boston, I, II: pp. 1–7.

KRUEGER, A. F. and WINSTON, J. S., 1974. A comparison of the flow over the tropics during two contrasting circulation regimes. *J. Atmos. Sci.*, 31: 358–370.

LAMB, H. H., 1967. On climate variations affecting the far south. In: *Polar Meteorology, WMO Tech. Note*, 87, pp. 428–453.

LAMB, H. H. and JOHNSON, A. I., 1966. Secular variations of the atmospheric circulation since 1750. *Geophys. Mem., Meteorol. Office, London*, 110, 125 pp.

LAMOND, M. H., PRICE, P. G., NEAL, A. B., and LAJOIE, F. A., 1972. GARP basic data set analysis project, the second experiment—June 1970. *Aust. Meteorol. Mag.*, 20: 193–204.

LANGFORD, J. C., 1957. Southern ocean analysis with special reference to the period December 1954 to March 1955. *Aust. Meteorol. Mag.*, 15: 1–22.

LANGFORD, J. C., 1960. Aspects of circulation and analysis of the Southern Ocean. In: *Antarctic Meteorology*. Pergamon, Oxford, pp. 256–273.

LEESE, J. A., BOOTH, A. L. and GODSHALL, F. A., 1970. Archiving and climatological application of meteorological satellite data. *ESSA Tech. Rep., NESC, 53, Section 3*. National Environmental Satellite Service, Washington, D.C., 23 pp.

LEIGH, R. M., 1969. A meteorological satellite study of a double vortex system over the western Pacific Ocean. *Aust. Meteorol. Mag.*, 17: 48–62.

LETTAU, H., 1976. Dynamic and energetic factors which cause and limit aridity along South America's Pacific coast. In: W. SCHWERDTFEGER (Editor), *Climates of Central and South America, World Survey of Climatology, Vol. 12.* Elsevier, Amsterdam, pp. 188–192.

LOURENSZ, R., 1977. Tropical cyclones in the Australian region, July 1909 to June 1975. *Bur. Meteorol., Melbourne, Meteorol. Summary,* 111 pp.

MACKINTOSH, N. A., 1946. The Antarctic convergence and the distribution of surface temperature in Antarctic waters. *Discovery Rep.,* 23: 177–212.

MACKINTOSH, N. A. and HERDMAN, H. F. P., 1940. Distribution of pack ice in the southern ocean. *Discovery Rep.,* 19: 285–296.

MAHER, J. V. and LEE, D. M., 1977. Upper air statistics: surface to 5 mb 1957 to 1975. *Bur. Meteorol., Melbourne, Meteorol. Summary,* 202 pp.

MAKSIMOV, I. V., 1961. The Antarctic Convergence front and long term changes of the northern boundary of iceberg occurrence in the Southern Ocean. *Problemy Arktiki i Antarktiki,* 8: 47–52.

MALKUS, J. S. and RIEHL, H., 1964. *Cloud Structure and Distributions over the Tropical Pacific Oceans.* Univ. Calif. Press, Berkeley, Calif., 229 pp.

MANABE, S., HOLLOWAY, J. L. and STONE, H. M., 1970. Tropical circulation in a time integration of a global model of the atmosphere. *J. Atmos. Sci.,* 27: 580–613.

MARAGOS, S. C., BAINES, G. B. and BEVERIDGE, P. J., 1973. Tropical cyclone Bebe creates a new land formation on Funafuti atoll. *Science,* 181: 1161–1163.

MARTIN, D. W., 1968. Satellite studies of cyclonic developments over the southern ocean. *IAMRC Tech. Rep. No. 9.* Bur. Meteorol., Melbourne, 64 pp.

MAWSON, D., 1915. *The Home of the Blizzard.* Heinemann, London, Vol. 1, 348 pp.

MCCREARY, J., 1976. Eastern tropical ocean responses to changing wind systems with applications to El Niño. *J. Phys. Oceanogr.,* 6: 632–645.

MCDONALD, W. F., 1938. Atlas of climatic charts of the oceans. *U.S. Weather Bur., Washington, D.C., Publ.,* 1247, 130 pp.

MCGUIRK, J. P. and REITER, E. R., 1976. A vacillation in atmospheric energy parameters. *J. Atmos. Sci.,* 33: 2079–2093.

MEINARDUS, W., 1923. *Meteorologische Ergebnisse der Deutschen Südpolar Expedition 1901–1923, Deut. Südpolar Exped. III.* Meteor. I. Reimer, Berlin, 578 pp.

MEINARDUS, W., 1929. Die Luftdruckverhältnisse und ihre Wandlungen südlich von 30° südl. Breite. *Meteorol. Z.,* 46: 41–49; 86–96.

MEINARDUS, W., 1934. Die Niederschlagsverteilung auf der Erde. *Meteorol. Z.,* 51: 345–350.

MEINARDUS, W., 1938. Klimakunde der Antarktis. In: W. KOEPPEN and R. GEIGER (Editors), *Handbuch der Klimatologie.* Bornträger, Berlin, IV(U), 133 pp.

MEINARDUS, W. and MECKING, L., 1911. Mittlere Isobarenkarten der höheren südlichen Breiten von Oktober 1901 bis März 1904. *Deutsche Südpolar Expedition 1901–1903, Meteorologischer Atlas.* G. Reimer, Berlin, 51 maps.

METEOROLOGICAL OFFICE, 1947. *Monthly Meteorological Charts of the Western Pacific Ocean.* M. O. 422, H. M. Stationery Office, London, 120 pp.

METEOROLOGICAL OFFICE, 1950. *Monthly Meteorological Charts of the Eastern Pacific Ocean.* M. O. 518, H. M. Stationery Office, London, 122 pp.

METEOROLOGICAL OFFICE, 1958. *Tables of Temperature, Relative Humidity, and Precipitation for the World, Part VI. Australia and the South Pacific Ocean.* M.O. 617f, H. M. Stationery Office, London, 54 pp.

MILLER, D. B. and FEDDES, R. G., 1971. *Global Atlas of Relative Cloud Cover 1967–1970 Based on Photographic Signals from Meteorological Satellites.* NOAA (NESS), USAF (Air Weather Service M.A.C.) joint production, Washington, D.C., 237 pp.

MÖLLER, F., 1951. Vierteljahrskarten des Niederschlags für die ganze Erde. *Petermanns Geogr. Mitt.,* 95: 1–7.

MOREL, P. and DESBOIS, M., 1974. Mean 200 mb circulation in the southern hemisphere deduced from EOLE balloon flights. *J. Atmos. Sci.,* 31: 394–407.

MUFFATTI, A. H. J., 1964. Aspects of the subtropical jet stream over Australia. In: J. W. HUTCHINGS (Editor), *Proc. Symp. Tropical Meteorology, Rotorua.* New Zealand Meteorol. Service, Wellington, pp. 72–88.

MURAKAMI, T., 1974. *Atmospheric Response to Heat Sources During July.* UHMET 74-04. Dept. Meteorol., Univ. Hawaii, Honolulu.

MURPHY, R. C., 1926. Oceanic and climatic phenomena along the west coast of South America during 1925. *Geograph. Rev.,* 16: 26–54.

MUSK, L. F., 1976. Rainfall variability and the Walker cell in the equatorial Pacific Ocean. *Weather,* 31: 34–47.

NAMIAS, J., 1950. The index cycle and its role in the general circulation. *J. Meteorol.*, 7: 130–139.

NAZAROV, V. S., 1962. L'dy antarkticheskikh Vod. Results of the IGY. Academy of Sciences, Soviet Geophysics Committee. *Okeanologiya Akad. Nauk SSSR, Moscow*, 6: 72 pp.

NEAL, A. B., 1972. The mean geostrophic flow at 200 mb in June 1970 between 20°S and 45°S. *Aust. Meteorol. Mag.*, 20: 231–236.

NEIBURGER, M., JOHNSON, O. S. and CHIEN, C. W., 1961. *Studies of the Structure of the Atmosphere over the Eastern Pacific Ocean in Summer, I. The Inversion over the Eastern North Pacific Ocean.* Univ. Calif. Press, Berkeley, Calif., 94 pp.

NEWELL, R. E., KIDSON, J. W., VINCENT, D. G. and BOER, G. J., 1972. *The General Circulation of the Tropical Atmosphere and Interactions with Extratropical Latitudes.* M.I.T. Press, Cambridge, Vol. 1, 258 pp.

NEWELL, R. E., KIDSON, J. W., VINCENT, D. G. and BOER, G. J., 1974. *The General Circulation of the Tropical Atmosphere and Interactions with Extratropical Latitudes.* M.I.T. Press, Cambridge, Vol. 2, 371 pp.

NEWTON, C. W., 1972. Southern hemisphere general circulation in relation to global energy and momentum balance requirements. In: C. W. NEWTON (Editor), *Meteorology of the Southern Hemisphere, Meteorological Monographs.* American Meteorol. Soc., Boston, 13: 215–240.

NEW ZEALAND METEOROLOGICAL SERVICE, 1973. Summaries of climatological observations to 1970. *N. Z. Meteorol. Serv. Misc. Pub.*, 143, 77 pp.

NEW ZEALAND METEOROLOGICAL SERVICE (undated). *Summaries of Climatological Observations (Table of Averages).* Meteorological Office, Laucala Bay, Suva, Fiji, 20 pp.

NICHOLLS, N., 1977. Tropical-extratropical interactions in the Australian region. *Mon. Weather Rev.*, 105: 826–832.

NILSSON, C. S., ANDREWS, J. C. and SCULLY, P. P,. 1977. Observations of eddy formation off east Australia. *J. Phys. Oceanogr.*, 7: 659–669.

NOWLIN, W. D., Jr., WHITWORTH,T. and PITTSBURG, R. D., 1977. Structure and transport of the Antarctic circumpolar current at Drake Passage from short term measurements. *J. Phys. Oceanogr.*, 7: 788–802.

OSTAPOFF, F., 1962. The salinity distribution at 200 metres and the Antarctic frontal zones. *Dtsch. Hydrograph. Z.*, 15: 133–142.

PALMÉN, E. and NEWTON, C. W., 1969. *Atmospheric Circulation Systems, Their Structure and Physical Interpretation.* Academic Press, New York, N.Y., 603 pp.

PALMER, C. E., 1942. Synoptic analysis over the Southern Oceans. *N. Z. Meteorol. Office, Prof. Note*, 1, 38 pp.

PALMER, C. E., 1951. Tropical meteorology. In: T. F. MALONE (Editor), *Compendium of Meteorology.* Am. Meteorol. Soc., Boston, pp. 859–880.

PALMER, C. E., 1952. Tropical meteorology. *Q. J. R. Meteorol. Soc.*, 78: 126–164.

PHILLPOT, H. R., 1967. *Selected Surface Climatic Data for Antarctic Stations.* Bur. Meteorol., Melbourne, 113 pp.

PHILLPOT, H. R., PRICE, P. G., NEAL, A. B. and LAJOIE, F. A., 1971. GARP basic data set analysis project, the first experiment November 1969. *Aust. Meteorol. Mag.*, 19: 48–81.

PICKARD, G. L., DONGUY J. R., HENIN, C. and ROUGERIE, F., 1977. A review of the physical oceanography of the Great Barrier Reef and Western Coral Sea. *Aust. Inst. Marine Sci., Monograph Ser.*, 2: 134 pp.

PIKE, A. C., 1968. A numerical study of the tropical circulations. *Air Force Cambridge Res. Lab. Sci. Rep.*, 68-0593, 129 pp.

PITTOCK, A. B., 1971. Rainfall and the general circulation. *Proc. Int. Conf. Weather Modification, 6–11 September 1971, Canberra, Australia*, pp. 330–338.

PITTOCK, A. B., 1975. Climatic change and the patterns of variation in Australian rainfall. *Search*, 6: 498–504.

POND, S., 1975. Exchanges of momentum, heat and moisture at the ocean-atmosphere interface. In: *Numerical Models of Ocean Circulation.* National Academy of Sciences, Washington, D.C., pp. 26–36.

POSNER, G. S., 1957. The Peru Current. *Bull. Bingham Oceanogr. Coll.*, 16: 106–155.

PRANTNER, G. D., 1962. *Cyclones and Anticyclones South of 50°S.* U.S. Navy Weather Research Facility, Norfolk, Va., 64 pp.

PREDOEHL, M. C., 1966. Antarctic pack ice boundaries established from Nimbus 1 pictures. *Science*, 153: 861–863.

PRIVETT, D. W., 1960. The exchange of energy between the atmosphere and the oceans of the southern hemisphere. *Meteorol. Office, London, Geophys. Mem.*, 13 (104), 61 pp.

QUINN, W. H., 1974. Monitoring and predicting El Niño invasions. *J. Appl. Meteorol.*, 13: 825–830.

QUINN, W. H. and BURT, W. V., 1972. Use of the Southern Oscillation in weather prediction. *J. Appl. Meteorol.*, 9: 20–28.

QUINN, W. H., BURT, W. V. and PAWLEY, W. M., 1969. A study of several approaches to computing surface insolation over tropical oceans. *J. Appl. Meteorol.*, 8: 205–212.

QUINN, W. H., ZOPF, D. O., SHORT, K. S. and KUO YANG, R. T. W., 1978. Historical trends and statistics of the Southern Oscillation, El Niño, and Indonesian droughts. *Fishery Bull.*, 76: 663–678.

RADOK, U., 1971. The Australian region and the general circulation of the Southern Hemisphere. In: J. GENTILLI (Editor), *Climates of Australia and New Zealand, World Survey of Climatology, Vol. 13.* Elsevier, Amsterdam, pp. 13–33.

RAMAGE, C. S., 1970. Meteorology of the South Pacific tropical and middle latitudes. In: W. S. WOOSTER (Editor), *Scientific Exploration of the South Pacific—Proceedings of a Symposium held during Ninth General Meeting of the Scientific Committee on Oceanic Research, 18–20 June 1968, Scripps Institution of Oceanography, La Jolla, Calif.* National Academy of Sciences, Washington, D.C., pp. 16–29.

RAMAGE, C. S., 1975. Preliminary discussion of the meteorology of the 1972–73 El Niño. *Bull. Am. Meteorol. Soc.*, 56: 234–242.

RAMAGE, C. S., 1977. Sea surface temperature and local weather. *Mon. Weather Rev.*, 105: 540–544.

REED, W. W., 1927. Climatological data for the tropical islands of the Pacific Ocean (Oceania). *Mon. Weather Rev.*, Suppl. No. 28, 28 pp.

REID, J. L., 1959. Evidence of a south equatorial countercurrent in the Pacific Ocean. *Nature*, 184: 209–210.

REID, J. L., 1961. On the geostrophic flow at the surface of the Pacific Ocean with respect to the 1,000-decibar surface. *Tellus*, 13: 489–502.

RIEHL, H., 1954. *Tropical Meteorology.* McGraw Hill, New York, N.Y., 392 pp.

RODEN, G. I., 1963. On sea level, temperature, and salinity variations in the central tropical Pacific and on Pacific Ocean islands. *J. Geophys. Res.*, 68: 455–472.

ROSENTHAL, S. L., 1960. A simplified linear theory of equatorial easterly waves. *J. Meteorol.*, 17: 484–488.

ROSSBY, C. G. and WILLETT, H. C., 1948. The circulation of the upper troposphere and lower stratosphere. *Science*, 108: 643–652.

ROWNTREE, P. R., 1972. The influence of tropical east Pacific Ocean temperatures on the atmosphere. *Q. J. R. Meteorol. Soc.*, 98: 290–321.

ROYAL AUSTRALIAN AIR FORCE, 1942. *Weather on the Australia Station: Local Information.* RAAF Publ. No. 252, Vol. II, Air Force Headquarters, Melbourne.

ROYAL METEOROLOGICAL SOCIETY, AUSTRALIAN BRANCH, 1977. *Proceedings of Two-Day Workshop on Fronts, 26–27 May 1977.* Bur. Meteorol., Melbourne.

ROYAL NEW ZEALAND AIR FORCE, 1943. *Climatological Notes: South Pacific Region.* New Zealand Meteorol. Office Ser. C, Directorate of Meteorological Services, Wellington.

RUBIN, M. J., 1955. An analysis of pressure anomalies in the southern hemisphere. *Notos*, 4: 11–16.

RUSSELL, H. C., 1893. Moving anticyclones in the southern hemisphere. *Q. J. R. Meteorol. Soc.*, 41: 1–11.

RUTHERFORD, G. T., 1966. The synoptic use of meteorological satellite data in sparse data regions. *Aust. Meteorol. Mag.*, 14: 133–151.

RUTHERFORD, G. T., 1969. Occlusion sequences south of Australia. In: *Satellite Meteorology. Proc. Inter-Regional Seminar on the Interpretation of Meteorological Satellite Data.* Bur. Meteorol., Melbourne, pp. 49–51.

SADLER, J. C., 1969. *Average Cloudiness in the Tropics from Satellite Observations.* East–West Center Press, Honolulu, 22 pp. and maps.

SADLER, J. C., 1972. *The Mean Winds of the Upper Troposphere over the Central and Eastern Pacific.* Tech. Rep., UHMET-72-04, Univ. of Hawaii, Honolulu, 29 pp.

SADLER, J. C., 1975. *The Upper Tropospheric Circulation over the Global Tropics.* Tech. Rep., UHMET-75-05, Dept. of Meteorol., Univ. of Hawaii, Honolulu, 35 pp.

SASAMORI, T., LONDON, J. and HOYT, D. V., 1972. Radiation budget of the southern hemisphere. In: C. W. NEWTON (Editor), *Meteorology of the Southern Hemisphere, Meteorological Monographs.* Am. Meteorol. Soc., Boston, 13: 9–23.

SCHELL, I. I., 1965. Origin and possible prediction of the fluctuations in the Peru Current and upwelling. *J. Geophys. Res.*, 70, 5529–5540.

SCHMITT, W., 1957. Synoptic meteorology of the Antarctic. In: M. P. VAN ROOY (Editor), *Meteorology of the Antarctic.* South African Weather Bureau, Pretoria, pp. 209–231.

SCHOTT, G., 1931. Der Peru-Strom und seine nördlichen Nachbargebiete unter normaler und abnormaler Ausbildung. *Ann. Dtsch. Hydrogr. U. Mar. Meteorol.*, 59: 161–169; 200–213; 240–253.

SCHOTT, G., 1932. The Humboldt Current in relation to land and sea condition on the Peruvian coast. *Geography*, 17: 87–98.

SCHOTT, G., 1935. *Geographie des Indischen und Stillen Ozeans.* C. Boysen, Hamburg, 413 pp.

SCHOTT, G., 1938. Klimakunde der Südsee-Inseln. In: W. KOEPPEN and R. GEIGER (Editors), *Handbuch der Klimatologie.* Bornträger, Berlin, Band IV (T), 114 pp.

SCHÜTTE, K., 1968. Untersuchungen zur Meteorologie und Klimatologie des El Niño Phänomens in Ecuador und Nord Peru. *Bonner Meteorol. Abhandl.*, 9, 152 pp.

SCHWEIGGER, E., 1942. Los fenomenos en el mar desde 1925 hasta 1941, en relacion con observaciones meteorologicas efectuadas en Puerto Chicama. [Phenomena in the sea from 1925 until 1941 in relation to meteorological observations recorded at Puerto Chicama.] In: *Tres estudios referentes a la oceanografica del Peru.* Jiron Junin, Lima, 64 pp.

SCHWERDTFEGER, W., 1970. The climate of the Antarctic. In: S. ORVIG (Editor), *Climates of the Polar Regions, World Survey of Climatology,* Vol. 14. Elsevier, Amsterdam, pp. 253 – 355.

SCHWERDTFEGER, W. (Editor), 1976. *Climates of Central and South America, World Survey of Climatology, Vol. 12.* Elsevier, Amsterdam, 532 pp.

SCHWERDTFEGER, W. and KACHELHOFFER, S. T., 1973. The frequency of cyclonic vortices over the southern ocean in relation to the extension of the pack ice belt. *Antarctic J. U.S.,* 8: 234.

SEELYE, C. J., 1943 – 1944. Climatic notes, South Pacific region. *N.Z. Meteorol. Office Ser. C,* Nos. 1 – 19.

SEELYE, C. J., 1950. Rainfall and its variability over the central and southwestern Pacific. *N.Z. J. Sci. Technol.,* 32: 11 – 14.

SEKIGUCHI, T., 1952. The rainfall distribution in the Pacific region. In: *Proc. Pacific Sci. Congr., Seventh Congr. III, Wellington,* pp. 101 – 102.

SIKDAR, D. N., YOUNG, J. A. and SUOMI, V. E., 1972. Time spectral characteristics of large scale cloud systems in the tropical Pacific. *J. Atmos. Sci.,* 29: 229 – 239.

SIMPSON, G. C., 1919. British Antarctic Expedition 1910 – 1913. *Meteorology, Vol. II, Weather Maps, and Pressure Curves.* Thacker, Spink & Co., Calcutta, 138 pp.

SOUTH AFRICAN WEATHER BUREAU, 1952 et seq. *Notos.*

SOUTH AFRICAN WEATHER BUREAU, 1962 – 66. *International Geophysical Year, World Weather Maps, Part III, Southern Hemisphere South of 20S: Daily Sea-level and 500 mb Charts.*

STRETEN, N. A., 1968a. Some aspects of high latitude southern hemisphere circulation as viewed by ESSA 3. *J. Appl. Meteorol.,* 7: 324 – 332.

STRETEN, N. A., 1968b. A note on multiple image photo mosaics for the southern hemisphere. *Aust. Meteorol. Mag.,* 16: 127 – 136.

STRETEN, N. A., 1968c. Characteristics of strong wind periods in coastal East Antarctica. *J. Appl. Meteorol.,* 7: 46 – 62.

STRETEN, N. A., 1969a. A case study of the winter circulation at 700 and 500 mb in middle and high southern latitudes. *Mon. Weather Rev.,* 97: 193 – 199.

STRETEN, N. A., 1969b. A note on the frequency of closed circulations between 50°S and 70°S in summer. *Aust. Meteorol. Mag.,* 17: 228 – 234.

STRETEN, N. A., 1970. A note on the climatology of the satellite observed zone of high cloudiness in the central South Pacific. *Aust. Meteorol. Mag.,* 18: 31 – 38.

STRETEN, N. A., 1973a. Some characteristics of satellite-observed bands of persistent cloudiness over the southern hemisphere. *Mon. Weather Rev.,* 101: 486 – 495.

STRETEN, N. A., 1973b. Satellite observations of the summer decay of the Antarctic sea ice. *Arch. Meteorol. Geophys. Bioklimatol.,* Ser. A, 22: 119 – 134.

STRETEN, N. A., 1975. Satellite derived inferences to some characteristics of the South Pacific circulation associated with the Niño event of 1972 – 73. *Mon. Weather Rev.,* 103: 989 – 995.

STRETEN, N. A., 1977. Seasonal climatic variability over the southern oceans. *Arch. Meteorol. Geophys. Bioklimatol.,* Ser. B, 25: 1 – 19.

STRETEN, N. A., 1978. A quasi periodicity in the motion of the South Pacific cloud band. *Mon. Weather Rev.,* 106: 1211 – 1214.

STRETEN, N. A., 1980. Some synoptic indices of the southern hemisphere mean sea level circulation 1972 – 1977. *Mon. Weather Rev.,* 108: 18 – 36.

STRETEN N. A. and KELLAS, W. R., 1973a. Aspects of cloud pattern signatures of depressions in maturity and decay. *J. Appl. Meteorol.,* 12: 23 – 27.

STRETEN, N. A. and KELLAS, W. R., 1973b. Some satellite observations of oceanic and continental contrasts in equatorial cloud features. *Sci. Rep., 19.* Commonwealth Meteorology Research Centre, Melbourne, 18 pp. and 11 figs.

STRETEN, N. A. and TROUP, A. J., 1973. A synoptic climatology of satellite observed cloud vortices over the southern hemisphere. *Q. J. R. Meteorol. Soc.,* 99: 56 – 72.

SULLIVAN, W., 1957. *Quest for a Continent.* Secker and Warburg, London, 372 pp.

SVERDRUP, H. U., JOHNSON, M. W. and FLEMING, R. H., 1942. *The Oceans: Their Physics, Chemistry and General Biology.* Prentice Hall, New York, 1087 pp.

TAFT, B. and JONES, J., 1973. Measurements of the equatorial undercurrent in the Eastern Pacific. In: B. A. WARREN (Editor), *Progress in Oceanography.* Pergamon Press, Oxford, 6: pp. 47 – 110.

TALJAARD, J. J., 1967. Development, distribution and movement of cyclones and anticyclones in the Southern Hemisphere during the IGY. *J. Appl. Meteorol.,* 6: 973 – 987.

TALJAARD, J. J., 1968. Climatic frontal zones of the Southern Hemisphere. *Notos*, 17: 23–34.

TALJAARD, J. J., 1972. Synoptic meteorology of the Southern Hemisphere. In: C. W. NEWTON (Editor), *Meteorology of the Southern Hemisphere, Meteorological Monographs*. Am. Meteorol. Soc., Boston, 13, pp. 39–213.

TALJAARD, J. J. and VAN LOON, H., 1962. Cyclogenesis, cyclones and anticyclones in the Southern Hemisphere during summer 1957–1958. *Notos*, 12: 37–50.

TALJAARD, J. J. and VAN LOON, H., 1964. Southern hemisphere weather maps for the International Geophysical Year. *Bull. Am. Meteorol. Soc.*, 45: 88–95.

TALJAARD, J. J., JENNE, R. L., VAN LOON, H. and CRUTCHER, H. L., 1968. Seasonal range, anomalies and other aspects of sea level pressure, isobaric height, temperature and dewpoint at selected levels in the Southern Hemisphere. *Notos*, 17: 63–140.

TALJAARD, J. J., SCHMITT, W. and VAN LOON, H., 1961. Frontal analysis with application to the Southern Hemisphere. *Notos*, 10: 25–58.

TALJAARD, J. J., VAN LOON, H., CRUTCHER, H. L. and JENNE, R. L., 1969. *Climate of the Upper Air, Southern Hemisphere*. NAVAIR, 50-1C-55, U.S. Navy, Washington, D.C., 1, 6 pp. and 134 figs.

TAYLOR, H. W., GORDON, A. L. and MOLINELLI, E., 1978. Climatic characteristics of the Antarctic polar front zone. *J. Phys. Oceanogr.*, 83: 4572–4578.

TAYLOR, R. C., 1973. *An Atlas of Pacific Islands Rainfall Data*. Rep. No. 25, HIG-73-9, Hawaii Institute of Geophysics, Univ. of Hawaii, Honolulu, 5 pp., figs., and tables.

TOLSTIKOV, E. I. (Editor), 1966. *Atlas Antarktiki, I*. G.U.C.K., Moscow (English translation: *Soviet Geography Reviews and Translations 1967*, 8: 5–6), American Geographical Society, New York, 225 pp.

TRELOAR, H. M. and NEWMAN, B. W., 1938. Weather conditions affecting aviation over the Tasman Sea. *Bur. Meteorol., Melbourne, Bull.*, 24, 34 pp.

TRENBERTH, K. E., 1975. A quasi biennial standing wave in the southern hemisphere and interrelations with sea surface temperature. *Q. J. R. Meteorol. Soc.*, 101: 55–74.

TRENBERTH, K. E., 1976a. Fluctuations and trends in indices of the southern hemisphere circulation. *Q. J. R. Meteorol. Soc.*, 102: 65–75.

TRENBERTH, K. E., 1976b. Spatial and temporal variations of the southern oscillation. *Q. J. R. Meteorol. Soc.*, 102: 639–653.

TRENBERTH, K. E. and NEALE, A. A., 1977. Numerical weather prediction in New Zealand. *Mon. Weather Rev.*, 105: 817–825.

TRESHNIKOV, A. F., 1967. The ice of the Southern Ocean. In: T. NAGATA (Editor), *Proc. Symp. on Pacific Antarctic Sciences*. National Science Museum, Tokyo, pp. 113–123.

TROUP, A. J., 1965. The southern oscillation. *Q. J. R. Meteorol. Soc.*, 91: 490–506.

TROUP, A. J. and STRETEN, N. A., 1972. Satellite-observed southern hemisphere cloud vortices in relation to conventional observations. *J. Appl. Meteorol.*, 11: 909–917.

TSUCHIYA, M., 1968. Upper waters of the intertropical Pacific Ocean. *Johns Hopkins Oceanogr. Stud.*, 4, 50 pp.

TSUCHIYA, M., 1970. Equatorial circulation of the South Pacific. In: W. S. WOOSTER (Editor), *Scientific Exploration of the South Pacific. Proceedings of a Symposium held during the Ninth General Meeting of the Scientific Committee on Oceanic Research, 18–20 June 1968, Scripps Institution of Oceanography, La Jolla, Calif.*, pp. 69–74.

TUCKER, G. B., 1961. Precipitation over the North Atlantic Ocean. *Q. J. R. Meteorol. Soc.*, 87: 147–158.

U.S. DEPARTMENT OF COMMERCE (monthly). *Monthly Climatic Data for the World*. Superintendent of Documents, Government Printing Office, Washington, D.C.

UNITED STATES NAVY, 1957. *Marine Climatic Atlas of the World, Vol. 3. Indian Ocean*. NAVAER 50-1C-530. Government Printing Office, Washington, D.C.

UNITED STATES NAVY, 1959. *Marine Climatic Atlas of the World, Vol. 5. South Pacific Ocean*. NAVAER 50-1C-532, Government Printing Office, Washington, D.C.

UNITED STATES NAVY, 1965. *Marine Climatic Atlas of the World, Vol. 7. Antarctic*. Government Printing Office, Washington, D.C.

UNITED STATES NAVY, 1969. *Marine Climatic Atlas of the World, Vol. 8. The World*. Government Printing Office, Washington, D.C.

UNITED STATES NAVY, 1979. *Marine Climatic Atlas, Vol. 5. South Pacific Ocean* (revised). NAVAIR 50-1C-532, Government Printing Office, Washington, D.C.

U.S.S.R. MINISTRY OF DEFENCE, 1974. *Ocean Atlas: Pacific Ocean*. USSR Ministry of Defence, Moscow, 14 pp. and 302 charts.

VAN HAMME, J. L., 1970. Mise en évidence de differences fondamentales entre la circulation de l'atmosphere arctique et celle de l'atmosphere antarctique. *Beitr. Phys. Atmos.*, 43: 150–170.

VAN LOON, H., 1955. Mean air temperature over the southern oceans. *Notos*, 4: 292–308.

VAN LOON, H., 1956. Blocking action in the Southern Hemisphere. *Notos*, 5: 171–175.

VAN LOON, H., 1960. Features of the atmospheric circulation in the South Pacific Ocean during the whaling seasons 1955–1956 and 1956–1957. In: *Antarctic Meteorology*. Pergamon, Oxford, pp. 477–487.

VAN LOON, H., 1962. On the movement of lows in the Ross and Weddell Seas sectors in summer. *Notos*, 11: 47–50.

VAN LOON, H., 1965. A climatological study of the atmospheric circulation in the Southern Hemisphere, Part I: 1 July 1957–31 March 1958. *J. Appl. Meteorol.*, 4: 479–491.

VAN LOON, H., 1966. On the annual temperature range over the southern oceans. *Geograph. Rev.*, 57: 497–515.

VAN LOON, H., 1967a. The half yearly oscillations in middle and high southern latitudes and the coreless winter. *J. Atmos. Sci.*, 24: 472–486.

VAN LOON, H., 1967b. A climatological study of the atmospheric circulation in the Southern Hemisphere during the IGY, Part II. *J. Appl. Meteorol.*, 6: 803–815.

VAN LOON, H., 1971. On the interaction between Antarctica and middle latitudes. In: L. QUAM (Editor), *Research in the Antarctic*. American Association for the Advancement of Science, Washington, D.C., pp. 477–487.

VAN LOON, H., 1972. (i) Pressure in the Southern Hemisphere. (ii) Cloudiness and precipitation in the Southern Hemisphere. In: C. W. NEWTON (Editor), *Meteorology of the Southern Hemisphere, Meteorological Monographs*. Am. Meteorol. Soc., Boston, 13, (i) pp. 59–86, and (ii) pp. 101–111.

VAN LOON, H. and JENNE, R. L., 1972. The zonal harmonic standing waves in the Southern Hemisphere. *J. Geophys. Res.*, 77: 992–1003.

VAN LOON, H. and JENNE, R. L., 1974. Standard deviations of monthly mean 500 and 100 mb heights in the Southern Hemisphere. *J. Geophys. Res.*, 79: 5561–5564.

VAN LOON, H. and MADDEN, R. A., 1981. The Southern Oscillation, Part I. Global associations with pressure and temperature in the northern winter. *Mon. Weather Rev.*, 109: 1150–1162.

VAN LOON, H. and ROGERS, J. C., 1981. The Southern Oscillation, Part II. Associations with changes in the middle troposphere in the northern winter. *Mon. Weather Rev.*, 109: 1163–1168.

VAN LOON, H., TALJAARD, J. J., JENNE, R. L. and CRUTCHER, H. L., 1971. *Climate of the Upper Air: Southern Hemisphere, Vol. II. Zonal Geostrophic Winds*. NCAR TN/STR-57 and NAVAIR 50-1C-56, National Center for Atmospheric Research, Boulder, Colorado, 43 pp.

VAN LOON, H., TALJAARD, J. J., SASAMORI, T., LONDON, J., HOYT, D. V., LABITZKE, K. and NEWTON, C. W., 1972. *Meteorology of the Southern Hemisphere, Meteorological Monographs*. American Meteorol. Soc., Boston, 13, 263 pp.

VIEBROCK, H., 1962. The transfer of energy between the ocean and the atmosphere in the Antarctic region. *J. Geophys. Res.*, 67: 4293–4302.

VISHER, S. S. and HODGE, D., 1925. Australian hurricanes and related storms. *Bur. Meteorol., Melbourne, Bull.*, 16, 54 pp.

VONDER HAAR, T. H. and ELLIS, J. S., 1974. *Atlas of Radiation Budget Measurements from Satellites (1962–70)*. Atmos. Sci. Pap., No. 231, Colorado State Univ., Dept. of Atmospheric Sciences, Fort Collins, Colo., 180 pp.

VONDER HAAR, T. H. and SUOMI, V. E., 1971. Measurements of the earth's radiation budget from satellites during a five-year period, Part I. Extended time and space means. *J. Atmos. Sci.*, 28: 305–314.

VOWINCKEL, E., 1957. Climate of the Antarctic Ocean. In: M. P. VAN ROOY (Editor), *Meteorology of the Antarctic*. Weather Bureau, Pretoria, pp. 91–110.

WALKER, G. T., 1923. A preliminary study of world weather-correlation in seasonal variations of weather, VIII. *Mem. Indian Meteorol. Dept.*, 24: 75–131.

WALKER, G. T., 1924. Correlation in seasonal variation of weather, IX. *Mem. Indian Meteorol. Dept.*, 24: 275–332.

WALKER, G. T., 1928a. Ceará (Brazil) famines and the general air movement. *Beitr. Phys. Freien Atmos.*, 14: 88–93.

WALKER, G. T., 1928b. World Weather III. *Mem. R. Meteorol. Soc.*, 2(17): 97–106.

WALKER, G. T., 1928c. World Weather—an address. *Q. J. R. Meteorol. Soc.*, 54: 79–87.

WALKER, G. T. and BLISS, E. W., 1932. World Weather V. *Mem. R. Meteorol. Soc.*, 4(36): 53–84.

WALKER, G. T. and BLISS, E. W., 1937. World Weather VI. *Mem. R. Meteorol. Soc.*, 4(39): 119–138.

WALLACE, J. M. and CHANG, C. P., 1969. Spectrum analysis of large scale disturbances in the tropical lower troposphere. *J. Atmos. Sci.*, 26: 1010–1025.

WARREN, B. A., 1970. General circulation of the South Pacific. In: W. S. WOOSTER (Editor), *Scientific Exploration of the South Pacific: Proceedings of a Symposium held during the Ninth General Meeting of the*

Scientific Committee on Oceanic Research, 18 – 20 June 1968, Scripps Institution of Oceanography, La Jolla, Calif. National Academy of Sciences, Washington, D.C., pp. 33 – 49.

WEARE, B. C., NAVATO, A. R. and NEWELL, R. E., 1976. Empirical orthogonal analysis of Pacific sea surface temperatures. *J. Phys. Oceanogr.,* 6: 671 – 678.

WEBSTER, P. J., 1973. Temporal variation of low latitude zonal circulations. *Mon. Weather Rev.,* 101: 803 – 816.

WEBSTER, P. J. and CURTIN, D. G., 1975. Interpretation of the EOLE experiment II. Spatial variation of transient and stationary modes. *J. Atmos. Sci.,* 32: 1848 – 1863.

WEBSTER, P. J. and KELLER, J. L., 1975. Atmospheric variations, vacillations and index cycles. *J. Atmos. Sci.,* 32: 1283 – 1300.

WEINERT, R. A., 1968. Statistics of the subtropical jet stream over the Australian region. *Aust. Meteorol. Mag.,* 16: 137 – 148.

WEXLER, H., 1959. The Antarctic convergence—or divergence. In: B. BOLIN (Editor), *The Atmosphere and the Sea in Motion.* (Scientific Contributions to the Rossby Memorial Volume). The Rockefeller Institute Press, New York, N.Y., pp. 107 – 120.

WHITE, W. B. and McCREARY, J. P., 1974. Eastern intensification of ocean spin-down: with application to El Niño. *J. Phys. Oceanogr.,* 4: 295 – 303.

WILLIAMS, K. T. and GRAY, W. M., 1973. A statistical analysis of satellite-observed trade wind cloud clusters in the western North Pacific. *Tellus,* 21: 323 – 336.

WINSTON, J. S., GRUBER, A., GRAY, T. I., VARNADORE, M. S., EARNEST, C. L. and MANNELLO, L. P., 1979. *Earth – Atmosphere Radiation Budget Analyses Derived from NOAA Satellite Data June 1974 – February 1978.* U.S. Department of Commerce, National Oceanic and Atmospheric Administration, National Environmental Satellite Service, Washington, D.C., August 1974, 8 pp. and diagrams.

WOOSTER, W. S., 1960. El Niño. *Calif. Coop. Oceanic Fish Invest. Rep.,* 7: 34 – 45.

WOOSTER, W. S., 1966. Peru Current. In: R. W. FAIRBRIDGE (Editor), *The Encyclopedia of Oceanography.* Reinhold, New York, N.Y., pp. 695 – 698.

WOOSTER, W. S., 1970. Eastern boundary currents in the South Pacific. In: W. S. WOOSTER (Editor), *Scientific Exploration of the South Pacific: Proceedings of a Symposium held during the Ninth General Meeting of the Scientific Committee on Oceanic Research, 18 – 20 June 1968, Scripps Institution of Oceanography, La Jolla, Calif.* National Academy of Sciences, Washington, D.C., pp. 60 – 68.

WOOSTER, W. S. and GUILLEN, O., 1974. Characteristics of El Niño in 1972. *J. Marine Res.,* 32: 387 – 404.

WOOSTER, W. S. and HEDGPETH, J. W., 1966. Oceanographic settings of the Galápagos. In: R. I. BOWMAND (Editor), *The Galápagos: Proc. Galápagos International Scientific Project.* Univ. of California Press, Berkeley, pp. 100 – 107.

WOOSTER, W. S. and JENNINGS, F., 1955. *Exploratory Oceanographic Observations in the Eastern Tropical Pacific—January to March 1953.* Univ. of Calif., Scripps Institution of Oceanography, La Jolla, Calif., pp. 163 – 174.

WOOSTER, W. S. and REID, J. L., 1963. Eastern boundary currents. In: M. N. HILL (Editor), *The Sea: Ideas and Observations on Progress in the Study of the Seas.* Interscience Publishers, New York, N.Y., pp. 253 – 276.

WRIGHT, A. D. F., 1974. Blocking action in the Australian region. *Bur. Meteorol., Melbourne, Tech. Rep.,* 10, 29 pp.

WRIGHT, P. B., 1975. *An Index of the Southern Oscillation.* Climatic Research Unit, Rep. CRU RP 4, Univ. of East Anglia, Norwich, 22 pp.

WYRTKI, K., 1963. The horizontal and vertical field of motion in the Peru Current. *Bull. Scripps Inst. Oceanogr.,* 8: 313 – 346.

WYRTKI, K., 1964. The thermal structure of the eastern Pacific Ocean. *Dtsch. Hydrograph. Z., Ergänzungsh.,* A6: 84 pp.

WYRTKI, K., 1965. Surface currents of the eastern tropical Pacific Ocean. *Inter-American Tropical Tuna Commission Bull.,* 9: 271 – 304.

WYRTKI, K., 1966a. Oceanography of the eastern equatorial Pacific Ocean. *Oceanogr. Mar. Biol. Ann. Rev.,* 4: 33 – 68.

WYRTKI, K., 1966b. *Seasonal Variation of Heat Exchange and Surface Temperature in the North Pacific Ocean.* HIG-66-3, Hawaii Institute of Geophysics, Univ. Hawaii, Honolulu, 8 pp. and figs.

WYRTKI, K., 1974. *The Dynamic Topography of the Pacific Ocean and its Fluctuations.* HIG-74-5, Hawaii Institute of Geophysics, Univ. Hawaii, Honolulu, 19 pp. plus figs. and tables.

WYRTKI, K., 1975a. Fluctuations of the dynamic topography in the Pacific Ocean. *J. Phys. Oceanogr.,* 5: 450 – 459.

WYRTKI, K., 1975b. El Niño—the dynamic response of the equatorial Pacific Ocean to atmosphere forcing. *J. Phys. Oceanogr.,* 5: 572 – 584.

WYRTKI, K., 1979. The response of sea surface topography to the 1976 El Niño. *J. Phys. Oceanogr.*, 9: 1223–1231.

WYRTKI, K. and BENNETT, E. B., 1963. Vertical eddy viscosity in the Pacific equatorial undercurrent. *Deep-Sea Res.*, 10: 449–455.

WYRTKI, K. and MEYERS, G., 1975. *The Trade Wind Field over the Pacific Ocean, 1. The Mean Field and the Mean Annual Variation.* HIG-75-1, Hawaii Institute of Geophysics, Univ. Hawaii, Honolulu, 26 pp. and figs.

WYRTKI, K., STROUP, E., PATZERT, W., WILLIAMS, R. and QUINN, W., 1976. Predicting and observing El Niño. *Science*, 191: 343–346.

YANAI, M. and NITTA, T., 1967. Computation of vertical motion and vorticity budget in a Caribbean easterly wave. *J. Meteorol. Soc. Jpn.*, 45: 444–466.

YASUNARI, T., 1977. Stationary waves in the Southern Hemisphere mid-latitude zone revealed from average brightness charts. *J. Meteorol. Soc. Jpn.*, 55: 274–285.

YOUNG, J. A. and SIKDAR, D. N., 1973. A filtered view of fluctuating cloud patterns in the tropical Pacific. *J. Atmos. Sci.*, 30: 392–407.

ZILLMAN, J. W., 1967. The surface radiation balance in high southern latitudes. In: *Polar Meteorology, WMO Tech. Note*, 87: 142–171.

ZILLMAN, J. W., 1969. Interpretation of satellite data over the Southern Ocean using the technique of Martin. In: *Satellite Meteorology. Proceedings of the Inter-regional Seminar on the Interpretation of Meteorological Satellite Data.* Bur. Meteorol. Melbourne, pp. 43–48.

ZILLMAN, J. W., 1970a. Sea surface temperature gradients south of Australia. *Aust. Meteorol. Mag.*, 18: 22–30.

ZILLMAN, J. W., 1970b. *The GARP Basic Data Set Analysis Project, 1. November Sea Surface Temperature Distribution in the Southern Hemisphere.* Working Paper No. 132, Bur. Meteorol., Melbourne, 33 pp.

ZILLMAN, J. W., 1972a. Solar radiation and sea-air interaction south of Australia. In: D. E. HAYES (Editor), *Antarctic Oceanology II: The Australian—New Zealand Sector. Antarctic Res. Ser., Vol. 19.* American Geophysical Union, Washington, D.C., pp. 11–40.

ZILLMAN, J. W., 1972b. *A Study of Some Aspects of the Radiation and Heat Budgets of the Southern Hemisphere Oceans. Meteorological Study 26.* Bur. Meteorol., Australian Government Publishing Service, Canberra, 562 pp.

ZILLMAN, J. W., 1972c. *Isentropically Time-averaged Mass Circulations in the Southern Hemisphere.* Ph.D. Thesis, Univ. Wisc., Madison, Wisc., 205 pp.

ZILLMAN, J. W., 1974a. *The Influence of Oceanic and Atmospheric Fronts on Sea—Air Energy Exchange South of Australia.* IAMAP/IAPSO Combined First Special Assemblies, Melbourne, January 1974. Abstracts p. AS-II-5.

ZILLMAN, J. W., 1974b. *Extension of the Equatorial Walker Circulation into High Latitudes of the Southern Hemisphere.* IAMAP/IAPSO Combined First Special Assemblies, Melbourne, January 1974, Abstracts p. US-VI-7.

ZILLMAN, J. W. and DINGLE, W. R. J., 1969. Southern Ocean sea-air energy exchange. *Aust. Meteorol. Mag.*, 17: 166–172.

ZILLMAN, J. W. and DINGLE, W. R. J., 1973. Meteorology. *Antarctic J. U. S.*, 8: 111–119.

ZILLMAN, J. W. and JOHNSON, D. R., 1983. Thermally forced mean mass circulations in the southern hemisphere. (Submitted for publication).

ZILLMAN, J. W. and MARTIN, D. W., 1968. A sharp cold frontal passage at Macquarie Is. in the Southern Ocean. *J. Appl. Meteorol.*, 7: 708–711.

ZILLMAN, J. W. and PRICE, P. G., 1972. On the thermal structure of mature southern ocean cyclones. *Aust. Meteorol. Mag.*, 20: 34–48.

Appendix — Climatic tables

The sources listed below have been used in preparing climatic tables for the various numbered marine and island stations shown in Fig. 2. These numbers are also used in the title of the tables. The source for each item is listed by an identifying letter. Because the periods of observation and the amount of data are quite variable, the climate tables, particularly those for the marine stations, may differ significantly from the map representations shown in the text.

A: U.S. Department of Commerce—Environmental Data Service, 1968.
World Weather Records 1951 — 60, Volume 6.
Antarctica, Australia, Oceanic Islands and Ocean Weather Stations.

A_1: U.S. Department of Commerce.
World Weather Records 1941 — 50, Antarctica, Australia, Oceanic Islands and Ocean Weather Stations.

B: World Meteorological Organization, 1971.
Climatological Normals CLINO for Climate and Climate Ship Stations for the Period 1931 — 1960, WMO/OMM No. 117 TP 52.

C: New Zealand Meteorological Service (Dept. Civil Aviation), 1966.
Summaries of Climatological Observations at New Zealand Stations to 1960. New Zealand Meteorological Service Misc. Pub. 122.

D: New Zealand Meteorological Service (Department of Transport), 1973. Temperature Normals 1941 to 1970—Stations in New Zealand, outlying islands, Cook Islands, Nive, and Western Samoa (mean daily maximum, mean daily minimum, and mean air temperatures). New Zealand Meteorological Service Misc. Pub. 149.

E: Taylor, R. C., 1973. *An Atlas of Pacific Islands Rainfall.*
Data Report No. 25, Dept. Meteorology, University of Hawaii.

F: Service Météorologique de la Polynésie Française, 1965.
Climatologie de l'Ile de Tahiti—surface—altitude.
Monographies de la Météorologie Nationale, No. 44.

G: Girard, J. and Rignot, D., 1971. *Climatologie de la Nouvelle Caledonie.* Monographie de la Météorologie Nationale, No. 82.

H: Haggard, W., personal communication, 1974. Tables assembled by Mr. Haggard from the records of the National Climatic Center, Asheville, North Carolina, U.S.A.

I: National Center for Atmospheric Research, Boulder, Colorado, U.S.A.

J: Bureau of Meteorology, Melbourne, Australia.

K: D'Hauteserre, M. A., 1960. *Les perturbations atmosphériques et le climat de la Polynésie Française.* Monographies de la Météorologie Nationale, No. 18.

TABLE I

CLIMATIC TABLE FOR MARINE AREA 1
Latitude 5.5°S, longitude 144.5°W, elevation 0 m, source H

Month	Mean press. (mbar)	Mean air temp. (°C)	Temp. extremes (°C)		Freq. precip. (%)	Most freq. wind dir.	Mean sea temp. (°C)
			max.	min.			
Jan.	1008.4	26.7	29.4	24.4	0.0	E	26.7
Feb.	1008.6	27.0	31.1	25.0	0.0	E	27.1
Mar.	1009.2	27.1	29.4	25.0	2.4	E	27.5
Apr.	1009.0	27.6	30.0	26.1	3.1	E	27.9
May	1009.6	27.5	28.3	26.1	0.0	E	27.4
June	1010.0	27.1	28.9	25.0	1.6	E	27.3
July	1009.6	26.8	28.3	25.6	2.2	E	27.0
Aug.	1010.5	26.4	27.8	23.3	1.4	E	26.5
Sep.	1010.2	26.3	28.3	25.0	0.0	E	26.2
Oct.	1010.8	26.2	27.2	24.4	0.0	E	26.4
Nov.	1009.5	26.3	28.9	25.0	0.0	E	26.5
Dec.	1008.9	26.2	29.4	22.2	0.0	E	26.4

TABLE II

CLIMATIC TABLE FOR MARINE AREA 2
Latitude 5.5°S, longitude 129.0°W, elevation 0 m, source H

Month	Mean press. (mbar)	Mean air temp. (°C)	Temp. extremes (°C)		Freq. precip. (%)	Most freq. wind dir.	Mean sea temp. (°C)
			max.	min.			
Jan.	1011.2	26.1	29.4	21.7	0.0	E	25.7
Feb.	1011.5	26.0	29.4	23.3	1.6	E	25.6
Mar.	1009.4	26.4	30.0	23.9	10.3	E	26.2
Apr.		26.7	28.9	25.0	7.4	SE	26.6
May	1011.8	27.0	29.4	25.6	3.7	E	27.4
June	1011.5	26.9	30.0	24.4	0.0	E	26.7
July	1011.8	26.3	28.9	23.9	0.0	E	25.9
Aug.	1012.8	25.2	27.8	23.3	3.6	E	25.4
Sep.	1012.8	25.5	27.8	23.9	0.0	E	25.3
Oct.	1012.9	25.5	28.3	23.3	0.0	E	25.3
Nov.	1011.9	25.6	29.4	23.3	0.0	E	25.2
Dec.	1011.6	25.4	29.4	23.3	0.0	E	25.0

TABLE III

CLIMATIC TABLE FOR MARINE AREA 3
Latitude 1.5°S, longitude 106.5°W, elevation 0 m, source H

Month	Mean press. (mbar)	Mean air temp. (°C)	Temp. extremes (°C)		Freq. precip. (%)	Most freq. wind dir.	Mean sea temp. (°C)
			max.	min.			
Jan.	1010.7	23.7	27.8	20.0	1.2	SE	23.2
Feb.	1010.5	25.4	29.4	22.8	14.5	SE	25.2
Mar.	1010.4	26.9	32.2	24.4	5.3	SE	26.2
Apr.	1009.8	26.6	28.9	21.7	11.9	SE	26.1
May	1011.2	25.0	27.8	22.8	0.0	SE	25.3
June	1012.0	24.4	27.8	22.8	0.0	SE	23.9
July	1012.3	22.3	25.6	20.0	0.0	SE	22.2
Aug.	1013.2	22.3	26.7	18.3	9.3	SE	21.6
Sep.	1012.9	22.1	27.2	18.9	8.2	SE	21.7
Oct.	1013.1	22.2	26.1	18.3	3.5	SE	21.7
Nov.	1012.3	22.2	25.6	19.4	2.4	SE	22.1
Dec.	1011.1	23.4	28.9	20.6	8.1	SE	22.7

TABLE IV

CLIMATIC TABLE FOR MARINE AREA 4
Latitude 7.5°S, longitude 97.5°W, elevation 0 m, source H

Month	Mean press. (mbar)	Mean air temp. (°C)	Temp. extremes (°C)		Freq. precip. (%)	Most freq. wind dir.	Mean sea temp. (°C)
			max.	min.			
Jan.	1011.6	23.7	26.7	21.1	1.9	SE	23.8
Feb.	1011.8	25.1	28.9	22.2	3.0	SE	25.3
Mar.	1010.6	26.1	30.6	22.8	6.3	SE	26.3
Apr.	1010.7	26.2	30.6	22.8	1.7	SE	26.6
May	1012.1	25.7	29.4	22.2	0.9	SE	25.8
June	1012.8	24.4	27.2	22.2	3.2	SE	24.8
July	1013.3	22.9	26.1	19.4	6.9	SE	23.3
Aug.	1013.8	21.6	24.4	18.9	1.9	SE	22.1
Sep.	1013.7	21.4	23.9	18.9	0.0	SE	21.8
Oct.	1013.5	21.6	25.6	18.9	3.0	SE	21.8
Nov.	1013.0	21.9	26.1	19.4	1.3	SE	22.1
Dec.	1012.8	22.6	26.1	20.0	0.5	SE	22.7

TABLE V

CLIMATIC TABLE FOR MARINE AREA 5
Latitude 13.5°S, longitude 163.5°E, elevation 0 m, source H

Month	Mean press. (mbar)	Mean air temp. (°C)	Temp. extremes (°C)		Freq. precip. (%)	Most freq. wind dir.	Mean sea temp. (°C)
			max.	min.			
Jan.	1008.8	28.2	30.6	26.1	0.0	SE	28.2
Feb.	1006.1	28.3	31.1	25.0			28.6
Mar.	1007.3	28.2	30.6	25.6	7.4	SE	28.6
Apr.	1009.5	27.4	29.4	23.3	0.0	E	28.2
May	1012.3	27.1	30.0	25.0	4.3	SE	27.6
June	1012.0	26.5	28.9	22.8	2.3	SE	26.6
July	1012.0	26.6	28.9	23.9	30.8	SE	26.9
Aug.	1012.6	26.2	28.3	23.9	12.5	SE	26.9
Sep.	1013.2	26.4	28.3	24.4	9.1	E	26.4
Oct.	1012.0	26.4	28.9	23.9	3.7	E	26.6
Nov.	1010.9	27.0	29.4	25.0	14.3	SE	27.6
Dec.	1010.4	27.5	29.4	23.9	12.1	SE	27.5

TABLE VI

CLIMATIC TABLE FOR MARINE AREA 6
Latitude 24.5°S, longitude 169.5°E, elevation 0 m, source H

Month	Mean press. (mbar)	Mean air temp. (°C)	Temp. extremes (°C)		Freq. precip. (%)	Most freq. wind dir.	Mean sea temp. (°C)
			max.	min.			
Jan.	1010.6	24.9	28.9	21.7	8.5	E	25.5
Feb.	1010.8	24.9	28.3	21.7	9.4	E	25.6
Mar.	1012.2	24.4	27.2	20.6	10.8	E	25.5
Apr.	1012.8	23.9	28.3	20.6	7.6	E	24.6
May	1013.6	22.4	26.1	17.2	6.6	SE	23.4
June	1015.0	21.9	26.1	17.2	6.6	E	22.9
July	1015.0	20.2	23.9	17.2	4.0	E	21.3
Aug.	1016.8	19.8	23.3	16.7	5.0	SE	21.1
Sep.	1017.0	20.1	23.3	17.2	3.1	E	21.1
Oct.	1015.7	21.7	27.2	17.8	3.8	SE	22.1
Nov.	1013.8	22.8	27.8	18.9	0.0	NE	22.8
Dec.	1012.2	23.5	26.7	20.6	4.9	E	24.1

TABLE VII

CLIMATIC TABLE FOR MARINE AREA 7
Latitude 12.5°S, longitude 177.5°W, elevation 0 m, source H

Month	Mean press. (mbar)	Mean air temp. (°C)	Temp. extremes (°C) max.	min.	Freq. precip. (%)	Most freq. wind dir.	Mean sea temp. (°C)
Jan.	1005.3	28.1	32.2	25.6	7.9	E	28.8
Feb.	1007.6	27.8	32.8	23.9	12.4	E	28.6
Mar.	1008.4	27.8	30.0	25.0	9.0	E	28.7
Apr.	1009.0	28.1	31.7	23.9	3.0	E	28.8
May	1010.1	27.8	32.2	23.9	7.9	E	28.6
June	1013.3	27.4	32.8	21.1	8.1	E	27.9
July	1010.1	27.1	29.4	23.9	1.3	E	27.6
Aug.	1010.8	26.9	30.0	23.9	5.1	E	27.4
Sep.	1011.3	27.3	31.7	24.4	7.0	E	27.5
Oct.	1011.7	27.3	32.8	22.2	7.1	E	27.9
Nov.	1008.7	27.8	32.8	23.9	8.9	SE	28.4
Dec.	1007.4	27.7	32.8	25.0	14.5	E	28.6

TABLE VIII

CLIMATIC TABLE FOR MARINE AREA 8
Latitude 24.5°S, longitude 170.0°W, elevation 0 m, source H

Month	Mean press. (mbar)	Mean air temp. (°C)	Temp. extremes (°C) max.	min.	Freq. precip. (%)	Most freq. wind dir.	Mean sea temp. (°C)
Jan.	1010.5	25.4	29.4	22.2	12.5	SE	25.4
Feb.	1010.9	25.1	29.4	22.2	18.6	SE	25.7
Mar.	1009.8	25.0	27.8	22.2	19.5	SE	25.4
Apr.	1016.1	24.3	27.8	22.2	8.1	E	24.7
May	1014.2	22.9	25.6	20.0	14.6	S	24.2
June	1015.3	21.7	25.0	18.9	13.0	SE	22.7
July	1015.0	20.4	25.0	17.2	16.4	SE	22.4
Aug.	1016.9	19.6	23.3	16.1	15.2	E	20.7
Sep.	1016.1	20.5	24.4	17.2	12.2	E	21.2
Oct.	1013.7	21.4	24.4	18.3	10.1	E	21.5
Nov.	1013.9	23.1	28.9	18.9	9.6	E	23.3
Dec.	1010.7	24.1	28.9	21.1	13.5	SE	24.3

TABLE IX

CLIMATIC TABLE FOR MARINE AREA 9
Latitude 10.5°S, longitude 164.0°W, elevation 0 m, source H

Month	Mean press. (mbar)	Mean air temp. (°C)	Temp. extremes (°C)		Freq. precip. (%)	Most freq. wind dir.	Mean sea temp. (°C)
			max.	min.			
Jan.		28.0	31.1	25.6	9.8	NE	29.3
Feb.	1009.2	28.7	31.1	26.7	8.3	E	28.8
Mar.		28.1	31.1	25.0	12.2	E	28.9
Apr.	1008.7	28.8	31.1	26.1	4.5	E	29.1
May		28.6	33.3	25.0	4.0	SE	29.7
June	1010.9	28.6	32.2	25.0	2.1	SE	28.6
July	1010.8	27.9	31.7	23.3	6.5	SE	28.4
Aug.	1009.8	27.6	30.0	25.0	8.1	E	27.7
Sep.	1011.6	28.2	31.7	25.0	3.8	E	28.6
Oct.	1009.3	28.0	30.6	26.1	4.8	E	28.4
Nov.		28.4	31.7	25.6	7.8	NE	28.7
Dec.	1009.4	28.2	32.8	24.4	8.6	NE	28.6

TABLE X

CLIMATIC TABLE FOR MARINE AREA 10
Latitude 18.5°S, longitude 154.0°W, elevation 0 m, source H

Month	Mean press. (mbar)	Mean air temp. (°C)	Temp. extremes (°C)		Freq. precip. (%)	Most freq. wind dir.	Mean sea temp. (°C)
			max.	min.			
Jan.	1010.0	27.0	30.6	23.9	13.7	NE	27.6
Feb.	1009.7	27.6	32.2	23.3	4.2	E	27.9
Mar.	1010.1	27.6	32.8	24.4	4.0	SE	27.9
Apr.	1010.8	26.4	31.1	23.3	3.2	SE	27.3
May	1012.6	25.9	30.6	22.8	6.5	SE	26.6
June	1013.3	25.1	29.4	21.7	8.0	SE	26.2
July	1014.0	24.6	30.6	20.6	4.0	E	25.4
Aug.	1014.7	24.6	29.4	20.6	3.0	SE	25.3
Sep.	1014.3	25.1	28.9	21.7	2.1	E	25.2
Oct.	1013.9	25.0	30.0	21.7	6.5	E	25.5
Nov.	1012.2	25.7	32.2	22.2	5.6	E	26.2
Dec.	1010.1	26.4	31.1	23.3	13.5	E	26.7

TABLE XI

CLIMATIC TABLE FOR MARINE AREA 11
Latitude 23.5°S, longitude 128.0°W, elevation 0 m, source H

Month	Mean press. (mbar)	Mean air temp. (°C)	Temp. extremes (°C)		Freq. precip. (%)	Most freq. wind dir.	Mean sea temp. (°C)
			max.	min.			
Jan.	1015.9	25.4	31.1	20.6	6.7	E	25.7
Feb.	1016.2	26.1	31.7	22.2	5.1	E	26.3
Mar.	1015.8	26.0	32.8	21.7	9.5	E	26.3
Apr.	1014.7	24.9	29.4	20.6	16.9	E	25.7
May	1016.0	23.9	28.3	16.1	14.1	E	24.7
June	1017.4	22.7	27.2	17.2	14.3	E	23.7
July	1018.2	22.3	28.3	17.8	12.9	E	23.0
Aug.	1018.5	21.8	27.2	17.2	6.0	E	22.5
Sep.	1018.2	21.7	27.2	17.8	9.4	SE	22.5
Oct.	1018.5	22.6	28.9	17.2	8.4	E	23.0
Nov.	1016.8	23.6	28.3	18.9	8.4	E	23.7
Dec.	1016.5	24.9	30.6	20.0	6.7	E	24.8

TABLE XII

CLIMATIC TABLE FOR MARINE AREA 12
Latitude 14.5°S, longitude 126°W, elevation 0 m, source H

Month	Mean press. (mbar)	Mean air temp. (°C)	Temp. extremes (°C)		Freq. precip. (%)	Most freq. wind dir.	Mean sea temp. (°C)
			max.	min.			
Jan.	1012.6	26.7	32.8	23.9	6.0	E	26.6
Feb.	1012.9	26.8	30.6	23.9	1.9	E	26.8
Mar.	1012.4	27.3	31.1	23.9	8.1	E	27.3
Apr.	1012.9	26.9	30.0	23.3	2.1	E	26.9
May	1013.1	26.9	32.8	23.9	6.0	E	26.4
June	1014.6	25.7	28.9	23.3	3.9	E	26.5
July	1015.1	25.6	30.6	23.3	3.3	E	25.4
Aug.	1014.8	24.8	29.4	22.8	5.7	E	25.1
Sep.	1015.0	25.1	30.0	22.2	4.0	E	25.2
Oct.	1015.0	25.1	29.4	21.7	5.2	E	25.4
Nov.	1013.6	25.9	30.6	21.7	3.6	E	25.8
Dec.	1012.9	26.3	30.6	23.9	2.8	E	26.3

TABLE XIII

CLIMATIC TABLE FOR MARINE AREA 13
Latitude 13.5°S, longitude 106.5°W, elevation 0 m, source H

Month	Mean press. (mbar)	Mean air temp. (°C)	Temp. extremes (°C)		Freq. precip. (%)	Most freq. wind dir.	Mean sea temp. (°C)
			max.	min.			
Jan.	1013.6	23.8	28.3	21.1	1.7	SE	23.8
Feb.	1013.7	24.6	28.3	22.2	0.0	E	24.6
Mar.	1012.7	24.8	28.9	23.3	0.6	E	25.1
Apr.	1013.2	24.9	29.4	21.7	1.9	E	25.3
May	1013.8	24.6	28.3	22.8	1.2	E	24.8
June	1015.0	23.4	26.1	21.1	6.8	E	24.2
July	1015.3	22.9	27.2	20.0	4.1	E	23.4
Aug.	1016.2	22.2	25.0	19.4	3.0	E	22.8
Sep.	1016.1	22.1	26.1	19.4	5.1	E	22.6
Oct.	1015.5	22.3	25.6	19.4	2.8	E	22.7
Nov.	1014.7	22.9	26.7	20.6	2.7	E	23.1
Dec.	1014.2	23.1	27.2	20.6	3.1	E	23.4

TABLE XIV

CLIMATIC TABLE FOR MARINE AREA 14
Latitude 21.5°S, longitude 107.0°W, elevation 0 m, source H

Month	Mean press. (mbar)	Mean air temp. (°C)	Temp. extremes (°C)		Freq. precip. (%)	Most freq. wind dir.	Mean sea temp. (°C)
			max.	min.			
Jan.	1016.7	23.7	27.8	21.7	0.0	SE	24.2
Feb.	1017.1	23.9	26.1	22.2	0.0	E	24.6
Mar.	1016.1	24.7	27.2	22.2	2.3	E	25.1
Apr.	1016.3	24.0	26.7	20.6	2.3	E	24.6
May	1017.1	22.8	26.7	20.0	10.3	NE	24.0
June	1019.9	22.3	24.4	21.1	0.0	E	22.7
July	1019.0	21.2	25.0	18.3	11.9	SE	22.2
Aug.	1021.2	21.2	22.8	19.4	2.8	N	21.7
Sep.	1020.8	20.7	25.0	18.9	0.0	SE	21.4
Oct.	1019.0	20.7	22.8	18.9	17.9	E	21.7
Nov.	1018.0	21.9	24.4	20.0	9.4	E	22.3
Dec.	1017.2	22.5	25.0	20.6	4.8	NE	22.4

TABLE XV

CLIMATIC TABLE FOR MARINE AREA 15
Latitude 23.5°S, longitude 92.0°W, elevation 0 m, source H

Month	Mean press. (mbar)	Mean air temp. (°C)	Temp. extremes (°C)		Freq. precip. (%)	Most freq. wind dir.	Mean sea temp. (°C)
			max.	min.			
Jan.	1020.5	22.5	25.6	18.3	7.4	SE	22.8
Feb.		22.7	25.0	20.0	17.5	E	22.7
Mar.	1017.7	23.3	29.4	20.6	0.0	SE	22.7
Apr.		21.4	23.3	20.0	16.7	SE	22.1
May		21.3	22.8	18.9	0.0	NW	21.5
June	1021.0	18.6	21.7	15.6	20.8	NW	19.8
July		18.4	21.1	16.7	5.0	E	18.1
Aug.		18.3	20.0	15.6	0.0	SE	19.1
Sep.		17.7	19.4	16.1	6.3	E	18.5
Oct.	1020.2	19.2	20.6	17.8	0.0	E	19.3
Nov.		19.2	22.8	17.2	0.0	E	19.4
Dec.	1019.3	19.7	21.7	18.3	5.3	SE	20.2

TABLE XVI

CLIMATIC TABLE FOR MARINE AREA 16
Latitude 25.5°S, longitude 80.0°W, elevation 0 m, source H

Month	Mean press. (mbar)	Mean air temp. (°C)	Temp. extremes (°C)		Freq. precip. (%)	Most freq. wind dir.	Mean sea temp. (°C)
			max.	min.			
Jan.	1015.7	20.4	24.4	17.2	0.0	SE	20.4
Feb.	1017.1				0.0	SE	20.7
Mar.	1015.9				0.0	SE	20.3
Apr.	1019.6	19.7	22.2	18.3	20.0	SE	20.2
May	1020.2				33.3	SE	20.4
June	1017.9	17.2	20.6	15.0	0.0	SE	18.2
July	1023.5	16.4	17.8	15.0	0.0	SE	17.6
Aug.	1019.2	16.8	20.6	13.3	12.5	SE	17.6
Sep.		16.8	24.4	13.9	7.1	SE	16.6
Oct.	1022.5	16.3	19.4	13.9	11.1	SE	16.9
Nov.	1021.6	17.7	19.4	16.1	0.0	SE	17.5
Dec.		18.7	21.1	17.2	0.0	SE	18.3

TABLE XVII

<small>CLIMATIC TABLE FOR MARINE AREA</small> 17
Latitude 13.5°S, longitude 79.5°W, elevation 0 m, source H

Month	Mean press. (mbar)	Mean air temp. (°C)	Temp. extremes (°C)		Freq. precip. (%)	Most freq. wind dir.	Mean sea temp. (°C)
			max.	min.			
Jan.	1013.1	22.9	26.1	18.9	3.8	SE	22.4
Feb.	1012.9	23.6	27.2	19.4	0.0	SE	22.6
Mar.	1011.3	24.2	28.3	20.6	2.6	SE	23.6
Apr.	1012.6	21.6	25.0	17.2	0.0	SE	21.4
May	1014.0	20.0	23.9	17.2	6.7	SE	20.3
June	1015.9	17.9	21.1	15.0	5.9	SE	18.2
July	1015.4	18.1	22.2	14.4	5.0	SE	18.4
Aug.	1015.3	17.1	20.0	14.4	5.6	SE	17.7
Sep.	1016.7	17.2	20.6	13.9	10.0	SE	17.4
Oct.	1015.4	18.8	23.3	15.0	0.0	SE	18.6
Nov.	1015.4	19.6	23.3	16.7	0.0	SE	19.3
Dec.	1012.6	21.9	27.8	18.9	0.0	SE	21.1

TABLE XVIII

<small>CLIMATIC TABLE FOR MARINE AREA</small> 18
Latitude 19.5°S, longitude 73.5°W, elevation 0 m, source H

Month	Mean press. (mbar)	Mean air temp. (°C)	Temp. extremes (°C)		Freq. precip. (%)	Most freq. wind dir.	Mean sea temp. (°C)
			max.	min.			
Jan.	1013.7	22.6	27.8	18.9	0.0	SE	22.8
Feb.	1013.4	23.9	28.9	18.9	0.0	SE	23.8
Mar.	1013.5	23.6	27.2	18.9	0.0	S	23.4
Apr.	1014.5	20.8	23.3	16.7	0.0	S	21.7
May	1015.0	19.4	25.0	19.4	0.0	SE	20.3
June	1017.2	16.8	20.0	14.4	0.0	SE	17.8
July	1017.3	16.3	21.1	13.3	0.0	SE	17.2
Aug.	1017.1	15.7	21.7	12.2	0.0	SE	16.5
Sep.	1018.1	16.6	21.7	12.8	0.0	SE	16.8
Oct.	1017.0	18.2	23.8	15.0	5.3	SE	18.1
Nov.	1015.3	19.1	23.3	15.6	0.0	S	19.1
Dec.	1013.2	21.3	26.1	17.8	0.0	S	21.0

TABLE XIX

Climatic table for marine area 19
Latitude 32.5°S, longitude 158.0°W, elevation 0 m, source H

Month	Mean press. (mbar)	Mean air temp. (°C)	Temp. extremes (°C)		Freq. precip. (%)	Most freq. wind dir.	Mean sea temp. (°C)
			max.	min.			
Jan.	1014.6	20.3	25.6	16.1	10.4	E	20.8
Feb.	1015.5	20.6	23.9	17.8	12.7	SE	21.2
Mar.	1014.0	20.7	23.9	17.8	14.7	SE	21.2
Apr.	1016.1	19.3	22.8	16.1	14.5	SE	20.3
May	1012.3	17.5	22.2	11.7	14.5	SW	18.6
June	1013.1	16.3	20.0	12.2	15.8	SW	17.4
July	1014.1	15.2	18.3	11.7	13.8	W	16.4
Aug.	1017.6	14.8	18.3	11.7	13.6	SW	16.1
Sep.	1018.7	15.3	18.9	11.1	19.1	SW	15.9
Oct.	1017.0	16.0	20.0	12.8	10.3	SW	16.3
Nov.	1015.2	16.8	20.6	13.9	9.6	SE	17.6
Dec.	1015.4	18.6	23.3	14.4	11.6	E	18.8

TABLE XX

Climatic table for marine area 20
Latitude 38.5°S, longitude 142.5°W, elevation 0 m, source H

Month	Mean press. (mbar)	Mean air temp. (°C)	Temp. extremes (°C)		Freq. precip. (%)	Most freq. wind dir.	Mean sea temp. (°C)
			max.	min.			
Jan.	1012.1	16.6	20.0	13.9	27.5	N	16.7
Feb.	1015.4	17.8	22.2	13.3	10.4	E	17.9
Mar.	1014.5	17.4	21.7	12.8	17.4	W	17.8
Apr.	1016.1	16.1	20.0	12.2	16.3	E	16.8
May	1013.2	14.3	17.8	10.0	9.5	SW	15.3
June	1012.6	13.2	16.7	10.0	7.5	NW	13.8
July	1012.3	12.2	16.1	9.4	24.1	NW	12.8
Aug.	1012.1	11.4	15.0	7.8	29.4	W	12.4
Sep.	1014.1	11.7	14.4	9.4	11.4	SW	12.1
Oct.	1015.0	11.9	15.0	9.4	17.3	SW	12.1
Nov.	1015.3	13.4	18.3	10.0	14.9	W	13.3
Dec.	1012.9	15.9	19.4	13.9	12.2	SW	15.4

TABLE XXI

<small>CLIMATIC TABLE FOR MARINE AREA</small> 21
Latitude 27.5°S, longitude 141.0°W, elevation 0 m, source H

Month	Mean press. (mbar)	Mean air temp. (°C)	Temp. extremes (°C) max.	Temp. extremes (°C) min.	Freq. precip. (%)	Most freq. wind dir.	Mean sea temp. (°C)
Jan.	1014.9	24.1	31.1	20.6	8.3	E	24.2
Feb.	1015.9	24.7	30.0	20.6	7.6	SE	25.1
Mar.	1015.1	24.4	29.4	20.0	10.7	E	24.6
Apr.	1013.5	22.7	27.8	18.9	18.0	SE	24.0
May	1016.5	21.0	25.6	17.2	17.0	SE	22.1
June	1016.4	20.1	26.1	15.6	18.5	SW	21.4
July	1016.4	19.0	22.8	15.6	13.9	SW	20.2
Aug.	1017.8	18.7	21.7	15.6	15.7	S	19.9
Sep.	1017.4	18.5	21.7	15.6	20.0	SE	19.7
Oct.	1019.4	19.5	25.0	15.6	14.4	SE	20.3
Nov.	1015.3	20.3	24.4	16.1	22.5	SE	20.9
Dec.	1014.3	22.3	27.8	18.9	10.0	E	22.5

TABLE XXII

<small>CLIMATIC TABLE FOR MARINE AREA</small> 22
Latide 35.5°S, Longitude 123.0°W, elevation 0 m, source H

Month	Mean press. (mbar)	Mean air temp. (°C)	Temp. extremes (°C) max.	Temp. extremes (°C) min.	Freq. precip. (%)	Most freq. wind dir.	Mean sea temp. (°C)
Jan.	1016.6	19.9	25.6	14.4	19.0	NW	19.5
Feb.	1017.8	20.4	27.2	16.1	6.6	S	20.7
Mar.	1019.6	19.9	24.4	16.7	10.3	SE	20.0
Apr.		18.8	23.9	15.6	12.2	E	19.3
May		16.4	19.4	11.7	32.3	SW	16.6
June	1010.3	14.5	18.3	12.2	4.5	SW	15.5
July	1012.0	14.4	17.2	11.7	27.3	NW	14.4
Aug.		12.4			25.9	SE	13.4
Sep.	1018.8	14.2	18.9	10.0	16.4	W	14.4
Oct.		14.4	18.3	8.3	29.2	W	15.1
Nov.	1020.9	16.9	21.1	12.8	32.3	NW	16.7
Dec.	1020.7	17.9	24.4	13.3	18.4	S	17.7

TABLE XXIII

CLIMATIC TABLE FOR MARINE AREA 23
Latitude 37.5°S, longitude 92.0°W, elevation 0 m, source H

Month	Mean press. (mbar)	Mean air temp. (°C)	Temp. extremes (°C)		Freq. precip. (%)	Most freq. wind dir.	Mean sea temp. (°C)
			max.	min.			
Jan.		18.4	22.2	13.9	13.6	NW	18.9
Feb.		19.1	23.9	15.0	2.2	S	19.6
Mar.		17.7	21.1	14.4	0.0	SW	17.8
Apr.		15.3	18.3	9.4	19.5	NW	16.3
May		14.3	17.2	11.7	3.1	S	15.5
June		12.7	16.7	8.3	18.7	S	13.3
July		12.4	16.1	7.2	16.2	W	13.2
Aug.	1017.3	11.7	16.7	9.4	12.9	N	12.8
Sep.		11.4	14.4	8.3	16.8	SW	12.3
Oct.	1021.4	11.7	18.3	10.0	20.3	NW	13.1
Nov.		14.6	17.8	11.1	0.0	W	14.1
Dec.		16.3	20.6	12.2	1.9	W	15.8

TABLE XXIV

CLIMATIC TABLE FOR MARINE AREA 24
Latitude 35.5°S, longitude 76.0°W, elevation 0 m, source H

Month	Mean press. (mbar)	Mean air temp. (°C)	Temp. extremes (°C)		Freq. precip. (%)	Most freq. wind dir.	Mean sea temp. (°C)
			max.	min.			
Jan.	1017.3	16.6	19.4	15.0		S	
Feb.	1015.0	17.2	18.9	13.9		S	16.9
Mar.	1015.4	17.1	20.0	13.3		S	17.9
Apr.	1016.9	16.6	20.6	13.3		S	16.9
May		14.0	15.6	12.8		SE	16.1
June	1017.2	12.6	15.6	10.6		S	13.8
July		12.9	15.6	9.4		NW	13.8
Aug.	1021.8	12.5	15.6	10.0		SE	13.3
Sep.	1019.1					S	13.4
Oct.		13.1	15.6	10.6		SE	13.1
Nov.	1021.0	13.5	17.8	10.6		S	14.1
Dec.	1017.0	15.8	17.8	13.9		S	15.7

TABLE XXV

<small>CLIMATIC TABLE FOR MARINE AREA 25</small>
Latitude 60.0°S, longitude 170.0°E, elevation 0 m, source H

Month	Mean press. (mbar)	Mean air temp. (°C)	Temp. extremes (°C)		Freq. precip. (%)	Most freq. wind dir.	Mean sea temp. (°C)
			max.	min.			
Jan.	998.9	6.2	11.1	1.7	15.8	N	5.8
Feb.	989.5	5.8	11.1	1.1	19.2	W	6.4
Mar.							
Apr.							
May							
June							
July							
Aug.							
Sep.	988.1	2.6	6.7	−2.2	17.3	W	3.3
Oct.	990.5	2.3	8.9	−6.7	23.1	W	2.4
Nov.	990.5	3.0	8.3	−3.9	20.7	W	2.8
Dec.	999.5	5.4	9.4	0.0	13.0	W	5.1

TABLE XXVI

<small>CLIMATIC TABLE FOR MARINE AREA 26</small>
Latitude 42.5°S, longitude 168.5°W, elevation 0 m, source H

Month	Mean press. (mbar)	Mean air temp. (°C)	Temp. extremes (°C)		Freq. precip. (%)	Most freq. wind dir.	Mean sea temp. (°C)
			max.	min.			
Jan.	1014.0	16.2	20.0	11.7	10.2	NW	16.1
Feb.	1015.3	16.3	20.6	12.8	10.7	W	16.8
Mar.	1017.8	15.7	19.4	12.2	12.6	SW	16.3
Apr.	1011.9	14.3	18.3	10.6	15.5	S	15.4
May		12.6	16.7	7.2	24.2	SW	14.1
June	1007.1	11.5	15.6	6.7	21.7	SW	12.9
July	1008.0	10.2	13.9	6.7	12.3	S	11.6
Aug.	1012.6	10.1	12.8	5.6	24.1	S	11.4
Sep.	1012.7	11.0	13.9	8.3	21.2	SE	11.7
Oct.	1015.0	11.7	15.0	7.8	11.9	SW	11.9
Nov.	1013.0	13.3	17.8	7.8	14.1	N	13.3
Dec.	1011.9	14.1	17.8	10.6	9.2	S	14.2

TABLE XXVII

CLIMATIC TABLE FOR MARINE AREA 27
Latitude 52.5°S, longitude 169.5°W, elevation 0 m, source H

Month	Mean press. (mbar)	Mean air temp. (°C)	Temp. extremes (°C)		Freq. precip. (%)	Most freq. wind dir.	Mean sea temp. (°C)
			max.	min.			
Jan.		11.3	16.1	6.1	21.5	W	10.2
Feb.		10.8	13.9	7.2	25.9	W	10.6
Mar.		9.8	13.3	5.0	23.3	SW	10.0
Apr.		9.4	12.2	7.2	41.2	SW	9.2
May		9.5	12.2	6.7	25.0	N	9.1
June		8.1	10.0	5.6	16.7	N	7.4
July		7.7	11.7	5.6	55.6	SW	6.8
Aug.		6.5	9.4	2.8	39.5	NW	6.9
Sep.		6.9	11.1	2.8	33.3	W	6.8
Oct.		8.3	12.8	4.4	20.6	W	7.8
Nov.		8.1	13.9	3.3	23.1	SW	7.7
Dec.		9.1	14.4	5.0	29.1	N	8.7

TABLE XXVIII

CLIMATIC TABLE FOR MARINE AREA 28
Latitude 52.5°S, longitude 156.5°W, elevation 0 m, source H

Month	Mean press. (mbar)	Mean air temp. (°C)	Temp. extremes (°C)		Freq. precip. (%)	Most freq. wind dir.	Mean sea temp. (°C)
			max.	min.			
Jan.		10.1	14.4	5.0	26.9	NW	9.7
Feb.		11.0	16.1	6.1	27.6	W	11.1
Mar.		9.2	13.3	4.4	24.5	SW	9.7
Apr.		8.6	12.2	6.1	45.2	W	8.7
May		7.0	10.6	3.3	26.0	S	7.8
June		5.9	8.9	1.1	29.4	SW	7.1
July		5.3	8.9	2.2	35.6	S	6.4
Aug.		6.3	8.9	2.2	28.6	SW	6.5
Sep.		6.1	8.9	1.7	14.3	NW	6.6
Oct.		7.0	10.6	1.7	39.1	NW	7.2
Nov.		8.0	12.2	4.4	26.3	SW	7.8
Dec.		9.2	13.3	3.9	22.6	NW	8.7

TABLE XXIX

CLIMATIC TABLE FOR MARINE AREA 29
Latitude 48.5°S, longitude 138.5°W, elevation 0 m, source H

Month	Mean press. (mbar)	Mean air temp. (°C)	Temp. extremes (°C)		Freq. precip. (%)	Most freq. wind dir.	Mean sea temp. (°C)
			max.	min.			
Jan.	1006.2	11.0	14.4	7.2	18.3	W	10.8
Feb.	1011.5	11.2	16.1	6.7	18.2	W	10.6
Mar.	1012.4	10.2	13.3	5.0	18.6	W	10.4
Apr.		9.0	12.2	2.8	29.7	W	9.7
May	1001.1	7.8	10.7	2.8	16.4	W	8.7
June	1001.2	7.2	11.1	2.8	24.7	W	8.4
July		7.0	11.7	1.1	30.5	SW	7.8
Aug.	1009.2	6.6	11.7	3.9	18.0	SW	7.3
Sep.		6.6	12.8	2.8	33.8	SW	7.4
Oct.	1006.8	7.3	12.8	3.9	26.1	SW	7.7
Nov.	1010.1	9.0	12.8	4.4	17.1	W	8.6
Dec.	1011.3	9.7	14.4	6.1	21.6	W	9.3

TABLE XXX

CLIMATIC TABLE FOR MARINE AREA 30
Latitude 54.5°S, longitude 138.5°W, elevation 0 m, source H

Month	Mean press. (mbar)	Mean air temp. (°C)	Temp. extremes (°C)		Freq. precip. (%)	Most freq. wind dir.	Mean sea temp. (°C)
			max.	min.			
Jan.	996.8	7.4	11.7	2.8	30.6	NW	7.2
Feb.	995.5	7.5	11.7	3.3	22.2	N	7.4
Mar.		6.6	12.2	2.8	54.5	W	7.1
Apr.	992.0	6.9	11.1	3.9	35.0	NW	6.8
May	992.9	3.9	9.4	−1.7	33.9	N	5.0
June			7.8	−1.7	41.2	N	
July	998.5	3.0	7.8	−2.8	55.0	S	4.6
Aug.	1000.0		7.8	−1.1	60.0	W	
Sep.		2.4	7.8	−1.7	33.3	NW	4.1
Oct.		5.4	9.4	1.1	26.9	N	5.2
Nov.		4.9	9.4	0.6	30.9	W	5.3
Dec.		6.9	10.6	1.1	30.8	NW	6.8

TABLE XXXI

CLIMATIC TABLE FOR MARINE AREA 31
Latitude 51.5°S, longitude 125.5°W, elevation 0 m, source H

Month	Mean press. (mbar)	Mean air temp. (°C)	Temp. extremes (°C)		Freq. precip. (%)	Most freq. wind dir.	Mean sea temp. (°C)
			max.	min.			
Jan.		9.5	13.9	5.6	31.5	W	8.9
Feb.	1004.7	9.3	14.4	3.9	26.6	W	9.1
Mar.	1002.7	8.4	13.9	2.2	32.3	NW	8.6
Apr.	1001.9	7.7	11.1	2.8	36.3	W	8.0
May	1002.4	6.9	10.6	1.7	33.6	NW	7.4
June		6.0	8.9	1.7	27.4	NW	6.9
July		5.4	8.9	0.0	31.9	NW	6.7
Aug.		5.0	8.3	1.1	32.7	W	6.4
Sep.		5.7	8.3	1.1	26.4	NW	6.3
Oct.	998.1	5.9	10.0	0.6	36.3	SW	6.6
Nov.	1004.3	6.9	10.6	0.6	25.0	W	7.2
Dec.		8.4	12.8	3.3	20.3	W	8.1

TABLE XXXII

CLIMATIC TABLE FOR MARINE AREA 32
Latitude 43.5°S, longitude 108.0°W, elevation 0 m, source H

Month	Mean press. (mbar)	Mean air temp. (°C)	Temp. extremes (°C)		Freq. precip. (%)	Most freq. wind dir.	Mean sea temp. (°C)
			max.	min.			
Jan.		13.1	17.2	9.4	16.7	W	12.9
Feb.		12.3	16.1	10.6	31.3		13.1
Mar.		13.3	17.8	9.4	18.3	S	13.3
Apr.		11.9	15.6	7.8	24.6	SW	12.2
May		10.4	13.3	7.8	5.3	NW	10.9
June		9.5	12.8	7.2	11.9	SW	10.6
July		8.3	9.4	6.7	21.4	SW	9.4
Aug.		7.5	8.3	5.6	11.1	N	8.7
Sep.		10.1	15.0	7.2	30.6	N	9.8
Oct.		10.9	13.3	8.3	45.2	NW	10.0
Nov.			12.2	7.2	13.5	SW	
Dec.		12.6	18.3	8.3	15.2	W	12.2

TABLE XXXIII

CLIMATIC TABLE FOR MARINE AREA 33
Latitude 52.5°S, longitude 106.5°W, elevation 0 m, source H

Month	Mean press. (mbar)	Mean air temp. (°C)	Temp. extremes (°C)		Freq. precip. (%)	Most freq. wind dir.	Mean sea temp. (°C)
			max.	min.			
Jan.	1000.2	8.6	14.4	5.6	25.4	W	8.1
Feb.	1001.5	8.4	12.8	4.4	24.4	W	8.4
Mar.	997.9	7.9	12.8	2.8	23.6	W	8.1
Apr.		6.7	11.7	1.7	27.3	W	7.5
May		6.2	9.4	1.1	34.3	W	6.8
June		5.4	8.3	2.8	29.3	W	6.3
July	1000.4	4.6	7.8	−0.6	41.4	W	5.8
Aug.	1004.2	4.3	6.7	0.6	29.8	W	5.6
Sep.		5.3	8.3	0.0	45.9	NW	5.8
Oct.	1005.2	5.8	9.4	0.6	33.9	NW	6.0
Nov.		6.4	8.9	2.2	40.8	W	6.6
Dec.	1003.5	8.0	12.2	3.9	22.1	W	7.7

TABLE XXXIV

CLIMATIC TABLE FOR MARINE AREA 34
Latitude 49.5°S, longitude 93.5°W, elevation 0 m, source H

Month	Mean press. (mbar)	Mean air temp. (°C)	Temp. extremes (°C)		Freq. precip. (%)	Most freq. wind dir.	Mean sea temp. (°C)
			max.	min.			
Jan.		9.4	13.3	6.1	16.4	NW	9.1
Feb.		10.7	13.9	8.3	15.1	NW	10.3
Mar.		9.0	12.2	6.1	15.5	SW	9.1
Apr.		8.1	11.7	4.4	36.1	W	8.5
May		7.3	8.3	5.0	30.4	W	8.2
June	1007.9	6.1	8.9	2.8	20.5	S	7.3
July		6.1	11.1	3.3	23.6	S	6.6
Aug.			11.1	4.4		SW	
Sep.		5.9	8.3	2.2	33.3	NW	6.4
Oct.		6.7	10.0	2.2	32.4	SW	6.4
Nov.		7.4	10.6	5.0	36.2	W	7.0
Dec.		8.1	11.7	5.0	13.5	W	8.1

TABLE XXXV

<small>CLIMATIC TABLE FOR MARINE AREA</small> 35
Latitude 57.5°S, longitude 92.0°W, elevation 0 m, source H

Month	Mean press. (mbar)	Mean air temp. (°C)	Temp. extremes (°C)		Freq. precip. (%)	Most freq. wind dir.	Mean sea temp. (°C)
			max.	min.			
Jan.	992.9	6.6	10.0	1.7	22.5	W	6.6
Feb.	997.4	7.2	11.7	1.1	35.5	W	6.9
Mar.	995.2	6.3	10.0	2.2	27.0	W	6.3
Apr.	993.6	5.3	7.8	1.7	34.6	W	5.5
May	993.8	4.5	6.7	0.6	34.5	NW	5.3
June		3.5	6.7	−2.2	57.3	W	4.4
July		3.1	5.6	0.6	33.8	W	4.4
Aug.	997.4		7.2	−0.6	42.1	NW	
Sep.	993.0	3.5	6.1	0.0	40.5	NW	4.3
Oct.		4.5	7.8	0.6	32.1	NW	4.6
Nov.	994.8	5.2	8.3	2.8	35.0	NW	5.1
Dec.		5.8	11.7	−0.6	38.4	SW	5.8

TABLE XXXVI

<small>CLIMATIC TABLE FOR MARINE AREA</small> 36
Latitude 43.5°S, longitude 81.5°W, elevation 0 m, source H

Month	Mean press. (mbar)	Mean air temp. (°C)	Temp. extremes (°C)		Freq. precip. (%)	Most freq. wind dir.	Mean sea temp. (°C)
			max.	min.			
Jan.		12.7	18.3	9.4		W	12.7
Feb.	1015.5	13.5	15.6	12.2		SW	14.1
Mar.		13.3	18.3	10.0		SW	13.8
Apr.		13.0	15.0	10.0			12.8
May	1011.8	10.9	14.4	7.2		NW	11.2
June						SW	
July		8.0	10.0	5.6		W	9.4
Aug.		8.1	10.6	5.6		W	8.8
Sep.		8.5	10.6	7.2		W	9.1
Oct.		8.9	11.7	5.0		W	8.4
Nov.		9.8	11.7	8.3		W	9.6
Dec.		10.8	13.3	8.3		W	10.7

TABLE XXXVII

CLIMATIC TABLE FOR MARINE AREA 37
Latitude 56.5°S, longitude 77.5°W, elevation 0 m, source H

Month	Mean press. (mbar)	Mean air temp. (°C)	Temp. extremes (°C)		Freq. precip. (%)	Most freq. wind dir.	Mean sea temp. (°C)
			max.	min.			
Jan.	994.1	8.1	13.9	3.9	25.0	W	7.3
Feb.	993.4	8.0	14.4	2.8	38.5	W	7.5
Mar.	991.9	7.1	12.2	2.2	32.0	W	7.0
Apr.	995.1	6.1	11.1	1.7	35.5	W	6.2
May	995.0	4.9	10.6	−2.2	34.5	W	5.6
June	996.8	3.7	8.9	−3.9	25.7	NW	4.8
July	998.2	3.5	8.9	−3.3	43.1	W	4.6
Aug.	997.6	3.8	9.4	−4.4	23.3	W	4.5
Sep.	995.8	4.4	10.0	−1.7	30.8	W	4.7
Oct.	997.2	5.1	11.7	−1.7	35.6	W	4.9
Nov.	994.1	5.9	12.8	0.6	22.2	W	5.4
Dec.	994.4	7.0	16.1	1.7	31.8	W	6.6

TABLE XXXVIII

CLIMATIC TABLE FOR MARINE AREA 38
Latitude 64.5°S, longitude 158.5°E, elevation 0 m, source H

Month	Mean press. (mbar)	Mean air temp. (°C)	Temp. extremes (°C)		Freq. precip. (%)	Most freq. wind dir.	Mean sea temp. (°C)
			max.	min.			
Jan.	991.2	0.6	7.8	−2.8	22.0	E	0.0
Feb.	983.3	−0.2	2.8	−5.0	28.8	W	−0.7
Mar.	990.1	−1.1	3.3	−11.7	29.6	W	−0.9
Apr.							
May							
June							
July							
Aug.							
Sep.							
Oct.							
Nov.							
Dec.		−0.9	2.8	−3.9	31.1	E	−0.5

TABLE XXXIX

CLIMATIC TABLE FOR MARINE AREA 39
Latitude 66.5°S, longitude 166.5°W, elevation 0 m, source H

Month	Mean press. (mbar)	Mean air temp. (°C)	Temp. extremes (°C)		Freq. precip. (%)	Most freq. wind dir.	Mean sea temp. (°C)
			max.	min.			
Jan.	987.7	−0.2	6.1	−3.3	27.6	S	−0.2
Feb.	984.1	−0.6	2.8	−3.3	27.8	SW	−0.6
Mar.	987.6	−1.1	1.1	−7.8	31.0	SW	−0.7
Apr.							
May							
June							
July							
Aug.							
Sep.							
Oct.							
Nov.							
Dec.		−1.3	0.6	−0.6	40.0	SE	−0.6

TABLE XL

CLIMATIC TABLE FOR MARINE AREA 40
Latitude 66.5°S, longitude 77°W, elevation 0 m, source H

Month	Mean press. (mbar)	Mean air temp. (°C)	Temp. extremes (°C)		Freq. precip. (%)	Most freq. wind dir.	Mean sea temp. (°C)
			max.	min.			
Jan.	992.8	1.3	3.9	−1.1	30.9	NE	1.3
Feb.		−0.2	2.2	−5.0	16.7	SW	0.7
Mar.		−0.1	3.3	−6.1	17.1	NE	0.6
Apr.							
May							
June							
July							
Aug.							
Sep.							
Oct.		−5.6	−2.2	−8.9	50.0	W	−1.6
Nov.							
Dec.	994.8	0.1	3.9	−3.3	24.3	E	−0.2

TABLE XLI

CLIMATIC TABLE FOR ALOFI, NIUE ISLAND (41)
Latitude 19°02'S, longitude 169°55'W, elevation 20 m

	Jan.	Feb.	Mar.	Apr.	May	June	July	Aug.	Sep.	Oct.	Nov.	Dec.	Year	Period and source
Mean sea level press. (mbar)	1009.2	1009.1	1009.5	1010.9	1012.4	1013.6	1014.0	1014.3	1014.6	1013.9	1011.4	1009.8	1011.9	1951–60 A
Mean monthly temp. (°C)	26.5	26.7	26.6	25.7	24.5	23.5	22.9	23.1	23.5	24.3	25.1	25.8	24.9	1907–60 C
Mean daily min. temp. (°C)	23.0	23.2	23.1	22.4	21.0	20.4	19.4	19.3	20.0	20.7	21.4	22.2	21.3	1941–70 D
Mean daily max. temp. (°C)	29.6	29.9	29.7	29.2	28.0	27.2	26.4	26.5	27.0	27.6	28.3	29.0	28.2	1941–70 D
Highest temp. (°C)	36.1	36.7	36.7	35.0	34.4	32.8	31.3	33.0	33.9	35.0	35.6	35.6	36.7	1907–60 C
Lowest temp. (°C)	16.1	15.6	13.3	12.8	13.3	13.9	12.2	12.8	13.3	13.3	14.4	15.0	12.2	1907–60 C
Mean monthly rainfall (mm)	277.9	263.5	308.7	193.9	126.9	82.2	108.7	101.2	101.4	145.1	148.1	189.4	2047.0	1921–71 E
Max. monthly rainfall (mm)	727.7	615.7	784.9	502.7	469.1	275.1	404.6	494.0	360.9	366.8	521.7	520.2	3185.4	1921–71 E
Min. monthly rainfall (mm)	23.1	54.9	24.9	34.0	7.9	4.3	5.3	13.2	19.8	5.1	9.1	22.1	1065.3	1921–71 E
No. of raindays	16	16	19	15	12	10	11	11	10	11	11	14	156	1905–60 C
Mean relat. humid. (%; 1800 GMT)	87	88	89	89	88	88	87	86	87	86	86	86	87	1928–60 C
Days with thunder (means)	1.5	1.7	1.7	1.1	0.8	0.3	0.4	0.5	0.4	0.6	1.0	1.2	11.2	1907–60 C

TABLE XLII

CLIMATIC TABLE FOR APIA, WEST SAMOA ISLAND (42)
Latitude 13°48'S, longitude 171°47'W, elevation 2 m

	Jan.	Feb.	Mar.	Apr.	May	June	July	Aug.	Sep.	Oct.	Nov.	Dec.	Year	Period and source
Mean sea level press. (mbar)	1008.2	1008.1	1009.2	1009.8	1010.7	1011.3	1011.8	1012.0	1012.0	1011.4	1009.2	1008.4	1010.2	1951–60 A
Mean monthly temp. (°C)	26.5	26.5	26.5	26.5	26.2	25.9	25.6	25.8	25.9	26.2	26.3	26.5	26.2	1890–1970 I
Mean daily min. temp. (°C)	23.7	23.7	23.7	23.7	23.4	23.1	22.6	22.7	22.8	23.2	23.3	23.5	23.3	1941–70 D
Mean daily max. temp. (°C)	30.2	30.1	30.2	30.5	30.2	29.8	29.3	29.4	29.6	29.7	30.0	30.0	29.9	1941–70 D
Mean monthly rainfall (mm)	424.2	369.3	356.4	246.3	170.6	137.6	96.1	103.4	139.9	197.3	259.1	369.5	2869.7	1890–1971 E
Max. monthly rainfall (mm)	1513.1	765.0	1297.4	528.3	599.2	341.6	275.0	392.0	402.6	434.8	847.0	933.0	4387.0	1890–1971 E
Min. monthly rainfall (mm)	84.2	90.0	36.1	45.0	10.0	16.5	4.0	2.0	7.0	31.0	25.0	77.0	1747.0	1890–1971 E
Bright sunshine (h; mean)	187	165	199	216	226	226	244	247	242	225	208	190	2575	1935–60 B
Most frequent wind direct.	E	E	E	E	E	E	E	E	E	E	E	E	E	H

397

TABLE XLIII

CLIMATIC TABLE FOR BORA BORA, LEEWARD ISLAND (43)
Latitude 16°31'S, longitude 151°45'W, elevation 3 m

	Jan.	Feb.	Mar.	Apr.	May	June	July	Aug.	Sep.	Oct.	Nov.	Dec.	Year	Period and source
Mean sea level press. (mbar)	1010.8	1010.7	1011.4	1011.5	1012.2	1013.5	1013.9	1014.0	1014.0	1013.6	1011.9	1011.1	1012.4	1951–60 A
Mean monthly temp. (°C)	26.9	26.9	27.2	27.0	26.2	25.6	25.1	25.3	25.2	25.9	26.3	26.6	26.2	1951–70 I
Mean monthly min. temp. (°C)	23.9	23.7	24.1	23.7	23.1	22.7	22.3	22.1	22.4	23.3	23.2	23.8	23.2	1951–57 K
Mean monthly max. temp. (°C)	29.8	30.1	30.5	30.1	29.2	28.5	28.2	28.3	28.3	29.2	29.3	29.8	29.3	1951–57 K
Highest temp. (°C)	32.0	32.8	32.9	32.5	34.4	30.9	30.9	30.7	30.7	32.0	31.6	32.4	34.4	1951–57 K
Lowest temp. (°C)	21.2	20.9	21.4	19.0	19.7	18.5	17.4	16.9	19.0	19.6	20.0	20.6	16.9	1951–57 K
Mean monthly rainfall (mm)	253	223	205	150	162	106	112	90	94	124	239	273	2031	1951–71 E
Max. monthly rainfall (mm)	678	443	520	279	358	262	316	320	233	255	588	782	2793	1951–71 E
Min. monthly rainfall (mm)	76	57	55	52	41	15	3	17	26	15	34	97	1386	1951–71 E
Max. daily rainfall (mm)	143.0	85.1	98.0	113.8	88.5	131.3	139.9	84.6	85.8	100.6	69.7	139.5	143.0	1951–57 K
No. of raindays	23	19	17	16	17	11	13	12	14	14	17	19	192	1951–57 K
Mean relat. humid. (%)	79	79	79	78	79	78	77	76	77	76	78	79	78	1951–60 B
Mean vap. press. (mbar)	28.3	28.5	29.1	28.9	27.9	26.0	25.3	24.8	25.6	26.2	27.6	28.0	27.2	1951–67 B

TABLE XLIV

CLIMATIC TABLE FOR CAMPBELL ISLAND (44)
Latitude 52°33'S, longitude 169°07'E, elevation 15 m

	Jan.	Feb.	Mar.	Apr.	May	June	July	Aug.	Sep.	Oct.	Nov.	Dec.	Year	Period and source
Mean sea level press. (mbar)	1004.3	1005.3	1005.7	1004.5	1005.0	1003.8	1007.5	1006.5	1003.3	1001.9	1000.7	1003.4	1004.3	1942–73 I
Mean monthly temp. (°C)	9.3	9.2	8.5	7.3	6.1	4.9	4.8	5.1	5.6	6.2	7.1	8.5	6.9	1941–70 D
Mean daily min. temp. (°C)	6.7	6.7	6.1	4.9	4.0	2.7	2.5	2.8	3.1	3.7	4.5	5.8	4.5	1941–70 D
Mean daily max. temp. (°C)	11.9	11.7	10.8	9.3	8.1	6.8	6.7	7.1	7.8	8.6	9.8	11.3	9.2	1941–70 D
Highest temp. (°C)	18.9	19.1	17.2	16.6	16.5	12.1	11.6	11.5	12.7	13.9	15.5	18.9	19.1	1941–64 C
Lowest temp. (°C)	−0.4	−0.2	−2.6	−2.8	−3.9	−6.2	−6.3	−5.1	−5.3	−4.3	−3.6	−1.8	−6.3	1941–64 C
Mean monthly rainfall (mm)	124.5	106.7	142.2	124.5	137.2	121.9	101.6	114.3	121.9	121.9	119.4	116.8	1452.9	1941–57 C
Max. daily rainfall (mm)	52.58	66.80	39.12	53.85	52.83	49.28	31.75	55.63	29.72	41.91	42.16	61.21	66.80	1941–64 C
No. of raindays	26	24	28	27	28	28	28	29	27	28	26	26	325	1941–57 C
Mean relat. humid. (%; 0900 GMT)	85	85	87	88	89	88	88	88	85	84	82	82	86	1941–57 C
Bright sunshine (h; mean)	92	80	61	36	21	12	15	27	54	67	89	99	653	1941–57 C
(% of possible)	19	20	16	12	8	6	6	9	16	16	20	20	15	1941–57 C
Days with snow (means)	0.2	0.2	0.8	1.8	3.6	7.5	6.4	5.6	6.6	4.6	2.6	0.7	40.6	1941–57 C
Days with hail (means)	2.7	2.9	5.7	4.9	5.0	7.7	5.8	5.8	7.1	7.2	4.8	3.1	62.7	1941–57 C

TABLE XLIV (*continued*)

	Jan.	Feb.	Mar.	Apr.	May	June	July	Aug.	Sep.	Oct.	Nov.	Dec.	Year	Period and source
Days with thunder (means)	0.1	0.1	0.1	0.1	0.1	0.2	0.1	0.1	0.2	0.2		0.1	1.4	1941–57 C
Days with fog (means)	1.7	1.8	2.1	1.6	0.6	0.6	0.8	0.8	0.8	0.6	0.6	1.3	12.6	1941–57 C
Days with ground frost (means)	0.3	0.4	0.7	1.7	3.6	7.6	9.5	7.1	5.2	3.5	2.2	1.7	43.5	1941–57 C
Days with frost in screen (means)	0.1	0.1	0.1	0.8	1.9	4.8	5.9	4.8	3.9	2.2	1.6	0.7	26.8	1941–57 C

TABLE XLV

CLIMATIC TABLE FOR CANTON ISLAND (45)
Latitude 02°46'S, longitude 171°43'W, elevation 3 m

	Jan.	Feb.	Mar.	Apr.	May	June	July	Aug.	Sep.	Oct.	Nov.	Dec.	Year	Period and source
Mean sea level press. (mbar)	1008.6	1008.6	1009.0	1009.2	1009.4	1009.6	1009.8	1009.8	1010.0	1009.9	1008.7	1008.3	1009.3	1941–60 B
Mean monthly rainfall (mm)	84.7	40.3	41.2	70.2	69.9	62.2	65.2	59.9	37.1	42.6	56.0	67.1	696.3	1942–72 E
Max. monthly rainfall (mm)	520.4	246.1	196.3	262.1	211.6	137.4	199.6	172.2	134.4	428.5	428.0	427.2	1596.6	1942–72 E
Min. monthly rainfall (mm)	0.0	0.5	0.5	20.1	8.9	5.1	5.8	4.6	1.3	0.8	0.0	1.0	197.9	1942–72 E
Relat. humidity (%)	74	74	74	76	73	74	72	72	70	72	70	70	73	1951–60 A
Mean monthly temp. (°C)	28.4	28.3	28.4	28.7	28.8	28.8	28.7	28.6	28.7	28.7	28.7	28.4	28.6	1931–60 A

TABLE XLVI

CLIMATIC TABLE FOR FUNAFUTI ISLAND (46)
Latitude 08°31'S, longitude 179°12'E, elevation 4 m

	Jan.	Feb.	Mar.	Apr.	May	June	July	Aug.	Sep.	Oct.	Nov.	Dec.	Year	Period and source
Mean sea level press. (mbar)	1007.0	1007.2	1007.8	1008.5	1008.8	1009.1	1009.4	1009.4	1009.5	1009.3	1007.9	1007.0	1008.4	1943–61 B
Mean monthly temp. (°C)	28.3	28.1	28.3	28.3	28.2	27.9	27.6	27.8	28.0	28.3	28.3	28.3	28.1	1932–70 I
Mean monthly rainfall (mm)	410.9	391.1	328.2	258.3	224.6	252.0	263.5	281.0	245.2	287.9	302.3	393.3	3638.3	1927–69 E
Min. monthly rainfall (mm)	167.1	148.1	92.5	92.7	53.8	56.6	57.7	40.9	47.0	59.7	75.4	129.8	2474.5	1927–69 E
Max. monthly rainfall (mm)	1141.5	1138.9	1293.1	618.2	669.8	566.4	617.2	1195.8	876.6	1206.2	677.4	806.4	6732.8	1927–69 E

TABLE XLVII

CLIMATIC TABLE FOR HONIARA, GUADALCANAL ISLAND (47)
Latitude 9°25'S, longitude 159°58'E, elevation 54.9 m

	Jan.	Feb.	Mar.	Apr.	May	June	July	Aug.	Sep.	Oct.	Nov.	Dec.	Year	Period and source
Mean stat. level pressure (mbar)	1000.3	1000.6	1001.2	1002.1	1002.7	1003.5	1003.5	1003.7	1003.6	1003.3	1002.0	1000.7	1002.3	1951–74 J
Mean daily max. temp. (°C)	30.7	30.5	30.2	30.5	30.7	30.4	30.1	30.4	30.6	30.7	30.7	30.5	30.5	1951–74 J
Mean daily min. temp. (°C)	23.0	23.0	23.0	22.9	22.8	22.5	22.2	22.1	22.3	22.5	22.7	23.0	22.7	1951–74 J
Highest temp. (°C)	33.9	34.1	33.9	33.4	33.6	32.8	33.3	33.5	33.4	33.3	33.4	34.8	34.8	1951–74 J
Lowest temp. (°C)	20.2	20.7	20.7	20.1	20.5	19.4	18.7	18.8	18.3	17.6	17.8	20.5	17.6	1951–74 J
Mean vap. press. (mbar)	28.3	28.5	28.7	28.3	28.0	27.2	26.7	26.4	26.9	27.2	27.7	28.2	27.7	1949–74 J
Mean monthly rainfall (mm)	277	287	362	214	141	97	100	92	95	154	141	217	2177	1949–74 J
No. of raindays	19	19	23	18	15	13	15	13	13	16	15	18	197	1949–74 J
Max. monthly rainfall (mm)	971	575	645	357	480	344	311	223	214	383	398	588	2962	1949–74 J
Min. monthly rainfall (mm)	33	82	119	110	24	1	16	7	11	25	39	65	1526	1949–74 J
Mean cloudiness (tenths)	6.3	6.5	6.5	6.1	5.5	5.5	5.7	5.4	5.5	5.8	5.7	6.1	5.8	1949–74 J
Mean fog freq. (days)	0.2	0.2	0.2	0	0	0	0.2	0	0	0	0.2	0.2	1.2	1959–63 J
Mean thunder freq. (days)	8.9	8.9	9.4	6.9	5.0	4.0	4.2	3.4	6.0	9.0	9.9	10.2	85.7	1951–74 J
Most freq. wind direction	SSW	CALM	CALM	SSW	SSW	SSW	SSW	SSW	SSW	SSW	SSW	SSW	SSW	1950–73 J

TABLE XLVIII

CLIMATIC TABLE FOR ISLA DE PASCUA, EASTER ISLAND (48)
Latitude 27°10′S, longitude 109°26′W, elevation 20 m

	Jan.	Feb.	Mar.	Apr.	May	June	July	Aug.	Sep.	Oct.	Nov.	Dec.	Year	Period and source
Mean sea level press. (mbar)	1019.3	1019.6	1019.3	1018.2	1017.9	1020.3	1020.9	1021.8	1021.8	1022.4	1020.8	1019.6	1020.2	1942–77 A₁
Mean monthly temp. (°C)	23.6	24.1	23.5	21.8	20.4	18.7	18.2	18.0	18.7	19.7	21.0	22.3	20.8	1942–50 A₁
Highest temp. (°C)	28.9	28.9	28.9	27.8	27.8	25.0	23.9	25.0	25.0	26.1	26.1	28.9	28.9	H
Lowest temp. (°C)	20.0	21.1	21.1	17.8	16.1	15.0	12.9	12.9	15.0	15.0	17.2	17.2	12.9	H
Mean monthly rainfall (mm)	121.4	94.0	115.6	105.9	117.2	109.3	90.0	77.4	68.0	93.6	116.9	124.8	1134.1	1941–50 A₁
Most freq. wind direction	E	E	E	E	NW	E	NW	NW	N	E	E	E	E	H

TABLE XLIX

CLIMATIC TABLE FOR JUAN FERNANDEZ ISLAND (49)
Latitude 33°37'S, longitude 78°50'W, elevation 6 m

	Jan.	Feb.	Mar.	Apr.	May	June	July	Aug.	Sep.	Oct.	Nov.	Dec.	Year	Period and source
Mean sea level press. (mbar)	1019.7	1019.0	1019.4	1018.6	1017.7	1019.1	1019.6	1020.4	1021.6	1022.1	1020.5	1019.9	1019.8	1941–50 A₁
Mean monthly temp. (°C)	18.8	19.2	18.5	16.8	15.2	13.8	12.8	12.1	12.9	13.5	15.3	17.5	15.5	1941–50 A₁
Mean monthly rainfall (mm)	17.0	31.4	36.5	98.6	173.0	172.6	213.0	140.9	75.2	61.4	30.7	17.5	1067.7	1941–50 A₁

Climate of the South Pacific Ocean

TABLE L

CLIMATIC TABLE FOR LORD HOWE ISLAND (50)
Latitude 31°32'S, longitude 159°4'E, elevation 0.0 m

	Jan.	Feb.	Mar.	Apr.	May	June	July	Aug.	Sep.	Oct.	Nov.	Dec.	Year	Period and source
Mean stat. level press. (mbar)	1008.9	1008.7	1010.2	1012.0	1011.1	1011.0	1011.5	1011.3	1012.3	1011.5	1009.4	1008.2	1010.5	1951–78 J
Mean daily max. temp. (°C)	25.1	25.4	24.8	23.1	21.0	19.3	18.4	18.4	19.3	20.5	22.1	23.9	21.8	1939–78 J
Mean daily min. temp. (°C)	20.0	20.4	19.7	18.0	16.1	14.6	13.5	13.3	14.0	15.4	16.9	18.6	16.7	1939–78 J
Highest temp. (°C)	29.3	29.4	29.5	27.3	25.2	23.1	22.1	22.4	24.4	24.9	27.8	28.2	29.5	1939–78 J
Lowest temp. (°C)	11.7	14.8	14.7	12.2	10.5	9.4	8.6	8.1	8.0	9.7	11.1	12.7	8.0	1939–78 J
Mean vapour press. (mbar)	21.2	22.2	21.1	18.3	16.2	14.7	13.5	13.5	14.3	16.1	18.0	19.8	17.4	1939–78 J
Mean monthly rainfall (mm)	116	106	128	165	159	185	186	137	140	126	116	121	1685	1886–1979 J
No. of raindays	11	12	15	17	20	21	22	20	17	13	12	11	191	1886–1979 J
Max. monthly rainfall (mm)	291	337	425	702	376	387	497	309	322	295	402	339	2865	1886–1979 J
Min. monthly rainfall (mm)	10	8	5	36	45	69	63	13	28	35	14	17	997	1886–1979 J
Mean cloudiness (tenths)	5.0	5.2	4.9	4.7	5.0	5.2	4.8	4.8	4.8	5.0	4.9	4.9	4.9	1939–78 J
Mean fog freq. (days)	0.1	0	0.1	0	0	0	0	0	0	0.1	0.2	0.2	0.7	1939–76 J
Mean thunder freq. (days)	1.3	0.9	1.4	1.6	2.9	3.0	3.0	3.7	3.3	3.3	3.3	2.6	30.2	1939–78 J
Most freq. wind direction	E	E	SE	SE	SW	SW	SW	SW	SW	SW	SW	E	SW	1939–78 J

407

TABLE LI

CLIMATIC TABLE FOR MACQUARIE ISLAND (51)
Latitude 54°30'S, longitude 158°57'E, elevation 6.1 m

	Jan.	Feb.	Mar.	Apr.	May	June	July	Aug.	Sep.	Oct.	Nov.	Dec.	Year	Period and source
Mean stat. level press. (mbar)	998.9	999.4	1000.1	998.9	1000.1	1000.7	1002.6	1000.9	997.1	996.4	996.2	998.4	999.1	1948–78 J
Mean daily max. temp. (°C)	8.5	8.3	7.8	6.7	5.7	4.9	4.7	4.9	5.1	5.5	6.3	7.7	6.3	1948–78 J
Mean daily min. temp. (°C)	5.2	5.0	4.5	3.4	2.4	1.4	1.4	1.5	1.3	1.7	2.6	4.2	2.9	1948–78 J
Highest temp. (°C)	12.6	12.2	10.9	10.1	9.4	11.1	8.3	8.2	8.6	10.3	10.7	12.1	12.6	1948–78 J
Lowest temp. (°C)	0.6	−0.6	−1.1	−4.5	−6.7	−7.0	−8.9	−8.9	−8.7	−4.6	−3.3	−1.7	−8.9	1948–78 J
Mean vapour press. (mbar)	8.8	8.7	8.5	8.0	7.5	7.0	6.9	7.0	6.9	6.9	7.3	8.3	7.7	1948–78 J
Mean monthly rainfall (mm)	79.3	74.6	87.6	87.3	75.5	70.6	61.9	57.8	68.3	66.6	64.2	68.8	862.5	1948–78 J
No. of raindays	25	23	25	26	27	25	25	25	25	24	23	24	297	1948–78 J
Max. monthly rainfall (mm)	151.9	127.8	139.4	136.5	113.2	122.7	114.0	128.3	106.5	108.1	108.0	114.6	1097	1948–78 J
Min. monthly rainfall (mm)	30.4	8.2	6.5	41.5	10.3	6.4	10.9	24.6	25.0	22.6	11.2	9.4	456	1948–78 J
Mean monthly sunshine (h)	100.5	95.1	76.6	51.9	29.3	16.7	24.6	43.5	58.1	79.1	90.9	90.9	757.3	1948–78 J
Mean cloudiness (tenths)	6.7	6.4	6.5	6.6	6.5	6.5	6.5	6.5	6.4	6.5	6.7	6.8	6.5	1948–78 J
Mean fog freq. (days)	6.5	5.4	5.5	5.8	6.0	5.1	6.1	5.3	3.4	4.0	5.3	5.7	64.1	1967–78 J
Mean thunder freq. (days)	0.2	0.2	0.4	0.2	0	0	0.2	0.4	0	0.2	0	0	1.8	1968–76 J
Most freq. wind direction	W	W	W	W	W	W	W	W	W	W	W	W	W	1948–78 J

TABLE LII

CLIMATIC TABLE FOR NANDI, VITI LEVU, FIJI ISLANDS (52)
Latitude 17°45'S, longitude 177°27'E, elevation 16 m

	Jan.	Feb.	Mar.	Apr.	May	June	July	Aug.	Sep.	Oct.	Nov.	Dec.	Year	Period and source
Mean sea level press. (mbar)	1007.7	1007.8	1008.1	1010.4	1011.9	1013.1	1013.6	1013.5	1013.5	1012.7	1010.5	1008.7	1011.0	1949–61 B
Mean monthly temp. (°C)	26.6	26.7	26.4	25.8	24.6	23.8	22.9	23.2	23.9	24.8	25.6	26.3	25.1	1951–70 I
Mean monthly rainfall (mm)	306.1	295.6	418.2	222.6	106.1	106.3	53.2	63.1	86.2	59.8	130.3	218.7	2066.2	1951–60 A
Bright sunshine (h; mean)	206	177	171	194	203	208	218	240	214	231	217	251	2530	1947–60 B

TABLE LIII

CLIMATIC TABLE FOR NAURU ISLAND (53)
Latitude 0°34'S, longitude 166°55'E, elevation 5.8 m

	Jan.	Feb.	Mar.	Apr.	May	June	July	Aug.	Sep.	Oct.	Nov.	Dec.	Year	Period and source
Mean stat. level press. (mbar)	1005.3	1005.5	1005.9	1006.3	1006.5	1006.9	1006.7	1006.9	1007.1	1007.0	1006.1	1005.3	1006.3	1953–77 J
Mean daily max. temp. (°C)	30.7	30.5	30.5	30.5	30.6	30.4	30.3	30.5	30.7	31.1	31.2	30.9	30.7	1951–77 J
Mean daily min. temp. (°C)	24.7	24.9	24.9	24.9	25.2	25.1	24.8	24.9	24.9	25.0	25.0	24.8	24.9	1951–77 J
Highest temp. (°C)	34.4	37.1	35.1	35.0	33.0	33.0	35.5	33.8	35.0	34.3	35.9	35.6	37.1	1951–77 J
Lowest temp. (°C)	21.1	21.7	21.1	21.7	20.4	21.1	20.3	21.1	20.6	21.4	21.2	21.0	20.3	1951–77 J
Mean vapour press. (mbar)	30.2	30.3	30.2	30.6	30.3	30.0	29.7	29.3	29.5	29.5	29.6	30.3	30.0	1953–77 J
Mean monthly rainfall (mm)	285	257	196	195	127	111	151	142	121	107	115	257	2064	1949–77 J
No. of raindays	17	13	13	14	12	13	14	13	10	8	9	15	151	1949–77 J
Max. monthly rainfall (mm)	781	641	548	474	450	300	557	427	526	411	414	597	3663	1949–77 J
Min. monthly rainfall (mm)	0	1	1	3	4	4	12	12	3	1	1	0	767	1949–77 J
Mean cloudiness (tenths)	5.6	5.5	5.1	5.1	4.6	4.7	4.8	4.6	4.5	4.5	4.9	5.5	4.9	1953–77 J
Mean thunder freq. (days)	0.2	0.2	0	0.1	0.3	0.1	0.1	0.4	0.1	0.2	0.1	0.4	2.3	1951–77 J
Most freq. wind direction	E	E	E	E	E	E	E	E	E	E	E	E	E	1953–77 J

TABLE LIV

CLIMATIC TABLE FOR NORFOLK ISLAND (54)
Latitude 29°4'S, longitude 167°58'E, elevation 108.8 m

	Jan.	Feb.	Mar.	Apr.	May	June	July	Aug.	Sep.	Oct.	Nov.	Dec.	Year	Period and source
Mean stat. level press. (mbar)	1000.5	1000.4	1001.3	1003.0	1003.0	1003.0	1003.3	1003.6	1004.6	1004.3	1002.6	1000.8	1002.5	1951–78 J
Mean daily max. temp. (°C)	24.4	24.8	24.1	22.6	20.8	19.1	18.2	18.2	18.8	20.1	21.8	23.2	21.3	1939–78 J
Mean daily min. temp. (°C)	18.9	19.5	19.1	17.6	15.8	14.6	13.3	13.1	13.4	14.5	15.9	17.5	16.1	1939–78 J
Highest temp. (°C)	28.1	28.0	35.9	26.6	24.3	22.9	22.0	23.6	22.3	23.7	26.5	27.9	35.9	1939–78 J
Lowest temp. (°C)	12.1	12.8	12.1	9.7	4.3	7.1	6.2	6.7	7.7	8.2	8.7	11.4	4.3	1939–78 J
Mean vapour press. (mbar)	20.4	21.6	20.9	18.4	16.4	15.3	13.9	13.9	14.2	15.5	17.2	18.9	17.2	1939–78 J
Mean monthly rainfall (mm)	88	98	105	128	140	150	147	136	94	89	67	81	1323	1890–1978 J
No. of raindays	10	12	15	16	18	19	21	19	15	13	9	11	178	1890–1978 J
Max. monthly rainfall (mm)	340	411	355	429	390	458	356	463	216	289	239	363	1934	1890–1978 J
Min. monthly rainfall (mm)	6	3	7	17	26	30	33	46	25	12	5	5	785	1890–1978 J
Mean monthly sunshine (h)	231.1	192.4	198.7	205.6	177.3	153.2	180.8	198.5	209.8	226.6	241.1	244.7	2459.7	1951–78 J
Mean cloudiness (tenths)	5.1	5.4	5.2	4.7	4.8	5.0	4.6	4.5	4.5	4.8	4.8	4.9	4.8	1939–78 J
Mean fog freq. (days)	0.9	1.0	0.8	0.5	0.2	0.2	0.0	0.1	0.4	0.5	0.6	0.6	5.8	1939–78 J
Mean thunder freq. (days)	0.5	0.3	0.4	0.6	1.3	1.4	1.4	1.8	1.1	1.1	0.8	0.6	11.2	1939–78 J
Most freq. wind direction	E	E	E	E	E	W	W	W	E	E	E	E	E	1939–78 J

TABLE LV

CLIMATIC TABLE FOR NOUMEA, NEW CALEDONIA ISLAND (55)
Latitude 22°16'S, longitude 166°27'E, elevation 72 m

	Jan.	Feb.	Mar.	Apr.	May	June	July	Aug.	Sep.	Oct.	Nov.	Dec.	Year	Period and source
Mean sea level press. (mbar)	1009.2	1008.9	1009.9	1012.9	1014.1	1015.7	1015.8	1016.3	1016.4	1015.2	1013.0	1010.7	1013.2	1952–66 G
Mean monthly temp. (°C)	25.7	26.3	25.4	23.8	22.3	21.0	19.8	19.9	20.6	22.1	23.7	25.0	23.0	1952–66 G
Mean monthly min. temp. (°C)	22.8	23.4	22.6	21.1	19.7	18.4	17.2	17.1	17.8	19.0	20.6	21.9	20.1	1952–66 G
Absolute min. temp. (°C)	19.4	19.8	18.8	17.6	16.0	13.6	13.5	13.2	13.3	14.9	15.2	17.8	13.2	1952–66 G
Mean monthly max. temp. (°C)	28.5	29.2	28.1	26.5	24.9	23.5	22.4	22.7	23.5	25.3	26.8	28.1	25.8	1952–66 G
Absolute max. temp. (°C)	35.2	34.0	33.8	33.2	31.8	29.8	27.2	29.8	28.8	30.8	32.3	35.3	35.3	1952–66 G
Mean monthly rainfall (mm)	110.6	94.6	114.4	132.3	113.2	94.6	78.0	77.3	44.7	39.7	45.1	66.1	1010.7	1952–66 G
Max. daily rainfall (mm)	219.6	107.0	96.6	135.2	120.7	59.7	58.5	91.1	44.7	58.8	36.2	83.9	219.6	1952–66 G
Max. monthly rainfall (mm)	351.6	183.9	269.6	263.6	339.0	207.6	209.9	226.7	112.6	89.9	95.6	273.7	1390.0	1952–66 G
Min. monthly rainfall (mm)	11.2	15.1	37.9	21.8	19.1	39.6	10.8	13.6	1.2	4.4	8.8	0.4	577.1	1952–66 G
Mean relat. humid. (%)	76	78	77	78	77	76	75	74	72	73	74	75	77	1952–66 G
Mean evap. (mm)	131.5	110.0	119.0	108.1	103.6	104.1	116.9	115.4	119.5	138.6	142.1	145.1	1453.9	1952–66 G
Mean vapour press. (mbar)	24.5	26.2	24.6	22.5	20.4	18.6	17.0	16.9	17.6	18.6	20.9	22.9	20.9	1952–66 G
Bright sunshine (h; mean)	236	231	203	204	184	167	188	206	204	270	254	267	2614	1951–60 B
Most freq. wind direction	E	E	E	E	E	E	E	E	E	E	E	E	E	1951–65 G

TABLE LVI

CLIMATIC TABLE FOR NUKU ALOFA, TONGATAPU ISLAND (56)
Latitude 21°08'S, longitude 175°12'W, elevation 3 m

	Jan.	Feb.	Mar.	Apr.	May	June	July	Aug.	Sep.	Oct.	Nov.	Dec.	Year	Period and source
Mean sea level press. (mbar)	1009.6	1009.6	1009.7	1011.6	1013.5	1014.5	1015.1	1015.1	1015.6	1014.8	1012.2	1013.3	1012.6	1951–60 A
Mean monthly temp. (°C)	25.6	25.8	25.8	25.1	23.4	22.3	21.2	21.4	21.9	22.8	23.9	25.0	24.4	1951–60 A
Mean monthly rainfall (mm)	200.6	219.8	227.2	150.1	113.1	92.0	106.3	112.8	112.5	113.2	109.4	129.8	1686.9	1926–69 E
Max. monthly rainfall (mm)	582.2	564.1	497.3	457.7	452.1	320.0	287.5	272.8	301.8	337.3	350.8	581.2	2435.6	1926–69 E
Min. monthly rainfall (mm)	9.7	11.4	83.8	27.7	17.3	7.1	26.4	16.5	11.4	2.8	6.6	3.3	988.8	1926–69 E

TABLE LVII

CLIMATIC TABLE FOR OCEAN ISLAND (57)
Latitude 00°52'S, longitude 169°35'E, elevation 65 m

	Jan.	Feb.	Mar.	Apr.	May	June	July	Aug.	Sep.	Oct.	Nov.	Dec.	Year	Period and source
Mean sea level press. (mbar)	1009.2	1009.3	1009.8	1010.3	1010.4	1010.6	1010.6	1010.3	1010.7	1010.9	1009.6	1009.1	1010.1	1951-60 A
Mean monthly temp. (°C)	27.8	27.8	27.7	27.9	28.0	28.0	27.8	28.0	28.3	28.4	28.3	28.0	28.0	1921-60 I
Mean monthly rainfall (mm)	295.9	220.6	182.2	152.2	112.8	110.1	133.1	110.9	93.4	95.2	136.5	205.5	1848.4	1905-69 E
Max. monthly rainfall (mm)	826.3	610.4	733.8	679.7	550.9	562.9	602.2	472.9	408.9	469.6	533.7	527.8	4487.7	1905-69 E
Min. monthly rainfall (mm)	3.0	0.3	2.5	0.3	2.0	4.1	12.7	14.0	3.3	0.0	0.0	0.0	360.9	1905-69 E

TABLE LVIII

CLIMATIC TABLE FOR PAPEETE, TAHITI ISLAND (58)
Latitude 17°32′S, longitude 149°35′W, elevataion 2 m

	Jan.	Feb.	Mar.	Apr.	May	June	July	Aug.	Sep.	Oct.	Nov.	Dec.	Year	Period and source
Mean sea level press. (mbar)*	1010.9	1011.1	1011.6	1011.7	1012.3	1013.6	1014.2	1014.5	1014.2	1013.8	1011.8	1011.3	1012.6	1949-62 F
Mean monthly temp. (°C)*	26.7	26.8	27.1	26.9	26.0	25.3	24.8	24.8	25.1	25.7	26.3	26.6	26.0	1935-70 I
Mean monthly min. temp. (°C)*	22.5	22.6	22.8	22.5	21.9	20.8	20.4	20.1	20.7	21.2	22.1	22.5	21.7	1949-62 F
Absolute min. temp. (°C)*	18.8	18.9	19.4	18.9	17.0	17.1	15.8	14.9	15.8	15.8	18.1	19.5	14.9	1949-62 F
Mean monthly max. temp. (°C)*	29.6	29.9	30.2	30.1	29.3	28.7	28.2	27.9	28.1	28.6	29.2	29.4	29.1	1949-62 F
Absolute max. temp. (°C)*	33.3	32.5	33.0	33.0	31.9	31.3	30.8	30.1	31.7	31.9	32.0	32.1	33.3	1949-62 F
Mean monthly rainfall (mm)*	335.1	262.9	157.2	134.4	121.4	65.5	64.6	49.0	74.0	83.9	181.6	287.2	1836.7	1922-71 E
Max. monthly rainfall (mm)*	1365.1	730.6	753.1	448.7	502.2	176.2	253.3	173.0	297.4	315.9	598.9	740.9	2908.9	1922-71 E
Min. monthly rainfall (mm)*	61.2	35.8	9.9	2.5	0.4	0.0	0.2	1.3	1.1	3.4	10.3	19.7	1020.3	1922-71 E
Max. daily rainfall (mm)*	273.0	231.0	125.5	143.9	275.0	114.6	79.6	77.5	127.2	99.3	168.2	157.7	275.0	1938-63 F
No. of raindays*	19	16	14	11	10	8	8	7	8	9	12	17	139	1938-63 F
Mean relat. humid. (%)	79	78	78	78	78	77	76	76	77	77	77	78	77	1951-60 A
Mean vapour press. (mbar)	27.2	27.4	27.6	27.1	26.1	24.3	23.5	22.8	23.7	24.7	25.9	26.6	25.6	1949-62 F
Most freq. wind direct.	NE	NE	NE	NE	NE	NE	NE	NE	NE	NE	NE	NE	NE	H

* Data from Tahiti F.A.A.A.

TABLE LIX

CLIMATIC TABLE FOR COOK ISLAND (59)
Latitude 09°00'S, longitude 158°03'W, elevation 1 m

	Jan.	Feb.	Mar.	Apr.	May	June	July	Aug.	Sep.	Oct.	Nov.	Dec.	Year	Period and source
Mean monthly temp. (°C)	27.7	27.9	28.2	28.4	28.4	27.9	27.6	27.3	27.7	27.7	27.8	27.9	27.9	1937-60 C
Mean daily min. temp. (°C)	25.6	25.6	25.8	26.1	26.3	25.8	25.4	24.3	25.4	25.5	25.5	25.7	25.7	1941-70 D
Mean daily max. temp. (°C)	29.8	29.9	30.1	30.4	30.3	29.6	29.4	29.1	29.2	29.5	29.8	29.8	29.7	1941-70 D
Highest temp. (°C)	33.9	34.3	33.9	35.1	33.3	32.9	33.6	33.9	33.3	33.6	34.2	34.7	35.1	1937-60 C
Lowest temp. (°C)	21.7	21.6	21.5	21.8	21.9	19.3	21.1	21.1	22.2	21.4	22.2	21.3	19.3	1937-60 C
Mean monthly rainfall (mm)	211.0	183.4	192.3	149.2	132.4	158.7	147.4	135.7	117.9	138.2	157.0	170.2	1893.4	1938-71 E
Max. monthly rainfall (mm)	645.2	539.5	495.3	346.7	437.4	569.7	659.1	612.4	277.1	396.2	451.1	565.7	3655.1	1938-71 E
Min. monthly rainfall (mm)	27.7	27.9	42.4	49.0	24.1	22.1	10.9	21.1	3.8	23.6	32.3	11.2	852.9	1938-71 E
Max. daily rainfall (mm)	102.11	111.25	74.93	101.85	140.46	84.58	144.02	153.16	99.06	119.89	191.01	143.26	191.01	1937-60 C
No. of raindays	14	13	18	16	14	16	15	16	14	16	14	14	180	1937-60 C
Mean relat. humid. (%)	82	83	84	84	83	83	82	82	82	81	80	80	82	1937-60 C
Days with thunder	0.6	1.0	2.1	2.2	0.4	0.8	0.4	0.2	0.1	0.9	1.1	0.9	10.7	1937-60 C

TABLE LX

CLIMATIC TABLE FOR PITCAIRN ISLAND (60)
Latitude 25°04'S, longitude 130°06'W, elevation 264 m

	Jan.	Feb.	Mar.	Apr.	May	June	July	Aug.	Sep.	Oct.	Nov.	Dec.	Year	Period and source
Mean monthly temp. (°C)	23.2	24.0	23.3	22.3	20.9	19.8	19.1	18.9	19.2	20.0	20.7	22.1	22.1	1956–60 I
Mean daily min. temp. (°C)	20.8	21.1	20.9	19.9	18.7	17.6	17.0	16.5	16.8	17.6	18.5	19.9	18.8	1941–70 D
Mean daily max. temp. (°C)	26.3	26.8	26.5	25.1	23.3	22.0	21.5	21.5	21.9	22.8	23.7	25.1	23.9	1941–70 D
Mean monthly rainfall (mm)	158.4	151.5	130.4	157.9	148.9	157.4	168.1	132.8	139.8	147.6	129.9	175.4	1798.0	1940–71 E
Max. monthly rainfall (mm)	318.3	353.3	420.4	315.2	314.2	339.1	419.6	325.1	311.4	362.7	364.7	339.6	2629.7	1940–71 E
Min. monthly rainfall (mm)	50.0	37.8	39.1	50.0	29.2	31.0	30.0	15.0	32.0	12.7	14.5	51.1	1030.7	1940–71 E

TABLE LXI

CLIMATIC TABLE FOR RAOUL ISLAND (61)
Latitude 29°15'S, longitude 177°55'W, elevation 38 m

	Jan.	Feb.	Mar.	Apr.	May	June	July	Aug.	Sep.	Oct.	Nov.	Dec.	Year	Period and source
Mean sea level press. (mbar)	1014.3	1014.7	1013.6	1016.1	1016.3	1016.8	1017.1	1017.1	1019.7	1017.7	1016.3	1014.0	1016.1	1951–60 A
Mean monthly temp. (°C)	21.5	22.3	21.8	20.5	18.7	17.2	16.1	15.9	16.2	17.3	18.7	20.4	18.9	1940–70 I
Mean daily min. temp. (°C)	19.0	20.0	19.4	18.1	16.2	14.8	13.6	13.4	13.6	14.7	16.0	17.9	16.4	1941–70 D
Mean daily max. temp. (°C)	24.2	24.8	24.3	23.2	21.4	19.8	18.7	18.6	19.0	20.2	21.5	23.1	21.6	1941–70 D
Highest temp. (°C)	27.9	27.7	27.4	27.3	25.3	24.0	23.2	22.3	22.3	23.8	25.4	26.2	27.9	1940–60 C
Lowest temp. (°C)	13.6	14.7	13.4	12.8	11.2	9.3	8.4	9.0	8.5	9.3	10.4	12.8	8.4	1940–60 C
Max. monthly rainfall (mm)	341.4	730.8	601.0	310.9	335.0	297.2	323.1	275.1	241.3	225.0	308.9	317.5	2448.6	1937–71 E
Min. monthly rainfall (mm)	2.0	25.7	6.1	14.7	51.1	69.3	39.1	'28.2	40.4	13.5	6.9	0.5	929.9	1937–71 E
Max. daily rainfall (mm)	127.25	176.02	125.73	137.41	162.81	122.17	114.81	70.10	108.20	88.39	81.28	125.22	176.02	1940–60 C
No of raindays	12	14	16	16	20	20	21	21	15	14	10	11	190	1940–60 C
Mean relat. humid. (%; 0900 GMT)	77	78	77	76	75	75	74	75	73	74	73	75	75	1937–60 C
Bright sunshine (h; mean)	205	170	168	165	154	131	155	170	166	192	203	215	2094	1940–60 C
(% of possible)	50	48	46	50	49	44	50	52	49	50	52	51	49	1940–60 C
Mean monthly rainfall (mm)	100.7	145.6	142.6	127.7	144.7	155.2	165.0	143.6	113.1	88.8	83.5	91.2	1501.7	1937–71 E

TABLE LXI *(continued)*

	Jan.	Feb.	Mar.	Apr.	May	June	July	Aug.	Sep.	Oct.	Nov.	Dec.	Year	Period and source
Days with hail						0.1	0.1	0.2					0.4	1937–60 E
Days with thunder	0.5	0.5	0.1	0.5	0.8	1.0	1.0	0.8	0.2	0.5	0.2	0.3	6.4	1937–60 C
Days with fog	0.1	0.1	0.1	0.3	0.1	0.1		0.1	0.3	0.1		0.2	1.4	1937–60 C
Most freq. wind direct.	E	E	E	E	E	W	W	W	E	E	E	E	E	H

TABLE LXII

CLIMATIC TABLE FOR RAPA, TUBUAI ISLANDS (62)
Latitude 27°25'S, longitude 144°17'W, elevation 6 m

	Jan.	Feb.	Mar.	Apr.	May	June	July	Aug.	Sep.	Oct.	Nov.	Dec.	Year	Period and source
Mean sea level press. (mbar)	1015.1	1015.4	1016.1	1015.9	1015.6	1017.3	1018.1	1017.6	1019.6	1019.9	1017.3	1016.7	1017.1	1952-60 A
Mean monthly temp. (°C)	23.1	23.7	23.1	22.2	20.3	18.8	18.4	17.7	18.1	19.1	20.3	22.0	20.6	1952-70 I
Mean monthly min. temp. (°C)	20.5	21.6	20.7	19.3	17.6	15.9	15.6	14.7	14.7	16.5	18.0	19.5	17.9	1951-57 K
Mean monthly max. temp. (°C)	25.9	26.6	26.1	24.7	23.0	21.7	20.6	20.2	20.1	21.8	23.3	25.0	23.3	1951-57 K
Highest temp. (°C)	29.1	30.0	30.0	28.6	27.4	25.2	24.0	24.1	24.5	25.8	28.4	29.9	30.0	1951-57 K
Lowest temp. (°C)	12.2	16.2	15.2	14.0	10.1	10.5	9.8	9.4	9.0	12.2	13.4	14.1	9.0	1951-57 K
Mean monthly rainfall (mm)	320	216	252	255	260	205	239	250	224	187	226	227	2860	1944-71 E
Max. monthly rainfall (mm)	696	368	520	735	578	412	496	567	597	521	592	510	3772	1944-71 E
Min. monthly rainfall (mm)	45	12	74	71	52	72	116	108	82	67	80	52	2194	1944-71 E
Max. daily rainfall (mm)	261.6	112.8	160.7	150.8	96.5	62.5	87.5	143.8	110.4	108.8	116.8	99.9	261.6	1951-57 K
Mean relat. humid. (%)	82	81	81	79	76	77	77	76	76	77	79	80	78	1951-60 B
No. of raindays	16	15	19	19	18	16	20	19	17	14	14	14	201	1951-57 K
Mean vapour press. (mbar)	23.4	24.2	23.5	21.7	18.7	17.1	16.6	15.7	16.2	17.2	19.1	21.2	19.6	1951-67 B

TABLE LXIII

CLIMATIC TABLE FOR RAROTONGA ISLAND, AIRPORT (63)
Latitude 21°12′S, longitude 159°49′W, elevation 5 m

	Jan.	Feb.	Mar.	Apr.	May	June	July	Aug.	Sep.	Oct.	Nov.	Dec.	Year	Period and source
Mean sea level press. (mbar)	1011.0	1011.3	1011.8	1012.7	1014.3	1015.4	1015.8	1015.8	1016.5	1016.0	1013.2	1011.6	1013.8	1951-60 A
Mean monthly temp. (°C)	25.7	25.9	25.6	24.8	23.3	22.2	21.6	21.6	22.1	22.9	23.8	24.8	23.7	1907-70 I
Mean daily min. temp. (°C)	22.7	22.8	22.4	21.8	20.1	19.1	18.3	18.1	18.8	19.9	20.9	21.7	20.6	1941-70 D
Mean daily max. temp. (°C)	28.8	29.1	28.9	28.3	26.7	25.7	25.0	24.9	25.3	26.1	27.1	27.8	27.0	1941-70 D
Highest temp. (°C)	31.6	31.8	32.8	31.3	32.1	28.3	28.3	28.1	27.9	29.8	30.2	30.5	32.8	1948-60 C
Lowest temp. (°C)	16.6	17.4	16.7	15.1	14.1	11.7	11.6	11.9	11.6	13.1	13.9	15.6	11.6	1948-60 C
Mean monthly rainfall (mm)	253.2	236.5	276.5	186.2	161.7	106.7	105.5	122.4	109.0	130.9	149.6	225.5	2063.8	1899-1971 E
Max. monthly rainfall (mm)	667.5	508.8	972.8	446.8	693.4	281.9	414.3	475.0	236.7	436.9	394.2	652.8	3132.8	1899-1971 E
Min. monthly rainfall (mm)	18.8	41.9	82.3	29.7	26.2	13.0	12.2	7.9	10.9	15.7	8.9	10.7	1169.4	1899-1971 E
Max. daily rainfall (mm)	172.21	154.69	91.69	94.23	129.54	156.97	163.83	76.45	69.34	107.95	140.97	189.23	189.23	1948-60 C
No. of raindays	19	19	20	17	16	13	14	12	14	14	14	19	191	1948-60 C
Mean relat. humid. (%; 1800 GMT)	85	86	90	86	83	85	84	83	80	79	82	84	84	1948-60 C
Bright sunshine (h; mean)	167	167	179	174	162	165	178	191	182	183	175	173	2096	1948-60 C
Most freq. wind direct.	E	E	E	E	E	E	E	E	E	E	E	E	E	H
Days with thunder	1.4	1.8	2.4	1.6	1.2	0.6	0.3	0.4	0.5	1.4	0.8	1.9	14.3	1948-60 C

TABLE LXIV

CLIMATIC TABLE FOR ROTUMA ISLAND, FIJI ISLANDS (64)
Latitude 12°30'S, longitude 177°03'E, elevation 26 m

	Jan.	Feb.	Mar.	Apr.	May	June	July	Aug.	Sep.	Oct.	Nov.	Dec.	Year	Period and source
Mean sea level press. (mbar)	1008.0	1008.0	1009.0	1010.2	1011.1	1011.9	1012.3	1012.3	1012.4	1011.8	1009.9	1008.6	1010.5	1951–60 A
Mean monthly temp.(°C)	27.4	27.2	27.3	27.5	27.3	26.8	26.4	26.5	26.8	26.9	27.1	27.3	27.0	1951–60 A
Mean monthly rainfall (mm)	359.1	398.8	412.9	251.8	299.0	266.4	208.0	202.6	258.1	320.3	345.2	326.7	3649.1	1947–69 E
Max. monthly rainfall (mm)	678.7	925.1	769.9	458.0	600.2	553.7	401.1	486.9	540.0	584.7	778.3	664.0	4286.2	1947–69 E
Min. monthly rainfall (mm)	104.9	116.3	97.3	96.3	55.9	96.5	33.5	96.3	76.2	138.9	182.1	27.2	2765.6	1947–69 E
Most freq. wind direct.	E	N	NNE	E	E	E	E	E	SE	E	E	N	E	H

TABLE LXV

CLIMATIC TABLE FOR SAN CRISTOBAL ISLAND (65)
Latitude 00°54′S, longitude 89°57′W, elevation 6 m

	Jan.	Feb.	Mar.	Apr.	May	June	July	Aug.	Sep.	Oct.	Nov.	Dec.	Year	Period and source
Mean monthly temp. (°C)	25.1	25.9	26.1	25.7	25.2	23.8	22.7	21.7	21.5	22.0	22.6	23.8	23.8	1951–60 A
Highest temp. (°C)	26.0	26.6	27.1	26.5	26.6	25.7	25.0	23.9	23.2	23.6	24.1	25.0	27.1	1951–60 A
Lowest temp. (°C)	24.4	25.1	25.2	24.0	23.8	22.0	20.6	19.9	19.6	20.5	21.1	22.3	19.6	1951–60 A
Mean monthly rainfall (mm)	70	126	145	138	64	49	51	42	52	37	36	42	852	1950–72 E
Max. monthly rainfall (mm)	215	487	651	805	438	416	172	224	164	169	102	136	2432	1950–72 E
Min. monthly rainfall (mm)	0	2	0	0	0	0	0	1	1	1	1	1	30	1950–72 E

TABLE LXVI

CLIMATIC TABLE FOR SANTA CRUZ ISLAND (66)
Latitude 10°43'S, longitude 165°48'E, elevation 22.8 m

	Jan.	Feb.	Mar.	Apr.	May	June	July	Aug.	Sep.	Oct.	Nov.	Dec.	Year	Period and source
Mean stat. level press. (mbar)	1003.9	1004.7	1005.5	1006.1	1007.0	1007.4	1007.5	1008.1	1008.2	1007.5	1006.0	1004.5	1006.3	1970–78 J
Mean daily max. temp. (°C)	30.4	30.7	30.4	30.5	30.1	29.8	29.1	29.0	29.3	29.8	30.6	30.7	30.0	1970–78 J
Mean daily min. temp. (°C)	23.9	23.9	23.9	23.9	23.8	23.7	23.6	23.5	23.6	23.7	23.8	23.7	23.7	1970–78 J
Highest temp. (°C)	33.1	35.3	33.0	33.4	32.8	32.1	31.7	31.1	31.4	32.2	33.1	33.0	35.3	1970–78 J
Lowest temp. (°C)	21.7	21.3	22.0	22.2	20.5	21.0	20.5	19.0	20.5	21.2	21.1	21.1	19.0	1970–78 J
Mean vapour press. (mbar)	30.7	30.9	30.8	31.1	30.5	30.1	29.2	28.8	28.7	29.4	30.2	30.3	30.1	1970–78 J
Mean monthly rainfall (mm)	516	397	481	310	333	280	325	396	329	356	320	316	4359	1970–78 J
No of raindays	24	22	25	23	23	22	24	25	21	23	21	22	275	1970–78 J
Max. monthly rainfall (mm)	1008	573	802	364	653	347	494	518	487	507	496	523	5479	1970–78 J
Min. monthly rainfall (mm)	232	287	300	208	148	214	247	296	204	170	168	157	3723	1970–78 J
Mean cloudiness (tenths)	6.0	5.8	5.8	5.4	5.4	5.2	5.4	5.4	5.1	5.3	5.4	5.7	5.4	1970–78 J
Days with thunder	3.8	4.8	4.5	3.6	4.0	3.6	3.0	4.1	3.1	3.9	7.0	5.7	51.0	1970–78 J
Most freq. wind direct.	CALM	CALM	CALM	CALM	CALM	CALM	SE	SE	SE	SE	CALM	CALM	CALM	1970–78 J

TABLE LXVII

CLIMATIC TABLE FOR SUVA, FIJI ISLANDS (67)
Latitude 18°08'S, longitude 178°26'E, elevation 5 m

	Jan.	Feb.	Mar.	Apr.	May	June	July	Aug.	Sep.	Oct.	Nov.	Dec.	Year	Period and source
Mean stat. level press. (mbar)	1008.3	1007.7	1009.6	1011.1	1012.2	1013.5	1014.5	1014.4	1014.2	1013.4	1010.9	1008.4	1011.5	H
Mean daily temp. (max. + min)/2 (°C)	26.6	26.7	26.7	26.0	24.7	23.9	22.9	23.2	23.5	24.2	24.9	26.0	24.9	H
Mean monthly rainfall (mm)	304.6	295.1	374.1	332.0	256.1	164.3	137.4	186.8	202.9	218.4	251.7	303.2	3026.6	1883–1969 E
Max. monthly rainfall (mm)	802.9	832.1	749.8	1154.4	917.7	638.8	469.6	948.2	456.7	900.2	752.6	1003.6	4561.6	1883–1969 E
Min. daily rainfall (mm)	42.9	96.8	82.8	81.3	45.7	18.5	7.4	22.4	2.8	13.0	13.5	51.8	1855.0	1883–1969 E
Most freq. wind direct.	E	E	E	E	SE	SE	SE	SE	SE	E	E	E	E	H

TABLE LXVIII

CLIMATIC TABLE FOR TAKAROA, TUAMOTU ARCHIPELAGO (68)
Latitude 14°30'S, longitude 145°05'W, elevation 2 m

	Jan.	Feb.	Mar.	Apr.	May	June	July	Aug.	Sep.	Oct.	Nov.	Dec.	Year	Period and source
Mean sea level press. (mbar)	1009.9	1009.9	1010.6	1010.4	1011.0	1011.9	1012.9	1012.9	1012.6	1012.3	1011.0	1010.2	1011.3	1951-60 B
Mean monthly temp. (°C)	27.4	27.8	28.0	27.9	27.5	26.7	26.2	25.9	26.1	26.5	26.9	27.3	27.0	1952-70 I
Mean monthly min. temp. (°C)	24.9	25.6	26.1	25.7	25.4	25.1	24.5	24.2	24.3	24.7	24.8	25.1	25.0	1952-57 K
Mean monthly max. temp. (°C)	29.9	30.5	30.2	29.8	29.6	28.1	27.6	27.7	27.7	28.5	28.8	29.4	29.0	1952-57 K
Highest temp. (°C)	32.7	33.2	32.7	33.0	32.8	31.0	30.9	31.0	31.6	30.9	32.0	33.0	33.2	1952-57 K
Lowest temp. (°C)	21.0	22.0	22.4	21.3	22.6	22.8	20.0	20.9	20.9	22.0	21.2	22.7	20.0	1952-57 K
Mean monthly rainfall (mm)	257	152	124	126	93	95	72	66	80	121	152	192	1530	1952-71 E
Max. monthly rainfall (mm)	681	388	394	292	274	233	299	270	220	279	432	525	2629	1952-71 E
Min. monthly rainfall (mm)	63	43	40	18	12	22	12	12	17	20	30	58	1065	1952-71 E
Max. daily rainfall (mm)	160.0	41.8	86.3	149.5	109.3	38.7	30.3	27.6	49.7	24.4	50.5	80.7	160.0	1952-57 K
Mean relat. humid. (%)	80	79	79	79	79	79	78	78	78	78	78	78	79	1951-60 B
No. of raindays	19	16	14	14	14	14	12	11	10	16	18	21	179	1952-57 K
Mean vapour press. (mbar)	28.7	28.9	29.3	29.3	28.4	27.3	26.2	26.1	26.5	27.1	27.8	28.2	27.8	1952-60 B

TABLE LXIX

CLIMATIC TABLE FOR VILA, EFATE ISLAND (69)
Latitude 17°44'S, longitude 168°19'E, elevation 20 m

	Jan.	Feb.	Mar.	Apr.	May	June	July	Aug.	Sep.	Oct.	Nov.	Dec.	Year	Period and source
Mean sea level press. (mbar)	1008.3	1007.9	1008.0	1011.1	1012.2	1014.3	1014.4	1014.5	1014.1	1013.8	1011.2	1009.6	1011.6	1954-60 A
Mean monthly temp. (°C)	26.3	26.6	26.2	25.2	24.1	23.2	22.3	22.5	23.2	24.0	24.9	25.9	24.5	1948-70 I
Mean monthly rainfall (mm)	247.4	240.9	351.1	213.1	145.0	133.6	117.0	80.3	148.6	90.7	138.9	180.1	2086.7	1954-60 A
Most freq. wind direct.	E	E	E	E	E	E	E	E	ESE	E	E	E	E	H

TABLE LXX

CLIMATIC TABLE FOR WAITANGI, CHATHAM ISLAND (70)
Latitude 43°57'S, longitude 176°34'W, elevation 44 m

	Jan.	Feb.	Mar.	Apr.	May	June	July	Aug.	Sep.	Oct.	Nov.	Dec.	Year	Period and source
Mean sea level press. (mbar)	1013.1	1016.0	1014.7	1012.5	1010.0	1007.4	1011.7	1013.9	1017.8	1013.1	1009.6	1010.5	1012.5	1951–60 A
Mean monthly temp. (°C)	14.5	14.9	14.2	12.4	10.4	8.7	7.8	8.1	9.0	10.1	11.6	13.3	11.3	1951–70 I
Mean daily min. temp. (°C)	11.3	11.9	11.4	9.7	8.0	6.5	5.3	5.7	6.6	7.6	8.7	10.5	8.6	1941–70 D
Mean daily max. temp. (°C)	17.4	17.7	17.0	15.1	12.8	10.8	10.1	10.4	11.5	12.7	14.4	16.2	13.8	1941–70 D
Highest temp. (°C)	22.5	22.8	22.2	19.4	16.6	14.6	14.0	14.5	16.0	17.1	19.1	20.4	22.8	1939–64 D
Lowest temp. (°C)	2.8	4.4	3.9	2.9	−0.7	−1.7	−1.2	−0.1	0.8	1.1	2.3	3.6	−1.7	1939–64 D
Mean monthly rainfall (mm)	71.1	61.0	66.0	66.0	88.9	99.1	91.4	81.3	63.5	48.3	55.9	58.4	850.9	1917–50 D
Max. daily rainfall (mm)	62.74	111.76	97.03	60.96	74.93	62.48	76.20	132.59	58.42	49.02	71.12	44.45	132.59	1917–50 C
No. of raindays	11	11	13	15	18	19	20	18	14	13	13	12	177	1917–50 C
Mean relat. humid. (%; 0900 GMT)	82	82	82	85	88	87	89	87	85	83	82	83	84	1939–58 C
Bright sunshine (h; mean)	200	145	128	103	88	67	79	100	109	132	161	158	1470	1957–67 D
Days with snow						0.2	0.3	0.2	0.1	0.1			0.9	1929–58 C
Days with hail	0.2	0.1	0.4	0.4	1.0	2.3	1.4	1.2	1.1	0.8	0.4	0.4	9.7	1929–58 C
Days with thunder	0.5	0.2	0.2	0.3	0.3	0.4		0.3	0.2	0.1	0.3	0.3	3.1	1929–58 C

TABLE LXX (*continued*)

	Jan.	Feb.	Mar.	Apr.	May	June	July	Aug.	Sep.	Oct.	Nov.	Dec.	Year	Period and source
Days with fog	1.9	2.5	2.0	2.5	1.5	1.2	1.6	1.2	1.7	1.8	2.1	1.9	21.9	1929-58 C
Days with ground frost		0.2	0.2	0.3	1.2	1.8	3.1	2.7	1.9	0.7	0.2	0.2	12.5	1939-58 C
Days with frost in screen (means)						0.2	0.4	0.2					0.8	1939-58 C

TABLE LXXI

CLIMATIC TABLE FOR WILLIS ISLAND (71)
Latitude 16°18'S, longitude 149°59'E, elevation 5.1 m

	Jan.	Feb.	Mar.	Apr.	May	June	July	Aug.	Sep.	Oct.	Nov.	Dec.	Year	Period and source
Mean stat. level press. (mbar)	1006.1	1005.7	1007.0	1009.8	1011.5	1013.0	1013.7	1013.6	1013.3	1011.9	1009.7	1007.5	1010.2	1952-78 J
Mean daily max. temp. (°C)	30.8	30.4	30.0	28.9	27.8	26.5	26.1	26.6	27.5	28.7	30.0	30.8	28.7	1939-78 J
Mean daily min. temp. (°C)	25.5	25.5	25.4	24.7	23.7	22.5	21.9	21.9	22.5	23.5	24.7	25.3	23.9	1939-78 J
Highest temp. (°C)	34.4	33.3	33.1	33.9	30.5	29.5	29.8	30.0	31.3	32.3	33.3	34.8	34.8	1939-78 J
Lowest temp. (°C)	20.7	22.2	21.5	21.2	16.1	17.5	17.2	17.9	18.6	18.9	21.5	21.1	16.1	1939-78 J
Mean vapour press. (mbar)	29.7	30.6	29.8	27.4	24.9	22.7	21.6	21.3	22.3	24.6	26.5	28.0	25.8	1939-78 J
Mean monthly rainfall (mm)	198	232	190	134	66	57	38	23	21	24	34	89	1106	1921-78 J
No. of raindays	15	16	17	15	12	11	9	7	6	5	6	10	129	1921-78 J
Max. monthly rainfall (mm)	645	685	605	511	359	320	253	216	206	166	278	451	2025	1921-78 J
Min. monthly rainfall (mm)	11	21	7	11	1	1	1	0	0	0	1	0	242	1921-78 J
Mean cloudiness (tenths)	4.9	5.2	5.0	4.8	4.4	4.5	4.4	4.0	3.8	3.5	3.5	4.0	4.3	1939-78 J
Days with fog	0.05	0.10	0.05	0	0.05	0	0.05	0	0	0.05	0	0	0.35	1950-78 J
Days with thunder	1.8	2.0	1.2	0.3	0.3	0.1	0	0	0	0.1	0.4	0.9	7.2	1939-78 J
Most freq. wind direc.	E	ESE	SE	SE	SE	SE	SE	SE	SE	ESE	ESE	ESE	SE	1939-78 J

Chapter 4

Climate of the North Pacific Ocean

K. TERADA* AND M. HANZAWA

Introduction

Geographic features

The Pacific Ocean is as big as the combined Atlantic and Indian oceans and amounts to almost 35% of the earth's surface. The North Pacific Ocean is bordered by North America on the east and Asia on the west. To the north, Siberia and Alaska almost join, and only the narrow and shallow Bering Strait connects the Pacific and Arctic oceans.

Along the northern and western boundaries lies a string of islands (Fig. 1) several hundred kilometres from the continent and nearly parallel to its coast. In the tropical North Pacific are many small islands, although fewer than in the South Pacific Ocean.

Early data summaries

It is practice in climatological analysis to rely on data at fixed points for long time intervals. This presents no overwhelming problem on land but, except for some islands and in recent years a few weather ships, most oceanic observations have been made by moving vessels. This raises many questions about the homogeneity of observations. Moreover, although the shipping lanes have enough observations for statistical treatment, the areas off the regular tracks have only sporadic observations and there are many such areas in the North Pacific. A bias is introduced in the data because observations are often not taken in areas with severe weather which ordinary ships (sometimes even ocean weather ships) avoid or leave.

The following early studies of marine data for the North Pacific Ocean may be mentioned. The U.S. HYDROGRAPHIC OFFICE (1878) published the *Meteorological Charts of the North Pacific Ocean*, covering the area from the Equator to 45°N and from the west coast of America to the 180th meridian. Based on data collected by Japanese merchant ships, in 1911 the Kobe Marine Observatory began to study the climate of the North Pacific and to publish current statistics of the meteorological elements. The DEUTSCHE SEEWARTE (1922) issued the *Dampferhandbuch für den Stillen Ozean*, a comprehensive manual on various meteorological conditions affecting navigation. The Netherlands Meteorological Institute (KONINKLIJK NEDERLANDSCH METEOROLOGISCH INSTITUUT, 1926) printed the *Monthly Meteorological Data for Ten-degree Squares in the Oceans*,

---

* Professor Kazuhiko Terada died on April 3, 1983 in Tokyo, Japan.

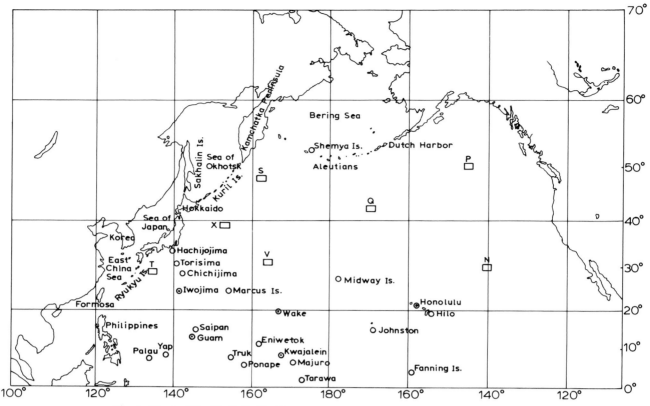

Fig. 1. Chart of the North Pacific Ocean.

which summarized data gathered from logbooks for the area 10—30°N and 140—160°W. Meteorological charts for the North Pacific were published by the United Kingdom (GREAT BRITAIN, 1928). Since these early efforts, many other nations have produced general or specialized charts of the climate of the North Pacific Ocean.

Ocean currents

In low latitudes, the South Equatorial Current reaches into the Northern Hemisphere (Fig. 2), the North Equatorial Current flows entirely within the Northern Hemisphere, and between the two the Equatorial Counter Current flows eastward across the North Pacific.

The North Equatorial Current flows from east to west and its volume increases as other water masses join from the north. It originates where the waters of the Equatorial Counter Current turn north off Central America. Farther west, waters of the California Current join the North Equatorial Current, and North Pacific waters are added west of the Sandwich Islands.

East of the Philippines, the North Equatorial Current divides into one branch turning south and a larger branch turning north. The latter closely follows the east side of the northernmost Philippines and Taiwan, and these warm waters continue as the well-known Kuroshio Current northeastward along the Ryukyu Islands and the south coast of Japan. A branch of the Kuroshio in the East China Sea and the Sea of Japan and

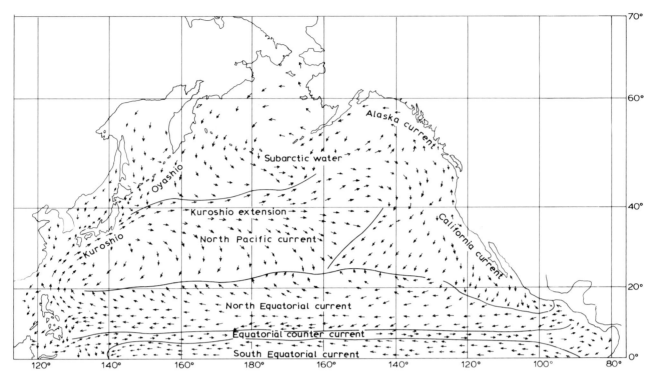

Fig. 2. Ocean currents of the North Pacific in winter.

the Kuroshio itself affect considerably the climate of east Asia, especially Japan.

After leaving Japan, one part of the Kuroshio curves to the east and as the Kuroshio Extension it becomes part of the western gyre and the North Pacific Current. Another part flows east-northeast where it meets the cold Oyashio Current and joins the North Pacific Current from the north.

The Oyashio Current is formed in the Sea of Okhotsk and the Bering Sea and moves southwestward past Kamchatka and along the Kurile Islands. The Oyashio is important in the climate of Japan, particularly in the north.

The North Pacific Current—Kuroshio Extension is a major part of the western gyre of the North Pacific. To the north is a marked transition to colder subarctic water in an extension of the Oyashio Current.

As the Kuroshio Extension flows eastward, one branch turns north along the coast of Alaska as the Alaska Current and another moves south along the American coast as the California Current.

The above is a general description of the various currents of the North Pacific Ocean. They vary with time and these variations affect appreciably the marine climates. Such variations were studied in some areas of the North Pacific by, e.g., REID (1973) in a report on the northwest Pacific Ocean in winter based on the *Boreas* Expedition of 1966 in Aleutian and Kurile waters, and in the USSR/USA Bering Sea experiment.

The ocean currents in winter are illustrated in Fig. 2; their intensity is qualitatively indicated by arrows; the thicker the arrow, the stronger the current. In summer, the South Equatorial Current becomes more predominant and the Equatorial Counter Current wider and stronger.

Mean pressure field

The seasonal extremes of the mean pressure distribution over the NPO are conspicuously different. In January (Fig. 3), the ocean is dominated by a deep extensive low, whereas in July (Fig. 4) it is completely covered by an anticyclone.

The pressure fields of the transition months are given in Figs. 5 and 6. The April map shows the expanding high and contracting low and the October map the rapidly developing mean low and shrinking high.

High pressure

The centre of the subtropical ridge of high mean pressure is found in all months in the southeastern North Pacific. In summer, the high extends over most of the ocean, and in winter a greatly reduced high lies over the southeastern part and is connected with the Siberian High by a ridge running along 25°N. The tropics, therefore, are under the influence of the south side of the high and its northeasterly trade winds throughout the year.

To illustrate the seasonal change of the mean subtropical high, the 1020-mbar isobar for every month is depicted in Fig. 7. Note that the area within the 1020-mbar isobar is similar from October to February and is confined to a region bounded by 22–38°N

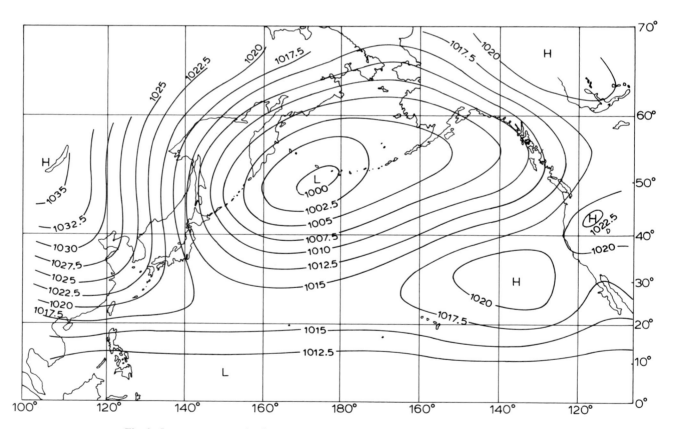

Fig. 3. January mean sea level pressure (mbar).

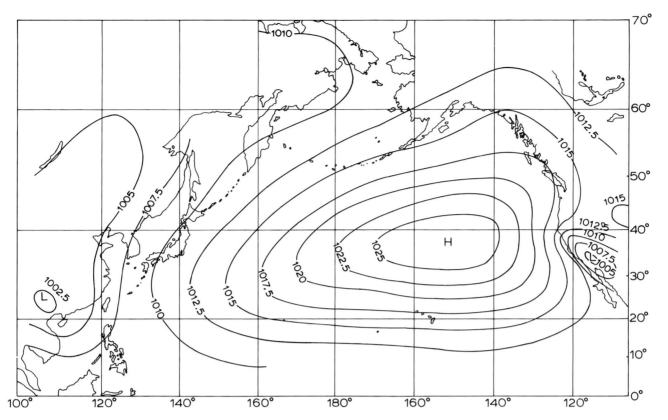

Fig. 4. July mean sea level pressure (mbar).

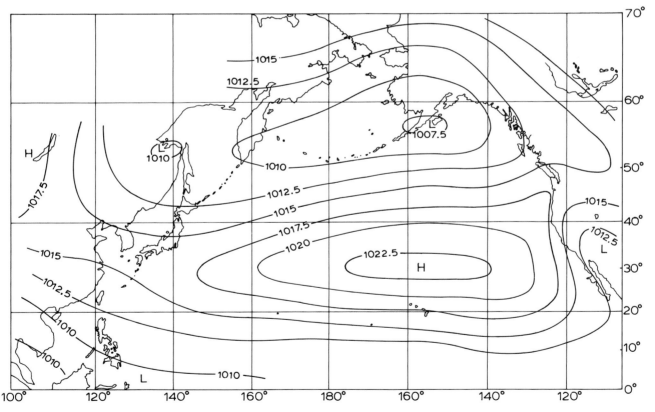

Fig. 5. April mean sea level pressure (mbar).

435

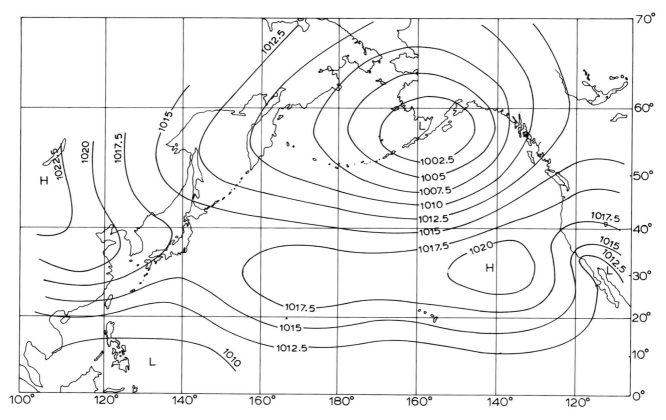

Fig. 6. October mean sea level pressure (mbar).

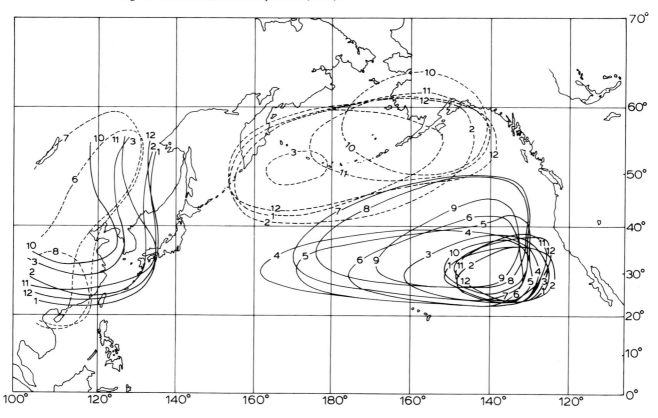

Fig. 7. Positions of monthly mean 1020-mbar isobars (solid lines) and 1005-mbar isobars (broken lines).

and 123−150°W. The size and location of the high in these months are closely related to the development of the Aleutian Low and Siberian High.

As summer approaches, the high expands and becomes more intense, first slowly but then rapidly between June and July. In summer, the 1020-mbar isobar extends to 50°N and west of the 180th meridian. From August to October, the high shrinks rapidly as the Aleutian Low deepens. The winter climate of the North Pacific is also closely related to the high pressure over Asia whose average central value exceeds 1040-mbar in winter. The 1020-mbar isobar of this high and the 1005-mbar isobar of the low are closest in December−February.

Low pressure

On maps of average monthly sea level pressure, the low pressure system around the Aleutian Islands is seen only during the colder part of the year. Its area is large as a result of the frequent passage of depressions through this region. These depressions—the source of the Aleutian Low—are formed between very different air masses. One originating in the region of the Siberian High is cold and dry; the other is a warmer, humid maritime air mass. Storms deepen rapidly over the vast area covered by the Aleutian Low, and as winter progresses the frequency of such developing depressions increases. An example of such a low is given in Fig. 8 with its path outlined from the region of origin toward the Aleutians.

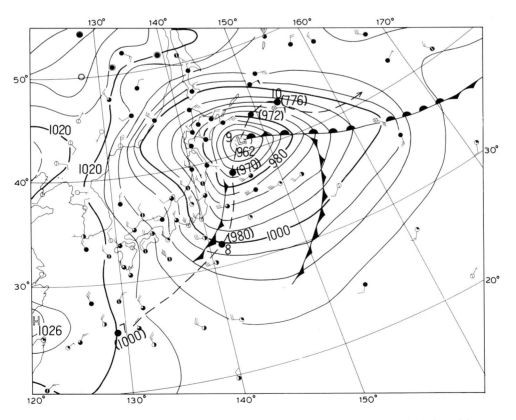

Fig. 8. Weather map for 9 January 1965 at 0000 GMT. Broken line is path of low with 12h positions.

Wind and depression tracks

Major wind systems

The major wind systems over the North Pacific are, briefly, as follows.

Trade winds, or tropical easterlies, are located between the doldrums of the Intertropical Convergence Zone and the subtropical belt of high pressure. The trades blow toward the west in a layer extending from the surface to 8 − 10 km in a limited latitude belt. Below 2 km the air is moist; above it is dry.

Antitrades are westerlies above the trades; the altitude is generally 10 − 16 km.

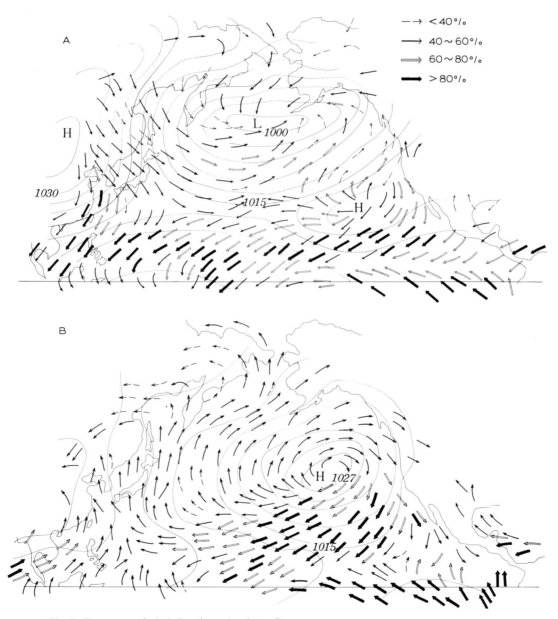

Fig. 9. Frequency of wind directions: A. winter; B. summer.

Equatorial westerlies are sometimes observed below 5 km; when they are observed at sea level, there are two zones of doldrums.

Prevailing westerlies are the westerlies of the middle latitudes which in winter extend into the stratosphere. In summer, their upper boundary is about 20 km. Their velocity is greatest in the jet stream, located about 25 – 40°N in winter and 40 – 55°N in summer, and they are stronger in winter than in summer.

Polar easterlies blow on the polar side of the Aleutian Low, are generally shallow (1 – 3 km deep), fluctuate much about the average, and sometimes are not observed in summer.

The *monsoons* in the North Pacific Ocean are examples of marked seasonal variation in the general circulation. In winter, strong northerly to westerly winds blow over the western part of the North Pacific between the Siberian High and the Aleutian Low, whereas in summer there is a weaker southeasterly flow from the ocean to the low pressure over the continent.

Fig. 9 was reproduced with some modifications from the *Morskoi Atlas* (NAVY MINIS-TRY, MOSCOW, 1953) and shows the frequency of wind direction in winter and summer. The wind systems associated with the large-scale mean pressure features—subtropical high, Aleutian Low, and Siberian High—are easy to identify, e.g., winter monsoon, westerlies, polar easterlies, and trades. In addition, the southeasterly trades, which blow from the Southern Hemisphere across the Equator in the eastern part of the ocean, are

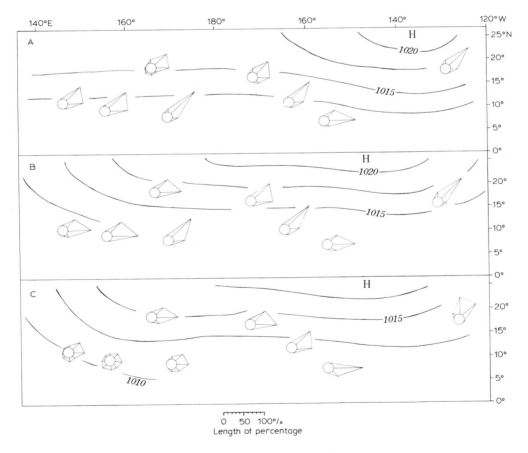

Fig. 10. Wind roses in the tropics: A. January; B. May; C. September.

439

visible in both seasons. The well-known fact that the mid-latitude westerlies are less constant than the tropical easterlies is also evident in the maps.

There are several areas where the wind direction reverses from summer to winter such as in the western, northern, and northeastern parts of the ocean. Here the summer winds diverge from the subtropical high and the winter winds converge into the Aleutian Low, or the reversals are monsoonal such as in the East and South China seas and adjacent areas.

In contrast, areas where the wind direction is nearly the same throughout the year are the middle latitudes southeast of the Aleutian Low in the westerly belt and that part of the tropics covered by the trades.

To illustrate the characteristics of the trades, frequencies of wind direction along with isobars for three months appear in Fig. 10. All year, pressure is high to the north and low to the south of the region shown and, consequently, the winds blow from northeast or east across the region. Examples of the annual march of monthly mean scalar wind speed in the centre of the trades are given in Table I.

TABLE I

MEAN MONTHLY SCALAR WIND AT TWO POINTS IN THE CENTRE OF THE TRADES (BEAUFORT SCALE)

	Jan.	Feb.	Mar.	Apr.	May	June	July	Aug.	Sept.	Oct.	Nov.	Dec.	Year
160°E	5.3	4.6	4.2	4.2	4.1	4.1	3.1	3.0	3.0	3.2	4.3	5.1	4.2
160°W	3.9	3.8	3.8	4.2	4.4	4.0	4.1	3.5	3.3	3.5	4.7	4.3	4.2

Another characteristic of the wind field is the strong wind in winter in the northwestern part of the North Pacific. The annual course of monthly mean scalar wind at ocean weather stations (OWS) X and Q is shown in Table II. Especially high frequencies (about 5%) of Beaufort 9 or more are experienced during winter in these areas.

TABLE II

MEAN MONTHLY SCALAR WIND AT OWS X (39°N 153°E) AND Q (42°N 173°W) (BEAUFORT SCALE)

	Jan.	Feb.	Mar.	Apr.	May	June	July	Aug.	Sept.	Oct.	Nov.	Dec.	Year
X	6.0	6.0	5.7	5.5	4.9	4.4	4.2	4.1	4.5	4.9	5.5	5.6	5.1
Q	5.3	5.7	5.1	5.1	4.5	4.3	4.2	3.8	4.5	4.9	5.2	5.5	4.8

Figs. 11 and 12 show wind roses at various points, together with mean isobars of January and July. Except for the trades and the winter monsoon, all directions are frequent in January; the summer monsoon is not evident beyond the South China Sea because its domain is the coastal areas and not the open ocean; but the winter monsoon is clearly seen over a wide area of the western Pacific, indicating the high frequency of its associated pressure pattern. The winter winds are by far stronger than those of summer. Table III shows the marked difference between seasons in the frequency of strong winds. Over the ocean north of a line connecting the Philippines and California, the wind is relatively strong for more than half the year. This area coincides with the Aleutian Low

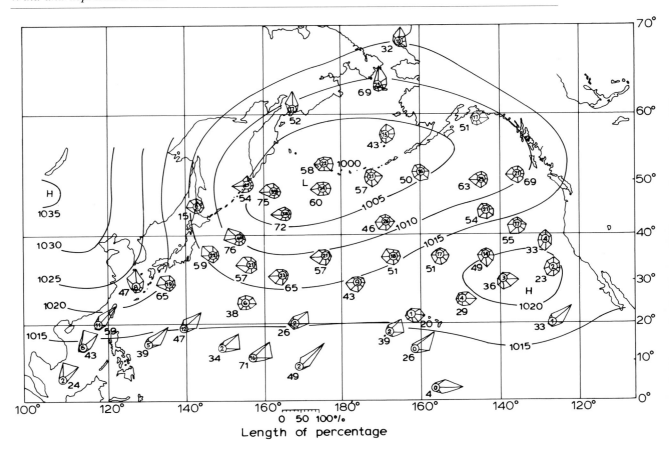

Fig. 11. Direction frequency of winds in January. Length of bars represents percentage frequency of wind observed from each direction. Figures in circles: percentage of all winds of ⩾ Bf 7. Adjacent figures: percentage of all winds ⩾ Bf 5.

and the winter monsoon. The winds in the trades region and the area covered by high pressure are comparatively weak so that winds stronger than Beaufort 7 are infrequent over most of the region south of a line connecting the Philippines and California.

Monsoons

The predominant feature of the winter monsoon is the relatively steady northerly flow over the South China Sea and the area near the Philippines. In the East China Sea, north and northwest winds dominate, but they are less steady than in the South China Sea because lows in the East China Sea often disturb the outflow from the continent. Over the Sea of Okhotsk and northeastern Japan, the direction of the monsoon is westerly to northwesterly, but due to the frequent passage of depressions the monsoon is greatly disturbed. Sometimes a low forms in the monsoon flow, and in the strong westerlies at its rear an instability line develops and becomes a danger to shipping.
The air in the winter monsoon is generally very dry, but as it flows across the ocean it accumulates water vapour. This is especially so when the monsoon blows over the Sea of Japan, where the main source of water vapour is the warm Tsushima Current along the eastern part of the Sea of Japan.

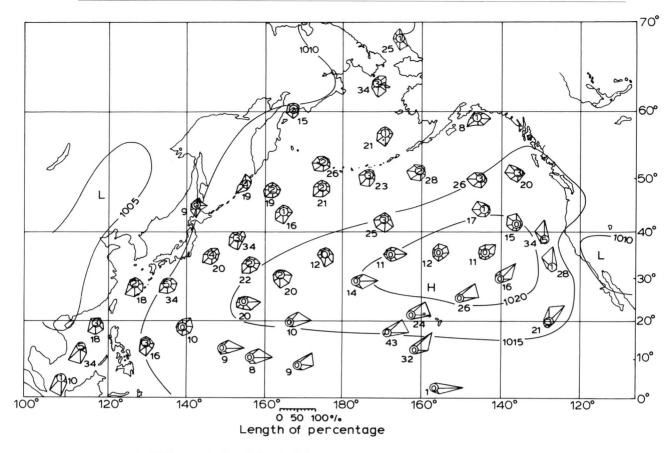

Fig. 12. Same as in Fig. 11, but for July.

TABLE III

PERCENTAGE OF WIND ABOVE 5 AND 7 BEAUFORT IN WINTER AND SUMMER*

Part of the ocean	Winter		Summer	
	> 7Bf	> 5Bf	> 7Bf	> 5Bf
East of northern Japan (X)	40	76	3	34
West of Aleutian Islands	39	75	2	19
South of Alaska (P)	16	50	2	28
South of Japan (T)	19	65	5	34
Western North Pacific (V)	6	38	0	20
Central North Pacific	18	51	0	11
Centre of high pressure (N)	3	26	0	16
South China Sea	6	43	5	34
Central equatorial area	16	71	0	8
Eastern equatorial area	0	26	0	32

* Letters in parentheses refer to the weather ships (see Fig. 1).

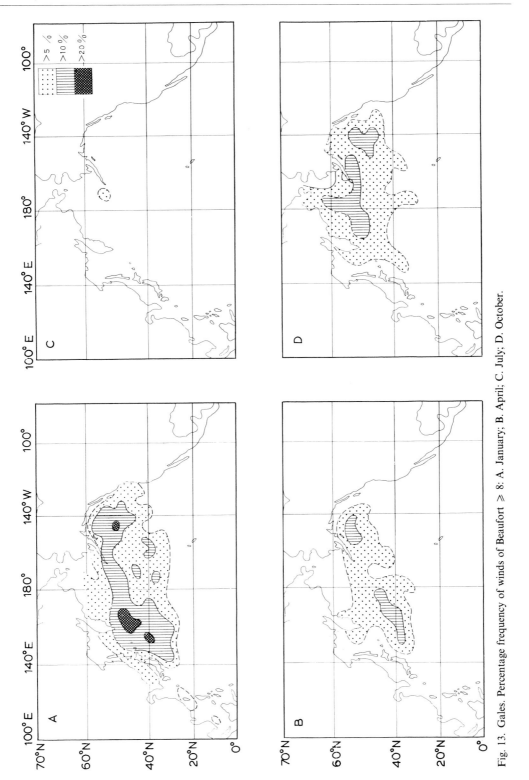

Fig. 13. Gales. Percentage frequency of winds of Beaufort $\geq$ 8: A. January; B. April; C. July; D. October.

The summer monsoon is weak, as evident from the surface pressure distribution. The direction of the summer monsoon varies more than 45° on either side of south; in the South China Sea it blows from the south and southwest.

If we compare summer and winter in the South China Sea, the contrast between the two monsoons is pronounced:

Winter: N/42%   NE/46%   E/3%   BF>5/43%

Summer: S/35%   SW/39%   SE/8%   BF>5/33%

There is, however, no such contrast between summer and winter at other places in the western Pacific.

Gale statistics and cyclogenesis

Over the northern half of the North Pacific, strong winds are frequent outside the summer season (Fig. 13). The area of frequency above 5% nearly covers the ocean north of 30°N in winter and north of 40°N in spring and autumn. In summer, the percentage remains below 2 or 3 over most of the ocean. The region enclosed by the 10% line coincides with the most frequent tracks of extratropical depressions, and the area just east of the Kurile Islands, where the frequency is above 20% in January, is strongly affected by the winter monsoon.

One might doubt that there is no area with a frequency of gales ⩾ 5% in the southwestern portion of the ocean where typhoons are frequent in summer and autumn. Despite the high density of tracks in the main typhoon areas (Fig. 16), the statistics show that gales are appreciably less frequent than in the northern part of the ocean. The reason is that depressions in the westerlies are accompanied by gales over a large area for a long period, whereas gales in typhoons cover small areas for shorter periods even though the winds near the centre of a typhoon may be much more violent than those anywhere in an extratropical low.

Table IV gives statistics from ocean weather stations on the duration of gales and the interval between them. In winter, the average duration of a gale is less than one day and the average interval is three days in the northern part of the Pacific.

Depressions originating over Asia in winter move eastward across the northern part of Japan, and those originating in the East China Sea or near Japan move northeastward. These depressions generally skirt southern Kamchatka and then pass eastward over the

TABLE IV

AVERAGE DURATION (DAYS) OF GALES AND AVERAGE INTERVAL (DAYS) BETWEEN AT OWS S AND X

	Duration		Interval	
	S	X	S	X
January	0.7	0.75	3	3
April	0.5	0.9	13	7.8
July	–	–	–	–
October	1.0	0.8	5	11

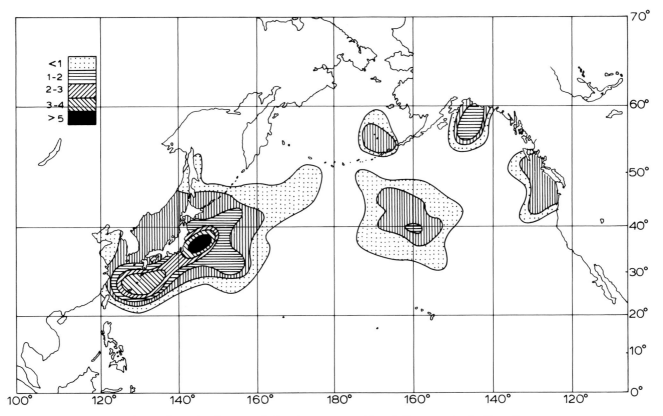

Fig. 14. Areas of genesis of extratropical cyclones expressed by relative amounts, for the entire year.

ocean near the Aleutians. Sometimes a high develops over southern Alaska and hinders the eastward movement of lows so that they change direction or become stationary. This is why strong winds in winter may be observed from most quarters near the Aleutians (Figs. 11 and 13).

The relative frequency of extratropical cyclogenesis is shown in Fig. 14. The area extending from Taiwan to south of Kamchatka is the primary area of cyclogenesis. Secondary areas appear in the mid-Pacific centred at 40°N 160°W, and along the North American coast.

Extratropical depression tracks

The Japan Meteorological Agency (formerly the Central Meteorological Observatory) has been publishing monthly weather reports since 1886 and among them are descriptions of main depressions. At first, such descriptions were confined to Japan and surroundings, but they were expanded to the entire North Pacific Ocean after 1910. Fig. 15 shows tracks of such main depressions from 1920 to 1940, from which we can grasp the general features of the movement of the depressions over the ocean.

The definition of a "main depression" is rather vague. Some might have considered a depression main by the amount of damage caused; others might have selected it from the level of its central pressure. Because of the uncertain criteria, the number of main extratropical depressions described is relatively small.

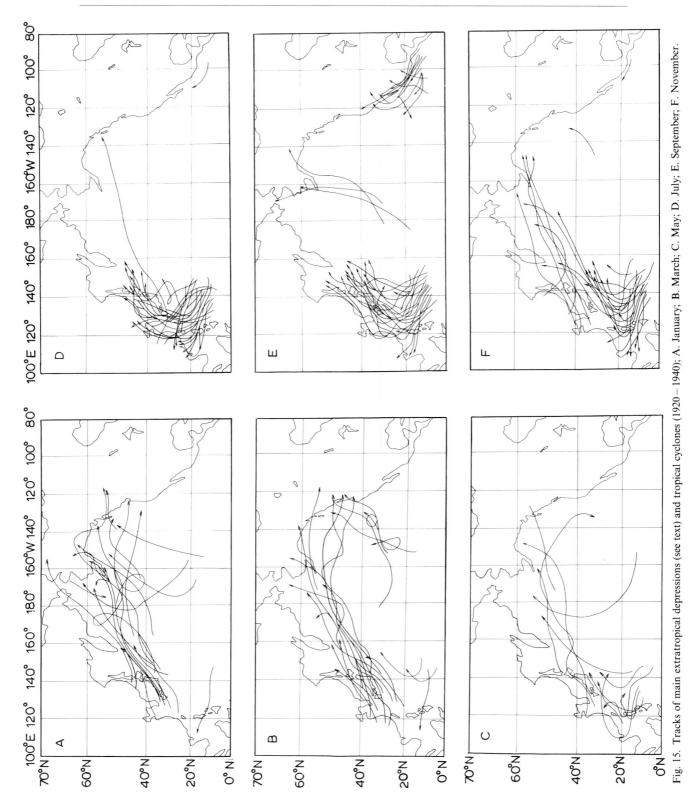

Fig. 15. Tracks of main extratropical depressions (see text) and tropical cyclones (1920–1940); A. January; B. March; C. May; D. July; E. September; F. November.

Japanese meteorologists previously had little personal experience with severe weather over the ocean, and it was not until after World War II that they encountered the harshness of extratropical depressions when serving on board ocean weather ships, as recounted by a meteorologist aboard the *Ikunamaru*: "...When I was at X (39°N 153°E) in winter, I experienced an extremely severe storm which was beyond my expectation. Our ship was unable to escape from this depression though she sailed more than 700 nautical miles. There were no islands to shelter us from the fury of the wind and we had to face gales or strong winds for a long period."

Winter. Depression tracks cover the northwestern part of the North Pacific (Fig. 15); most of them start southwest of Japan, point northeastward, and reach the Aleutians and surroundings; some extend as far as North America. Fewer lows originate in the central part of the North Pacific and move northward to Alaska or western Canada. Near the Aleutians, the tracks are often complex.

Spring. Main depressions are more infrequent, but the distances covered are longer than in winter.

Summer. Main extratropical depressions are rare.

Autumn. The number of extratropical depressions increases.

The above statistics show that the western and northwestern portions of the North Pacific are the seat of major depressions, but the area from 160°E to 120°W south of 40°N is rarely touched by such depressions. The latter area is usually occupied by the Pacific High.

Tropical and subtropical cyclones

Typhoons, the most intense depressions known, develop from weak tropical lows over an area between 5°N and 25°N west of about 170°E. Another, less important, area is the ocean off lower California and Central America. From the extensive research on the typhoon, it has been pointed out that three stages are necessary for its formation—wave, warming, and developing. For example, in the development of typhoon Doris in 1958, a slow amplification of an existing easterly wave was observed in the first stage. The warming stage was characterized by a gradual increase in temperature over the area of ascent and a slight decrease over the surrounding area of descent. From a study of the moisture budget, it was confirmed that these temperature changes were associated with the liberation of latent heat of condensation and with cooling due to evaporation. Once a weak tropical depression has formed, it gains energy from the convergent maritime air through release of latent heat.

In early summer, typhoons generally move westward across the Philippines, but in late summer and autumn they often turn northwest farther east and then north or northeast (Fig. 16). Typhoons are most frequent during summer and autumn and comparatively rare in winter and spring. The average number of typhoons per year and the frequency by months appear in Table V. In the typhoon's centre, the pressure is very low, sometimes lower than 920 mbar; the pressure gradients are steep, the winds violent, and the rain heavy. Table VI shows maximum wind velocity ranked by decreasing value of some typhoons observed in Japan.

On 26 September 1959, typhoon Vera attacked central Japan (Fig. 17) causing an extraordinarily high storm surge and killing more than 5,000 persons. In this example, the

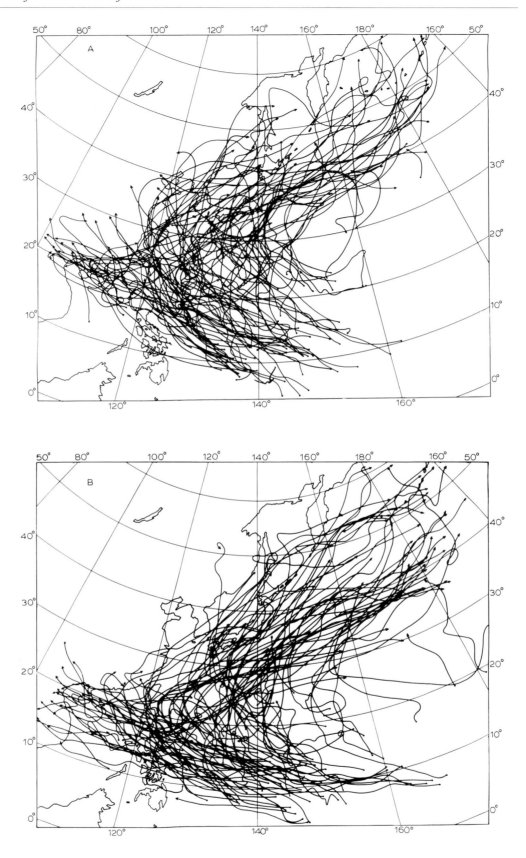

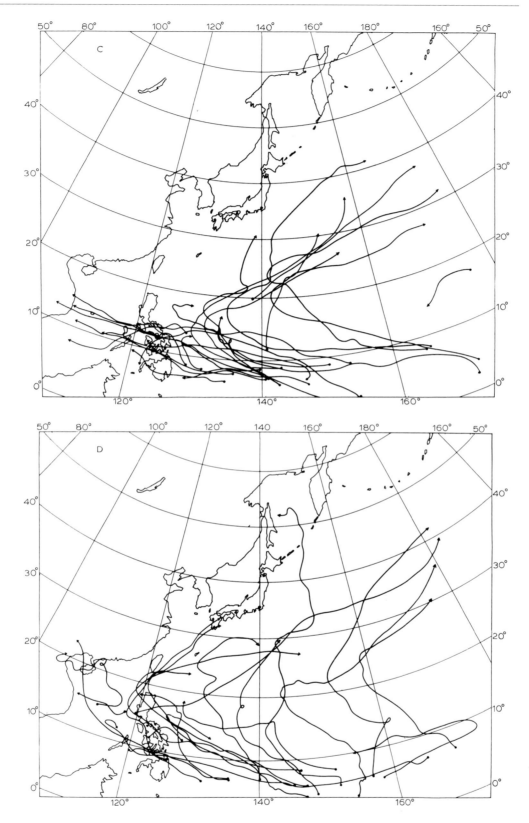

Fig. 16. Typhoon tracks: A. summer; B. autumn; C. winter; D. spring.

TABLE V

MONTHLY AND YEARLY AVERAGE NUMBER OF TYPHOONS

Jan.	Feb.	Mar.	Apr.	May	June	July	Aug.	Sept.	Oct.	Nov.	Dec.	Year
0.5	0.5	0.3	0.9	0.8	1.8	4.3	6.5	4.8	3.9	2.2	1.6	28.1

extreme velocity of 45.4 m s⁻¹ was observed at Irako. As seen in Fig. 17, severe gales in a typhoon cover in most cases an area with a radius of only 100 km, whereas the storm area of extratropical depressions, as noted above, is frequently much larger.

It is obvious (Fig. 18) that tropical storms form where the sea surface temperature is high, but this does not imply that the storms are particularly frequent in years of high

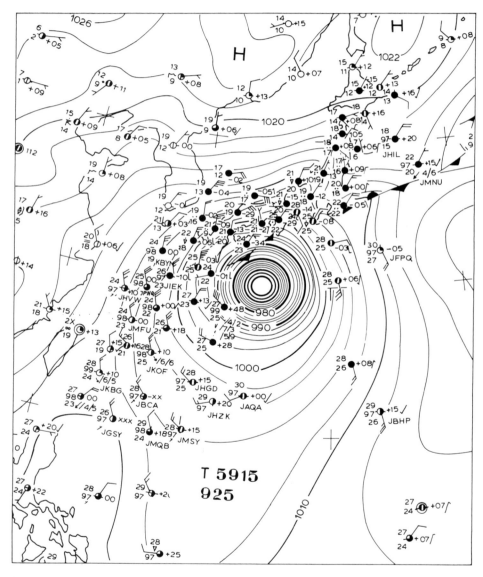

Fig. 17. Weather map for 26 September 1959 0000 GMT, showing typhoon Vera.

TABLE VI

MAXIMUM WIND VELOCITIES OBSERVED IN TYPHOONS

Observatory	Lat. (N)	Long. (E)	Direction	Speed (m s^{-1})	Date
Miyakojima	24°47′	125°17′	NE	85.3	5 Sept. 1966
Murotomisaki	33°15′	134°11′	WSW	>84.5	16 Sept. 1961
Miyakojima	24°47′	125°17′	NW	79.8	22 Sept. 1968
Naze	28°23′	129°30′	ESE	78.9	13 Aug. 1970
Miyakojima	24°47′	125°17′	NNE	78.0	23 Sept. 1968
Naha	26°14′	127°41′	S	73.6	8 Sept. 1956
Uwajima	33°14′	132°33′	W	72.3	25 Sept. 1964
Miyakojima	24°47′	125°17′	NNE	70.0	23 June 1950
Yakushima	30°27′	130°30′	ENE	68.5	24 Sept. 1964
Hachijojima	33°06′	139°47′	S	67.8	5 Oct. 1975
Tokushima	34°04′	134°35′	SSE	>67.0	10 Sept. 1965
Ishigakijima	24°20′	124°10′	SSE	66.9	22 Sept. 1971
Yakushima	30°27′	130°30′	SW	65.0	6 Sept. 1956
Yonakunijima	24°28′	123°01′	SSW	64.0	22 Sept. 1971

sea surface temperature. The important factors are the temperature of the surface water *and* the stability determined by the vertical distribution of temperature.

The Kona cyclones, which are subtropical systems, have been described by SIMPSON (1952) who investigated 76 from the winters of 20 years. They are large, cold-core lows, usually apart from the mainstream of polar westerlies, and originating and moving principally in the central and eastern subtropical parts of the ocean. The Kona lows

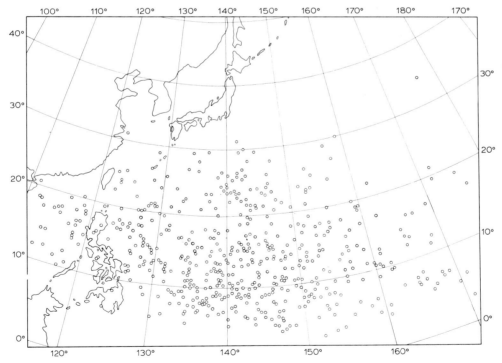

Fig. 18. Birthplaces of typhoons (1940–1959).

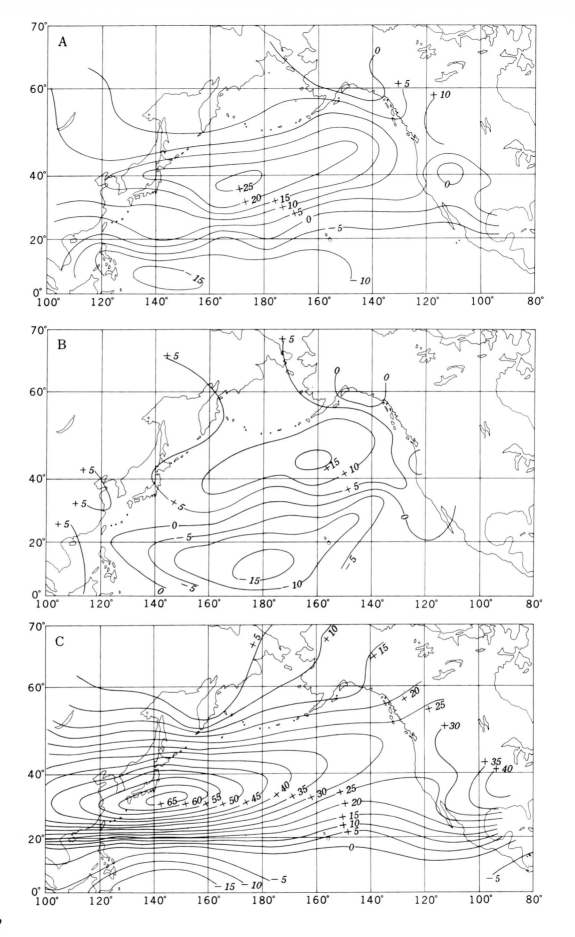

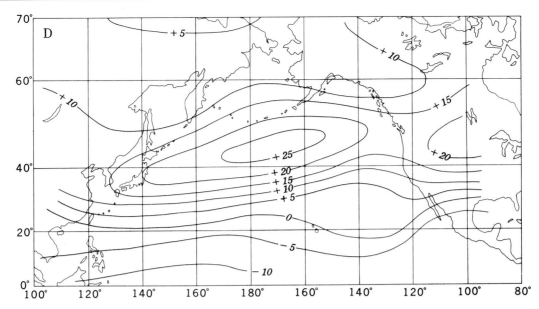

Fig. 19. Upper wind. Zonal mean component (knots). A. 850 mbar, December, January, February; B. 850 mbar, June, July, August; C. 500 mbar, December, January, February; D. 500 mbar, June, July, August.

move erratically, westward as well as eastward, and some have moved more than 2500 km before dissipating or recurving.

Upper winds. The mean upper winds over the NPO have a large zonal component (Fig. 19). Westerlies predominate in higher latitudes and easterlies at lower latitudes. In summer, the easterlies extend somewhat farther north and the westerlies weaken. The highest mean wind speed in the westerlies is shown in Table VII.

TABLE VII

HIGHEST MEAN SPEED OF THE ZONAL WESTERLY WIND COMPONENT AT 850 AND 500 MBAR

Season	850 mbar	500 mbar
Winter	25 kt., centred at 38°N 170°E	65 kt., centred at 33°N 145°E
Summer	15 kt., centred at 45°N 160°W	25 kt., centred at 45°N 170°W

Temperature and humidity

Temperature distribution

Air temperature is highly dependent on insolation and were it not for other influences, isotherms would parallel the latitude circles. Over the North Pacific Ocean, they come nearest to doing so in summer outside the areas with meridional ocean currents (Fig. 20). The distribution of air temperature in winter (Fig. 21) is quite different from that in summer as the isotherms cross the Pacific from west-southwest to east-northeast; in summer they run east-southeast south of the zone where they are parallel to the latitude

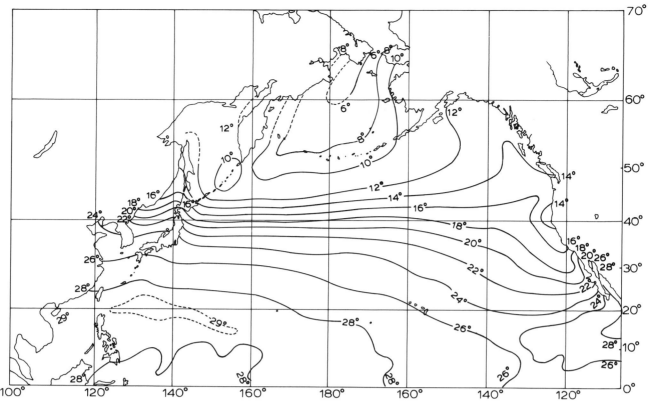

Fig. 20. Air temperature, July.

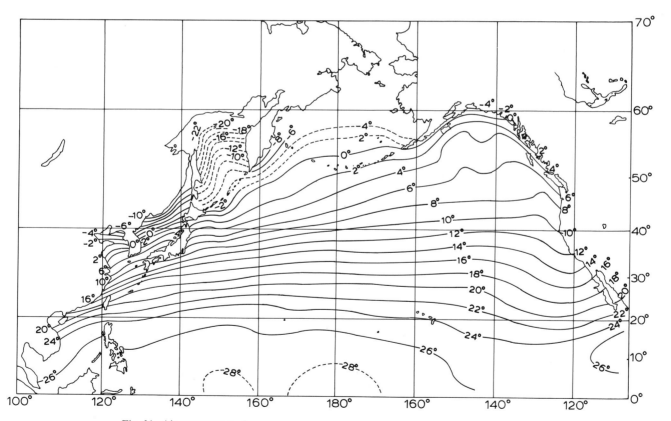

Fig. 21. Air temperature, January.

454

circles. The temperature distribution in winter is strongly influenced by the winter monsoon in the northwestern parts, and in the east the effect of the Alaska and California currents is obvious in both seasons.

In July, the strongest latitudinal temperature gradient, 10°C per 1000 km, is in the middle latitudes; in other areas the gradient is 2−3°C per 1000 km, but there is no measurable latitudinal gradient in the equatorial region. By contrast, in winter large temperature gradients are common outside the tropics. Gradients such as 10°C per 1000 km are found over a very wide area from 20° to 40°N, and even larger gradients, about 20°C per 1000 km, are observed in the Sea of Japan, the East China Sea, and the Sea of Okhotsk.

A characteristic of low latitudes in summer is the existence of wide areas where the temperature deviates little from 28°C. There is a region with temperatures above 29°C, but not exceeding 30°C, near the Caroline Islands where typhoons often form. In winter, the temperature of the equatorial zone remains about 28°C in the west and centre, and 26°C in the east. This is related not only to the small seasonal deviation of the sun's height from the perpendicular but also the uniformity of the trades through the year.

To demonstrate the change of temperature throughout the year over the North Pacific, the 0°C, 10°C, and 20°C isotherms for each month are shown in Fig. 22. Fig. 22C illustrates the monthly variation in the position of the 20°C isotherm. Note that these isotherms run parallel to latitude circles west of the California Current. From January to April, the isotherm is farthest south at 27−28°N. Then it moves north and reaches about 42°N in August after which it moves south again. Therefore, this isotherm moves 5 degrees of latitude per month on the average. As the isotherms of 26°C and 28°C stay in the tropics, the change in the location of the 20°C isotherm with season means that the latitudinal temperature gradient is steep in the subtropics and northern parts of the tropics in winter and weak in summer.

The 20°C isotherms have a southward bulge throughout the year over the easternmost part of the Pacific which is associated with the cold California Current. Over the western part of the ocean, the 20°C isotherms are forced south from October to May by the winter monsoon. In summer, the same isotherms generally lie at a more northerly latitude off Asia than in the middle part of the ocean, showing that the air temperature over the Sea of Japan and adjacent regions is highly influenced by Asia and land areas such as Japan. The same is evident from the positions of the 10°C isotherms.

The 10-degree isotherms are almost all oriented west-southwest−east-northeast in agreement with the orientation of the Kuroshio and its extension and, as with the 20-degree isotherms, the 10-degree isotherms in winter remain within a narrow latitude zone. In several summer months, the 10-degree lines turn north over the Alaska Current. The zero-degree isotherm appears only for the period from September to April, as the air over the North Pacific is warmer than 0°C in the summer season. This isotherm extends northeastward from the Sea of Japan, east of Hokkaido and across the Aleutians to the Gulf of Alaska. Its position coincides with the boundary between the Kuroshio and Oyashio off the Kuriles and with the north side of the Alaska Current.

Summarizing Figs. 20−22, we note the marked difference of temperature distribution between the east and west sides of the North Pacific. In the eastern North Pacific, one must go from California to the Gulf of Alaska, a distance of 40° latitudes, to cover a difference of 20°C, whereas such a temperature difference is observed within only 15

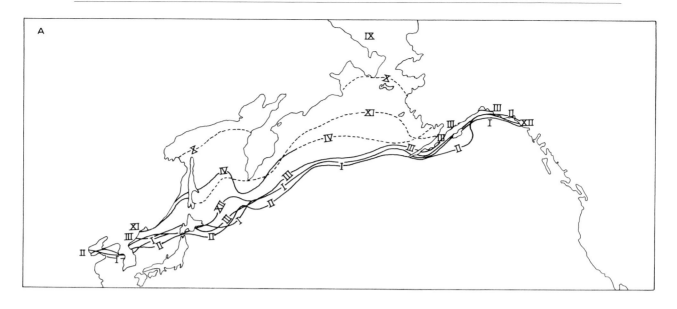

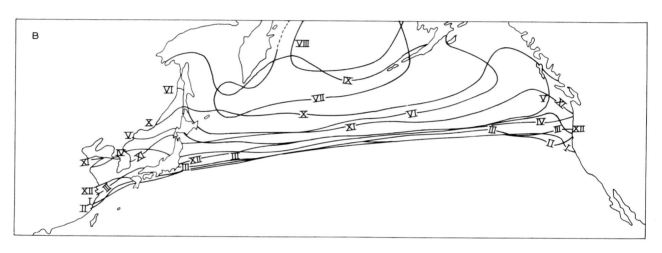

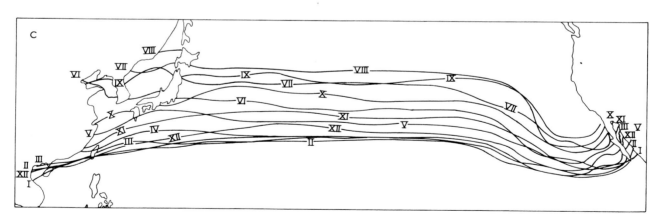

Fig. 22. Variation of air temperature by month illustrated by selected isotherms. A. 0°C isotherms; B. 10°C isotherms; C. 20°C isotherms.

degrees of latitude in the western Pacific, especially in winter. The climates over the eastern part of the Pacific, therefore, change less over the same distance than in the west, and on the same parallel they are by far milder in the east than in the west in winter. This is related to the position of the high in the eastern part of the Pacific and the northeastward tracks of the frequent depressions.

Difference between air and sea surface temperature

The sea surface temperature maps for February and August (Figs. 23, 24) resemble those of the air temperature for winter and summer and need no further discussion, but the main features of the difference between air and sea temperatures are described below. In summer, air and sea surface temperatures do not differ much over most of the North Pacific. The air temperature over the tropics is slightly lower than the sea surface temperature, and north of about 30°N the air is slightly warmer than the sea. Northwest of a line connecting Japan, the Aleutians, and Alaska, the air—sea temperature difference is generally less than 1.5°C, the air being warmer. Over the ocean close to land, such as the western part of the Sea of Japan, the waters next to Sakhalin, Kamchatka, the Kuriles, and west of Alaska, the air is about 2°C warmer than the sea, but it is nowhere warmer than 4°C. Higher air than sea temperatures are found over the cold ocean currents, except over the California Current where the air is colder than the sea, but the difference is small.

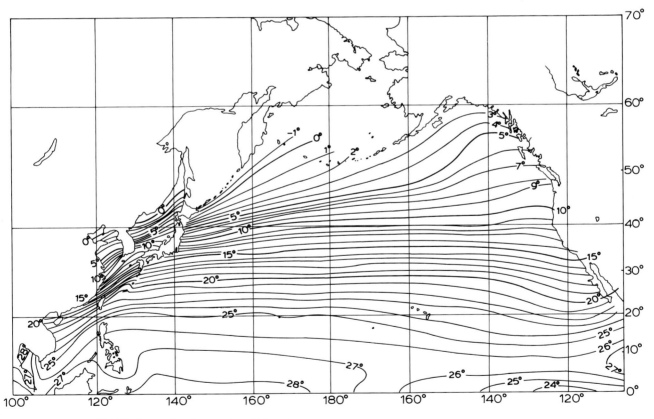

Fig. 23. Sea surface temperature, February.

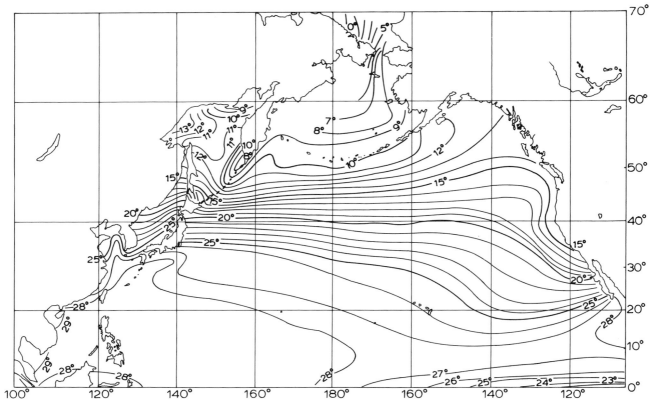

Fig. 24. Sea surface temperature, August.

The air temperature in winter is generally lower than the sea surface temperature over the entire North Pacific, especially over the northwestern half in the domain of the winter monsoon. The area where the sea is warmer than the air by more than 2°C includes all of the northern and northwestern parts and extends into the central part; over the Kuroshio and Tsushima currents, the difference is 4°C or more. North of the Sea of Japan and in the Okhotsk Sea, the difference exceeds 10°C, but as the ocean here is covered by ice and the air temperature is very low, such a difference between air and sea should not be considered in the same sense.

The vast areas where the water is warmer than the air by more than 2°C are traversed by frequent depressions which develop rapidly owing to the release of energy within them in the form of latent heat.

Humidity

From monthly mean air temperature and wet bulb temperature at 33 points in the *Marine Climatic Atlas of the World* (U.S. NAVY, 1956), monthly means of relative humidity are obtained. As expected, the relative humidity is higher in summer than in winter (Table VIII).

Monthly mean relative humidity below 70% is observed only in winter over the ocean east of Japan, near the Aleutians, and south of the Gulf of Alaska. The winter monsoon is very dry, but this does not emerge from the monthly means of relative humidity. The

TABLE VIII

MEAN AND RANGE OF RELATIVE HUMIDITY OVER THE NORTH PACIFIC (IN PERCENTAGES)

	Jan.	Apr.	July	Oct.
Mean	82.4	82.8	88.0	82.5
Range	59–91	74–91	76–94	72–100

vapour pressure at three points in the monsoon region makes this dryness evident (Table IX). In January and April, the vapour pressure at these points is low, but it increases greatly in summer and autumn. Other places do not show such marked seasonal change of vapour pressure.

TABLE IX

RELATIVE HUMIDITY % AND VAPOUR PRESSURE (MBAR)

Month	S. of Japan, OWS T		E. of Japan, OWS X		SE. of Japan, 30°N 163°E	
	(%)	(mbar)	(%)	(mbar)	(%)	(mbar)
Jan.	77	12.8	69	9.1	77	12.3
Apr.	75	15.8	83	13.7	83	14.8
June	88	27.0	91	26.8	91	26.8
Oct.	74	23.4	81	20.3	82	23.4

Cloudiness, visibility, fog, and precipitation

Cloudiness

Fig. 25 shows percentage frequencies for January and July of total cloud amount of 2/10 or less; only frequencies ≥ 20% are shown. Where the percentage is high, fair weather prevails and where it is small, cloudy or rainy weather predominates. Over the subtropics in the position of the ridge, the weather is fair about one-third of the time; this area of comparatively high frequency moves northwestward from winter to summer. The values are high on either side of the ocean in both January and July. The high frequencies in the east are associated in part with the California Current, in part with outflow from the continent. The high values near Asia in winter are explained by frequent outbursts of continental polar air.

Fig. 26 shows the frequency of total cloud amount of 6/10 or more in January and July; only contours of 50% and more are shown. The areas outlined by these frequencies coincide with some of less than 20% frequency in Fig. 25. We have already noted that many depressions travel across the northern North Pacific in winter, but only a few in summer. This might suggest that cloudiness would be higher in winter than in summer; yet the statistics show a higher occurrence of total cloudiness in summer. This is, however, due to frequent low stratus and fog in summer. Mean values of total cloud amount

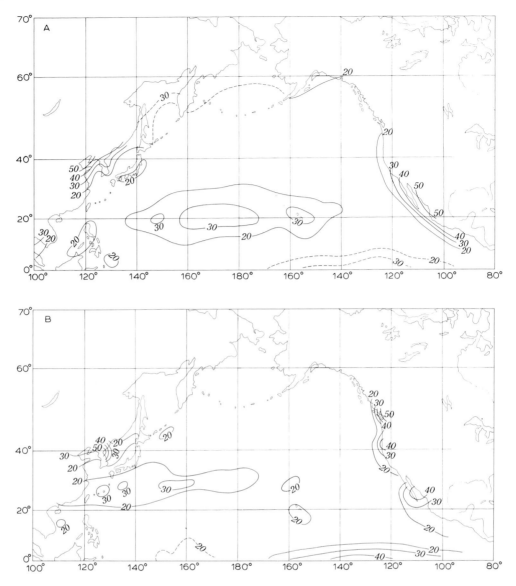

Fig. 25. Percentage frequency of total cloud amount ≤ 2/10: A. January; B. July.

> 85% in both seasons are denoted by large dots; in these regions overcast skies are frequent.

Since 1967, cloud photographs by meteorological satellites have yielded valuable information which could not be deduced from surface observations. In the following, some typical results are mentioned.

Clouds in the tropics are often organized into groups called "cloud clusters." These clusters whose horizontal dimensions range from 2 to 12 degrees longitude usually appear within cloud bands extending over several thousand kilometres and move from day to day. The area of a cluster may be as large as $5 \cdot 10^5$ km² and over the western Pacific occasionally as large as $15 \cdot 10^5$ km².

The low-latitude cloud band north of the Equator, which is farthest south over the

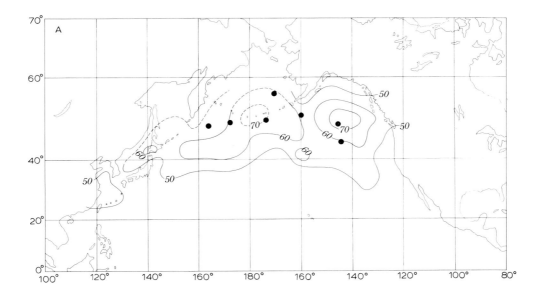

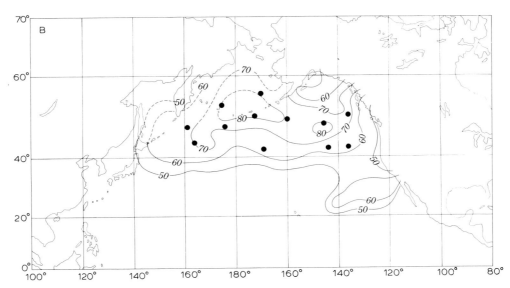

Fig. 26. Percentage frequency of total cloud amount ⩾ 6/10: A. January; B. July. Dots denote an average coverage >85%.

central North Pacific and farthest north over the western and eastern parts, is prominent throughout the year, although its intensity varies with season. The cloud clusters in the western side of the ocean are frequently well-developed vortices and often become typhoons, particularly in late northern summer; those in the central Pacific rarely show a vortex structure. Those in the eastern Pacific often show a vortex structure and occasionally develop into hurricanes. The clusters in the central Pacific often appear to be associated with waves and sometimes there are pairs of clusters with one on either side of the Equator.

461

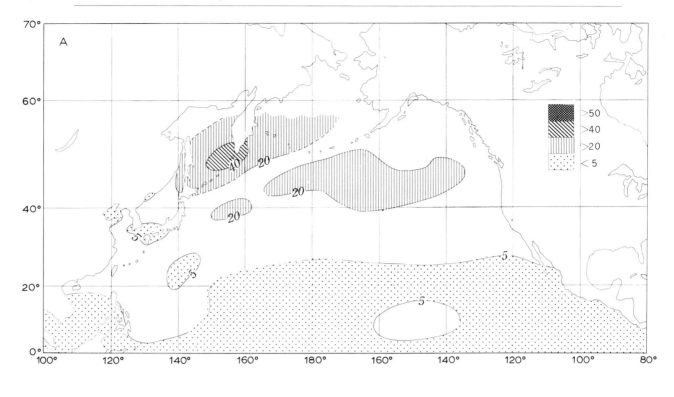

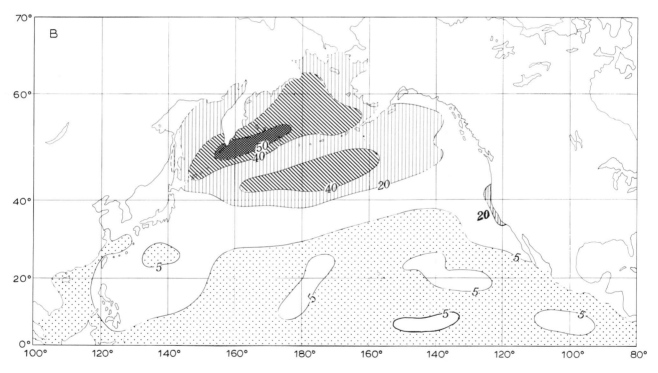

Fig. 27. Frequency (percent) of visibility less than 5 nautical miles. A. January; B. July.

Visibility

Observation of visibility on board ship is difficult because suitable targets are lacking. Fig. 27 shows the frequency of visibility of less than 5 nautical miles. The region over which the frequency is less than 5% changes little throughout the year. Much of it is within the trades where the cloud amount is relatively small and rain infrequent.

In January, frequencies of above 20% dominate north of 40°N, exceeding 40% south of Kamchatka. These high-frequency regions roughly coincide with major depression tracks. In July, the region of poor visibility is much larger and the frequency of low visibility within that region is greater than in January.

TABLE X

MAXIMUM DURATION (DAYS) OF VISIBILITY OF < 2 NAUTICAL MILES

	OWS X (39°N 153°E)	S (48°N 162°E)	Q (43°N 167°W)	P (50°N 145°W)
Jan.	1.0	1.0	0.8	1.3
Mar.	0.7	1.5	0.3	2.0
May	1.0	1.4	1.7	1.0
July	2.1	2.3	2.8	1.8
Sept.	3.0	1.5	1.8	1.7
Nov.	0.7	0.9	0.7	1.0

Table X summarizes the maximum duration of visibility < 2 nautical miles and clearly shows that poor visibility lasts longer in summer than in winter, suggesting that fog is more frequent during the warmer part of the year.

The mean value of visibility lies between 10 and 25 nautical miles south of 40°N, being especially good south of 30°N, where there are almost no cases of mean visibility less than 5 nautical miles outside the easternmost parts of the ocean. The mean value lies between 5 and 10 nautical miles within the 20% isopleth in Fig. 27, but near Kamchatka there are a few places with smaller means. The poorest mean visibility is recorded at Harmukotan, an island in the Kuriles just south of Kamchatka where the value falls

TABLE XI

PROBABILITY (PERCENT) OF VISIBILITY OF < 0.25 NAUTICAL MILE AT HARMUKOTAN

Jan	Feb.	Mar.	Apr.	May	June	July	Aug.	Sept.	Oct.	Nov.	Dec.
20	13	12	12	19	31	30	40	19	4	4	6

below 2 nautical miles in summer, the season of poorest visibility (Table XI), and is as low as 1/4 to 1/2 mile in August. Observations of visibility less than 1/4 mile are numerous between 42° and 50°N, 155° and 165°E, southeast of Kamchatka. The frequencies of various visibility intervals at Harmukotan are given in Table XII, together with the frequencies at other localities for comparison.

TABLE XII

FREQUENCY (%) OF VISIBILITY (N MILES) AT HARMUKOTAN, OWS X AND S, AND 11°N 156°E

Place	Month	<1/10	1/10–1/4	1/4–1/2	1/2–1	1–2	2–5	5–10	10–25	>25
Harmukotan	Jan.	10	10	6	4	8	12	28	15	7
Harmukotan	Mar.	6	5	8	3	6	12	24	26	10
Harmukotan	May	9	10	6	1	3	15	24	21	11
Harmukotan	June	15	15	12	3	6	13	13	12	11
Harmukotan	July	8	22	14	2	4	16	14	10	10
Harmukotan	Aug.	10	30	11	6	10	10	10	10	13
Harmukotan	Sept.	7	12	6	2	5	18	20	20	10
OWS X	Jan.	–	–	–	–	6	14	42	36	2
OWS X	Aug.	–	–	–	–	–	8	17	70	5
OWS X	Sept.	–	–	–	–	–	7	13	75	5
OWS S	Jan.	–	6	4	2	14	25	31	17	5
OWS S	July	10	16	9	4	4	10	35	10	2
OWS S	Aug.	5	6	8	5	4	10	50	10	5
OWS S	Sept.	–	2	3	4	2	10	48	31	–
U.S. Navy Atlas	Jan.	–	–	–	–	–	2	38	60	–
(11°N 156°E)	May	–	–	–	–	–	–	50	50	–

Fog

Of the three types of fog occurring over the ocean, sea fog is the most frequently observed. Frontal fog may be encountered, e.g., when a warm front is stationary near Japan, but steam fog is not often reported over the North Pacific.

The east side of Hokkaido is notorious for foggy weather. Statistics of the duration of fog at Nemuro (42°20′N 145°35′E) appear in Table XIII. From late spring to autumn, poor visibility at Nemuro is due to fog which is observed then as often as half the time (August), with visibility of less than 200 m 5% of the month. The reason for the high frequency of fog is the combination of the cold Oyashio Current just east and north of

TABLE XIII

DURATION (H/MONTH) OF VISIBILITY LESS THAN 200, 500, AND 1000 M AT NEMURO (43°20′N 145°35′E)

Month	Visibility			Cause of poor visibility (%)		
	<200 m	<500 m	<1000 m	fog	snow	rain
Jan.	4	12	26	10	85	
Feb.	3	7	26	27	73	
Mar.	9	27	40	65	35	
Apr.	11	49	83	92	8	
May	20	50	108	99		
June	35	92	175	100		
July	31	102	202	100		
Aug.	40	110	208	100		
Sept.	12	30	50	100		
Oct.	2	4	7	80		20
Nov.	4	6	8	100		
Dec.		2	7	30	70	

Nemuro and frequent warm air advection from the south or southeast over this cold water. The sea surface temperature map for August (Fig. 24) shows the cold water extending southeastward along the Kuriles. In summer, when southerly winds blow over this area, the warm, humid air is cooled over the cold water and fog frequently forms. Off the southwestern part of Kamchatka and over the northern Kuriles, fog may be observed even with northerly wind.

In summer, late spring, and early autumn, a high of moderate intensity is often observed over the area from east of Hokkaido to south of the Aleutians. In such a situation, moderate southerly winds blow over a surface with lower temperature and fog forms northwest of the ridge line. Such fog continues for several hours, sometimes for more than one day, over a wide area.

Two of the most notorious places in the world for fog are the northwestern Pacific and the northwestern Atlantic. The area of foggy weather in summer is far greater in the Pacific than in the Atlantic. In both, fog is linked to the warm and cold ocean currents, but the difference in the size of the fog areas depends on the features of the warm currents—the Kuroshio and the Gulf Stream. The Kuroshio and its extension have more

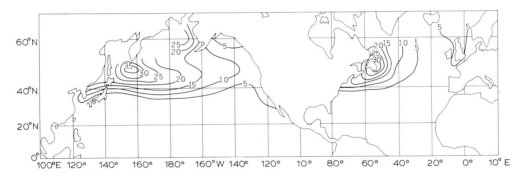

Fig. 28. Percentage frequency of fog in summer.

of an eastward component than the Gulf Stream where the northward component is stronger than in the Pacific. Therefore, the area of the Atlantic where fog is common is the western part, whereas that in the Pacific extends across the ocean (Fig. 28).

Precipitation

It is difficult to obtain on board ship reliable measurements of rain and snow, but the precipitation noted in the ship's weather log provides a means to compile statistics on the frequency of precipitation over the ocean. Precipitation in the westerlies, of course, is mainly observed in depressions. Therefore, the areas of high frequency will coincide with the major storm tracks. This is evident in Fig. 29 where the highest frequencies are in the northernmost part of the ocean in all seasons. In the subtropical ridge, the values sink below 10%.

Another region with comparatively high frequency is the Intertropical Convergence Zone, particularly its eastern side. The precipitation there is principally convective and squally.

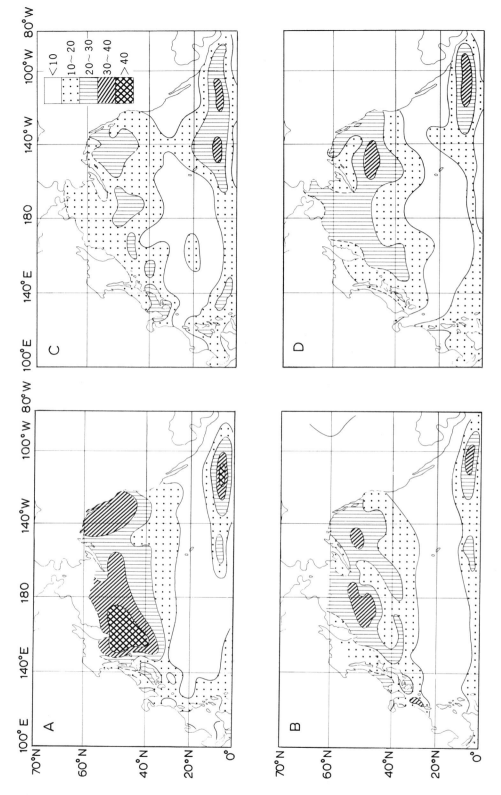

Fig. 29. Percentage frequency of observations reporting precipitation: A. January; B. April; C. July; D. October.

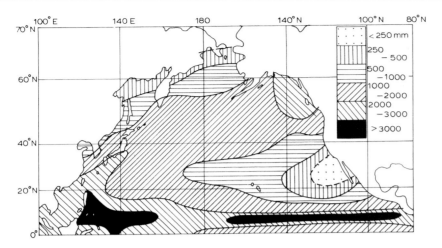

Fig. 30. Estimated annual amount of precipitation.

From precipitation observed on islands and coastal stations, one can estimate the annual amount of precipitation over the North Pacific (Fig. 30), but precipitation at neither of such places is representative of the amounts over the open sea. ELLIOTT and REED (1973) demonstrated that the amount at sea within 55 km of the Washington and Oregon coasts was only one-half to one-third that found at coastal stations; and DORMAN et al. (1974) found that the mean rainfall over 20 years at OWS N (30°N 140°W), 23 cm per year, was less than half previous estimates. The values of Fig. 30 should be read with these observations in mind.

The high amount of precipitation at low latitudes does not necessarily agree with areas of maximum frequency in Fig. 29; e.g., the equatorial parts near the Philippines and Carolines have the largest annual precipitation, but the frequency is not high. This naturally means that the intensity of precipitation in these areas is great. In contrast, there is a very high percentage of days with precipitation over the northern part in winter, but the estimated amount is not large because in winter rain and snow may continue for a long time at weak intensity. A belt of little precipitation from California to the central part of the ocean is in the region dominated by the Pacific High.

Snow and ice

Snow is observed only north of a line from south of Japan to the Canadian south coast. Fig. 31 shows the yearly variations of the area where snow occurs; the 1 percent frequency contour was chosen as the boundary as it indicates the limit of substantial snowfall. From Fig. 31D, it is seen that this boundary is nearly stationary from December to March. From May to June it shifts rapidly to the Bering Sea, and from September to November it moves just as quickly southward again. The highest frequency in winter, more than 40%, is to the south of Kamchatka in the region of poor visibility. The highest frequencies are located mainly on the rear side of the Aleutian Low.

In winter, the Sea of Japan is under the influence of the cold and dry northwest monsoon air which absorbs a large amount of water vapour when passing over the warm Tsushima Current. Much of this is deposited as snow as the monsoon ascends the mountain ranges of Japan.

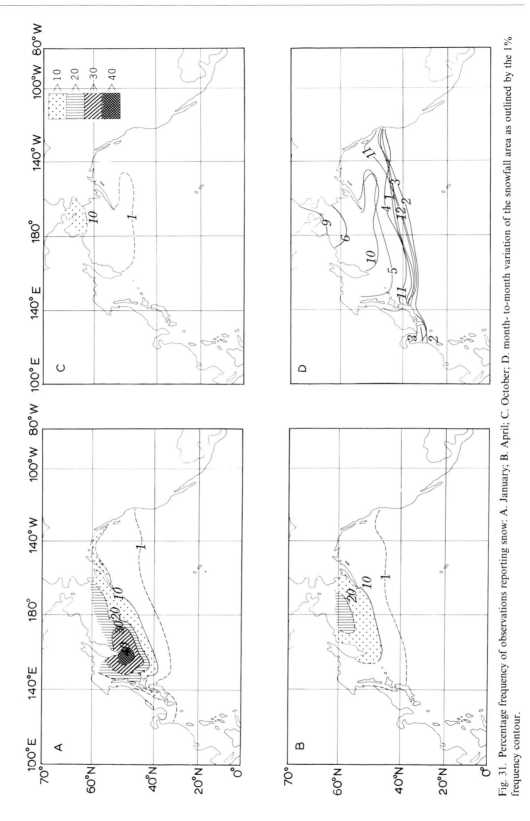

Fig. 31. Percentage frequency of observations reporting snow: A. January; B. April; C. October; D. month-to-month variation of the snowfall area as outlined by the 1% frequency contour.

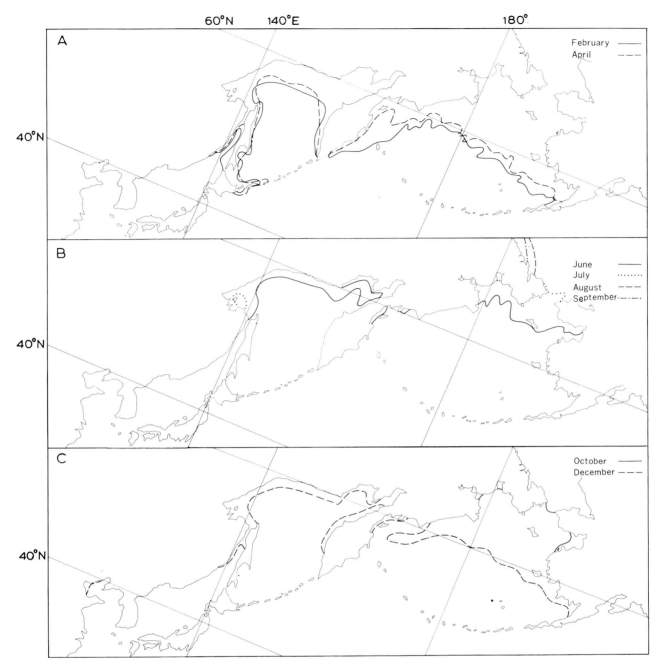

Fig. 32. Seasonal variation of sea ice.

The northern waters, such as the Sea of Okhotsk or the Bering Sea, are covered by ice during winter and early spring, limiting the sea traffic and thus the number of observations. Therefore, the data of the climatic elements in winter are not as representative as in other seasons or areas. We have indicated uncertainty by dotted lines in the figures. The seasonal variation of sea ice is illustrated in Fig. 32. During winter, the ice conditions change little from month to month, and the ice covers the coastal waters in the Sea of Okhotsk and the northern half of the Bering Sea. In May and June, the ice withdraws to the northern parts of the two seas, and in summer it lies in the Arctic Sea beyond the Bering Strait. In October, sea ice reappears in the northernmost Bering Sea. The seasonal time lag of sea ice is longer than that of the boundary of snowfall (cf. Fig. 31).

One characteristic of the North Pacific Ocean is the lack of icebergs. The separation of the Bering Sea from the Arctic Sea by the narrow and shallow Bering Strait prevents any appreciable flow of ice from the Arctic. Over the North Pacific only drift ice, which usually consists of small, thin floes, is observed.

Large-scale energy exchange between ocean and atmosphere

Early computations of seasonal or monthly heat exchange between the ocean and atmosphere were carried out by, e.g., JACOBS (1951) and ALBRECHT (1960). Heat exchange studies may be classified as: (*a*) clarification of the variability of short-term heat exchange patterns; (*b*) long-term patterns; and (*c*) consideration of the problem as part of the radiative balance of the earth's surface.

The first approach uses heat exchange information between the ocean and the atmosphere for daily or short-term weather forecasting. In this field, LAEVASTU (1965) studied extensively several North Pacific weather situations. LAEVASTU (1960) also reviewed the formulae used in the computations of energy exchange between the ocean and atmosphere.

The second approach is associated with long-range weather forecasting and is represented by NAMIAS (e.g., 1959, 1962) who stressed the relation between ocean and atmosphere anomalies and the importance of teleconnections.

The third approach was carried out mainly by BUDYKO (e.g., 1963) who treated the energy exchange between the ocean and atmosphere from the viewpoint of global radiation balance.

Evaporation

Evaporation is one of the most important topics in the study of the interaction between air and sea, not only for meteorology but also for oceanography. That portion of solar radiation which reaches the surface penetrates into the sea, although only to a comparatively shallow depth. The energy partly warms the water and is partly used for evaporation. Because our observations of the amount of insolation reaching the ocean surface are inadequate, we must calculate it indirectly by means of other elements such as sea surface temperature, air temperature, humidity, etc.

Starting from the equation:

$$Q_e = 8.3(e_w - e_a) \, w_a \, (\text{cal.cm}^{-2} \, \text{day}^{-1})$$

where e_w and e_a are vapour pressure (mbar) at the ocean surface and in the atmosphere at about 10 m, and where the wind speed w_a is in m s^{-1}, JACOBS (1942) calculated values for every 5-degree block in the North Pacific and North Atlantic oceans. His results showed that the regions of large energy exchange in the North Pacific are on the western

TABLE XIV

AMOUNTS OF EVAPORATION AT SELECTED LOCATIONS (LY/DAY)

Place	Jan.	Apr.	June	Oct.	Remarks
12°N 162°W	470	190	160	190	Eastern part of tropics
19°N 177°W	480	290	270	240	
14°N 130°E	360	310	410	410	Western part of tropics; high values in
13°N 148°E	610	250	420	430	summer and autumn relate to typhoon formation
29°N 127°E	580	360	170	370	East China Sea
35°N 141°E	800	380	130	520	In the Kuroshio
33°N 156°E	630	250	50	470	West side of Pacific ridge
31°N 164°E	690	300	150	230	
51°N 173°W	120	90	10	90	In the westerlies
43°N 162°E	340	70	40	340	
50°N 145°W	170	140	70	290	Gulf of Alaska
39°N 129°W	80	160	170	120	Off California
25°N 150°W	420	370	330	390	Central part of Pacific ridge;
36°N 168°W	260	200	220	260	nearly uniform values

side and that the value of Q_e is generally small north of 45°N and over the California Current.

The senior author made a similar study for this chapter, based on the *Marine Climatic Atlas of the World* (U.S. NAVY, 1956). The results were nearly the same as Jacobs', although the effect of the Kuroshio was more marked and seasonal changes in many places were more evident. Thirty-three locations were used and some results are presented in Table XIV.

The value of Q_e is generally larger in winter than in other seasons owing to the large difference between e_s and e_a, especially in the western part of the North Pacific. Evaporation in the tropics supplies a large amount of energy to the atmosphere, especially in the western part of the ocean. The latent heat is the energy base of tropical cyclones, and values above 400 cal.cm^{-2} day^{-1} are found in summer and fall when most tropical cyclones form. Very high values, about 800 cal.cm^{-2} day^{-1}, were obtained over the Kuroshio Current in winter, and even in autumn Q_e is higher there than at the other locations, indicating the large influence of the Kuroshio on marine climate. Large values are characteristic of regions with warm ocean currents and small values of regions with cold ocean currents.

Marine climate studies

It is possible to prepare average sea surface temperature charts for single months over wide areas of the North Pacific Ocean and, consequently, also deviations from the average. These anomalies should be associated with the fluctuations of climate, but it

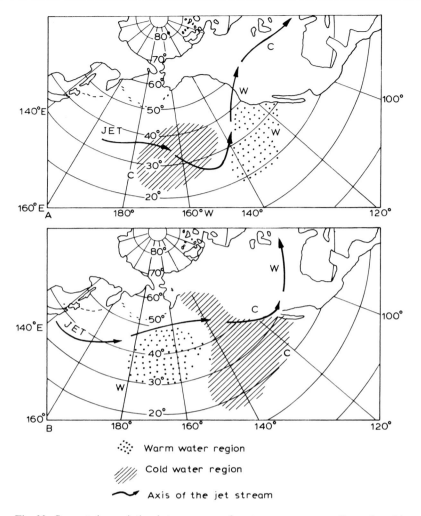

Fig. 33. Suggested association between sea surface temperature anomalies and position of the jet stream: A. winters of the 1960s; B. winter of 1971–1972. Hatched area is cold water region; dotted area warm water region; *C* = cold air; *W* = warm air.

is very difficult to establish if particular sea surface temperature anomalies correspond to anomalies of the climate of the same area or an adjacent area or remote areas, either simultaneously or with a time lag. In the following, some studies of marine climate are mentioned.

HANZAWA and INOUE (1952) noticed the seesaw of sea surface temperatures across the North Pacific and associated it with the atmospheric circulation over the area.

NAMIAS (1959) investigated a pronounced warming of the water in the eastern North Pacific and related it to the overlying atmospheric circulation. Fig. 33 is a drawing of the association between temperature anomalies in the sea surface and the position of the jet stream axis. Namias' several studies of this subject have been invaluable in piecing together the description of large-scale interaction between air and sea over the North Pacific.

BJERKNES (1966) showed that large positive sea surface temperature anomalies in 1957–1958 (3° or 4°C) extending across the equatorial zone from South America to the mid-Pacific had the effect of strengthening the zonal wind systems within the same longitude sector in the Northern Hemisphere winter, as the anomalies were associated with a negative pressure anomaly in the extratropical cyclone belt over the sector considered.

Similar large-scale remote effects, presumably also attributable to an extensive equatorial warming in the eastern Pacific, occurred in 1940—1941 and 1952—1953.

NAMIAS (1970) studied the marine climate of the North Pacific using eight million sea surface temperature observations taken over 20 years and corresponding sea level pressure data, and determined macroscale relationships with lags from 0 to 8 seasons. Evidence for transport of water masses in the North Pacific gyre was found in the sea surface temperature lag correlations. The magnitude of the indicated transport (0.2— 0.3 kt.) agrees well with mean surface current charts. Namias inferred that the Pacific gyre may be accelerated, decelerated, or otherwise modified in certain years by longlasting anomalous wind systems.

Fig. 34 shows the seasonal average North Pacific sea surface temperature departures from the 1947—1966 mean. A difference of more than 1°F (between 1958 and 1965) has been observed in the annual average sea surface temperature over the entire North Pacific north of 20°N. If we assume that all parameters (except sea surface temperature) responsible for evaporation were equal to their normal values and take the sea surface temperature as 1°F above normal, we can estimate that the latent heat of evaporation would increase by $47.8 \cdot 10^{20}$ cal. from the colder to the warmer year.

Fig. 34 shows that anomalies may persist for as long as three years. This conservatism, to which no counterpart exists in the atmosphere, not only shows the great heat capacity of the mixed layer in the ocean but may also imply the existence of processes whereby anomalous water masses frequently redevelop.

BJERKNES (1969) has also studied large-scale, almost global, interactions between air and sea. He found a link between equatorial sea surface temperatures in the central Pacific, the extent and intensity of the subtropical anticyclones, and the extratropical westerlies. He showed that changes in equatorial rainfall are associated with changes in the temperature of the underlying sea surface and reasoned that varying liberation of latent heat of condensation results in variations in the strength of the Hadley cell, which then influences higher latitudes through angular momentum transport. Bjerknes used the precipitation records from Canton Island (2°48′S 171°43′W) to demonstrate large, year-to-year fluctuations in the upward flux of water vapour. Large monthly totals of

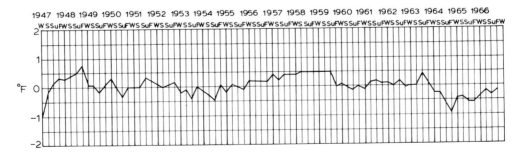

Fig. 34. Seasonal average North Pacific sea surface temperature departures from the 1947—1966 mean (NAMIAS, 1970).

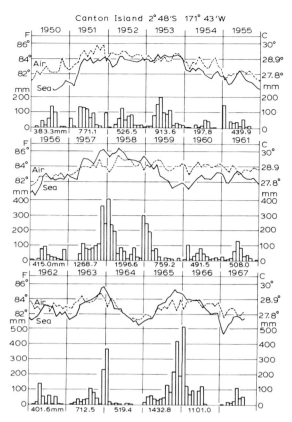

Fig. 35. Time series 1950–1967 of monthly sea and air temperatures and monthly precipitation at Canton Island (BJERKNES, 1969).

rain at this island occur only when the ocean is warmer than the atmosphere (Fig. 35). The maximum upward transfer of moisture takes place under these conditions.

The concept of zonal vertical cells (Walker circulation) implied in Bjerknes' study, and their changes with variations in the location of energy supply are illustrated by Bjerknes

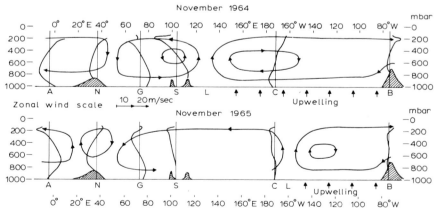

Fig. 36. Schematic zonal profiles of air flow near the Equator, in November 1964 and November 1965, based on rawin measurements at Abidjan 5°15′N 3°56′W; Nairobi 1°18′S 36°45′E; Gan 0°42′S 73°10′E; Singapore 1°21′N 103°54′E; Canton Island 2°48′S 171°43′W; and Bogotá 4°42′N 74°9′W. *L* marks position of minimum sea level pressure at the Equator. Arrows denote upwelling. (BJERKNES, 1969.)

in Fig. 36, which shows that the cells can change from year to year. In November 1964, the centre of lowest pressure was located at western New Guinea (130°E), and in November 1965 the low was near Canton Island (170°W). In November 1964, the equatorial Pacific was in a state of extensive upwelling which in November 1965 had vanished over the western and central regions.

Acknowledgements

We acknowledge extensive use of the *U.S. Marine Climatic Atlas of the World* for our North Pacific illustrations. Many figures and tables were compiled with the support of the Japan Meteorological Agency and with assistance from former colleagues in the Agency.

For the material in the many tables, we are much indebted to Mr. W. H. Haggard of the National Weather Records Center, Asheville, North Carolina. Dr. R. C. Gentry and Mr. H. Thomsen scrutinized the first draft written in 1966, and following their suggestions we redrafted the manuscript to include some recent studies. For this purpose, we received valuable suggestions from Drs. H. E. Landsberg, J. Bjerknes, and J. Namias.

We express our hearty thanks to the above organizations and gentlemen.

References

ALBRECHT, F., 1960. Jahreskarten des Wärme- und Wasserhaushaltes der Ozeane. *Ber. Dtsch. Wetterdienstes,*9: 1 – 19.

BJERKNES, J., 1959. The recent warming of the North Atlantic. In: B. BOLIN (Editor), *The Atmosphere and Sea in Motion.* Rockefeller, New York, N.Y., pp. 65 – 73.

BJERKNES, J., 1962. Synoptic survey of the interaction of sea and atmosphere in the North Atlantic. *Geofys. Publ.,* 24: 115 – 145.

BJERKNES, J., 1966. A possible response of the atmospheric Hadley circulation to equatorial anomalies of ocean temperature. *Tellus,* 18: 820 – 829.

BJERKNES, J., 1969. Atmospheric teleconnections from the equatorial Pacific. *Mon. Weather Rev.,* 97: 163 – 172.

BJERKNES, J., 1973. Rhythmic variations of the Hadley circulation over the Pacific during 1963 – 67. *NORPAX Tech. Rep.,* 24 pp.

BJERKNES, J., 1974a. *Atmospheric Teleconnections from the Equatorial Pacific During 1963 – 67.* Final Rep., National Science Foundation, Washington, D.C., Grant GA 27754, 38 pp.

BJERKNES, J., 1974b. *Preliminary Study of the Atmospheric Circulation During the Period Preceding the 1972 – 1973 El Niño.* A NORPAX contribution, Department of Meteorology, University of California at Los Angeles.

BUDYKO, M. I., 1956. *Teplovogo Balansa Zemnogo Shara (The Heat Balance of the Earth's Surface).* Gidromet. Izdatel'stvo, Leningrad, 259 pp. (English translation by N. A. Stepanova, U.S. Dept. of Commerce, Washington, D.C., 1958).

BUDYKO, M. I., 1963. *Atlas Teplovogo Balansa Zemnogo Shara (Atlas of Heat Balance of the Earth's Surface).* Glavnaia Geofiz. Observ., Moscow, 69 pp. plus charts.

CENTRAL METEOROLOGICAL OBSERVATORY, 1942. *Far Eastern Weather Records.* Tokyo, Vol. 2: 560 pp; Vol. 3: 544 pp. (in Japanese).

CENTRAL METEOROLOGICAL OBSERVATORY, 1950. *The Results of Marine Meteorological and Oceanographical Observations, Jan. 1947 – Dec. 1950.* Tokyo, Nos. 1 – 8.

DEUTSCHE SEEWARTE, 1922. *Steamship Handbook for the Pacific Ocean (Dampferhandbuch für den Stillen Ozean).* Eckhardt and Messtorff, Hamburg, 624 pp.

DORMAN, C. E., PAULSON, C. A. and QUINN, W. H., 1974. An analysis of 20 years of meteorological and oceanographic data from ocean station N. *J. Phys. Oceanogr.*, 4: 645–653.

ELLIOTT, W. P. and REED, R. K., 1973. Oceanic rainfall of the Pacific northwest coast. *J. Geophys. Res.*, 78: 941–948.

GREAT BRITAIN, 1928. *Meteorological Charts for the World, Charts 2917–18; 2930–34.* H. M. Stationery Office, London.

HANZAWA, M. and INOUE, T., 1952. Relation between oceanic and atmospheric states in the North Pacific Ocean for the year of bad and good rice crops. *Oceanogr. Rep., Central Meteorol. Observ.*, 2: 283–288.

HANZAWA, M. and TOURNIER, T. H., 1968. System for the collection of ships' weather reports. *World Weather Watch Planning Reports 25.* World Meteorological Organization, Geneva, 32 pp., xx annexes.

HAYASHI, R. and ENDO, M., 1971. Large-scale ocean/atmosphere interaction resulting from variable heat transfer at the Equator. *Tenki*, 18: 221–226. (Summary, in Japanese, of lecture given by J. Bjerknes in Tokyo, 25 September 1970.)

JACOBS, W. C., 1942. On the energy exchange between sea and atmosphere. *J. Marine Res.*, 5: 37–66.

JACOBS, W. C., 1951. The energy exchange between sea and atmosphere and some of its consequences. *Univ. Cal. Bull. Scripps Inst. Oceanogr.*, 6: 27–122.

JAPAN METEOROLOGICAL AGENCY, 1961. *Report of the Ise Bay Typhoon (Vera) in September 1959.* Japan Meteorol. Agency, Tokyo, 899 pp. plus data, 482 pp. (in Japanese).

JAPAN METEOROLOGICAL AGENCY, 1962. *Proceedings of the Interregional Seminar on Tropical Cyclones, Tokyo, 18–31 January 1962.* Japan Meteorological Agency, Tokyo, 315 pp.

JAPAN METEOROLOGICAL AGENCY, 1963. *Marine Climatological Tables of the North Pacific Ocean, 1942–1960.* Japan Meteorological Agency, Tokyo, Part 1: 322 pp.; Part 2: 298 pp.

JAPAN METEOROLOGICAL AGENCY, 1965. *Tracks of Tropical Cyclones, 1940–1964.* Japan Meteorological Agency, Tokyo, 58 pp.

KASAHARA, A., 1961. A numerical experiment on the development of a tropical cyclone. *J. Meteorol.*, 18: 171–180.

KOBE IMPERIAL MARINE OBSERVATORY, 1925, 1929, 1932. *The Mean Atmospheric Pressure, Cloudiness and Sea Surface Temperatures of the North Pacific Ocean and Surroundings for the Lustra, 1916–1920, 1921–1925, 1926–1930.* Supplemental Tables for the period from 1911–1925, 581 pp., 623 pp., 161 pp., 234 pp.

KONINKLIJK NEDERLANDSCH METEOROLOGISCH INSTITUUT, 1926. *Monthly Meteorological Data for Ten-degree Squares in the Oceans.* Computed from International logs and observations. Meteorol. Inst., De Bilt, No. 107b.

LAEVASTU, T., 1960. Factors affecting the temperatures of the surface layer of the sea. *Soc. Sci. Fenn., Comm. Physico-Mathematicae*, 25: 1–136.

LAEVASTU, T., 1965. Daily heat exchange in the North Pacific, its relations to weather and its oceanographic consequences. *Soc. Sci. Fenn., Comm. Physico-Mathematicae*, 31: 1–53.

MALONE T. F. (Editor), 1951. *Compendium of Meteorology.* American Meteorol. Soc., Boston, 1334 pp.

MURAKAMI, T., 1951. Tsuyu ake no Kikô (Mechanism of the ending of Tsuyu (rainy season). *J. Meteor. Soc. Japan, Ser. B*, 29: 160–175.

NAMIAS, J., 1959. Recent seasonal interactions between North Pacific waters and the overlying atmospheric circulation. *J. Geophys. Res.*, 64: 631–646.

NAMIAS, J., 1962. Influences of abnormal surface heat sources and sinks on atmospheric behavior. *Proceedings of International Symposium on Numerical Weather Prediction.* Meteorological Society of Japan, Tokyo, pp. 615–627.

NAMIAS, J., 1963. Large-scale air–sea interactions over the North Pacific from summer 1962 through the subsequent winter. *J. Geophys. Res.*, 68: 6171–6186.

NAMIAS, J., 1969. Seasonal interactions between the North Pacific Ocean and the atmosphere during the 1960's. *Mon. Weather Rev.*, 97: 173–192.

NAMIAS, J., 1970. Macroscale variations in sea-surface temperatures in the North Pacific. *J. Geophys. Res.*, 75: 565–582.

NAVY MINISTRY, MOSCOW, 1953. *Morskoi Atlas (Marine Atlas).* Glavnyi Shtab Voenna-Morshikh Sil., Moscow, II: 76 plates.

OKADA, T., 1931. *The Climate of Japan.* Central Meteorol. Observatory, Tokyo, 338 pp. plus 35 plates.

OOYAMA, K., 1969. Numerical simulation of the life cycle of tropical cyclones. *J. Atmos. Sci.*, 26: 3–40.

REID, J. L., 1973. North Pacific Ocean waters in winter. *The Johns Hopkins Oceanographic Studies No. 5*, The Johns Hopkins University Press, Baltimore.

RIGBY, M., 1953. Selective annotated bibliography on the climatology and marine meteorology of the Pacific. *Meteorol. Abstr. Bibl.*, 4: 1036–1082.

SIMPSON, R. H., 1952. Evolution of the kona storm, a subtropical cyclone. *J. Meteorol.,* 9: 24—35.

SVERDRUP, H. N., JOHNSON, M. S. and FLEMING, R. H., 1941. *The Oceans.* Prentice-Hall, Englewood Cliffs, N.J., 1087 pp. plus 7 charts.

TERADA, K. and OSAWA, K., 1953. On the energy exchange between sea and atmosphere in the adjacent seas of Japan. *Geophys. Mag.,* 24: 155—170.

UNESCO, 1963. *Changes of Climate. Proceedings of the Rome Symposium, October 1961.* Organized by UNESCO, Paris, and World Meteorological Organization, Geneva, 448 pp.

U.S. HYDROGRAPHIC OFFICE, 1878. *Meteorological Charts of the North Pacific Ocean from the Equator to Latitude 45°N and from the American Coast to the 180th Meridian.* Giving Information Regarding Winds, Calms, Fog, Rain, Squalls, Weather, Barometer, and Temperature of the Air and of Sea Water at the Surface. Washington, D.C., 3 pp., tables and charts.

U.S NAVY, 1956. *Marine Climatic Atlas of the World, NAVAIR 50-1C-529, Vol. II. North Pacific Ocean.* U.S. Government Printing Office, Washington, D.C., 18 pp. plus 275 charts.

WADATI, K., 1953. *Encyclopedia of Meteorology.* Toykodo Co., Tokyo, 640 pp. plus 14 charts (in Japanese).

WADATI, K., 1958. *Climate of Japan.* Tokyodo Co., Tokyo, 492 pp. plus 71 charts (in Japanese).

WORLD METEOROLOGICAL ORGANIZATION, 1961. *WMO Bibliography of Climatic Fluctuations.* World Meteorological Organization, Geneva, 90 pp.

WORLD METEOROLOGICAL ORGANIZATION, 1963. *International List of Selected and Supplementary Ships.* World Meteorological Organization, Geneva, 47, TP. 18, 185 pp.

WORLD METEOROLOGICAL ORGANIZATION, 1964a. *Commission for Maritime Meteorology.* World Meteorological Organization, Geneva, Fourth Session 164, RP 61, 160 pp.; Third Session 101, RP 41, 150 pp.

WORLD METEOROLOGICAL ORGANIZATION, 1964b. Survey on the effectiveness of the ship reporting scheme and dissemination of the reports. *Commission for Maritime Meteorology, (CMM-IV), Doc. 31.* World Meteorological Organization, Geneva.

YAMANOUCHI, Y. and OGAWA, A., 1970. *Statistical Diagrams on the Winds and Waves of the North Pacific Ocean.* Ship Research Institute, Tokyo, 144 figures.

YANAI, M., 1961. A detailed analysis of typhoon formation. *J. Meteor. Soc. Japan, Ser. 2,* 39: 187—214.

YANAI, M., 1964. Formation of tropical cyclones. *Rev. Geophys.,* 2: 367—414.

YANAI, M., 1969. Reports of the JOC Study Group on Tropical Disturbances. *Tenki,* 16 (3) (in Japanese).

YANAI, M. and TOKIOKA, T., 1969. Axially symmetric meridional motions in the baroclinic circular vortex, a numerical experiment. *J. Meteorol. Soc. Jpn.,* 47: 183—197.

Appendix – Climatic tables for ocean weather ships and land or island stations

TABLE XV

CLIMATIC TABLE FOR OWS P
Latitude 50°N, longitude 145°W

Month	Mean press. (mbar)	Mean temp. (°C)	Mean dewpoint temp. (°C)	Mean sea surface temp. (°C)	Number of days with				Mean cloudiness (tenths)	Most freq. wind direct.	Mean wind speed (m s^{-1})
					precip.	thunderstorm	fog	gale			
Jan.	1004.0	5.1	2.7	5.6					9.1	W	
Feb.	1011.0	5.1	2.9	5.0					9.1	W	
Mar.	1012.5	4.6	2.7	5.0					9.3	W	
Apr.	1013.0	5.3	3.3	5.0					9.3	SW	
May	1013.5	6.9	5.6	6.1					9.6	S	
June	1017.2	8.8	7.1	7.8					9.5	S	
July	1021.6	11.3	9.7	10.6					9.5	W	
Aug.	1020.7	13.3	12.3	13.3					9.5	W	
Sep.	1015.6	13.1	11.5	13.3					9.1	W	
Oct.	1008.1	10.7	8.2	11.1					8.8	W	
Nov.	1005.5	8.1	5.4	8.3					8.6	W	
Dec.	1005.6	5.9	4.1	6.7					9.1	W	
Annual	1012.4	8.2									

TABLE XVI

CLIMATIC TABLE FOR OWS S
Latitude 48°N, longitude 162°E

Month	Mean press. (mbar)	Mean temp. (°C)	Mean dewpoint temp. (°C)	Mean sea surface temp. (°C)	Number of days with				Mean cloudiness (tenths)	Most freq. wind direct.	Mean wind speed (m s^{-1})
					precip.	thunderstorm	fog	gale			
Jan.	1000	−0.1	−2.4	2.2					9.2	W	
Feb.	1001	−1.9	−3.6	2.2					8.4	W	
Mar.	1008	0.2	−1.2	1.1					9.5	W	
Apr.	1010	0.8	−0.2	1.7					9.4	W	
May	1011	2.6	1.6	1.7					9.6	W	
June	1008	5.0	3.6	5.0					9.6	W	
July	1014	7.8	7.1	7.8					9.9	SW	
Aug.	1012	11.4	10.8	11.7					9.9	SW	
Sep.	1014	11.3	9.8	12.8					9.2	W	
Oct.	1012	8.5	6.9	10.0					8.4	W	
Nov.	1008	3.1	1.2	6.1					8.8	W	
Dec.	1005	0.4	−1.3	3.3					8.5	W	
Annual											

TABLE XVII

CLIMATIC TABLE FOR OWS Q
Latitude 43°N, longitude 167°W

| Month | Mean press. (mbar) | Mean temp. (°C) | Mean dewpoint temp. (°C) | Mean sea surface temp. (°C) | Number of days with | | | | Mean cloudi- ness (tenths) | Most freq. wind direct. | Mean wind speed (m s⁻¹) |
					precip.	thunder- storm	fog	gale			
Jan.	1003	7.7	6.6	8.9					8.5	W	
Feb.	997	7.9	6.4	–					7.3	SW	
Mar.	1020	8.2	6.8	8.9					9.5	W	
Apr.	1015	8.4	6.8	9.4					9.0	W	
May	1017	9.1	7.6	10.0					9.8	W	
June	1017	11.1	8.4	10.6					9.5	SW	
July	1022	14.7	13.7	12.8					9.5	S	
Aug.	1022			17.2					9.6	S	
Sep.	1018	16.8	15.0	17.2					9.3	W	
Oct.	1018	12.8	11.4	13.9					8.9	W	
Nov.	1013	10.8	9.3	11.7					9.0	W	
Dec.	1007	8.7	8.2	10.6					8.6	W	
Annual											

TABLE XVIII

CLIMATIC TABLE FOR OWS X
Latitude 39°N, longitude 153°E

| Month | Mean press. (mbar) | Mean temp. (°C) | Mean dewpoint temp. (°C) | Mean sea surface temp. (°C) | Number of days with | | | | Mean cloudi- ness (tenths) | Most freq. wind direct. | Mean wind speed (m s⁻¹) |
					precip.	thunder- storm	fog	gale			
Jan.	1008.9	6.6	3.7	12.9	29	1	–		8.4	NW	13.6
Feb.	1011.2	5.4	1.7	11.4	29	3	1		8.5	NW	12.9
Mar.	1012.6	6.3	2.0	10.5	27	–	–		8.2	NW	11.8
Apr.	1014.6	9.6	6.6	11.6	24	–	3		7.7	N	10.9
May	1014.8	12.9	10.4	13.9	19	1	5		7.9	SW	8.6
June	1011.7	16.0	14.0	16.8	22	1	6		9.0	SSW	8.0
July	1013.4	20.3	18.9	20.5	22	–	12		8.5	SSW	7.2
Aug.	1013.6	23.3	20.7	24.0	20	1	2		7.0	SW	7.4
Sep.	1014.6	21.9	18.8	23.3	22	–	2		7.3	S	8.5
Oct.	1016.4	17.4	12.9	19.8	22	–	1		7.9	N	9.5
Nov.	1017.2	12.8	8.0	17.3	26	1	–		7.5	WNW	11.4
Dec.	1012.5	9.3	5.5	15.3	31	4	1		8.1	NW	12.4
Annual	1013.5	13.5	10.3	16.4	24	1	3		8.0		10.2

TABLE XIX

CLIMATIC TABLE FOR OWS V
Latitude 34°N, longitude 164°E

Month	Mean press. (mbar)	Mean temp. (°C)	Mean dewpoint temp. (°C)	Mean sea surface temp. (°C)	Number of days with				Mean cloudiness (tenths)	Most freq. wind direct.	Mean wind speed (m s⁻¹)
					precip.	thunder-storm	fog	gale			
Jan.	1012.6	14.9	13.3	19.4				–	7.0	W	
Feb.	1014.1	14.1	13.1	18.3				–	7.5	W	
Mar.	1015.4	14.4	13.8	18.3				6	7.6	S	
Apr.	1021.2	16.1	14.6	18.9				6	8.0	SE	
May	1018.0	18.3	17.9	21.1				0	8.5	S	
June	1014.5	19.9	–	22.8				0	8.0	S	
July	1017.0	23.8	23.2	25.0				1	6.8	S	
Aug.	1015.8	25.3	23.6	26.1				0	5.9	S	
Sep.	1016.1	24.2	21.7	25.6				3	4.4	E	
Oct.	1018.9	22.5	21.4	25.0				2	4.8	S	
Nov.	1018.3	19.8	18.6	23.3				–	7.9	E	
Dec.	1014.8	17.1	15.9	21.1				–	7.5	W	
Annual	1016.4	19.2									

TABLE XX

CLIMATIC TABLE FOR OWS N
Latitude 31°N, longitude 141°W

Month	Mean press. (mbar)	Mean temp. (°C)	Mean dewpoint temp. (°C)	Mean sea surface temp. (°C)	Number of days with				Mean cloudiness (tenths)	Most freq. wind direct.	Mean wind speed (m s⁻¹)
					precip.	thunder-storm	fog	gale			
Jan.	1021.5	17.7	14.9	18.9				–	8.4	E	
Feb.	1020.1	17.2	13.8	18.9				–	8.4	E	
Mar.	1023.2	16.9	13.2	18.9				4	7.8	NE	
Apr.	1023.5	17.3	13.4	18.9				2	8.3	E	
May	1023.1	18.0	14.2	19.4				1	8.5	NE	
June	1023.8	19.4	15.8	20.6				0	8.4	NE	
July	1023.3	20.3	17.3	21.7				0	8.1	NE	
Aug.	1022.8	21.1	17.3	22.8				0	7.5	NE	
Sep.	1020.9	21.7	18.2	23.3				1	5.5	NE	
Oct.	1021.2	21.2	16.9	22.8				0	7.3	NE	
Nov.	1022.7	20.2	16.1	21.7				6	8.0	E	
Dec.	1021.9	18.1	14.5	20.0				6	8.5	E	
Annual	1022.3	19.1									

TABLE XXI

CLIMATIC TABLE FOR OWS T
Latitude 29°N, longitude 135°E

Month	Mean press. (mbar)	Mean temp. (°C)	Mean dewpoint temp. (°C)	Mean sea surface temp. (°C)	Number of days with				Mean cloudiness (tenths)	Most freq. wind direct.	Mean wind speed (m s^{-1})
					precip.	thunder-storm	fog	gale			
Jan.	1018.9	16.2	10.4	19.8	24	–	–	–	6.8	NW	10.0
Feb.	1019.1	16.1	11.1	18.9	21	–	–	–	6.8	NW	9.6
Mar.	1016.0	17.3	12.7	19.4	19	–	–	7	7.1	NW	9.0
Apr.	1016.0	19.3	14.4	20.6	15	1	1	8	7.0	SE	8.7
May	1012.8	21.5	18.1	22.2	18	–	2	5	7.7	E	8.7
June	1010.8	24.5	23.1	24.4	21	2	1	4	7.7	SW	7.0
July	1010.1	27.6	25.2	27.6	15	2	1	5	6.0	WSW	7.0
Aug.	1009.2	28.1	24.6	28.6	19	4	–	8	5.4	ESE	6.7
Sep.	1011.0	27.3	23.4	28.0	18	2	–	3	5.3	E	6.7
Oct.	1014.0	21.6	19.8	26.1	20	1	–	7	6.3	NE	8.0
Nov.	1017.0	21.9	17.3	28.8	19	–	–	7	6.6	ENE	8.8
Dec.	1019.7	17.7	11.9	21.3	18	–	–	–	6.1	N	8.6
Annual	1014.6	21.6	17.7	23.8	19	1	0		6.6		8.2

TABLE XXII

CLIMATIC TABLE FOR DUTCH HARBOR
Latitude 53°53′N, longitude 166°32′W, elevation 4 m

Month	Mean stat. press. (mbar)	Temperature (°C) daily mean	mean daily range	extremes max.	min.	Mean vapour press. (mbar)	Precipitation (mm) mean	24-h max.	Relat. humid. (%)
Jan.	1001.4	0.3	5.0	13.4	−15.0		150	66	
Feb.	1001.2	−0.3	5.0	15.6	−14.0		142	94	
Mar.	1005.8	1.1	5.6	20.6	−15.0		125	64	
Apr.	1006.5	2.2	5.6	16.1	−10.5		99	89	
May	1006.1	4.7	6.1	18.9	−6.6		107	69	
June	1010.2	8.1	6.1	22.8	−1.1		69	76	
July	1014.2	10.6	6.7	26.7	2.2		48	41	
Aug.	1013.6	11.7	6.7	26.7	−2.2		61	53	
Sep.	1007.8	9.3	6.1	26.7	−2.2*		135	64	
Oct.	1003.4	5.6	5.6	18.4	−4.4		185	81	
Nov.	1002.4	2.5	5.0	14.5	−10.0		145	81	
Dec.	999.3	0.9	5.0	13.4	−12.2		178	53	
Annual	1005.9	4.7		26.7	−15.0		1443	94	

Month	Mean evap. (mm)	Number of days with precip. (⩾0.25 mm)	thunder-storm	heavy fog	Mean cloudiness (tenths)	Mean sunshine duration (h)	Most freq. wind direct.	Mean wind speed (m s⁻¹)	Sun-shine (% of possible)
Jan.		22	0	0.8	7.7		SE, S	2.4	
Feb.		18	0	0.7	7.5		NW	3.8	
Mar.		20	0	0.9	7.4		SE	5.8	
Apr.		19	0	0.2	7.9		SE, NW	4.0	
May		19	0	2.2	8.1		SE	4.5	
June		13	0	3.7	8.0		SE	4.9	
July		13	0	3.1	7.7		SE	5.0	
Aug.		12	0	2.9	7.5		–	4.5	
Sep.		18	0	1.9	8.0		SE	5.2	
Oct.		23	0	1.9	7.9		NW	4.7	
Nov.		21	0	0.0	7.5		SW	3.8	
Dec.		21	0	0.6	7.3		SE, SW	3.6	
Annual		219	0	19.4	7.7		SE	4.3	

TABLE XXIII

CLIMATIC TABLE FOR SHEMYA ISLAND
Latitude 52°43′N, longitude 174°06′E, elevation 37 m

Month	Mean stat. press. (mbar)	Temperature (°C) daily mean	mean daily range	extremes max.	min.	Mean vapour press. (mbar)	Precipitation (mm) mean	24-h max.	Relat. humid. (%)
Jan.	998.8	−0.4		7.2	−8.9		64	33	
Feb.	999.4	−0.7		5.6	−7.8		58	54	
Mar.	1005.8	0.5		5.6	−8.3		65	61	
Apr.	1010.5	1.9		7.8	−6.1		52	39	
May	1007.4	3.3		10.6	−2.2		61	41	
June	1012.1	5.4		13.9	−1.1		34	45	
July	1011.2	8.0		17.2	3.9		55	43	
Aug.	1011.3	9.8		15.6	3.3		54	57	
Sep.	1012.8	9.1		14.4	2.8		55	42	
Oct.	1006.2	4.3		12.2	−3.3		70	133	
Nov.	999.6	2.1		8.9	−5.6		69	44	
Dec.	999.2	−0.3		7.8	−7.8		54	29	
Annual	1006.2	3.6		17.2	−8.9		693	133	

Month	Mean evap. (mm)	Number of days with precip. (≥0.25 mm)	thunder-storm	heavy fog	Mean cloudi-ness (tenths)	Mean sunshine duration (h)	Most freq. wind direct.	Mean wind speed (m s⁻¹)	Sun-shine (% of possible)
Jan.		21	0	1	8.6		ENE	9.0	
Feb.		19	0	1	8.8		NNE	9.2	
Mar.		18	0	3	8.8		ESE	9.1	
Apr.		15	0	4	9.0		NW	8.2	
May		15	0	6	9.3		NW	7.5	
June		13	0	14	9.5		NW	6.2	
July		14	0	24	9.8		WSW	5.9	
Aug.		16	0	16	9.5		WSW	6.2	
Sep.		16	0	6	8.3		WSW	7.1	
Oct.		22	0	2	8.2		WSW	9.0	
Nov.		22	0	2	8.4		WNW	9.6	
Dec.		20	0	2	8.7		WSW	9.1	
Annual		211	0	80	8.9			8.0	

TABLE XXIV

CLIMATIC TABLE FOR HACHIJOJIMA
Latitude 33°06′N, longitude 139°47′E, elevation 80.9 m

Month	Mean stat. press. (mbar)	Temperature (°C) daily mean	mean daily range	extremes max.	min.	Mean vapour press. (mbar)	Precipitation (mm) mean	24-h max.	Relat. humid. (%)
Jan.	1005.3	10.2	5.6	13.2	7.6	8.5	182.3		
Feb.	1006.1	10.2	5.9	13.4	7.5	8.7	202.8		
Mar.	1005.4	12.3	6.2	15.6	9.4	10.4	243.2		
Apr.	1006.4	16.1	5.9	19.2	13.3	13.9	236.0		
May	1002.9	19.2	5.5	22.1	16.6	18.1	289.0		
June	1000.6	21.9	4.7	24.4	19.7	23.2	350.4		
July	1000.7	25.3	4.6	27.9	23.3	28.3	181.2		
Aug.	1001.0	26.6	5.3	29.6	24.3	29.5	231.0		
Sep.	1002.7	25.0	5.4	28.0	22.6	25.9	336.1		
Oct.	1005.4	20.9	5.4	23.8	18.4	19.2	484.4		
Nov.	1007.9	17.2	5.5	20.1	14.6	14.6	373.3		
Dec.	1007.4	12.9	5.5	15.8	10.3	10.3	173.9		
Annual	1004.3	18.1	5.5	21.1	15.6	17.6	3283.6		
	1951–70	1941–70	1941–70	1941–70	1941–70	1941–70	1941–70		

Month	Mean evap. (mm)	Number of days with precip. (≥0.25 mm)	thunder-storm	heavy fog	Mean cloudi-ness (tenths)	Mean sunshine duration (h)	Most freq. wind direct.	Mean wind speed (m s⁻¹)	Sun-shine (% of possible)
Jan.		14.8	2.7	0.0	7.5	117.2	W	8.3	
Feb.		15.1	2.0	0.0	7.5	114.8	W	7.5	
Mar.		15.2	1.6	0.3	7.9	151.3	W	7.4	
Apr.		14.0	0.9	1.0	7.2	163.2	W	6.6	
May		15.0	0.9	1.7	8.0	160.3	W	5.9	
June		16.2	0.8	2.3	8.7	113.8	W	5.3	
July		10.4	1.8	1.5	8.0	177.5	SW	5.3	
Aug.		12.7	3.9	0.1	6.7	225.3	W	5.1	
Sep.		15.9	2.8	0.0	6.9	180.9	W	5.5	
Oct.		18.2	1.5	0.0	7.2	133.6	NE	6.9	
Nov.		15.4	1.9	0.0	7.1	117.2	W	6.3	
Dec.		12.4	2.2	0.0	6.9	123.8	W	7.2	
Annual		174.9	23.1	7.0	7.4	1778.8		6.4	
		1951–70	1941–70	1941–70	1941–70	1941–70	1951–70	1961–70	

TABLE XXV

CLIMATIC TABLE FOR TORISHIMA
Latitude 30°29′N, longitude 140°18′E, elevation 82.5 m

| Month | Mean stat. press. (mbar) | Temperature (°C) | | | | Mean vapour press. (mbar) | Precipitation (mm) | | Relat. humid. (%) |
| | | daily mean | mean daily range | extremes | | | mean | 24-h max. | |
				max.	min.				
Jan.	1006.1	14.4	4.5	16.7	12.2	11.4	101.5		
Feb.	1006.3	14.1	4.6	16.5	11.9	11.3	102.1		
Mar.	1005.9	15.4	4.7	18.0	13.3	11.9	108.0		
Apr.	1006.4	18.1	4.2	20.4	16.2	16.4	144.9		
May	1003.3	20.5	4.1	22.7	18.6	20.9	259.9		
June	999.2	23.3	4.1	25.5	21.4	25.7	263.1		
July	1001.4	26.5	4.7	29.1	24.4	29.5	98.8		
Aug.	1001.0	27.1	4.8	30.1	25.3	30.0	118.5		
Sep.	1002.4	26.8	4.9	29.7	24.8	28.1	135.9		
Oct.	1004.5	23.7	4.3	26.2	21.9	22.8	296.4		
Nov.	1007.5	20.3	4.1	22.6	18.5	17.5	225.6		
Dec.	1007.5	17.0	4.1	19.2	15.1	13.2	129.8		
Annual	1004.2	20.6	4.4	23.0	18.6	20.0	1984.6		
	1950–64	1947–64	1947–64	1947–64	1947–64	1947–64	1947–64		

| Month | Mean evap. (mm) | Number of days with | | | Mean cloudi-ness (tenths) | Mean sunshine duration (h) | Most freq. wind direct. | Mean wind speed (m s^{-1}) | Sun-shine (% of possible) |
		precip. (⩾0.25 mm)	thunder-storm	heavy fog					
Jan.		14.9	1.1	–	6.8	113.1	WNW	9.3	
Feb.		12.3	0.5	0.3	6.9	109.6	WNW	8.4	
Mar.		10.4	0.6	0.2	7.2	143.4	NNW	8.4	
Apr.		9.5	0.3	4.1	7.4	144.6	S	7.2	
May		14.5	0.2	5.8	8.2	129.9	S	7.0	
June		12.8	0.7	6.8	8.3	132.7	S	6.9	
July		6.6	1.3	1.2	6.7	221.6	WSW	5.6	
Aug.		9.3	1.2	0.2	6.0	234.7	S	5.5	
Sep.		10.1	1.4	–	5.9	211.2	S	5.9	
Oct.		13.5	1.2	0.1	7.0	149.2	NE	6.6	
Nov.		13.0	0.6	0.1	7.1	119.1	N	7.2	
Dec.		10.4	0.6	0.1	6.4	121.2	NNW	7.8	
Annual		138.0	9.6	18.4	7.1	1830.2		7.2	
		1950–64	1950–64	1950–64	1950–64	1950–64	1950–64	1951–60	

TABLE XXVI

CLIMATIC TABLE FOR MIDWAY ISLAND
Latitude 28°13′N, longitude 177°22′W, elevation 13 m

Month	Mean stat. press. (mbar)	Temperature (°C)				Mean vapour press. (mbar)	Precipitation (mm)		Relat. humid. (%)
		daily mean	mean daily range	extremes			mean	24-h max.	
				max.	min.				
Jan.		18.9	6.7	26.1	7.8		107	86	
Feb.		18.4	6.7	26.7	8.9		94	43	
Mar.		19.2	6.1	27.2	10.0		107	97	
Apr.		19.7	7.2	27.8	10.0		130	142	
May		21.7	7.8	30.6	11.7		91	155	
June		24.2	7.2	31.7	13.4		86	102	
July		25.3	7.2	32.2	18.9		79	104	
Aug.		25.9	7.2	32.8	17.8		89	79	
Sep.		25.6	7.8	32.2	15.0		127	150	
Oct.		23.6	7.2	31.7	13.4		132	160	
Nov.		21.7	6.7	28.9	11.7		43	36	
Dec.		19.2	6.1	26.7	10.6		91	53	
Annual		22.0		32.8	7.8		1176	160	

Month	Mean evap. (mm)	Number of days with			Mean cloudi-ness (tenths)	Mean sunshine duration (h)	Most freq. wind direct.	Mean wind speed (m s⁻¹)	Sun-shine (% of possible)
		precip. (≥0.25 mm)	thunder-storm	heavy fog					
Jan.		16	1	1	6		W	6.8	
Feb.		16	(<0.5)*	1	6		W	6.2	
Mar.		13	(<0.5)	1	7		E	5.8	
Apr.		11	1	1	6		ENE	5.2	
May		10	(<0.5)	1	6		E	4.6	
June		10	(<0.5)	(<0.5)*	6		E	4.4	
July		15	1	(<0.5)	5		E	5.0	
Aug.		14	1	(<0.5)	5		E	4.5	
Sep.		15	1	(<0.5)	5		E	4.2	
Oct.		12	1	(<0.5)	5		E	5.1	
Nov.		12	(<0.5)	1	6		E	5.3	
Dec.		18	(<0.5)	1	6		E	5.2	
Annual		162	6	7	6		E	5.2	

* Less than 0.5 but more than 0.

TABLE XXVII

CLIMATIC TABLE FOR CHICHIJIMA
Latitude 27°05'N, longitude 142°11'E, elevation 3 m

Month	Mean stat. press. (mbar)	Temperature (°C)				Mean vapour press. (mbar)	Precipitation (mm)		Relat. humid. (%)
		daily mean	mean daily range	extremes			mean	24-h max.	
				max.	min.				
Jan.	1017.6	17.5	6.2	26.1	7.1		99.2	100.0	70.2
Feb.	1016.7	17.3	6.2	27.0	7.0		79.3	77.8	70.9
Mar.	1017.0	18.2	6.4	26.7	7.6		106.0	73.3	74.2
Apr.	1015.7	20.5	6.1	29.3	9.6		122.6	169.8	80.1
May	1013.0	22.7	5.6	32.0	12.2		204.6	171.1	84.2
June	1012.3	25.5	6.0	33.0	14.7		128.7	115.4	84.5
July	1011.8	27.1	6.8	34.9	19.1		102.7	177.7	81.0
Aug.	1008.8	27.1	5.8	34.3	18.8		171.2	182.5	82.8
Sep.	1011.7	26.8	6.0	33.8	17.5		137.2	113.8	82.5
Oct.	1013.4	25.4	5.8	32.9	15.9		140.7	232.4	82.1
Nov.	1016.2	22.7	5.8	30.1	12.4		145.9	163.9	78.7
Dec.	1017.2	19.4	5.9	27.5	8.0		131.6	118.4	73.0
Annual	1014.4	22.5	6.0	34.9	7.0		1569.7	232.4	78.7
		1907–29	1907–29	1907–29	1907–29		1907–29	1907–29	1907–29

Month	Mean evap. (mm)	Number of days with			Mean cloudiness (tenths)	Mean sunshine duration (h)	Most freq. wind direct.	Mean wind speed (m s⁻¹)	Sunshine (% of possible)
		precip. (≥0.25 mm)	thunderstorm	heavy fog					
Jan.		16.5	0.2		6.3	150	N	2.6	
Feb.		15.1	–		6.7	150	N	2.8	
Mar.		15.4	0.4		6.8	170	N	2.5	
Apr.		14.9	0.4		7.2	168	S	2.2	
May		17.6	0.8		7.8	169	NE	2.2	
June		11.9	0.2		6.9	209	S	2.1	
July		13.8	0.4		5.9	261	SE	1.7	
Aug.		16.6	0.8		6.6	237	E	2.1	
Sep.		17.0	1.0		5.8	230	E	2.1	
Oct.		18.4	1.8		8.7	199	NE	2.1	
Nov.		17.1	0.4		6.3	154	NE	2.2	
Dec.		17.0	0.2		6.5	135	N	2.3	
Annual		191.3	6.6		6.5	2239	S	2.2	
		1907–29	1925–29		1907–29	1907–29		1907–29	

TABLE XXVIII

Latitude 24°18′N, longitude 153°58′E, elevation 16.7 m, period 1951–1963

Month	Mean stat. press. (mbar)	Temperature (°C)				Mean vapour press. (mbar)	Precipitation (mm)		Relat. humid. (%)
		daily mean	mean daily range	extremes			mean	24-h max.	
				max.	min.				
Jan.	1016.3	22.2	4.2	29.7	15.6	19.2	62.5	43.1 ⎫	
Feb.	1016.8	21.7	4.2	28.5	15.1	19.2	53.2	36.3 ⎬ *1	
Mar.	1016.7	22.2	4.2	29.9	15.5	20.2	49.8	40.5 ⎭	
Apr.	1017.6	23.9	3.8	31.9	17.9	23.4	39.9	86.0	
May	1016.1	25.6	4.0	33.3	19.5	25.7	38.5	50.4	
June	1015.2	27.8	4.7	35.0	22.0	28.2	57.8	139.2	
July	1013.4	28.0	4.9	35.6	21.6	29.0	186.0	176.9 ⎫	
Aug.	1013.0	28.0	4.5	34.7	21.8	29.8	174.7	417.6 ⎪	
Sep.	1013.4	28.3	4.6	35.3	21.7	29.6	89.3	56.1 ⎬ *2	
Oct.	1013.8	27.5	4.2	33.5	21.9	28.5	89.6	53.5 ⎪	
Nov.	1015.5	26.3	3.8	34.2	19.8	26.2	57.7	36.8 ⎪	
Dec.	1015.9	24.3	3.9	31.6	18.0	22.7	66.7	91.6 ⎭	
Annual	1015.3	25.5	4.3	35.6	15.1	25.1	963.4	417.6*3	

Month	Mean evap. (mm)	Number of days with			Mean cloudi- ness (tenths)	Mean sunshine duration (h)	Most freq. wind direct.	Mean wind speed (m s⁻¹)	Sun- shine (% of possible)
		precip. (≥0.25 mm)	thunder- storm	heavy fog					
Jan.		9.1	0.3		5.5		NE	7.1	
Feb.		7.8	0.0		5.7		ENE	7.3	
Mar.		7.2	0.0		5.3		E	7.5	
Apr.		5.9	0.1		5.0		E	7.8	
May		6.0	0.4		4.8		ESE	5.8	
June		7.6	0.7		4.5		E	5.0	
July		14.0	2.2		6.4		E	5.6	
Aug.		13.4	2.9		6.3		ESE	6.5	
Sep.		12.0	2.3		5.5		E	6.8	
Oct.		10.8	1.0		5.2		E	7.5	
Nov.		8.4	0.4		5.0		ESE	7.5	
Dec.		8.9	0.3		5.5		E	7.9	
Annual		111.1	10.6		5.4		E	6.9	

*1 1952–63; *2 1951–62; *3 August 17, 1957.

TABLE XXIX

CLIMATIC TABLE FOR HONOLULU OAHU
Latitude 21°20′N, longitude 157°55′W, elevation 5 m

Month	Mean stat. press. (mbar)	Temperature (°C)				Mean vapour press. (mbar)	Precipitation (mm)		Relat. humid. (%)
		daily mean	mean daily range	extremes			mean	24-h max.	
				max.	min.				
Jan.	1015.6	22.5	3.9	28.9	12.2		96	114	71
Feb.	1015.9	22.4	5.0	28.9	11.1		84	163	69
Mar.	1017.0	22.7	5.6	28.9	11.7		73	343	67
Apr.	1017.7	23.4	5.6	30.0	15.0		33	203	66
May	1017.5	24.4	5.6	30.6	15.6		25	119	64
June	1017.2	25.5	5.0	31.1	17.2		8	76	63
July	1016.7	26.0	5.0	31.1	17.2		11	31	64
Aug.	1016.1	26.3	5.0	31.1	17.2		23	53	64
Sep.	1015.4	26.2	5.0	31.1	17.2		21	152	64
Oct.	1015.5	25.7	5.6	32.2	17.2		47	117	67
Nov.	1015.7	24.4	5.6	30.0	15.0		55	140	68
Dec.	1015.9	23.1	5.0	29.5	12.8		76	155	70
Annual	1016.4	24.1		32.2	11.1		556	343	66

Month	Mean evap. (mm)	Number of days with			Mean cloudiness (tenths)	Mean sunshine duration (h)	Most freq. wind direct.	Mean wind speed (m s^{-1})	Sunshine (% of possible)
		precip. (⩾0.25 mm)	thunderstorm	heavy fog					
Jan.		14	1	0	5.4		ENE	4.6	65
Feb.		11	2	0	5.7		ENE	4.8	66
Mar.		13	2	0	5.8		ENE	5.1	71
Apr.		12	–	0	6.3		ENE	5.3	70
May		11	0	0	6.2		ENE	5.6	71
June		12	0	0	5.5		ENE	5.7	74
July		14	0	0	5.2		ENE	6.0	77
Aug.		13	0	0	5.2		ENE	6.0	78
Sep.		13	0	0	5.0		ENE	5.2	76
Oct.		13	1	0	5.6		ENE	4.8	69
Nov.		13	1	0	5.6		ENE	5.0	62
Dec.		15	1	0	5.6		ENE	4.8	60
Annual		154	8	0	5.6		ENE	5.2	70

TABLE XXX

CLIMATIC TABLE FOR HILO HAWAII
Latitude 19°44′N, longitude 155°04′W, elevation 11 m

Month	Mean stat. press. (mbar)	Temperature (°C)				Mean vapour press. (mbar)	Precipitation (mm)		Relat. humid. (%)
		daily mean	mean daily range	extremes			mean	24-h max.	
				max.	min.				
Jan.	1015.8	21.6	8.9	31.7	13.4		300	252	76
Feb.	1016.0	21.4	9.4	31.1	12.8		329	340	75
Mar.	1017.1	21.4	8.3	31.1	12.8		373	234	76
Apr.	1017.7	22.0	8.3	30.6	13.4		303	239	76
May	1017.6	22.8	8.9	29.5	14.5		237	150	77
June	1017.5	23.5	8.9	31.1	16.1		172	58	76
July	1017.0	23.8	8.9	30.6	16.7		249	137	77
Aug.	1016.1	24.3	8.3	33.9	17.2		291	132	77
Sep.	1015.4	24.2	8.9	33.4	16.7		216	81	76
Oct.	1015.5	23.9	8.3	31.1	16.7		274	226	77
Nov.	1015.7	22.9	8.3	30.0	14.5		340	241	79
Dec.	1016.0	21.9	8.3	29.5	13.4		386	267	78
Annual	1016.5	22.8		33.9	12.8		3470	340	77

Month	Mean evap. (mm)	Number of days with			Mean cloudi-ness (tenths)	Mean sunshine duration (h)	Most freq. wind direct.	Mean wind speed (m s⁻¹)	Sun-shine (% of possible)
		precip. (⩾0.25 mm)	thunder-storm	heavy fog					
Jan.		20	–	0	6.6		SW	3.4	47
Feb.		19	2	0	6.8		SW	3.6	44
Mar.		25	2	0	7.8		SW	3.6	38
Apr.		25	1	0	8.2		SW	3.5	35
May		25	–	0	8.3		NE	3.3	29
June		24	0	0	7.5		WSW	3.3	41
July		28	–	0	7.6		NE	3.2	43
Aug.		27	0	0	7.7		SW	3.2	39
Sep.		23	0	0	7.2		WSW	3.2	39
Oct.		25	1	0	7.3		WSW	3.1	39
Nov.		23	1	0	7.5		SW	3.1	33
Dec.		23	1	0	7.2		SW	3.3	36
Annual		287	8	0	7.5		SW	3.3	39

TABLE XXXI

CLIMATIC TABLE FOR WAKE ISLAND
Latitude 19°17′N, longitude 166°39′E, elevation 4 m

Month	Mean stat. press. (mbar)	Temperature (°C)				Mean vapour press. (mbar)	Precipitation (mm)		Relat. humid. (%)
		daily mean	mean daily range	extremes			mean	24-h max.	
				max.	min.				
Jan.	1014.9	25.2		30.6	18.3		29	32	69
Feb.	1015.3	25.1		30.0	18.3		34	44	70
Mar.	1015.7	25.4		31.1	19.4		37	69	73
Apr.	1016.4	25.8		30.6	20.0		47	41	73
May	1015.7	26.8		31.7	21.1		52	57	76
June	1015.0	27.4		32.2	21.7		48	94	76
July	1014.1	27.8		32.2	20.6		117	97	77
Aug.	1013.4	28.0		33.3	20.0		180	113	77
Sep.	1013.0	28.1		32.8	21.1		133	380	76
Oct.	1013.5	27.6		32.8	20.0		134	86	77
Nov.	1014.3	26.9		31.1	18.3		78	159	76
Dec.	1014.5	26.1		31.1	17.8		46	46	74
Annual	1014.7	26.7		33.3	17.8		936	380	75

Month	Mean evap. (mm)	Number of days with			Mean cloudi- ness (tenths)	Mean sunshine duration (h)	Most freq. wind direct.	Mean wind speed (m s⁻¹)	Sun- shine (% of possible)
		precip. (≥0.25 mm)	thunder- storm	heavy fog					
Jan.		10	0	0	4.7		ENE	5.8	
Feb.		9	0	0	4.2		ENE	5.9	
Mar.		12	0	0	4.6		ENE	6.2	
Apr.		15	(<0.5)*	0	5.1		ENE	6.8	
May		15	(<0.5)	0	5.4		ENE	6.4	
June		16	1	0	5.3		E	5.2	
July		19	1	0	6.6		E	5.5	
Aug.		19	1	0	6.5		E	5.4	
Sep.		18	1	0	6.3		E	5.3	
Oct.		20	2	0	6.1		ENE	6.0	
Nov.		14	(<0.5)	0	4.8		ENE	6.6	
Dec.		13	(0.5)	0	4.4		ENE	6.0	
Annual		179	7	0	5.3		ENE	5.9	

* Cf. Table XXVI.

TABLE XXXII

CLIMATIC TABLE FOR JOHNSON ISLAND
Latitude 16°44′N, longitude 169°31′W, elevation 2 m

Month	Mean stat. press. (mbar)	Temperature (°C)				Mean vapour press. (mbar)	Precipitation (mm)		Relat. humid. (%)
		daily mean	mean daily range	extremes			mean	24-h max.	
				max.	min.				
Jan.	1012.6	25.0		30.0	18.3		99	68	
Feb.	1013.8	24.9		29.4	17.8		39	16	
Mar.	1014.0	25.1		29.4	19.4		59	66	
Apr.	1014.6	25.5		29.4	20.0		58	98	
May	1014.8	26.2		30.6	20.0		25	116	
June	1014.3	26.8		30.6	20.6		21	29	
July	1014.1	27.2		31.1	21.1		33	13	
Aug.	1013.4	27.5		31.1	21.7		57	202	
Sep.	1012.9	27.6		31.1	21.7		60	60	
Oct.	1013.3	27.2		31.1	18.9		83	243	
Nov.	1013.0	26.4		31.1	17.2		52	23	
Dec.	1012.9	25.3		30.0	16.7		76	234	
Annual	1013.6	26.3		31.1	16.7		663	243	

Month	Mean evap. (mm)	Number of days with			Mean cloudi- ness (tenths)	Mean sunshine duration (h)	Most freq. wind direct.	Mean wind speed (m s⁻¹)	Sun- shine (% of possible)
		precip. (≥0.25 mm)	thunder- storm	heavy fog					
Jan.		10	0	0	4.7		NE	5.9	70
Feb.		13	(<0.5)*	0	4.9		NE	6.9	75
Mar.		14	(<0.5)	0	5.1		NE	6.6	68
Apr.		15	(<0.5)	0	6.8		NE	6.5	64
May		13	(<0.5)	0	7.1		NE	6.9	63
June		14	0	0	6.1		E	6.9	75
July		13	0	0	6.1		E	6.8	78
Aug.		14	0	0	6.9		E	6.7	69
Sep.		15	(<0.5)	0	6.3		E	6.1	67
Oct.		16	(<0.5)	0	7.1		E	6.0	61
Nov.		14	0	0	6.4		E	6.8	56
Dec.		16	1	0	5.4		E	6.5	59
Annual		165	2	0	6.1			6.5	68

* Cf. Table XXVI.

492

TABLE XXXIII

CLIMATIC TABLE FOR GUAM
Latitude 13°27'N, longitude 114°39'E, elevation 19 m

Month	Mean stat. press. (mbar)	Temperature (°C)				Mean vapour press. (mbar)	Precipitation (mm)		Relat. humid. (%)
		daily mean	mean daily range	extremes			mean	24-h max.	
				max.	min.				
Jan.	1011.4	26.4	5.0	31.7	20.0		76	114	
Feb.	1011.7	26.1	5.6	31.7	19.5		69	86	
Mar.	1011.3	26.7	5.6	32.2	20.0		76	105	
Apr.	1011.1	27.5	6.1	33.4	21.7		51	58	
May	1010.7	28.1	6.1	34.5	21.7		109	114	
June	1010.8	28.1	6.1	34.5	22.2		142	112	
July	1010.0	27.2	5.6	33.4	21.1		361	267	
Aug.	1009.1	27.2	5.6	32.8	21.7		376	244	
Sep.	1009.6	27.2	5.6	32.8	21.1		361	168	
Oct.	1009.6	27.2	5.6	32.8	20.6		333	406	
Nov.	1009.3	27.5	5.0	32.2	20.6		196	226	
Dec.	1009.9	27.0	5.0	33.4	21.1		125	76	
Annual	1010.4	27.2		34.5	19.5		2273	406	

Month	Mean evap. (mm)	Number of days with			Mean cloudi-ness (tenths)	Mean sunshine duration (h)	Most freq. wind direct.	Mean wind speed (m s⁻¹)	Sun-shine (% of possible)
		precip. (≥0.25 mm)	thunder-storm	heavy fog					
Jan.		14	(<0.5)*	(<0.5)*	7		E	3.6	52
Feb.		12	0	(<0.5)	7		NE	4.4	58
Mar.		12	0	(<0.5)	7		E	3.8	71
Apr.		10	1	0	7		E	3.8	67
May		14	1	0	7		E	3.7	62
June		18	2	0	7		E	2.5	63
July		24	5	0	8		E	2.2	44
Aug.		23	5	0	8		E	2.2	46
Sep.		23	8	0	8		E	2.0	42
Oct.		23	4	(<0.5)	8		E	2.6	46
Nov.		20	2	0	7		E	3.3	55
Dec.		18	(<0.5)	(<0.5)	7		E	3.8	52
Annual		211	28	(<0.5)	7		E	3.2	55

* Cf. Table XXVI.

TABLE XXXIV

CLIMATIC TABLE FOR ENIWETOK
Latitude 11°20'N, longitude 162°20'E, elevation 3 m

Month	Mean stat. press. (mbar)	Temperature (°C)				Mean vapour press. (mbar)	Precipitation (mm)		Relat. humid. (%)
		daily mean	mean daily range	extremes max.	min.		mean	24-h max.	
Jan.	1010.5	27.3		31.1	21.7		26	12	
Feb.	1010.5	27.2		31.1	22.2		47	11	
Mar.	1010.5	27.4		31.7	22.8		47	14	
Apr.	1011.0	27.8		32.2	22.2		33	78	
May	1010.6	28.0		32.8	21.1		116	73	
June	1010.6	28.2		32.2	21.7		86	70	
July	1010.2	28.2		33.3	21.7		164	47	
Aug.	1009.5	28.3		32.8	21.1		173	53	
Sep.	1009.8	28.5		33.9	21.2		158	96	
Oct.	1009.5	28.3		33.3	21.7		231	98	
Nov.	1009.1	28.1		32.2	22.2		160	63	
Dec.	1009.3	27.8		31.1	21.7		67	103	
Annual	1010.1	27.9		33.9	21.1		1307	103	

Month	Mean evap. (mm)	Number of days with			Mean cloudiness (tenths)	Mean sunshine duration (h)	Most freq. wind direct.	Mean wind speed (m s⁻¹)	Sunshine (% of possible)
		precip. (≥0.25 mm)	thunderstorm	heavy fog					
Jan.		11	0	0	5.5		ENE	7.6	
Feb.		10	0	0	6.0		ENE	8.3	
Mar.		12	0	0	6.5		ENE	8.0	
Apr.		13	0	0	6.9		ENE	7.1	
May		17	(<0.5)*	0	7.5		ENE	6.8	
June		17	(<0.5)	0	7.0		ENE	5.5	
July		22	1	0	7.2		E	5.1	
Aug.		21	1	0	7.4		E	4.5	
Sep.		21	1	0	6.9		E	4.7	
Oct.		22	1	0	6.9		E	4.7	
Nov.		21	(<0.5)	0	6.1		ENE	6.3	
Dec.		16	(<0.5)	0	5.7		ENE	7.7	
Annual		202		0	6.6		ENE	6.4	

* Cf. Table XXVI.

TABLE XXXV

CLIMATIC TABLE FOR YAP, CAROLINE IS.
Latitude 09°31′N, longitude 138°08′E, elevation 17 m

Month	Mean stat. press. (mbar)	Temperature (°C)				Mean vapour press. (mbar)	Precipitation (mm)		Relat. humid. (%)
		daily mean	mean daily range	extremes max.	min.		mean	24-h max.	
Jan.	1010.0	26.9	6.1	34.5	20.0		200	117	
Feb.	1010.7	26.9	6.1	33.9	20.6		118	163	
Mar.	1010.4	27.2	6.7	35.6	20.0		137	140	
Apr.	1010.0	27.6	6.1	36.7	19.5		162	76	
May	1009.9	27.7	6.1	35.6	21.1		242	224	
June	1010.1	27.7	6.7	35.0	22.2		272	147	
July	1009.6	27.4	7.2	34.5	21.1		350	320	
Aug.	1009.1	27.4	7.2	34.5	21.7		373	125	
Sep.	1009.5	27.4	7.2	34.5	20.6		356	137	
Oct.	1009.5	27.5	6.7	34.5	20.6		335	150	
Nov.	1008.7	27.5	6.1	34.5	21.1		283	135	
Dec.	1009.1	27.2	5.6	34.5	21.1		258	150	
Annual	1009.7	27.4		36.7	19.5		3086	320	

Month	Mean evap. (mm)	Number of days with			Mean cloudi-ness (tenths)	Mean sunshine duration (h)	Most freq. wind direct.	Mean wind speed (m s⁻¹)	Sun-shine (% of possible)
		precip. (≥0.25 mm)	thunder-storm	heavy fog					
Jan.	20	(<0.5)*	0	9.3		NE	4.6	54
Feb.	17	(<0.5)	0	9.4		NE	4.8	60
Mar.	18	(<0.5)	0	9.3		NE	4.5	61
Apr.	18	1	0	9.1		NE	3.6	70
May	22	1	0	8.9		NE	3.4	51
June	24	1	0	9.3		NE	2.6	49
July	25	1	0	9.4		SW	2.4	46
Aug.	24	2	0	9.6		SW	2.4	47
Sep.	23	2	0	9.4		SW	2.6	46
Oct.	23	2	0	9.3		W	2.9	35
Nov.	22	2	0	9.4		NE	3.3	57
Dec.	23	2	0	9.4		NE	4.0	50
Annual	259	15	0	9.3		NE	3.5	51

* Cf. Table XXVI.

TABLE XXXVI

CLIMATIC TABLE FOR KWAJALEIN, MARSHALL IS.
Latitude 08°45′N, longitude 167°46′E, elevation 3 m

Month	Mean stat. press. (mbar)	Temperature (°C)				Mean vapour press. (mbar)	Precipitation (mm)		Relat. humid. (%)
		daily mean	mean daily range	extremes			mean	24-h max.	
				max.	min.				
Jan.	1009.2	26.7		32.8	21.7		92	103	
Feb.	1009.4	26.7		33.3	21.7		55	85	
Mar.	1009.8	26.9		32.8	22.2		164	50	
Apr.	1010.2	27.1		33.3	21.7		128	104	
May	1010.0	27.3		33.9	22.2		208	116	
June	1010.2	27.4		33.3	21.7		220	69	
July	1010.2	27.6		34.4	21.1		226	94	
Aug.	1009.3	27.7		35.0	21.7		242	85	
Sep.	1009.9	27.7		33.9	22.2		259	88	
Oct.	1009.9	27.7		36.1	22.2		276	99	
Nov.	1009.0	27.4		33.3	21.7		308	83	
Dec.	1009.1	27.1		32.2	20.6		228	170	
Annual	1009.7	27.3		36.1	20.6		2407	170	

Month	Mean evap. (mm)	Number of days with			Mean cloudi-ness (tenths)	Mean sunshine duration (h)	Most freq. wind direct.	Mean wind speed (m s⁻¹)	Sun-shine (% of possible)
		precip. (⩾0.25 mm)	thunder-storm	heavy fog					
Jan.		15	(<0.5)*	0	7.2		ENE	8.1	
Feb.		14	(<0.5)	0	7.1		ENE	8.0	
Mar.		14	(<0.5)	0	7.2		ENE	7.5	
Apr.		17	0	0	8.0		ENE	6.6	
May		21	1	0	8.2		ENE	6.1	
June		23	1	0	8.0		ENE	5.2	
July		24	2	0	8.5		E	4.7	
Aug.		24	1	0	8.1		ENE	4.4	
Sep.		23	1	0	8.3		E	4.2	
Oct.		23	1	0	8.5		E	4.2	
Nov.		24	1	0	8.6		ENE	5.6	
Dec.		18	1	0	8.3		ENE	7.5	
Annual		241	10	0	8.0			6.0	

* Cf. Table XXVI.

TABLE XXXVII

CLIMATIC TABLE FOR TRUK CAROLINE IS.
Latitude 07°27'N, longitude 151°50'E, elevation 2 m

Month	Mean stat. press. (mbar)	Temperature (°C)				Mean vapour press. (mbar)	Precipitation (mm)		Relat. humid. (%)
		daily mean	mean daily range	extremes			mean	24-h max.	
				max.	min.				
Jan.	1009.6	27.1	3.8	31.4	23.1		213	158	
Feb.	1010.0	27.0	4.1	31.1	22.0		100	114	
Mar.	1009.9	27.1	3.8	33.0	22.2		197	131	
Apr.	1010.0	27.1	4.3	31.3	22.3		313	128	
May	1010.3	27.1	4.5	31.1	22.4		359	168	
June	1010.2	27.1	4.8	32.0	22.7		301	108	
July	1010.2	26.9	5.0	32.1	22.3		313	256	
Aug.	1009.6	26.9	5.2	32.4	22.3		325	125	
Sep.	1010.2	26.9	5.4	32.5	22.5		320	131	
Oct.	1010.0	27.0	5.3	33.0	23.0		342	128	
Nov.	1008.9	27.1	4.9	32.0	22.0		314	264	
Dec.	1009.0	27.1	4.3	31.4	20.8		336	379	
Annual	1009.8	27.1		33.0	20.8		3493	379	

Month	Mean evap. (mm)	Number of days with			Mean cloudi- ness (tenths)	Mean sunshine duration (h)	Most freq. wind direct.	Mean wind speed (m s⁻¹)	Sun- shine (% of possible)
		precip. (≥0.25 mm)	thunder- storm	heavy fog					
Jan.		19	1	0	8.9		NNE	4.9	57
Feb.		17	(<0.5)*	0	9.2		NNE	4.8	48
Mar.		18	1	0	9.1		NNE	4.7	61
Apr.		22	1	0	9.1		NNE	3.1	54
May		28	2	0	9.0		NNE	2.2	49
June		24	2	0	9.0		SE	1.9	47
July		26	2	0	9.0		S	1.9	53
Aug.		26	2	0	8.8		S	1.6	48
Sep.		21	2	0	8.9		SW	2.0	44
Oct.		26	3	0	9.0		S	2.0	44
Nov.		26	3	0	8.9		NNE	2.8	57
Dec.		25	2	0	9.2		NNE	4.0	48
Annual		277	22	0	9.0		NNE	3.2	51

* Cf. Table XXVI.

TABLE XXXVIII

CLIMATIC TABLE FOR KOROR, PALAU IS.
Latitude 07°21′N, longitude 134°29′E, elevation 33 m

Month	Mean stat. press. (mbar)	Temperature (°C)				Mean vapour press. (mbar)	Precipitation (mm)		Relat. humid. (%)
		daily mean	mean daily range	extremes			mean	24-h max.	
				max.	min.				
Jan.	1009.4	26.8	5.0	31.7	20.6		298	213	
Feb.	1009.9	26.8	5.6	31.7	21.7		181	145	
Mar.	1009.8	27.1	5.6	31.7	22.2		194	84	
Apr.	1009.3	27.4	5.6	32.2	22.8		264	127	
May	1009.2	27.4	5.0	32.2	22.2		372	137	
June	1009.6	27.3	5.0	32.8	22.2		330	104	
July	1009.3	27.1	5.0	31.7	21.7		385	203	
Aug.	1009.0	27.2	5.0	32.8	21.7		400	122	
Sep.	1009.4	27.2	5.6	32.8	21.7		364	142	
Oct.	1009.3	27.3	5.6	31.7	21.7		332	125	
Nov.	1008.4	27.3	5.6	31.7	22.2		328	213	
Dec.	1008.5	27.2	5.0	31.7	22.2		298	165	
Annual	1009.3	27.2		32.8	20.6		3746	213	

Month	Mean evap. (mm)	Number of days with			Mean cloudi-ness (tenths)	Mean sunshine duration (h)	Most freq. wind direct.	Mean wind speed (m s⁻¹)	Sun-shine (% of possible)
		precip. (≥0.25 mm)	thunder-storm	heavy fog					
Jan.		25	1	0	9.3		NE	3.5	53
Feb.		21	1	0	9.1		ENE	3.7	55
Mar.		23	1	0	9.0		NE	3.5	65
Apr.		19	2	0	9.0		ENE	2.9	67
May		25	4	0	9.1		E	2.5	48
June		25	5	0	9.4		E	2.4	42
July		26	4	0	9.4		NW	2.4	50
Aug.		24	5	0	9.5		SW	2.7	51
Sep.		24	6	0	9.4		W	2.7	52
Oct.		24	5	0	9.0		W	3.2	52
Nov.		24	4	0	9.1		NE	2.7	58
Dec.		26	3	0	9.5		ENE	3.1	48
Annual		286	40	0	9.2		ENE	2.9	53

TABLE XXXIX

CLIMATE TABLE FOR MAJURO, MARSHALL IS.
Latitude 07°06′N, longitude 171°24′E, elevation 3 m

Month	Mean stat. press. (mbar)	Temperature (°C)				Mean vapour press. (mbar)	Precipitation (mm)		Relat. humid. (%)
		daily mean	mean daily range	extremes			mean	24-h max.	
				max.	min.				
Jan.	1008.9	26.8		31.1	20.6		177	243	
Feb.	1008.9	26.9		31.1	22.2		217	160	
Mar.	1009.5	26.9		31.1	21.1		298	111	
Apr.	1009.7	26.9		31.1	21.1		313	99	
May	1009.7	26.9		31.7	21.7		324	149	
June	1010.1	26.8		31.1	21.1		326	146	
July	1009.9	26.7		31.7	22.2		318	137	
Aug.	1009.3	26.7		32.2	21.7		297	85	
Sep.	1009.8	26.8		31.7	22.2		306	130	
Oct.	1009.8	26.9		32.8	21.7		385	183	
Nov.	1008.9	26.9		31.1	22.2		404	254	
Dec.	1008.5	26.8		31.1	21.1		282	105	
Annual	1009.4	26.8		32.8	20.6		3648	254	

Month	Mean evap. (mm)	Number of days with			Mean cloudiness (tenths)	Mean sunshine duration (h)	Most freq. wind direct.	Mean wind speed (m s^{-1})	Sunshine (% of possible)
		precip. (⩾0.25 mm)	thunderstorm	heavy fog					
Jan.		16	(<0.5)*	0	8.6		ENE	5.8	64
Feb.		17	1	0	8.6		ENE	6.5	55
Mar.		17	(<0.5)	0	8.5		ENE	5.9	63
Apr.		21	1	0	8.7		ENE	5.5	55
May		23	1	0	8.6		ENE	5.1	56
June		24	2	0	8.7		ENE	4.2	51
July		25	2	0	8.8		ENE	3.8	51
Aug.		23	2	0	8.6		ENE	3.5	59
Sep.		23	3	0	8.8		E	3.3	53
Oct.		25	3	0	8.9		E	3.3	52
Nov.		24	3	0	8.9		E	4.1	50
Dec.		21	1	0	8.8		ENE	5.7	56
Annual		259	19	0	8.7		ENE	4.7	56

* Cf. Table XXVI.

TABLE XL

CLIMATIC TABLE FOR PONAPE, CAROLINE IS.
Latitude 06°58′N, longitude 158°13′E, elevation 46 m

Month	Mean stat. press. (mbar)	Temperature (°C)				Mean vapour press. (mbar)	Precipitation (mm)		Relat. humid. (%)
		daily mean	mean daily range	extremes			mean	24-h max.	
				max.	min.				
Jan.	1009.3	26.9	4.4	31.1	21.1		281	168	
Feb.	1009.8	26.9	5.0	31.1	20.6		247	239	
Mar.	1009.6	26.9	5.0	31.1	21.7		370	128	
Apr.	1010.0	26.9	5.0	31.7	21.1		509	150	
May	1010.0	26.9	5.6	31.7	20.6		516	142	
June	1010.2	26.7	6.1	31.1	20.6		424	142	
July	1010.0	26.5	6.7	31.7	20.6		412	163	
Aug.	1009.8	26.3	7.2	32.2	20.6		415	109	
Sep.	1010.0	26.4	6.7	32.2	20.0		402	170	
Oct.	1009.9	26.4	7.2	32.2	20.6		406	170	
Nov.	1008.8	26.6	6.1	32.2	20.6		428	137	
Dec.	1008.7	26.7	5.0	31.1	20.0		466	155	
Annual	1009.7	26.7		32.2	20.0		4875	239	

Month	Mean evap. (mm)	Number of days with			Mean cloudi-ness (tenths)	Mean sunshine duration (h)	Most freq. wind direct.	Mean wind speed (m s⁻¹)	Sun-shine (% of possible)
		precip. (≥0.25 mm)	thunder-storm	heavy fog					
Jan.		24	2	0	9.1		NE	3.4	49
Feb.		21	1	0	9.0		NE	4.1	45
Mar.		24	2	0	9.1		NE	3.4	50
Apr.		27	1	0	9.3		NE	2.7	38
May		29	3	0	9.0		NE	2.4	38
June		28	2	0	8.9		NE	1.9	48
July		28	4	0	8.8		E	1.6	49
Aug.		27	3	0	8.7		ESE	1.4	51
Sep.		26	4	0	8.9		S	1.4	47
Oct.		25	4	0	8.8		ESE	1.6	45
Nov.		25	4	0	8.9		NE	2.0	49
Dec.		27	2	0	8.9		NE	2.8	47
Annual		311	31	0	8.9		NE	2.4	46

TABLE XLI

CLIMATIC TABLE FOR FANNING ISLAND
Latitude 3°54′N, longitude 159°23′W, elevation 3 m

Month	Mean stat. press. (mbar)	Temperature (°C)				Mean vapour press. (mbar)	Precipitation (mm)		Relat. humid. (%)
		daily mean	mean daily range	extremes			mean	24-h max.	
				max.	min.				
Jan.		27.5	6.1	35.0	21.7		274	152	
Feb.		27.5	6.1	36.1	20.0		267	178	
Mar.		27.8	5.6	36.1	21.2		272	196	
Apr.		27.8	5.6	36.7	21.7		358	196	
May		27.8	5.6	36.1	21.7		307	152	
June		28.1	6.1	37.2	21.7		254	109	
July		28.1	6.1	36.7	21.1		208	229	
Aug.		28.4	6.7	36.7	22.2		112	102	
Sep.		28.1	7.2	37.8	20.6		81	79	
Oct.		28.4	6.7	37.2	21.1		91	102	
Nov.		28.4	6.7	37.8	20.0		74	79	
Dec.		27.8	6.7	36.1	20.6		203	112	
Annual		28.1		37.8	20.0		2515	229	

Month	Mean evap. (mm)	Number of days with			Mean cloudi- ness (tenths)	Mean sunshine duration (h)	Most freq. wind direct.	Mean wind speed (m s⁻¹)	Sun- shine (% of possible)
		precip. (⩾0.25 mm)	thunder- storm	heavy fog					
Jan.		13	0	0	5.9		SE	5.7	
Feb.		15	0	0	5.9		E	6.4	
Mar.		18	0	0	5.9		E	6.0	
Apr.		20	(<0.5)*	0	6.3		E	5.7	
May		19	1	0	5.9		E	4.8	
June		16	2	0	5.8		SE	4.3	
July		14	(<0.5)	0	5.4		SE	4.8	
Aug.		10	0	0	5.0		SE	5.8	
Sep.		8	0	0	4.9		SE	4.8	
Oct.		8	0	0	5.0		SE	5.8	
Nov.		8	0	0	4.9		SE	5.8	
Dec.		11	0	0	5.5		SE	5.8	
Annual		160	3	0	5.5		SE	5.8	

* Cf. Table XXVI.

TABLE XLII

CLIMATIC TABLE FOR TARAWA
Latitude 01°25′N, longitude 172°56′W, elevation 4 m

Month	Mean stat. press. (mbar)	Temperature (°C)				Mean vapour press. (mbar)	Precipitation (mm)		Relat. humid. (%)
		daily mean	mean daily range	extremes			mean	24-h max.	
				max.	min.				
Jan.	1007.7	27.7	5.6	33.9	21.7		290	91	
Feb.	1007.8	27.7	5.6	32.8	22.8		203	191	
Mar.	1008.4	27.6	5.6	32.8	22.8		140	94	
Apr.	1008.7	27.7	5.6	34.5	22.8		196	125	
May	1008.9	27.9	5.6	34.5	21.1		142	132	
June	1009.0	27.8	5.6	33.4	21.1		122	64	
July	1009.2	27.8	6.1	33.4	21.1		102	43	
Aug.	1009.0	27.9	6.7	34.5	21.7		198	69	
Sep.	1009.4	28.1	6.7	33.4	22.8		158	76	
Oct.	1009.4	28.0	6.1	33.9	22.2		71	48	
Nov.	1008.3	27.9	6.7	33.9	22.8		58	28	
Dec.	1007.5	27.9	5.6	33.9	23.4		318	94	
Annual	1008.6	27.8		34.5	21.1		1996	191	

Month	Mean evap. (mm)	Number of days with			Mean cloudiness (tenths)	Mean sunshine duration (h)	Most freq. wind direct.	Mean wind speed (m s⁻¹)	Sunshine (% of possible)
		precip. (≥0.25 mm)	thunderstorm	heavy fog					
Jan.		14	0		6.0		E		
Feb.		12	0		5.0		E		
Mar.		14	0		5.4		E		
Apr.		15	0		5.1		E		
May		15	0		4.8		E		
June		14	1		5.0		E		
July		16	0		4.1		E		
Aug.		18	0		4.1		E		
Sep.		15	0		3.8		E		
Oct.		11	0		3.8		E		
Nov.		10	0		4.3		E		
Dec.		17	0		5.4		E		
Annual		171	1				E		

TABLE XLIII

CLIMATIC TABLE FOR SEYMOUR ISLAND (GALÁPAGOS)
Latitude 0°28′S, longitude 90°18′W, elevation 11 m

Month	Mean stat. press. (mbar)	Temperature (°C) daily mean	mean daily range	extremes max.	min.	Mean vapour press. (mbar)	Precipitation (mm) mean	24-h max.	Relat. humid. (%)
Jan.	1011	26.1	7.8	32.2	18.9		20	8	
Feb.	1008	27.0	6.1	32.2	21.1		36	25	
Mar.	1009	27.5	7.2	33.9	21.1		28	64	
Apr.	1009	27.2	6.7	32.2	21.7		18	64	
May	1010	26.4	7.2	31.7	20.6		<3	<3	
June	1010	25.0	6.7	30.6	20.0		<3	<3	
July	1011	23.9	6.7	31.1	19.5		<3	3	
Aug.	1010	23.4	7.8	30.6	15.6		<3	3	
Sep.	1010	22.8	7.8	30.0	15.0		<3	<3	
Oct.	1012	23.4	7.8	30.0	14.5		<3	3	
Nov.	1010	23.6	7.2	30.0	16.7		<3	<3	
Dec.	1009	24.7	7.2	31.1	16.1		<3	<3	
Annual		25.0		33.9	14.5		112	64	

Month	Mean evap. (mm)	Number of days with precip. (≥0.25 mm)	thunder-storm	heavy fog	Mean cloudi-ness (tenths)	Mean sunshine duration (h)	Most freq. wind direct.	Mean wind speed (m s⁻¹)	Sun-shine (% of possible)
Jan.						4.5	E		
Feb.						3.3	E		
Mar.						3.0	E		
Apr.						3.3	E		
May						3.6	E		
June						5.8	SE		
July						7.0	SE		
Aug.						6.6	SE		
Sep.						6.4	SE		
Oct.						5.9	SE		
Nov.						5.8	SE		
Dec.						5.8	SE		
Annual									

Chapter 5

Climate of the Indian Ocean South of 35°S

J. J. TALJAARD AND H. VAN LOON

Introduction

General

The area to be discussed is the Indian Ocean extending from 20°E to 115°E between 35°S and the coast of Antarctica (Fig. 1). The intersection of 35°S and 20°E almost coincides with Cape Agulhas, the southernmost tip of Africa, whereas the intersection of 35°S and 115°E is only about 40 km from Cape Leeuwin, which is the southwestern-most point of Australia. The longitudinal span of the area is one quarter of the circumference of the hemisphere. The average position of the axis of the subtropical high pressure belt is only slightly equatorward of 35°S and, therefore, the area can also be described as the southern Indian Ocean between the subtropical ridge and the Antarctic coast. However, because the core of the subtropical ridge shifts seasonally and periodically through several degrees of latitude, a description of the synoptic climatology and the upper-air circulation of the region will require occasional reference to phenomena north of the ridge axis.

The relative paucity of meteorological data for the middle and high latitude zones of the Southern Hemisphere tends to raise doubts in the minds of climatologists and meteorologists regarding the acceptability of conclusions reached for the region. Conceding that for routine daily weather map analysis and weather forecasting the existing observation networks, in addition to information from weather satellites, have been and still are inadequate, nevertheless many conclusions and results are qualitatively correct, although they can, of course, be refined with the expected improved conventional and satellite information.

Geographical setting and bottom topography

Fig. 1 shows the setting of the southern Indian Ocean between South Africa, Australia, and Antarctica, and the locations of New Amsterdam, Marion, Crozet, Kerguelen, and Heard Islands, as well as antarctic coastal stations that provide control data for the boundary conditions in the extreme south. The boundaries of Fig. 1 extend slightly beyond those of the area under discussion (indicated by interrupted thick lines) for showing the orientation of coastlines that act as barriers for sea currents, as well as for indicating the topography of the land masses.

The topography of South Africa and Antarctica adjacent to the Indian Ocean and the bottom topography of the ocean are indicated by 1000-m contour intervals. Only the

505

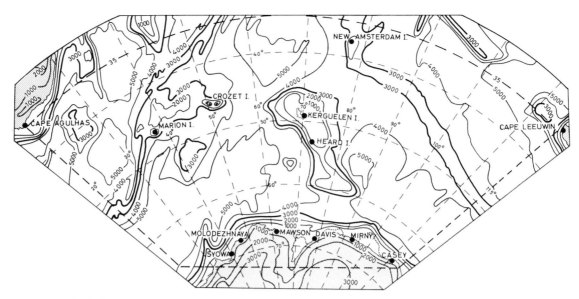

Fig. 1. Geographical setting and bottom topography of the Southern Indian Ocean. Contours and isobaths are in meters.

extreme southwestern portion of Australia is shown since the land here does not rise above about 300 m.

The southernmost part of South Africa consists of dissected land with several E — W oriented mountain ranges rising to between 1000 and 1500 m. The main escarpment along 32°S rises mostly to between 1500 and 1800 m, but to well over 2000 m in the extreme northeastern part of the included territory. On the whole, the land rises steeply to either the southern mountain ranges or the main escarpment, thus obviously providing barriers for guiding or influencing the eastward movement of pressure troughs and ridges over the sea at the latitude of the subtropical high pressure belt.

The massive ice cap of East Antarctica faces the Indian Ocean and rises steeply along the entire distance between 20° and 115°E. From the point of intersection of longitude 20°E with the Antarctic coast at 70°S, the coastline extends ENE to Cape Batterbee at 66°S, 54°E, then ESE to 68°S at 70°E where its smooth trend is interrupted over a distance of 10° longitude by the major indentation of the Lambert Glacier, Amery Ice Shelf, and the Amundsen and Prydz bays. From about 80°E to the eastern boundary of the area at 115°E, the coastline deviates less than one degree of latitude from the polar circle at 66.5°S.

The contours of the ice cap are closely parallel to the coastline, the slope being steepest near the sea, such that the 1000-m contour is reached within 100 km, or one degree of latitude, from the coast. The 2000-m contour is located within 250 km from the sea in most areas, but the 3000-m contour is located at about double this distance on the average.

The bottom topography of oceans has recognizable effects on the surface currents if the currents are of great vertical extent, such as the West Wind Drift (or Antarctic Circumpolar Current) of the southern oceans. Examples of such effects exist in the southern Indian Ocean and, because sea and air temperature distributions are related to currents, these features are described below.

The Agulhas Bank adjoining South Africa extends to about 400 km south of Cape Agulhas but narrows to almost zero width off the Natal coast, so that the steep continental slope stretches NE−SW, forming a steering barrier for the core of the Agulhas Current. The Agulhas Plateau, located in the block bounded by 37−42°S and 24−29°E, rises to a minimum depth of less than 2500 m but evidently has a less spectacular effect on the surface flow than the mid-oceanic ridges and plateaus located farther east. The Atlantic−Indian Ocean Ridge enters the area at 53° S, extending NE-ward to 55°E at 35°S. The Crozet Plateau, from which the Prince Edward and Crozet Island groups arise, extends eastward along 45°S from the Atlantic−Indian Ridge over more than 1000 km and has a marked influence on the West Wind Drift. The Kerguelen Plateau, from which Kerguelen and Heard Islands rise, is an extensive block situated centrally in the southern Indian Ocean and stretching over 15° of latitude. It has a definite but apparently less spectacular influence on the West Wind Drift than the Crozet Plateau. The only remaining major feature of the bottom topography is the Mid-Indian Ocean Ridge stretching to ESE from New Amsterdam Island.

Currents, convergences, and pack ice

The climate of an ocean area is closely dependent on the temperatures and the flow patterns of the water masses that circulate in the area. The high heat capacity of water and the vertical transport of heat through turbulent exchange in the upper 50−100 m cause surface temperatures over the sea to be conservative to such an extent that the annual range is less than 2°C in parts of the Southern Ocean. It is evident, therefore, that the current regime should be studied and relationships established with physical barriers, such as coastlines, subsurface ridges and plateaus, as well as circulation barriers, such as convergences and divergences in the surface flow.

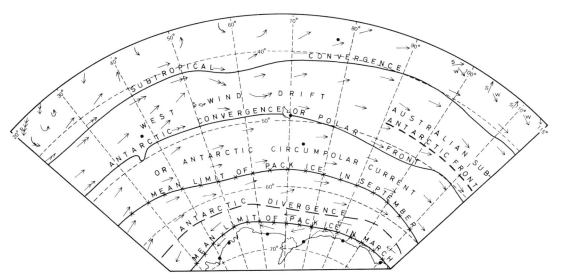

Fig. 2. Currents, convergences, and mean northern limits of pack ice in March and September. One arrowhead indicates current speed of 0.1−0.5 m s⁻¹; two arrowheads 0.6−1.0 m s⁻¹; etc.

Fig. 2 shows the currents, oceanic convergences (or fronts), and the mean northern boundaries of the Antarctic pack ice in March and September. In spite of the appreciable research activity that has been going on in some areas since the International Geophysical Year (IGY), no comprehensive updated exposition of these features exists as yet. The southern Indian Ocean has received particularly little attention from oceanographers except over the extreme eastern and northwestern sectors. Therefore, information has to be gleaned from climatic atlases published two decades or more ago, or from more recent oceanographical textbooks containing generalized diagrams on scales too small to show details for limited regions.

Concerning currents, the most detailed and useful information for the Indian Ocean equatorward of 50°S is contained in KONINKLIJK NEDERLANDS METEOROLOGISCH INSTITUUT (1952) and in METEOROLOGICAL OFFICE, GREAT BRITAIN (1949). The current directions and speeds indicated in Fig. 2 are based on these charts, but account was also taken of the generalized expositions in SVERDRUP (1942), SVERDRUP et al. (1954), DEFANT (1961), DIETRICH (1963), and NEUMANN (1968). The study by VAN LOON (1971) has drawn attention to the striking current maximum north of the Crozet Islands. New information on water movements has been forthcoming recently from drifting buoys south of South Africa (GRUNDLINGH, 1977, 1978; HARRIS and STAVROPOULOS, 1978).

The surface currents for the Antarctic zone of the Indian Ocean, i.e., south of 50°S, are based on a summary by TALJAARD (1957) of information in DEACON (1937), SVERDRUP (1942), GREAT BRITAIN (1948), and UNITED STATES NAVY HYDROGRAPHIC OFFICE (1957b), as well as the discussions in the oceanographic textbooks cited above.

A consideration in constructing surface current charts is the fact that water masses as a rule are not supposed to flow across the oceanic convergences such as the Subtropical and Antarctic convergences. However, due to ascent or descent of the water masses at the convergences, streamlines can start or end there. Also, the convergences might be zones of mixing due to eddies.

The currents of the southern Indian Ocean can be considered separately for the zone north of the Subtropical Convergence, the zone with Subantarctic Surface Water between the Subtropical and Antarctic convergences, and the zone with Antarctic Surface Water south of the Antarctic Convergence.

The swift-flowing, SW-setting Agulhas Current is found in the extreme northwestern corner of the area, off the South African coast, where GRUNDLINGH (1978) reported speeds of $2-2.5$ m s^{-1} by a buoy tracked by NIMBUS 6. Part of this current continues into the Atlantic Ocean and part is deflected southward and finally eastward in the zone between 39° and 42°S, i.e., immediately north of the Subtropical Convergence. The region between the Agulhas Current and its recurved east-setting continuation is one with numerous eddies altogether forming an anticyclonic gyre between about 23° and 33°E. Farther east in the subtropical zone to about 65°E, the flow is slow and irregular except in the 2° or 3° wide zone adjacent to the Subtropical Convergence where the current sets eastward reaching speeds up to 1 m s^{-1}. The southward flexure of the Subtropical Convergence northeast of the Crozet Islands is reflected in a local southward component of the flow of the subtropical water, but from 65° to about 90°E, the northward component of the flow is well marked. In the region between 90°E and the Australian coast are seasonal differences, with fluctuating weak currents in summer mostly setting N, but during autumn, winter, and spring the current sets predominantly E by ESE.

In the zone between the convergences, the West Wind Drift sets easterly on the whole, but with some significant features apparent in the Marion—Crozet—Kerguelen region. There is evidence that the current exceeds 0.5 m s^{-1} quite frequently in the west close to the Subtropical Convergence, but it is believed that the current is strongest within 1° or 2° north of the Antarctic Convergence (HARRIS and STAVROPOULOS, 1978). From about the longitude of Marion Island to the Crozet Islands, a consistent ENE set of the current is evident and the speed increases close to the Subtropical Convergence, where it sets due easterly to a position north of the latter islands. At this position, a marked southward flexure of the Subtropical Convergence and the current is found (VAN LOON, 1971), but 10—15° farther east an opposite flexure in the streamlines and the convergence occurs. At the longitude of Kerguelen, the strongest current is found halfway between the convergences but farther east its speed diminishes and the flow pattern is diffluent.

There is very little doubt that the flow pattern in this zone from 35° to 75°E is influenced by the existence of the Mid-Atlantic—Indian Ocean Ridge and the Crozet and Kerguelen plateaus. Some indication exists that the Australian Subantarctic Front at 47—49° S eastward of 90°E is due to a similar process over the Mid-Indian Ocean Ridge, but current observations are few in this area.

In the zone between the Antarctic Convergence and the Antarctic Divergence, the currents are believed to set mostly ENE or E by ENE. Speeds decrease from north to south. Vertical N/S sections of the water circulation in this zone invariably show that the Antarctic Surface Water flows northward and sinks at the Antarctic Convergence. The view has been aired by some writers (MEYER, 1923; SVERDRUP, 1942), however, that a southward transport occurs in the region from 20° to 40°E where the pack-ice limit is oriented WNW—ESE.

The Antarctic Divergence is defined as the axis of a zone in which upwelling occurs and from which, therefore, surface water is transported slowly northward to the Antarctic Convergence and southward to the Antarctic coast. It is assumed to be more or less coincident with the circumpolar low-pressure trough of the atmospheric sea-level circulation.

The westward-setting East Wind Drift is found between the Antarctic Divergence and the coast of Antarctica. It is a weak but persistent current in a zone two to five degrees wide.

The position of the Subtropical Convergence shown in Fig. 2 is taken from DEACON (1937, 1945, 1964) and after 30 years there still seems to be no new version to present. Considerable attention has been given to the structure and position of the Antarctic Convergence during the past two decades, but the Subtropical Convergence, a considerably stronger discontinuity in certain areas, has evidently not been studied seriously. It is the southernmost position reached by water of tropical or subtropical origin, so that considerable discontinuities or gradients of temperature develop where the warm water meets the cool Subantarctic Surface Water of the West Wind Drift. This happens at 39—42°S in the area south and southwest of Cape Agulhas where the swift-flowing Agulhas Current impinges on much cooler water masses. The current recurves eastward at this position and flows side by side with Subtropical and Subantarctic Surface Water. From about 15°E 41°S to a position north of the Crozet Islands, the Subtropical Convergence is marked by strong temperature discontinuities or gradients of more than 5°C and occasionally even 10°C over distances of less than two degrees of latitude. It is evident, however, from many ship crossings that appreciable meandering and mixing occur in eddies along the zone because the temperature contrasts and the positions at which they occur vary over short periods.

DEACON (1945) maintains that the average temperature discontinuity along the Subtropical Convergence around the hemisphere is 5°C. This is more than justified in the South African area, but from about 60°E toward Australia the convergence diminishes in intensity so that it cannot be readily discerned from the records of most ships crossing the indicated position. Instead, another zone with appreciable temperature gradient, viz., 3.6°C over an average width of about 1.3° latitude, appears at 47—48°S from 90°E eastward (ZILLMAN, 1970a). This zone has come to be called the Australian Subantarctic Front and might be related to the current pattern across the Mid-Indian Ocean Ridge over which it is found.

The Antarctic Convergence, formerly known as the Meinardus Line and at present being preferably called the Polar Front by oceanographers, is located at the average latitude of 50°S in the southern Indian Ocean. The position indicated in Fig. 2 is from MACKINTOSH (1946). ZILLMAN (1970b) shows a mean position for the convergences and a band about 4° wide within which the convergence wanders. According to Zillman, its mean position at 20°E is about 48°S, but east of Kerguelen it is found at increasingly higher latitudes so that at 115°E its position is given as 53°S. However, the latter position is about 3° south of that indicated by MACKINTOSH (1946), who made the first comprehensive study of the convergence.

As already mentioned, the Antarctic Convergence is a discontinuity between cold Antarctic Surface Water and less cold Subantarctic Surface Water. According to Mackintosh, the temperature jump across the convergence is about 2°C on the average, but it is most readily recognized by marked differences in salinity and marine life of the two water masses. The temperature profiles by VAN LOON (1966), based upon 12 cruises between Cape Town and Syowa, and many other individual ship crossings, show that in the western part of the Indian Ocean the convergence, or more specifically its southern boundary, is located along the line where the weak meridional temperature gradient of the Antarctic Surface Water reverts to the strong gradient characteristic of the convergence itself and of the Subantarctic Surface Water. In other areas and in the eastern Indian Ocean, these characteristics are not as marked as in the western part of the ocean. A summary of the theories of development of the convergence and a survey of relevant literature are given by ZILLMAN (1970b).

The Antarctic Divergence, indicated in Fig. 2 (after UNITED STATES NAVY HYDROGRAPHIC OFFICE, 1965), is by way of its formation not a spectacular narrow zone with contrasting sea surface temperatures but, rather, a phenomenon contributing to the characteristic homogeneity and weak temperature gradients of the water poleward of the Antarctic Convergence. The existence of the divergence, with the expectation of relatively warm upwelling deep water along a circumpolar belt coinciding roughly with the atmospheric low-pressure belt, is not obvious or spectacular at the sea surface. Its reality is indicated in the southwestern part of the southern Indian Ocean, as will be mentioned below in connection with the pack ice.

The mean northern limits of the pack ice in March and September (according to the UNITED STATES NAVY HYDROGRAPHIC OFFICE, 1957b) are indicated in Fig. 2. These limits agree to within less than a degree of latitude from those given by MACKINTOSH and HERDMAN (1940). The pack-ice extent around Antarctica is at its minimum in March with little difference between the outermost limits of the ice in September and October. This is in accord with the annual variation of sea surface temperature in southern middle and high latitudes. The ice limit should coincide theoretically with the $-1.8°C$ sea isotherm.

Melting of the ice starts in November, but the boundary actually retreats less than 1° latitude during this month. Rapid melting and retreat occur mainly during December but continue during January so that by the end of January the ice edge is less than five degrees from the coast in the Indian Ocean sector. Further slow retreat occurs during February and March when large parts of the coast are ice-free and the outer limit at which ice is found is one to four degrees from the coast, depending on the positions of promontories and bays. The growth period of the pack ice is from April to September, but the ice increase in April is quite minimal. Rapid growth occurs from May to July when the outer limit of the ice attains a position only about one degree from the September—October maximum limit. Therefore, during five months (July—November) there is little change of the outer limit, whereas during the four summer and autumn months, January—April, the ice edge is close to the coast.

The ice edge is tortuous in individual years or specific periods so that the limits shown in Fig. 2 represent long-term most likely positions. The marked southward flexure of the September limit between 30° and 50°E is a reflection of ice tongues that frequently extend from the Atlantic to the Indian Ocean along 56°—58°S, being transported there by the Weddell or Bouvet Current flowing ENE from the Weddell Sea. It is possible, as indicated in Fig. 2, that a current with a southward component exists in this area, but recent studies of pack-ice distribution by satellite (GLOERSEN et al., 1974; ZWALLY and GLOERSEN, 1977) have shown that in winter a belt of relatively thin ice, or with incomplete coverage, often extends from the region of the flexure of the ice boundary towards an extensive polynya located farther SW, straddling the Greenwich meridian close to the Antarctic coast. The polynya can be evidence for the existence of an active zone of divergence, i.e., part of the Antarctic Divergence, in the extreme southwestern part of the Indian Ocean.

Historical survey and data used

A survey of meteorological observations and climatic studies for the southern Indian Ocean cannot be isolated from that pertaining to the whole of the Southern Ocean because very few studies and publications refer to this sector only.

A detailed general survey of pre-IGY scientific expeditionary cruises to high southern latitudes and the stations established south of 45°S for the purpose of carrying out meteorological and other observations is given in VENTER (1957). These observations constitute only minor sources of data compared to the routine meteorological observations taken by ships since about 1850. Sailing ships dominated the scene until about 1920 when steamers and motor vessels became universal. The sailing ships availed themselves of the prevailing winds to reach their destinations so that more than half a century after their demise the main routes they followed in the Southern Ocean are still clearly indicated by a relative abundance of historical data. In the southern Indian Ocean, they preferably sailed from west to east in the zone 37—47°S where the Roaring Forties assisted their speedy passage. From about 90°E, the main track turned ENE toward the southwest and south coasts of Australia. Few ships sailed south of 50°S so that Antarctic waters remained ill-provided with observations. The steamers and motor vessels which came into use early in the present century preferred to sail north of 40°S so that since about 1920 the middle latitude zone has been starved of observations.

The advent of motorised vessels made pelagic whaling possible so that for nearly 40 years between 1920 and the early 1960s (the war years 1939—45 excluded) a fair amount of data for the Antarctic zones became available for the summer season from late November to early April. Whaling activities were most concentrated in the South Atlantic and Indian oceans, but on the whole the eastern third of the Antarctic sector of the Indian Ocean had fewer observations than the central and western parts. The whaling ships (at times more than 20 fleets) sailed south to reach the pack ice in late November or early December. The pack-ice limit (Fig. 2) is located along 57—60°S in November so that whaling was active in the zone 50—60°S at this time, but with rapid melting in December and January the operations could be carried out farther south step by step, and thus in February and March the activities were mostly in the 60—70°S zone. The peculiar habits of the whaling fleets and the consequent effects on climatic studies are described by Vowinckel (1956).

The South African Weather Bureau collected over 100,000 whaling ship reports from 1920 to 1955 which Vowinckel (1956, 1957) and Vowinckel and Van Loon (1957) used for studying the climate of the Antarctic Ocean (defined as the ocean south of about 50°S). It would appear that an appreciable amount of these data, having been taken partly from special logs maintained by certain fleets, has never been incorporated in the data decks of marine meteorological services.

An upswing in meteorological observations other than those of whaling ships in the southern ocean commenced in 1955 with the preparations for the International Geophysical Year (IGY, 1957—58). However, the supply ships for maintaining the Antarctic stations during and since the IGY sailed southward and returned northward as quickly as possible, so that the total amount of data from them over a period of two decades is insufficient for studying representative climatic conditions, except in a few areas located outside the Indian Ocean. Cruises for oceanographic studies are better suited for obtaining data over extended periods (e.g., the cruises of the U.S. *Eltanin* in the Pacific, southwestern Atlantic, and extreme eastern Indian oceans), but considering the enormous area of the Southern Ocean, all these observations provide too sparse a coverage in middle and high latitudes to satisfy the criteria of some climatologists.

The overall picture is well illustrated in a printout of the total numbers of observations and average values of several meteorological elements in 2 × 2 degree blocks supplied for this investigation by the United States Environmental Data Center in 1976. This particular data set has been the basis for deriving the sea and air temperature charts north of 50°S (as given on pp. 515, 516 and 521), and thus also for the diagrams partly derived from these. Incidentally, this data set clearly illustrates the preponderance of observations along the preferred route of the sailing ships in lower middle latitudes. Further confirmation of this situation can be found in the daily synoptic charts prepared by Meldrum (1861) for January, February, and March 1861 for a large part of the South Atlantic and Indian oceans and by Meinardus and Mecking (1911) for southern middle and high latitudes during 1901—1904.

An important development in understanding the meteorology and climatology of the Southern Hemisphere was the establishment of several stations on middle-latitude islands and on Antarctica during and after World War II. As regards the Indian Ocean, Australia established Heard Island in 1948 and maintained the station until 1954. Also in 1948 South Africa established Marion Island as a permanent station. The French stations on New Amsterdam and Kerguelen islands were established in 1949 and converted to full surface meteorological stations in 1951. Upper-air soundings were carried out when conditions allowed at Heard Island during the five years of its existence. Soundings were introduced at Marion Island in 1950, whereas at New Amsterdam soundings commenced in 1955 and at Kerguelen in 1968 (upper-wind soundings started at Kerguelen in 1961). Along the Antarctic coast, Australia established Mawson (68°S 63°E) as a full surface and upper-air station in 1954. The IGY (July 1957 until December 1958) was the incentive for establishing several more stations along the Indian Ocean sector of the Antarctic coast, viz., Syowa (69°S 40°E) by Japan, Davis (69°S 78°E) by Australia, Mirnyi (67°S 93°E) by the USSR (1956), and Wilkes (66°S 110°E) by the USA. The latter station was taken over by Australia after the IGY and renamed Casey. In 1962 the USSR established Molodezhnaya at 68°S 46°E. These coastal stations are mentioned because their surface and upper-air data have been used for evaluating conditions at the southern boundary of the Indian Ocean.

From a synoptic climatological point of view, the demise of the whaling ships has been partly compensated since 1965 by the advent of satellite visible and infrared cloud pictures on which the locations of cold fronts and vortices are indisputably indicated. Cloudiness, pack-ice limits, the approximate distribution of upper-air temperature and, therefore, trough/ridge patterns and to a certain extent sea temperature distribution are other quantities realized by satellite observations. Although climatological values have already been obtained for some of these quantities, improvement of techniques and the routine hemispherical evaluation of some will probably be forthcoming. A promising development for obtaining improved ground truth coverage for surface pressure and sea temperature for the Southern Hemisphere is the drifting buoy programme which began in 1979 as part of the Global Weather Experiment.

The following is an incomplete sequential list of publications containing climatic charts or discussions for the Southern Oceans as a whole or the Indian Ocean in particular. For more detailed references, the reader should consult SCHMITT (1957), VAN LOON (1972a,b,c,d), and TALJAARD (1972b).

BUCHAN (1869). First sea-level mean monthly pressure maps, published about 20 years after the introduction of routine weather observations by ships.

TEISSERENC DE BORT (1893). Pressure maps for 1.5, 2.9, and 4.0 km.

HANN (1901). Mean monthly sea-level pressure and air temperature maps.

MEINARDUS and MECKING (1911). Mean monthly pressure maps for October 1901 to March 1904, based on daily synoptic maps for the region south of about 30°S. The coverage (sailing ships) was fair for middle latitudes while several scientific expeditions operated in the Antarctic.

SHAW (1927). Mean monthly sea-level pressure and temperature maps and isobars for the 4000 m level. The charts are incomplete for the region south of 60°S.

MEINARDUS (1928, 1929, 1938, 1940). Expositions of the meteorology and climatology of southern middle and high latitudes based primarily on the synoptic charts of MEI-

NARDUS and MECKING (1911) but also on data which became available in the interim periods. Considered altogether, these studies constitute an overview of the basics of the surface weather and climate of the hemisphere which has been improved upon only by the establishment of over 30 stations on the islands and on Antarctica during and since World War II, as well as by the considerable amounts of data from whaling and expeditionary ships since 1920.

VOWINCKEL (1956, 1957) and VOWINCKEL and VAN LOON (1957). Discussions of the climate of the Antarctic Ocean based on over 100,000 ship reports. Part I of the discussion elaborated the late-spring, summer, and early-autumn conditions in the sector 20–40°E between 46° and 70°S. Part II of the study discusses conditions for the same latitude interval but from 50°W to 150°E. VOWINCKEL and VAN LOON (1957) completed the study for the Antarctic Ocean round the hemisphere, summarizing at the same time the results of VOWINCKEL (1956). VOWINCKEL (1957) is mainly also a summary of the material contained in the preceding two references. This discussion appears in VAN ROOY (1957) in which contributions by Fabricius on the climate of the subantarctic islands, Hofmeyr on the sea-level pressure and upper-air conditions, Loewe on precipitation and evaporation, and Schmitt on the synoptic meteorology were also included. Meanwhile, several climatic atlases of the Indian Ocean were published, viz., METEOROLOGICAL OFFICE, GREAT BRITAIN (1949), KONINKLIJK NEDERLANDS METEOROLOGISCH INSTITUUT (1952), UNITED STATES NAVY HYDROGRAPHIC OFFICE (1957b, revised 1976). These atlases contain additional details, e.g., on wind and precipitation which are not discussed in this chapter. The climate atlas of McDONALD (1938) is also a valuable source of climate data, as well as the *Meteorological Atlas of the International Indian Ocean Expedition* (RAMAGE et al., 1972; RAMAGE and RAMAN, 1972).

The post-IGY publications on aspects of the climate (including synoptic climatology) of the Southern Hemisphere are too many to be enumerated here. The most important ones are mentioned in the appropriate sections below. Suffice it to say that the daily sea-level and upper-air synoptic charts for the IGY published by the SOUTH AFRICAN WEATHER BUREAU (1962–1966), the climate atlas of TALJAARD et al. (1969), the chapters by Van Loon on the climate and by Taljaard on the synoptic meteorology of the Southern Hemisphere in VAN LOON et al. (1972), the printout of ship data supplied by the United States Environmental Data Center already mentioned, and the surface and upper-air climate tables (see the Appendix, pp. 592–601) supplied by the Australian, French, and South African Weather Services and by the National Center for Atmospheric Research in Boulder, Colorado, USA, are the main sources of information utilized. As regards satellite data, STRETEN and TROUP (1973) and CARLETON (1979) are the most useful publications.

Temperature

Sea surface temperature

In most parts of the southern Indian Ocean, the mean sea surface temperature (abbreviated to sea temperature below) is very slightly higher in March than in February and slightly lower in September than in August. Therefore, March and September are selected for presenting charts of the warmest and coldest months and for determining the maximum annual range of the sea temperature. For this purpose, the data set supplied by the United States National Environmental Data Center mentioned on p. 512 was used for obtaining the sea temperature distribution in the zones 35−50°S in March and 35°−48°S in September. The data set contains average sea temperatures and numbers of observations in 2 × 2-degree blocks for the southern Indian Ocean to 60°S for each month. As a rule, more than 50 observations per month per 2 × 2-degree block are available for the zones mentioned above, whereas for the 10-degree zone centred along 41°S in the western and central parts of the ocean the observations mostly vary between 100 and 300 per 2 × 2-degree block. Therefore, fair reliance can be attached to the charts derived for the regions equatorward of about 50°S.

For the region between 50°S and the pack ice, the charts of TALJAARD (1972a) for February and August were used after applying minor adjustments to convert them to March and September values. It was assumed that the pack-ice limit represents the −1.8°C isotherm in each case. The charts of TALJAARD (1972a) were based on the collection of ship data used by VOWINCKEL and VAN LOON (1957).

The charts for March and September are given in Figs. 3 and 4. The most striking feature is the strong temperature gradient north of about 50°S and the weak gradient south of that latitude, the dividing line coinciding approximately with the Antarctic Convergence. The temperature gradient is particularly strong between 40° and 50°S in the western half of the area.

According to MACKINTOSH (1946), the temperature at the northern boundary of the Antarctic Convergence should be between 5.0° and 5.2°C along its entire length in the southern Indian Ocean in the warmest month. Also, the temperature discontinuity across the convergence should be about 0.8°C in the east, centre, and extreme west of the area and 2.3°C south of Marion Island. Fig. 3 shows that the temperature along the indicated middle position of the convergence is 4.5−5.5°C in March.

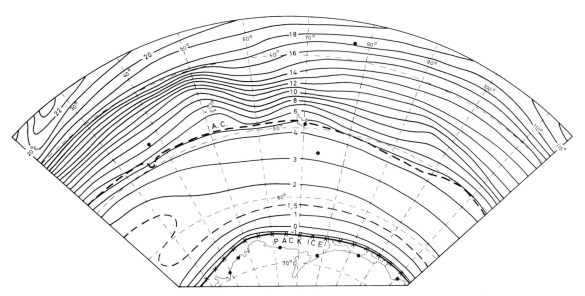

Fig. 3. Sea temperature in March. *A.C.* = Antarctic Convergence.

In the northwestern corner of Fig. 3, the Agulhas Current is apparent as a tongue of high temperature pointing in a southwesterly direction. East of this warm intrusion the isotherms in the 35—40°S zone extend WSW—ENE to 100°E, resulting in a slow eastward decrease of temperature, whereafter an increase sets in from 105°E to the Australian coast. In the zone between 40° and 45°S, the temperature gradient along 20°E is about 2.4°C per degree of latitude, but this average gradient is frequently exceeded along narrow zones between 40° and 43°S where the Subtropical Convergence is located. Proceeding eastward to the region between 35° and 50°E, we see a remarkable concentration of isotherms between 40° and 43°S. The temperature remains practically constant at 17°C along 40°S, whereas farther south the isotherms are oriented WSW—ENE. From 50° to 60°E, a sharp flexure of the isotherms to SE occurs, reverting to an ENE alignment from 60°E to the longitude of Kerguelen. Thus the resulting wave pattern of the isotherms is found only in the 10-degree zone between 40° and 50°S. It is evidently produced by the deflecting effects of the bottom topography on the West Wind Drift. VAN LOON (1971) describes the associated current maximum in this area.

Eastward of 70°E, the isotherms are diffluent so that they are aligned WSW—ENE in the north and WNW—ESE in the south. However, east of 90°E the data suggest a concentration of isotherms over the Mid-Indian Ocean Ridge along 45° to 48°S, thereby outlining the Australian Subantarctic Front.

A notable feature of the region south of the Antarctic Convergence is the looped shape of the 1.5°C isotherm in the west of the area. This is evidently produced by the transport of cold water from the Weddell Sea by the Bouvet Current along 55—60°S and the upwelling of slightly warmer water along the Antarctic Divergence between 60° and 65°S. In the central and eastern parts of the chart area, the isotherms slowly trend southward across the latitude circles as the longitude increases. The temperature gradient increases in the extreme south in the vicinity of the pack ice.

The sea temperature distribution in September (Fig. 4) is mainly similar to that in March, but some differences exist. The gradient in the west between 40° and 50°S is

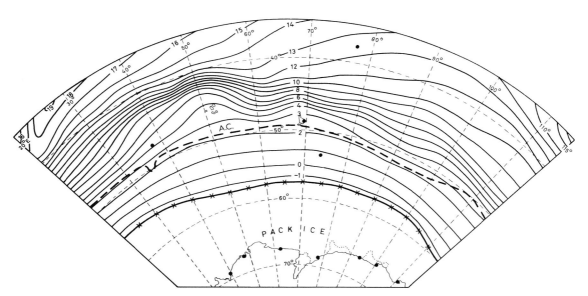

Fig. 4. Sea temperature in September. *A.C.* = Antarctic Convergence.

reduced, particularly between 40° and 45°S. In the zone 35−40°S and also in the zone 40−45°S east of the Crozet−Kerguelen wave pattern, the gradient is reduced to about half its value in March. The longitude of lowest temperature at 35−40°S is now found at 90−95°E, i.e., about 10 degrees farther west of Australia than in March. The zone between the Antarctic Convergence and the pack ice is now half its width in March, whereas the temperature along the convergence is only 2.5°C lower than in March. Thus the meridional gradient in this zone is doubled, although still remarkably lower than north of the convergence. The WNW−ESE alignment of isotherms close to the Australian coast remains the same as in summer and so does the concentration of isotherms along the Australian Subantarctic Front.

March−September range of sea temperature

The range of sea temperature from March to September, based on Figs. 3 and 4, is shown in Fig. 5. The greatest range of more than 6°C is found along 35°S from 70° to 90°E. It decreases westward to 4°C near Cape Agulhas and eastward to less than 4°C near Cape Leeuwin. In the east, the range decreases progressively southward to reach a minimum of less than 1.5°C between 45° and 50°S. A tongue of low range stretches WNW from this minimum while a zone with range less than 2.5°C stretches westward along 50−53°S to join a second area of very low range around 47°S in the extreme west. A wedge with intermediate range stretches ESE from two patches with ranges of more than 4°C at 38−42°S, 35−55°E. The range increases to over 3°C in the vicinity of the pack-ice limit in September.

It should be noted that the range of sea temperature determined from Figs. 3 and 4 is sensitive to small errors in the drafting of isotherms in regions of strong gradient, such as between 40° and 50°S, so that some of the irregularities of the isopleths in this zone might be overemphasized.

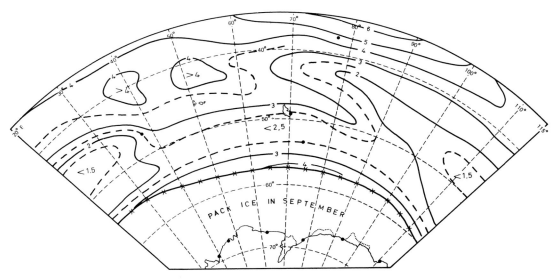

Fig. 5. Sea temperature range (°C): March minus September.

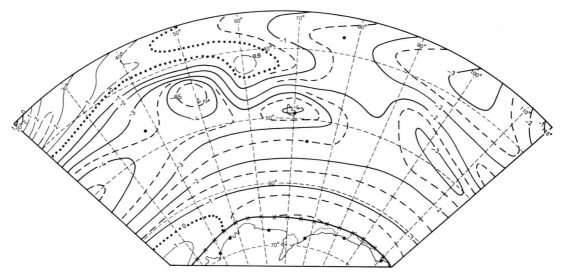

Fig. 6. Deviations of sea temperature from hemispheric zonal means in March.

Anomalies of sea temperature in March and September

The deviations of sea temperature in March and September from the hemispheric means along circles of latitude (as derived from the adjusted February and August charts of TALJAARD, 1972a) are given in Figs. 6 and 7. The anomaly patterns are on the whole similar in the two months, and there is little reason to expect markedly different patterns during the remaining 10 months.

The tropical water transported southward by the Agulhas Current in the extreme northwest of the area is reflected as positive anomalies of about 3°C in March and more than 4°C in winter. As the land is approached, however, the anomaly decreases to such an extent that in March a negative anomaly of −1°C appears at the intersection of 35°S and 20°E. A broad area of negative anomaly exists off the Australian west coast, cul-

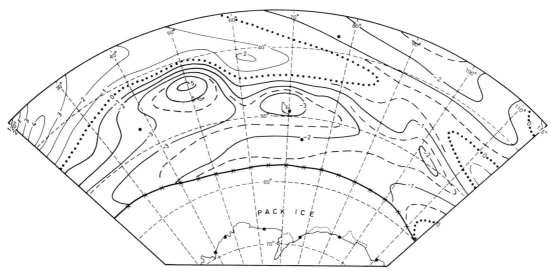

Fig. 7. Deviations of sea temperature from hemispheric zonal means in September.

minating in deviations of more than −3°C at 37°S 107°E in March and more than −2°C at 35°S 95°E in September. The center of this anomaly therefore shifts westward from summer/autumn to winter/spring. Positive deviations elongated along WNW − ESE axes have their highest values at 42°S 61°E.

Marked negative anomalies stretch across both charts at middle latitudes. The anomalies north of the Crozet and Kerguelen Islands are associated with the wave patterns of the isotherms in that region (Figs. 3 and 4). The negative anomalies aligned WNW − ESE at 50°S near the eastern boundaries of the charts reflect the concentration of isotherms along the Australian Subantarctic Front which is located immediately north of the axes of these anomalies. In March a centre with maximum negative anomaly evidently lies outside the western boundary of the chart at 50°S. Poleward of these negative centres the deviations decrease toward the pack-ice boundaries, so that in March a positive anomaly appears south of 62°S from 20° to 45°E. Small positive anomalies are also apparent in September at the northeastern and southeastern extremities of the charts.

The annual variation of sea temperature along 43°S

Fig. 8 shows profiles of the monthly mean sea temperature along 43°S from 20° to 90°E, based directly on the data in 2 × 2-degree blocks supplied by the United States National Environmental Data Center. A limited amount of smoothing was necessary and certain unlikely bulges in the curves indicate that more smoothing would not have invalidated the main properties. The curve for August has been omitted because it practically coincides with the September curve.

Fig. 8A illustrates conditions during the half year with falling temperatures, i.e., from March to September, while Fig. 8B illustrates conditions during the half year with rising

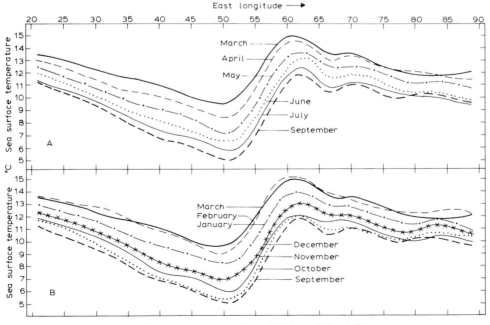

Fig. 8. Profiles of mean monthly sea temperature along 43°S from 20° to 90°E.

temperatures. The period with most rapidly falling temperature is March — June and the period with rapidly rising temperature is November — February.

The most striking feature of the temperature profiles is the gradual eastward drop in sea temperatures in the area west of 50°E and east of 60°E and the abrupt rise of 6°C in the interval between 50° and 60°E. This phenomenon in the Crozet — Kerguelen area has already been discussed in the preceding subsections.

The minor wave pattern north of Kerguelen at 67 — 73°E is also reflected in the profiles. Between 80° and 90°E, most of the monthly profiles show a rise and fall of about 0.5°C and, therefore, it has to be accepted as a true feature of the sea temperature distribution. The Crozet — Kerguelen wave pattern of the sea isotherms is a feature practically confined to the 40 — 50°S zone.

The profiles of Fig. 8 also indicate a steady increase of the annual sea temperature range along 43°S from about 2.5°C at 20°E to about 4.5°C at 50°E and then a gradual decline to about 2.5°C again at 90°E, in accordance with Fig. 5.

Sea temperature extremes

The range of extreme sea temperature in March and September is given in Fig. 9 (METEOROLOGICAL OFFICE, GREAT BRITAIN, 1949). The two diagrams and, in fact, the diagrams for all the months of the year show a zone with remarkable fluctuations in sea temperature centred between 40° and 45°S. The highest range in practically all months is found at 41°S, 40 — 50°E which coincides with the region north of Crozet Island where the maximum gradient of sea temperature is located. It can be expected that even small meridional displacements of the water masses to either side of an area with strong temperature gradient will result in large fluctuations. The possibility should be mentioned that relatively small errors in position determination by ships, which might particularly have occurred in sailing ship times, would result in an apparent high range of sea temperature in areas with strong gradients.

Although the location of the band with strong sea temperature fluctuations differs slightly in March and September and indeed also during the remaining months, it obviously

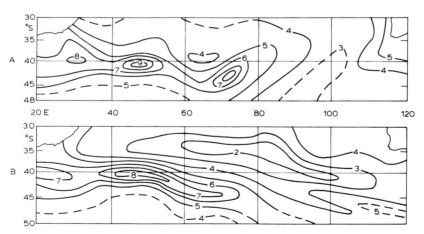

Fig. 9. Range of extreme sea temperatures: A. in March, and B. in September (after METEOROLOGICAL OFFICE, GREAT BRITAIN, 1949).

520

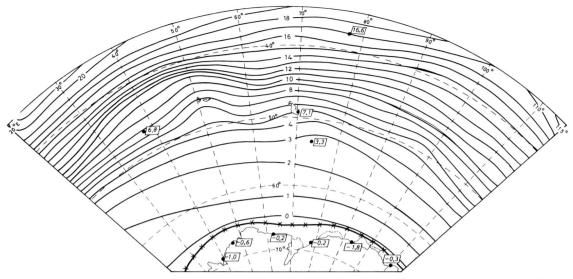

Fig. 10. Mean air temperature in January, with mean temperatures at the islands and Antartic stations inserted.

must be associated with the Subtropical Convergence and the Crozet – Kerguelen isotherm wave pattern.

Surface air temperature

Surface air temperature in January and July

For convenience, surface air temperature is abbreviated to air temperature below. The distributions of monthly mean air temperature over the southern Indian Ocean are shown in Fig. 10 for January and in Fig. 11 for July. It was shown above (p. 517) that mean sea temperature is highest in early March and lowest in early September, and it is shown below (p. 523) that mean air temperature is highest in February and lowest in

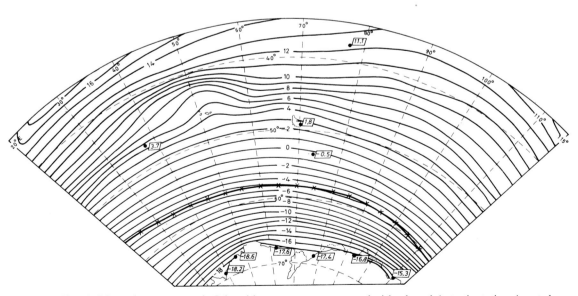

Fig. 11. Mean air temperature in July, with mean temperatures at the islands and Antartic stations inserted.

521

August in most areas, so that the annual course of sea temperature lags behind that of air temperature by a few weeks. March and September were chosen to illustrate the distribution of sea temperature (Figs. 3 and 4) and to derive the maximum range of sea temperature (Fig. 5). January and July illustrate air temperature distribution in accordance with normal practice. Actually the main features of sea and air temperature distributions change little throughout the year.

Few comments are needed on the air temperature distribution because it is more expedient to discuss the prominent features by means of derived diagrams. Also, mean air and sea temperatures differ very little and sea temperature distribution is discussed in detail above.

Similar to sea temperature, the air temperature in summer, represented in Fig. 10 by the January distribution, shows a marked contrast in meridional gradient north and south of the Antarctic Convergence located along about 50°S. Along the western boundary of the area, the temperature difference between 35° and 50°S is 16°C, whereas between 50° and 65°S it is only 4°C. However, due to fanning out of the isotherms as the longitude increases, the corresponding temperature differences at 105°E near the eastern boundary are 11°C and 7°C. Superimposed on this eastward spreading of the isotherms and smoothing of the meridional gradient are regional variations that are reflected, for instance, in the Marion—Crozet—Kerguelen wave pattern and the practically isothermal condition between 55° and 65°S at 20°E. Quantitatively, the eastward fanning out of the isotherms results in a 6°C drop in air temperature from 25° to 105°E along 35°S and a 3°C rise from 20° to 115°E along 50°S.

Apart from a weaker meridional air temperature gradient north of the Antarctic Convergence in July, conditions in this zone are mainly similar to that of January, but south of the convergence the distribution is spectacularly different. This is naturally due to the presence of the pack ice between the Antarctic coast and the average latitude of about 58°S in winter.

The true distribution of air temperature between the Antarctic Convergence and the pack ice in winter is uncertain because of the dearth of observations and a complete lack of observations over the pack itself. Therefore, the indicated distributions of air temperature in these zones are mostly hypothetical. According to Mackintosh (1946), the middle sea temperature along the Antarctic Convergence in the Indian Ocean is 2.5—3.0°C in July, so that if the assumption is made that air and sea temperatures are close along the convergence, then a fix is available for this position. Air temperatures are also available for the Antarctic coast (Schwerdtfeger, 1970) so that the problem remains of filling in the isotherms over the intermediate zones. The mean pack-ice limit in July supplies a clue but a rather uncertain one. The mean sea temperature at the pack-ice boundary must theoretically be −1.8°C, but it is problematical what the mean air temperature would be.

According to Fig. 11, the −5°C isotherm should coincide with the ice edge. This is in agreement with the finding of Vowinckel (1956, his fig. 2 of Part I). This could be argued about at length on the basis of surface temperatures over the pack ice, the stability and lapse rates of air masses flowing northward from the ice and southward from the warmer seas, the rates of warming and cooling in the two air masses, etc. The outcome of such considerations, right or wrong, is that the average air temperature over the sea close to the ice edge would be lower than the sea temperature by a few degrees.

Therefore, for drafting Fig. 11 it was assumed that at the ice edge the air temperature is 3°C lower than the sea temperature and that the air–sea temperature difference diminishes to zero at the Antarctic Convergence.

The next problem is the gradation of temperature between the ice edge and the Antarctic coast. It was simply assumed that the meridional gradient is uniform between the ice edge and the coast. This assumption, however, results in a marked increase in the gradient from west to east, viz., from 0.8°C per degree of latitude in the west to 1.7°C per degree of latitude in the east of the area. The truth is probably that the gradient on the ice is strongest near Antarctica and south of the circumpolar trough but finality can probably only be reached by means of satellite measurements.

The annual variation of air temperature

The annual course of air temperature is indicated in Figs. 12, 13, and 14. In Fig. 12, the mean monthly air temperatures recorded at the weather stations on New Amsterdam, Marion, Kerguelen, and Heard islands (Appendix, Tables XI–XIV) are compared to the temperatures based on ship observations contained in the climatic atlas of TALJAARD et al. (1969) at the positions of the islands. At New Amsterdam, Kerguelen, and Heard islands, the highest monthly means clearly occur in February, but at Marion Island the February and March temperatures are almost identical. As noted above, sea temperatures are almost identical in February and March, with March slightly in the lead at many places.

As regards the lowest monthly temperatures there is considerable disparity. At Marion and Heard Island stations, the air temperatures are lowest in September, at New Amsterdam in August, and at Kerguelen in July. Over the adjacent sea, the air temperatures are almost identical in August and September.

It should be recognized that air temperature, even the 24-h mean, at island stations does not necessarily reflect the temperature over the adjacent sea. This is very obvious at New Amsterdam, Marion, and Kerguelen. At New Amsterdam and Marion Island

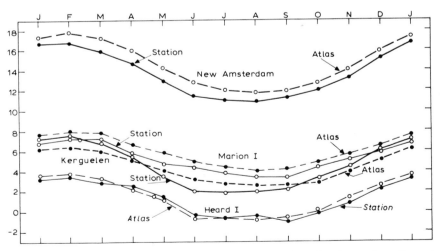

Fig. 12. Annual course of air temperature recorded at Marion, New Amsterdam, Kerguelen, and Heard Island meteorological stations and the mean air temperatures over the adjacent sea (after TALJAARD et al., 1969).

stations, the 24-h means are 0.5 – 1.5°C lower than over the sea throughout the year, a phenomenon not easy to explain. The difference is greatest in summer and autumn. At Kerguelen the station, Port-aux-Français, is situated near the east coast, about 100 km from the west coast near which mountains and plateaus rise to about 1000 m. Therefore, the degree of continentality reflected at the station is to be expected. In summer, the station temperature is up to 1°C higher than the sea temperature, whereas in June the station temperature is 1°C lower. At Heard Island, the station was located on a narrow isthmus north of the main island, where ventilation from the prevailing westerlies may have resulted in practically no difference between the station and the adjacent sea temperatures.

Fig. 13 shows the annual variation of mean monthly air temperature for 12 selected blocks measuring 2 degrees of latitude by 10 degrees of longitude, as determined from the printout of ship data from which the mean sea and air temperature distributions of Figs. 3, 4, 10, and 11 were derived. The curves of Fig. 13 show unsmoothed values, whereas for the charts (Figs. 10 and 11) a small amount of smoothing was necessary. The curves show the annual variation at 4-degree intervals of latitude in the western (20 – 30°E), central (70 – 80°E), and eastern (110 – 120°E) parts of the ocean. The areas at 110 – 120°E actually transgress into the sea area discussed in Chapter 3. The curves for 35°S are based on the data for the zone between 34° and 36°S, etc.

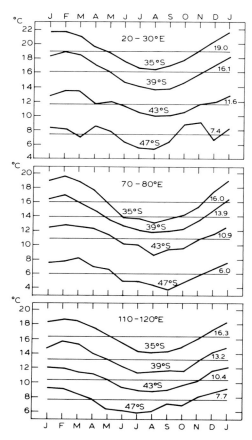

Fig. 13. Annual course of mean monthly air temperature in selected ocean blocks measuring 2 degrees latitude by 10 degrees longitude.

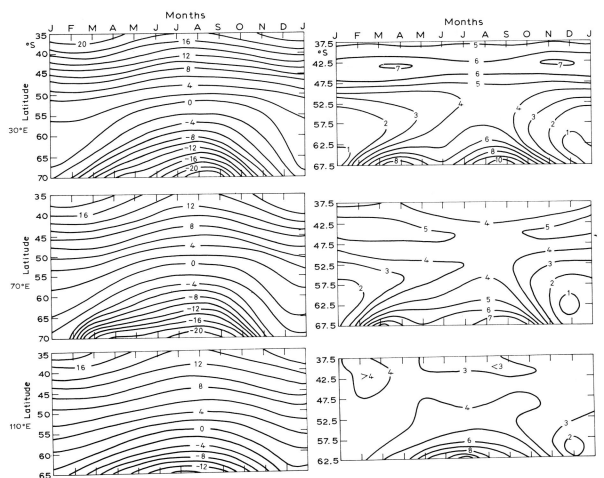

Fig. 14. Profiles of mean monthly air temperature and the corresponding meridional gradients of the temperature (°C per 5 degrees latitude) along 30°E, 70°E, and 110°E.

Being unsmoothed curves, irregularities in some cases are due to insufficient (less than 100) observations in the blocks. The block centred along 47°S between 20° and 30°E is particularly bad, but surprisingly smooth curves are obtained in most of the blocks at 35°, 39°, and 43°S, for several of which more than 1,000 observations are available.

It is clear that air temperature is highest in February in most areas and lowest in August. It is also notable that the annual range of air temperature is highest at 35°S, where it is about 5°C, and lowest at 47°S, where it is about 3°C on the average.

The annual variation of surface air temperature along 30°, 70°, and 110°E from 35° to 65° or 70°S, based on the gridpoint values in TALJAARD et al. (1969), is shown in Fig. 14. The corresponding meridional gradients (°C per 5 degrees of latitude) are also shown in the same diagram. It has to be pointed out that the more detailed analyses of air temperature shown in Figs. 10 and 11 are slightly at variance with those of the atlas in a few areas.

Features catching attention are the following.

(*1*) The strong gradient north of 50°S in the west exceeding 7°C per 5 degrees of latitude at 42.5°C in some months, decreasing progressively eastward to the extent that at 110°E it is not apparent anymore.

(2) The quasi-isothermal condition south of 50°S in the west in summer, this condition being less obvious in the center and hardly apparent at all in the east.

(3) The poleward increasing gradient south of about 50°S in autumn, winter, and spring, with double maxima in autumn and late winter at 30° and 70°E. For 110°E, the diagrams stop at 65° and 62.5°S, which are 5 degrees equatorward of the southern limits for 30° and 70°E, resulting from the more northerly position of the continental coast, viz., 66°S at 110°E.

Difference between air and sea temperature

Mean monthly sea temperatures were determined for the same 2 × 10-degree blocks as for air temperatures, with the exclusion of those for 47°S. The annual course of air-minus-sea temperature for the blocks is shown in Fig. 15. Erratic fluctuations are more in evidence than for air temperature alone (Fig. 13). However, it is clear that air temperature is predominantly lower than sea temperature, more so in winter than in summer. In half of the blocks, the air temperature is slightly higher (mostly less than 0.5°C) than sea temperature during 3 or 4 summer months. The annual variation is from the most positive, or least negative, differences in January to the greatest negative differences in the cold months May to August.

It is noteworthy that at 39°S in the west the mean air temperature is 1° to 2°C lower than the sea temperature throughout the year, indicating predominant cold air advection from west and south over the warm tropical water of the Agulhas Current. The situation is rather similar along 35°S at 110−120°E, i.e., just off the Australian southwest coast

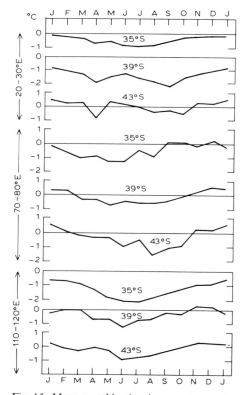

Fig. 15. Mean monthly air-minus-sea temperatures in nine 2-degree latitude by 10-degree longitude blocks.

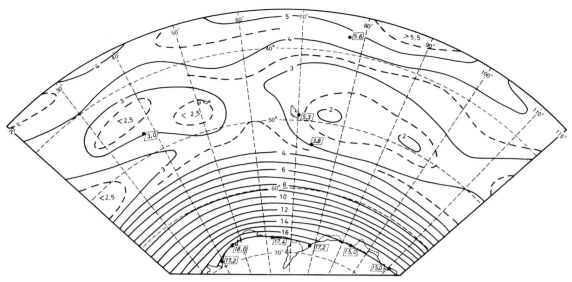

Fig. 16. Range of mean monthly air temperature: January minus July.

where the sea temperature is several degrees higher than along the temperature "trough" located 10 — 20 degrees farther west.

Summer to winter air temperature range

The January to July range of mean air temperature (Fig. 16) is based on Figs. 10 and 11. Along 35°S, the range increases from 4.5°C near Cape Agulhas to almost 6°C at 90°E and then decreases to less than 4°C off Cape Leeuwin. This relatively high range is part of the subtropical zone of high annual variation (VAN LOON, 1966, 1972a). The range decreases southward to 3°C or less along 45 — 50°S, with values lower than 2°C in patches east of Kerguelen Islands. From 50°S to the Antarctic coast, the range increases spectacularly, this being due to the presence of the pack ice south of about 58°S

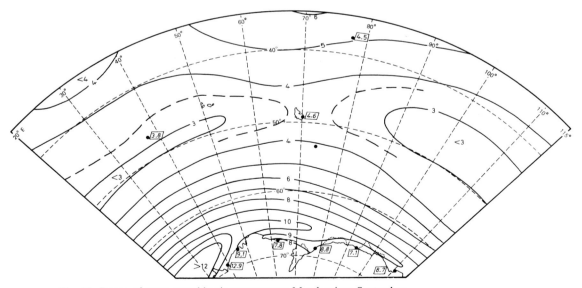

Fig. 17. Range of mean monthly air temperature: March minus September.

in winter, as well as to the extremely low temperatures over Antarctica in this season. It is well to remember that the low winter temperatures over the ice are confined to a surface layer of air only a couple of hundred metres deep.

Because sea temperature is higher in March than in January and lower in September than in July (Fig. 8), it should be interesting to compare the March-to-September air temperature range (Fig. 17) with that of January-to-July (Fig. 16). Fig. 17 was derived from the gridpoint values of air temperature in TALJAARD et al. (1969) in which atlas the surface temperature charts were analysed in less detail than for Figs. 10 and 11. Therefore, the March-to-September range shows a smoother pattern than the January-to-July range, but the main features agree, viz., a high range in the north, a zone with low range between 45° and 50°S, and strongly increasing range as Antarctica is approached. However, with March temperatures at Antarctic coastal stations 8−10°C lower than January temperatures, whereas the air temperatures are practically unchanged at 60°S, and considering also that air temperature in the far south changes very little from July to September (e.g., SCHWERDTFEGER, 1970), the pattern close to Antarctica for the two periods differs markedly. Fig. 17 shows a belt with maximum range a few degrees north of the Antarctic coast.

January and July air temperature anomalies

Using the hemispheric zonal mean monthly air temperatures of TALJAARD et al. (1969) for 5-degree latitude intervals from 35° to 65°S and the mean temperatures for the Indian Ocean for January and July (Figs. 10 and 11), the anomalies shown in Figs. 18 and 19 are obtained. Because the distributions of temperature given in Figs. 10 and 11 for the southern Indian Ocean are not strictly identical with those for January and July contained in the atlas, there will be discrepancies in the anomalies determined from the zonal means according to the atlas, but these discrepancies are very small.

In spite of possible atmospheric and oceanic circulation changes from January to March and from July to September and in spite of the variation of air-minus-sea temperature from month to month (Fig. 15), there is appreciable similarity between the pairs of deviation maps for air and sea temperatures for January and March (Figs. 18 and 6) and for July and September (Figs. 19 and 7).

In both summer and winter, the region off the South African coast, being flooded by the warm water of the Agulhas Current, is anomalously warm; the anomalies exceeding 2°C in January and 4°C in July. Following 35°S eastward, the anomalies decrease and change sign to become about −2°C at 100°E in January and −1°C in July. Apart from some irregularities over the eastern part of the ocean, the anomalies are increasingly negative southward to the zones astride 50°S in summer and about 60°S in winter. The wave patterns of the sea and air isotherms in the Crozet/Kerguelen area are well reflected in the anomaly patterns. The latitudinal shift of the strongest negative anomalies in middle latitudes is obviously related to the absence in summer and presence in winter of the pack ice in the Antarctic zone. The pack ice covers a broader zone in the Atlantic and western Indian Oceans than in the relatively warm Pacific Ocean and the ice boundary is located 5−10° farther north in the former regions than in the latter region. This results in a considerably greater annual range of air temperature in the zone bounded by approximately 55° and 65°S in the Atlantic and Indian oceans than in the Pacific

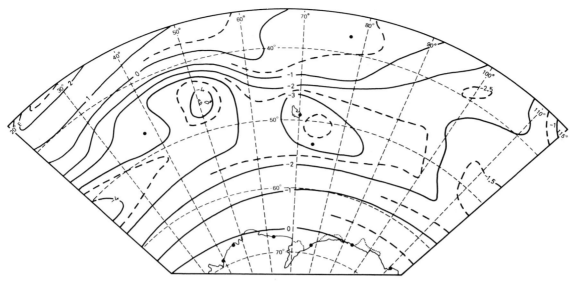

Fig. 18. Deviation of January air temperature from hemispheric zonal mean air temperature over the sea.

Ocean. It is this disparity in the annual range of temperature that leads to the change in the latitude of the negative anomaly.

The eastward spreading out of the air isotherms (Figs. 10 and 11) in the southern Indian Ocean is reflected by the change from positive to negative anomalies going eastward at 35° and 40°S and the decreasing negative anomalies following 50° and 60°S in the same direction.

Air temperature extremes

The ranges of extreme air temperature in January and July (METEOROLOGICAL OFFICE, GREAT BRITAIN, 1949) are shown in Fig. 20. Similar to the ranges of extreme sea temperature in March and September (Fig. 9), the ranges of air temperature are greatest

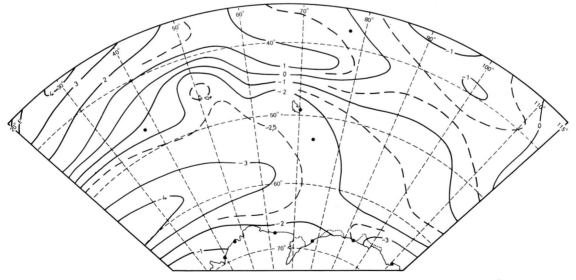

Fig. 19. Deviation of July air temperature from hemispheric zonal mean air temperature over the sea.

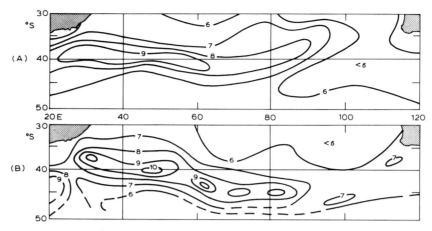

Fig. 20. Range of extreme air temperatures: A. in January, and B. in July (after METEOROLOGICAL OFFICE, GREAT BRITAIN, 1949).

along 38 – 45°S which is roughly the position of the Subtropical Convergence, or at least the zone with the strongest meridional sea and air temperature gradients. The fluctuations decrease sharply east of about 90°E where the sea and air temperature gradients are much weaker than in the western and central parts of the ocean.

The comments made on p. 520 on possible errors in sea temperature data due to position determination are also applicable to air temperature.

TABLE I

MEAN MONTHLY, SEASONAL, AND ANNUAL SURFACE AIR TEMPERATURE (°C) AT MARION, NEW AMSTERDAM, AND KERGUELEN ISLANDS FROM 1951 TO 1975[*1]

	Marion Island				New Amsterdam Island				Kerguelen Island			
	(a)	(b)	(c)	(d)	(a)	(b)	(c)	(d)	(a)	(b)	(c)	(d)
Jan.	6.8	0.58	8.1	5.8	16.7	1.15	18.9	14.8	7.2	0.82	8.9	5.9
Feb.	7.3	0.63	8.5	6.2	16.8	0.84	18.6	15.2	7.5	0.94	9.5	5.5
Mar.	7.2	0.50	8.4	6.5	15.9	0.94	17.7	14.6	6.9	0.63	8.3	5.2
Apr.	5.8	0.62	7.4	4.5	14.8	0.93	16.2	12.1	5.6	0.84	7.2	3.6
May	4.8	0.70	6.2	3.7	13.0	0.84	15.3	11.8	3.5	0.90	5.2	1.8
June	4.4	0.66	5.8	3.3	11.5	0.67	12.5	9.9	2.1	0.77	3.6	1.0
July	3.8	0.54	4.9	2.6	11.1	0.61	12.1	9.9	1.9	0.82	3.0	0.7
Aug.	3.4	0.60	4.7	2.1	10.9	0.77	12.3	9.0	2.0	0.87	3.5	0.7
Sep.	3.4	0.47	4.1	2.3	11.4	0.46	12.1	10.5	2.3	0.83	4.0	0.0
Oct.	4.4	0.55	5.2	3.4	12.1	0.58	13.0	10.6	3.3	0.67	4.6	1.8
Nov.	5.1	0.62	6.8	4.1	13.3	0.63	14.4	12.1	4.5	0.68	5.6	2.9
Dec.	5.9	0.52	6.9	4.7	15.3	0.58	16.6	14.2	6.1	0.67	7.4	5.1
DJF	6.7	0.42	7.3	5.8	16.2	0.67	17.3	14.9	6.9	0.64	8.1	5.8
MAM	5.9	0.45	7.0	5.1	14.6	0.78	16.4	13.0	5.3	0.62	6.4	4.1
JJA	3.9	0.36	4.6	3.6	11.2	0.53	11.9	10.1	2.0	0.63	3.2	0.5
SON	4.3	0.39	5.1	3.6	12.3	0.43	13.1	11.5	3.4	0.53	4.2	1.6
Ann.	5.2	0.23	5.6	4.8	13.6	0.49	14.5	12.6	4.4	0.41	5.2	3.5

[*1] (a) Mean monthly, seasonal, and annual; (b) standard deviation; (c) highest mean monthly; and (d) lowest mean monthly temperature. (Information supplied by Roy Jenne of the National Center for Atmospheric Research, Boulder, Colo., U.S.A.)

TABLE II

MEAN MONTHLY TEMPERATURES (°C) AT HOYVIK ON THE FAROE ISLANDS (62°N 7°W; height 20 m above MSL)
(After LYSGAARD, 1969)

Jan.	Feb.	Mar.	Apr.	May	June	July	Aug.	Sept.	Oct.	Nov.	Dec.	Year
3.9	3.7	4.6	5.4	7.3	9.2	11.0	11.1	10.0	7.9	6.1	5.9	7.1

Temperature variability at Marion, New Amsterdam, and Kerguelen

The mean monthly, seasonal, and annual air temperatures and the corresponding standard deviations and the highest and lowest values for the period 1951—1975 are given in Table I (see also Fig. 12 for the graphical presentation of the mean monthly temperatures and pp. 523—524 for a discussion thereof).

The standard deviations are notably smaller for Marion Island than for the other two islands. It might be surprising that the deviations at New Amsterdam at 37°S, i.e., between the subtropical high pressure zone and the middle latitude westerlies, are appreciably greater than at Marion Island at 47°S, where fronts and airmass changes are more frequent and more pronounced. Then again, Kerguelen is situated only 3 degrees poleward of Marion Island. The same tendency is reflected in the ranges of the highest and lowest mean monthly, seasonal, and annual values for the 25-year period.

It is instructive to compare the mean monthly temperatures at Marion Island (47°S 38°E) and at Kerguelen (50°S 70°E) with those at Hoyvik in the Faroes (62°N 7°W) given in Table II. Equating January at the Faroes with July at Marion and Kerguelen Islands, we see that the lowest mean monthly temperatures at Hoyvik are almost the same as at Marion Island (15 degrees closer to the Equator), but that Kerguelen is 2° colder in spite of being 12 degrees closer to the Equator. In summer the temperature at Hoyvik rises almost 4°C higher than at the two Southern Hemisphere stations.

This clearly illustrates the tremendous difference of an oceanic area washed by a meridional warm current (the eastern part of the far North Atlantic Ocean) and that in which a zonal cold current flows in relatively low latitudes (the middle latitudes of the South Indian and Atlantic oceans).

Upper-air temperature

General

As this chapter is mainly concerned with surface climate, the description of upper-air temperature will be brief with emphasis on conditions at 500 mbar. Considerably more detail can be found in TALJAARD et al. (1968, 1969), CRUTCHER et al. (1971), and VAN LOON (1972a).

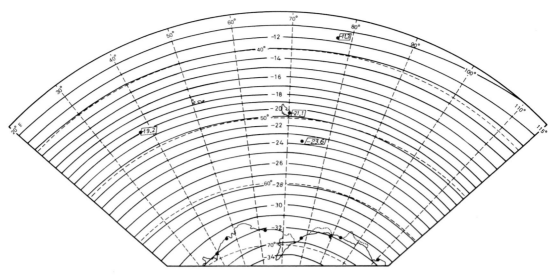

Fig. 21. Mean temperature for January at the 500-mbar constant pressure surface (after TALJAARD et al., 1969).

Mean monthly, seasonal, and annual upper-air temperatures and standard deviations at Marion, New Amsterdam, Kerguelen, and Heard islands

Mean monthly temperatures and standard deviations for four island stations are given in the appendix for the standard pressure levels from 850 to 100 mbar. The length of records used are 23−25 years for Marion Island, 15−19 years for New Amsterdam, 4−6 years for Kerguelen Island, and 2−5 years for Heard Island (not all soundings reached the 200−100 mbar levels). These long-term monthly means have been taken into account for drafting Figs. 21 and 22, but otherwise the diagrams in this section are based upon the grid-point data in TALJAARD et al. (1969).

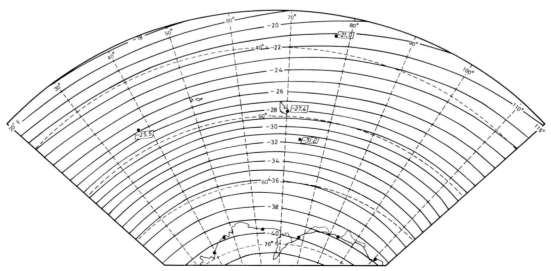

Fig. 22. Mean temperature for July at the 500-mbar constant pressure surface (after TALJAARD et al., 1969).

532

The temperature distribution at 500 mbar in January and July

In both months, the isotherms of the 500-mbar constant pressure surface appear superficially as concentric circles around the South Pole, with only minor variations in the meridional gradients along both the longitudes and latitudes (Figs. 21 and 22). This is also apparent for the remaining 10 months of the year. However, close scrutiny and evaluation of the gradients reveal notable zonal and meridional differences.

Following latitude 35°S from west to east, the drop in temperature from 20° to 70°E in January is zero and from 70° to 110°E it is only 1°C, but in July the decrease is 4°C from 45° to 115°E. Also in January the temperature is approximately constant along 60°S, whereas in July it rises about 1.5°C from 20° to 115°E. These changes of opposite sign at 35° and 60°S in winter produce a clear diffluent pattern of the isotherms toward the east and a decrease of the meridional gradient from 0.7°C at 20°E to 0.5°C per degree of latitude at 115°E. The temperature gradients are discussed on pp. 537, 539. Other notable features of the temperature distribution are best discussed by using derived diagrams.

Temperature change from January to July at 500 mbar

Using Figs. 21 and 22, we give the January-minus-July temperature distribution in Fig. 23. The range is greatest near Australia where it reaches 11°C, whereas the smallest range of about 7°C is found near Marion Island. The Marion Island minimum extends to E and ESE past Kerguelen and Heard islands to 60°S at 115°E. Farther south the range increases slightly to just over 9°C at Syowa and Molodezhnaya and about 8°C at Casey. Near South Africa the highest range of just over 8°C is found along about 38°S. Because upper-air temperature over southern middle latitudes (exemplified by Marion and Kerguelen islands) is highest in February and lowest in August, the temperature change from February to August in this zone is somewhat greater than for January to July. However, at lower and higher latitudes, the extremes occur in January and July,

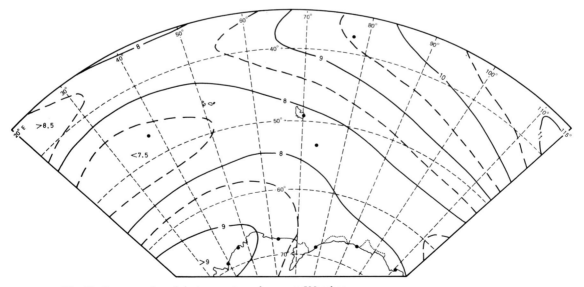

Fig. 23. January-minus-July temperature change at 500 mbar.

implying that the regional difference in a February-minus-August map would be slightly smaller than indicated in Fig. 23.

Temperature deviations from the hemispheric zonal means at 500 mbar

Temperature deviations from the hemispheric zonal means (i.e., the isanomals of temperature) at 500 mbar are shown in Fig. 24. The hemispheric zonal mean temperatures used are based on TALJAARD et al. (1969), and the temperatures are those for January and July shown in Figs. 21 and 22.

In January, the deviation is positive over the western part of the subtropical zone, the highest deviation being about 1.5°C. Over the eastern part of the Antarctic zone the negative deviation is more than −1°C. The zero anomaly line runs from 44°S in the west to 51°S in the east.

In July, the deviation along 35°S west of the 60th meridian is more than 2°C and then it sharply drops to −1.5°C at 105°E. Over the western part of the ice-covered Antarctic zone, the anomaly exceeds −2°C. The meridional change of over 4°C in the zonal anomalies in the western half of the southern Indian Ocean contrasts sharply with the almost constant value of about −1.5°C at 100° to 115°E.

Comparison of Fig. 24 with the corresponding surface air temperature anomalies (Figs. 18, 19) shows that: (*1*) the marked irregularities of the middle latitude surface air tem-

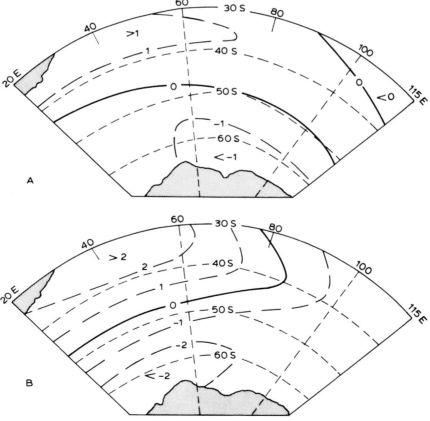

Fig. 24. Deviation of 500 mbar temperature from the hemispheric zonal means in: A. January; and B. July.

peratures virtually disappear at 500 mbar; (2) the strongest positive anomalies exist in the northwestern (African) sector of the ocean at both levels and in both months, the anomalies in July being strongest; and (3) the strongest negative anomalies are near 60°S in the west in winter. The anomalies over the eastern third of the ocean are weak and rather irregular.

Annual variation of upper-air temperature

The annual variation of the mean temperature at 500 mbar along the 30°, 70°, and 115°E meridians is shown in Fig. 25A. Maxima occur in February equatorward of 60°S and in January farther south. North of 55°S, minima occur in July but the data up to 1966 (TALJAARD et al., 1969) suggest that at 55°S and poleward thereof a secondary minimum with about the same value as in July occurs in September. The rise in temperature in August and the cooling in September might be explained in terms of enhanced meridional exchange in the zone south of 50°S, but the phenomenon has not yet been thoroughly investigated for the southern Indian Ocean. The following question can be posed: Do

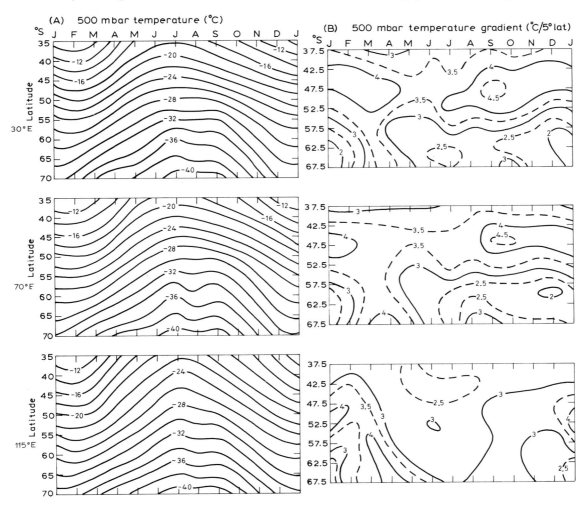

Fig. 25. Annual variation of: A. 500-mbar temperature, and B. 500-mbar meridional temperature gradient (°C per 5 degrees latitude) along 30°E, 70°E, and 115°E.

the shifting trough-ridge patterns south of Australia in winter/spring described by VAN LOON (1967a, 1972b) also affect the southern Indian Ocean?

In the zone 35° — 50/55°S, the profiles show a more rapid drop in temperature in autumn (February — May), viz., 6.5°C, than a rise in spring (August — November), viz., 4.5°C. This skew distribution indicates that temperatures drop during about five months and rise during the remaining seven months. From 50/55° to 70°S, which is over the continent, the skewness is reversed with temperatures falling during about six months (January — July), then rising evidently in August and falling again in September, whereafter a rapid rise sets in during four months (October — January).

The annual course of the 500-mbar temperature at the four island stations and at four continental stations along or close to the boundaries of the southern Indian Ocean is shown in Fig. 26. Although Perth (32°S) is located two degrees equatorward of Cape Town (34°S), its 500-mbar temperature is mostly 1 — 2°C lower than at Cape Town, which illustrates the cooler conditions over the eastern than over the western subtropical Indian Ocean. At New Amsterdam Island at 37.6°S, the temperature is naturally another 1 — 2°C lower than at Perth. At this group of stations, the steep autumnal drop in temperature, the clear minimum in July and the relatively slow rise of temperature from July to February, mentioned above, are clear.

At Marion and Kerguelen islands, the peak in February is pronounced, the winter minimum around August is flat and the rate of fall and rise in autumn and spring is roughly equal. At Heard Island, the rise in spring is stronger than the rate of fall in autumn, while the indicated rise in August followed by a drop in September might be due to the small number of months of observations available (3 for July, 5 for August, and 3 for September).

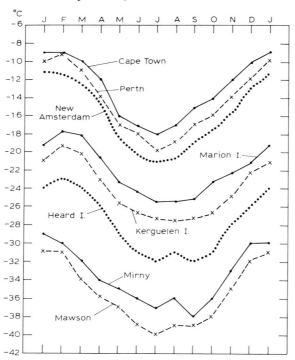

Fig. 26. The annual variation of temperature at 500 mbar at Cape Town, Perth, Mawson, Mirny (from TALJAARD et al., 1969) and at Marion, New Amsterdam, Kerguelen, and Heard Islands (from Table XV).

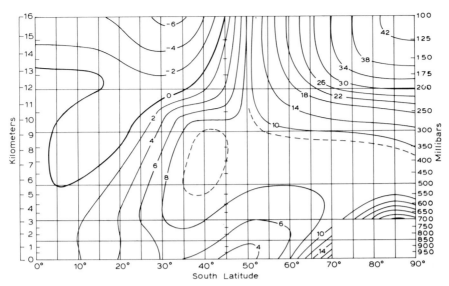

Fig. 27. Profile of January-minus-July temperature change from the surface to 100 mbar along 70°E.

At Mirny and Mawson on the Antarctic coast, the slow fall in temperature from summer to winter contrasts markedly with the rapid rise in spring. Again there is evidence for a secondary minimum of temperature in September in the central part of the far southern Indian Ocean.

The January-minus-July temperature difference, or range, from the surface to 100 mbar (about 16 km) along 70°E between the Equator and the Pole is shown in Fig. 27. At sea level, the range increases from about zero at the Equator to 6°C at 30°S, then decreases to about 3°C near 50°S (cf. Fig. 16), whereafter it increases markedly to about 20°C at the Antarctic coast. At the 500—300-mbar levels, the range increases from about zero at the Equator and 10°S to over 9°C at 40°S, whereafter it remains practically constant at 8—9°C right to the Pole. In the upper tropical troposphere, i.e., from 300 to 100 mbar, the temperature mostly rises from summer to winter north of 45°S, but at middle and high latitudes the temperature drops remarkably at these levels, the range being 43°C at the Pole!

Meridional temperature gradient at 500 mbar

Profiles of the meridional temperature gradient, expressed in °C per 5-degree latitude interval, are shown in Fig. 25B for 30°, 70°, and 115°E. At 30° and 70°E, the gradient is strongest along 45—50°S being slightly weaker in winter than in summer. This situation is not evident at 115°E. On the average, the gradient at 60—65°S is only roughly half of its value at 45—50°S. It is noteworthy that the gradient varies semiannually in the Antarctic zone, reaching peaks in autumn (March—April) and spring (September-October-November). The autumn peak is pronounced at 115°E.

In Fig. 28, the difference of temperature at 500 mbar along 45° and 70°S, for the interval 20—115°E, and for the whole hemisphere (based on TALJAARD et al., 1969), also shows two maxima, viz., a very marked peak in March/April and a less marked peak in September/October, thus proving the existence of a clear semiannual variation.

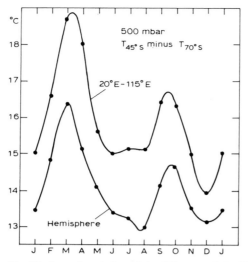

Fig. 28. Annual variation of the difference of the 500-mbar temperature at 45° and 70°S for the hemisphere and for the sector 20 – 115°E.

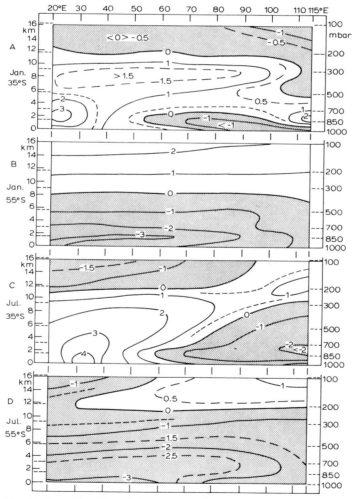

Fig. 29. Vertical profiles of the January and July temperature deviations from the hemispheric zonal means along 35°S (A and C) and along 55°S (B and D) between 20°E and 115°E.

538

Zonal temperature anomaly profiles along 35° and 55°S in January and July

The profiles along 35°S (Fig. 29, A and C) show the expected positive anomalies close to South Africa below about 12 km in both January and July. In the eastern half of the ocean, the anomalies along 35°S revert to negative values, but close to Australia a remarkable contrast occurs in the two seasons in the layer between the surface and 700 mbar. In January, the anomaly increases almost 4°C from the surface to 800 mbar, whereas in winter it decreases about 3°C in this layer, thus indicating marked stability in summer and relative instability in winter. In the upper troposphere or lower stratosphere (12−16 km), the anomalies are relatively weak, being negative in both seasons, except near Australia in July.

At 55°S (Fig. 29, B and D), the anomalies in the troposphere below about 10 km are negative, with values of −3° to −4°C at 850 mbar in January and at the surface in July thus illustrating the relative coldness of the midlatitude Indian Ocean. The anomalies decrease eastward. In the stratosphere the anomalies are mostly positive, values of over 2°C occurring at 16 km in January.

Pressure

Sea-level pressure

Mean sea-level pressure in January and July

Superficially, the mean pressure distributions for January and July (Figs. 30 and 31) differ very little. The main features are high pressure along the northern boundary of the area and low pressure along the circumpolar trough at 61−63°S, separated by closely spaced isobars in middle latitudes (40−60°S). The meridional pressure gradient is greatest in the zone 45−50°S where it is approximately 2 mbar per degree of latitude. The evidence is strong for the existence of two statistical low pressure centers along the circumpolar trough, viz., at 90−100°E and somewhat to the west of 20°E. The average pressure difference between 35°S and the trough axis is about 7 mbar more in July than in January, the greatest increase in the difference being in the centre and west (about 10 mbar), with very little change in the east (about 2 mbar). Apart from these general aspects, the monthly mean pressure patterns for January and July are by no means representative of conditions in all of the intervening months. The change of pressure is composed of annual and semiannual fluctuations and there are changes affecting both the meridional gradients as well as the locations of the subtropical high pressure cell and the circumpolar trough as described below.

The subtropical ridge

Aspects of the high-pressure zone along the northern boundary of the area are the location and movement of the axis of highest pressure from month to month and of the center of highest pressure, i.e., the Indian Ocean anticyclone. It should be kept in mind that the discussion here concerns only the statistical means of the markedly fluctuating daily positions and intensities of the individual anticyclones.

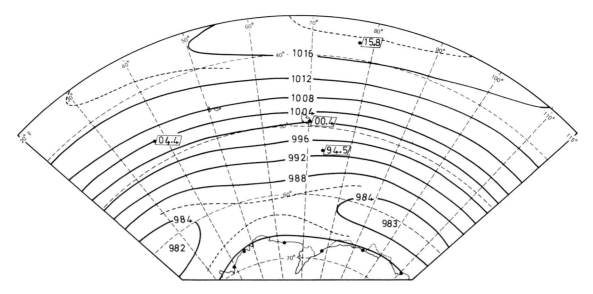

Fig. 30. Mean sea-level pressure in January (after TALJAARD et al., 1969). New long-term means (1951 – 1976) are inserted for Marion, New Amsterdam, and Kerguelen.

The anticyclonic center is displaced quite remarkably during the year (Fig. 32), a phenomenon to which attention was drawn by VOWINCKEL (1955) and VAN LOON (1972b). The centre is farthest south in March (35°S at 80°E), whereafter it is displaced westward and northward until June when it is located at 30°S 53°E. During the next three months, it starts retracing its previous path but then suddenly in October it moves northward through two degrees of latitude. Then in November it is displaced rapidly eastward to attain its easternmost position at 85°E, which is 32 degrees of longitude from its position in June. Subsequently, the displacement occurs southward step by step to the above-mentioned position in March.

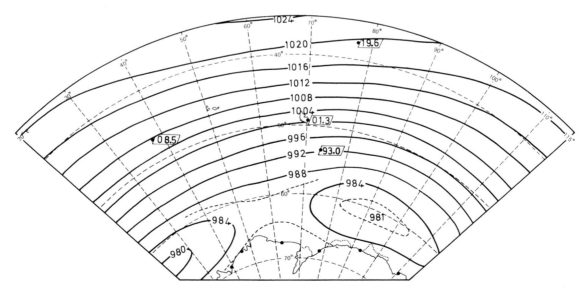

Fig. 31. Mean sea-level pressure in July (after TALJAARD et al., 1969). New long-term means (1951 – 1976) are inserted for Marion, New Amsterdam, and Kerguelen.

The remarkable westward displacement of the anticyclone in winter explains the fresh-to-strong SE trade-wind current of the western central part of the subtropical southern Indian Ocean in this season, whereas in summer a broad easterly current sets in, transporting humid warm air to Malagasy and southern Africa. The westward displacement of the high from March to June occurs at the same time as the enhancement of the skew trough which characterises middle latitudes southwest of Australia in winter.

The axis of the subtropical ridge, followed eastward and westward from the anticyclonic center trends ESE and WSW to positions near the south coast of Australia and Africa throughout the year, except near South Africa from May to September. In fact, during the warmer half of the year the axis is found along 37 – 39°S, i.e., several degrees south of the South African and Australian coasts.

Fig. 32(B) shows the variation of the central pressure of the anticyclone during the year. The annual pressure wave is obviously predominant, but in conjunction with the positional change of the high the effects on the pressure distribution in the subtropical and middle latitude zones are substantial.

The circumpolar trough

On charts of mean pressure, the most obvious feature along the circumpolar trough is the col at 60°E separating two areas of lower pressure at about 20° and 90° to 100°E. The location of the col is probably determined by the Enderly Land promontory, which projects northward to 65.8°S, flanked on the east by the deep indentation of the Amery Ice Shelf and the Lambert Glacier. The offshore SE winds are likely to be steered farther north and with a stronger southerly component in this region than elsewhere.

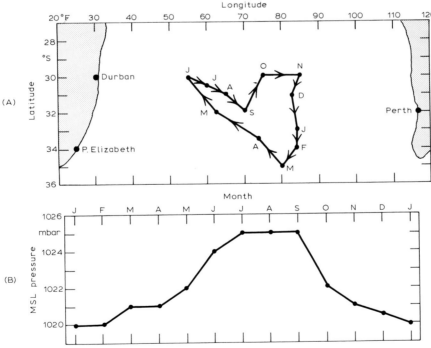

Fig. 32. A. Location, and B. central pressure of the subtropical anticyclone for each month of the year (based on TALJAARD et al., 1969).

The outstanding characteristic of the trough in the Indian Ocean, as elsewhere around Antarctica, is its predominant semiannual fluctuation in both position and intensity (VAN LOON, 1967a, 1972b). The semiannual wave in the position of the trough in the southern Indian Ocean (Fig. 33) accounts for 74% of its total yearly variance, whereas the annual wave accounts for a mere 5% (the third harmonic actually accounts for 14% of the variance). At the time of the equinoxes in late March and September, the position of the trough is about 4 degrees of latitude south of the position it occupies during the solstices in June and December. Similarly, there is a marked, though less prominent, semiannual wave in the mean pressure along the trough. In this case, the annual wave accounts for 37% of the variance, compared to 48% of the semiannual wave.

The remarkable characteristic of the semiannual fluctuation is that the pressure along the trough is highest when it is located farthest north and lowest when located closest to the Antarctic coast. A puzzling aspect is that pressure not only along the trough, but also over the Antarctic coast and Antarctica, is higher in summer (December–February) than in winter (June–August), whereas according to the temperatures and densities of the air over the high latitude regions, the pressure at the surface, being hydrostatically a reflection of the total mass overhead, should have been low in summer and high in winter.

A striking property of the pressure along the trough is its particularly low value in September and October, when it is 3–7 mbar lower than during the remainder of the year.

Annual course of pressure at the islands and at Mawson

The annual course of pressure at the island stations (Table III) and at Mawson at 63°E on the Antarctic coast (Fig. 34), shows marked differences from one position to another

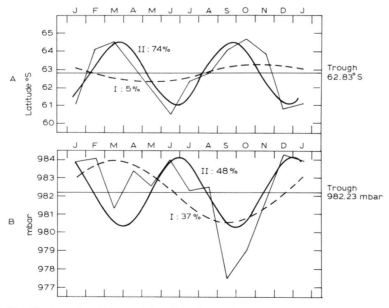

Fig. 33. Annual variation and first (*I*) and second (*II*) harmonics of: A. the mean latitude of the circumpolar trough; and B. the mean pressure along the trough between 20° and 115°E.

TABLE III

MEAN MONTHLY, SEASONAL, AND ANNUAL SEA-LEVEL PRESSURES, STANDARD DEVIATIONS, AND EXTREMES AT MARION, NEW AMSTERDAM AND KERGUELEN ISLANDS[*1]

	Marion Island				New Amsterdam Island				Kerguelen Island			
	(a)	(b)	(c)	(d)	(a)	(b)	(c)	(d)	(a)	(b)	(c)	(d)
Jan.	04.1	3.1	08.0	−04.0	15.8	3.0	21.3	10.0	00.4	4.5	07.0	06.0
Feb.	07.3	4.0	15.1	−02.0	18.3	3.5	24.7	11.0	05.8	5.2	20.0	−04.0
Mar.	09.8	2.9	15.0	03.0	19.8	3.3	24.2	12.0	07.8	4.6	17.0	−02.0
Apr.	07.2	3.5	14.0	02.0	19.1	3.0	23.0	12.0	04.7	3.0	10.0	−01.0
May	06.8	3.1	14.4	01.0	16.6	3.2	23.0	11.0	02.4	3.9	11.0	−04.0
June	08.3	4.6	17.0	−01.0	16.0	4.3	21.0	02.0	00.2	3.8	08.0	−08.0
July	08.5	4.2	15.0	00.0	19.6	3.5	24.0	12.0	01.3	4.3	10.0	−06.0
Aug.	08.6	3.4	14.0	02.0	20.0	4.8	28.0	04.0	02.0	5.5	13.0	−09.0
Sept.	08.4	2.6	15.0	04.0	19.8	4.0	26.0	12.0	00.8	4.6	10.0	−07.9
Oct.	08.3	3.5	17.0	02.0	18.6	2.4	23.0	15.0	01.5	3.4	08.0	−03.7
Nov.	05.2	4.0	12.2	−02.0	16.3	3.7	23.0	09.0	−01.6	4.7	06.0	−09.0
Dec.	01.5	2.7	07.5	−03.0	16.1	3.1	23.5	10.0	−01.5	4.0	06.5	−09.0
DJF	04.3	2.2	07.5	−01.3	16.8	2.5	23.2	12.0	01.6	3.4	09.0	−04.7
MAM	08.0	2.0	14.1	04.3	18.5	2.2	22.7	11.7	05.0	2.4	10.3	00.0
JJA	08.4	2.6	12.7	04.7	18.5	3.2	22.7	08.7	01.2	2.9	06.3	−04.7
SON	07.3	2.3	13.0	03.0	18.3	2.4	22.3	13.3	00.2	3.1	06.0	−06.1
Ann.	07.0	1.3	10.3	05.2	18.0	1.9	21.5	11.3	02.0	1.5	05.5	−01.3

[*1] (a) Mean monthly, seasonal, and annual pressures (1000 + ... mbar); (b) standard deviations (mbar); (c) highest mean monthly, seasonal, and annual pressures (1000 + ... mbar); (d) lowest mean monthly, seasonal, and annual pressures (1000 + ... mbar). (Information supplied by Roy Jenne of the National Center for Atmospheric Research, Boulder, Colorado, U.S.A.)

in the ocean. At New Amsterdam Island in the subtropics, there is a clear semiannual wave of the pressure, with peaks occurring in March and August/September. At Mawson, the semiannual fluctuation is exactly opposite to that at New Amsterdam. At the three mid-latitude islands, pressure is clearly highest in March, but during subsequent months the similarity fades, though a semiannual variation and a weaker annual variation can be discerned. Heard and Mawson close to the middle longitude of the ocean, but on opposite sides of the Antarctic trough, show marked inverse variations from January until August, but this relationship breaks down from September to December. Pressure is lowest in November/December at Marion and Kerguelen islands.

Mid-seasonal fluctuations of pressure

VAN LOON (1972b) discussed the annual and semiannual components of pressure over the Southern Hemisphere in detail, but it should also be illuminating to discuss the total seasonal fluctuations of the sea-level pressure distribution over the southern Indian Ocean (Figs. 35−38). Fig. 35 shows the December-to-March pressure rise of the middle latitudes and the associated decrease along and south of the circumpolar trough. This distribution implies that the meridional pressure gradient increases between the largest rise and fall from December to March and reflects the poleward shift of the trough through 4 degrees of latitude.

From March to June (Fig. 36), the trough moves north from 64.5° to 60.5°S, reflected

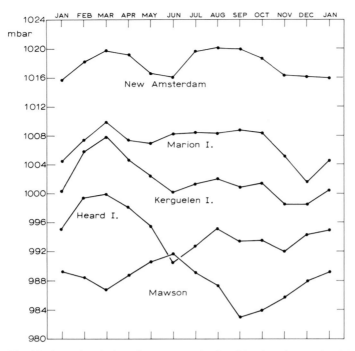

Fig. 34. Annual variation of pressure at the four island stations and at Mawson on the Antarctic coast.

in a pressure rise of up to 8 mbar along the Antarctic coast. The most spectacular change is over the eastern middle latitudes where pressure drops more than 10 mbar, as against a change of less than 2 mbar at the same latitude south of Africa, thus indicating the development of a trough southwest of Australia, as well as increasing pressure gradient north of the anomaly minimum and decreasing gradient to the south. At the same time, the pressure rises several millibars over the western subtropical part of the ocean, this being in agreement with the westward displacement and intensification of the anticyclonic center.

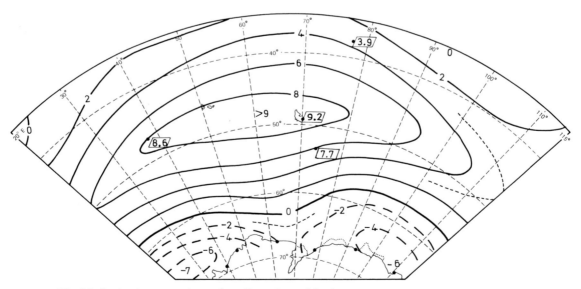

Fig. 35. Sea-level pressure change from December to March.

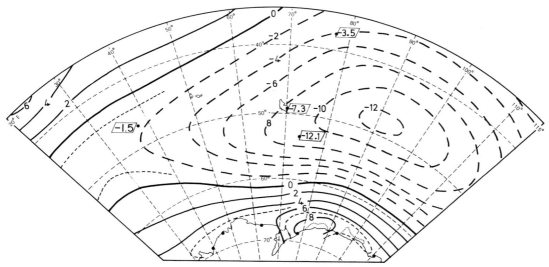

Fig. 36. Sea-level pressure change from March to June.

From June to September (Fig. 37), the trough off southwestern Australia weakens and the circumpolar trough shifts south to 64°S, where the average pressure drops to a pronounced minimum of 978 mbar. The meridional pressure difference now increases between middle and high latitudes.

September to December (Fig. 38) signals the second northward shift of the circumpolar trough and a pressure rise of 6—10 mbar along and near the Antarctic coast. Pressure falls over the western and central parts of the middle latitudes and subtropical zones concomitant with the eastward shift and weakening of the subtropical anticyclone, and the meridional pressure gradients again decrease between middle and high latitudes.

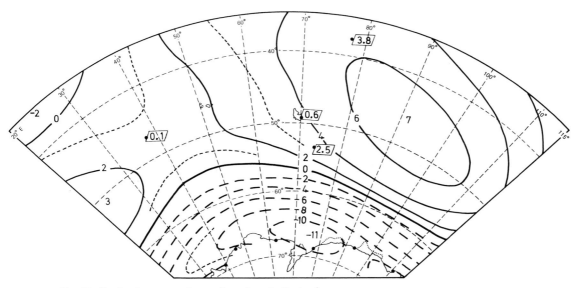

Fig. 37. Sea-level pressure change from June to September.

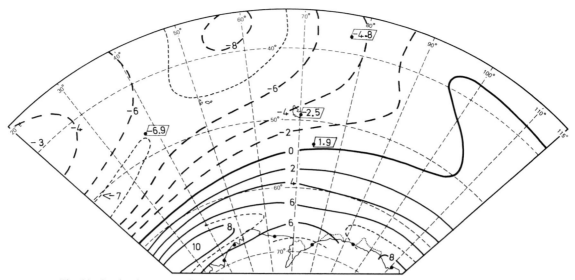

Fig. 38. Sea-level pressure change from September to December.

Variability of daily, monthly, seasonal, and annual mean sea-level pressures

The annual course of mean monthly pressure at the four island stations is shown in Fig. 34 and discussed on p. 543. The actual values of the mean pressure, including those for the seasons and the year, as well as the corresponding standard deviations and the extreme values recorded at Marion, New Amsterdam, and Kerguelen islands during 1951–1975, are given in Table III. It should be pointed out that the individual monthly data for New Amsterdam for 1961 used for compiling Table III are doubtful, for the mean annual pressure for this year is about 7 mbar below the long-period average, such that for most months, seasons, and the year the lowest minima given in the table are for this year.

The standard deviations for New Amsterdam and Kerguelen islands are very similar and higher than those for Marion Island. It might have been expected that the fluctuations of mean pressure at New Amsterdam, being a subtropical station at 37°S, ought to be less than those for the two middle-latitude stations. Assuming that the 1961 pressures for New Amsterdam are too low (let us say by 6 or 7 mbar), the altered standard deviation and the extreme fluctuations for this island would be slightly lower than given in the table. For instance, the extreme range for August is 24 mbar, viz., 1028 mbar in 1968 to 1004 mbar in 1961. The normal and acceptable maximum fluctuations are in the range 10–20 mbar.

The standard deviation of daily sea-level pressure, based on the daily weather maps for the IGY, are discussed by TALJAARD (1966), from which the example for summer, shown in Fig. 39, is taken. The standard deviation increases from about 5 mbar along 35°S to over 12 mbar at 56–61°S in the western part of the area. There is a gradual decrease eastward in middle and subantarctic latitudes, which phenomenon might be ascribed to probably less reliable analysis in the region southwest of Australia where ship observations were more sparse than elsewhere, but the same feature also appears on the maps for the other seasons when no bias existed as regards the scanty observations everywhere.

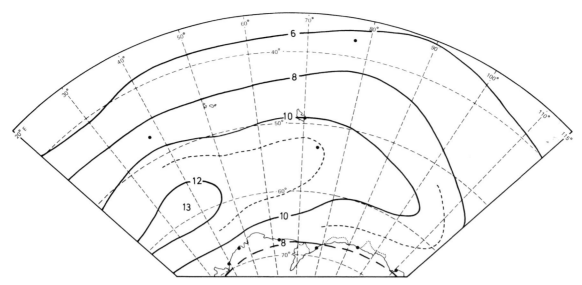

Fig. 39. Standard deviation of daily sea-level pressure in summer (after TALJAARD, 1966).

Upper-air pressure

General

As for temperature, the upper-air pressure will be discussed only briefly with emphasis on conditions at the 500-mbar level. Again, considerable detail is to be found in TAL-JAARD et al. (1968, 1969), CRUTCHER et al. (1971), VAN LOON et al. (1971), JENNE et al., (1971), and VAN LOON (1972b).

Mean monthly, seasonal, and annual heights of constant-pressure surfaces and standard deviations

Monthly, seasonal, and annual mean geopotential heights for constant-pressure surfaces from 850 to 100 mbar for Marion, New Amsterdam, Kerguelen, and Heard islands are given in the Appendix (Table XVI). The corresponding standard deviations are included in the table. The values are based on 23–25 and 15–19 years' observations at Marion and New Amsterdam islands, respectively, but for Kerguelen and Heard islands the periods are only 4–6 and 2–5 years, respectively. Hence the values for the latter two stations cannot be considered as very representative.
The data can be most profitably discussed by using derived diagrams.

Contour charts for 500 mbar in January and July

The 500-mbar charts for January and July (Figs. 40, 41) are taken from TALJAARD et al. (1969). Slight adjustments were made to accommodate the new values up to 1974 for Marion and New Amsterdam islands, as well as for the Kerguelen values which were not available for the original charts.
In January, the height gradient is clearly strongest along 45–50°S, the isohypses being almost parallel to the latitude circles. The gradient decreases sharply south of 50°S and

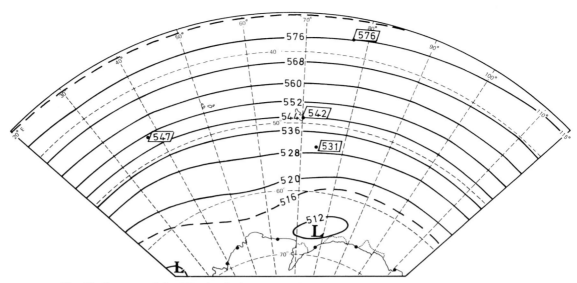

Fig. 40. Contours of the 500-mbar isobaric surface in January (after TALJAARD et al., 1969). Heights are in geopotential dekametres.

the evidence indicates the existence of a very weak zonal trough along about 67°S, with possibly closed centres at 80°E and at a position west of 20°E. Along 35°S, the height decreases slowly east of 80°E, reflected in a weakening of the meridional gradient near Australia.

In July, the W—E differences are amplified such that a weak ridge appears at 40—50°E while the western flank of a trough is indicated in the east of the area. The meridional height gradient decreases markedly from summer to winter east of the area.

These features of the height patterns are well reflected in the geostrophic wind profiles at 50° and 110°E (Figs. 51 and 52 in the section on wind).

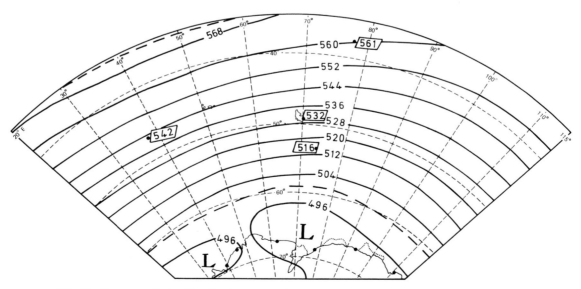

Fig. 41. Contours of the 500-mbar isobaric surface in July (after TALJAARD et al., 1969). Heights are in geopotential dekametres.

January—July height change and zonal anomalies at 200 mbar

Instead of using Figs. 40 and 41 for deriving the January—July 500-mbar height changes, the 200-mbar charts in TALJAARD et al. (1969) were used for obtaining Fig. 42, because the changes in the middle and upper troposphere and lower stratosphere are very similar. Fig. 42 shows that quite a notable relative ridge, with axis along 45°E, develops equatorward of 50°S in winter. Therefore, the 200-mbar isohypses must necessarily have a marked northward flexure between 50°E and Australia in winter, as can be verified from the original charts. The same phenomenon is well illustrated by the zonal anomaly charts for 200 mbar in January and July (Fig. 43). Only a weak trough is indicated by the anomalies at 95°E in January, whereas in July the 200-mbar height is 190 gpm lower at the southwestern tip of Australia than at 35°E, 45°E. This situation is a reflection of the relatively cold troposphere in subtropical and middle latitudes over the eastern Indian Ocean and southern half of Australia in winter (Figs. 22, 23). Although a discussion of the subtropical jet stream in the Southern Hemisphere falls outside the scope of this chapter, it should be noted that the tropospheric temperature and constant-pressure height distribution over the eastern part of the southern Indian Ocean point to the existence of a strong jet stream at 25—30°S over Australia. Furthermore, the positive anomalies in the subtropics and negative anomalies in middle latitudes over the western half of the area reflect the existence of a stronger-than-average polar frontal zone, or polar jet stream at 45—50°S in this sector, extending also into the South Atlantic Ocean (p. 558 and Fig. 51A and C).

Variability of pressure heights

The values of the standard deviation of pressure heights given in the Appendix (Table XVI) do not show marked regional differences or obvious seasonal variations. The deviations naturally increase with height at all stations and this is true even for the

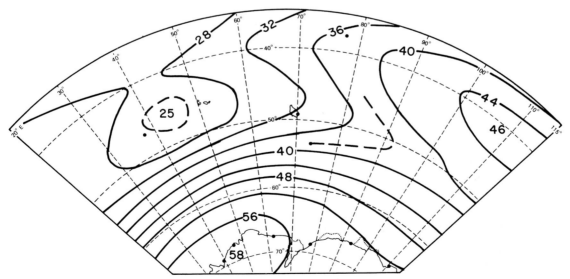

Fig. 42. January-minus-July height change of the 200-mbar isobaric surface.

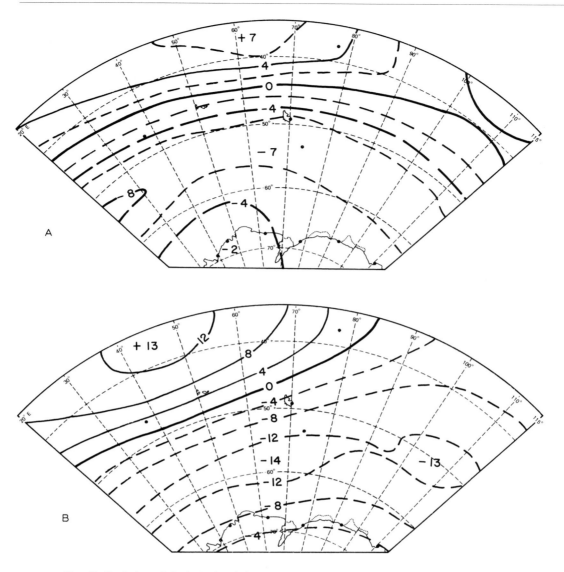

Fig. 43. Deviation of the 200-mbar height from the hemispheric zonal mean heights for January and July (after TALJAARD et al., 1968).

100-mbar surface which is well into the polar stratosphere. It would appear, however, that the monthly fluctuations at Heard Island often are lower than at the remaining three stations, this being in accordance with the known fact that temperature and pressure-height fluctuations are highest in the core of the west wind belt. In this respect, the magnitude of the standard deviations at New Amsterdam, located at the southern flank of the subtropics, is lower than that at Marion Island located in the core of the strongest westerlies.

The standard deviation of the daily 500-mbar heights over the southern Indian Ocean for 5 summer months (December—March) of the IGY is shown in Fig. 44 (from TALJAARD, 1966). The maximum variability clearly exists along 50—55°S and decreases from west to east.

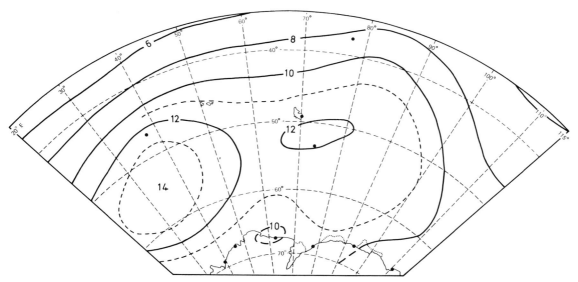

Fig. 44. Standard deviation (gpm) of the daily 500-mbar heights during 5 summer months of the IGY (after TALJAARD, 1966).

Wind

Surface wind

North of about 45°S, the mean resultant wind has a northerly component over the western parts of our area and a southerly component over the eastern parts, and the same distribution is found south of about 55°S; although there are seasonal variations in the meridional components, they are not big. Between these latitudes, the mean resultant wind is almost due west (see JENNE et al., 1971). The resultant wind speed, averaged across the southern Indian Ocean is shown in Fig. 45. The speed increases with increasing latitude to a peak in 45–50°S and decreases farther south to a minimum at 62°S. The latitudinal shear of the wind is larger in January and the speed at the peak is higher in January than in July. The summer maximum which owes its existence to the extensive water surface of the hemisphere will be further described below.

The difference between mid-season months of the west-to-east component (u) of the geostrophic wind is given in Fig. 46. At lower latitudes, u grows, i.e., becomes more westerly, from January to April and from April to July and it then diminishes during the rest of the year. Near 50°S, the net change from January to July is negative, an indication of the summer maximum mentioned above. The seasonal change in the sub-antarctic has a dominant half-yearly component as the wind increases from January to April, decreases from April to July, increases again from July to October, and drops from October to January. Fig. 46 shows that the annual course of the wind near 45°S contains a half-yearly component too, which in contrast with the one farther south has its maxima in the extreme seasons. The half-yearly oscillations of the wind are described in VAN LOON (1967a, 1971). They are related to the seasonal changes of the subantarctic trough discussed on p. 542.

The observational material over the southern oceans south of the subtropical high is so

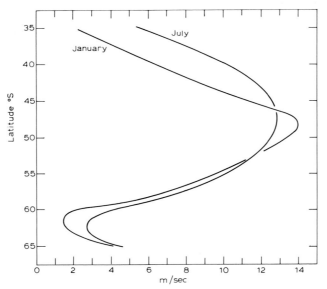

Fig. 45. The zonal geostrophic wind at sea level (m s^{-1}) between 35° and 65°S, averaged between 20° and 115°E. From VAN LOON et al., 1971.

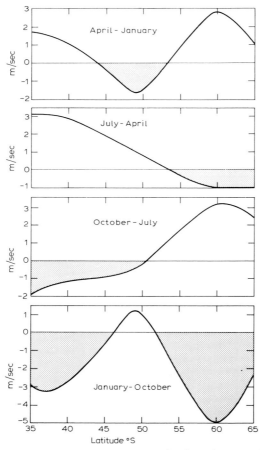

Fig. 46. Mid-season changes (m s^{-1}) of zonal geostrophic wind at sea level between 35° and 65°S, averaged between 20° and 115°E.

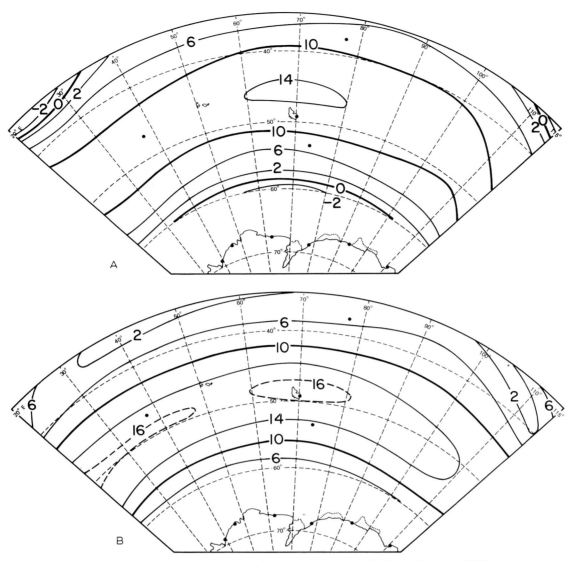

Fig. 47. The zonal geostrophic wind (m s⁻¹): A. for December 1957; B. for December 1958.

scant during most of the year that it cannot be used to describe the wind in single months of individual years, nor can reliable frequency distributions be made of wind direction and speed from the few ship observations. Wind data from islands are not representative of the surrounding ocean, and one must therefore turn to daily maps of sea-level pressure to obtain such descriptions as are contained in Figs. 47—50. The following statistics are based on the daily analyses for the International Geophysical Year 1957/1958.

The mean zonal geostrophic wind between 35° and 60°S for two early summer months, December 1957 and 1958, has the same general pattern which consists of a continuous maximum across the ocean in 45° to 50°S, somewhat stronger in 1958 than in 1957 (Fig. 47). In 1958, however, the westerlies extended over all of our area, whereas in 1957 the polar easterlies reached north of 60°S and the subtropical easterlies came south of 35°S in the northern corners. In December 1957, the strength of the maximum is close to the

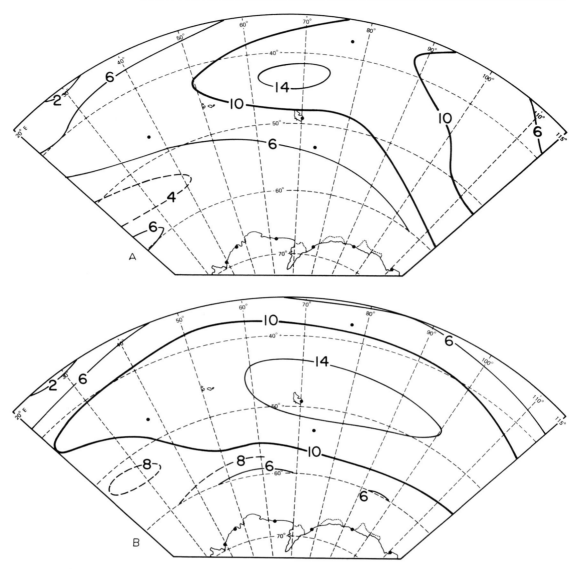

Fig. 48. As Fig. 47 but for August 1957 and 1958, resp.

long-term average (cf. VAN LOON et al., 1971, fig. 2); in 1958, it was somewhat above average. The long-period mean maps of the summer months all contain such a maximum in middle latitudes, with only small latitudinal displacements from one month to another. In winter, August 1957 and 1958 (Fig. 48), the general pattern is also similar in both months with a major maximum whose axis runs WNW – ESE across the area and a minor maximum in the southwestern corner, but the pattern is less simple than that in summer.

The frequency distributions of resultant geostrophic wind speeds and directions in Figs. 49 and 50 are based on values for every 24 h in December 1957, January, February, March, and December 1958 for summer (152 days), and on June – September 1957 and 1958 for winter (244 days). The nine data points are placed at 40°, 50° and 60°S in the western, central and eastern parts of the region. The directions are for 30-degree sectors, 345 – 15°, etc., and they are plotted at the midpoint of the sector: 360°, 30°, 60°, etc.

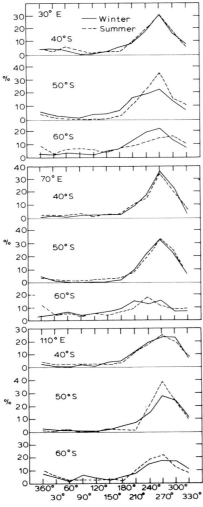

Fig. 49. Frequencies of geostrophic wind directions at 30°E, 70°E, and 110°E, at 40°S, 50°S, and 60°S, during the winters of 1957 and 1958, and the summer of 1957/58 and December 1958.

There is remarkably little difference between the two seasons in frequency of direction, no more, in fact, than one might ascribe to the small sample. Winds from westerly directions dominate, as expected, at all nine locations, more so at 40° and 50°S than at 60°S.

The frequencies of daily resultant wind speeds (Fig. 50) are given for five categories. In all three latitudes, calms and winds ≤ 1.4 m s^{-1} are rare in both seasons, rarer than winds ≥ 21.5 m s^{-1}. Indeed, the latter category is frequent both summer and winter in the central and western parts at 50°S. At 40° and 60°S, the distributions in summer are skewed toward the lower values; in winter at the same latitudes they are closer to a normal distribution or skewed slightly toward the higher values. These seasonal differences agree with the latitudinal expansion of stronger winds from summer to winter in Fig. 45. At 50°S, the seasonal differences in the central and western regions are small; if anything, the stronger values in the central parts are a little more frequent in summer than in winter.

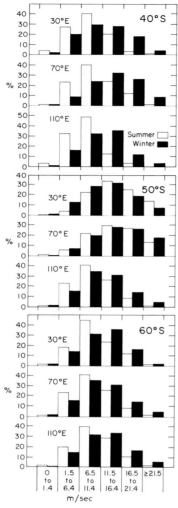

Fig. 50. Frequencies of resultant geostrophic wind speeds for the same periods and places as in Fig. 49.

Wind in the troposphere

Despite the preponderance of ocean in the Southern Hemisphere—just over 80% is covered with water—there are marked longitudinal differences in the pattern of the wind (see, e.g., VAN LOON, 1964, 1972a). There are appreciable differences even in such a short span of longitude as we are dealing with, especially during the colder part of the year. The meridians of 50° and 110°E are used to illustrate the mean zonal geostrophic wind in the troposphere and lowest stratosphere. In summer (Fig. 51,A,B), the pattern is uncomplicated: the subtropical ridge (the zero line) which slopes toward the Equator with height in the troposphere marks the border between westerlies and easterlies, from about 35°S at sea level to 14−17°S at 10−12 km. The westerly maximum, which is 5 m s^{-1} stronger at 50°E than at 110°E, lies 2−3° closer to the Pole at 110°E, and at the latter meridian a subtropical maximum appears which extends eastward beyond the Tasman Sea. The polar easterlies above the violent katabatic winds of the surface layer are in the mean shallow and weak.

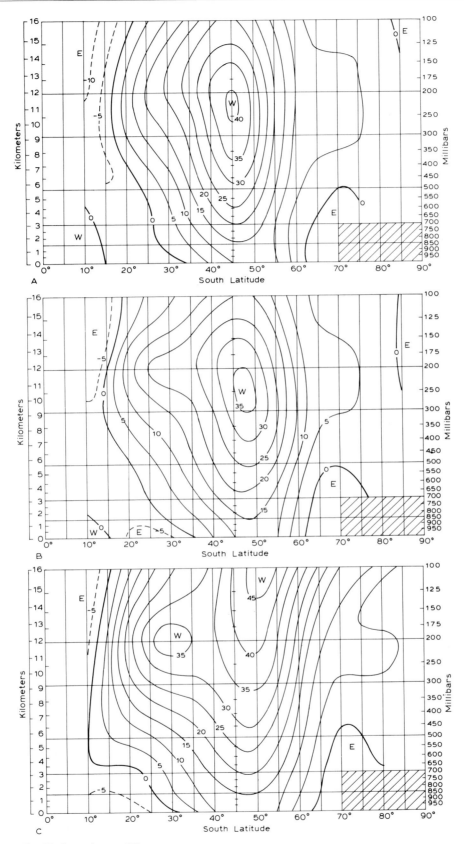

Fig. 51. Legend on p. 558.

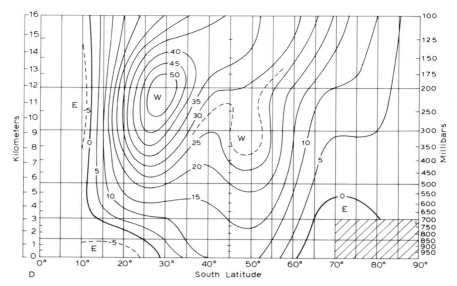

Fig. 51. Latitudinal vertical sections of zonal geostrophic wind (m s⁻¹): A. at 50°E for January; B. at 110°E for January; C. at 50°E for July; D. at 110°E for July. From VAN LOON et al., 1971.

The subtropical maximum in the winter sections (Fig. 51,C,D) is much stronger at 110°E than at 50°E, and it extends downward at 110°E to 4 km above the surface whereas at 50°E it is clearly limited to the upper troposphere. The westerlies at middle latitudes, the Polar Front westerlies, are almost equally strong at the surface at both meridians but are considerably stronger in the troposphere in the western part—a testimony to the stronger latitudinal temperature gradients there (cf. Fig. 14). As in summer, the polar easterlies are shallow and weak, but in contrast with summer the equatorward slope of the border between the westerlies and low-latitude easterlies is accomplished in little more than 3 km.

The differences between January and July (Fig. 52) are very illuminating. The westerly component increases from January to July, or the easterly component decreases, over most of the region (January minus July < 0). The increase of the westerlies has three distinct maxima: one in the subtropics, a second in the stratosphere, and a third in the subantarctic in the troposphere. In the last instance, the mean westerlies are, however, only slightly stronger in winter. The westerlies decrease from summer to winter, or easterlies increase, in the trades (January minus July > 0), in the polar region, and over the Polar Front zone of middle latitudes, which is of particular interest since it is the opposite of what happens in the Northern Hemisphere.

The amplitude of the changes from summer to winter in the subtropics is by far larger over the eastern part of the sector.

The stronger westerlies in summer over the Polar Front zone is a feature peculiar to an ocean hemisphere (VAN LOON, 1966, 1972b). WILHELM MEINARDUS (1929, 1940) had already noted that the pressure gradient in these latitudes is steeper in summer than in winter and that cyclones there move eastward more quickly in summer. Meinardus also pointed out that the latitudinal temperature gradient at the surface in southern middle latitudes is steeper in summer than in winter, and he related it to his observation that the annual temperature range decreased with increasing latitude between the subtropics

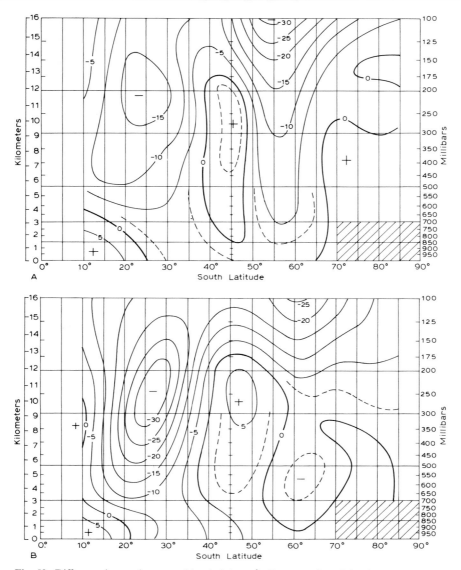

Fig. 52. Difference in zonal geostrophic wind (m s⁻¹), January minus July: A. at 50°E; B. at 110°E.

and middle latitudes. This characteristic of the annual range is discussed by VAN LOON (1966, 1972b) who showed that it is a property of a planetary ocean surface and that it also appears in the central to eastern parts of the oceans in the Northern Hemisphere. There are two major factors associated with the poleward decreasing annual range of temperature over the open ocean in middle latitudes. The first is that cloudiness increases poleward in the latitudes where the annual temperature range diminishes poleward. Although the possible annual range of shortwave radiation which reaches the surface is larger in middle latitudes than in the subtropics, the larger amount of clouds in middle latitudes makes the actual range of short-wave radiation at the surface smaller in middle latitudes.

The second factor is that the energy reaching the ocean surface at middle latitudes is distributed as heat through a much deeper layer of water in middle latitudes than in the subtropics, owing to the mixing by the stronger winds and currents of middle latitudes. An estimate of the annual variation of heat storage in the mixed layer of the ocean at 35° and 50°S, and of the resulting seasonal change in the temperature of the layer (VAN LOON, 1966), does suggest that the combined effect of the reduction of insolation by clouds and of the mixing of heat over a deeper layer in the ocean in the west wind zone than in the subtropics may adequately account for the diminishing annual temperature range as one goes toward the Pole.

This feature of the annual range of both sea surface temperature and surface air temperature is seen in Figs. 5, 16, and 17, and its vertical extent in the troposphere is shown in Fig. 27. The subtropical peak in the annual range slopes poleward over 10 degrees of latitude from the surface to the upper troposphere, and the minimum at middle latitudes extends into the subantarctic over a vertical distance of 7–8 km (Fig. 27). Since the temperature falls less from summer to winter at higher than at lower latitudes, it is obvious that the temperature contrast between the two latitude zones in the troposphere must be smaller in winter than in summer. The west wind will therefore increase faster with increasing height in summer than in winter over the region between the tropics and subantarctic.

As mentioned earlier, the annual march of the wind has an appreciable half-yearly component in the subantarctic, where the maxima occur in the transition seasons. The half-yearly oscillation wanes where the wind has its summer maximum in middle latitudes, but it reappears farther north where its maxima then are in the extreme seasons (see, e.g., VAN LOON, 1967b, fig. 11). The half-yearly oscillations are discussed in VAN LOON (1967a) and their implications for surface currents in the Indian Ocean in VAN LOON (1971).

Precipitation and clouds

Precipitation

Although the amounts of precipitation measured at the four island stations in the southern Indian Ocean (Appendix, Tables XI–XIV) are not representative of the amounts which fall over the open ocean, it is likely that one can use the monthly or seasonal means to describe how the precipitation is distributed over the year. The seasonal percentages in Table IV show that most of the precipitation at Marion and Heard islands falls in summer and autumn, whereas at Kerguelen and New Amsterdam the largest amounts fall in autumn and winter. Where the cyclonic activity is unabated through the year, the heaviest precipitation may be expected during the warmer part of the year and the annual distribution would be as it is at Marion and Heard islands. It is understandable that the largest amounts fall in autumn and winter at New Amsterdam where the subtropical high is nearest the island during the warmer part of the year, December–March. It is not clear why Kerguelen does not have the same annual distribution as Marion and Heard islands. JACOBS (1969) has the following seasonal apportionment in the Indian Ocean between 40° and 50°S; summer 22%; fall 26%; winter 28%; and spring 24%, which is the same distribution as at Kerguelen and New Amsterdam islands. MÖLLER (1951) also has the highest amount in winter and fall between 45° and 50°S for

TABLE IV

SEASONAL PERCENTAGES OF THE ANNUAL PRECIPITATION

	Dec.–Feb.	Mar.–May	June–Aug.	Sept.–Nov.
Heard Island	*30.1*	*32.7*	17.9	19.3
Marion Island	*26.8*	*27.9*	22.4	22.9
Kerguelen Island	22.7	*27.1*	28.4	21.8
New Amsterdam Island	20.8	*26.0*	27.9	25.3

the whole Southern Hemisphere, and gives a distribution such as the one at Marion and Heard islands only south of 55°S. The precipitation rates derived from satellite observations for two years (RAO et al., 1976) show a seasonal distribution south of 40°S similar to the one at Marion and Heard islands. It is possible that the record at Kerguelen, which began before those at the two other islands, has influenced early investigators to place the peak in winter.

The percentage of all observations with precipitation in whatever form is given for all seasons by McDONALD (1938). His maps for summer and winter are shown, redrawn, as our Fig. 53. The frequency is higher in winter than in summer north of 45°S, but there is an indication that in winter the highest frequency, 30%, is reached at 42°S and that the number decreases farther south. The UNITED STATES NAVY (1957b, revised 1976) published similar maps, but does not have such a peak in winter, since in all months the frequencies increase poleward as far as the maps go: 55°S. At 48°S, the

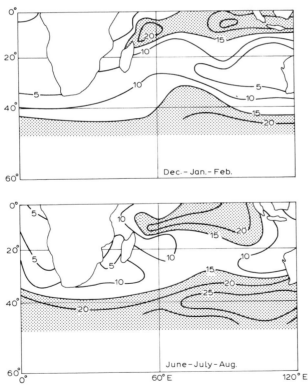

Fig. 53. Percent of all ship observations with rain in whatever form. Adapted from McDONALD, 1938.

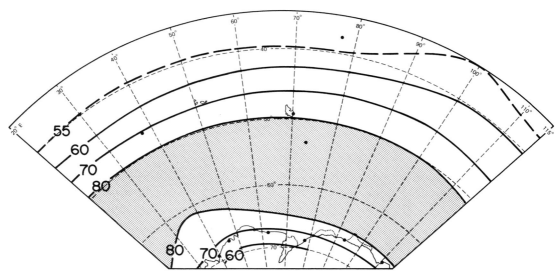

Fig. 54. Total cloudiness in January in percent. From VAN LOON, 1972d.

percentage of observations with precipitation increases from 25 in summer to 35 in winter in the Navy atlas. VOWINCKEL and OOSTHUIZEN (1953) reported that in summer south of about 55°S about 20% of the observations in the western end of our area and about 40% of those in the eastern end reported precipitation. Their results are based on whaling ship observations from four summers and are extended over more years for the far south by VOWINCKEL (1956). Further information about precipitation, particularly on the islands, is given by LOEWE (1957).

Total cloudiness

The mean total cloudiness increases from the northern border to a maximum between 50° and 60°S and decreases from there toward Antarctica. The summer picture (Fig.

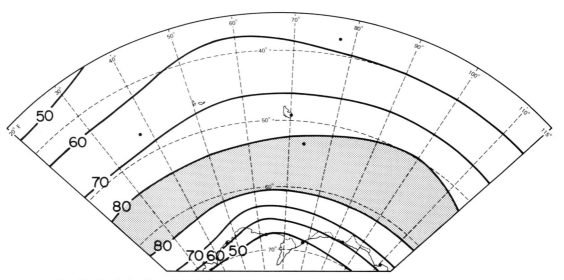

Fig. 55. Total cloudiness in July in percent. From VAN LOON, 1972d.

54), which is based on a fair amount of whaling ship observations (VOWINCKEL and VAN LOON, 1957), is reliable, but the winter picture (Fig. 55) is based on very little evidence. If the difference between the two months is as given in the illustrations, there are somewhat more clouds on the average in summer than in winter over most of the region south of 45°S.

Synoptic climatology

General

Synoptic climatology may be defined as the climate, or characteristic frequencies and ranges of weather elements associated with each of the circulation patterns, or synoptic map types, which occur in an area. For this purpose, a classification of the surface and preferably also the upper-air weather maps of an area is required, supplemented by compilation of the weather elements associated with the various sectors of each map type.

For the southern Indian Ocean, this could be done by using the routine data of the island stations in conjunction with, for instance, the daily synoptic maps of the Australian or South African Weather Services for the past few decades. The coverage could be improved by using ship data for selected blocks, mainly for summer. A suitable period would be the 1951−1960 decade and the early years of the 1961−1970 decade when whaling fleets and expeditionary ships operated in the antarctic and subantarctic zones. However, this is an onerous task which has not been tackled yet.

One aspect of synoptic climatology is the study of the occurrence, development, movement, and general behaviour of weather systems in an area. This excludes the study of the long-term averages, such as the mean monthly or seasonal pressure distributions, already discussed.

The source material for discussing this aspect of synoptic climatology is contained mainly in TALJAARD and VAN LOON (1962, 1963), TALJAARD (1964, 1965, 1967a), VAN LOON (1965), and STRETEN and TROUP (1973). Taljaard and Van Loon used the daily hemispheric maps for the IGY published by the SOUTH AFRICAN WEATHER BUREAU (1962, and on), whereas Streten and Troup used satellite cloud mosaics for the hemisphere for the summer and intermediate season months of the period November 1966 to March 1969. The IGY maps were based on an optimum of observations gathered by the World Meteorological Organization, as well as on many thousands of unpublished observations contained in ship logbooks. They can be relied upon for the summer months for most regions, but for the months April to October they give only an approximate indication of the systems of the Antarctic Ocean between 50° and 65°S.

Satellite photos are advantageous for the determination of cyclonic vortices and cold fronts, but anticyclones and some weak lows and warm fronts cannot be determined very well. STRETEN and TROUP (1973) used only the satellite pictures obtained in the visible spectral range and therefore did not study winter conditions. One comprehensive investigation using infrared cloud pictures has, however, recently appeared (CARLETON, 1979).

As the discussion below will show, the anticyclones of the Southern Hemisphere occur predominantly in the subtropical and lower middle latitude zones, viz., equatorward of 45°S. In the Indian Ocean, the observations at the three long-standing island stations, as well as those made at Heard Island (1950—1954) and in recent years at Crozet Islands, in conjunction with scattered ship data north of 45°S on the South Africa—Australia route throughout the year, have ensured that synoptic maps for the lower middle latitude and subtropical zones are essentially reliable, except for relatively small systems such as cyclones in their formative stages. Anticyclones are mostly extensive systems and therefore their positions and frequencies ought to be fairly well determined from the existing weather map series.

Anticyclones

Anticyclogenesis

Anticyclogenesis may be defined as the first appearance of a high-pressure cell with closed isobars or as the intensification of an existing system. Evidently no study has been made thus far of the latter aspect of southern anticyclones and so it will not be attempted to present statistics. Only a few brief remarks can be made.

Many new high-pressure cells which cannot be traced as having passed as closed cells across South Africa or within a few degrees of latitude off its south coast, appear on synoptic maps off the east coast of this country. The usual sequence of events is found to be the eastward ridging of a South Atlantic anticyclone across the land or along its south coast and then the appearance of closed isobars in the leading part of the ridge some distance from the east coast. This leap forward has been called "budding" and also occurs across the Andes and off the Australian southwest coast. It is probably caused by topographic influences and need not be true anticyclogenesis affecting a deep tropospheric layer. However, those anticyclones whose centers pass eastward at distances of more than three or four degrees south of the land advance as normal closed cells all the way.

A phenomenon which is evidently also confined to the western part of the southern Indian Ocean is the rather sudden intensification of anticyclones at latitudes 40—50°S due south or southeast of South Africa in the Marion—Crozet Island area. These anticyclones are often not evident at Gough Island (40°S 10°W) in the Atlantic Ocean and yet they reach Marion Island with central pressures of 1025—1040 mbar. It can be argued that due to lack of regular observations the eastward passage of some anticyclones at 40—50°S in the Atlantic Ocean might proceed unheeded, in which case the sudden appearance of strong highs in the vicinity of Marion Island might not be cases of anticyclogenesis at all. However, the finding by VAN LOON (1956) that the Marion—Crozet Island region is one of three favoured areas for the occurrence of blocking anticyclones in the Southern Hemisphere lends support to the impression that this might be a region of more than the average occurrence of anticyclogenesis.

Anticyclone movement

The existence of above-average pressure in certain positions along the subtropical high-pressure belt might give the wrong impression that anticyclones will be found in roughly these positions every day. Daily weather maps show that very few anticyclones are actually stationary for more than a few days. They move eastward in harmony with waves in the subtropical upper westerlies and the frontal troughs of the middle latitudes, but the eastward displacement of the trough-ridge systems in the two zones is not necessarily always in phase so that the progress of the surface anticyclones is often irregular. Many examples of anticyclone tracks for IGY months are given by TALJAARD and VAN LOON (1962, 1963) and TALJAARD (1964, 1965, 1967a, 1972b). A midsummer and a midwinter month have been chosen to illustrate anticyclone and cyclone tracks for the Indian Ocean (Figs. 56, 57). Due to well-known deviations of samples for short periods from normal conditions, the given examples are not necessarily representative of what is to be found during many months and years. For instance, the clustering of anticyclones around 37°S 55°E instead of around 35°S 95°E in January 1958 was somewhat

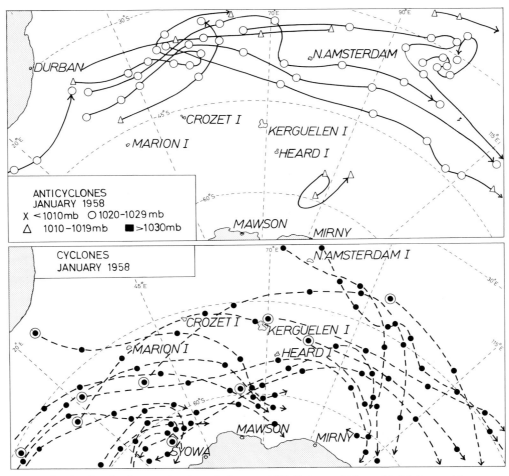

Fig. 56. Tracks of anticyclones and cyclones in January 1958. Systems which persisted 3 days and longer are taken into account. Black dots are daily positions, of 24-h intervals, of the cyclones; the symbols for anticyclones are according to their intensity.

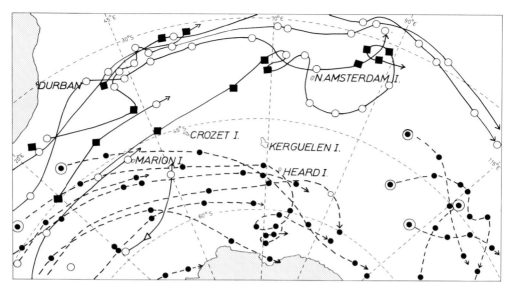

Fig. 57. Tracks of anticyclones and cyclones in July 1958. Systems which persisted 3 days or longer are taken into account. See Fig. 56 for the legend.

unusual. So also was the eastward passage of several strong anticyclones at 42−47°S south of South Africa in July 1958.

It will be evident from the examples that the anticyclones generally move eastward with a slight equatorward component. The northward component is appreciable in the west of the region, but close to Australia this tendency is reversed in both summer and winter. Very few centres ever occur in the region between 105°E and the Australian west coast, even in winter when cells approaching Australia either perform a leap forward on to the land or proceed via the sea south of the land. Due to the usual slow northward movement, it is along the northern fringe of the anticyclonic zone that the systems fade out. Retrograde motions are quite common north of the axis of highest frequency.

Anticyclone frequency

The distribution of anticyclones over the Southern Hemisphere during the IGY are given in TALJAARD (1967a, 1972b), but their seasonal frequencies for the six years 1959−1964 for the Indian Ocean south of 20°S are given in Fig. 58 and in Table V, based on daily hemispheric charts published by the SOUTH AFRICAN WEATHER BUREAU (1963−1970). Summer is defined as December to March, winter as June to September, and the intermediate months as April, May, October, and November.

Fig. 58 was derived by counting all the anticyclone centers which occurred in blocks measuring 5 degrees latitude by 10 degrees longitude during summer, winter, and the intermediate season months and then converting these numbers to those which would have occurred per season of four months in blocks with an area of 438,000 km². This is the area of the "unit block" between 42.5 and 47.5 degrees latitude taken over 10 degrees of longitude. Thus the conversion provides a measure giving the density of systems per season per equal area unit, so that the diagrams are exactly comparable to those of TALJAARD (1967a, 1972b).

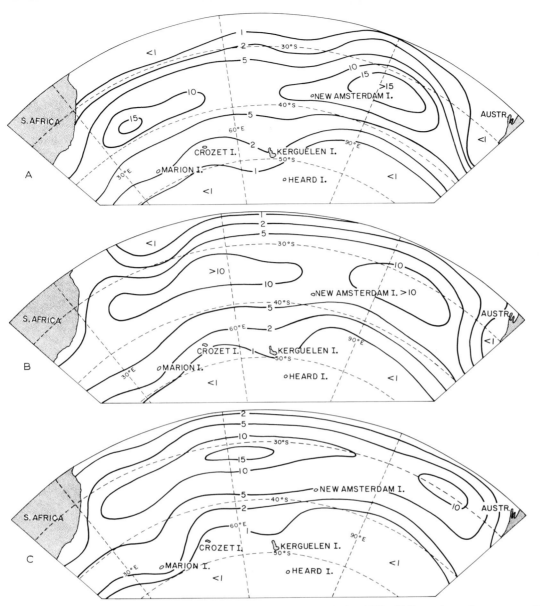

Fig. 58. Numbers of anticyclone centres which occurred during 1959—1964 per 4-month season per block of 438,000 km² in: A. summer; B. the intermediate season months; and C. winter.

The area where most anticyclones occur in summer is clearly centred around 35°S 95°E, but a secondary centre with almost equal frequency is found in the west around 37°S 40°E. The latter maximum was also evident during the IGY, but not as well marked. It might be expected that the mean pressure distribution in summer over the ocean should show two maxima, but this is not the case because Fig. 30 shows that during summer the single maximum is found at about 85°E. Therefore, it can be deduced that there must be a longitudinal variation of the central pressures of systems. It is logical to deduce that in the area around 37°S 40°E the anticyclones are less intense than farther east. However, it is well known that the area 95—100°E at 25—30°S is a "graveyard" of many systems. In the central part of the ocean, the systems probably advance rela-

TABLE V

NUMBERS OF ANTICYCLONES WITH CENTRAL PRESSURES IN 5-MBAR INTERVALS WHICH OCCURRED DURING 1959–1964 IN 5-DEGREE LATITUDE ZONES FROM 20° TO 115°E[*1]

Zone (°S)	1040 mbar	1035–40 mbar	1030–35 mbar	1025–30 mbar	1020–25 mbar	1015–20 mbar	1015 mbar	Total
(A)								
20–25	–	–	–	–	1(0.2)	–	1(0.2)	2(0.3)
25–30	–	–	–	9(1.2)	46(6.1)	21(2.8)	3(0.4)	79(10.5)
30–35	–	–	11(1.8)	137(19.1)	238(33.2)	59(8.3)	–	445(62.1)
35–40	–	1(0.2)	109(16.2)	271(40.3)	185(27.5)	24(3.6)	5(0.8)	595(88.4)
40–45	3(0.5)	1(0.2)	48(7.7)	95(15.2)	44(7.1)	9(1.5)	2(0.3)	202(32.3)
45–50	–	1(0.2)	5(0.8)	17(2.6)	16(2.8)	2(0.4)	2(0.4)	43(7.5)
50–55	–	–	–	4(0.8)	1(0.2)	8(1.6)	2(0.4)	15(2.9)
55–60	–	–	–	–	–	1(0.2)	–	1(0.2)
(B)								
20–25	–	–	–	–	7(0.9)	16(2.1)	2(0.3)	25(3.2)
25–30	–	–	7(1.0)	67(8.9)	110(14.7)	39(5.2)	2(0.3)	225(30.0)
30–35	–	10(1.4)	95(13.3)	293(40.8)	141(19.7)	24(3.4)	–	563(78.6)
35–40	1(0.2)	26(3.9)	143(21.3)	205(30.5)	71(10.6)	17(2.5)	2(0.3)	465(69.2)
40–45	–	3(0.5)	41(6.6)	168(26.9)	26(4.2)	13(2.1)	6(1.0)	257(41.4)
45–50	–	2(0.4)	5(0.9)	19(3.3)	14(2.5)	12(2.1)	4(0.7)	56(9.8)
50–55	–	–	1(0.2)	2(0.4)	2(0.4)	3(0.6)	3(0.6)	11(2.2)
55–60	–	–	–	1(0.2)	2(0.5)	2(0.5)	1(0.2)	6(1.3)
(C)								
20–25	–	–	–	–	22(2.8)	18(2.3)	–	40(5.1)
25–30	–	–	37(4.9)	153(20.4)	137(18.2)	14(1.9)	–	341(45.4)
30–35	3(0.4)	65(9.1)	237(33.1)	217(30.3)	46(6.4)	1(0.2)	1(0.2)	570(79.5)
35–40	15(2.3)	66(9.8)	129(19.2)	62(9.2)	17(2.5)	3(0.5)	1(0.2)	293(43.5)
40–45	3(0.5)	14(2.3)	28(4.5)	16(2.6)	18(2.9)	6(0.9)	1(0.2)	86(13.8)
45–50	3(0.5)	1(0.2)	7(1.2)	10(1.8)	5(0.9)	5(0.9)	2(0.4)	33(5.8)
50–55	–	–	–	–	1(0.2)	1(0.2)	2(0.4)	4(0.8)
55–60	–	–	–	–	1(0.2)	–	–	1(0.2)

[*1] Numbers in parentheses are the corresponding reduced frequencies of systems per season of 4 months per "unit zone", as defined in text. (A) summer (DJFM); (b) intermediate months (AMON); and (c) winter (JJAS). Data from SOUTH AFRICAN WEATHER BUREAU, 1963–1970.

tively rapidly, but no speed determinations have been made.

It is seen that the anticyclonic belt terminates sharply in the north, i.e., the tradewind belt, this boundary being farther north in winter than in summer and intermediate seasons. Anticyclones are also extremely sparse south of 50°S, or even south of 45°S in the east of the ocean. The observations at Marion and Kerguelen islands ensures that no anticyclones are missed when they occur at middle latitudes west of 70°E, but southwest of Australia some systems might exist unheeded before they affect the land or Macquarie Island.

It is striking to see the southward flexures of the frequency isopleths past South Africa and Australia.

In the winter, the anticyclonic frequencies increase in the central part of the ocean so that a clear maximum is found at 32°S 65°E. This is in accord with the westward shift of the center of average highest pressure. The axis of highest frequency of systems shifts 4.5 degrees equatorward from summer to winter. Again, there are evidently very few

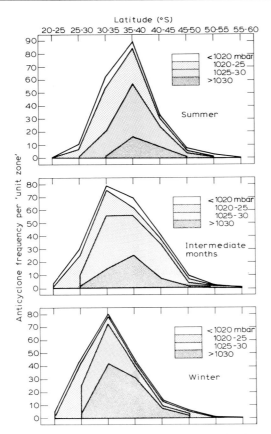

Fig. 59. Meridional profiles of numbers of anticyclones which occurred per 4 month season per "unit zone" of 5 degrees width (see text) in summer, the intermediate season months and winter over the sea between 20° and 115°E during 1959—1964.

anticyclones to be found in the 40—50°S zone eastward of Kerguelen, as well as poleward of 50°S in all areas.

During the intermediate months, there is only a slight break at 80°E in the almost uniform longitudinal spread of the anticyclones.

Meridional profiles of the frequency of anticyclones in four categories of intensity for the period 1959—1964 are shown in Fig. 59. These diagrams are based upon the statistics in Table V, giving the counts of anticyclones in 5-degree latitude zones from 20—25°S to 55—60°S over the sea from 20° to 115°E. The data in the table also distinguish systems with central pressures below 1015 mbar and above 1040 mbar and in 5-mbar intervals between 1015 and 1040 mbar. The values for the 5-degree zones, adjusted to one season of four months and to the area of the "unit zone" of 95 degrees longitude between 42.5° and 47.5°S, are included in the table. The diagrams of Fig. 59 are based on the latter values, i.e., they reflect the average frequencies over all longitudes for single seasons for areas of 4,161,000 km².

Several significant features are apparent from the diagrams and statistics. First, the relatively weak anticyclones with central pressures lower than 1025 mbar occur predominantly on the equatorward side of the anticyclonic belt. This is the zone where they fade out and where strong systems naturally cannot exist without giving rise to very strong winds. Second, strong systems, viz., those with central pressures over 1030 mbar, are predominantly found a few degrees poleward of the core of the anticyclonic belt. If the curves had been smoothed, the cores of highest frequency of the strong systems would have been 2 − 3 degrees poleward of systems with central pressures in the range 1025 − 1030 mbar. In winter the strong systems, viz., those with central pressure over 1035 mbar, mostly occur 5 degrees poleward of the core of all systems combined. Considering that the average pressure rises about 5 mbar from summer to winter along the high pressure axis, one should perhaps define strong anticyclones in winter as those with central pressures over 1035 mbar, because more than half of the systems found in the zones 30 − 35°S and 35 − 40°S have central pressures higher than 1030 mbar in this season.

The seasonal shift of the cores of highest anticyclonic frequency is clear, amounting to about 4 degrees northward from summer to winter. This, of course, is in accord with the seasonal shift of the axis of the highest pressure. It might be noted, however, that the high pressure axis is located about 3 degrees equatorward of the core of highest anticyclonic frequency, this being due to the fact that frontal or intercell troughs and occasional cyclones also occur in the anticyclonic belt, in which cases the highest pressures (in the cols) are found well to the north.

Cyclones

Cyclogenesis

Cyclogenesis may be defined as the first appearance of a closed low-pressure system which then persists for several days. This is equivalent to vortex formation which can be identified on satellite cloud photos. This definition excludes the intensification of already existing low-pressure systems which can, of course, also be taken as cyclogenesis. It also excludes the appearance of shallow heat or lee lows, which for the ocean area under investigation do not come into consideration.

Cyclogenesis in the Southern Hemisphere was studied by TALJAARD and VAN LOON (1962, 1963), TALJAARD (1964, 1965), and VAN LOON (1965) using the IGY maps, which were based on surface observations and upper-air soundings. The advent of polar orbiting meteorological satellites eight years after the IGY supplied visual confirmation of cyclonic vortices and of their first appearances. STRETEN (1968a,b, 1969) made the first efforts to study cyclogenesis and cyclones south of 40°S for a 3-month period (December 1966 to February 1967). This was followed by STRETEN and TROUP (1973) who studied satellite cloud mosaics for the summer and transitional season months for the period November 1966 to March 1969.

The studies of Taljaard and Van Loon are summarized in TALJAARD (1967a, 1972b). Table VI shows the numbers of systems which developed during 5 summer months, 7 winter months and 6 intermediate-season months of the IGY in the sector between 20°E and 115°E. As in the remaining southern oceans, most cyclogeneses occurred between latitudes 35° and 55°S, but in summer a considerable number occurred north of 35°S, several of the cases having been tropical cyclones which originated north of 25°S. According to the analysis, only eight cases of cyclogenesis occurred poleward of 55°S during the entire period. During winter and the intermediate seasons, most systems developed between 35° and 45°S, but during summer the peak occurred between 50° and

TABLE VI

(a) FREQUENCIES (ACTUAL NUMBERS OF CASES) OF CYCLOGENESIS IN 5-DEGREE LATITUDE ZONES OF THE SOUTHERN INDIAN OCEAN BETWEEN 20° AND 110°E, i.e., along one quarter of the hemispheric circumference, compared to: (b) one quarter of all cases of cyclogenesis in the same zones around the hemisphere during the IGY (From the IGY daily charts, SOUTH AFRICAN WEATHER BUREAU 1962–1966)

Latitude zone (°S)	Summer		Winter		Intermediate seasons		Year	
	(a)	(b)	(a)	(b)	(a)	(b)	(a)	(b)
15–20	4	3.5	0	1.2	2	1.5	6	6.2
20–25	7	4.7	2	5.0	1	3.5	10	13.2
25–30	3	3.2	3	8.5	5	6.5	11	18.2
30–35	7	6.7	5	11.0	6	8.2	18	25.9
35–40	5	7.5	12	11.2	8	9.0	25	27.6
40–45	5	9.7	9	14.0	12	11.7	26	35.4
45–50	7	9.7	6	11.0	8	10.7	21	31.4
50–55	9	6.0	7	8.0	12	8.0	28	22.0
55–60	2	5.0	2	3.7	1	4.7	5	13.4
60–65	1	1.7	1	4.2	1	3.0	3	8.9

55°S. However, conditions differed remarkably between the same months of the two successive years, pointing to the fact that a much longer period is required for reaching finality on seasonal and regional characteristics. For instance, the distributions of cyclogeneses in January 1958 (Fig. 56B) and July 1958 (Fig. 57) differ appreciably from those of some other summer and winter months of 1957/58.

Because the first appearances of vortices on satellite photos must be assumed as convincing evidence, the results of STRETEN and TROUP (1973) are given in more detail here. These authors considered the local bulging of an existing cold frontal cloud band with or without a change in the orientation of the band, and the inverted comma type cloud formations in the rear of cold fronts as evidence for cyclogenesis. The frequency distribution of such developments for twelve summer months is shown in Fig. 60A. The isopleths indicate the numbers of cases per unit block of 438,000 km^2. The highest frequency of over three systems per unit block per season was found at 45°S to the southwest of Australia. At this position and farther west, except in the extreme west of the ocean, the axis of most frequent cyclogenesis coincides with the Polar Front (TALJAARD, 1968). Between 45° and 70°E, cyclogenesis was less frequent than on either side. This was also found by TALJAARD (1972b), but this author found a band of higher frequency along 50°S, as did STRETEN (1968a).

Fig. 60B shows the axis of the zones of highest frequency of cyclogenesis for summer and the intermediate seasons. So-called "late vortex developments," included in Fig. 60A, are omitted for this diagram. In summer, the highest frequency is along 50°S with a branch from lower latitudes aligned NW–SE across the central and eastern parts of the ocean. In the intermediate seasons, the highest frequency of these early vortex developments is along 44–45°S throughout the ocean.

Streten and Troup used only the satellite photos obtained in the visible spectrum so that no study could be carried out for winter.

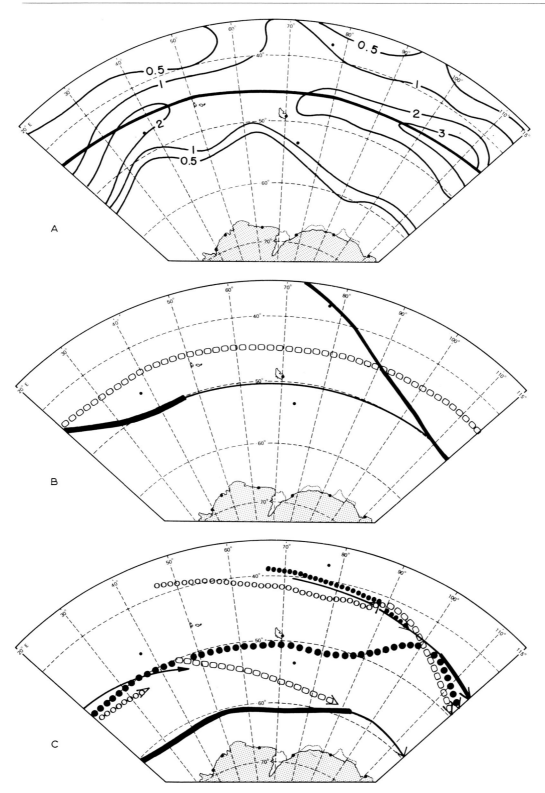

Fig. 60. A. Frequency of developing vortices per unit block (438,000 km²) per summer of 4 months. B. Axis of zones of highest frequency of "early development" of vortices for summer (open dots) and the main intermediate season months (full lines). C. Main tracks of vortices; heavy lines are "major" tracks; finer lines are "minor" tracks. November: open-dotted; January: full-dotted; March: full line. (From STRETEN and TROUP, 1973.)

Cyclone movement

The tracks followed by cyclones with a lifespan of more than 3 days during a mid-summer and a mid-winter month of the IGY are shown in Fig. 56B and Fig. 57. The tracks of systems during January 1958 were well substantiated by the presence of several whaling fleets and scientific expedition ships in the area. It is seen that most systems moved eastward with a southward component, but several systems first moved ENE and then ESE, which is not characteristic for summer months and for the Indian Ocean. The systems slowed down within 5–10 degrees from the circumpolar trough at about 65°S and quite often they performed retrograde motions in the vicinity of the coast of Antarctica.

The tracks for July 1958 at first seem questionable, for they crowd together over the western and central parts of the ocean south of 42°S and terminate near the Antarctic coast between 70° and 110°E, where many short-lived systems were also found. Cyclogenesis occurred several times southwest of Australia and these systems moved off to the SE, but none of the western systems joined these "Australian" systems. The analysis between Kerguelen, Australia, and Antarctica in winter might be questioned. However, concerning the absence of cyclones north of Marion Island and Kerguelen in this particular month there can be no question. The position in July 1957 was almost identical (TALJAARD and VAN LOON, 1962) and broadly similar conditions prevailed during most other IGY winter months. Conversely, it cannot be stated that the events of January 1958 were perfectly representative of what happened with cyclones during the remaining four IGY summer months. It should be added that during summer tropical cyclones develop in the zone 10–20°S, most of them moving westward at first and subsequently SW, S, and SE in the zone 20–35°S. Therefore, the tracks of five to ten systems in a normal summer will be found crossing the high-pressure belt from tropical to subtropical latitudes (e.g., three systems in January 1958). But there are no obvious preferred major paths for such systems, at least nothing comparable to what happens in the western Atlantic and central Pacific Oceans.

The situation as found by STRETEN and TROUP (1973) for November, January, and March, based on three months' data for each track, is shown in Fig. 60C. Except for a northward deflection of the January main track at 90–100°E, most systems advance eastward along 50°S in this month. In November, two tracks can be identified—one from subtropical latitudes at 50°E and curving to SE west of Australia and the other trending ESE from the vicinity of Marion Island. In March, the main track is very far to the south along 60°S, while a secondary track coincides more or less with that of November. However, these major tracks cannot be considered as well established as, or identical with, those found during the IGY.

Comparing with the track diagrams produced by various authors (e.g., DUBENTSOV and DAVIDOVA, 1964; BAKAYEV, 1966), one is tempted to say that the topic of main cyclone arteries in the Southern Hemisphere has not been resolved to general satisfaction, apart from the two major tracks in the Pacific and Atlantic oceans.

Cyclone frequencies

TALJAARD (1972b) cites more than a dozen publications dealing with the distribution and frequencies of cyclones in various sectors and zones of the Southern Hemisphere. For the present purpose, it will suffice to summarize only some of the results and conclusions of TALJAARD (1967a, 1972b) for the IGY, and of STRETEN (1969) and STRETEN and TROUP (1973) who used synoptic maps derived from conventional data and satellite cloud mosaics for the period November 1966 to March 1969.

Fig. 61 shows the distribution of cyclone frequencies in the Indian Ocean south of 35°S during summer as derived by TALJAARD (1967a) and by STRETEN (1969). The isopleths indicate the numbers of systems per unit block of 438,000 km² per 4-month season. Summer is taken as December—March, but Streten used only December—February, the data of which were then converted to a 4-month season. The results correspond well from 60°S to the Antarctic coast but not so along 50°S. In the Antarctic zone, the three centres of maximum cyclone frequency at 20—40°E, 60—75°E, and 95° to beyond 115°E are in agreement with a clear break north of Enderby Land (55°E), but during the IGY there were no separate maxima at 30—40°E and 60—80°E along 50°S. In the latter areas, Streten actually found twice as many cyclones as during the IGY. Since the conventional analysis here is controlled by the island observations, this difference is indicative of marked circulation differences in the two periods. In fact, the charts of PRANTNER (1962), based on six years of pre- and post-IGY conventionally analysed maps, show closer agreement with the IGY distribution than with those of the 1966—69 period.

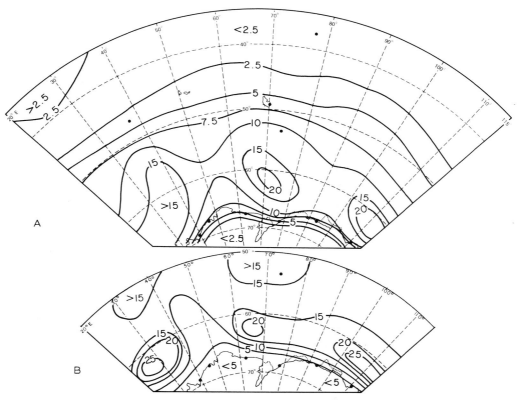

Fig. 61. A. Frequency of cyclone centres per 4 month summer per unit block (438,000 km²) during the IGY (after TALJAARD, 1967a). B. Same as A but only for the region south of 50°S and based mainly on satellite photos for three summers 1966—67, 1967—68, and 1968—69 (after STRETEN, 1969).

North of 50°S, the IGY maps indicate a gradual decrease of cyclone frequencies toward 35°S but equatorward of this latitude there is again an increase (see fig. 7 of TALJAARD, 1967a; VAN LOON, 1966, fig. 8) to more than five systems per unit block in the western half of the ocean. This must be due to the tropical cyclones and weaker lows which frequent the region north of the subtropical ridge.

A point worthy of note is that during all seasons of the IGY there was no clear evidence for a predominance of cyclones along a belt stretching from Malagasy or South Africa toward SE, a belief which is widely held, beginning with PALMER's (1942) postulate of a branch of the Polar Front to be found in the western subtropical Indian Ocean. The vortex track diagram of Streten and Troup shown in Fig. 60C does, however, indicate weak tracks at latitude 40°S starting in the middle of the ocean.

Speeds of cyclones

MEINARDUS (1928, 1929), VAN LOON (1967b), and STRETEN and TROUP (1973) discussed the speeds of cyclones but they either discussed the speed distributions in various latitude zones for the entire hemisphere or used sectors which do not coincide with the Indian Ocean. Therefore, no statistics for this ocean alone are available. Suffice it to say that average speeds are highest (24 – 29 kt.) in the zone 40 – 60°S, that they are probably of the order of 30 kt. along 45 – 50°S and that in the zone 30 – 40°S they are higher by about 5 kt. in winter than in summer. This increase is in response to the northward broadening of the west wind belt in winter with consequent stronger westerlies in the 30 – 40°S zone.

Meinardus and Van Loon found that in the Atlantic – Indian Ocean sector the cyclone speeds are somewhat higher than in the Pacific sector. The probability exists that cyclone speeds are on the average greatest in the western Indian Ocean and eastern Atlantic Ocean sectors where the Polar Front jetstream is well defined in both summer and winter, such that in middle latitudes the average wind speed at 500 mbar is appreciably higher here than over the remainder of the hemisphere. Thus, GILBERT et al. (1953) found that the average speed of fronts between Marion and Kerguelen islands is 38 kt. and the associated cyclones, therefore, probably advance at speeds of over 30 kt. Attention has been drawn to the fact that subtropical anticyclones are confined to a narrower belt of latitude in the Indian Ocean than in the other oceans, giving the impression that in middle latitudes conditions are more favoured here for the unimpeded eastward and poleward movement of cyclones.

Fronts

VAN LOON (1965) determined the frequency distribution of fronts for the hemisphere for three summer months (January – March, 1958) and three winter months (July – September, 1957) using the IGY maps. The distributions for the Indian Ocean are shown in Fig. 62. The periods used, being short, might not be very well representative of long-term conditions, but they are qualitatively sound. He used both cold and warm fronts analysed according to the principles described by TALJAARD et al. (1961). Modern maps based on cloud patterns alone would probably contain fewer fronts because warm fronts, whether they actually exist as sloping hyper-baroclinic layers or not, are only

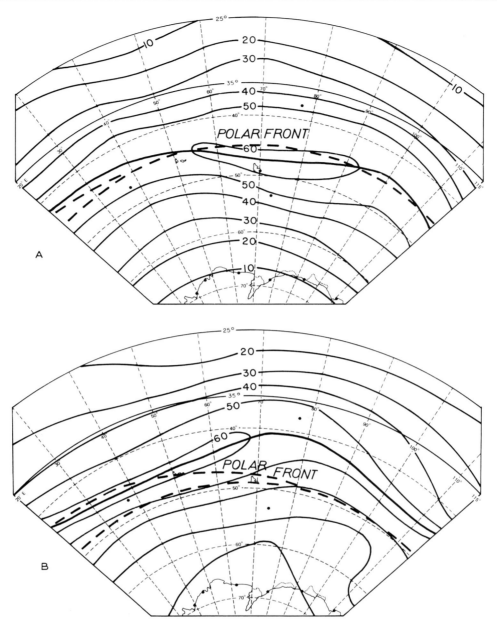

Fig. 62. Frequency of surface fronts per block of 400,000 km²: A. during January, February, and March 1958; B. during July, August, and September 1957. (From VAN LOON, 1965.)

seldom clearly displayed by cloud patterns.

Fronts are just about equally frequent in summer and winter, with the maxima reaching 60 systems per 100 days in blocks measuring 400,000 km² along parts of the zone 40−45°S. Frequencies along 25°S and along the Antarctic coast are about twice as high in winter as in summer. The slight southward flexure of the core with highest frequency in summer and northward flexure thereof in winter are probably due to the short periods of data used. The mean positions of the Polar Front in summer and winter, based on the positions of the maximum 1000−500 mbar and 500−300 mbar thickness gradients (TALJAARD et al., 1968) are also indicated in Fig. 62 and do not show these flexures but

where c = factor varying with latitude (Table X).

$$I_1 = 4s\sigma T_a{}^3(T_s - T_a)$$

where T_s = sea temperature.

The rate of effective longwave radiation then becomes:

$$Q_b = s\sigma T_a{}^4(0.39 - 0.05\sqrt{e_a})\,(1 - cN^2) + 4s\sigma T_a{}^3(T_s - T_a)$$

The difference between the effective shortwave and the effective longwave radiation is called the radiation balance:

$$R = Q_r - Q_b$$

A major component of the heat balance is the latent heat of evaporation supplied by the sea to the atmosphere. This is given by the expression:

$$Q_e = L\rho k_1(e_s - e_a)u$$

where L = latent heat of evaporation = $2.48 \cdot 10^6$ J kg^{-1}; ρ = density of sea water = 10^3 kg m^{-3}; k_1 = evaporation factor = $0.1044 \cdot 10^{-8}$ (after Sverdrup, 1951; adapted for S.I. units); e_s = saturation vapour pressure at sea temperature (mbar); u = scalar mean wind speed.

Sensible heat, being the heat conducted from the sea to the air or vice versa, and then transported by turbulence into the two media, is given by:

$$Q_c = C_p k_2(T_s - T_a)u$$

where C_p = specific heat of air at constant pressure = $1.0 \cdot 10^3$ J kg^{-1} K^{-1}; k_2 = constant = $2.5 \cdot 10^{-3}$ kg m^{-3} (Budyko, 1963).

Results and discussion

The results of the computations of the various components of the heat balance and of the heat balance itself for January and July are depicted in Figs. 64–75.

The rate of effective shortwave radiation absorbed by the uppermost layers of the sea Q_r is shown in Fig. 64 for January and in Fig. 65 for July. In January, Q_r decreases from 230–240 W m^{-2} along 35°S to just below 160 W m^{-2} along 56–59°C, and then increases to about 190 W m^{-2} close to the antarctic coast. The zone with the lowest values coincides with that of the maximum cloudiness. The cloudiness particularly determines the total albedo at this time of the year when the pack ice is confined to a narrow coastal zone south of 65°S. Along 35–40°S, Q_r is about 5% lower near Australia than off South Africa.

In July, Q_r decreases from 90 W m^{-2} along 35°S to near zero along the Antarctic coast.

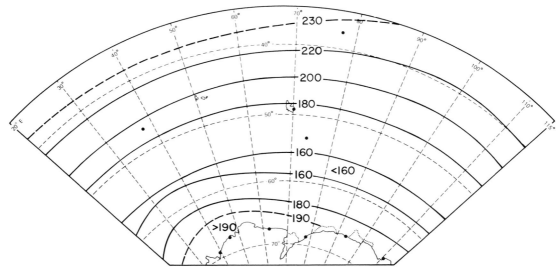

Fig. 64. Effective shortwave radiation in January (W m^{-2}).

The decrease is rapid north of 50°S, but south of this latitude the high cloudiness and the pack ice raise the albedo to such an extent as to allow very little of the slanting rays of the sun to be absorbed at the surface.

The effective longwave radiation Q_b from the sea surface in January (Fig. 66) amounts to about 60 W m^{-2} along 35°S, with a peak value of 67 W m^{-2} close to the Australian coast. At 20°E 40°S the warm Agulhas Current also produces a peak value of 67 W m^{-2}. South of 40°S, the isopleths of Q_b are sinuous, a zone of values below 40 W m^{-2} being located between 50° and 55°S over the western half of the ocean and between 55° and 60—65°S over the eastern half of the ocean. Values of more than 50 W m^{-2} are evidently to be found in the extreme southwestern part of the area.

In July (Fig. 67), Q_b shows values of 70—77 W m^{-2} along 35—40°S over the western half of the ocean and near Australia. The 55 W m^{-2} isopleth is sinuous, indicating a

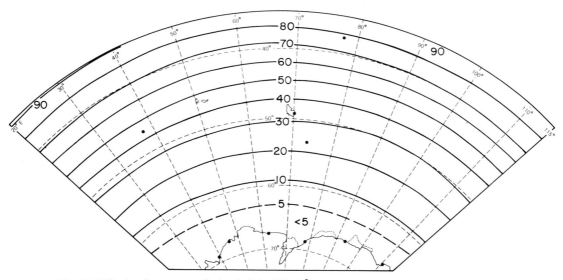

Fig. 65. Effective shortwave radiation in July (W m^{-2}).

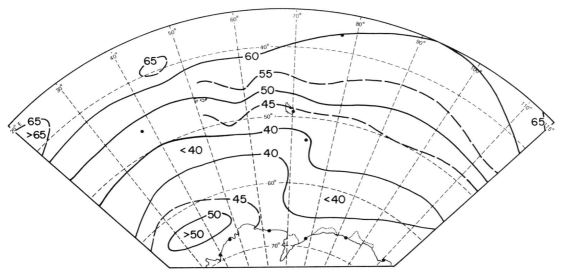

Fig. 66. Effective longwave radiation in January (W m^{-2}).

weak minimum in the Marion and Crozet Islands region. A zone with Q_b below 45 W m^{-2} straddles 60°S, but values increase again to over 55 W m^{-2} north of the Prydz Bay indentation.

The radiation balance R in January (Fig. 68) shows values 170 – 180 W m^{-2} along 35°S, the higher values being found in the west and the center. Southward R decreases to less than 125 W m^{-2} along roughly the zone 53 – 59°S, but then increases again to just over 150 W m^{-2} near the antarctic coast. In July (Fig. 69), R is weakly positive north of 40°S, but negative southward to reach a value of –60 W m^{-2} in the Prydz Bay area. However, the meridional gradient is variable, being very small along 50 – 60°S west of 70°E.

The heat loss due to evaporation Q_e shows a maximum of more than 130 W m^{-2} over the Agulhas Current in January (Fig. 70) with a belt of relatively high values stretching

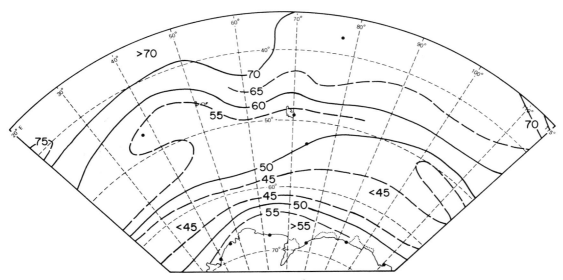

Fig. 67. Effective longwave radiation in July (W m^{-2}).

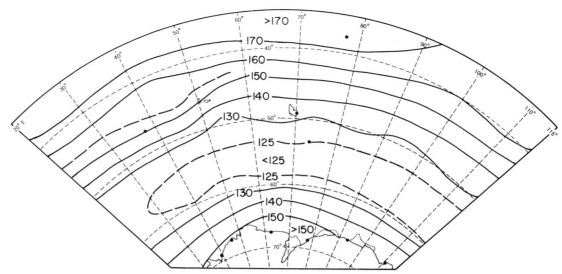

Fig. 68. Radiation balance in January (W m^{-2}).

eastward to 75°E. Except close to the coast, Q_e is notably low, viz., lower than 70 W m^{-2}, in the subtropical zone southwest of Australia. The anomalously cold sea in the Marion/Crozet area is reflected in the minimum of lower than 40 W m^{-2}. South of 50°S, values decrease poleward very slowly to become lower than 20 W m^{-2} near the Antarctic coast. In July (Fig. 71), Q_e is mostly 10−20% higher than in January in the zone 35−50°S. South of 50°S, the situation is the opposite with values dwindling to near zero close to the Antarctic coast.

The sensible heat loss of the sea to the air Q_c presents an apparently confused pattern of small values in January (Fig. 72). The sea loses or gains heat according to whether it is warmer or colder than the air and these temperature differences are mostly small and difficult to determine from small samples of observations. An area where the sea is decidedly warmer than the air is over the Agulhas Current where the rate of sensible

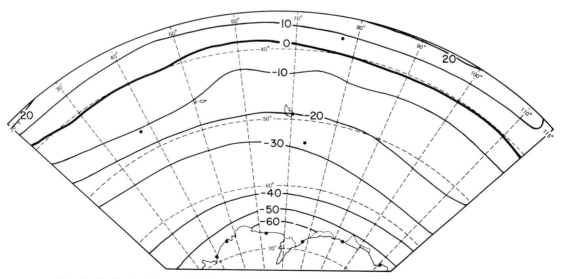

Fig. 69. Radiation balance in July (W m^{-2}).

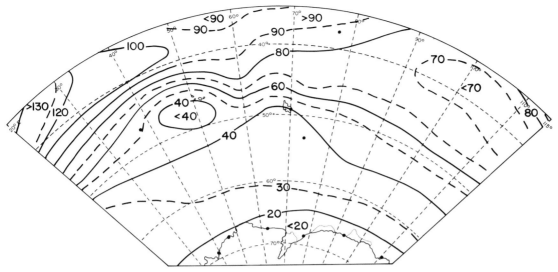

Fig. 70. Heat loss due to evaporation in January (W m^{-2}).

heat loss is shown to be over 20 W m^{-2} in January and over 50 W m^{-2} in July (Fig. 73). An area where the air is warmer than the sea throughout the year is located along 40 – 45°S between 20°E and the Crozet Islands. With a narrow interruption at 55 – 60°E, this belt of heat gain stretches eastward to reach a secondary maximum of over 20 W m^{-2} in January at about the position of the Australian Subantarctic Front. The Antarctic Ocean is clearly also an area of heat gain in January.

In July, heat loss prevails everywhere except in the Marion/Crozet area already mentioned and apparently also over the Australian Subantarctic Front. The debatable feature of Fig. 73 is the indicated zone of strong heat loss immediately north of the pack-ice edge at 55 – 60°S. Although the sea close to the edge of the ice will undoubtedly lose heat quite rapidly to the very cold air masses which flow northward in the rear of low-pressure systems, the loss might be restricted to a narrower zone than indicated.

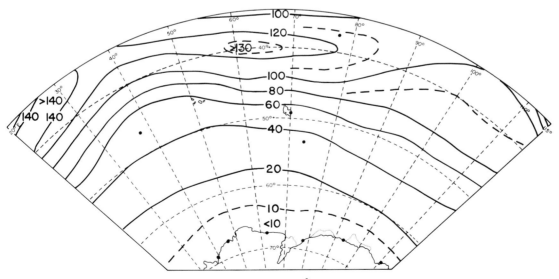

Fig. 71. Heat loss due to evaporation in July (W m^{-2}).

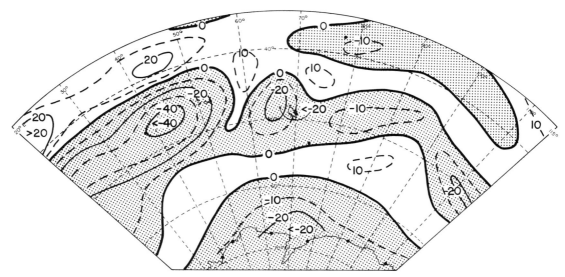

Fig. 72. Loss of sensible heat, sea to air, in January (W m^{-2}). Stippled areas: regions of sensible heat gain; unstippled areas: loss of heat.

The question of the sea-minus-air-temperature difference in the zone between the pack ice and the Antarctic Convergence is discussed on p. 522. Over the pack ice itself, the rate of heat loss is probably well below 10 W m^{-2} due to the very low conductivity of the ice.

The heat balance in January and July is given in Figs. 74 and 75. These diagrams show great resemblance to Figs. 72 and 73 because the isopleths of the radiation balance are mainly zonal upon which the patterns of the latent and sensible heat fluxes are then superimposed. In January, the tortuous isopleths between 40° and 55°S reflect the wave pattern of the sea and air isotherms, with maxima of heat intake between Marion and Crozet Islands and near Kerguelen, areas where the sea is anomalously cold. The heat gain is relatively small in the zone centred along 55°S, where cloudiness and the albedo are high. The fairly high heat intake in the Antarctic Ocean is also notable.

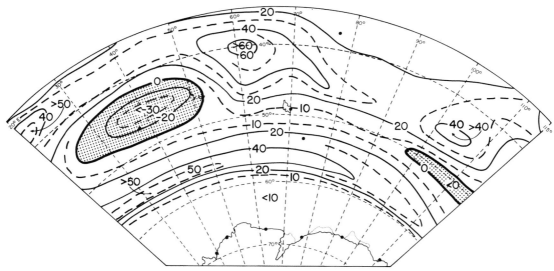

Fig. 73. Loss of sensible heat, sea to air, in July (W m^{-2}). See also Fig. 72.

586

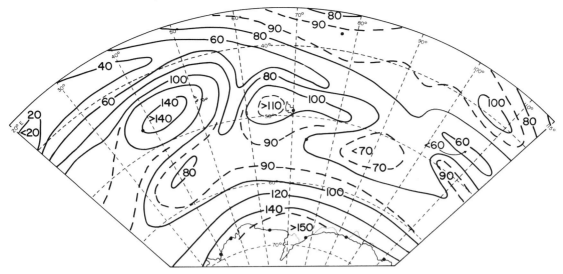

Fig. 74. Heat balance in January (W m^{-2}).

In July, a centre of pronounced heat loss is located at 40°S 60°E, whereas between Marion and Crozet there is very little heat loss. The heat loss of less than 60 W m^{-2} at 53°S in the extreme east of the area is surprisingly low.

The strong heat loss over the Agulhas Current at 20°E 40°S diminishes rapidly eastward, as can be expected because the uptake of sensible heat and the rapid loss due to evaporation into the cold air normally flowing from west to east in that area must diminish after a couple of hundred kilometres. The zone of relatively strong heat loss straddling 55°S is a consequence mainly of the strong uptake of sensible heat north of the pack ice boundary as discussed above. A zone of relatively weak heat loss evidently exists along 60–62°S, whereas the heat loss increases gradually farther southward to the Antarctic coast.

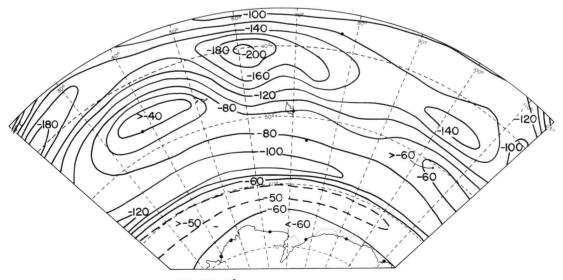

Fig. 75. Heat balance in July (W m^{-2}).

Acknowledgements

Mrs. M. Luus of the South African Weather Bureau was responsible for drafting most of the diagrams of this chapter, and Mr. T. I. Potgieter of the same Bureau performed the programming and calculations for the heat budget discussed in the last section.

References

BAKAYEV, V. G. (Editor), 1966. *Atlas Antarktiki*. Moscow, Glavnoe Upravlenie Goedezii i Kartografii MG SSSR, xxiii plus 225 pp.

BUCHAN, A., 1869. The mean pressure of the atmosphere and the prevailing winds over the globe for the months and for the year. *Trans. R. Soc.*, Edinburgh, 25: 575–637.

BUDYKO, M. I., 1956. *Teplovogo Balansa Zemnogo Shara (The Heat Balance of the Earth's Surface)*. Gidromet., Leningrad, Izdatel'stvo, 259 pp. (English translation by N. A. Stepanova, U.S. Department of Commerce, Washington, D.C., 1958.)

BUDYKO, M. I., 1963. *Atlas Teplovogo Balansa Zemnogo Shara (Atlas of the Heat Balance of the Earth's Surface)*. Main Geophysical Observatory, Moscow, V plus 69 charts.

CARLETON, A. M., 1979. A synoptic climatology of satellite-observed ultratropical cyclone activity for the Southern Hemisphere winter. *Arch. Meteorol. Geophys. Bioklimatol.*, B, 27: 265–279.

CRUTCHER, H. L., JENNE, R. L., TALJAARD, J. J. and VAN LOON, H., 1971. *Climate of the Upper Air: Southern Hemisphere, Vol. IV. Selected Meridional Cross Sections of Temperature, Dew Point and Height*. NAVAER 5-1C-58, 62 pp.

DEACON, G. E. R., 1937. The hydrology of the Southern Ocean. *Discovery Rep.*, 15: 124 pp.

DEACON, G. E. R., 1945. Water circulation and surface boundaries in the oceans. *Q. J. R. Meteorol. Soc.*, 71: 11–25.

DEACON, G. E. R., 1964. The Southern Ocean. In: R. PRIESTLEY, R. J. ADIE and G. DE Q. ROBIN (Editors), *Antarctic Research*. Butterworths, London, pp. 292–307.

DEFANT, A., 1961. *Physical Oceanography*. Pergamon, Oxford, 745 pp.

DIETRICH, G., 1963. *General Oceanography*. Interscience, New York, N.Y., XV plus 588 pp.

DUBENTSOV, V. R. and DAVIDOVA, N. G., 1964. *Atlas Aeroklimaticheskikh Kart i Kart Putei i Povtoriaemosti Tsiklonov i Antisiklonov Yuzhnogo Polushariia*. (Atlas of aeroclimatological maps and maps of the paths and frequencies of cyclones and anticyclones in the Southern Hemisphere.) Central Forecasting Institute, Moscow, 102 pp.

GILBERT, J., LAMBERT, J. and GOYER, G., 1953. Quelques caractéristiques des perturbations observées à Kerguelen en 1951. *Météorologie*, 4th Series, 32: 245–252.

GLOERSEN, P., WILHEIT, T. T., CHANG, T. C., NORDBERG, W. and CAMPBELL, W. J., 1974. Microwave maps of the polar ice on the earth. *Bull. Am. Meteorol. Soc.*, 55: 1442–1448.

GREAT BRITAIN, 1948. *Antarctic Pilot*. 2nd ed., Hydrographic Office, London.

GRUNDLINGH, M. L., 1977. Drift observations from NIMBUS 6 satellite-tracked buoys in the southwestern Indian Ocean. *Deep-Sea Res.*, 24: 903–913.

GRUNDLINGH, M. L., 1978. Drift of a satellite-tracked buoy in the southern Agulhas Current and Agulhas Return Current. *Deep-Sea Res.*, 25: 1209–1224.

HANN, J., 1901. *Lehrbuch der Meteorologie*. Tauchnitz, Leipzig, 805 pp.

HARRIS, T. F. W. and STAVROPOULOS, C. C., 1978. Satellite-tracked drifters between Africa and Antarctica. *Bull. Am. Meteorol. Soc.*, 59: 51–59.

JACOBS, W. C., 1969. The seasonal apportionment of precipitation over the ocean. In: *Yearbook of the Association of Pacific Coast Geographers*. Oregon State University Press, Corvallis, pp. 63–78.

JENNE, R. L., CRUTCHER, H. L., VAN LOON, H. and TALJAARD, J. J., 1971. *Climate of the Upper Air: Southern Hemisphere, Vol. III. Vector Mean Geostrophic Winds*. NCAR TN/STR-58 and NAVAER 50-1C-57, National Center for Atmospheric Research, Boulder, Colo., 68 pp.

KONINKLIJK NEDERLANDS METEOROLOGISCH INSTITUUT, 1952. *Indian Ocean Oceanographic and Meteorological Data*. No. 135, De Bilt.

LOEWE, F., 1957. Precipitation and evaporation in the Antarctic. In: M. P. VAN ROOY (Editor), *Meteorology of the Antarctic*. Weather Bureau, Pretoria, 71–89 plus 5 charts.

LYSGAARD, L., 1969. Klimaet paa Faerøerne. *Dan. Meteorol. Inst., Medd.*, 20: 29 pp.

MACKINTOSH, N. A., 1946. The Antarctic Convergence and the distribution of the surface temperatures in antarctic waters. *Discovery Rep.*, 23: 172−212.

MACKINTOSH, N. A. and HERDMAN, H. F. P., 1940. Distribution of the pack ice in the Southern Ocean. *Discovery Rep.*, 19: 285−296.

McDONALD, W. F., 1938. *Atlas of Climatic Charts of the Oceans*. U.S. Weather Bureau No. 1247, Government Printing Office, Washington, D.C., 130 charts.

MEINARDUS, W., 1928. *Deutsche Südpolar-Expedition 1901−1903. Meteorologie, Band III*. Walter de Gruyter, Berlin, pp. 133−307.

MEINARDUS, W., 1929. Die Luftdruckverhältnisse und ihre Wandlungen südlich von 30° südlichen Breite. *Meteorol. Z.*, 46: 41−49; 89−96.

MEINARDUS, W., 1938. Klimakunde der Antarktis. In: W. KÖPPEN and R. GEIGER (Editors), *Handbuch der Klimatologie*, Band IV. Berlin, Bornträger, 133 pp.

MEINARDUS, W., 1940. Die interdiurne Veränderlichkeit der Temperatur und verwandte Erscheinungen auf der Südhalbkugel. *Meteorol. Z.*, 57: 165−176; 219−233.

MEINARDUS, W. and MECKING, L., 1911. *Tägliche synoptische Wetterkarten der höheren südlichen Breiten, von Oktober 1901 bis März 1904*. Deutsche Südpolar Expedition 1901−1903. Meteorologischer Atlas, Berlin.

MELDRUM, C., 1861. *Synoptic Weather Charts for the Indian Ocean for the Months January, February and March, 1861*. Edinburgh.

METEOROLOGICAL OFFICE, GREAT BRITAIN, 1949. *Monthly Meteorological Charts of the Indian Ocean*. M.O. 519, London, 122 pp.

MEYER, H. H. F., 1923. Die Oberflächenströmungen des Atlantischen Ozeans im Februar. *Veröff. Inst. Meereskunde*, Reihe A, Heft XI, Berlin, 35 pp.

MÖLLER, F., 1951. Vierteljahrskarten des Niederschlags für die ganze Erde. *Petermanns Geograph. Mitt.*, 95: 1−7.

NEUMANN, G., 1968. *Ocean Currents*, Elsevier, Amsterdam, 352 pp.

PALMER, C. W., 1942. Synoptic analysis over the southern oceans. *N. Z. Meteorol. Office, Prof. Note*, 1, 38 pp.

PRANTNER, G. D., 1962. *Cyclones and Anticyclones south of 50°S*. U.S. Navy Weather Research Facility, Norfolk, Va., 16-0962-067, 64 pp.

RAMAGE, C. S., MILLER, F. R. and JEFFERIES, C., 1972. *Meteorological Atlas of the International Indian Ocean Expedition, Vol. 1. The Surface Climate of 1963 and 1964*. National Science Foundation, Washington, D.C., 144 charts.

RAMAGE, C. S. and RAMAN, C. V. R., 1972. *Meteorological Atlas of the International Indian Ocean Expedition, Vol. 2. Upper Air*. National Science Foundation, Washington, D.C., 121 charts.

RAO, M. S. V., ABBOT, W. V. and THEON, J. S., 1976. *Satellite Derived Global Oceanic Rainfall Atlas (1973) and 1974)*. NASA SP-410, 31 pp plus maps.

SCHMITT, W., 1957. Synoptic meteorology of the antarctic. In: M. P. VAN ROOY (Editor). *Meteorology of the Antarctic*. Weather Bureau, Pretoria, pp. 209−231.

*SCHOTT, G., 1935. *Geographie des Indischen und Stillen Ozeans*. C. Boysen, Hamburg, XX plus 413 pp.

SCHWERDTFEGER, W., 1970. The climate of the Antarctic. In: S. ORVIG (Editor), *Climates of the Polar Regions. World Survey of Climatology, Vol. 14*. Elsevier, Amsterdam, pp. 253−355.

SHAW, N., 1927. *Manual of Meteorology*, Cambridge University Press, Cambridge, England, 472 pp.

SOUTH AFRICAN WEATHER BUREAU, 1962−1966. *International Geophysical Year World Weather Maps, Part III. Southern Hemisphere, South of 20°S*. Daily sea-level and 500-mb maps. Pretoria.

SOUTH AFRICAN WEATHER BUREAU, 1963−1970. *Notos*. Pretoria, 12−19.

STRETEN, N. A., 1968a. Some aspects of high latitude southern hemisphere summer circulation as viewed by ESSA. *J. Appl. Meteorol.*, 7: 324−332.

STRETEN, N. A., 1968b. A note on multiple image photo-mosaics for the Southern Hemisphere. *Aust. Meteorol. Mag.*, 16: 127−136.

STRETEN, N. A., 1969. A note on the frequency of closed circulations between 50°S and 70°S in summer. *Aust. Meteorol. Mag.*, 17: 228−234.

STRETEN, N. A. and TROUP, A. J., 1973. A synoptic climatology of satellite-observed cloud vortices over the Southern Hemisphere. *Q. J. R. Meteorol. Soc.*, 99: 56−72.

SVERDRUP, H. U., 1942. *Oceanography for Meteorologists*. Prentice Hall, Englewood Cliffs, N. J., 246 pp.

SVERDRUP, H. U., JOHNSON, M. W. and FLEMING, R. H., 1954. *The Oceans, Their Physics, Chemistry and General Biology*. Prentice Hall, Englewood Cliffs, N. J., 2nd ed., 1087 pp.

---

* Useful reference beyond those mentioned in text.

TALJAARD, J. J., 1957. Geographical and hydrological features of the antarctic. In: M. P. VAN ROOY (Editor), *Meteorology of the Antarctic*. Weather Bureau, Pretoria, pp. 1—16.

TALJAARD, J. J., 1964. Cyclogenesis, cyclones and anticyclones in the Southern Hemisphere during autumn, 1958. *Notos*, 13: 31—36.

TALJAARD, J. J., 1965. Cyclogenesis, cyclones and anticyclones in the Southern Hemisphere during the period June to December 1958. *Notos*, 14: 73—84.

TALJAARD, J. J., 1966. Standard deviation of daily sea-level pressure and 500 mb height over the Southern Hemisphere during the IGY. *Notos*, 15: 29—36.

TALJAARD, J. J., 1967a. Development, distribution and movement of cyclones and anticyclones in the Southern Hemisphere during the IGY. *J. Appl. Meteorol.*, 6: 973—987.

TALJAARD, J. J., 1967b. The behaviour of 1000—500 mb thickness anomalies in the Southern Hemisphere. *Notos*, 16: 3—20.

TALJAARD, J. J., 1968. Climatic frontal zones of the Southern Hemisphere. *Notos*, 17: 23—34.

TALJAARD, J. J., 1972a. Physical features of the Southern Hemisphere. In: C. W. NEWTON (Editor), *Meteorology of the Southern Hemisphere. Am. Meteorol. Soc. Monograph*, 13: 1—8.

TALJAARD, J. J., 1972b. Synoptic meteorology of the Southern Hemisphere. In: C. W. NEWTON (Editor), *Meteorology of the Southern Hemisphere. Am. Meteorol. Soc. Monograph*, 13: 139—214.

TALJAARD, J. J. and VAN LOON, H., 1962. Cyclogenesis, cyclones and anticyclones in the Southern Hemisphere during the winter and spring of 1957. *Notos*, 11: 3—20.

TALJAARD, J. J. and VAN LOON, H., 1963. Cyclogenesis, cyclones and anticyclones in the Southern Hemisphere during summer 1957—1958. *Notos*, 12: 37—50.

TALJAARD, J. J., SCHMITT, W. and VAN LOON, H., 1961. Frontal analysis with application to the Southern Hemisphere. *Notos*, 10: 25—58.

TALJAARD, J. J., JENNE, R. L., VAN LOON, H. and CRUTCHER, H. L., 1968. Seasonal range, anomalies and other aspects of sea-level pressure, isobaric height, temperature and dew point at selected levels in the Southern Hemisphere. *Notos*, 17: 63—140.

TALJAARD, J. J., VAN LOON, H., CRUTCHER, H. L. and JENNE, R. L., 1969. *Climate of the Upper Air: Southern Hemisphere, Vol. I. Temperatures, Dew Points and Heights at Selected Pressure Levels*. NAVAER 50-1C-55, Washington, D.C., 135 pp.

TEISSERENC DE BORT, L., 1893. *Report on the Present State of Our Knowledge Respecting the General Circulation of the Atmosphere*. E. Stanford, London, 20 pp. plus maps.

UNITED STATES NAVY HYDROGRAPHIC OFFICE, 1957a. *Oceanographic Atlas of the Polar Seas, Part 1. Antarctic.* H. O. Publ. No. 705, Washington, D.C., 70 pp.

UNITED STATES NAVY HYDROGRAPHIC OFFICE, 1957b. *Marine Climatic Atlas of the World, Vol. III. Indian Ocean.* NAVAER 50-1C-530, Washington, D.C., 267 pp. (revised 1976).

UNITED STATES NAVY HYDROGRAPHIC OFFICE, 1965. *Marine Climatic Atlas of the World, Vol. VII. Antarctic.* NAVWEPS 50-1C-50, Washington, D.C.

VAN LOON, H., 1956. Blocking action in the Southern Hemisphere. *Notos*, 5: 171—177.

VAN LOON, H., 1964. Mid-season average zonal winds at sea level and at 500 mb south of 25°S and a brief comparison with the Northern Hemisphere. *J. Appl. Meteorol.*, 5: 554—563.

VAN LOON, H., 1965. A climatological study of the atmospheric circulation in the Southern Hemisphere during the IGY, Part I. *J. Appl. Meteorol.*, 4: 479—491.

VAN LOON, H., 1966. On the annual temperature range over the southern oceans. *Geograph. Rev.*, 56: 497—515.

VAN LOON, H., 1967a. The half-yearly oscillations in middle and high southern latitudes and the coreless winter. *J. Atmos. Sci.*, 24: 472—486.

VAN LOON, H., 1967b. A climatological study of the atmospheric circulation in the Southern Hemisphere during the IGY, Part II. *J. Appl. Meteorol.*, 6: 803—815.

VAN LOON, H., 1971. A half-yearly variation of the circumpolar drift in the Southern Hemisphere. *Tellus*, 23: 511—516.

VAN LOON, H., 1972a. Temperature in the Southern Hemisphere. In: C. W. NEWTON (Editor), *Meteorology of the Southern Hemisphere. Am. Meteorol. Soc. Monograph*, 13: 25—58.

VAN LOON, H., 1972b. Pressure in the Southern Hemisphere. In: C. W. NEWTON (Editor), *Meteorology of the Southern Hemisphere. Am. Meteorol. Soc. Monograph*, 13: 59—86.

VAN LOON, H., 1972c. Wind in the Southern Hemisphere. In: C. W. NEWTON (Editor), *Meteorology of the Southern Hemisphere. Am. Meteorol. Soc. Monograph*, 13: 87—100.

VAN LOON, H., 1972d. Cloudiness and precipitation in the Southern Hemisphere. In: C. W. NEWTON (Editor), *Meteorology of the Southern Hemisphere. Am. Meteorol. Soc. Monograph*, 13: 101—112.

VAN LOON, H., 1976. On the interaction between Antarctica and middle latitudes. In: L. QUAM (Editor), *Research in the Antarctic*. American Association for the Advancement of Science, Washington, D.C., 477−487.

VAN LOON, H., TALJAARD, J. J., JENNE, R. L. and CRUTCHER, H. L., 1971. *Climate of the Upper Air: Southern Hemisphere, Vol. II. Zonal Geostrophic Winds*. NAVAER 50-1C-56 and NCAR TN/STR-57, National Center for Atmospheric Research, Boulder, Colo., 43 pp.

VAN LOON, H., TALJAARD, J. J., SASAMORI, T., LONDON, J., HOYT, D. V., LABITZKE, K. and NEWTON, C. W., 1972. *Meteorology of the Southern Hemisphere*. Am. Meteorol. Soc. Monograph, 13, No. 35: 263 pp.

VAN ROOY, M. P. (Editor), 1957. *Meteorology of the Antarctic*. Pretoria, 240 pp.

VENTER, R. J., 1957. Sources of meteorological data for the Antarctic. In: M. P. VAN ROOY, (Editor), *Meteorology of the Antarctic*. Pretoria, pp. 17−38.

VOWINCKEL, E., 1955. Southern Hemisphere weather map analysis. 5-year mean pressures. *Notos*, 4: 17−50; 204−216.

VOWINCKEL, E., 1956. Das Klima des Antarktischen Ozeans: I Nord-Süd-Schnitt zwischen 20°E und 40°E; II West-Ost-Schnitt zwischen 50°W und 150°E. *Arch. Meteorol. Geophys. Bioklimatol.*, B, 7: 317−341; 341−369.

VOWINCKEL, E., 1957. Climate of the Antarctic Ocean. In: M. P. VAN ROOY (Editor), *Meteorology of the Antarctic*, Pretoria, 91−108.

VOWINCKEL, E. and OOSTHUIZEN, C. M., 1953. Weather types and weather elements over the Antarctic Ocean during the whaling season. *Notos*, 2: 157−182.

VOWINCKEL, E. and VAN LOON, H., 1957. Das Klima des Antarktischen Ozeans, III. Die Verteilung der Klimaelemente und ihre Zusammenhang mit der allgemeinen Zirkulation. *Arch. Meteorol. Geophys. Bioklimatol.*, B, 8: 75−102.

ZILLMAN, J. W., 1967. The surface radiation balance in high southern latitudes. *WMO Tech. Note, No. 87, Polar Meteorology*, pp. 142−171.

ZILLMAN, J. W., 1970a. Sea surface temperature gradients south of Australia. *Aust. Meteorol. Mag.*, 18: 22−30.

ZILLMAN, J. W., 1970b. *The GARP Basic Data Set Project. November Sea Surface Temperature Distribution in the Southern Hemisphere*. Working Paper No. 132, Bureau of Meteorology, Australia.

ZWALLY, H. J. and GLOERSEN, P., 1977. Passive microwave images of the polar regions and research applications. *Polar Rec.*, 18: 431−450.

Appendix – Climatic tables

TABLE XI

CLIMATIC TABLE FOR MARION ISLAND[1]
Latitude 46°53'S, longitude 37°52'E, elevation 22.6 m, period 1950–77

Month	Mean sta. press. (mbar)	Temperature (°C)				Mean relat. humid (%)	Precipitation (mm)		Mean daily total rad.[3] (cal. cm⁻²)	Mean sea surf. temp. (°C)
		mean daily	mean daily range	extreme[2] max.	min.		mean	24-h max.		
Jan.	1001.2	6.8	5.7	22.2	−1.8	82	226	88	474	5.5
Feb.	1004.4	7.3	5.6	22.3	−1.4	84	201	87	400	6.0
Mar.	1007.2	7.2	5.6	(20.4)	(−0.8)	84	216	63	294	5.9
Apr.	1004.3	5.9	5.1	(17.8)	(−2.2)	84	230	103	184	5.5
May	1004.2	4.9	4.9	(19.6)	(−3.0)	85	237	81	129	4.9
June	1005.5	4.3	4.8	(14.7)	(−6.0)	86	214	70	85	4.5
July	1005.7	3.7	4.7	(13.3)	(−6.0)	85	214	71	102	4.2
Aug.	1005.6	3.4	4.9	16.2	−5.5	84	187	51	167	4.1
Sept.	1005.7	3.4	5.1	15.9	−6.8	83	181	95	261	4.0
Oct.	1005.5	4.3	5.5	17.7	−4.4	81	170	66	386	4.2
Nov.	1002.6	5.1	5.9	21.4	−2.5	81	196	79	478	4.7
Dec.	998.4	5.9	5.7	18.5	−1.5	82	227	69	499	5.1
Year	1004.2	5.2	5.3	(22.3)	(−6.8)	83	2499	103	288	4.9

Month	Mean evap. (mm)	Mean number of days with			Mean cloudiness (%)	Mean sunshine (h)	Wind		Mean no. of days with winds[4] > 16 m s⁻¹
		Precip. ≥ 0.1 mm	thunder-storm	fog			preval. direct.	mean speed (m s⁻¹)	
Jan.		26	0.1	3.7	78	157	NW	7.2	14.1
Feb.		22	0.1	4.7	74	136	NW	6.8	14.2
Mar.		24	0.3	5.4	73	116	NW	6.7	12.1
Apr.		26	0.4	4.2	75	85	NW	7.2	13.4
May		27	0.3	4.0	75	74	NW	7.2	16.6
June		28	0.3	3.8	77	55	NW	7.7	22.6
July		28	0.5	4.5	74	63	WNW	7.9	21.8
Aug.		27	0.7	3.6	81	81	NW	7.8	21.6
Sept.		26	0.3	2.5	78	96	NW	7.6	16.1
Oct.		25	0.1	2.6	78	138	NW	7.6	18.3
Nov.		25	0.1	2.5	79	154	NW	7.6	18.1
Dec.		27	0.1	3.0	81	157	NW	7.5	21.7
Year		311	3.3	44.5	77	1312	NW	7.4	210.6

[1] Source: Weather Bureau, Department of Transport, Pretoria, Republic of South Africa.
[2] ( ) No readings March to June 1966 and only 20 readings for July 1966.
[3] Measured with Robitzsch Actinograph for 9−11 years, and with the Hartman & Braun solarimeter for 6−7 years.
[4] Measured for 14−15 years.

TABLE XII

CLIMATIC TABLE FOR NEW AMSTERDAM[*1]
Latitude 37°50′S, longitude 77°34′E, elevation 27 m, period 1951–70

Month	Mean sta. press. (mbar)	Temperature (°C)				Mean vap. press. (mbar)	Precipitation (mm)		Mean sea surf. temp. (°C)
		mean daily	mean daily range	extreme			mean	24-h max.	
				max.	min.				
Jan.	1014.2	16.6	5.0	24.3	7.0	15.5	60	111	16.4
Feb.	1015.8	16.8	5.2	24.0	6.4	15.5	47	90	17.2
Mar.	1017.3	16.0	4.9	23.8	6.0	13.8	59	100	16.8
Apr.	1016.9	14.8	4.3	22.2	4.3	13.8	71	78	15.8
May	1014.0	13.0	4.3	20.4	3.8	12.3	91	66	14.5
June	1012.5	11.4	4.3	18.3	3.0	11.1	79	60	13.2
July	1016.6	10.9	4.0	16.3	1.8	11.0	81	57	12.8
Aug.	1017.3	10.8	4.2	16.1	1.9	10.6	78	54	12.5
Sept.	1017.4	11.4	4.2	16.2	3.1	10.9	82	34	12.6
Oct.	1015.6	12.2	4.5	18.0	3.1	11.5	66	39	13.0
Nov.	1013.4	13.3	4.5	20.5	4.7	12.5	68	43	13.7
Dec.	1014.1	15.3	4.6	22.3	6.6	14.5	71	45	15.3
Year	1015.4	13.5	4.5	24.3	1.8	12.8	853	111	14.5

Month	Mean evap. (mm)	Mean number of days with			Mean cloudi-ness (%)	Mean sun-shine (h)	Wind		Mean no. of days with winds > 16 m s⁻¹
		precip. ⩾ 0.1 mm	thunder-storm	fog			preval. direct.	mean speed (m s⁻¹)	
Jan.	92	16	0.3	1.2	73	172	W	6.5	7
Feb.	85	14	0.1	1.2	74	145	W	6.1	6
Mar.	86	16	0.3	1.1	74	132	W	6.0	8
Apr.	81	18	0.4	1.1	75	105	W	6.5	11
May	89	21	0.5	0.5	71	108	W	7.9	17
June	91	25	1.0	0.6	69	96	W	9.2	21
July	87	25	0.7	0.5	70	104	W	8.9	21
Aug.	90	24	0.5	0.3	71	120	W	8.8	20
Sept.	89	21	0.4	0.7	74	122	W	9.0	19
Oct.	86	20	0.2	0.8	75	141	W	8.2	16
Nov.	82	19	0.1	1.2	75	146	W	7.9	14
Dec.	77	16	0.3	3.2	75	162	W	6.8	8
Year	1035	235	4.8	12.4	73	1553	W	7.6	168

[*1] Source: Météorologie Nationale, Paris, France.

TABLE XIII

CLIMATIC TABLE FOR HEARD[*1]
Latitude 53°06'S, longitude 72°31'E, elevation 5 m, period Feb. 1948–Dec. 1954

Month	Mean sta. press. (mbar)	Temperature (°C)				Mean vap. press. (mbar)	Precipitation (mm)	
		mean daily	mean daily range	extreme			mean	24-h max.
				max.	min.			
Jan.	994.5	3.3	3.7	12.6	−0.9	6.7	139	37
Feb.	999.7	3.6	3.5	14.4	−1.3	6.8	148	45
Mar.	1001.4	3.1	3.6	13.3	−1.9	6.7	143	34
Apr.	997.8	2.7	3.6	14.0	−4.4	6.5	161	39
May	995.1	1.5	3.6	10.6	−5.5	5.7	157	41
June	989.3	−0.4	3.6	6.9	−2.9	5.0	98	28
July	993.0	−0.5	3.7	8.9	−9.1	5.0	97	31
Aug.	995.9	−0.4	3.9	11.6	−10.6	5.0	63	23
Sept.	991.8	−1.1	4.1	4.8	−8.9	4.7	67	25
Oct.	993.3	−0.1	3.3	6.7	−8.6	5.1	104	36
Nov.	991.7	0.8	3.2	6.2	−4.4	5.5	104	45
Dec.	993.7	2.4	3.4	11.7	−1.8	6.3	135	33
Year	994.8	1.2	3.6	14.4	−10.6	5.7	1416	45

Month	Mean number of days with			Mean cloudiness (%)	Wind	
	precip. ≥ 0.1 mm	thunderstorm	fog		preval. direct.	mean speed (m s⁻¹)
Jan.	27	0	4.3	77	SW	7.2
Feb.	24	0	3.0	75	SW	7.6
Mar.	27	0	2.0	75	SW	7.6
Apr.	27	0	1.5	73	W	8.1
May	25	0	1.0	73	W	8.4
June	21	0	1.0	71	SW	8.3
July	20	0	0.7	71	WNW	8.8
Aug.	21	0	0.5	72	SW	9.3
Sept.	19	0	0.3	71	WNW	9.3
Oct.	19	0	1.2	71	WNW	8.5
Nov.	20	0	0.8	75	SW	8.9
Dec.	23	0	3.7	77	SW	7.2
Year	273	0	20.0	73	SW	8.3

[*1] Source: Bureau of Meteorology, Melbourne, Australia.

TABLE XIV

CLIMATIC TABLE FOR KERGUELEN (PORT-AUX-FRANÇAIS)[1]
Latitude 49°21′S, longitude 70°15′E, elevation 29 m, period 1951–70

Month	Mean sta. press. (mbar)	Temperature (°C)				Mean vap. press. (mbar)	Precipitation (mm)		Days with snow (mean)
		mean daily	mean daily range	extreme			mean	24-h max.	
				max.	min.				
Jan.	997.5	7.2	7.0	23.0	−0.8	7.2	94	40	4
Feb.	1001.4	7.6	7.1	20.6	−0.6	7.4	78	36	2
Mar.	1003.1	7.0	6.8	20.5	−1.6	7.4	99	34	4
Apr.	1001.3	5.9	6.3	18.0	−4.4	7.3	99	54	6
May	997.9	3.7	5.6	16.2	−7.2	6.4	104	101	11
June	995.7	1.9	5.3	13.5	−9.4	5.7	121	47	15
July	997.2	1.8	5.5	13.2	−8.9	5.5	106	42	17
Aug.	998.4	1.9	5.7	15.0	−8.4	5.6	91	47	17
Sept.	997.5	2.1	5.8	13.9	−6.3	5.3	72	53	16
Oct.	997.7	3.3	6.5	16.8	−5.1	5.7	78	32	14
Nov.	994.9	4.6	6.7	19.9	−4.4	6.1	94	44	12
Dec.	994.8	6.3	6.8	22.1	−3.3	6.8	81	41	6
Year	998.1	4.4	6.3	23.0	−9.4	6.4	1117	101	124

Month	Mean evap.[2] (mm)	Mean number of days with			Mean cloudi-ness (%)	Mean sun-shine (h)	Wind		Days with gale (mean)
		precip. ⩾ 0.1 mm	thunder-storm	fog			preval. direct.	mean speed (m s⁻¹)	
Jan.	123	19	0.1	0.5	75	181	W	9.3	9
Feb.	100	16	0.1	0.4	74	161	W	9.0	9
Mar.	85	18	0.0	0.3	71	154	W	9.4	11
Apr.	54	19	0.0	0.6	70	121	W	9.4	12
May	40	22	0.0	0.4	69	91	W	9.5	12
June	28	23	0.0	0.6	70	68	W	9.6	12
July	33	23	0.0	0.6	68	86	W	10.3	16
Aug.	44	23	0.0	0.3	69	108	W	10.8	16
Sept.	56	22	0.0	0.1	73	127	W	11.0	16
Oct.	81	22	0.0	0.3	74	160	W	10.5	14
Nov.	102	21	0.0	0.2	75	169	W	10.0	12
Dec.	118	19	0.1	0.1	76	178	W	9.5	9
Year	864	247	0.3	4.4	72	1604	W	9.9	148

[1] Source: Météorologie Nationale, Paris, France.
[2] Penman period after 1954

TABLE XV

AVERAGE (AVE) MONTHLY, SEASONAL, AND ANNUAL TEMPERATURES AND STANDARD DEVIATION (STD) OF THE MEAN MONTHLY VALUES FOR MARION, NEW AMSTERDAM, KERGUELEN, AND HEARD ISLANDS

(Data supplied by National Center for Atmospheric Research, Boulder, Colo., U.S.A.)

	850 mbar		700 mbar		500 mbar		300 mbar		200 mbar		150 mbar		100 mbar	
	AVE	STD	AVE	STD	AVE	STD	AVE	STD	AVE	STD	AVE	STD	AVE	STD
Marion Island (1950–1974)														
J	0.4	1.4	−5.4	1.6	−19.2	1.6	−42.4	1.7	−50.3	2.5	−51.9	1.7	−53.4	2.0
F	2.1	1.7	−3.6	1.6	−17.6	1.4	−41.4	1.5	−51.4	2.2	−52.9	2.3	−54.6	2.2
M	2.1	1.3	−3.8	1.3	−18.0	1.3	−42.2	1.3	−53.0	2.0	−54.5	1.8	−56.1	1.9
A	0.5	1.2	−6.0	1.4	−20.6	1.8	−44.5	1.7	−54.5	2.3	−54.9	1.7	−55.9	1.5
M	−1.0	1.5	−8.1	1.7	−23.3	1.6	−47.5	1.3	−56.3	2.2	−55.2	1.4	−55.4	1.6
J	−1.5	1.8	−8.9	1.8	−24.3	1.8	−49.1	1.4	−58.4	2.7	−56.6	2.0	−57.0	1.6
J	−2.5	1.1	−9.9	1.4	−25.5	1.5	−49.6	1.3	−58.4	1.9	−57.2	1.4	−57.9	1.5
A	−3.0	1.3	−10.2	1.3	−25.5	1.3	−49.3	1.4	−56.9	2.0	−55.9	2.2	−56.9	2.2
S	−3.2	1.0	−10.3	1.2	−25.0	1.6	−48.1	1.6	−55.0	1.3	−54.5	1.4	−55.1	2.0
O	−2.5	1.2	−8.8	1.3	−23.2	1.4	−46.4	1.8	−53.5	1.7	−53.7	1.7	−54.1	1.7
N	−1.6	1.4	−7.9	1.8	−22.3	1.9	−45.5	1.5	−52.4	1.8	−52.9	1.3	−54.1	1.1
D	−0.8	1.1	−7.0	1.2	−21.0	1.4	−44.2	1.4	−51.1	1.5	−51.8	1.4	−53.0	1.6
DJF	0.5	0.9	−5.4	1.0	−19.3	1.0	−42.7	1.2	−50.9	1.7	−52.2	1.5	−53.7	1.7
MAM	0.5	0.9	−6.0	1.0	−20.7	1.0	−44.7	1.2	−54.6	1.8	−54.9	1.2	−55.8	1.3
JJA	−2.3	0.8	−9.7	0.9	−25.1	1.0	−49.3	1.0	−57.9	1.7	−56.5	1.4	−57.3	1.4
SON	−2.4	0.7	−9.0	0.9	−23.5	1.3	−46.7	1.4	−53.7	1.0	−53.8	0.7	−54.5	0.8
ANN	−0.9	0.5	−7.5	0.6	−22.1	0.7	−45.9	1.0	−54.3	1.2	−54.4	0.9	−55.3	1.0
New Amsterdam (1955–1974)														
J	8.9	1.4	3.5	1.1	−11.3	2.0	−36.1	1.2	−53.2	1.4	−58.7	1.2	−62.0	1.7
F	8.7	1.0	3.2	2.0	−11.4	1.7	−36.9	1.3	−53.8	1.2	−59.0	1.0	−62.2	1.1
M	7.4	0.9	2.4	0.9	−12.6	1.4	−38.2	1.1	−54.3	1.6	−58.8	1.7	−62.0	2.1
A	6.5	1.2	0.9	1.1	−14.3	1.3	−40.2	1.5	−56.3	1.6	−59.4	1.6	−60.9	1.4
M	4.0	1.0	−2.6	1.0	−18.2	1.3	−44.3	2.0	−56.5	2.1	−56.1	1.9	−57.4	1.7

TABLE XV (continued)

	850 mbar		700 mbar		500 mbar		300 mbar		200 mbar		150 mbar		100 mbar	
	AVE	STD	AVE	STD	AVE	STD	AVE	STD	AVE	STD	AVE	STD	AVE	STD
J	2.7	0.7	−4.4	0.8	−20.3	0.9	−45.6	1.3	−55.4	2.0	−54.6	1.6	−56.7	1.7
J	1.8	1.1	−4.9	1.1	−21.0	1.1	−46.6	1.0	−55.0	1.7	−54.6	1.5	−57.1	1.8
A	1.5	1.5	−4.9	1.2	−20.6	1.4	−45.7	1.5	−53.5	2.2	−54.2	2.1	−56.5	2.3
S	2.4	0.9	−3.3	1.3	−18.8	1.1	−43.9	1.5	−53.5	2.0	−54.2	1.8	−55.8	1.9
O	3.2	0.9	−2.5	1.3	−17.4	1.3	−42.6	1.5	−53.5	1.7	−54.3	2.0	−55.4	1.9
N	4.7	1.1	−1.2	1.2	−15.7	1.1	−40.9	1.2	−54.8	1.9	−56.0	1.7	−57.1	2.2
D	7.9	1.2	1.9	1.0	−12.8	1.2	−38.4	1.6	−54.6	1.4	−57.8	1.8	−60.6	2.5
DJF	8.5	0.8	2.9	1.0	−11.8	1.1	−37.1	1.0	−53.8	0.9	−58.4	0.9	−61.5	1.3
MAM	6.0	0.7	0.2	0.8	−15.0	1.1	−40.9	1.4	−55.7	1.3	−58.1	1.3	−60.1	1.4
JJA	2.0	0.8	−4.8	0.8	−20.6	0.9	−46.0	1.0	−54.7	1.4	−54.5	1.5	−56.8	1.8
SON	3.4	0.6	−2.3	0.9	−17.3	0.9	−42.5	1.2	−53.9	1.5	−54.8	1.6	−56.1	1.6
ANN	4.9	0.5	−1.0	0.7	−16.2	0.7	−41.7	0.9	−54.6	0.9	−56.5	1.1	−58.7	1.3
Kerguelen (1968–1974, but several years incomplete)														
J	−0.7	1.8	−6.6	1.4	−21.1	1.6	−43.8	1.8	−49.8	0.3	−50.4	0.8	−51.5	1.0
F	0.1	1.9	−4.7	2.2	−19.2	1.9	−42.9	1.1	−51.7	2.0	−53.2	2.4	−54.3	2.0
M	−0.0	1.8	−5.9	1.1	−20.3	1.0	−44.2	1.2	−53.8	1.5	−54.6	1.2	−56.0	0.6
A	−1.8	1.5	−8.2	1.3	−23.2	1.5	−46.3	1.4	−54.8	0.2	−54.0	0.9	−55.4	0.4
M	−3.6	0.7	−10.2	1.0	−25.7	1.5	−49.2	1.0	−56.5	1.6	−54.1	2.2	−54.6	2.3
J	−4.3	1.5	−11.0	1.3	−26.7	1.6	−51.8	2.3	−58.3	2.5	−56.1	2.6	−56.4	2.2
J	−5.1	0.9	−11.5	1.3	−27.4	1.5	−51.1	0.6	−58.8	1.5	−56.4	1.0	−57.3	1.3
A	−4.6	1.4	−11.8	1.2	−27.7	1.1	−50.6	0.6	−57.7	1.6	−56.1	2.2	−56.8	2.6
S	−5.0	0.7	−11.4	1.3	−27.3	1.6	−50.5	0.5	−57.0	2.1	−54.9	2.1	−54.0	2.3
O	−5.5	0.7	−11.8	1.3	−26.9	1.0	−48.9	0.8	−53.0	0.7	−51.4	1.1	−51.3	0.9
N	−4.3	1.2	−10.5	1.7	−24.9	2.0	−46.8	0.7	−50.5	1.1	−50.0	0.7	−49.6	0.6
D	−0.9	0.4	−7.4	1.3	−22.3	1.5	−46.0	1.4	−51.4	1.4	−51.0	1.5	−51.4	1.8
DJF	−0.4	1.2	−6.1	1.4	−20.7	1.5	−44.2	1.0	−51.3	0.9	−51.9	1.1	−52.7	1.3
MAM	−1.8	0.9	−8.1	0.7	−23.1	1.0	−46.6	1.0	−55.0	0.7	−54.2	0.9	−55.4	1.1
JJA	−4.8	0.4	−11.6	0.5	−27.3	0.8	−50.8	0.4	−58.1	1.4	−56.2	1.7	−56.8	1.8

TABLE XV (continued)

	850 mbar		700 mbar		500 mbar		300 mbar		200 mbar		150 mbar		100 mbar	
	AVE	STD	AVE	STD	AVE	STD	AVE	STD	AVE	STD	AVE	STD	AVE	STD
SON	−4.9	0.5	−11.2	0.9	−26.3	0.9	−48.7	0.5	−53.6	0.6	−52.0	0.4	−51.5	0.8
ANN	−3.0	0.5	−9.3	0.4	−24.4	0.5	−47.6	0.6	−54.5	0.6	−53.6	0.7	−54.1	0.7
Heard Island (1950–1954)														
J	−3.5	1.5	−9.6	2.0	−23.6	1.9	−45.9	2.0	−48.1	1.6				
F	−3.3	0.5	−7.0	0.1	−22.1	0.5	−45.5	1.0	−51.9	0.5				
M	−2.8	0.6	−8.0	1.0	−22.4	0.9	−45.1	1.4	−51.5	1.9				
A	−3.7	1.3	−10.4	1.5	−25.4	0.6	−48.2	0.9	−52.3	2.8				
M	−5.8	1.4	−12.7	1.2	−27.9	1.2	−51.0	0.3	−55.9	2.4				
J	−7.4	0.9	−15.4	1.0	−30.3	1.1	−52.5	0.3	−57.0	0.5				
J	−7.5	0.6	−15.7	1.1	−31.2	1.3	−54.2	0.0	−59.0	1.5				
A	−8.0	1.1	−15.6	2.0	−30.4	1.2	−53.8	1.0	−60.4	2.0				
S	−9.1	0.3	−16.8	0.1	−31.6	0.8	−53.8	0.6	−57.2	1.7				
O	−8.5	0.7	−15.7	1.3	−31.1	0.7	−52.4	0.8	−55.3	1.8				
N	−6.8	0.5	−13.7	0.5	−27.8	1.3	−48.6	1.4	−49.4	1.9				
D	−4.2	1.3	−10.7	1.5	−25.0	2.0	−47.2	1.0	−50.7	2.6				
DJF	−3.7	0.8	−9.5	0.9	−23.8	1.2	−46.3	0.9	−49.9	1.4				
MAM	−4.1	0.5	−10.3	0.6	−25.2	0.6	−48.3	0.4	−53.5	1.8				
JJA	−7.5	0.5	−15.3	0.6	−30.5	1.0	−53.4	0.4	−59.0	0.9				
SON	−8.2	0.3	−15.6	0.2	−30.2	0.4	−51.6	0.6	−53.9	0.8				
ANN	−5.8	0.4	−12.6	0.3	−27.3	0.4	−49.7	0.1	−54.1	0.4				

TABLE XVI

AVERAGE MONTHLY, SEASONAL, AND ANNUAL HEIGHTS (AVE, GPM) AND STANDARD DEVIATIONS (STD, GPM) FOR CONSTANT PRESSURE SURFACES AT FOUR INDIAN OCEAN ISLANDS

(Computations by National Center for Atmospheric Research, Boulder, Colorado, U.S.A. using "Monthly Climatic Data for the World")

	850 mbar		700 mbar		500 mbar		300 mbar		200 mbar		150 mbar		100 mbar	
	AVE	STD	AVE	STD	AVE	STD	AVE	STD	AVE	STD	AVE	STD	AVE	STD
Marion Island (1950–1974)														
J	1355	29	2897	36	5471	49	9091	68	11770	65	13631	53	16236	37
F	1385	37	2937	45	5528	59	9172	79	11847	81	13688	58	16290	36
M	1409	25	2951	32	5549	43	9183	56	11847	64	13675	43	16261	38
A	1380	30	2920	34	5482	47	9081	71	11722	77	13552	60	16129	58
M	1369	32	2899	39	5438	55	8991	72	11600	73	13435	58	16022	57
J	1379	39	2904	47	5434	63	8970	80	11560	76	13383	50	15962	37
J	1378	38	2898	44	5418	55	8939	72	11525	76	13330	57	15889	54
A	1376	32	2893	39	5411	50	8935	64	11531	72	13357	57	15930	67
S	1371	23	2888	27	5409	38	8943	61	11557	72	13377	60	15970	66
O	1375	29	2899	33	5435	44	8997	61	11630	69	13477	71	16078	81
N	1356	35	2884	43	5430	58	9004	82	11649	86	13490	74	16100	67
D	1327	25	2860	30	5417	40	9010	56	11672	59	13534	52	16147	48
DJF	1356	20	2898	25	5472	33	9091	45	11763	48	13618	34	16224	26
MAM	1386	20	2927	25	5490	33	9085	45	11723	53	13554	35	16136	35
JJA	1378	23	2899	27	5421	35	8948	46	11538	50	13357	26	15927	28
SON	1368	20	2891	24	5425	33	8982	54	11612	63	13450	55	16049	55
ANN	1372	12	2904	14	5452	20	9026	30	11659	40	13494	25	16085	27
New Amsterdam (1955–1974)														
J	1499	26	3098	31	5758	40	9497	56	12201	65	14033	71	16551	79
F	1529	38	3125	36	5783	35	[9498]		12210	60	14036	60	16512	72
M	1527	20	3115	24	5760	33	9469	42	12155	43	13974	48	16494	58
A	1515	17	3099	20	5726	27	9421	61	12069	59	13885	65	16398	80
M	1477	27	3042	31	5635	40	9253	57	11883	70	13695	70	16261	80
J	1466	26	3020	26	5593	31	9178	36	11808	38	13641	44	16233	55

TABLE XVI (*continued*)

	850 mbar		700 mbar		500 mbar		300 mbar		200 mbar		150 mbar		100 mbar	
	AVE	STD	AVE	STD	AVE	STD	AVE	STD	AVE	STD	AVE	STD	AVE	STD
J	1491	27	3043	31	5608	39	9186	48	11811	46	13648	47	16227	54
A	1487	42	3039	43	5609	51	9195	61	11833	60	13677	60	16263	73
S	1488	37	3045	41	5630	49	9245	59	11899	61	13740	63	16328	78
O	1474	46	3036	49	5633	54	9269	65	11930	68	13785	56	16374	54
N	1479	33	3050	33	5657	33	9321	41	11985	45	13817	51	16381	61
D	1497	23	3086	28	5730	36	9435	54	12118	65	13940	73	16478	83
DJF	1509	16	3104	20	5757	26	9475	57	12178	48	14005	53	16512	78
MAM	1506	13	3085	16	5707	25	9381	37	12036	50	13851	50	16385	67
JJA	1481	22	3034	23	5603	28	9186	35	11816	36	13655	41	16239	51
SON	1484	18	3048	19	5644	25	9282	33	11940	37	13781	43	16361	52
ANN	1495	9	3067	11	5677	16	9329	27	11989	32	13820	37	16368	50
Kerguelen, (1968–1974)														
J	1315	25	2853	37	5421	44	8996	69	11675	79	13558	82	16191	86
F	1367	45	2909	53	5492	75	9100	97	11772	89	13627	80	16229	65
M	1379	39	2924	40	5489	43	9074	65	11726	56	13584	49	16158	52
A	1330	11	2854	7	5394	17	8953	32	11583	43	13421	38	16005	33
M	1313	26	2831	33	5352	44	8870	60	11470	59	13312	56	15909	60
J	1317	26	2830	35	5342	52	8841	63	11423	58	13244	35	15814	49
J	1315	37	2824	44	5325	59	8815	77	11402	75	13219	70	15781	64
A	1327	40	2835	40	5328	47	8826	56	11412	69	13239	79	15809	98
S	1311	27	2821	32	5322	44	8817	69	11408	57	13249	40	15840	43
O	1296	26	2801	27	5305	34	8818	45	11444	51	13300	54	15922	58
N	1279	39	2792	42	5313	60	8850	79	11503	75	13385	73	16030	63
D	1303	42	2833	48	5381	63	8950	83	11598	83	13475	75	16104	63
DJF	1336	30	2873	39	5440	51	9026	73	11682	74	13560	68	16179	61
MAM	1341	17	2869	21	5412	25	8966	39	11593	39	13439	39	16026	46
JJA	1318	15	2827	16	5328	20	8824	29	11410	26	13233	35	15801	47
SON	1297	21	2807	23	5317	30	8833	38	11455	39	13315	38	15936	33
ANN	1322	9	2843	11	5373	15	8911	25	11536	28	13386	32	15984	35

TABLE XVI (*continued*)

Heard Island (1950-1954)

	850 mbar		700 mbar		500 mbar		300 mbar		200 mbar		150 mbar		100 mbar	
	AVE	STD	AVE	STD	AVE	STD	AVE	STD	AVE	STD	AVE	STD	AVE	STD
J	1268	44	2789	52	5322	71	8864	11	11539	81				
F	1338	19	2861	22	5407	19	8978	15	11637	11				
M	1328	20	2845	25	5401	33	8969	44	11612	43				
A	1286	13	2799	3	5313	3	8840	21	11457	36				
M	1272	19	2772	26	5263	38	8761	55	11345	49				
J	1219	13	2711	19	5176	22	8639	39	11214	31				
J	1231	30	2718	31	5197	64	8640	78	11202	76				
A	1264	37	2748	43	5218	55	8671	70	11217	83				
S	1243	6	2723	5	5174	14	8626	30	11178	19				
O	1231	18	2717	25	5179	35	8629	41	11207	31				
N	1223	15	2719	17	5212	13	8708	29	11345	40				
D	1271	32	2783	39	5302	55	8849	94	11482	77				
DJF	1292	26	2810	31	5341	39	8892	68	11549	60				
MAM	1298	12	2811	14	5329	21	8861	27	11474	26				
JJA	1242	19	2731	21	5204	33	8659	46	11220	47				
SON	1231	9	2717	11	5184	11	8649	9	11239	12				
ANN	1267	7	2768	9	5268	12	8770	18	11377	18				

Climate of the Indian Ocean North of 35°S

C. S. RAMAGE

Introduction

For the purpose of this chapter, the Indian Ocean extends from Africa in the west to the Timor Sea and Australia in the east, and from southern Asia and southern Indonesia in the north to 35°S in the south. The land-enclosed Persian Gulf and Red Sea are not included. The Indian Ocean south of 35°S is discussed in Chapter 5.

Volume 4 of the WSC, *Climate of the Free Atmosphere,* does not obviate the need to discuss conditions above the surface layer over the Indian Ocean, but I shall emphasize the surface layer and refer to it without specification.

Although no oceanic climate is independent of the influences of surrounding land masses, none is so massively affected as the climate of the Indian Ocean. As a result, most of the ocean north of 10°S is dominated by monsoons.

The monsoons, which blow in response to annual change in the difference in pressure over land and sea, result from the temperature difference between land and sea, and where great continents border an ocean, temperature differences are large. When the sun moves north of the Equator in the northern summer, the land mass of Asia is rapidly warmed because of its relatively low heat capacity. On the other hand, the northern Indian Ocean stores the sun's heat within its deep surface layer. Consequently, the land gives off heat more readily than the sea, and the air over land becomes warmer and air pressure lower than over the neighbouring ocean.

Thus, during summer air flows from the Indian Ocean toward lower pressure over southern Asia, ascending as it is heated over the land until it reaches a level at which the pressure gradient is reversed, whereupon it flows on a return trajectory from land to sea where, descending, it is once more taken up by the landward-directed pressure gradient (HARWOOD, 1924). As long as the land is significantly warmer than the sea, this great circulation persists.

In winter the reverse occurs. The low heat capacity of Asia relative to the northern Indian Ocean ensures that air over the land is colder than over the sea. We observe then the typical winter monsoon in which at low levels air flows out from the continent over the sea where it rises and returns in the middle and higher layers of the troposphere to the land, sinks to the surface, and resumes the cycle.

Over Africa the monsoons have a somewhat different character from those of southern Asia. During the Northern Hemisphere summer, the desert areas of northern Africa heat rapidly and pressure there falls in the same way as over Asia. South of the Equator over Africa during the Southern Hemisphere winter, cooling occurs, establishing a pressure gradient across Africa which, in turn, sets up a massive flow of air from south to north across the Equator.

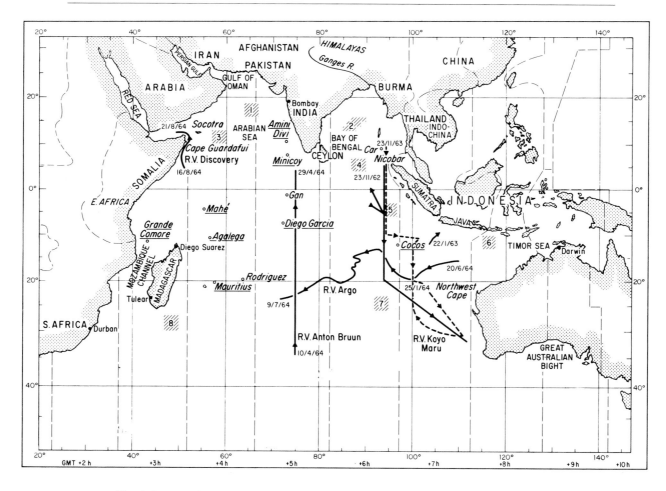

Fig. 1. Locator chart showing places and the tracks of research vessels mentioned in the text. Stations with climatic tabulations underlined. Ocean areas with climatic tabulations hatched.

Because the deflecting force due to the earth's rotation reverses its direction at the Equator and is weak in equatorial regions, air flow is more directly from high to low pressure than is the case in higher latitudes. The influence of Africa on the atmospheric circulation extends 800 km east of the continent and merges with the influence of Asia farther north. In the Northern Hemisphere summer, wind circulates in a huge gyre from the southeast around the northern edge of the southern Indian Ocean anticyclone and toward the coast of Africa near the Equator, swinging into the south and then southwest to parallel the African, Arabian, and Asian coasts and finally sweeping across India, Burma, and the Indochina-Thailand peninsula as the southwest or summer monsoon. Six months later a complete reversal takes place. Northern Africa is cold and southern Africa is warm and so the winds blow from the north across the Equator in the western Indian Ocean. With the southward passage of the sun, Asia cools much more rapidly than the ocean, and winds now blow out from high pressure over Asia toward relatively low pressure over the Indian Ocean. This winter monsoon blows generally from the northeast and is much weaker than the summer monsoon. Climate in the vicinity of Australia is less intensely monsoonal.

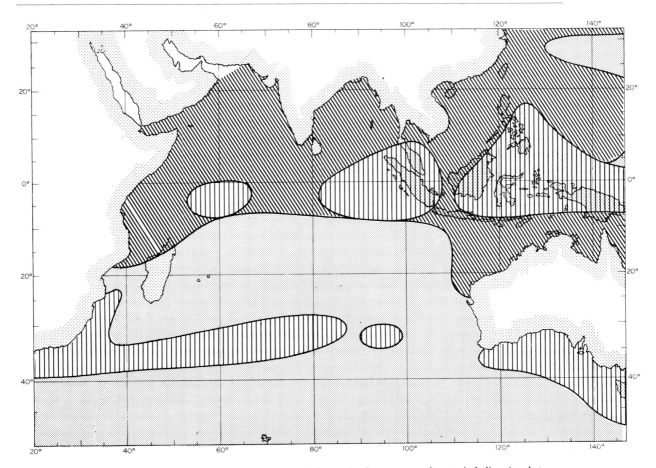

Fig. 2. Wind direction as a monsoon criterion. A change in the mean resultant wind direction between January and July of more than 90° indicates that the ocean area possesses a monsoonal climate (diagonally hatched). Direction change less than 90° (stippled); winds in both January and July less than 2.5 m s^{-1} (vertically hatched).

The extent of the monsoon regime can be gauged from Figs. 2 and 3. Fig. 2 shows that in terms of the direction of the circulation, monsoons dominate the Indian Ocean north of about 10°S. Fig. 3 confirms that most rain falls during the moist landward flow of the summer monsoon.

Although monsoonal influences are slight over the open ocean south of 10°S, the low-latitude pressure trough, the trade winds, and the subtropical high-pressure ridge do shift northward in winter and southward in summer. Subsequent discussion will take account of these two sharply differing climatic zones.

Sea surface currents

Just as the Himalayas prevent severe outbursts of polar continental air from reaching the Indian Ocean, so Asia blocks any flow of cold water from the north into it. Where winds undergo a monsoon reversal (Fig. 2), surface currents correspondingly reverse, whereas over the open ocean south of the Equator annual variability is slight.

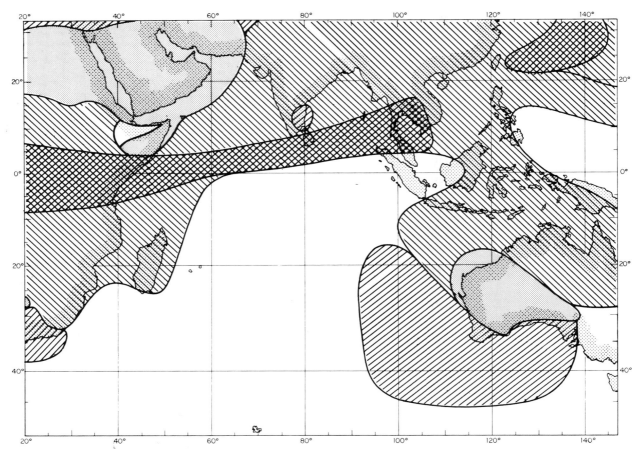

Fig. 3. Rainfall as a monsoon criterion. Periods during which more than 75% of the annual rain falls are regionally delineated. Regions with winter rainfall maximum (diagonally hatched, upward to right); regions with summer rainfall maximum (diagonally hatched, upward to left); regions with double rainfall maximum (cross-hatched); regions with less than 250 mm yr^{-1} are classed as deserts (stippled); regions accumulating 75% of the annual rainfall in over seven months are considered to have no seasonal maximum.

Monsoon region

In winter over the Arabian Sea and Bay of Bengal, northeast winds and coastline orientation combine to produce counterclockwise surface currents. As summer advances, the coastal circulations reverse first and then as the southwest monsoon becomes established, the open sea circulations also reverse. During summer, the coastal current off eastern Africa and Somalia is one of the strongest in the world. Currents flowing parallel to the coast of southern Indonesia also reverse direction with the change of season.

Non-monsoon region

South of about 7° S, the circulation reflects the influence of the subtropical anticyclone. A great anticlockwise gyre brings cool water to the ocean west of Australia and warm water to the coast of Africa. Madagascar produces a local distortion forcing the westward flowing current to split. The equatorial zone is dominated by the eastward flowing

Equatorial Counter Current. The current is strongest between October and December when depressions north and south of the Equator cause westerly winds to prevail. It is also well-marked between May and September when it merges with the southwest monsoon current farther north.

Data

The most important data sources utilized are listed in the references at the end of the chapter. Their contributions are summarized in Table I (Appendix, p. 659).

During recent years and particularly in 1963 and 1964, scientists associated with the International Indian Ocean Expedition have been studying the meteorology of the region intensively (RAMAGE, 1965). I have drawn heavily on the data resources and research results of this project.

The series of mean charts for January and July (Figs. 12–25 and 27–40) and tabulations of ocean area climates (Tables III–X) stem primarily from the following five sources.

(*1*) United Kingdom meteorological atlas of the Indian Ocean (METEOROLOGICAL OFFICE, 1949).

(*2*) U.S. Navy meteorological atlas of the Indian Ocean (U.S. WEATHER BUREAU, 1957).

(*3*) Dutch meteorological atlas of the Indian Ocean (VAN DUIJNEN MONTIJN, 1952).

(*4*) United Kingdom meteorological atlas of the Pacific Ocean (METEOROLOGICAL OFFICE, 1947).

(*5*) International Indian Ocean Expeditions collections.

Table II lists the sources for each element (Appendix, p. 659). Frequencies of precipitation obtained from the primary source are usually significantly lower than from the secondary sources, but otherwise variability is not great. Material in other atlases was used to supplement and check.

Two sets of derived fields were also computed:

(*1*) *Divergence of the mean resultant surface winds.* The Dutch Atlas depicts mean resultant winds for 2-degree squares. However, when VAN DIJK (1956) computed horizontal divergence directly from these data, inhomogeneities in sample sizes between adjacent squares resulted in an unrealistically detailed picture. To avoid this, I used the 2-degree square winds to obtain 5-degree square averages from which the divergence was computed (see also HANTEL, 1971).

(*2*) *Net heat balance at the sea surface.* Many investigators have tried to derive empirical equations relating the elements of standard marine meteorological observations to the heat budget at the ocean's surface. All the equations suffer from lack of accurate direct measurements of the budget components, particularly during periods of strong winds and high seas. The equations used for this chapter are based chiefly on the work of BUDYKO (1956) and RODEN (1959). Although the patterns of heat exchange derived from the equations are probably realistic, absolute values and steepness of gradients qualify only as approximations (GARSTANG, 1965).

As outlined earlier, although general features of Indian Ocean climate can be accounted for by the annual march of the sun and by varying degrees of interaction between ocean and continents, only by noting the synoptic distributions and their frequencies which are the components of climate can we understand climatological details. Assimilating the information contained in a multitude of synoptic charts and distilling the essence and meaning of climate are impossible within the compass of a brief survey, but the attempt must be made, for mean charts used uncritically may be both uninformative and misleading.

I have chosen to discuss three synoptic situations somewhat peculiar to the Indian Ocean and thereafter to attempt physical explanations for the patterns of atmospheric elements and the interrelations among elements as portrayed on mean charts for January and July. Tropical cyclones need separate treatment, for in the northern Indian Ocean they are most frequent in the transition seasons of spring and autumn. Finally, with the aid of time-latitude sections along 73°E and climatological tables for ocean areas and small islands, the salient features are interconnected in space and time.

Typical synoptic situations

Numerous published case studies have familiarized the reader with the meteorology of cold fronts, tropical cyclones, and subtropical anticyclones in the Indian Ocean (see, for example, DEUTSCHES HYDROGRAPHISCHES INSTITUT, 1960). Since these phenomena are in no way peculiar to the Indian Ocean and their descriptions would contribute nothing new, I have chosen to concentrate on three types of situations which, besides being better developed over the Indian than over other oceans, have received relatively little attention until now.

Subtropical cyclones in the northeastern Arabian Sea account for most of the heavy summer monsoon rains of western India; east — west oriented *troughs on either side of the Equator* are persistent low-latitude features throughout most of the year; off the coasts of Arabia and Somalia, *summertime interaction between the southwest monsoon and cold upwelling water* helps maintain a remarkably strong and stable circulation.

Each case illustrates the power of data from weather satellites in supplementing point--sampling by conventional observations.

Subtropical cyclone

PETTERSSEN (1956) has suggested that persistent, quasi-stationary circulation systems such as the trade winds or the low-level polar anticyclone may significantly modify conditions in peripheral regions through thermodynamic mechanisms leading to vorticity export. Using this concept, I have tried to evaluate the effect of the summer heat low on its environment and on the monsoon rains of western India (RAMAGE, 1966). Sequentially:

(*1*) Intense solar heating over the deserts of West Pakistan, Arabia, and Somalia develops a heat low and its accompanying southwest monsoon circulation.

(*2*) Monsoon depression systems in the Bay of Bengal entering India near the Ganges Delta (see p. 630) bring summer monsoon rains to northeastern India. The resultant release of latent heat increases the north — south temperature gradient and hence strengthens the upper tropospheric easterlies above and south of the rains (RAMAN and RAMANATHAN, 1964). This, in turn, by causing convergence downstream in the easterlies, produces subsidence and warming over the heat low to the west further intensifying the heat low. The monsoon depressions, which normally move west-northwestward across central India, are probably the commonest vehicles for transporting moist air in considerable depth from the Bay of Bengal to the northeastern part of the Arabian Sea, displacing a low-level subsidence-limited moist layer.

(*3*) The export of middle and upper tropospheric cyclonic vorticity from the intensified heat low acts on the moist air over the northeastern Arabian Sea to trigger a middle tropospheric or "subtropical" cyclone (SIMPSON, 1952; RAMAGE, 1962). This initiates the summer monsoon rains of western India.

(*4*) The subtropical cyclone, by releasing latent heat of condensation in the upper troposphere, in turn, further increases subsidence above the heat low intensifying it and its associated low-level monsoon circulation over the Arabian Sea.

(*5*) The subtropical cyclone strengthens and then slowly weakens as it acts on its environment to draw progressively drier air into the circulation from the north and northwest (MILLER and KESHAVAMURTHY, 1968). Eventually, the dry air destroys the buoyancy of the system in much the same way that buoyancy of hurricanes may be destroyed; the cyclone fills and rain stops.

(*6*) The break persists until conditions to the east favour renewed intensification of the heat low and moist air once again becomes available for supporting cyclogenesis over the northeastern Arabian Sea.

In a normal summer, feedback between heat lows, depressions over central and northeastern India, and subtropical cyclones maintains both the reliable rhythmic monsoon rains of western India, and over the Arabian Sea the steadiest strong wind circulation in the world.

Subtropical cyclones may develop elsewhere over the Indian Ocean (RAMAGE, 1964) but are not the dominant rain producers.

Subtropical cyclone of 12 and 13 August 1964 (MILLER and KESHAVAMURTHY, 1968)

From 12 to 13 August, TIROS VII photographed dramatic changes in the cloud structure of the middle and high troposphere over the northeastern Arabian Sea and west coast of India. On 12 August, when a mid-tropospheric cyclone was active with its center near Bombay and heavy rain was falling along the west coast, the photographs reveal a great shield of altostratus and cirrus extending southwestward from the west coast (Fig. 4). To the south, a break is visible between 10° and 15°N. Also clearly evident are elongated layers of clouds extending southwestward from the southern tip of India. These layers probably consisted of middle- and high-level stratiform clouds which were associated with cumulus congestus or cumulonimbus embedded in the deep, unstable southwesterly flow.

Even at this time, signs of decay were evident at the 500-mbar level, for incursion of moist air from the Bay of Bengal had ceased and the subtropical cyclone circulation was incorporating dry air from north.

On 13 August, after the cyclone had moved northward to 25°N and weakened, TIROS VII photographed a marked decrease in cloudiness over the northeastern Arabian Sea with the cloud-fringed outline of the west coast clearly visible (Fig. 5). Cloudiness further decreased as the subtropical cyclone dissipated and a monsoon lull supervened.

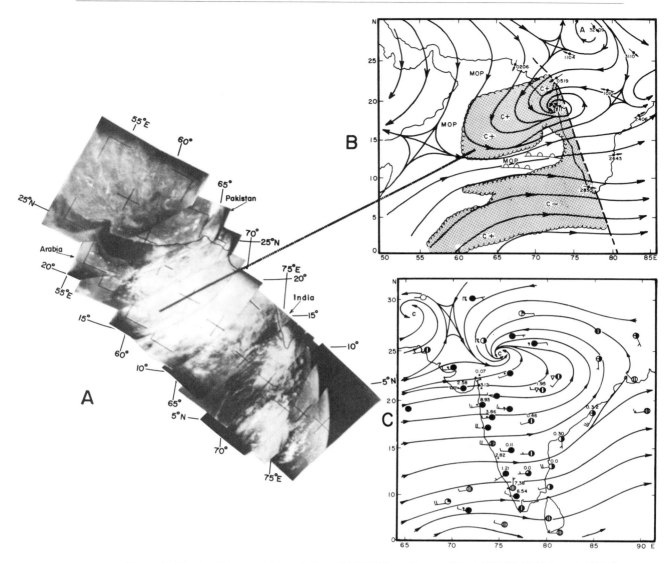

Fig. 4. A. Mosaic of photographs made from TIROS VII weather satellite at 1040 GMT 12 August 1964. B. Corresponding nephanalysis and 500 mbar streamline analysis for 1200 GMT 12 August. C. 1500 m streamline analysis and total cloud amounts for 1200 GMT 12 August and rainfall totals in cm for the 24-h period ending 0300 GMT 12 August. (From MILLER and KESHAVAMURTHY, 1968)

Near-equatorial troughs

Many students of Indian Ocean meteorology (FLETCHER, 1945; WEICKMANN, 1964; RA-MAN, 1967; and others) have called attention to the existence throughout the year of an east—west oriented pressure trough in the tropics of each hemisphere. The troughs, separated by persistent equatorial west winds, move in response to the sun's march and participate in a double oscillation through the year. In July the Northern Hemisphere trough is found across Arabia, Pakistan, and northern India, while the Southern Hemisphere trough lies just south of the Equator. Between July and January, the Northern Hemisphere trough does not move south, but rather dissipates and redevelops just north

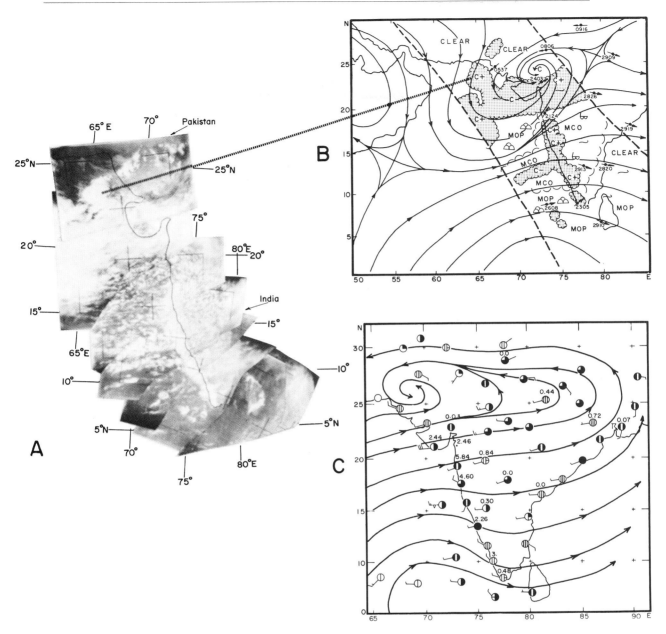

Fig. 5. Same as for Fig. 4, but: A. 0950 GMT 13 August 1964; B. 1200 GMT 13 August; C. 1200 GMT 13 August and 24-h period ending 0300 GMT 13 August.

of the Equator (see p. 651). The Southern Hemisphere trough off Africa and Australia likewise dissipates and redevelops near about 20°S, whereas over the central ocean, movement southward is much less and probably continuous. After January, the cycle reverses until the troughs once more occupy their July positions. Cyclones of varying intensities frequently develop in the extensive and vigorous summer hemisphere trough with consequent severe weather. However, depressions may and do develop in the winter hemisphere trough and, although generally weak, on rare occasions they may reach hurricane force and do unexpected widespread damage. It is not uncommon, especially in the spring and autumn transition seasons, to observe tropical cyclones developing or coexisting on both sides of the Equator at more or less the same time.

Near-coincidence of trough and thermal equator which facilely explains the trough in the summer hemisphere certainly cannot account for the trough in the winter hemisphere. Studies (GORDON, 1967; GORDON and TAYLOR, 1975) hint that the large gradient of the Coriolis force between the parallels of 5° and 10° predisposes the lower troposphere to develop cyclonic vorticity. If this is so, then the shift of the summer hemisphere trough to higher latitudes represents a thermally induced distortion of the norm.

SADLER (1964) reports that weather satellite photographs taken over the tropical northeast Pacific show the near-equatorial trough to be relatively cloud-free except when tropical storms develop in it. Maximum cloudiness and surface convergence occur 150−300 km north and south of the trough line.

Over the Indian Ocean, careful study of the mean January charts (Figs. 12−25) and the mean July charts (Figs. 27−40) likewise reveals that the troughs as defined by pressures and winds seldom coincide with the maxima of surface convergence, cloudiness or rain.

However, they lie close to the axes of maximum air temperature and maximum ocean surface heating. It appears then that trough location, even in the winter hemisphere, may be partially under thermal control. Air converging into a trough would ascend and clouds, rain, and occasionally vigorous cyclonic circulations would develop and, by interrupting incoming solar radiation, sharply reduce the surface heating needed to maintain the trough. A new trough would form where insolation was much greater—on the edge of the rain area. Presumably, the trough would tend to reform toward the sun's zenithal latitude. In the winter hemisphere, on the average it should lie equatorward of the rain belt and in the summer hemisphere poleward (cf. Figs. 46 and 50). Thus in the mean, maximum rains and thermal troughs cannot quite coincide (RIEHL, 1954, pp. 78−79).

Near-equatorial troughs of April−May 1964

Between 10 and 29 April 1964, the research vessel <i>Anton Bruun</i> sailed northward along the 75°E meridian from 37°S to 10°N (Fig. 1). Besides standard weather observations, daily radiosoundings were made from the ship. An analysis of these soundings and corresponding soundings from Diego Garcia, Gan, and Minicoy is reproduced in Fig. 6 as a time/space section.

For the first three days, the ship experienced moderate trade winds and generally fair weather beneath a sharp subsidence inversion, indicating that the subtropical ridge lay south of its normal April latitude.

The winds gradually backed into the northeast and at about 23°S a front was traversed. Winds veering from NNE to SW and becoming strong, and replacement of the subsidence inversion by a deep moist layer, indicated cyclogenesis was in progress on the front just to the east of the ship. This depression subsequently intensified further and moved slowly southeastward. From 25° to 19°S, skies remained overcast and occasional rain (not showers) fell. Between 18° and 9°S, light or moderate trades prevailed. The associated subsidence inversion, higher to the north than to the south, could not be detected north of 14°S. Between 9° and 6°S, the ship encountered the Southern Hemi-

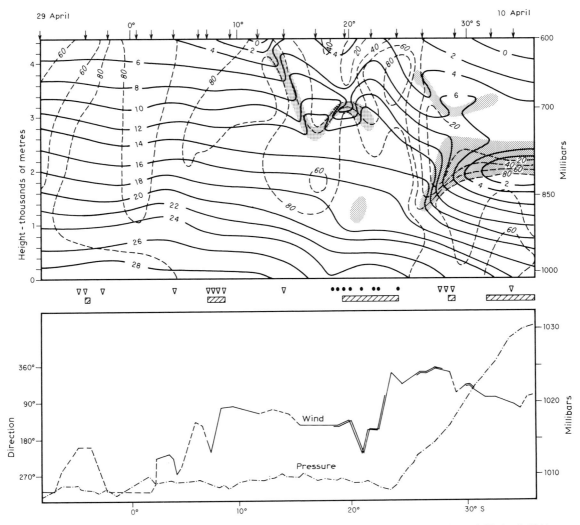

Fig. 6. Analysis of measurements made along 75°E by R.V. *Anton Bruun* between 10 and 29 April 1964. Upper: isotherms (full lines); relative humidity isopleths (dashed lines); inversion layers (stippled); periods of low cloud ⩾ 75% (hatched). Precipitation symbolically shown; arrows denote locations of radiosonde ascents. Lower: mean sea level pressure, compensated for diurnal variation (dot-dashed line); surface wind, 0–5 m s⁻¹ (dashed line), 6–10 m s⁻¹ (full line), ⩾ 11 m s⁻¹ (double line).

sphere near-equatorial trough. The trough appears to have been rather complex, as twice the winds veered from SE to SW. Relatively dry westerlies predominated from 4°S to 1°N while from 1° to 5°N the Northern Hemisphere near-equatorial trough affected both winds and weather at the ship. A temporary shift from NW to SW, a deepening of the moist layer, and showers all stemmed from a small depression east of the track. TIROS photo sweeps made at 0424, 0532, and 0603 GMT on 2 May 1964 (Fig. 7) and the surface weather chart analysis for 06 GMT 2 May (Fig. 8) show that the major features encountered by the *Anton Bruun* over the previous three weeks were still present and, therefore, may not be uncommon at this time of year.

The front along about 25°S appears rather weak in the photographs lacking the hard bright appearance typical of deep cumulonimbus. Clouds are scarce to the north of the

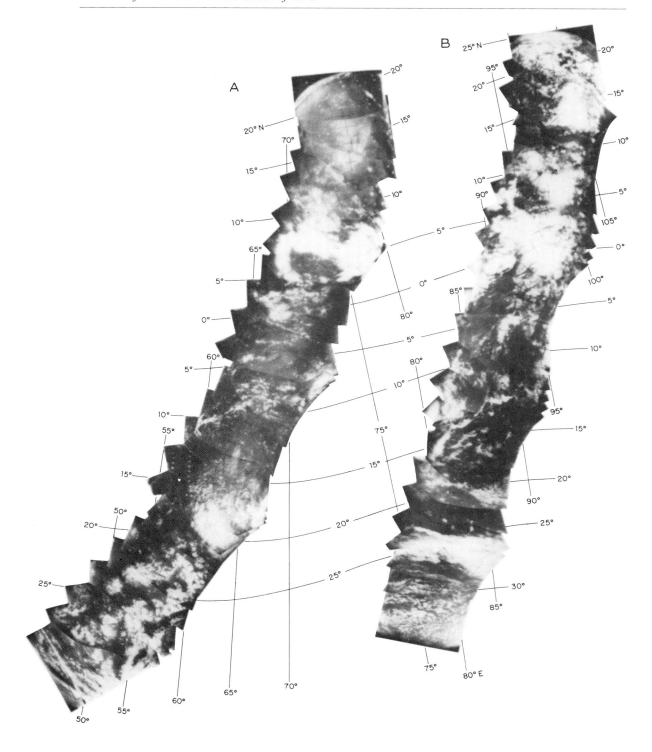

front where the air is being slightly cooled and stabilized by the sea, and immediately to the south of the front where subsidence is effective. Farther south the cellular, cumuliform appearance of the clouds betokens vigorous addition of heat and moisture from ocean to air.

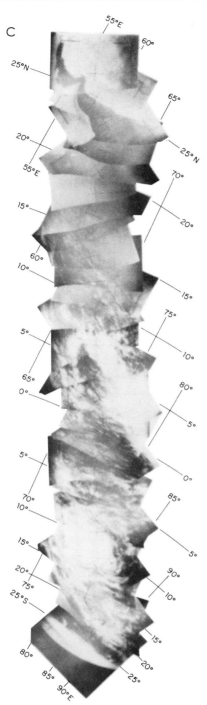

Fig. 7. Mosaics of photographs made from weather satellites on 2 May 1964: A. TIROS VIII orbit 1932 at 0630 GMT; B. TIROS VIII orbit 1925 at 0424 GMT; C. TIROS VII orbit 4701 at 0532 GMT.

To the west, the front apparently merges into the circulation around a depression centered near 17°S 66°E, at the western extremity of the *Southern Hemisphere near-equatorial trough*. The trough, oriented WSW — ENE, contains another and vigorous depression in which winds exceeding 15 m s^{-1} are observed. The centre, located near 11°S

615

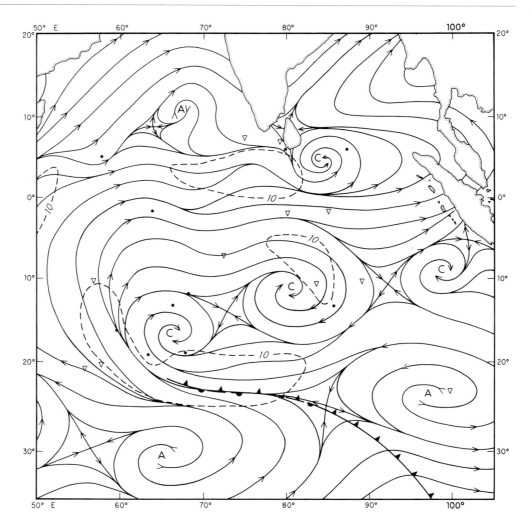

Fig. 8. Kinematic analysis of surface winds for 0600 GMT 2 May 1964. Streamlines (full lines); isotachs in m s^{-1} (dashed lines); cyclones denoted by *C*, anticyclones by *A*; precipitation, where observed, symbolically shown.

81°E, is surrounded by a mass of deep, strongly reflecting cloud. Within 24 h, this depression intensified into a severe tropical storm. Visible on the very edge of the TIROS mosaic for 0424 GMT, near 10°S 95°E, is the western edge of a cloud mass surrounding another depression south of Sumatra.

For about 5—10 degrees north of the trough, where westerly winds predominate, skies are partly cloudy.

The *Northern Hemisphere near-equatorial trough* dominates the southern Bay of Bengal and the southeastern Arabian Sea. Numerous ship reports confirm the existence of a depression in the trough near 5°N 85°E. The satellite views depict rather more cloud to north and south than at the trough line.

The strongly reflecting cloud mass apparent over the southeastern Arabian Sea on Fig. 7, A and C probably owes its strange shape to a complex interaction between the east—west oriented, near-equatorial trough and the north—south oriented, heat trough over peninsular India.

Along and north of the ridge line at 12°N, weather is fine.

Air—sea interaction over the western Arabian Sea

Although mean charts lack sufficient detail to show it, summer surveys of the western Arabian Sea by ships and aircraft confirm the persistence of a narrow low-level southwesterly jet (e.g., FINDLATER, 1969). The jet is maintained by a feedback mechanism linking ocean and atmosphere (BUNKER, 1967). When the heat low first develops, in response to the sun's zenithal march, south-westerlies set in off Somalia and Arabia, forcing surface waters away from the shore (STOMMEL and WOOSTER, 1965). The ensuing upwelling of colder water cools the adjacent air producing a local pressure ridge and a large thermal gradient. These two effects combine, further strengthening the winds by protecting the heat low from inflowing maritime air and creating a low-level jet.

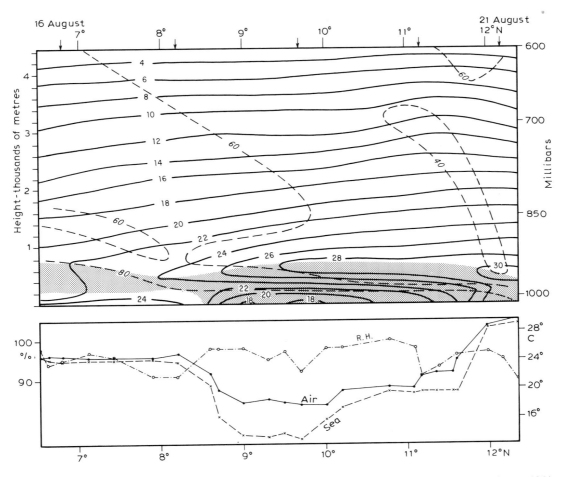

Fig. 9. Analysis of measurements made along about 50°E by R.V. *Discovery* between 16 and 21 August 1964. Upper: isotherms (full lines); relative humidity isopleths (dashed lines); inversion layer (stippled). Arrows denote locations of radiosonde ascents. Lower: air temperature at deck level (full line); relative humidity at deck level (dot-dashed line); sea surface temperature (dashed line).

Off Somalia to the south of Socotra where upwelling is most intense, rain seldom falls, clouds are scanty, and for a third of the time visibility is reduced below 10 km by mist or haze (Figs. 36 – 39). However, *fog is practically unknown*, despite the fact that the air passing over the cold water has its temperature lowered well below the dew point it possessed a day or two previously. The causes for this combination of phenomena can best be sought in discussion of the following example.

Weather off Somalia in August 1964

Between 16 and 21 August 1964, the research vessel *Discovery* sailed northward along about 50°E from 6.5° to 12.3°N (Fig. 1) making numerous sea surface and air temperature measurements as well as five aerological soundings (Fig. 9). The synoptic situation, which changed little during this period, is illustrated by the surface analysis for 06 GMT 20 August (Fig. 10) and by weather satellite photos taken at 0907 GMT 20 August (Fig. 11).

From 16 to 21 August, *Discovery* never recorded more than 10 percent of low cloud nor visibility at less than 10 km. Winds blew persistently from between S and SW at from 10 to 15 m s⁻¹, diminishing only after the ship sailed to the lee of Cape Guardafui. At the outset, air and sea temperatures were the same and a weak inversion extending from about 200 to 800 m inhibited low cloud development. Then when the ship encountered cold water near 8.5°N, the inversion extended to the surface being intensified both from below and above to greater than 10°C. Although relative humidity of the

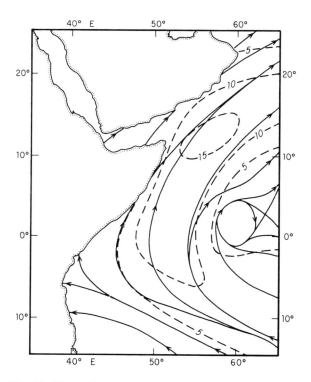

Fig. 10. Kinematic analysis of surface winds for 0600 GMT 20 August 1964. Streamlines (full lines), isotachs (dashed lines) in m s⁻¹.

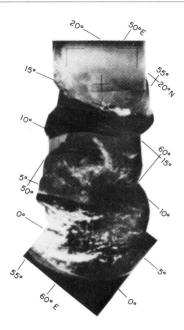

Fig. 11. Mosaic of photographs made from the weather satellite TIROS VIII, orbit 3523 at 0907 GMT 20 August 1964.

surface air exceeded 90%, no fog developed despite the fact that only 24 h before the air moving over the cooling surface had possessed a dew point 9°C higher than the temperature of the upwelled water.

EMMONS and MONTGOMERY (1947) point out that when air is cooled from beneath, moisture as well as heat is transported downward to condense on the underlying surface. Thus, unless the air is further cooled by some independent agency such as radiation, saturation might never be attained.

Discovery's soundings reveal a rapid decrease in relative humidity accompanied by a slight increase in mixing ratio with height in the inversion confirming the existence of a downward moisture flux (ROLL, 1965). The strong surface winds no doubt facilitated heat and moisture transport despite the great stability of the air.

A mystery remains—why does fog never form here, whereas off California apparently similar conditions frequently result in fog? The answer can be found in the character of the summer circulation over the Arabian Sea (RAMAGE, 1966). As compensation for the great upward motion accompanying the monsoon rains of India into which the surface southwesterlies feed, convergent easterlies dominate the upper troposphere over the western Arabian Sea north of 7° or 8°N and cause massive and persistent subsidence (see p. 630). Because the surface winds accelerate downstream and diverge under the influence of the heat low (Figs. 31 and 33) and because (in contrast to the trade winds) no heat-induced convection can develop in the lowest layers, the subsidence inversion almost reaches the ocean surface. Warming due to subsidence at the top of the inversion overpowers cooling by both radiation above and upwelled water at the base to prevent fog forming, a conclusion borne out by the *Discovery* measurements. At the inversion top, the temperature at 7°N south of both the main subsidence and the upwelled water was 6°C below the temperature at 10°N where maximum surface cooling took place.

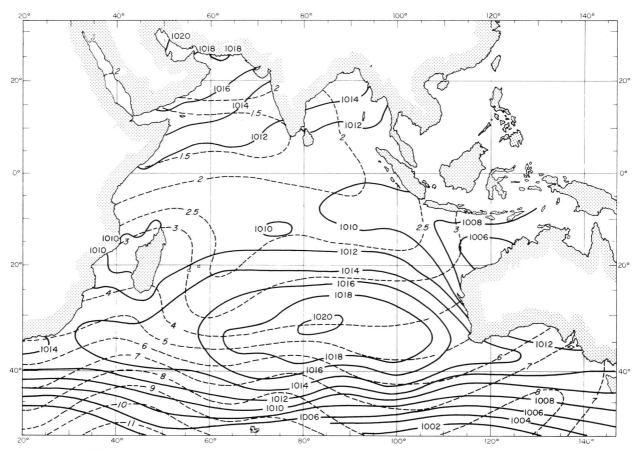

Fig. 12. Sea-level pressure, January. Mean (full lines) and standard deviation (dashed lines) in mbar.

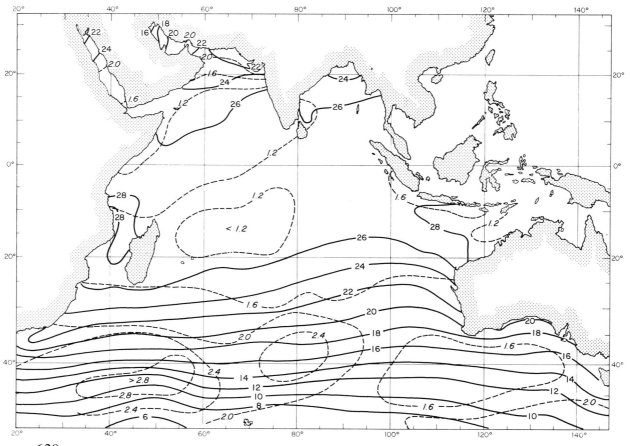

Fig. 13. Air temperature, January. Mean (full lines) and standard deviation (dashed lines) in °C.

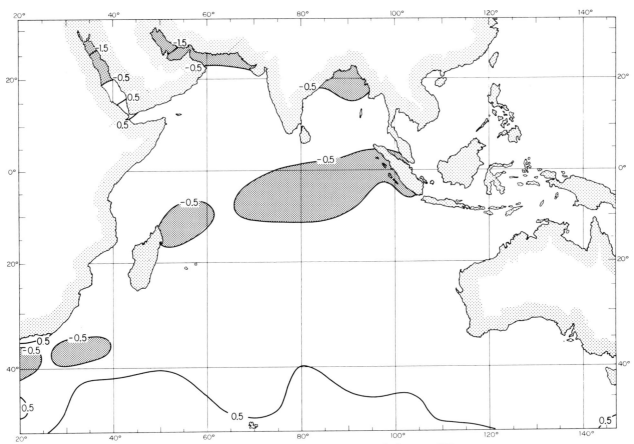

Fig. 14. Mean values of air temperature minus sea temperature in January (°C).

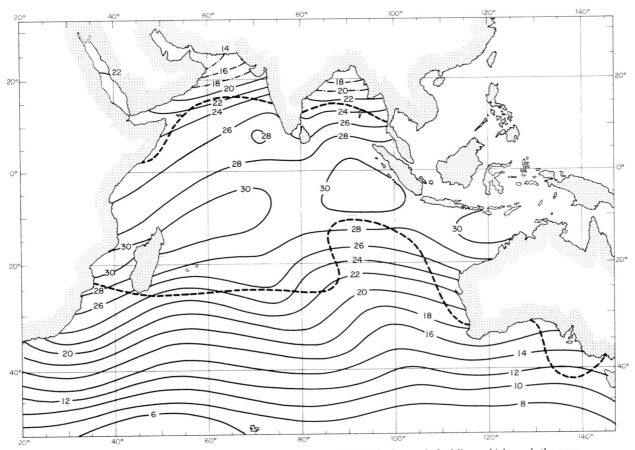

Fig. 15. January, mean vapour pressure in mbar (full lines). The heavy dashed lines which mark the equatorward limits of regions where the median 60% of air temperature observations ranges over more than 2°C also roughly delimit frontal occurrences.

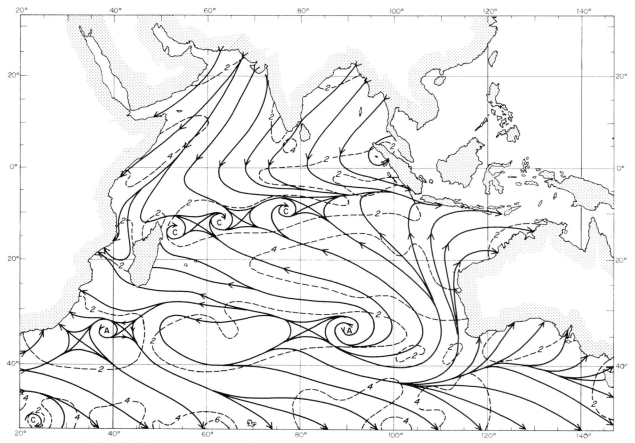

Fig. 16. Mean resultant surface winds in January. Streamlines (full lines), isotachs (dashed lines) in Beaufort force numbers.

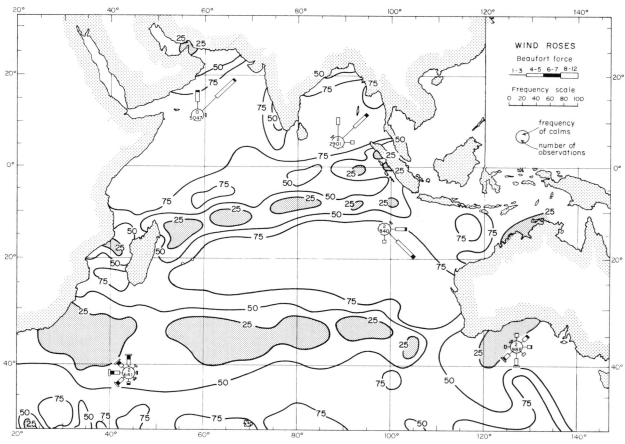

Fig. 17. Surface winds in January. Steadiness (full lines) in percent $\left(\dfrac{\text{mean resultant wind} \times 100}{\text{mean scalar wind}}\right)$ and representative eight-point wind roses.

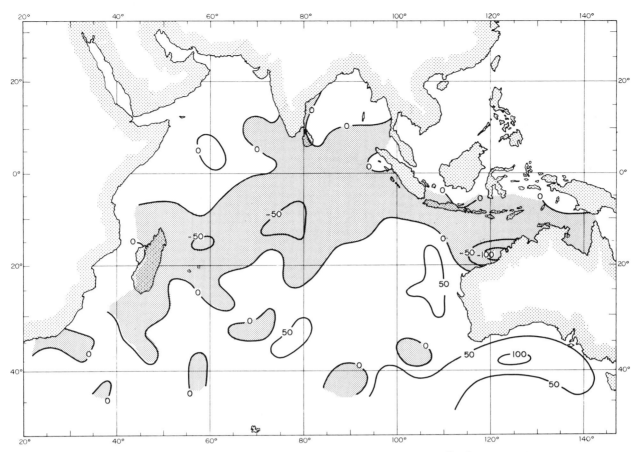

Fig. 18. Divergence of the mean resultant surface winds for January, in 10^{-7} s^{-1}. Regions of convergence stippled.

Fig. 19. Mean resultant winds for January at 700 mbar (3 km). Streamlines (full lines), isotachs (dashed lines) in m s^{-1}.

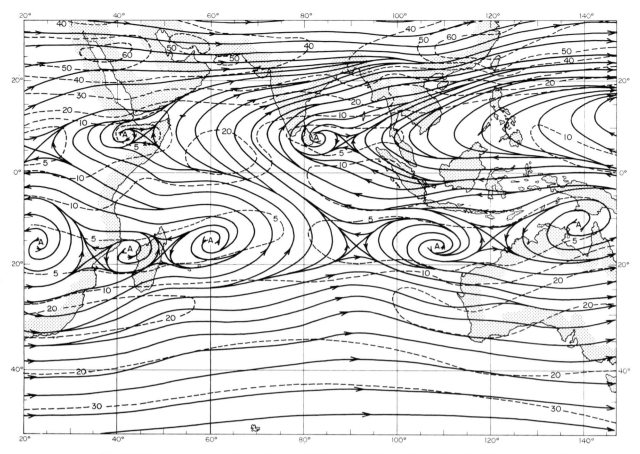

Fig. 20. Mean resultant winds for January at 200 mbar (12 km). Streamlines (full lines), isotachs (dashed lines) in m s^{-1}.

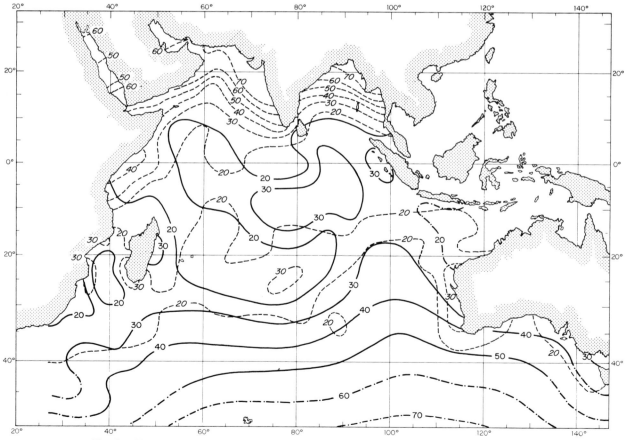

Fig. 21. Cloudiness for January. Percentage frequency of low cloud cover equal to or exceeding 60% (full lines); percentage frequency of total cloud cover equal to or less than 20% (dashed lines).

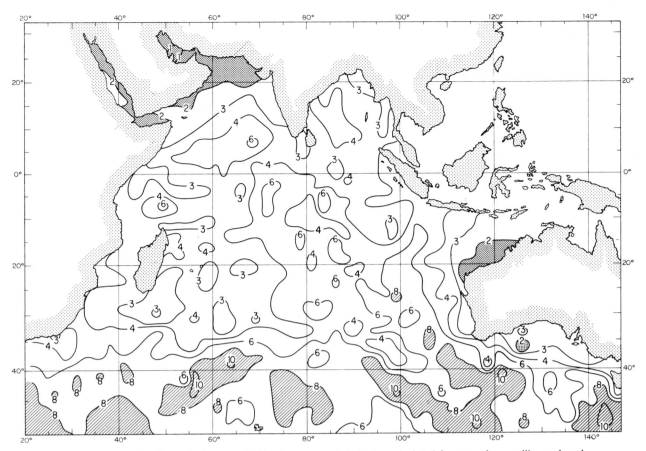

Fig. 22. Cloudiness for January, 1966. Mean amounts in tenths calculated from weather satellite nephanalyses.

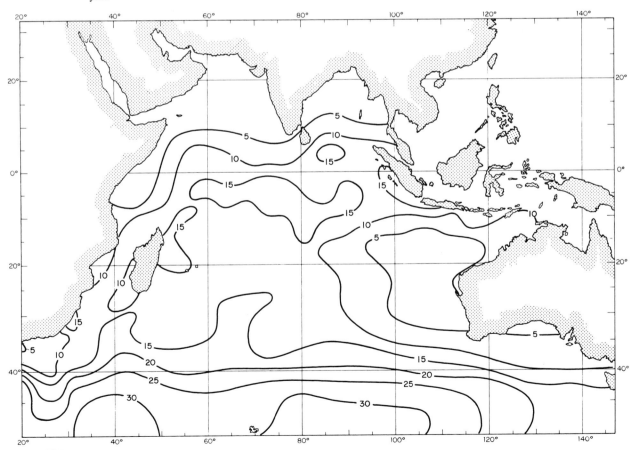

Fig. 23. Precipitation for January. Percentage of observations reporting precipitation.

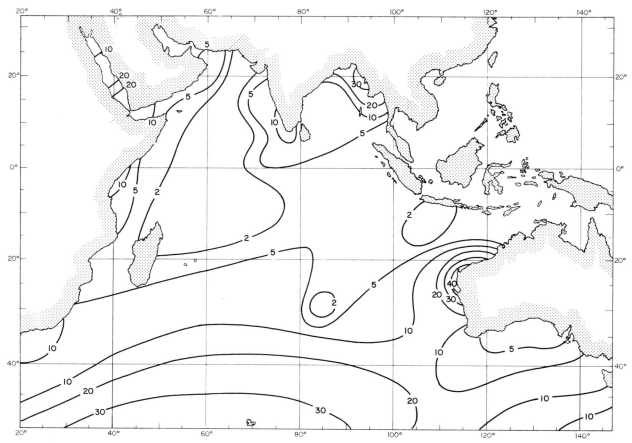

Fig. 24. Visibilities less than 8 km for January. Percentage frequency of fog, mist, and haze.

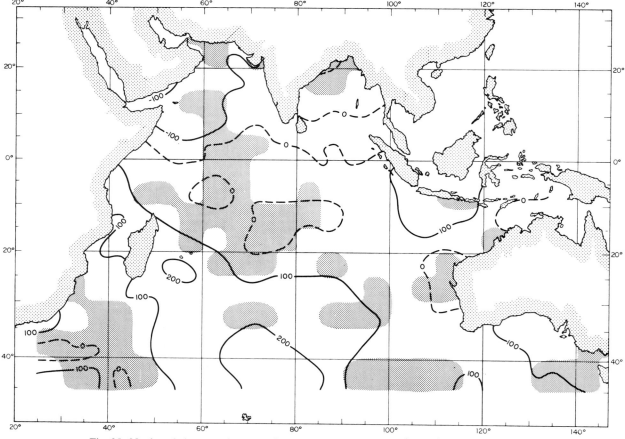

Fig. 25. Net heat balance at the sea surface for January in cal. cm^{-2} day^{-1}. Average of 1963 and 1964. In the stippled areas, the difference between 1963 and 1964 exceeded 100 cal. cm^{-2} day^{-1}.

Conditions during January and July

Study of the region dominated by the monsoons and the region relatively unaffected by the monsoons reveals many distortions in the simple circulation model outlined in the introduction. Significant anomalies will be delineated and explained through mean charts for January (Figs. 12–25) and July (Figs. 27–40) and by climatic tables for ocean areas and island stations (Appendix), pp. 660–671.

January

Monsoon region

Northeast monsoon. Radiational cooling over Asia exceeding that over the neighbouring ocean has established shallow continental highs and a seaward-directed pressure gradient. The huge Siberian polar anticyclone, source of chilling gale-force outbreaks which sweep southward across the China Seas, is effectively cut off from the Indian Ocean by the Himalayas and the contiguous mountain ranges of Afghanistan and Iran. Thus, the winter monsoon is a gentle phenomenon in the Indian Ocean—prevailing moderate northeast winds are temporarily freshened in the rear of depressions moving eastward south of the massif. Air is dry over the continent but as it flows out across the Arabian Sea and Bay of Bengal, it absorbs both heat and moisture from the ocean surface. Thus, temperature, vapour pressure, cloudiness, and rainfall increase along the trajectory, whereas atmospheric stability and the incidence of mist or haze decrease. Changes between Amini Divi and Minicoy and between areas 1 and 3 conform to this pattern. Generally, clear skies and light winds so combine to favour insolational heating over evaporative cooling that even in midwinter only the western Arabian Sea loses significant heat.

Over Africa, cooling in the north and heating in the south have established a significant trans-equatorial pressure gradient affecting the circulation 800 km to the east and extending the northeast monsoon well into the Southern Hemisphere.

(*1*) Along the coast, periods of dry outflow from Arabia and Somalia and the stabilizing effect of general divergence, as the airstream splits to spiral into the Kalahari heat low or into the trough east of Madagascar (Fig. 16), combine to reduce the rate of air mass transformation. Latitude for latitude, vapour pressure, cloudiness, and rain are lowest immediately off the African coast while turbidity frequency varies little between the Gulf of Oman and the Mozambique Channel.

(*2*) Farther east the north–south pressure gradient, large in the north and smaller near to and south of the Equator, results, according to GORDON (1967), in air trajectories diverging north of the Equator and turning anticlockwise and converging between the Equator and the Southern Hemisphere near-equatorial trough.[*1] High January rainfall at Mahé, Grande Comore, Agalega, and Diego Garcia (see the tables in the Appendix) lends support to Gordon's ideas, for all of them lie in a region of anticlockwise turning of the mean resultant winds north of the trough.

[*1] That convergence does not dominate this region on the mean divergence chart (Fig. 18) might be accounted for by the paucity of ship observations.

627

Northern Hemisphere near-equatorial trough. Over the southern Bay of Bengal (Appendix, Table VI), the distribution of pressure and, in particular, of the winds at 700 mbar (Fig. 19) indicates the presence of a trough just north of the Equator.

Convergence, cloudiness, and squally, thundery showers are relatively frequent north of the trough line, approximately beneath the subtropical upper-tropospheric ridge. It is here that much of the moisture evaporated earlier from the sea surface by the northeast monsoon is condensed and the released latent heat goes to warm the troposphere. The rain-shadowing of India and Ceylon weakens the effect over the Arabian Sea.

Over the central equatorial region, weak ephemeral counterclockwise eddies (extensions of the near-equatorial trough to the east) expand and contract in complex response to changes in higher latitudes. Stronger winds on the western than on the eastern sides of the eddies reflect a net momentum transport southward. As quasi-stationary areas of bad weather develop, intensify, and dissipate, frequent showers fall from moist, conditionally unstable air.

Southern Hemisphere near-equatorial trough. Anchored by heat lows over the Kalahari and Australian deserts, the pressure trough arcs northward to near 10°S over the central ocean. The trough which may shift appreciably from day to day is the birthplace of tropical cyclones (see p. 643). In the mean, it combines convergence and high uniform temperature and vapour pressure. However, the axes of maximum cloudiness and precipitation lie somewhat equatorward of the trough axis (see p. 612). In the trough region the ocean gains heat, for although clouds may significantly reduce incoming radiation and the ocean loses sensible heat to the atmosphere, light winds and high humidity minimize evaporational cooling.

The Northern and Southern Hemisphere near-equatorial troughs and the intermediate zone of light, predominantly westerly winds comprise the *doldrums*.

Australian summer monsoon

The heat low over Australia draws in trade wind air deflected from the southwest (see below). Appreciable rain falls in the area of anticlockwise-turning surface winds west of Sumatra (Table IX). Farther south, however, the tropical maritime air converging into the Australian heat low is surprisingly deficient in cloud and rain, a situation resembling that occurring over the northern Arabian Sea in July and discussed in detail below (see p. 630).

A puzzling feature (which is, however, not nearly so marked in all ocean atlases) is the high incidence of mist and haze (fogs are rare) off Northwest Cape (Fig. 24), but without the associated cold upwelling water many oceanographers expected. Since clouds are scanty here, the frequent poor visibility may be due to dust. Inland over the desert, dust is convectively mixed through a depth of at least 3 km and then often blown to sea by the predominant easterlies on the south side of the heat low.

Non-monsoon region

The non-monsoon region, south of the summer latitude of the pressure trough (just south of Agelega and just north of Cocos), includes the trade winds and the subtropical ridge at about 35°S.

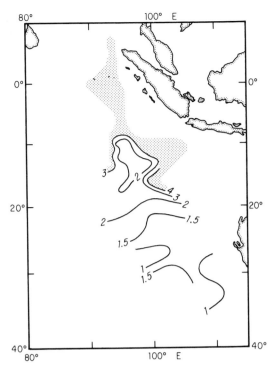

Fig. 26. Height of the base of the trade wind inversion (in km) determined from 62 soundings made by R.V. *Koyo Maru* between 23 November 1962 and 22 January 1963 and between 23 November 1963 and 25 January 1964. In the stippled area, soundings revealed no inversion.

Between the trough and ridge where the trade winds prevail (Table IX), temperature is uniform and poor visibility rare. In the east between the heat low over Australia and the subtropical ridge to the southwest, the pressure gradient is large and winds are fresh and divergent, with an eastern branch spiralling in to the heat low and a western branch sweeping around the subtropical ridge. The divergent southerlies advect and cause cold water to upwell (although apparently not along the Australian west coast), which, in turn, affects the distribution of temperature and vapour pressure[*1] and presumably accounts for recognizable cold fronts penetrating farther equatorward than in other longitudes (see also TALJAARD, 1967). On most soundings made by the research vessel *Koyo Maru* (Figs. 1 and 26), the base of a sharp trade wind inversion is found between 1 and 3 km.

Winds blowing up the vapour pressure gradient beneath an inversion cap result in considerable cloud, little rain, and small ocean heat gain.

In contrast, the trade winds farther west (Tables X and XVIII) nearly parallel the vapour pressure isopleths. Cloudiness is less than in the east, but because the air is moister and flow converges toward the pressure trough, rain is more frequent.

A stable layer is found about 2 to 3 km above the surface on 88% of occasions at Mauritius but on only 53% of occasions at Tulear farther west (EMON, 1949).

After developing in the trough to the north, tropical cyclones usually travel southwest-

[*1] Van Dijk (1956) conversely suggests that the presence of cold water causes the surface wind to diverge.

ward and then recurve toward the southeast, completely disrupting the normal trade wind weather with hurricane winds and torrential rains (Fig. 44, p. 643).

The *subtropical ridge*, roughly coinciding with the axis of maximum wind divergence, lies about 1000 km *south* of the axis of minimum rainfall frequency. That this is a significant displacement is borne out by measurements made during the *Meteor* expedition to the South Atlantic (FICKER, 1936), where the base of the trade wind inversion was closest to the surface 1000 to 1500 km north of the ridge. Presumably, divergence acting for some time on air flowing out from the ridge produces a downstream lowering of the inversion and decrease of precipitation.

Beneath the ridge because of light winds and fair weather, the net heat gain by the sea is greatest.

July

Monsoon region

Southwest monsoon. Intense heating over southern Asia, particularly the desert arc extending from Somalia to northwest India, maintains vigorous heat lows and a landward-directed pressure gradient.

As described earlier, southwestern winds off the coasts of Arabia and Somalia (Table V) force cold water to upwell which, in turn, modifies the air above (see p. 617).

Strong winds and adjacent deserts account for the high incidence of mist and haze over the western Arabian Sea. However, the apparent anomaly of surface convergence and fair weather over the northern Arabian Sea can only be explained in terms of middle tropospheric subsidence. Air converging into the heat lows certainly rises but spreads out beneath the subsidence inversion before significant condensation can occur; thus, deep precipitating clouds cannot develop. Frequent heavy rains off the west coast of India, where in the mean surface winds neither converge nor diverge, can be accounted for by development of subtropical cyclones (see p. 608).

"Monsoon depressions" (having the thermal properties of tropical cyclones but seldom reaching storm intensity) develop near the head of the Bay of Bengal and moving slowly westnorthwest merge with and reinforce the trough lying along the Ganges Valley. Over the northern Bay of Bengal (Table IV), these depressions maintain southwesterlies that are weaker than those over the Arabian Sea. Upper tropospheric easterlies attain maximum strength along about 70°E in response to the combined condensation above and radiational heating of the Himalayas and Tibet (RAMAGE, 1971). East of 70°E, the easterlies are divergent and combine with convergent lower tropospheric southwesterlies to produce upward motion, cloudiness, and considerable rain. Over the southern Bay of Bengal, the rain-shadowing effect of peninsular India and Ceylon extends a considerable distance downstream, while over the central Arabian Sea and the southern Bay of Bengal, evaporational cooling by strong monsoon winds and reduction of incoming solar radiation by clouds surprisingly result in a net cooling of the surface (COLÓN, 1964).

To the west, high pressure in the cool south and low pressure in the warm north extend the monsoon south over Africa to 20°S. Blowing from SSE rather than SW, this monsoon branch belongs, in effect, to the wintertime trade winds.

630

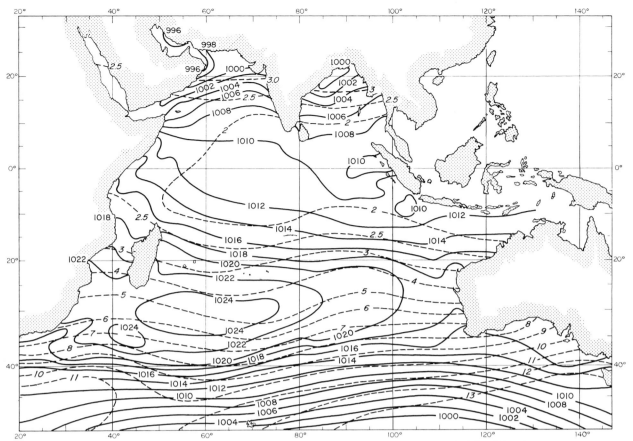

Fig. 27. Sea level pressure for July. Mean (full lines) and standard deviation (dashed lines) in mbar.

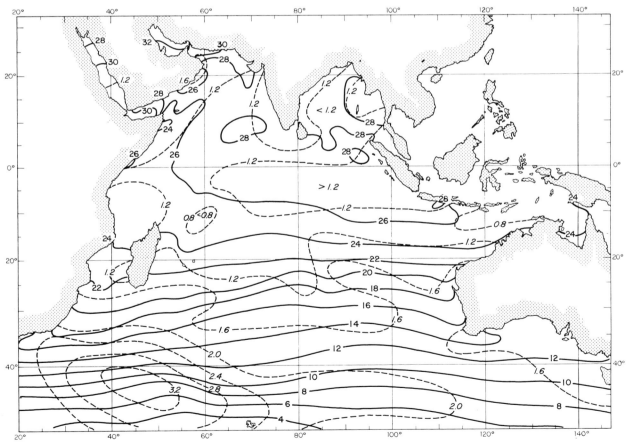

Fig. 28. Air temperature for July. Mean (full lines) and standard deviation (dashed lines) in °C.

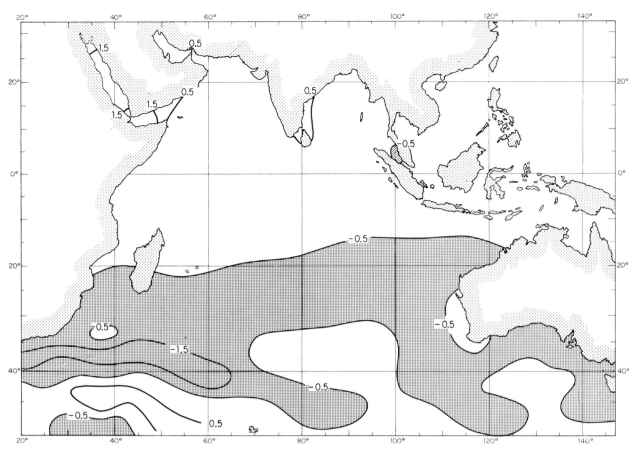

Fig. 29. Mean values of air temperature minus sea temperature for July, in °C.

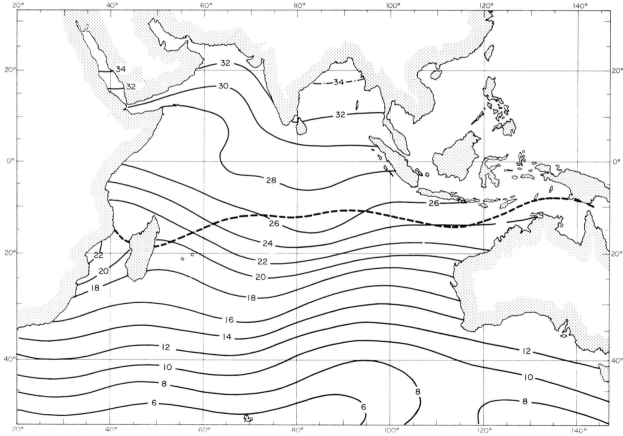

Fig. 30. Mean vapour pressure for July, in mbar (full lines). The heavy dashed line which marks the equatorward limit of the region where the median 60% of air temperature observations ranges over more than 2°C also roughly delimits frontal occurrences.

632

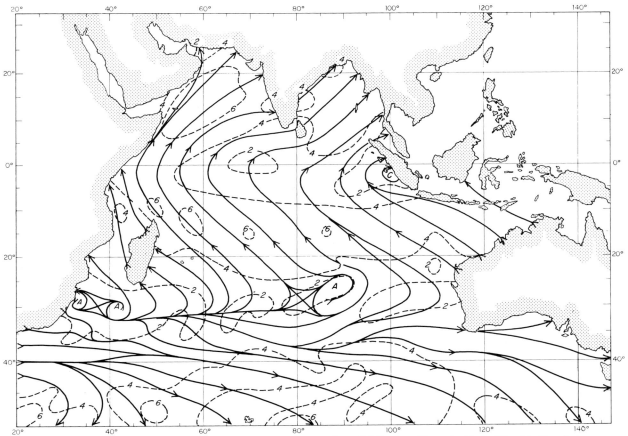

Fig. 31. Mean resultant surface winds for July. Streamlines (full lines) and isotachs (dashed lines) in Beaufort force numbers.

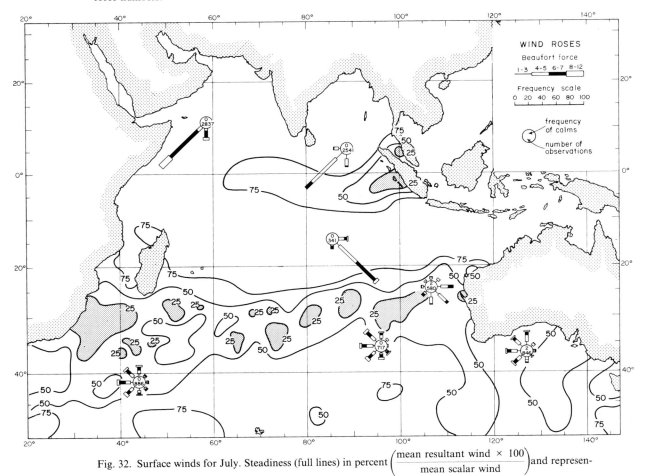

Fig. 32. Surface winds for July. Steadiness (full lines) in percent $\left(\dfrac{\text{mean resultant wind} \times 100}{\text{mean scalar wind}}\right)$ and represen-

tative eight-point wind roses.

633

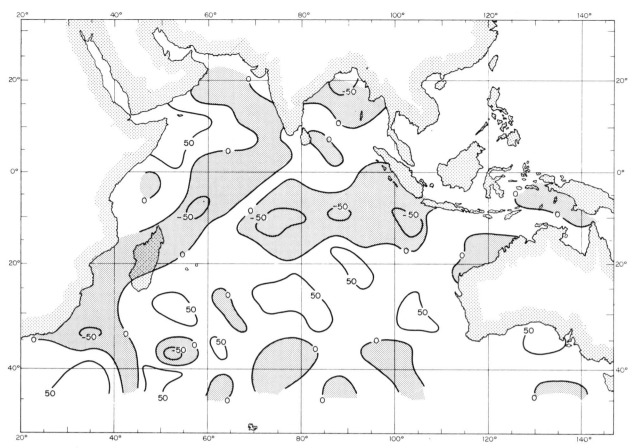

Fig. 33. Divergence of the mean resultant surface winds for July, in 10^{-7} s^{-1}. Regions of convergence stippled.

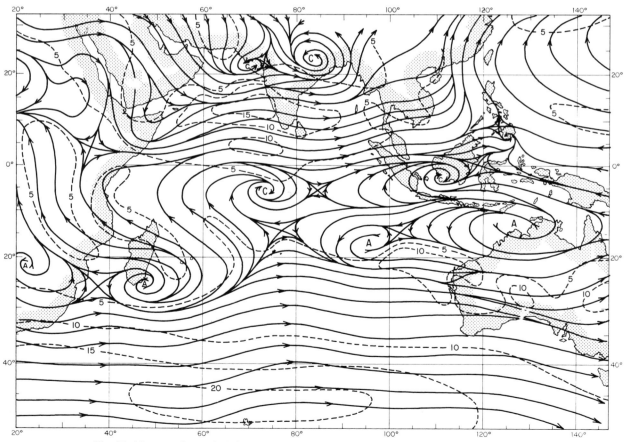

Fig. 34. Mean resultant winds for July at 700 mbar (3 km). Streamlines (full lines) and isotachs (dashed lines) in m s^{-1}.

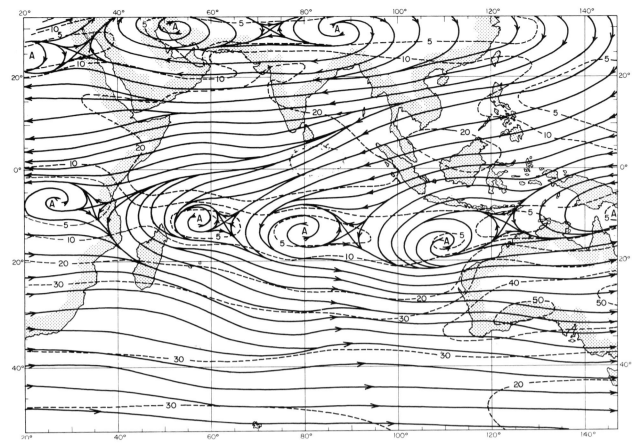

Fig. 35. Mean resultant winds for July, at 200 mbar (12 km). Streamlines (full lines) and isotachs (dashed lines) in m s^{-1}.

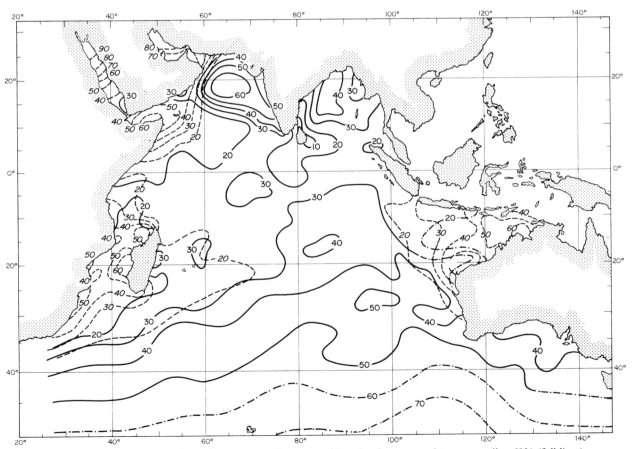

Fig. 36. Cloudiness for July. Percentage frequency of low cloud cover equal to or exceeding 60% (full lines) and percentage frequency of total cloud cover equal to or less than 20% (dashed lines).

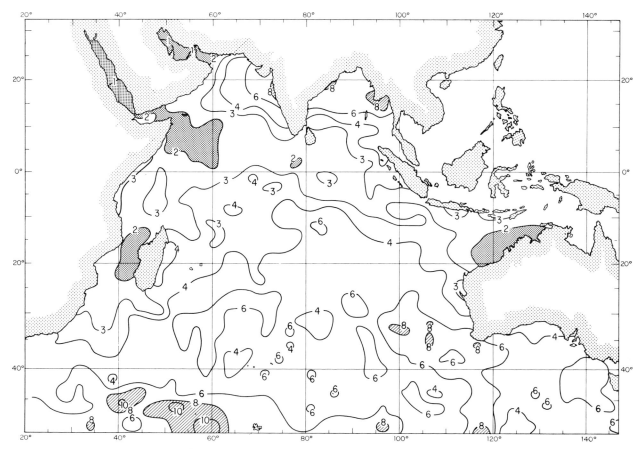

Fig. 37. Cloudiness for July, 1965. Mean amounts in tenths calculated from weather satellite nephanalyses.

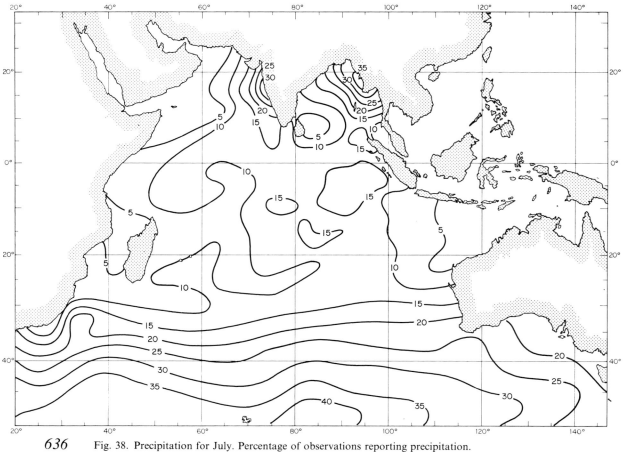

Fig. 38. Precipitation for July. Percentage of observations reporting precipitation.

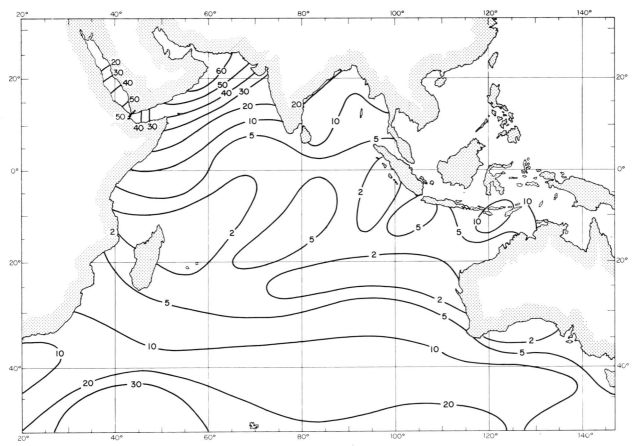

Fig. 39. Visibilities less than 8 km for July. Percentage frequency of fog, mist, and haze.

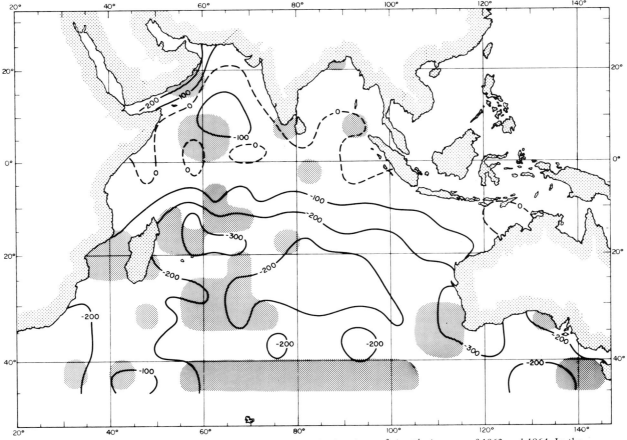

Fig. 40. Net heat balance at the sea surface for July, in cal. cm^{-2} day^{-1}. Average of 1963 and 1964. In the stippled areas, the difference between 1963 and 1964 exceeded 100 cal. cm^{-2} day^{-1}.

Off the African coast, the low rainfall at Mahé, Agalega, and possibly Grande Comore*¹ is to be expected (GORDON, 1967). With the pressure gradient now directed from south to north, trajectories would be expected to diverge south of the Equator and rain would be inhibited.

Southern Hemisphere near-equatorial trough. Although the mean pressure distribution is uninformative, the wind field at both the surface and 700 mbar indicates a trough just south of the Equator (Table VII). The axes of bad weather, convergence, and the ridge at 200 mbar all lie south of the trough which probably stems from causes akin to those responsible for the January near-equatorial trough in the Northern Hemisphere. Doubtless, the orientation of the Somali coast, the effect of upwelling in restricting inflow to the heat low, and the circulation around the subtropical ridge to the south contribute to extending the trough across the width of the ocean, although west of 55°E, where the winds are strong and steady from the south, cyclonic circulations are unknown.

Over the central and eastern ocean in the doldrums, weak clockwise eddies prevail, appearing and disappearing and occasionally drifting across the Equator. As in January, the eddies respond to changes in higher latitudes. For example, when the Arabian Sea branch of the southwest monsoon is strong, the eddy to the south is small and confined to the equatorial zone; when the monsoon weakens, the eddy expands northward as a weak ridge. Stronger winds on the western sides than on the eastern sides of the eddies reflect a net transport northward.

Australian winter monsoon. Between Indonesia and western Australia (Table VIII), outflow from the continental anticyclones contributes to upwelling south of Java (WYRTKI, 1962). Cloudiness and rainfall are scanty but increase downstream as warmer waters modify the air mass. As in the west, the monsoon between 5° and 25°S is, in effect, an extension of the wintertime trade winds.

Over the northeast Timor Sea and northwest of Madagascar, surface heat exchange is small, probably because usually clear skies allow a significant fraction of the solar radiation to reach the surface. The effect is similar to, but because of the stronger winds is not as marked as, the January heating of the northern Indian Ocean.

Non-monsoon region

Trade winds. Although the near-equatorial trough is farther from Cocos in July than in January, the peculiar influence of the Australian heat low is absent, the island lies in convergent southeasterlies, and appreciably more rain falls than in midsummer.

The trade winds are characterized by even temperature and good visibility. Weather is fair in the east where the air is predominantly of continental origin and in the west (Mauritius and area 8) where divergence develops as air accelerates northward under the influence of the south — north pressure gradient across Africa. Between 2 and 3 km above the surface at Mauritius and Diego Suarez, a stable layer exists on more than 90% of occasions (EMON, 1949).

In July, evaporative cooling of the sea surface by the trade winds is only slightly greater

---

*¹ The secondary maximum of rainfall in July at Grande Comore (Table XVI) is unlikely to reflect long-term reality in view of the shortness of record, the single peaked curve of rain day frequencies, the rain frequencies in the area as determined from ship reports, and the absence of any July peak in the records from other islands in the Comore group.

than in January but because of the change in season, incoming solar radiation is much less. Thus, surface cooling exceeding 150 cal. cm^{-2} day^{-1} contrasts with a slight net warming in the summer.

From 20 June to 9 July 1964, the research vessel *Argo* on passage from Darwin to Mauritius (Fig. 1) made 13 radiosonde ascents between 15°S 110°E and 25°S 70°E. The soundings are analyzed in Fig. 41, which also shows cloud cover and weather observed

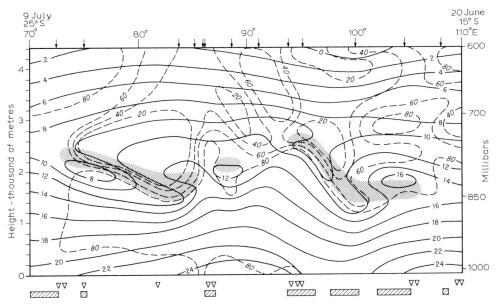

Fig. 41. Analysis of measurements made in the Southern Hemisphere trade wind zone by R.V. *Argo* between 20 June and 9 July 1964. Isotherms (full lines); relative humidity isopleths (dashed lines); inversion layers (stippled); periods of low cloud ⩾ 75% (hatched). Showers symbolically shown; arrows denote locations of radiosonde ascents.

from the ship. The sequence is typical of the trade winds, in particular the degree of variability it reveals. Showers fell when the subsidence inversion was high or absent, but until the ship reached 78°E winds, while fluctuating in speed, remained constant from SE. Although there were few supporting data, it seems unlikely that easterly waves could have accounted for the unsettled periods. Below 1.5 km, relative humidity changed little, ranging from 80 to 95%. Above 1.5 km, however, there were wide fluctuations— from less than 20% in subsiding layers to more than 80%. Showers were associated with a deep moist layer or with more than one moist layer. Of 21 observations of showers, 12 occurred when surface pressures were rising and 9 when they were falling. West of 78°E, *Argo* traversed a weak, almost stationary depression; winds at the ship backed through NE to SW and finally to SE. The sounding made at 72.5°E just west of the depression centre shows no inversion and a deep moist layer.

It seems likely then, except for rare depressions, that rain in the trade wind regime of the southern Indian Ocean may well fall in response to changes in the middle as well as in the lower troposphere.

The *subtropical ridge*, 500 km farther north but closer to the belt of minimum rainfall than in January, is marked by fair weather.

Winter — summer differences

Comparison of the January and July mean charts reveals marked seasonal asymmetry between the largely monsoonal Northern Hemisphere and the largely non-monsoonal Southern Hemisphere, and a less marked asymmetry between west and east in the Indian Ocean.

Heat balance

Fig. 42 shows the difference in net heat balance at the ocean surface between January and July. The south-central Arabian Sea gains more heat in January than in July. Strong winds and cloudy skies in July and light winds and clear skies in January outweigh greater solar radiation in July. In the monsoon regions of the Southern Hemisphere, the effect is scarcely perceptible. However, somewhat cloudier skies and more rain in the west and stronger winds in the east in January than in July reduce the summer-winter difference in the net heat balance. Differences are also small in the doldrums between the equator and 10°S where skies are generally cloudy and winds light throughout the year.

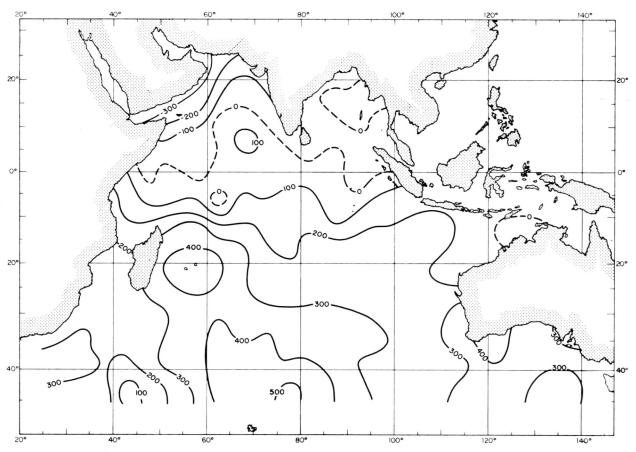

Fig. 42. Net heat exchange at the sea surface as a monsoon criterion. The difference, January minus July, is shown in cal. cm^{-2} day^{-1}. Away from the Equator, small changes between January and July are confined to regions with wet or cloudy summers and dry sunny winters.

The largest differences occur in the regions of trade winds reflecting the constancy of winds and weather and the change in solar radiation.

Trough and ridge shifts

Since the subtropical ridge inclines with height toward higher temperatures, its position at 200 mbar is controlled in varying degrees by the locations of: (*1*) surface ridge; (*2*) maximum surface heating (the trough); and (*3*) maximum release of latent heat by condensation. Hence, most of the time the 200 mbar ridge lies equatorward of location 1 and poleward of locations 2 and 3.

In the Northern Hemisphere, the surface trough and the ridge at 200 mbar move more than 2000 km northward between January and July in response to the development of heat lows over the great mass of southern Asia.

Between July and January in the Southern Hemisphere, the surface trough moves only 600 km southward in the central ocean as compared with 2000 km in the east over Australia and 1200 km in the west over Africa where continental effects prevail. At 200 mbar, the ridge shifts little over the central ocean *and over Australia*, but in the west it is displaced over 1000 km to the south.

Over Australia, although a heat low develops along about 20°S by January and the *surface* ridge shifts more than 1200 km south of its July latitude, very heavy rains and daily thunderstorms over Indonesia convert such large quantities of latent heat to sensible heat and transport it to the high troposphere that the upper tropospheric ridge moves only a little south.

Trans-equatorial flow

On the average over the Indian Ocean, surface pressure is lower and rain is heavier in the summer hemisphere than in the winter hemisphere. Consequent with the greater release of latent heat, upper tropospheric pressure is higher in the summer hemisphere. Thus, mean trans-equatorial flow is directed toward the summer hemisphere at low levels and away from it at high levels (RAO, 1960).

Tropical cyclones

Because tropical cyclones possess a bimodal frequency distribution over much of the northern Indian Ocean with peaks in spring and autumn, and because in their relative rarity they make little impression on mean monthly conditions, they could not be treated adequately in the preceding section.

Figs. 43 and 44 show the frequency distribution and track envelopes for tropical cyclones. The length of record (Appendix, Table I) and the fact that the northern Indian Ocean and the western and eastern parts of the southern Indian Ocean are crossed by busy shipping lanes make it unlikely that many storms went undetected in these areas. However, over the southern Indian Ocean between 70° and 100°E where few ships voyage, information from the meteorological satellites suggests that storms may have gone unseen in the past (SADLER, 1967). Therefore, the sharp frequency discontinuities along 70° and 110°E may be partly fictitious.

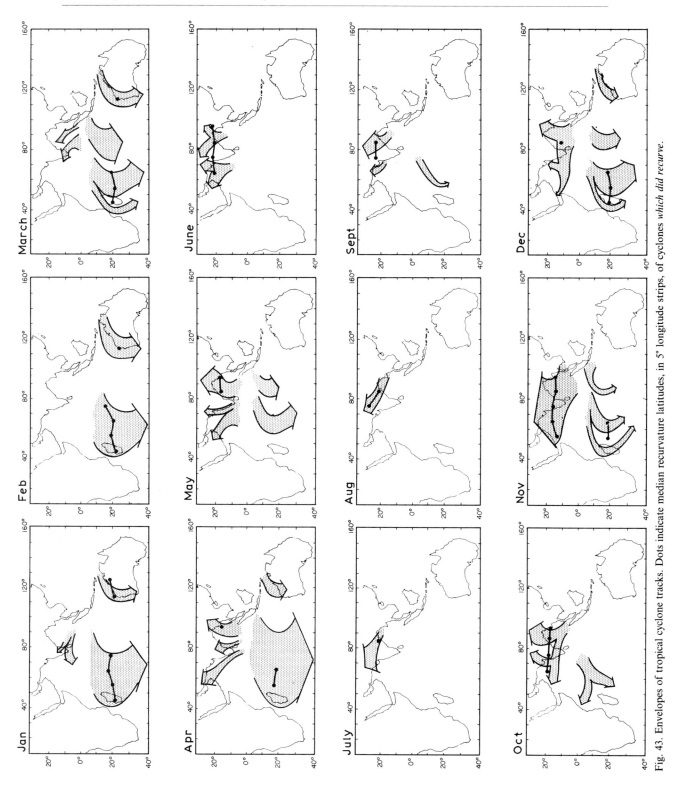

Fig. 43. Envelopes of tropical cyclone tracks. Dots indicate median recurvature latitudes, in 5° longitude strips, of cyclones *which did recurve*.

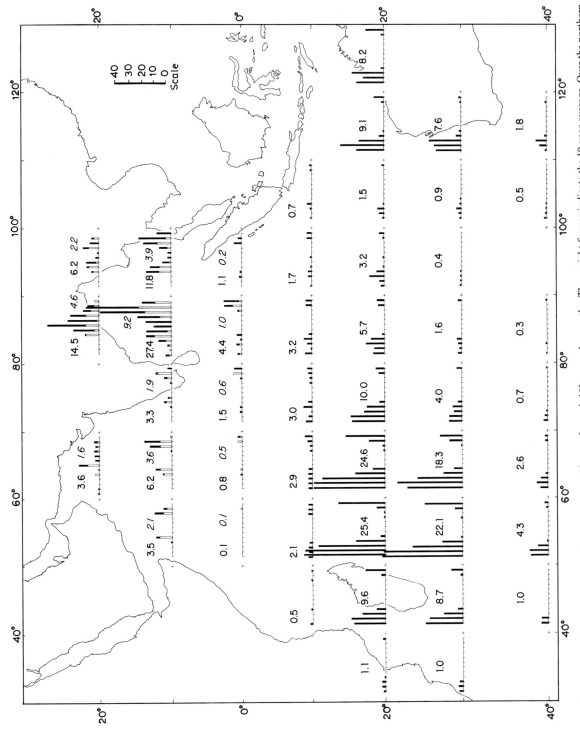

Fig. 44. Tropical cyclones. Average numbers of occurrences per 10 years for each 10°-square by months. The upright figures indicate the 10-year average. Over the northern Indian Ocean, the unblacked segments of the histograms and the sloping figures give corresponding information for severe tropical storms.

643

Tropical cyclones over the northern Indian Ocean

Variations in total frequency and in the frequency distribution of tropical cyclones over the northern Indian Ocean might be related to three interacting factors: (*1*) seasonal shifts of the surface trough (GALLÉ, 1924); (*2*) effect of the Indian subcontinent on the tropospheric circulation; and (*3*) influence of the arc of deserts to the north and west of the Arabian Sea.

In spring and again in autumn, when the trough lies near 10°N and wind shear between 850 and 200 mbar averages less than 10 m s^{-1}, conditions favour cyclone development. Lower frequencies in the Arabian Sea stem from buoyancy-inhibiting intrusions of desert air from the north and west. Cyclones are extremely rare in the Gulf of Aden (RONDELEUX, 1916; METEOROLOGICAL OFFICE, 1943), only seven having been reported between 1850 and 1965.

By July when the trough lies almost entirely over land, pressure distribution favours cyclone development only in the extreme north of the Arabian Sea and Bay of Bengal. The summer monsoon seriously restricts tropical cyclone development (RAMAGE, 1959). The preceding section described a persistent speed maximum in the upper tropospheric easterlies in Indian longitudes (Fig. 35). Thus, upper-level divergence over the northern Bay of Bengal is associated with rising tropospheric air while upper-level convergence over the northern Arabian Sea is associated with sinking tropospheric air. Since beneath the easterlies strong southwesterlies prevail in the lower troposphere, wind shear between 850 and 200 mbar averages more than 25 m s^{-1}. Thus, in July and August over the northern Arabian Sea, sinking motion and shear combine to prevent any tropical cyclone from developing and make tropical storms or depressions extremely rare. Over the northern Bay of Bengal, although the shear inhibits tropical cyclones, the general upward motion favours less intense warm-cored lows, known locally as "monsoon depressions" (see p. 630).

The fact that rainfall off the west coast of India is comparable to rainfall at the head of the Bay of Bengal can be accounted for by the occurrence of subtropical cyclones over the former region (see p. 608). In other words, the prime rain producers during summer over the northern Bay of Bengal are warm-cored monsoon depressions and over the northern Arabian Sea middle tropospheric subtropical cyclones (Fig. 45).

Tropical cyclones over the southern Indian Ocean

Throughout the region, the frequency distribution shows a single seasonal peak during January—March. Although tropical cyclones appear generally to develop over the warmest ocean areas, insufficient data prevent determination of actual causes of development. The only noticeable monsoon effect is observed northwest of Madagascar where cyclones are almost unknown. Incursions of dry air from eastern Africa over this part of the ocean during the height of the cyclone season prevent development and weaken westward-moving cyclones.

Tropical cyclone movement

The charts showing the envelopes of tropical cyclone tracks contain no surprises. Cy-

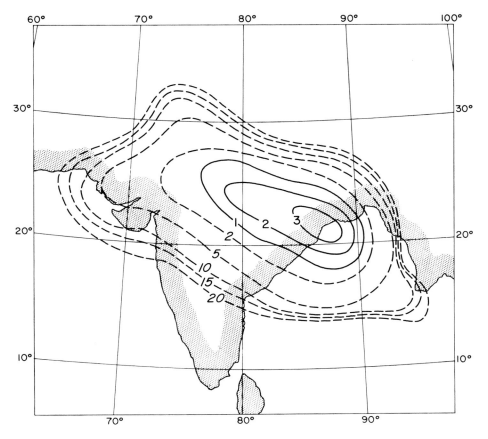

Fig. 45. Mean frequency of surface depression development during summer. Isopleths labelled in times per season (solid) and in years between occurrences (dashed).

clones tend to move toward the west when they are located in low latitudes and then as they drift poleward they usually recurve, probably in the latitude of the subtropical ridge aloft (HARWOOD, 1924), as observed over other ocean areas.

March of climate through the year (Figs. 46–52; Tables III–XXIV)

SCHOTT (1935) plotted the distribution of monthly means on time-latitude sections along the eastern and western borders of the Indian Ocean to help him describe in the compass of a few diagrams salient features of the annual variation in climate. Unfortunately, the sections were more representative of continental than oceanic conditions. In recent years, stations have been established on several small, above-water protuberances on the Central Indian Ocean Rise along about 73°E and so now it becomes possible to use Schott's powerful aid along a truly oceanic meridian, comparing and relating observations from seven stations with data derived from meteorological atlases.

I use the time-latitude sections and climatological tables for small islands and ocean areas to link the earlier discussions of conditions in January, July, and the tropical cyclone seasons into the continuous chain of the year, while calling attention to peculiar features of the transition months.

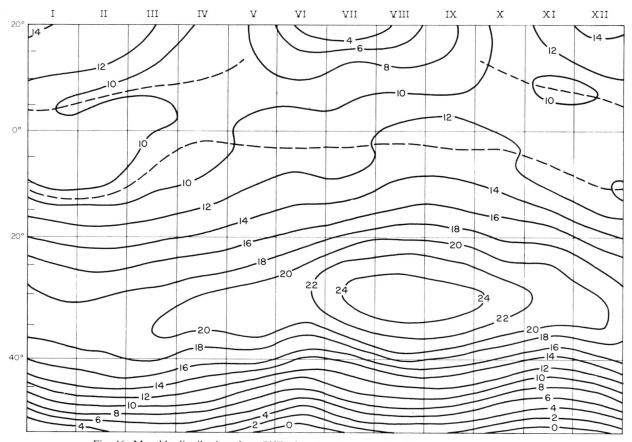

Fig. 46. Monthly distribution along 73°E of sea level pressure in 1000+ mbar. Dashed lines show climatological trough locations.

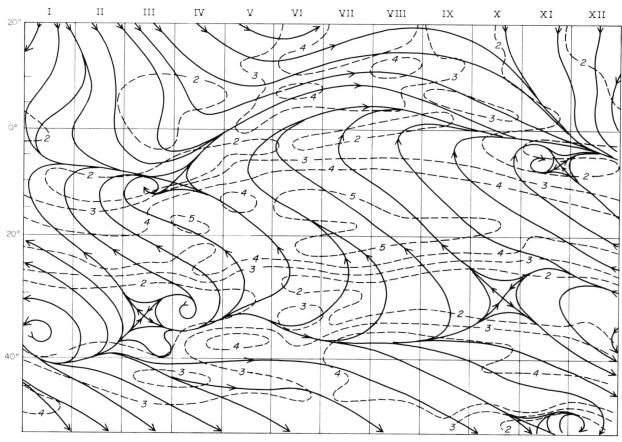

Fig. 47. Monthly distribution along 73°E of resultant surface winds. The full and dashed lines show the sequence of wind direction and wind speed (Beaufort force numbers) and should not be confused with streamlines or isotachs.

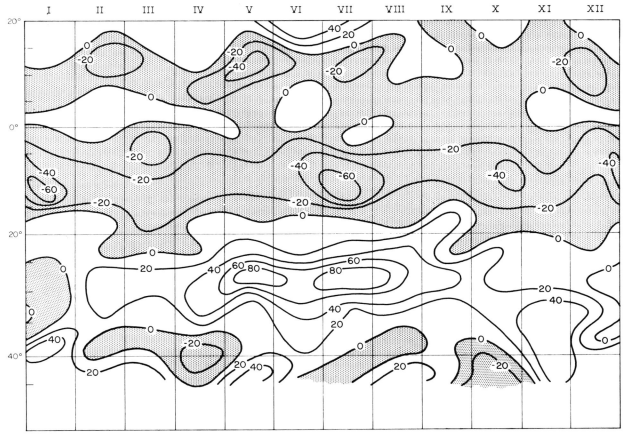

Fig. 48. Monthly distribution along 73°E of the divergence of the resultant surface winds in $10^{-7}\,s^{-1}$. Regions of convergence stippled.

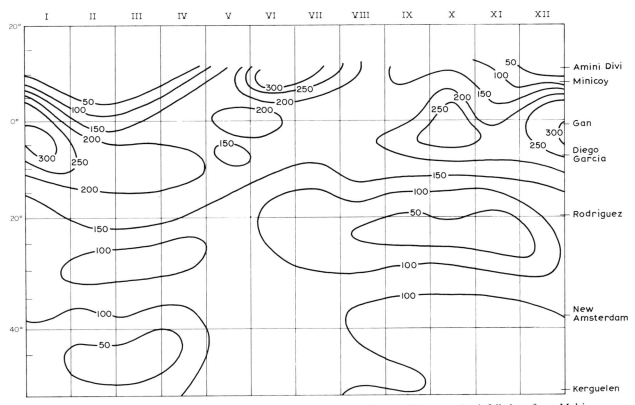

Fig. 49. Monthly distribution along 73°E of precipitation in mm. Four years' rainfall data from Mahé (4°12′N) (WELLS, 1948) fit the analysis quite well.

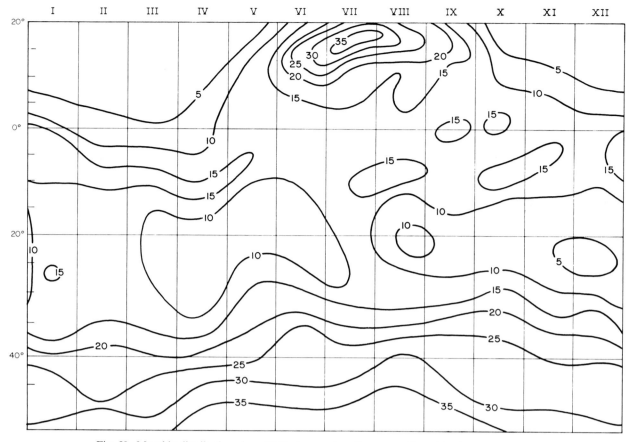

Fig. 50. Monthly distribution along 73°E of percentage frequency of precipitation.

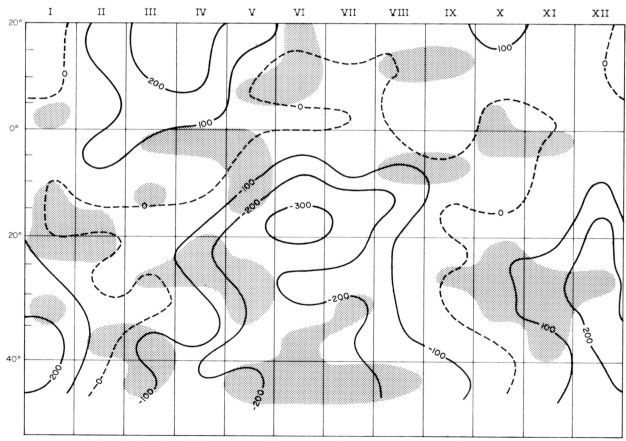

Fig. 51. Monthly distribution along 73°E of net heat balance at the sea surface in cal. cm^{-2} day^{-1}. Average of 1963 and 1964. In the stippled areas, the difference between 1963 and 1964 exceeded 100 cal. cm^{-2} day^{-1}.

During transition months, the circulation over the Indian Ocean often closely resembles circulations over the other oceans. Small land−ocean temperature gradients muffle monsoonal effects.

Means for a transition month are seldom widely observed on any day, for they average complex combinations and sequences in which winterlike situations alternate with summerlike situations. For example, twice between 1 and 15 April 1963, circulations typical of winter developed over the western Indian Ocean on *both* sides of the Equator. The southeast trades of the Southern Hemisphere and the northeast monsoon of the Northern Hemisphere merged into equatorial easterlies. Westerlies prevailed in the upper troposphere. This situation strikingly resembled normal conditions over the east central Pacific.

Although a day in March is more likely to resemble a day in January than a day in July, and the converse applies to May, nevertheless, in some years over some parts of the ocean the resemblances may be reversed.

Northern Hemisphere monsoon region

November to April

North of 6° to 8°N, rainfall decreases through the period, paralleling a decrease in tropical depression frequency. Northern or northeastern winds flow outward from the continental anticyclone. In February, however, warming over Burma, peninsular India, and Arabia weakens the ridge and anticyclonic cells appear over the northern parts of the Arabian Sea and Bay of Bengal and shift southward as the season advances.

In the upper troposphere, convergent westerlies to the south of the subtropical jet stream complete the picture of widespread subsidence and nearly cloudless skies.

In April, rain increases as tropical depressions reappear. On an average in this month, the high cells are located near 12° to 13°N over the Arabian Sea and Bay of Bengal while a weak heat low prevails over central India. North of the highs, southwesterlies have set in although farther south northerlies still prevail.

Although the westerlies have decreased considerably from March, the pattern remains unchanged in the upper troposphere.

Until January, the sea surface gains slightly more heat by radiation through nearly cloudless skies than it loses to the dry cold northerlies. Thereafter, net heat gain to the surface increases to a maximum in April as the northeast monsoon weakens, sea and air temperatures come into equilibrium, and insolation strengthens.

South of 6° to 8°N, the Northern Hemisphere near-equatorial trough is well-marked over the Bay of Bengal, but after November it is very weak in the eastern Arabian Sea and non-existent in the western Arabian Sea where air flows across the Equator into the summertime low pressure over south central Africa. Weather satellite pictures (SADLER, 1967) hint at a relative cloud maximum just north of the trough line. During March, heating over northeast Africa reverses the trans-equatorial pressure gradient and winds off Somalia veer from NE to SE in response.

May to mid-September

During May, heating increases over the continental arc from the Sudan to Burma, intensifying the heat trough. By the middle of the month, the anticyclones only occasionally appear in the Arabian Sea and Bay of Bengal and flow typical of the southwest monsoon prevails. However, the heat trough is very shallow and *west of 90°E* the circulation aloft shows little change from April. A remnant of the near-equatorial trough also comes and goes between 5° and 10°N; tropical storms develop in it and move northwest; and cloud and rainfall are greatest along about 5°N.

The considerable rain-shadowing effect of the southern Indian Peninsula and Ceylon, a feature of the summer monsoon, first appears in May. *East of 90°E,* conditions throughout the troposphere differ from those in April. In the mean, the subtropical ridge at 500 mbar is replaced by a well-marked north—south trough, while above 400 mbar convergent southwesterlies give way to a ridge. These average changes reflect development of subtropical cyclones and tropical storms bringing onset of monsoon rains over the southeastern half of the Bay of Bengal. For example, average rainfall in May at Car Nicobar (9°15′N 92°48′E), 317 mm, greatly exceeds the May averages for Minicoy (199 mm) and Amini Divi (175 mm).

In May and June, interaction between the advancing tropical maritime air of the southwest monsoon and modified continental air, which is rather less stable in the middle and upper troposphere, probably accounts for the fact that thunderstorms are commonest in these months.

In June, the trend established in May continues until by the last week the summer monsoon dominates and persists without significant interruption until mid-September. Although the average surface circulation changes little from May except for strengthening winds, the changes aloft, in the incidence of depressions and in the distribution of rain are remarkable. The sequence appears as follows.

The northward march of the sun and increasingly frequent overland thunderstorms combine to heat the air to produce a thermal equator throughout the troposphere just south of the Asian mountain massif (ANANTHAKRISHNAN and RANGARAJAN, 1964). At the thermal equator, a trough in the lower troposphere is overlain by the subtropical ridge at 200 mbar; the subtropical westerly jet stream has moved north of the mountains and a broad belt of strong easterlies prevails in the upper troposphere south of 20°N. Onset of heavy rains stems from monsoon depressions over the northern Bay of Bengal or subtropical cyclones over the northeastern Arabian Sea which make their appearance some weeks after the moist surface southwesterlies have set in (METEOROLOGICAL OFFICE, 1943). Along the west coast of India, the rains may start any time from early May in the south to the second half of June in the north with an average standard deviation of six days (RAMDAS et al., 1954).

Beginning in May, strengthening southwesterlies combine with increased cloud to counteract heating by a sun at its zenith, even leading to a net cooling of the sea surface where the winds are strongest. In area 1, the sea loses more heat in August than in any other month.

Mid-September to October

The weather of the second half of September resembles that of May. Although southwest winds prevail, they are becoming weaker and less steady as the continental heat trough weakens and the monsoon rains of the eastern Arabian Sea and northeastern

Bay of Bengal also diminish. Small anticyclones appear over the central Arabian Sea and Bay of Bengal and net heat gain at the ocean surface increases to a secondary maximum. The near-equatorial trough between 5° and 10°N once more begins to effect an increase in cloud and rain. In the upper troposphere, westerlies appear south of the Himalayas as the subtropical ridge shifts equatorward and the easterlies south of the ridge decrease. Monsoon depressions give way to tropical cyclones in the Bay of Bengal while a rare surface cyclone may develop in the Arabian Sea.

Transition to the winter monsoon is rapidly completed in October. A weak secondary maximum of thunderstorm frequency results from interaction between maritime and continental air masses. By the end of the month, anticyclonic cells over the Arabian Sea and Bay of Bengal give way to the continental high and northerlies prevail. In the upper troposphere, the subtropical ridge lies along 15°N and the subtropical westerly jet stream is established south of the Himalayas before midmonth (YEH, 1950). The near-equatorial trough between 8° and 10°N is the birthplace of depressions which travel northward and occasionally intensify into cyclones. Maximum precipitation is found somewhat south of the trough.

Although the dominant pressure trough is located near the Equator in spring and autumn and near 30°N in midsummer, Fig. 3 reveals that the double rainfall maximum is confined to latitudes below 15°N. This may be accounted for by the fact that *not one trough moves continuously between latitudinal extremes,* but rather that two distinct troughs are involved (THOMPSON, 1965; RAMAGE, 1968; C.R.V. Raman, personal communication, 1968). The continental heat trough predominates during the summer months, and the two troughs coexist in the transition months. Between May and June and September and October, weakening of one trough and intensification of the other may give the erroneous impression of major latitudinal displacement (see, for example, RAMAN and DIXIT, 1964).

In spring, as the near-equatorial trough intensifies, rainfall in the vicinity increases but diminishes when the continental heat trough becomes dominant. Then redevelopment of the near-equatorial trough in autumn is accompanied by a second rain enhancement. A second diminution follows in winter as the Southern Hemisphere trough draws air across the Equator. Latitudes above 10°−15°N are not traversed by a trough (in the climatological sense) and, therefore, experience only a single rainfall maximum as the continental heat trough develops and intensifies to the north.

Fig. 52A depicts average soundings at Minicoy for a dry month (March, 1964) and a wet month (June, 1964). Between 900 and 600 mbar, the dry month sounding is drier and slightly warmer than the wet month sounding and is more stable between 900 and 800 mbar; the differences are probably accounted for by subsidence during the dry month.

Equatorial region

During the transition seasons, a bewildering variety of circulations and weather occurs in the equatorial doldrums region which extends across the width of the ocean. Clockwise and counterclockwise eddies may coexist or, if the near-equatorial troughs are active in both hemispheres, moderate or fresh westerlies prevail. Little net trans-equatorial exchange takes place either at the surface or in the upper troposphere where winds are almost due easterly.

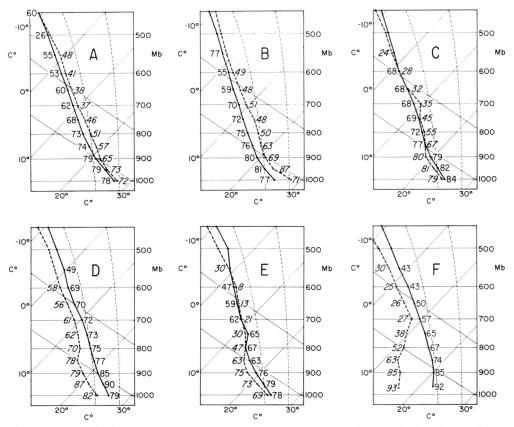

Fig. 52. Mean aerological soundings in wet (full lines) and dry months (dotted lines). Numbers denote relative humidities (%). A. Minicoy, 8°18′N 73°00′E, 1200 GMT; June 1964 294 mm rainfall, March 1964 10 mm. B. Gan, 00°41′S 73°09′E, 1200 GMT; October 1963 368 mm, March 1964 17 mm. C. Mahé, 4°37′S 55°27′E, 00 GMT; January 1964 806 mm, June 1964 75 mm. D. Diego Garcia, 7°14′S 72°26′E, 00 GMT; January 1964 408 mm, July 1964 79 mm. E. Cocos, 12°05′S 96°53′E, 00 GMT; May 1964 300 mm, November 1963 10 mm. F. Mauritius, 20°18′S 57°30′E, 00 GMT; March 1964 346 mm, October 1963 76 mm.

At Gan (eight years of record), rainfall is rather uniform apart from a relatively dry spell in February and March. Charts of rainfall frequencies show that Gan is typical of the equatorial regime between 65° and 90°E.

Fig. 52B, which compares average soundings at Gan for a dry month (March, 1964) and a wet month (October, 1963), resembles Fig. 52A for Minicoy. Again the difference probably stems from subsidence in the dry month.

Southern Hemisphere monsoon region

Western Indian Ocean

RIEHL (1954) remarks that Mahé has a single sharp rainfall peak during the southern summer, whereas stations in equatorial Africa exhibit a well-defined double maximum (see Fig. 3). Curves for Grande Comore and Agalega resemble Mahé's.

By December, typical summer conditions prevail with trans-equatorial flow becoming most intense in January. In March, as the heat low over the Kalahari Desert and the high over the Sahara weaken, a near-equatorial trough develops over Africa but becomes diffuse during May as the Northern Hemisphere heat lows take over, southerlies set in along the Equator, and trade winds extend to the coast of Africa between 5° and 25°S. The sequence is reversed in the second half of the year.

Thus Agalega and Grande Comore, which lie between the dominant mid-summer heat trough and the near-equatorial trough of the transition seasons, and Mahé, always lying north of the trough until becoming enveloped in the trans-equatorial flow of the southwest monsoon, experience only a single rainfall maximum.

At Mahé, average temperatures in the lower troposphere for a wet month differ little from those for a dry month (Fig. 52C). However, in the drier month, the layer between 900 and 700 mbar is more stable and levels above 850 mbar contain less moisture. The island remains throughout the year under the influence of a single tropical maritime air mass which subsides in the dry season (see WRIGHT and EBDON, 1968).

Eastern Indian Ocean

Although data are sparse, discontinuous trough development may also occur here. However, because the Equator traverses an oceanic environment, development of vigorous near-equatorial troughs in the transition seasons is not well-marked.

As in the Arabian Sea and Bay of Bengal, the curve of net heat gain at the sea surface is bimodal. An insignificant secondary minimum in summer reflects the relative weakness of the Australian summer monsoon.

Central Indian Ocean

Over the central Indian Ocean, the near-equatorial trough appears to move more or less continuously between 13°S (in February) and 3°S (in July and August). Most of the displacement takes place in the transition months. The trough does not usually coincide with the areas of maximum rainfall (see p. 612). That there is good coincidence between the rainfall and mean resultant wind convergence maxima indicates a likely preponderance of warm-cored, rain-producing disturbances. Fig. 52D reveals that Diego Garcia is affected by the trough during the wet season. The average sounding for January 1964 is moist and conditionally unstable. In July 1964, however, with the trough to the north, the mean sounding, having a stable layer between 850 and 800 mbar, typifies weak trade wind conditions. Despite the effect of subsidence, because the dry air is colder at all levels than the moist air, two different air masses are involved—equatorial maritime in summer and tropical maritime (cf. Fig. 52C) in winter.

Southern Hemisphere non-monsoon region

Trade winds

Between 14° and 24°S and 60° and 110°E, the trade winds prevail throughout the year. They are most extensive and strongest in winter, dominating the width of the ocean between 5° and 25°S, and least extensive and weakest in summer when the near-equatorial trough and the African and Australian monsoons restrict them to north, west, and east.

Although some interseasonal differences can be detected in the trades, the changes are gradual. The most significant is the annual variation of net heat exchange at the ocean surface; its range averaging about 300 cal. cm^{-2} day^{-1} is accounted for almost entirely by the march of the sun.

In the *western trade winds,* conditions described for January (see pp. 629–630) prevail from October to April and July conditions (see pp. 638–639) typify the period May–September.

Average soundings at Mauritius (Fig. 52F) depict a difference in air mass between equatorial maritime in the wet month of March 1964 (cf. Fig. 52D) and relatively cool strongly subsident tropical maritime (trade wind) in the dry month of October 1963.

In the *eastern trade winds,* the combination of considerable cloudiness, subsidence inversion, and little rain persists from November to March. From then until August, the rainfall minimum shifts northeast toward the Timor Sea and the cloud maximum shifts westward to coincide with a rainfall maximum along about 90°E, while probably changing from predominantly stratiform to predominantly cumuliform types. During September and October, the sequence is rapidly reversed. Cocos Island lies within the trade winds. However, Fig. 52E shows that on an average subsidence is stronger in a dry month such as November 1963 than in a wet month such as May 1964.

Acknowledgements

I thank the many meteorological services which have provided me with essential data, my colleagues in Honolulu and Bombay for their encouragement and help, and the U.S. National Science Foundation for support through Grant GA-386.

This paper constitutes Contribution No. 73-2 of the Department of Meteorology, University of Hawaii.

References

ANANTHAKRISHNAN, R. and RANGARAJAN, S., 1964. Some features of the thermal structure of the atmosphere over India and neighbourhood. In: J. W. HUTCHINGS (Editor), *Proceedings of the Symposium on Tropical Meteorology.* N.Z. Meteorological Services, Wellington, pp. 62–71.

BARGMAN, D. J. (Editor), 1960. *Tropical Meteorology in Africa.* Munitalp Foundation, Nairobi, 446 pp.[*1,2]

BUDYKO, M. I., 1956. *The Heat Balance of the Earth's Surface.* Gidrometeorologischeskoe izdatel'stvo, Leningrad, 255 pp. Translated by Nina A. Stepanova and distributed by the U.S. Weather Bureau, Washington, D.C., 1958.[*1]

BUNKER, A. F., 1967. Interaction of the summer monsoon air with the Arabian Sea. In: P. R. PISHAROTY (Editor), *Proceedings of the Symposium on Meteorological Results of the International Indian Ocean Expedition.* India Meteorological Department, New Delhi, pp. 3–16.

CHAUSSARD, A. and LAPLACE, L., 1964. Les cyclones du sud-ouest de l'Ocean Indien. *Mem. Météor. Nationale, Paris,* 49: 155 pp.[*1]

[*1] Publications providing more detailed information on the subject discussed.
[*2] Publications with extensive literature references and/or bibliographies.

COLÓN, J. A., 1964. On interactions between the southwest monsoon current and the sea surface over the Arabian Sea. *Indian J. Meteorol. Geophys.*, 15: 183 – 200.

DEUTSCHES HYDROGRAPHISCHES INSTITUT, 1960. *Monatskarten für den Indischen Ozean*. Hamburg, 50 pp.

EMMONS, G. and MONTGOMERY, R. B., 1947. Note on the physics of fog formation. *J. Meteorol.*, 4: 206.

EMON, J., 1949. L'inversion de l'alizé dans l'Ocean Indien sud-ouest. *La Météorologie*, Ser. 4, 14: 73 – 83.

FICKER, H. V., 1936. Die Passatinversion. *Veröff., Meteorol. Inst., Berlin*, 1, (4): 33 pp.

FINDLATER, J., 1969. A major low-level current near the Indian Ocean during northern summer. *Q. J. R. Meteorol. Soc.*, 95: 362 – 380.

FLEAGLE, R. G., BADGLEY, F. I. and HSUEH, Y., 1967. Calculation of turbulent fluxes by integral methods. *J. Atmos. Sci.*, 24: 356 – 373.

FLETCHER, R. D., 1945. The general circulation of the tropical and equatorial atmosphere. *J. Meteorol.*, 2: 167 – 174.

FLOHN, H., 1963. Comments on a synoptic climatology of southern Asia. In: *Meteorology and the Desert Locust. W. M. O., Geneva, Tech. Note*, 69: 245 – 252.

GALLÉ, P. H., 1907. Cyclone in the Arabian Sea. October 18th – November 4th 1906. *K. Ned. Meteorol. Inst., Meded. Verh.*, 5: 8 pp.

GALLÉ, P. H., 1924. Climatology of the Indian Ocean. *K. Ned. Meteorol. Ins., Meded. Verh.*, No. 102, 29, a: 87 pp; b: 38 pp; c: 31 pp.

GARSTANG, M., 1965. *Distribution and Mechanism of Energy Exchange Between the Tropical Oceans and Atmosphere*. Department of Meteorology, Florida State University, Tallahassee, 110 pp.

GERMAIN, H. and CHAUSSARD, A., 1964. Aperçu sur le temps à Madagascar. *La Météorologie*, Ser. 4, 75: 232 – 240.

GORDON, A. H., 1967. A Lagrangian approach to problems in tropical meteorology. *Weather*, 22: 455 – 468.

GORDON, A. H. and TAYLOR, R. C., 1975. Computations of surface layer air parcel trajectories, and weather, in the oceanic tropics. *Int. Indian Ocean Exped. Meteorol. Monographs, No. 7*. University Press of Hawaii, Honolulu, 112 pp.[1]

HANTEL, M., 1971. Surface wind vergence over the tropical Indian Ocean. *J. Appl. Meteorol.*, 10: 875 – 881.

HARWOOD, W. A., 1924. The free atmosphere in India-upper air movement in the Indian monsoons and its relation to the general circulation of the atmosphere. *Mem. India Meteorol. Dept.*, 24: 249 – 273.

IYER, V. D. and FRANCIS, K. A., 1941. The climate of Seychelles with special reference to its rainfall. *Mem. India Meteorol. Dept.*, 27: 45 – 59.

MEINARDUS, W., 1893. Beiträge zur Kenntniss der klimatischen Verhältnisse des nordöstlichen Theils des Indischen Ozeans. *Arch. Dtsch. Seewarte*, 16 (7): 48 pp.

METEOROLOGICAL OFFICE, 1943. *Weather in the Indian Ocean, Vol. I. General Information* (52 pp.); *Vol. II, Local Information* (Parts 1 – 9). H. M. Stationery Office, London.

MILLER, F. R. and KESHAVAMURTHY, R. N., 1968. Structure of an Arabian Sea summer monsoon system. *Int. Indian Ocean Exped. Meteorol. Monographs, No. 1*. East-West Center Press, Honolulu, 94 pp.[1]

NEWNHAM, E. V., 1922. Hurricanes and tropical revolving storms. *Geophys. Mem.*, 19: 333 pp.[1,2]

PETTERSSEN, S., 1956. *Weather Analysis and Forecasting, Vol. I. Motion and Motion Systems*. McGraw-Hill, New York, N.Y., 2nd ed., 428 pp.

RAMAGE, C. S., 1959. Hurricane development. *J. Meteorol.*, 16: 227 – 237.[2]

RAMAGE, C. S., 1962. The subtropical cyclone. *J. Geophys. Res.*, 67: 1401 – 1411.

RAMAGE, C. S., 1964. Some preliminary research results from the International Meteorological Centre. In: J. W. HUTCHINGS (Editor), *Proceedings of the Symposium on Tropical Meteorology*. N. Z. Meteorological Service, Wellington, pp. 403 – 408.

RAMAGE, C. S., 1965. *Meteorology in the Indian Ocean*. World Meteorological Organization, 166 T.P., 181: 31 pp.

RAMAGE, C. S., 1966. The summer atmospheric circulation over the Arabian Sea. *J. Atmos. Sci.*, 23: 144 – 150.[1]

RAMAGE, C. S., 1968. Problems of a monsoon ocean. *Weather*, 23: 28 – 37.

RAMAGE, C. S., 1969. Indian Ocean surface meteorology. In: H. BARNES (Editor), *Oceanography and Marine Biology: An Annual Review, Vol. 7*. Allen and Unwin, London, pp. 11 – 30.

RAMAGE, C. S., 1971. *Monsoon Meteorology*. Academic Press, New York and London, 296 pp.[1,2]

RAMAN, C. R. V., 1967. Cyclonic vortices on either side of the equator and their implications. In: P. R. PISHAROTY (Editor), *Proceedings of the Symposium on Meteorological Results of the International Indian Ocean Expedition*. India Meteorological Department, New Delhi, pp. 155 – 163.

RAMAN, C. R. V. and DIXIT, C. M., 1964. Analyses of monthly mean resultant winds for standard pressure levels over the Indian Ocean and adjoining continental areas. In: J. W. HUTCHINGS (Editor), *Proceedings of the Symposium on Tropical Meteorology*. N. Z. Meteorological Service, Wellington, pp. 107 – 118.

RAMAN, C. R. V. and RAMANATHAN, Y., 1964. Interaction between lower and upper tropical tropospheres. *Nature*, 204: 31–35.

RAMDAS, L. A., JAGANNATHAN, P. and GOPAL RAO, S., 1954. Prediction of the date of establishment of southwest monsoon along the west coast of India. *Indian J. Meteorol. Geophys.*, 5: 1–10.

RAO, Y. P., 1960. Interhemispherical features of the general circulation of the atmosphere. *Q. J. R. Meteorol. Soc.*, 86: 156–166.

RIEHL, H., 1954. *Tropical Meteorology*, McGraw-Hill, New York, N.Y., 392 pp.[*2]

RODEN, G. I., 1959. On the heat and salt balance of the California Current region. *J. Mar. Res.*, 18: 36–61.

ROLL, H. U., 1965. *Physics of the Marine Atmosphere*. Academic Press, New York, N.Y., 426 pp.[*2]

RONDELEUX, M., 1916. *Cyclones théorie succincte prévision et manoevre préventive d'après les travaux météorologique récents*. Imprimerie Nationale, Paris, 80 pp.

SADLER, J. C., 1964. Tiros observations of the summer circulation and weather patterns of the eastern North Pacific. In: J. W. HUTCHINGS (Editor), *Proceedings of the Symposium on Tropical Meteorology*. N. Z. Meteorological Service, Wellington, pp. 553–571.

SADLER, J. C., 1967. Satellite meteorology in the International Indian Ocean Expedition. In: P. R. PISHAROTY (Editor), *Proceedings of the Symposium on Meteorological Results of the International Indian Ocean Expedition*. India Meteorological Department, New Delhi, pp. 269–284.

SCHICK, M., 1953. Die geographische Verbreitung des Monsuns. *Abh., Nova Acta Leopoldina*, 16: 129–157.

SCHOTT, G., 1935. *Geographie des Indischen und Stillen Ozeans*. C. Boysen, Hamburg, 413 pp.[*2]

SIMPSON, R. H., 1952. Evolution of the kona storm, a subtropical cyclone. *J. Meteorol.*, 9: 24–35.

STOMMEL, H. and WOOSTER, W. S., 1965. Reconnaissance of the Somali Current during the southwest monsoon. *Proc. Nat. Acad. Sci.*, 54: 8–13.

TALJAARD, J. J., 1967. The behaviour of 1000-500 mb thickness anomalies in the Southern Hemisphere. *Notos*, 16: 3–20.

THOMPSON, B. W., 1965. *The Climate of Africa*. Oxford University Press, Nairobi, 15 pp., 132 maps.

USMANOV, R. F., 1957. Vertikal'nyĭ nazrez atmosfery nad Indiiskim okeanom, 1957. *Meteorol. Gidrol.*, 4: 24–28.

VAN DIJK, W., 1956. An investigation of the vergence field of the wind and ocean currents of the Indian Ocean. *Archiv. Meteorol. Geophys., Bioklimatol., Ser. A*, 9: 158–177.

VENKATESWARAN, S. V., 1956. On evaporation from the Indian Ocean. *Indian J. Meteorol. Geophys.*, 7: 265–284.

VERPLOEGH, G., 1960. On the annual variation of climatic elements of the Indian Ocean. *K. Ned. Meteorol. Inst., Meded. Verh.*, 77: 64 pp., 28 charts.

WALLACE, J. A., Jr., 1961. *An Annotated Bibliography on the Climate of the South Indian Ocean Islands*. U.S. Weather Bureau, Washington, D. C., 55 pp.[*2]

WEICKMANN, L., Jr., 1963. Mittlere Luftdruckverteilung im Meeresniveau während der Hauptjahreszeiten im Bereiche um Afrika, in dem Indischen Ozean und den angrenzenden Teilen Asiens. *Meteorol. Rndsch.*, 16: 89–100.

WEICKMANN, L., Jr., 1964. Mittlere Lage und vertikale Struktur grossräumiger Diskontinuitäten im Luftdruck- und Strömungsfeld der Tropenzone zwischen Afrika und Indonesien. *Meteorol. Rndsch.*, 17: 105–112.

WRIGHT, P. B. and EBDON, R. A., 1968. Upper air observations at the Seychelles, 1963–64. *Geophys. Mem.*, 111: 85 pp.

WYRTKI, K., 1962. The upwelling in the region between Java and Australia during the south-east monsoon. *Aust. J. Marine Freshwater Res.*, 13: 217–225.

YEH, T. C., 1950. The circulation of the high troposphere over China in the winter of 1945–46. *Tellus*, 2: 173–183.

Data sources

ANANTHAKRISHNAN, R., 1964. *Tracks of Storms and Depressions in the Bay of Bengal and the Arabian Sea*. India Meteorological Department, New Delhi, 13 pp., 167 charts.

BUNKER, A. F. and CHAFFEE, M., 1969. Tropical Indian Ocean clouds. *Int. Indian Ocean Exped. Meteorol. Monographs, No. 4*. East-West Center Press, Honolulu, 193 pp.

DEUTSCHES HYDROGRAPHISCHES INSTITUT, 1960. *Monatskarten für den Indischen Ozean*. Hamburg, 50 pp.

ELIOT, J., 1906. *Climatological Atlas of India*. Bartholemew, Edinburgh, 32 pp., 120 plates.

FINDLATER, J., 1971. Mean monthly airflow at low levels over the western Indian Ocean. *Geophys. Mem.*, 115: 53 pp.

FROST, R. and STEPHENSON, P. M., 1965. Mean streamlines and isotachs at standard pressure levels over the Indian and west Pacific Oceans and adjacent land areas. *Geophys. Mem.*, 109: 24 pp.

GALLÉ, P. H., 1914. Luft- und Wassertemperatur im Indischen Ozean. *K. Ned. Meteorol. Inst., Meded. Verh.*, 18: 29 pp.

HOGAN, J., 1948. Meteorology of the Indian Ocean area between western Australia and India. *Bur. Meteorol., Aust., Bull.*, 40: 88 pp.

MANI, A., CHACKO, O., KRISHNAMURTHY, V. and DESIKAN, V., 1967. Distribution of global and net radiation over the Indian Ocean and its environments. *Arch. Meteorol. Geophys, Bioklimatol., B.*, 15: 82–98.

MCDONALD, W. F., 1938. *Atlas of Climatic Charts of the Oceans*. U.S. Weather Bureau, Washington, D. C., 138 charts.

METEOROLOGICAL OFFICE, 1917–1938. *Réseau Mondial, 1910–1931*. H. M. Stationery Office, London (issued in annual volumes).

METEOROLOGICAL OFFICE, 1943. *Weather in the Indian Ocean, Vol. 1. General Information* (52 pp.), Vol. II. Local Information (Parts 1–9). H.M. Stationery Office, London.

METEOROLOGICAL OFFICE, 1947. *Monthly Meteorological Charts of the Western Pacific*. H.M. Stationery Office, London, 120 pp.

METEOROLOGICAL OFFICE, 1949. *Monthly Meteorological Charts of the Indian Ocean*. H.M. Stationery Office, London, 98 pp.

METEOROLOGICAL OFFICE, 1958. *Tables of Temperatures, Relative Humidity and Precipitation for the World, Part IV. Africa, the Atlantic Ocean South of 35°N and the Indian Ocean*. H.M. Stationery Office, London, 220 pp.

MILLER, F. R. and JEFFERIES, C., 1967. *Mean Monthly Sea Surface Temperatures of the Indian Ocean During the International Indian Ocean Expedition*. Hawaii Institute of Geophysics Rept. No. HIG-67-14, 6 pp., 25 charts.

NEWNHAM, E. V., 1949. The climates of Addu Atoll, Agalega Islands and Tristan da Cunha. *Prof. Notes, Meteorol. Office, Lond.*, 7, 101: 20 pp.

NICHOLSON, J. R., 1969. Meteorological data catalogue. *Int. Indian Ocean Exped. Meteorol. Monographs, No. 3*. East-West Center Press, Honolulu, 59 pp.

PORTMAN, D. J. and RYZNAR, E., 1969. An investigation of heat exchange. *Int. Indian Ocean Exped. Meteorol. Monographs, No. 5*. East-West Center Press, Honolulu, 78 pp.

PRIVETT, D. W., 1959. Monthly charts of evaporation from the N. Indian Ocean including the Red Sea and the Persian Gulf. *Q. J. R. Meteorol. Soc.*, 85: 424–428.

RAMAGE, C. S., MILLER, F. R. and JEFFERIES, C., 1972. *Meteorological Atlas of the International Indian Ocean Expedition, Vol. 1. The Surface Climate of 1963 and 1964*. National Science Foundation, Washington, D. C., 144 charts.

RAMAGE, C. S. and RAMAN, C. R. V., 1972. *Meteorological Atlas of the International Indian Ocean Expedition, Vol. 2. Upper Air*. National Science Foundation, Washington, D. C., 121 charts.

RAMAN, C. R. V. and DIXIT, C. M., 1964. Analyses of monthly mean resultant winds for standard pressure levels over the Indian Ocean and adjoining continental areas. In: J. W. HUTCHINGS (Editor), *Proceedings of the Symposium on Tropical Meteorology*. N.Z. Meteorological Service, Wellington, pp. 107–118.

RAMANATHAN, K. R. and RAMAKRISHNAN, K. P., 1939. The general circulation of the atmosphere over India and its neighbourhood. *Mem. India Meteorol. Dept.*, 26: 189–245.

RAMANATHAN, K. R. and VENKITESHWARAN, S. P., 1948. *Climatological Charts of the Indian Monsoon Area*. India Meteorological Dept., New Delhi, 3 pp., 72 charts.

ROMANOV, Y. A., 1965. Mean resultant wind divergence and vorticity charts for the Indian Ocean. *Acad. Sci., USSR, Inst. of Oceanology*, 78: 119–127 (in Russian).

SADLER, J. C., 1969. Average cloudiness in the tropics from satellite observations. *Int. Indian Ocean Exped. Meteorol. Monographs, No. 2*. East-West Center Press, Honolulu, 22 pp., 12 charts.

SCHOTT, G., 1902. *Wissenschaftliche Ergebnisse der Deutschen Tiefsee-Expedition auf dem Dampfer "Valdivia" 1898–1899, Vol. I. Oceanographie und Maritime Meteorologie*. Gustav Fischer, Jena, 404 pp.

SEEWETTERAMT HAMBURG, 1964. *International Geophysical Year 1957–58, World Weather Maps, Part II. Tropical Zone 25°North–25°South*. Deutscher Wetterdienst, Offenbach am Main, 18 months of daily surface and 500 mbar charts.

THOMPSON, B. W., 1965. *The Climate of Africa*. Oxford University Press, Nairobi, 15 pp., 132 maps.

U.S. HYDROGRAPHIC OFFICE, 1944. *Atlas of Surface Currents Indian Ocean*. H. O. Pub. No. 566, U.S. Government Printing Office, Washington, D. C., 12 charts.

U.S. HYDROGRAPHIC OFFICE, 1950. *Sailing Directions for the South Coast of Australia*. H. O. Pub. No. 167, U.S. Government Printing Office, Washington, D. C., 246 pp.

U.S. HYDROGRAPHIC OFFICE, 1951. *Sailing Directions for the Bay of Bengal*. H. O. Pub. No. 160, U.S. Government Printing Office, Washington, D. C., 276 pp.

U.S. HYDROGRAPHIC OFFICE, 1952. *Sailing Directions for the North and West Coasts of Australia.* H. O. Pub. No. 170, U.S. Government Printing Office, Washington, D. C., 372 pp.

U.S. HYDROGRAPHIC OFFICE, 1952. *Sailing Directions for the South Indian Ocean.* H. O. Pub. No. 161, U.S. Government Printing Office, Washington, D. C., 364 pp.

U.S. WEATHER BUREAU, 1957. *U.S. Navy Marine Climatic Atlas of the World, Vol. III. Indian Ocean.* Chief of Naval Operations, Washington, D. C., 7 pp., 267 charts.

VAN DUIJNEN MONTIJN, J. A., 1952. *Indian Ocean Oceanographic and Meteorological Data.* Royal Netherlands Meteorological Institute, De Bilt, No. 135, 31 pp., 24 charts (2nd ed.).

VAN LOON, H., TALJAARD, J. J., SASAMORI, T., LONDON, J., HOYT, D. V., LABITZKE, K. and NEWTON, C. W., 1972. Meteorology of the Southern Hemisphere. In: C. W. NEWTON (Editor), *Meteorological Monographs,* No. 35, Vol. 13. American Meteorological Society, Boston, 263 pp.

WALTER, A., 1927. Results of observations on the direction and velocity of the upper air current over the south Indian Ocean. *Geophys. Mem.,* 39: 32 pp.

WEATHER BUREAU, SOUTH AFRICA, 1962—1966. *International Geophysical Year (1957—58) World Weather Maps, Part III, Southern Hemisphere South of 20°S.* Daily sea level and 500 mb charts. Department of Transport, Pretoria.

WELLS, A. J., 1948. The weather of the Maldive Islands. *Weather,* 3: 310—313.

WRIGHT, P. B. and EBDON, R. A., 1968. Upper air observations at the Seychelles, 1963—64. *Geophys. Mem.,* 111: 85 pp.

WRIGHT, P. B. and STUBBS, M. W., 1971. Circulation patterns at 850, 700, 500 and 200 millibars over the eastern hemisphere from 40°N to 40°S during May and June. *Geophys. Mem.,* 114: 101 pp.

Appendix — Climatic tables

TABLE I

LENGTH OF RECORD OF CLIMATIC TABLES

Type of observation	Length of record (years)
Surface marine	Over 100
Upper air marine	2–4
Surface land station	Up to 30
Upper air land station	2–14
Aircraft	5
Weather satellite	2–4
Tropical cyclones	
Northern Indian Ocean	1891–1960
Southern Indian Ocean	
a) West of 110°E	1849–1948
b) East of 100°E	1907–1964

TABLE II

PRINCIPAL DATA SOURCES IN ORDER OF IMPORTANCE FOR MEAN CHARTS AND OCEAN CLIMATIC TABLES

Element	Sources
Surface	
Pressure	3, 2
Standard deviation of pressure	2
Air temperature	3, 2
Standard deviation of air temperature	2
Sea temperature	2
Air–sea temperature difference	2
Vapour pressure	2
Resultant winds	3, 4
Mean wind speeds	2, 3
Wind steadiness	3, 4
Wind roses	2
Gales	2, 1, 4
Cloud	2, 5
Fog, mist and haze	2, 4
Precipitation	2, 1, 4
Lightning	1, 4
Net heat balance at sea surface	
(a) Charts	5
(b) Tables	2
Upper air	
Resultant winds—incorporating soundings from land and ship stations and flying level winds from aircraft	5

1 = METEOROLOGICAL OFFICE 1949.
2 = U. S. WEATHER BUREAU 1957.
3 = VAN DUIJNEN MONTIJN 1952.
4 = METEOROLOGICAL OFFICE 1947 (western Pacific).
5 = International Indian Ocean Expedition Collections.

TABLE III

CLIMATIC TABLE FOR OCEAN AREA 1, CENTRED NEAR 17°30′ N 65°30′E

Month	Mean press. (mbar)	Mean temp. (°C)	Mean vapour press. (mbar)	Mean sea surf. temp. (°C)	% Freq. of			Mean (%) cloudiness	Most freq. wind dir.	Mean wind vel. (m s⁻¹)	Net heat bal. at sea surf. (cal. cm⁻² day⁻¹)
					pre-cip.	light-ning	gale				
Jan.	1015.2	23.6	20.8	25.0	0	1	0	16	NE	4.7	-24
Feb.	1014.7	23.8	23.2	25.0	0	1	0	10	NE	2.8	235
Mar.	1012.6	25.2	26.9	26.1	0	1	0	4	N	3.6	332
Apr.	1010.7	27.4	28.5	27.8	0	1	0	7	NW	3.3	347
May	1008.4	28.5	33.1	28.9	0	2	0	19	W	4.4	345
June	1004.2	28.3	32.7	28.9	11	2	2	46	SW	9.0	-12
July	1004.4	27.4	30.7	27.2	8	1	5	76	SW	11.9	-77
Aug.	1006.8	26.2	29.0	26.7	8	0	4	83	SW	9.9	-175
Sept.	1008.8	26.7	27.7	26.7	4	0	0	46	W, SW	5.4	128
Oct.	1011.4	26.7	27.3	27.2	1	2	0	15	N	3.8	198
Nov.	1013.0	26.7	25.1	27.2	0	2	0	10	NE	4.5	75
Dec.	1014.5	25.0	22.5	26.1	1	1	0	19	NE	4.7	-81
Year	1010.4	26.3	27.3	26.9	2.8	1.2	0.9	29	NE	5.7	104

TABLE IV

CLIMATIC TABLE FOR OCEAN AREA 2, CENTRED NEAR 14°30′N 87°30′E

Month	Mean press. (mbar)	Mean temp. (°C)	Mean vapour press. (mbar)	Mean sea surf. temp. (°C)	% Freq. of			Mean (%) cloudiness	Most freq. wind dir.	Mean wind vel. (m s⁻¹)	Net heat bal. at sea surf. (cal. cm⁻² day⁻¹)
					pre-cip.	light-ning	gale				
Jan.	1013.6	25.3	22.8	26.1	3	2	0	25	NE	5.7	-26
Feb.	1012.7	25.4	26.5	26.1	0	2	0	15	NE	3.0	283
Mar.	1011.5	26.6	27.8	27.2	0	2	0	12	NE	3.0	323
Apr.	1008.9	28.3	32.6	28.9	2	7	0	20	S	3.5	310
May	1006.3	29.3	33.5	29.4	6	10	0	38	SW	5.6	221
June	1003.6	28.9	33.5	28.9	10	12	2	74	SW	8.9	-25
July	1003.9	28.2	33.2	28.3	13	9	3	78	SW	9.3	50
Aug.	1004.3	27.8	33.2	27.8	13	8	1	60	SW	8.3	116
Sept.	1006.6	27.7	31.7	28.3	12	9	0	60	SW	6.6	83
Oct.	1009.0	27.7	31.7	28.3	12	9	0	46	SW	4.7	160
Nov.	1010.0	27.2	28.1	27.8	10	6	0	48	NE	6.0	-53
Dec.	1012.7	25.8	25.3	26.7	7	3	0	49	NE	6.8	-186
Year	1008.6	27.4	30.0	27.8	7.3	6.6	0.8	44	SW	6.0	101

TABLE V

CLIMATIC TABLE FOR OCEAN AREA 3, CENTRED NEAR 11°30′N 58°30′E

Month	Mean press. (mbar)	Mean temp. (°C)	Mean vapour press. (mbar)	Mean sea surf. temp. (°C)	% Freq. of			Mean (%) cloudi-ness	Most freq. wind dir.	Mean wind vel. (m s⁻¹)	Net heat bal. at sea surf. (cal. cm⁻² day⁻¹)
					pre-cip.	light-ning	gale				
Jan.	1014.0	25.1	24.1	25.6	1	1	0	31	NE	6.0	−12
Feb.	1013.4	25.0	24.0	25.6	1	1	0	29	NE	3.7	157
Mar.	1012.4	25.8	26.5	26.7	1	1	0	23	NE	3.6	248
Apr.	1011.3	27.6	27.1	27.8	0	2	0	20	NE	2.1	319
May	1009.3	28.6	30.8	28.9	3	5	0	28	SW	3.5	238
June	1007.7	27.7	30.9	27.8	3	2	16	48	SW	12.8	−50
July	1007.4	26.3	28.0	25.6	1	1	25	40	SW	14.6	160
Aug.	1008.7	25.4	28.4	25.0	0	1	9	30	SW	12.2	215
Sept.	1010.2	25.6	28.5	25.6	2	1	1	36	SW	7.3	189
Oct.	1012.2	26.6	27.1	26.7	3	1	0	29	NE	3.1	212
Nov.	1012.3	26.6	26.1	26.7	5	2	0	30	NE	4.9	108
Dec.	1013.6	25.6	24.7	26.1	3	2	0	33	NE	6.3	−3
Year	1011.0	26.3	27.2	26.5	1.9	1.7	4.2	31	NE	6.7	145

TABLE VI

CLIMATIC TABLE FOR OCEAN AREA 4, CENTRED NEAR 5°30′N 88°30′E

Month	Mean press. (mbar)	Mean temp. (°C)	Mean vapour press. (mbar)	Mean sea surf. temp. (°C)	% Freq. of			Mean (%) cloudi-ness	Most freq. wind dir.	Mean wind vel. (m s⁻¹)	Net heat bal. at sea surf. (cal. cm⁻² day⁻¹)
					pre-cip.	light-ning	gale				
Jan.	1011.0	27.2	29.1	27.2	14	8	0	45	NE	4.9	175
Feb.	1010.6	27.4	29.3	27.8	8	10	0	41	NE	3.0	228
Mar.	1010.3	27.7	30.1	28.3	10	11	0	45	NE	3.0	236
Apr.	1009.2	28.3	31.3	28.9	13	15	0	49	SW	2.9	207
May	1008.4	28.2	31.7	28.3	15	15	0	59	SW	6.3	98
June	1008.2	28.1	31.6	28.3	13	6	0	63	SW	7.2	48
July	1008.5	27.8	31.0	27.8	9	4	0	63	SW	7.6	42
Aug.	1009.2	27.6	30.7	27.8	12	4	0	63	SW	7.3	−18
Sept.	1009.6	27.4	30.5	27.8	13	4	0	63	SW	6.8	−2
Oct.	1010.1	27.2	29.2	27.8	18	5	0	62	SW	4.8	76
Nov.	1010.3	27.1	29.2	27.8	17	4	0	63	SW	4.0	85
Dec.	1010.6	27.2	29.4	27.8	16	7	0	52	NE	3.9	109
Year	1009.7	27.6	30.2	28.0	13.2	7.8	0	56	SW	5.1	104

TABLE VII

CLIMATIC TABLE FOR OCEAN AREA 5, CENTRED NEAR 4°30′S 95°30′E

Month	Mean press. (mbar)	Mean temp. (°C)	Mean vapour press. (mbar)	Mean sea surf. temp. (°C)	% Freq. of			Mean (%) cloudiness	Most freq. wind dir.	Mean wind vel. (m s⁻¹)	Net heat bal. at sea surf. (cal. cm⁻² day⁻¹)
					pre-cip.	light-ning	gale				
Jan.	1009.8	27.6		27.8	14	7	0	56	W	4.0	
Feb.	1010.1	27.7		27.8	14	7	0	55	W	3.2	
Mar.	1009.8	27.5		28.3	14	11	0	56	NW	3.2	
Apr.	1009.2	28.4		28.3	15	8	0	50	W	3.0	
May	1008.9	28.4		28.3	14	8	0	51	SE	3.3	
June	1009.6	27.7		28.3	13	7	0	56	SE	3.5	
July	1009.8	27.7		27.8	16	7	0	50	SE	4.3	
Aug.	1010.2	27.5		27.8	13	8	0	63	SE	4.7	
Sept.	1010.7	26.7		27.8	16	7	0	78	SE	5.2	
Oct.	1010.5	26.8		27.8	22	5	0	79	SE	4.5	
Nov.	1010.3	26.7		27.8	22	3	0	72	E	3.9	
Dec.	1010.0	26.9		27.8	18	6	0	63	W, SE	3.8	
Year	1009.9	27.5		28.0	15.9	7.0	0	61	SE	3.9	

TABLE VIII

CLIMATIC TABLE FOR OCEAN AREA 6, CENTRED NEAR 11°30′S 116°30′E

Month	Mean press. (mbar)	Mean temp. (°C)	Mean vapour press. (mbar)	Mean sea surf. temp. (°C)	% Freq. of			Mean (%) cloudiness	Most freq. wind dir.	Mean wind vel. (m s⁻¹)	Net heat bal. at sea surf. (cal. cm⁻² day⁻¹)
					pre-cip.	light-ning	gale				
Jan.	1007.5	27.6	30.6	28.3	12	8	4	38	SW	6.5	117
Feb.	1007.8	28.4	31.7	28.9	10	8	6	41	SW	4.4	194
Mar.	1008.9	28.4	30.9	28.9	10	10	0	35	SW	3.9	225
Apr.	1010.0	28.5	30.0	28.3	4	4	0	41	E	4.8	140
May	1010.8	27.5	28.9	27.8	6	3	0	29	E	7.5	−65
June	1011.3	26.7	27.7	26.7	0	2	0	30	E	7.4	−5
July	1012.8	25.3	25.7	25.6	0	1	0	31	E	8.0	−81
Aug.	1012.5	25.5	24.5	26.1	0	1	0	21	E	7.0	26
Sept.	1013.3	25.7	25.1	26.1	0	1	0	21	E	5.5	155
Oct.	1011.7	27.2	28.1	27.8	3	3	0	25	SE	4.3	210
Nov.	1009.9	28.3	29.4	28.9	5	7	0	33	S	4.7	157
Dec.	1007.5	28.6	31.1	29.4	8	11	5	45	SW	6.6	71
Year	1010.3	27.3	28.6	27.7	4.8	4.9	1.2	33	E	5.9	92

TABLE IX

CLIMATIC TABLE FOR OCEAN AREA 7, CENTRED NEAR 24°30's 93°30'E

Month	Mean press. (mbar)	Mean temp. (°C)	Mean vapour press. (mbar)	Mean sea surf. temp. (°C)	% Freq. of			Mean (%) cloudi-ness	Most freq. wind dir.	Mean wind vel. (m s⁻¹)	Net heat bal. at sea surf. (cal. cm⁻² day⁻¹)
					pre-cip.	light-ning	gale				
Jan.	1016.4	22.8	20.3	22.8	7	0	2	45	E	10.1	− 28
Feb.	1015.2	22.8	21.7	23.3	7	0	2	45	E	7.5	99
Mar.	1016.7	22.7	24.3	23.3	7	1	1	48	E	7.5	150
Apr.	1017.6	22.2	22.8	23.3	8	1	0	50	E	5.8	40
May	1018.2	20.7	19.4	22.2	8	1	0	53	E	6.7	−151
June	1018.7	19.9	18.9	21.1	10	1	0	55	SE	5.4	− 79
July	1020.6	17.9	16.2	20.0	12	1	0	60	SE	7.7	−266
Aug.	1021.5	18.3	16.8	20.0	8	0	0	59	SE	5.9	− 90
Sept.	1022.2	18.3	15.3	19.4	8	0	1	56	SE	7.1	−124
Oct.	1019.6	19.1	17.8	20.0	7	1	1	49	SE	6.1	105
Nov.	1019.3	20.6	18.3	21.7	4	0	1	48	E	5.9	60
Dec.	1017.9	21.5	19.1	22.2	3	1	0	52	E	8.3	14
Year	1018.6	20.6	19.2	21.6	7.4	0.6	0.7	52	E	7.0	− 23

TABLE X

CLIMATIC TABLE FOR OCEAN AREA 8, CENTRED NEAR 28°30's 48°30'E

Month	Mean press. (mbar)	Mean temp. (°C)	Mean vapour press. (mbar)	Mean sea surf. temp. (°C)	% Freq. of			Mean (%) cloudi-ness	Most freq. wind dir.	Mean wind vel. (m s⁻¹)	Net heat bal. at sea surf. (cal. cm⁻² day⁻¹)
					pre-cip.	light-ning	gale				
Jan.	1013.6	24.0	21.9	24.4	12	3	2	43	E	7.7	76
Feb.	1013.7	24.2	23.4	24.4	10	4	3	41	E	7.0	130
Mar.	1015.1	23.9	23.8	24.4	8	5	3	39	E	7.0	50
Apr.	1017.8	22.6	20.9	23.9	9	7	3	39	E	7.3	−190
May	1019.6	21.6	15.1	22.2	13	10	4	39	E	7.7	−375
June	1022.1	20.6	14.4	21.1	12	5	0	38	NE	6.6	−308
July	1024.0	19.3	16.3	20.0	7	6	3	34	E	7.2	−191
Aug.	1023.2	20.0	17.5	20.0	9	2	2	35	NE	7.0	− 62
Sept.	1022.3	20.1	19.0	20.6	8	2	1	36	NE	7.1	38
Oct.	1018.8	21.1	19.0	21.1	6	3	2	43	NE	8.0	81
Nov.	1017.1	21.7	18.1	21.7	6	2	3	39	NE	7.5	71
Dec.	1015.2	22.8	21.6	23.3	10	3	0	41	NE	7.3	178
Year	1018.5	21.8	19.2	22.2	9.2	4.3	2.2	39	E, NE	7.3	− 42

TABLE XI

Latitude 11°07′N, longitude 72°44′E, elevation 4 m, record 1931–1960

| Month | Mean sta. press.*[2] (mbar) | Temperature (°C) | | | | Mean vapour press.*[2] (mbar) | Precip. (mm) | |
| | | mean daily*[3] | mean daily range | extremes | | | mean | 24-h max. |
				max.	min.			
Jan.	1012.6	27.6	7.4	36.7	18.9	26.5	15	74
Feb.	1012.2	28.5	7.1	35.7	19.4	26.8	2	27
Mar.	1011.4	29.2	7.0	37.2	20.6	28.7	2	32
Apr.	1010.2	30.0	5.9	37.7	20.0	30.4	24	122
May	1008.5	29.8	5.9	37.5	21.7	31.2	175	161
June	1007.8	28.0	4.2	35.9	21.1	32.1	353	211
July	1008.4	27.3	3.9	33.3	21.7	31.6	321	180
Aug.	1009.0	27.4	3.9	33.3	22.2	31.0	203	242
Sept.	1010.0	27.6	4.3	33.9	21.7	30.1	139	218
Oct.	1010.5	27.8	5.2	36.9	20.6	29.8	149	134
Nov.	1011.0	28.0	6.5	35.0	18.3	29.0	87	89
Dec.	1012.1	27.6	7.6	35.1	18.9	27.0	23	119
Year	1010.3	28.2	5.7	37.7	18.3	29.5	1493	242

| Month | Mean evap. (mm) | Mean number of days | | | Mean cloudi-ness*[2] (%) | Mean sun-shine (h) | Most freq. wind direct.*[2] | Mean wind speed*[2] (m s⁻¹) |
		precip. ≥ 0.3 mm	thunder-storm	fog				
Jan.		1	0.5	0	38		N	1.6
Feb.		0.4	0.1	0	38		N	2.2
Mar.		0.6	0.7	0	35		NW	2.4
Apr.		2	1.3	0	45		NW	2.8
May		9	3	0	62		NW	3.8
June		21	3	0	81		W	5.5
July		22	0.7	0	80		W	6.4
Aug.		18	0.2	0	74		W	6.3
Sept.		15	0.8	0	65		NW	5.1
Oct.		11	3	0	65		NW	2.8
Nov.		7	3	0	53		N	1.5
Dec.		3	0.7	0	40		N	1.4
Year		111	17	0	56		NW	3.5

*[1] Source: Meteorological Department, Poona, India.
*[2] Observations at 0830 IST.
*[3] (Max. + min.)/2.

TABLE XII

CLIMATIC TABLE FOR MINICOY[1]

Latitude 08°18′N, longitude 73°00′E, elevation 2 m, record 1931–1960

Month	Mean sta. press.[2] (mbar)	Temperature (°C)				Mean vapour press.[2] (mbar)	Precip. (mm)	
		mean daily[3]	mean daily range	extremes			mean	24-h max.
				max.	min.			
Jan.	1011.1	26.2	6.8	32.8	17.8	27.5	36	126
Feb.	1010.8	26.6	6.3	32.2	17.2	28.1	25	57
Mar.	1010.0	27.6	5.9	32.8	19.4	29.4	14	55
Apr.	1009.0	28.7	5.0	35.6	21.1	30.8	54	121
May	1007.9	28.8	5.0	36.7	21.7	31.7	199	213
June	1008.3	27.8	5.2	33.9	22.2	31.1	296	149
July	1008.8	27.3	4.5	31.7	21.1	30.3	224	155
Aug.	1009.0	27.3	4.4	31.7	21.1	30.0	200	201
Sept.	1009.6	27.3	4.4	32.2	21.1	29.5	144	108
Oct.	1009.8	27.1	5.0	33.3	19.4	29.2	180	128
Nov.	1010.1	26.4	5.6	32.2	17.2	29.0	141	132
Dec.	1010.8	26.5	6.5	32.2	18.3	27.8	76	188
Year	1009.6	27.3	5.4	36.7	17.2	29.5	1589	213

Month	Mean evap. (mm)	Mean number of days			Mean cloudi-ness[2] (%)	Mean sun-shine (h)	Most freq. wind direct.[2]	Mean wind speed[2] (m s^{-1})
		precip. ⩾0.3 mm	thunder-storm	fog				
Jan.		4	1	0	49		NE	1.7
Feb.		3	1	0	49		N	1.8
Mar.		3	2	0	46		N	1.9
Apr.		6	4	0	55		NW	2.1
May		14	6	0	74		NW	3.2
June		22	3	0	86		W	4.8
July		21	2	0	81		W	4.5
Aug.		18	0.5	0	80		W	4.1
Sept.		17	0.5	0	70		NW	3.6
Oct.		15	1	0	69		NW	2.8
Nov.		11	2	0	61		NW	1.9
Dec.		6	1	0	53		NE	1.7
Year		140	24	0	64		NW	2.8

[1] Source: Meteorological Department, Poona, India.

[2] Average of 0830 and 1730 IST observations.

[3] (Max. + min.)/2.

TABLE XIII

CLIMATIC TABLE FOR MAHÉ PORT VICTORIA[1]
Latitude 04°37′S, longitude 55°27′E, elevation 3 m, record 1944–1962

Month	Mean sta. press.[2] (mbar)	Temperature (°C)				Mean vapour press. (mbar)	Precip. (mm)[3]	
		mean daily	mean daily range	extremes			mean	24-h max.
				max.	min.			
Jan.	1010.3	26.7	4.4	32.3	19.3	28.9	358	172
Feb.	1010.0	27.0	4.6	31.9	20.1	29.0	242	475
Mar.	1009.9	27.6	5.0	33.3	19.4	29.7	204	134
Apr.	1009.7	27.9	5.1	33.1	20.8	30.1	170	164
May	1010.2	27.7	4.7	32.9	21.2	29.3	161	164
June	1011.3	26.5	3.7	31.8	20.4	27.5	85	95
July	1011.9	25.8	3.6	29.7	19.8	26.3	83	58
Aug.	1011.8	25.8	3.4	29.4	20.0	26.3	78	94
Sept.	1011.7	26.1	3.4	30.2	20.7	27.1	116	103
Oct.	1011.7	26.3	4.3	31.5	19.3	27.5	167	104
Nov.	1011.1	26.6	5.2	32.2	20.4	28.1	206	119
Dec.	1010.7	26.7	4.9	32.4	21.2	28.8	332	129
Year	1010.9	26.7	4.4	33.3	19.3	28.2	2202	475

Month	Mean evap.[4] (mm)	Mean number of days			Mean cloudiness (%)	Mean sunshine (h)	Most freq. wind direct.[5]	Mean wind speed[5] (m s⁻¹)
		precip. ≥0.1 mm	thunderstorm	fog				
Jan.	121	19	3	0	74	174	W	3.7
Feb.	121	16	3	0	71	179	W	3.5
Mar.	151	15	3	0	61	223	NE	2.8
Apr.	143	17	2	0	58	243	E	2.9
May	156	15	2	0	59	257	SE	4.3
June	138	18	0	0	71	210	SE	5.7
July	145	17	0	0	71	229	SE	5.9
Aug.	171	16	0	0	74	223	SE	6.4
Sept.	177	15	1	0	69	207	SE	5.9
Oct.	167	14	1	0	61	223	SE	4.3
Nov.	150	16	1	0	61	210	VAR.	3.3
Dec.	149	21	4	0	69	189	W	3.5
Year	1789	199	20	0	67	2567	SE	4.3

[1] Source: East African Meteorological Department, Nairobi, Kenya.
[2] Period of record 12 years.
[3] Period of record 30 years.
[4] Period of record 4 years.
[5] Average of 0600 and 1200 GMT observations.

TABLE XIV

CLIMATIC TABLE FOR DIEGO GARCIA[*1]
Latitude 07°14'S, longitude 72°26'E, elevation 2.1 m, record 1951–1960

Month	Mean sta. press. (mbar)	Temperature (°C)					Mean vapour press. (mbar)	Precip. (mm)	
		mean daily	mean daily range	extremes				mean	24-h max.
				max.	min.				
Jan.	1009.9	28.2	5.3	33.4	22.0	30.1		322	135
Feb.	1009.7	28.0	5.5	32.4	22.0	30.7		233	83
Mar.	1010.1	29.2	5.4	33.6	22.1	30.4		229	115
Apr.	1009.6	28.6	4.8	33.3	23.3	31.1		212	121
May	1009.8	27.9	4.3	32.9	21.8	29.9		141	105
June	1010.8	26.8	3.7	31.4	21.9	28.5		183	128
July	1011.3	26.2	3.8	29.6	21.9	27.5		140	199
Aug.	1011.6	26.1	3.2	30.3	21.5	27.2		161	329
Sept.	1011.6	26.4	4.0	30.2	21.4	27.8		180	123
Oct.	1011.3	26.8	4.2	32.0	21.7	28.8		233	190
Nov.	1011.1	27.5	4.7	32.0	21.4	30.1		218	152
Dec.	1010.6	28.0	5.4	32.5	22.2	29.9		211	88
Year	1010.6	27.5	4.5	33.6	21.4	29.3		2463	329

Month	Mean evap. (mm)	Mean number of days			Mean cloudi-ness (%)	Mean sun-shine (h)	Most freq. wind direct.	Mean wind speed (m s⁻¹)	Mean number of days with gale
		precip. ⩾0.1 mm	thunder-storm	fog					
Jan.	21	2.6	0		75		W	2.1	0
Feb.	20	3.0	0		78		W	2.3	0
Mar.	18	1.9	0		68		W	1.8	0
Apr.	18	2.1	0		70		W	2.0	0
May	17	0.9	0		69		SE	2.7	0
June	17	0.6	0		74		SE	3.5	0
July	18	0.4	0		74		SE	3.7	0
Aug.	17	0	0		76		SE	3.9	0
Sept.	17	0.1	0		78		SE	3.6	0
Oct.	20	0.9	0		78		SE	3.1	0
Nov.	18	0.7	0		71		SE	2.0	0
Dec.	18	2.0	0		71		W	1.5	0
Year	219	15.2	0		74		SE	2.7	0

[*1] Source: Meteorological Services, Vacoas, Mauritius.

TABLE XV

CLIMATIC TABLE FOR AGALEGA*1
Latitude 10°33'S, longitude 56°45'E, elevation 3.1 m, record 1951–1960

Month	Mean sta. press. (mbar)	Temperature (°C)				Mean vapour press. (mbar)	Precip. (mm)	
		mean daily	mean daily range	extremes			mean	24-h max.
				max.	min.			
Jan.	1009.0	27.6	5.7	33.8	21.3	30.3	272	123
Feb.	1008.5	27.6	5.8	32.3	21.0	30.2	198	136
Mar.	1008.9	27.9	6.1	33.6	21.4	30.8	195	98
Apr.	1009.4	27.9	5.7	33.6	22.2	30.1	168	104
May	1010.9	26.3	5.2	32.6	20.5	28.4	175	157
June	1012.5	25.9	4.9	31.1	20.2	26.3	82	54
July	1013.3	25.1	4.9	29.6	19.4	24.9	90	154
Aug.	1013.5	24.9	4.9	28.8	20.1	24.4	87	119
Sept.	1013.3	25.4	4.8	29.1	19.9	24.9	91	152
Oct.	1012.7	25.9	5.0	30.1	20.2	25.8	143	79
Nov.	1011.8	26.8	5.3	31.0	20.1	26.6	113	318
Dec.	1010.2	27.5	5.8	32.3	20.8	28.9	225	366
Year	1011.2	26.6	5.3	33.8	19.4	27.6	1839	366

Month	Mean evap. (mm)	Mean number of days			Mean cloudi-ness (%)	Mean sun-shine (h)	Most freq. wind direct.	Mean wind speed (m s⁻¹)	Mean number of days with gale
		precip. ≥0.1 mm	thunder-storm	fog					
Jan.	21	3.9	0	72		W	2.7	0	
Feb.	20	3.0	0	66		W	2.1	0	
Mar.	21	3.4	0	64		NW	2.1	0	
Apr.	18	3.0	0	58		SE	2.6	0	
May	18	1.3	0	56		SE	4.0	0.1	
June	20	0	0	59		SE	5.1	0	
July	21	0.1	0	60		SE	5.6	0	
Aug.	21	0	0	62		SE	5.9	0	
Sept.	18	0	0	55		SE	5.4	0	
Oct.	16	0.4	0	58		SE	4.7	0.1	
Nov.	14	0.6	0	54		SE	3.8	0.1	
Dec.	20	2.4	0	53		SE	2.2	0.1	
Year	228	18.1	0	60		SE	3.9	0.4	

*1 Source: Meteorological Services, Vacoas, Mauritius.

TABLE XVI

CLIMATIC TABLE FOR GRANDE COMORE (MORONI)[1]
Latitude 11°41'S, longitude 43°15'E, elevation 59 m, record 1951–1960

Month	Mean sta. press. (mbar)	Temperature (°C)					Mean vapour press. (mbar)	Precip. (mm)	
		mean daily	mean daily range	extremes				mean	24-h max.
				max.	min.				
Jan.	1009.2	27.1	6.3	33.5	20.2	27.8		347	220
Feb.	1008.9	26.8	6.3	32.7	20.8	28.1		291	148
Mar.	1009.1	26.9	6.8	32.7	20.3	28.1		289	262
Apr.	1010.4	26.6	7.3	33.8	20.5	27.5		270	437
May	1012.5	25.5	7.4	32.3	17.5	24.8		200	296
June	1014.9	24.1	7.6	31.9	14.1	22.4		227	261
July	1015.6	23.3	7.8	30.7	14.8	21.1		245	445
Aug.	1015.4	23.2	8.1	31.0	13.9	21.8		133	157
Sept.	1014.7	23.7	7.9	31.2	15.9	22.7		129	391
Oct.	1013.5	24.8	7.8	31.3	16.5	24.3		89	99
Nov.	1011.3	26.2	8.1	34.2	18.4	25.8		104	146
Dec.	1010.0	26.8	7.3	35.6	19.6	27.0		238	273
Year	1012.1	25.4	7.4	35.6	13.9	25.1		2562	445

Month	Mean evap. (mm)	Mean number of days			Mean cloudi-ness (%)	Mean sun-shine (h)	Most freq. wind direct.	Mean wind speed (m s⁻¹)	Mean number of days with gale
		precip. ≥0.1 mm	thunder-storm	fog					
Jan.		18	9	0	74		N	3.0	
Feb.		16	10	0	67		N	2.8	
Mar.		18	10	0.1	69		N	1.9	
Apr.		17	7	0	63		S	2.2	
May		12	1	0	54		S	2.3	
June		12	0.3	0	54		S	3.3	
July		12	0.1	0	51		S	3.7	
Aug.		11	0.1	0	55		S	3.5	
Sept.		12	0	0	56		S	2.8	
Oct.		13	0.4	0	57		SSW	2.1	
Nov.		11	2	0	57		SW	2.5	
Dec.		16	8	0	63		N	2.4	
Year		168	48	0.1	60		S	2.7	

[1] Source: Météorologie Nationale, Paris, France.

TABLE XVII

CLIMATIC TABLE FOR COCOS[1]
Latitude 12°05'S, longitude 96°53'E, elevation 5 m, record 1906–1962[2]

Month	Mean sta. press. (mbar)	Temperature (°C)				Mean vapour press. (mbar)	Precip. (mm)	
		mean daily[3]	mean daily range	extremes			mean	24-h max.
				max.	min.			
Jan.	1010.4	27.6	5.3	33.8	21.1	29.7	155	178
Feb.	1009.9	27.8	5.1	33.7	20.8	30.9	187	183
Mar.	1009.6	27.8	4.8	32.8	20.4	31.1	218	152
Apr.	1010.4	27.5	4.3	34.5	21.1	31.2	247	218
May	1011.1	27.2	3.9	32.8	21.6	30.6	193	158
June	1011.5	26.5	3.9	32.4	20.6	29.3	221	187
July	1012.1	26.1	3.8	31.8	20.6	28.5	220	259
Aug.	1012.8	26.0	4.1	31.7	18.3	27.6	131	287
Sept.	1013.4	26.2	4.4	32.1	20.4	27.4	101	213
Oct.	1013.2	26.6	4.8	32.3	20.8	28.1	94	323
Nov.	1011.9	26.9	4.9	32.6	21.1	28.6	109	365
Dec.	1010.6	27.2	4.9	32.8	20.1	29.3	115	269
Year	1011.4	26.9	4.5	34.5	20.1	29.4	1991	365

Month	Mean evap. (mm)	Mean number of days[4]			Mean cloudi-ness[4] (%)	Mean sun-shine (h)	Most freq. wind direct.[4]	Mean wind speed[4] (m s⁻¹)
		precip. ⩾0.1 mm	thunder-storm	fog				
Jan.		14	0.6	0	43		SE	2.6
Feb.		14	1.3	0	50		SE	2.9
Mar.		15	0.6	0	50		SE	2.9
Apr.		19	1.0	0	49		SE	3.6
May		16	0.1	0	47		SE	4.3
June		17	0.5	0	51		SE	4.9
July		19	0	0	51		SE	5.1
Aug.		19	0.1	0.1	47		E	4.6
Sept.		12	0.1	0	45		E	4.9
Oct.		13	0	0	45		SE	4.3
Nov.		12	0	0	46		SE	3.6
Dec.		12	0	0.1	46		SE	3.6
Year		182	4.3	0.2	47		SE	3.9

[1] Source: Bureau of Meteorology, Melbourne, Australia.
[2] Records not maintained from July 1946 through January 1952.
[3] (Max. + min.)/2.
[4] Period of record 5–13 years.

TABLE XVIII

CLIMATIC TABLE FOR MAURITIUS (PLAISANCE)[1].
Latitude 20°26'S, longitude 57°40'E, elevation 57.3 m, record 1951–1960

Month	Mean sta. press. (mbar)	Temperature (°C)				Mean vapour press. (mbar)	Precip. (mm)	
		mean daily	mean daily range	extremes			mean	24-h max.
				max.	min.			
Jan.	1004.4	26.0	7.2	35.9	18.7	27.7	237	262
Feb.	1004.3	25.8	6.8	33.0	18.3	27.6	241	190
Mar.	1004.6	25.6	6.6	32.9	15.9	28.0	383	296
Apr.	1007.1	24.3	6.8	31.3	15.0	25.1	206	171
May	1009.4	23.0	6.6	29.7	13.5	22.9	175	127
June	1012.4	21.3	6.3	30.4	12.5	20.2	117	67
July	1013.2	20.7	6.5	27.7	11.0	19.3	130	91
Aug.	1013.8	20.6	6.8	27.9	11.4	19.2	85	32
Sept.	1013.0	21.1	7.2	28.6	12.1	19.8	84	124
Oct.	1012.0	22.0	7.6	29.4	13.3	20.6	56	41
Nov.	1009.4	23.5	8.0	31.2	12.2	22.7	80	84
Dec.	1006.9	25.0	7.8	33.0	16.9	25.6	172	292
Year	1009.2	23.2	7.0	35.9	11.0	23.2	1966	296

Month	Mean evap. (mm)	Mean number of days			Mean cloudi-ness (%)	Mean sun-shine (h)	Most freq. wind direct.	Mean wind speed (m s⁻¹)	Mean number of days with gale
		precip. ≥0.1 mm	thunder-storm	fog					
Jan.	192	23	4.1	0	66	235	E	2.9	0.1
Feb.	155	22	3.8	0	66	207	E	3.1	0.1
Mar.	152	25	6.1	0	70	188	E	2.8	0
Apr.	138	21	2.3	0	59	201	SE	2.6	0
May	127	24	0.8	0	59	191	SE	2.5	0
June	114	22	0	0	61	167	SE	3.3	0
July	115	25	0	0	60	175	SE	3.2	0
Aug.	130	26	0	0	61	188	E	3.3	0
Sept.	138	21	0	0	59	197	E	3.3	0
Oct.	180	16	0.4	0	58	237	E	3.1	0
Nov.	189	16	0.9	0	59	237	E	2.7	0
Dec.	201	19	3.0	0	59	247	E	2.6	0
Year	1831	260	21.4	0	61	2470	E	2.9	0.2

[1] Source: Meteorological Services, Vacoas, Mauritius.

Chapter 7

Climate of Iceland

MARKÚS Á. EINARSSON

Introduction

Iceland is situated in the North Atlantic close to the Arctic Circle between latitudes 63°23'N and 66°32'N and longitudes 13°30'W and 24°32'W. The shortest distance to Greenland is about 290 km, to Scotland 800 km, and to Norway 970 km. The total area of the country is 103,100 km².

Because of Iceland's latitude, the solar altitude is never large and there is a great difference in the length of the day between summer and winter (Table I). In the northern-most part of the country, midnight sun is seen in midsummer and the nights are light in other parts at the same time.

Iceland is mountainous with an average height of 500 m above sea level, the highest peak Öræfajökull being 2,119 m. Only a quarter of the country lies below 200 m. The biggest lowland regions are in southern Iceland where the coasts are also sandy with a smooth outline. In most other parts, numerous fjords cut into the rocky landscape giving the coasts an irregular outline. In the innermost parts of the fjords are usually small lowland areas. This type of landscape is especially found in the basalt areas in the northwestern and eastern parts. Between them, in the middle zone, from southwest to northeast, the landscape is somewhat smoother and plateau-like with occasional, steep mountains.

Glaciers cover about 11,800 km² or 11.5% of the total area of Iceland. Vatnajökull with an area of 8,400 km² is by far the largest. The glaciers respond to climatic variations by retreating in warm periods and advancing in cold ones.

The history of meteorological observations in Iceland is not long. The first instrumental observations were carried out from 1749—1751 near Reykjavík and later similar observations were made temporarily at several locations. The first meteorological station with systematic and continuous weather observations was established at Stykkishólmur in 1845 and has been in operation ever since.

TABLE I

SOLAR ALTITUDE AT NOON AND THE LENGTH OF DAY AT THE SOLSTICES

Station	Latitude	Solar altitude at noon		Length of day	
		summer solstice	winter solstice	summer solstice	winter solstice
Vestmannaeyjar	63°27'N	50°00'	3°07'	20h37	4h30
Reykjavík	64°08'N	49°18'	2°25'	21h09	4h08
Akureyri	65°40'N	47°46'	0°53'	23h32	3h05
Grímsey	66°32'N	46°55'	0°01'	24h	2h13

The Danish Meteorological Institute was founded in 1872 and assumed in 1873 operation of the weather station at Stykkishólmur, along with the new station at Berufjördur (Teigarhorn). The Institute was responsible for all observations in Iceland until the Icelandic Meteorological Office was established in 1920. The number of stations increased gradually from 2 in the beginning to 17 in 1901 and 48 in 1931. In 1971, 127 stations functioned in Iceland—40 synoptic stations, 38 climatological stations, 42 precipitation stations, and 7 irregular stations. Agrometeorological observations were made at 7 stations, duration of sunshine was measured at 9 stations, global radiation was recorded at 2 stations, and there was 1 aerological station at Keflavík airport. Of all stations, 61% were located between 0 and 50 m above sea level, 18% between 51 and 100 m, 14% at 101 to 200 m, and 7% above 200 m, the highest being Hveravellir at 642 m. Fig. 1 shows synoptic and climatological stations in Iceland in 1971.

Climatic factors

Several meteorological as well as geographical factors influence weather and climate in Iceland to a great extent. At its latitudes, there is a considerable annual deficit in the total radiation balance of earth and atmosphere. Consequently, a transfer of heat from lower latitudes is carried out by oceanic and atmospheric circulations.

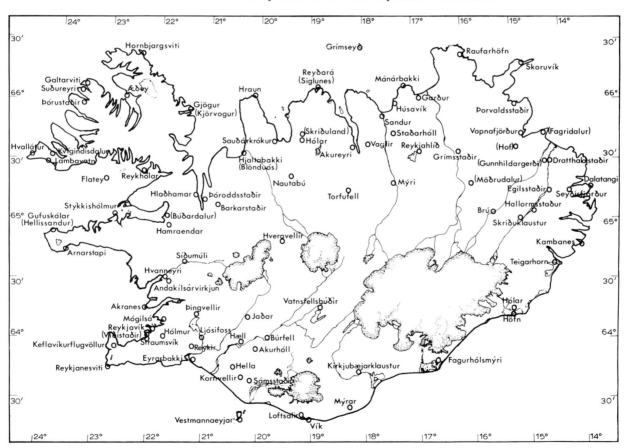

Fig. 1. Synoptic and climatological stations in Iceland in 1971(EINARSSON, 1976).

Iceland is situated near the border between warm and cold ocean currents. The North Atlantic Drift passes just to the south on its course northeastwards, and one of its branches—the Irminger Current—encircles the south, west, and north coasts. On the other hand, a branch of the cold East Greenland Current, known as the East Iceland Current, flows in a southerly and southeasterly direction along the east coast. Off the northwest and southeast coasts, a temperature front is found between these two different currents. It is obvious that the oceanographic conditions just described must influence weather and climate considerably, both directly at the coasts and also because all air masses arrive in Iceland after having passed over the sea.

A concomitant of climatic deterioration in Iceland and its vicinity is an increased extension of *sea ice*, which then may reach the coasts of Iceland in winter, particularly late winter. This sea ice is part of the main ice flow in the East Greenland Current. The extent of the ice flow varies from year to year and also with the time of year. Fig. 2 shows the limits of sea ice in October when it is near minimum and in March—May, the months of largest extent. It can be seen that under normal conditions the main ice edge does not reach the coast of Iceland, whereas in severe ice years the ice may extend along the northwest, north, and east coasts, and in extreme cases is even carried westward along the south coast. When sea ice is present near or at the coasts, the temperature of the coastal region falls appreciably and the largest negative deviations as a rule are then found at the north and east coasts.

As with ocean currents, warm and cold air masses often meet near Iceland. The polar front can almost always be found somewhere over the North Atlantic. Cyclones which form as disturbances on this front often intensify and pass close to Iceland and irregular

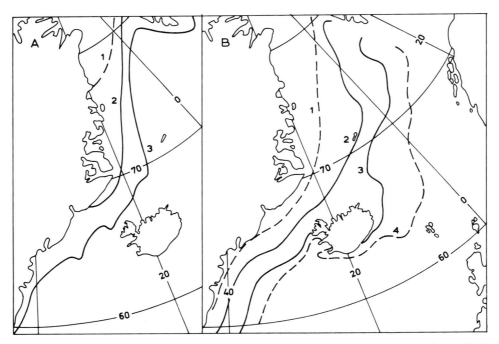

Fig. 2. A. Recent limits of sea ice in early October: *1* = minimum; *2* = normal; *3* = maximum. B. Limits of sea ice in March—May: *1* = recent minimum; *2* = recent normal; *3* = recent maximum; *4* = estimated maximum in historical times. (EYTHÓRSSON and SIGTRYGGSSON, 1971.)

and large pressure variations are therefore common. The lowest pressure ever measured in Iceland, reduced to mean sea level, was 920 mbar; the highest was 1,054 mbar.

Maps of annual mean pressure over the North Atlantic confirm that cyclones must be frequent near Iceland, as a mean low pressure centre—the *Icelandic Low*—is found a short distance southwest of the country. The travelling cyclones bring precipitation and strong winds, and rapid changes in weather may occur in their path. For instance, in winter it may change in a few hours from strong southerly winds with rain or drizzle and temperatures 5–10°C above freezing to northerly gale with temperatures 5–10°C below freezing. The temperature contrasts between tropical and polar air masses are largest in winter and, consequently, the lows are then more intense than in summer.

From mean pressure maps it may be seen that Greenland, over which a pronounced high often is situated, can have considerable influence on weather in Iceland, mainly by a strengthening of the northeasterly air flow.

The fact that Iceland is mountainous is important in many respects. The temperature conditions depend on the height above sea level and on the terrain; in addition, mountains may form barriers in places preventing maritime air from reaching the interior. Their influence on cloudiness and precipitation is obvious, as both elements increase windward of the mountains, but decrease leeward with resulting rain shadow and even clear weather.

The frequency distribution of wind direction depends considerably on the form and direction of valleys and fjords; local conditions also affect wind speed.

According to Köppen's classification, Iceland lies in a border region between two climatic types. In southern and western Iceland, a temperate rainy climate with cool and short summers (Cfc) exists, but in large parts of northern Iceland and the highlands the type is snow climate (ET).

Weather types

When describing the climatic conditions of Iceland, it is necessary to have in mind the general characteristics of circulation and weather. It has been mentioned that cyclones are frequent in the vicinity of Iceland, especially in winter, and the weather depends to a great extent on the position of their track. At a given time, the weather type differs considerably from one part of the country to another because of wind direction and topography. The main weather types are as follows.

Southeastern

Frequently, cyclones approach Iceland from the southwest and ahead of them southeast winds increase. Occluded fronts or warm fronts cause precipitation in most parts except perhaps in northeastern Iceland, although the sky there will also be overcast. As a rule, maximum precipitation is found in southern or southeastern Iceland or along the west coast; in this weather type, the temperature is relatively high—even above freezing in winter.

For considerable periods, Iceland may experience a series of cyclones from the southwest or even south. This happens when a blocking anticyclone over the British Isles or Scandinavia steers all cyclones northward just west of Iceland. The weather then alternates between the southeastern and southwestern types.

Southwestern or western

Following the passage of a cold front or an occlusion, the wind usually turns from southeast to southwest or sometimes even to west. The air mass which then invades the country is cold, often originating in the polar regions of North America. While passing the North Atlantic, it becomes unstable so that gusty winds and showers are common in southern and western Iceland when the air mass arrives. In the northeast and east, the clouds dissolve and fair weather prevails. When the air comes directly from the Greenland ice cap, the showers abate and even dry weather may occur.

Southern with warm air mass

Low-pressure systems are sometimes almost stationary for days east or southeast of Cape Farewell in Greenland, and at the same time an anticyclone is usually situated over western Europe. Then tropical air reaches Iceland from the south. It is stable in the lower layers after flowing over a relatively cool sea surface and thus fog and drizzle are common in southern Iceland. If the wind is strong, orographic rain may also result. When the air descends on the leeward side in northern and eastern Iceland, clouds disappear and föhn will bring locally warm and dry weather.

Warm air mass originating in Europe

This weather type is often associated with southeasterly winds which carry air laden with industrial dust from the British Isles or the continent. In most parts of Iceland, the weather is hazy with rather poor visibility and when the air blows over the cold East Iceland Current, fog frequently forms off the north and east coasts. In this type of weather, the temperature is usually relatively high in western Iceland.

Eastern

This weather type is caused by cyclones to the south of Iceland and can be rather persistent if they are almost stationary. New lows may arrive from the west and join the primary low.
East winds then prevail on the south coast. Occluded fronts lying from west to east are often found near the south coast and bring precipitation to the south and east. Fog occasionally forms at the northeast and north coasts, but in western Iceland the weather is usually favourable.

Northeastern

When the high over Greenland is strong, northeast winds prevail in Iceland. The cyclone tracks then lie south of Iceland, pointing east and later often northeast. If the cyclones are deep and close to the country when they reach the ocean between Iceland and

Norway, the northeast wind will be strong with snow (rain or drizzle in summer) in northern and eastern Iceland. At other times, the wind is so gentle that the weather may be tolerable in the inner parts of these regions, even if some precipitation falls in the outer parts. In southern Iceland, fair weather prevails in this situation.

Northern

In winter, deep lows east or northeast of Iceland cause strong northerly winds with heavy snowfall in northern Iceland which may last for days. The snowfall often extends southwards along the west and east coasts, but the weather will be dry in southern Iceland. Usually, such weather first abates farthest west when the low moves off, so that in northern Iceland the wind may be weak and the weather fair in the west while it is still snowing on the northeast coast. In many instances, the northeastern and the northern weather types are much alike.

A high over Iceland

The monthly mean pressure in Iceland is highest in spring. In all seasons, however, it may happen that a high over Iceland, often connected to the Greenland high, rules the weather. The winds are then generally light and variable and the weather fair with occasional fog at the coasts, especially in summer.

Climatic variations

Some evidence of climatic variations in geological as well as historical times can be found in Iceland, and the observations at Stykkishólmur have provided a continuous record of temperature variations from 1845 on.

It is considered (TH. EINARSSON, 1969) that the temperature rise after the last glacial peak started some 15,000–18,000 years ago. Then the glaciers generally retreated and the sea level rose. During this period of generally improving climate, however, there is evidence of two periods of deterioration—one a little more than 12,000 years ago, called the *Álftanes stage*, and the other, *the Búda stage,* some 11,000 years ago.

About 10,000 years ago, temperature suddenly began to rise and it is believed that the Ice Age glacier completely disappeared 2,000 years later. The vegetation which had survived began to spread; the birch advanced rapidly over most of the country about 9,000 years ago, indicating a warm and dry climate. This period has been named the *earlier birch stage*. Later, about 6,000–7,000 years ago, precipitation increased while temperature remained high so that the birch retreated and bogs expanded. This is the *earlier bog stage*. Approximately 5,000 years ago, the climate became drier again and the birch increased (*later birch stage*).

It is generally believed that the period from 9,000 to 2,500 years ago was as a whole several degrees warmer than now. But, 2,500 years ago, a sudden deterioration of the climate resulted in a second bog period (*later bog stage*), and the climate gradually gained the characteristics it had when Iceland was settled between 874 and 930 A.D.

No contemporary accounts from the first centuries following settlement can be found. The Icelandic sagas deal with this period, but because they were written down much later, their accounts of weather are unreliable. Nevertheless, such indirect information as mention of barley cultivation indicates relatively favourable conditions.

During the 13th century, the writing of annals began and gradually increased, except during the 15th century (after 1430), a period of almost no reliable records.

THORODDSEN (1916–1917) collected all available information on weather and sea ice in annals and sagas from settlement until 1900. On the basis of Thoroddsen's work and other sources, several authors (see, for instance, BERGTHÓRSSON, 1969) discussed the main characteristics of climatic changes from the time of settlement. The principal conclusions are that the climatic conditions have been subject to the following changes.

It is believed that during the first centuries after settlement (9th–12th centuries) conditions were at least as favourable as in the warm part of the present century (1920–1964). At that time, glaciers were small and barley was cultivated. About 1200, the climate deteriorated, temperature decreased and from that time the climatic conditions were more or less unfavourable until about 1920, when a marked improvement was experienced. Nevertheless, appreciable temperature variations took place during these cold centuries. A temperature minimum occurred around 1300, but some increase followed in the last part of the 14th century. The 15th century is thought to have been rather mild, but records of that time are sparse. After 1500, the climate deteriorated again and the 17th, 18th, and 19th centuries were very cold.

In 1845, systematic meteorological observations started in Iceland; a weather station still in operation was set up at Stykkishólmur in western Iceland. Temperature observations from this station give a fairly good picture of the conditions for the country as a whole because temperature variations tend to be in phase in different parts of the country. In winter, temperature variations are more extreme than in summer. It is possible to divide the entire period from 1846 into several distinct shorter periods according to the mean temperature of that season (December–March) at Stykkishólmur, as shown in Table II (SIGFÚSDÓTTIR, 1970).

The table indicates that low temperatures prevailed from 1853 to 1920, especially during the first part, whereas a relatively warm period followed in 1921–1965 (1921–1964 in northern Iceland). A dramatic change took place around 1965; the annual temperature dropped to a level comparable with the cold period before 1920, and 1966–1971 were all much below normal values. This decrease in temperature was most pronounced in northeastern and eastern Iceland and was associated with the presence of sea ice, one of several serious attendants of climatic deterioration in Iceland.

Fig. 3, showing 10-year overlapping means of temperature for six stations in Iceland, illustrates the above. At Stykkishólmur, the curve has a minimum in 1859/68, but it then rises slowly until 1889/98. The following 25 years it is nearly horizontal with some irregular variations, but from 1916/25 it rises suddenly until 1925/34. A maximum is found for 1932/41 and 1933/42. Later, the curve descends in general, although a minimum occurs in 1947/56 to 1949/58 and a maximum after 1951/60.

Iceland represents an environment in many respects very sensitive to climatic changes. A deterioration in the climate as a rule is accompanied by increased sea ice near the coasts, which in extreme cases can obstruct navigation and hinder the fisheries. A lowering of temperature often causes winter-killing of grasses and retards growth during

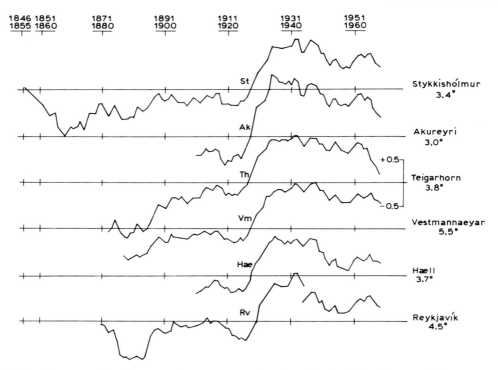

St — Stykkishólmur 3.4°

Ak — Akureyri 3.0°

+0.5

Th — Teigarhorn 3.8°

−0.5

Vm — Vestmannaeyar 5.5°

Hae — Hæll 3.7°

Rv — Reykjavík 4.5°

Fig. 3. Ten-year overlapping means of temperature at six stations in Iceland (EYTHÓRSSON and SIGTRYGGSSON, 1971).

TABLE II

MEAN TEMPERATURE (°C) IN WINTER AT STYKKISHÓLMUR

Period	Temperature
1846–1852	−0.7
1853–1892	−2.3
1893–1920	−1.7
1921–1965	−0.1
1966–1971	−2.0

the growing season. In both cases the result is a failure of hay, the most important crop in Iceland.

Temperature

The climate of Iceland is maritime with cool summers and mild winters. *Annual mean temperature* ranges from 2.0°C to 5.7°C in the lowlands. It is above 5°C at several stations along the south coast, but generally 4 − 5°C elsewhere in southern Iceland and 3 − 4°C in the interior of western Iceland. Nowhere in northern Iceland does it reach 4°C, but it is 3 − 4°C in the outlying areas and 2 − 3°C in some inland areas. In the highlands, of course, it is lower.

The *annual range of temperature,* i.e., the difference between the average temperatures of the warmest and coldest month, is rather small. Lowest values are found at the coasts, usually 9 – 11°C, except in the west where it is 11 – 12°C. The lowest single value is 8.7°C at Dalatangi on the east coast. In most inland regions, the range is 12 – 13°C, but 13 – 14°C in the interior of southwestern and northeastern Iceland, and in the latter are the highest values—above 15°C. *July* is the warmest month in most parts except at the north and east coasts where *August* is a little warmer. In southwestern Iceland, *January* is the coldest month, but in all other parts it is *February*. The difference between these two months is always very small, however.

Figs. 4 and 5 show the mean temperatures of *January* and *July* for 1931 – 1960 based on all available temperature data.

The January map shows that only in a few places does the temperature of this month exceed 0°C—at Reykjanes and some parts of the south and east coasts. The highest single value is 1.4°C at Vestmannaeyjar. At the west and partly also at the north coast, the temperature is 0° to − 1°C, but already in the fjords it decreases to − 2°C. Temperature generally decreases towards the interior, partly because of an increase in altitude and partly because temperature in winter decreases with increasing distance from the shore. This decrease is on the average approximately 2°C/100 km. In the highlands, except for the highest mountains and glacier caps, the January temperature is from − 4° to − 8°C.

In Fig. 5, which shows the mean temperature for July, one should first note the regions

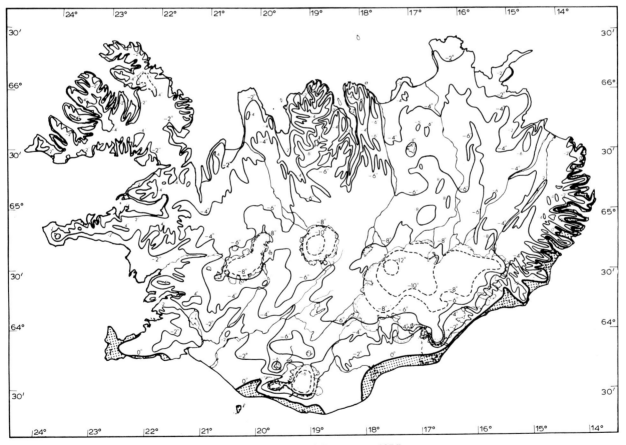

Fig. 4. Mean temperature for January in Iceland (EINARSSON, 1976).

681

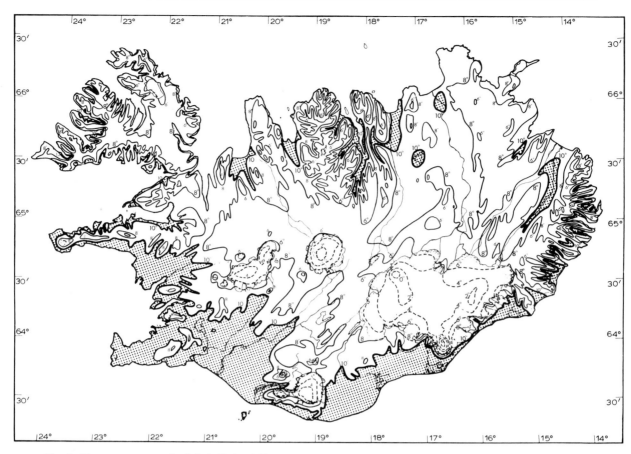

Fig. 5. Mean temperature for July in Iceland (Einarsson, 1976).

where the temperature exceeds 10°C, the temperature which separates temperate rainy climates (C) and snow climates (E) according to Köppen's classification. These regions are found in all lowland areas from Snaefellsnes southward and eastward to southeastern Iceland. In the Reykjavík area and inland of Borgarfjördur and southwestern Iceland, the mean temperature even exceeds 11°C. Other regions over 10°C include the lowlands in inner parts of the fjords in western, northern and eastern Iceland. At the coasts of Vestfirdir and northern and eastern Iceland, the July temperature is 8−10°C, and in the highlands it is generally 6−9°C. Excluding the effect of elevation, the temperature in summer increases inland.

Monthly mean temperatures are quite variable from year to year, especially in winter. At Reykjavík, the highest mean value in January in this century was 3.6°C (1964) and the lowest −7.3°C (1918). In July, the highest value was 13.3°C (1917) and the lowest 9.5°C (1970). Corresponding figures for Akureyri in northern Iceland are: January, highest 3.2°C (1947) and lowest −13.5°C (1918); July, highest 13.6°C (1927) and lowest 6.6°C (1915).

The mean diurnal variation of temperature is, on the whole, small in Iceland. In December and January, it is normally less than 1°C but increases towards spring and summer. In extreme coastal areas such as Dalatangi in eastern Iceland, it is less than 2°C even during summer. It is 3−5°C at Reykjavík and 4−6°C at inland stations in summer.

It may be noted that the range between the mean maximum and the mean minimum temperatures of each month shows very little variation during the year, contrary to the diurnal amplitude. Irregular temperature variations caused by frontal passages and air mass changes are responsible for the range in winter, whereas the influence of the regular diurnal cycle and insolation prevails in summer, even though air mass changes are also rather frequent during that season.

Temperature changes caused by advection of air masses are often sudden and large. A notable example of such an abrupt change occurred on 9 April 1963 when the temperature at Reykjavík was 7°C at noon, but had dropped to −7° to −8°C the same night. Similar drops in temperature were experienced in all parts of the country, and because this occurred in spring considerable damage was caused, especially to trees.

In spite of the maritime character of the climate, extreme values of temperature deviate considerably from the averages. At Reykjavík, for instance, the highest maximum in 1931−1960 was 23.4°C and the lowest minimum −17.1°C. At Akureyri, corresponding figures are 28.6°C and −22.1°C. There is, of course, a significant difference in the extremes between coasts and inland. The maximum temperature also shows a distinct difference between the north and east and the south and west. In summer, fair weather in northern and eastern Iceland is associated with warm air of a southerly origin, and a föhn may add to the temperature, which may thus rise to 20−25°C in the afternoon. In the south and west, fair weather, in contrast, occurs with cool, northerly air so that here the maximum temperature will be lower as a rule.

The highest reliable temperature measured in Iceland is 30.5°C at Teigarhorn in June 1939, and the lowest −37.9°C at Grímsstadir in January 1918.

Frost is frequent in Iceland, but does not normally last long as winter thaws are common; as a matter of fact, thaws are a peculiarity of Icelandic weather.

Table III shows the average number of *frost days* for five weather stations during 1951−1960, i.e., days during which the minimum temperature at 2 m height was below freezing. Frost days are rare in June−August except at highland stations (Grímsstadir). In December−February, the number of frost days per month is about 20 or more except at Vestmannaeyjar. The annual number is highest, commonly above 150, in northern and eastern Iceland, but lowest at Vestmannaeyjar, an extreme maritime station off the south coast.

In Table IV, the average dates of last frost in spring and first frost in autumn are shown

TABLE III

AVERAGE NUMBER OF FROST DAYS AT 2 M HEIGHT, 1951–1960

	Jan.	Feb.	Mar.	Apr.	May	June	July	Aug.	Sept.	Oct.	Nov.	Dec.	Year
Reykjavík	24	20	16	10	4	0.1	–	0.1	2	6	12	21	115
Akureyri	26	22	22	17	8	0.9	–	0.3	2	12	16	25	151
Grímsstadir	29	26	25	24	15	6	0.5	1	9	20	22	28	206
Hólar i Hornaf.	22	19	16	11	5	0.3	–	–	0.8	6	10	19	109
Vestmannaeyjar	14	13	11	6	3	–	–	–	0.6	3	6	13	70

TABLE IV

AVERAGE DATE OF LAST FROST IN SPRING AND FIRST FROST IN AUTUMN AT 2 M HEIGHT, 1951–1960

	Last frost	First frost
Reykjavík	10/5	1/10
Akureyri	31/5	15/9
Grímsstadir	20/6	20/8
Hólar í Hornaf.	15/5	9/10
Vestmannaeyjar	5/5	16/10

for the same five stations. Usually, the frost-free period in summer is shorter in the interior than at the coasts. The period is also shorter in northern Iceland than in the south, as cold air from the north causes frost there later in spring and earlier in autumn.

Precipitation

A large part of precipitation in Iceland falls with winds between east and south which prevail in the forward part of cyclones arriving from the southwest. Fig. 6 shows that the annual precipitation is distributed accordingly and that topography influences the pattern. The highest precipitation is in the southeast, with estimated maximum annual values of more than 4,000 mm on the ice caps Vatnajökull and Mýrdalsjökull and values above 1,600 mm in the lower areas. The highest annual value for a weather station is over 3,300 mm at Kvísker. In southwestern and western Iceland, the amount in the lowlands is 1,000–1,600 mm at the coasts, but 700–1,000 mm farther inland. The northern and northeastern districts have the least precipitation, with values of 400–600 mm in lower areas and less than 400 mm in an extensive area north of the huge glacier Vatnajökull, which creates a rain shadow for the southeasterly winds.

Precipitation depends to a great extent on local conditions and can vary much within a short distance. A good example is the Reykjavík area. At Reykjavík airport, the annual precipitation is about 800 mm; in the eastern part of town, it is 900 mm; on a small hill less than 10 km southeast of town, it is 1,040 mm; and in the mountains farther southeast, estimated values reach 3,000 mm.

Rain gauges, which in Iceland have openings about 1.5 m above the ground, give too low precipitation values, especially where wind velocities are high and where considerable precipitation falls as snow. No systematic investigations have been made in Iceland to quantify this point, but preliminary figures indicate that measured values for rain on the average may be approximately 25% too low. This figure is highly dependent on wind velocity and a higher value is to be expected for snow (F. H. Sigurdsson, pers. comm., 1972). Unfortunately, no reliable corrections could be applied when this was written and this fact must be considered when interpreting the precipitation measurements.

Autumn and early winter represent the seasons of greatest precipitation in most of Iceland, with a maximum usually occurring in *October*. An exception is interior northeastern Iceland where maximum precipitation occurs in July or August. *May* and *June*

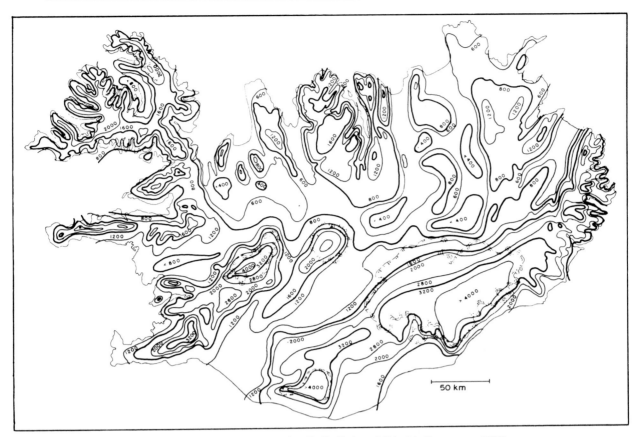

Fig. 6. Annual precipitation in Iceland (after Sigfúsdóttir, published in EINARSSON, 1976).

are the driest months of the year. As an average for the country as a whole, October's share of the annual precipitation is about 12% and that of May about 5%.

In most parts, some 50−65% of the annual precipitation falls during the months of October to March, except in some places in northeastern Iceland where the winter share is 40−50%.

Monthly values of precipitation show large year-to-year variations, and months with no precipitation at all exist. Very dry months are more likely to occur from January to August than in autumn. At Reykjavík, for example, the mean precipitation in October is 97 mm; the highest value for this month from 1931−1960 is 181 mm and the lowest 33 mm. In May, the mean value is 42 mm, the highest 96 mm and the lowest 0 mm. As a rule, maximum values deviate much more from the mean than the minimum values, especially for wet regions.

The largest monthly amount of precipitation measured in Iceland was 769 mm at Kvísker in southeastern Iceland in October 1965. The maximum precipitation measured in 24 h was 242.7 mm at Kvísker in October 1979.

As convective precipitation during summer is not an essential part of the total amount, it follows that intense rain of short duration is rare. Intensity measurements are sparse, but as an example, the highest 10-min value for Reykjavík is 4.6 mm and the highest 1-h value 17.3 mm. Thunderstorms are accordingly very infrequent; they may occur 1−2 days per year on the average in southern and southwestern Iceland, usually in winter, and are connected with frontal passages or unstable cold air masses. In northern Iceland, they are observed even more rarely.

The annual number of days with precipitation of 0.1 mm or more is quite variable—200 or more in southern Iceland and at points along the east and west coasts, with a maximum value of 235 days at Vestmannaeyjar. In northern and northeastern Iceland, the number is less even at the coasts, and farther inland it is 130—140 days. There is considerable difference between summer and winter in the monthly number of days with precipitation. In autumn and winter, it is 16—23 days at the south, west and east coasts, but 12—16 days elsewhere. In western Iceland, the lowest number is in one of the months from May to August, 11—15 days, and in northern and eastern Iceland it is 7—10 days in May or June in the inland regions. In southern Iceland, the number is variable in summer, but in general it is lowest in May or June, 13—17 days. August, on the contrary, resembles the wet autumn months.

Dry periods in summer are of great importance in Iceland for curing hay and there is no doubt that the interior of northern and eastern Iceland is the most favourable region in this respect. Here 66% of all days during June—August is dry; the corresponding figure for southern Iceland is 47%.

Snow

In northern Iceland, more than half the winter precipitation falls as snow and a complete snow cover may persist for weeks or even months, whereas in southern Iceland the snow cover is more variable. Winter thaws are frequent and may alter the snow cover, especially in the south where the thaws are usually accompanied by rain. It is difficult to classify precipitation as rain or snow because a considerable part of precipitation falling between observation times is mixed, or of both types, which is also classified as mixed. According to average values for several weather stations during 1965—1972, snow alone is only about 5—10% of the total precipitation in southern Iceland during December through March. In northern Iceland, in contrast, snow amounts to 50—70% of total precipitation from November to March and only 3—15% is classified as rain.

The snow cover, expressed in percent of total cover and taken as the average for October—May, is largest in northern and northeastern Iceland and in the northernmost part of Vestfirdir, where it is 50—70%, with the highest monthly mean values of 80—90%. In the south and in most of the west, the values are much lower, 15—38%, with minimum values at the south coast.

Table V shows the average dates for 1951—1970 of first and last snowfall, first and last day of complete snow cover, and first day of no snow cover in spring for several weather stations.

The first snowfall in autumn usually occurs in October in southern and western Iceland, but in September in the northern districts. At Hornbjargsviti in the extreme northwest and at Grímsstadir in the northeastern highlands, the first snowfall is in late August.

The last snowfall in spring is in early May in southern and western Iceland, in late May or early June in the northern and eastern districts, and even later at Hornbjargsviti and Grímsstadir.

The first day of complete snow cover in autumn is on the average in the first part of

TABLE V

SNOWFALL AND SNOW COVER, AVERAGE DATES 1951–1970

	First snowfall	Last snowfall	Complete snow cover		First day of no snow cover
			first day	last day	
Reykjavík	14/10	30/4	4/11	9/4	19/4
Sídumúli	5/10	9/5	2/11	25/4	2/5
Hornbjargsviti	27/8	13/6	23/10	20/5	4/6
Akureyri	25/9	27/5	23/10	22/4	13/5
Grímsstadir	23/8	16/6	7/9	30/5	2/6
Raufarhöfn	10/9	4/6	23/10	12/5	1/6
Hallormsstadur	6/10	29/5	30/10	22/4	4/5
Hólar (Höfn)	23/10	29/4	10/11	8/4	22/4
Kirkjubaejarklaustur	19/10	9/5	8/11	10/4	29/4
Vestmannaeyjar	8/10	7/5	19/11	8/4	22/4

November in southern and western Iceland, but in the last days of October in other parts.

The last day of complete snow cover in spring is rather variable, but it generally occurs in the first part of April in the south and late April or May elsewhere. The date of no snow cover varies from a few days to one month after the last day of complete snow cover—on the average about 15 days later. Although the last date of snow cover is on the average in April or May, shorter or longer periods of no snow cover are common in all winter months in southern Iceland.

Snow depth is quite variable and difficult to measure. The deepest snow will be found in the north, and several times since 1931 the coastal stations Hornbjargsviti and Raufarhöfn have reported depths in excess of 200 cm, the highest known values for that period.

Wind

The distribution of annual mean pressure over the North Atlantic with the *Icelandic Low* to the southwest indicates that, where unobstructed, wind directions between northeast and southeast should be frequent. Fig. 7, showing the annual frequency of the different wind directions for 1965–1971 confirms this, at least at the coastal stations in southern and northwestern Iceland and at Grímsey off the north coast. Apart from these places, the wind roses are rather irregular. Local conditions, landscape, and direction of fjords or valleys control the frequencies, as seen for instance at Sídumúli, Thóroddsstadir, Akureyri, and Egilsstadir. Although west and northwest winds are rare, especially in the west, some are found in the east, particularly at Raufarhöfn in the northeast corner.

The frequency of calms is given at each station. Unfortunately, the figures are hardly comparable between stations as estimates of calms differ, especially between stations with and without anemometers. Nevertheless, one can conclude that calms are uncommon on the south coast and in the highlands, but are rather frequent in lowland regions of the interior.

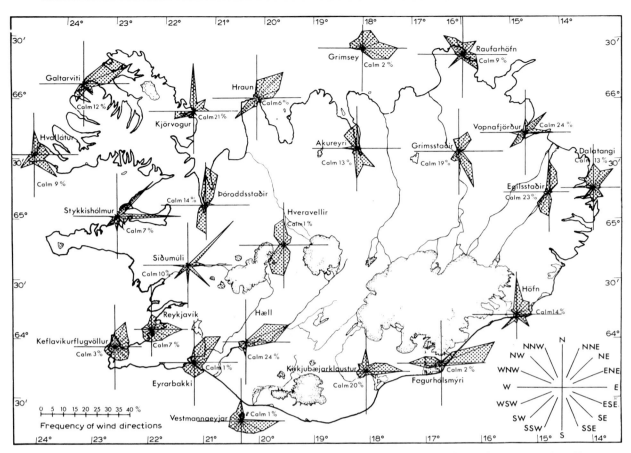

Fig. 7. Annual frequency of different wind directions at several Icelandic weather stations, 1965—1971. (EINARSSON, 1976).

In summer, sea breezes influence the frequencies of wind directions considerably at the coasts. At Reykjavík, for instance, the average frequency of the sea breeze directions, west to north, is 44% in July but 11% in January. At Akureyri, the frequency of the directions northwest to north is 54% in July but 21% in January.

As expected in a country where low pressure systems pass frequently, wind velocities are generally high, especially at the coasts and in winter when the cyclones are most intense. Typical winter values of mean monthly wind velocity are $6-7$ m s^{-1} at the coasts, while summer values are $4-6$ m s^{-1}. The southernmost station, Vestmannaeyjar, has exceptional wind conditions with a highest monthly mean value of about 13 m s^{-1}, partly caused by the height (120 m) of the station and its position on a cape. Otherwise, wind speeds are without doubt highest at the south coast. At lowland stations in the interior, the mean wind speed is $3-4$ m s^{-1}, but in the highlands it is rather high. The annual number of days with gale force winds ($\geqslant 8$ Beaufort) is in general highest at the coasts and in the highlands, as seen in Table VI, where the numbers $\geqslant 9$, $\geqslant 10$, $\geqslant 11$, and $\geqslant 12$ Beaufort are also given for several weather stations. The number for Vestmannaeyjar is extremely high for the reason described above.

The highest measured 10-min average of wind speed in Iceland is 55.6 m s^{-1} at Vestmannaeyjar in October 1963 and 51.5 m s^{-1} in January 1957. In January 1949, the wind force was estimated 17 Beaufort ($56.1-60.7$ m s^{-1}) at the same place. A few values

TABLE VI

ANNUAL MEAN NUMBER OF DAYS, 1965—1971 WITH DIFFERENT BEAUFORT WIND VALUES
(after ÓLAFSSON and BERGTHÓRSSON, 1972)

	≥8	≥9	≥10	≥11	≥12
Reykjavík	43	16	5	0.7	0.1
Galtarviti	40	18	7	2.3	1.6
Saudárkrókur	46	15	9	0.6	0.2
Akureyri	18	7	3	0.6	0.3
Raufarhöfn	45	16	4	0.6	
Dalatangi	27	8	1	0.1	
Höfn	33	12	5	1.1	0.4
Vestmannaeyjar	199	135	82	41	17
Hveravellir	92	45	18	4	0.5

between 36.0 and 49.4 m s^{-1} are known from other stations. The highest measured value for Reykjavík is 39.6 m s^{-1} in February 1981.

Measurements of wind gusts have only recently become available from a few stations, in addition to the airports at Reykjavík and Keflavík where older records are available. The highest gust velocity measured till now is 61.8 m s^{-1} at Thyrill in Hvalfjördur, beneath a mountain, but further measurements are likely to yield higher values. The occurrence of gusts depends greatly on topography. Extreme gusts probably do occur near mountains, but this has not been sufficiently confirmed owing to lack of measurements.

Humidity

Because the climate of Iceland is maritime, the humidity of the air is comparatively high. Monthly mean vapour pressure for 1958—1967 (computed from mean relative humidity and saturation vapour pressure at monthly mean temperature) is generally lowest in December and January (4—5 mbar) and highest in July (9—11 mbar), with maximum values in southern Iceland. The annual means are on the order of 6—7 mbar. For the same period, the monthly mean relative humidity is mostly in the interval 75—90%. The variation from month to month is rather small and irregular, and the same applies to the variation between different parts of the country. It is common to find the lowest monthly means in one of the months from January to June, frequently in May. The highest monthly means, on the other hand, are found in the latter half of the year, usually in August or October. In Iceland, relative humidity is measured by a psychrometer which is often unreliable when the temperature is below freezing.

Cloudiness

The mean cloud cover is high in Iceland with monthly mean values (1931—1960) of 5.0—7.0 oktas. In western and northern Iceland, the highest monthly means are most

TABLE VII

AVERAGE MONTHLY AND ANNUAL NUMBER OF OVERCAST DAYS, 1961–1970

	Jan.	Feb.	Mar.	Apr.	May	June	July	Aug.	Sept.	Oct.	Nov.	Dec.	Year
Reykjavík	14	13	15	16	13	19	15	16	15	15	11	13	174
Galtarviti	21	17	20	17	17	18	18	20	17	22	20	19	226
Grímsey	20	18	23	18	20	17	22	22	19	23	22	23	247
Akureyri	16	13	15	14	18	15	18	19	16	18	17	17	196
Grímsstadir	13	10	12	12	12	11	11	14	13	14	13	13	148
Hallormsstadur	15	11	13	14	15	14	15	16	16	16	12	13	170
Dalatangi	17	13	16	16	19	17	18	19	18	17	16	14	200
Hólar (Höfn)	15	12	14	15	16	19	17	18	15	16	10	11	178
Kirkjubaejarkl.	12	12	13	15	16	17	12	14	13	12	9	11	156
Vestmannaeyjar	17	16	17	16	16	20	15	16	17	16	13	14	193

often found in one of the months August – December, but in July – September in eastern and southern Iceland. Lowest monthly values are found in January or February in the south and in May or June in the north. The variation between months is very small. The cloud cover is lowest in the interior of northeastern and eastern Iceland and in the south. The highest values are, on the other hand, found in the northwest.

Tables VII and VIII give the average number of overcast and clear days at several weather stations for 1961 – 1970. The definition of an overcast day is that the sum of the cloud cover values from 8 daily observations is 52 oktas or more. A day is considered clear when the sum of the cloud cover from 8 daily observations is not more than 12 oktas. The 10-year period used here is hardly representative in the southwest, as the cloud conditions in June were exceptional there during the period with almost no clear but many overcast days.

The monthly number of overcast days is rather variable, mainly in the range 15 – 23 at

TABLE VIII

AVERAGE MONTHLY AND ANNUAL NUMBER OF CLEAR DAYS, 1961–1970

	Jan.	Feb.	Mar.	Apr.	May	June	July	Aug.	Sept.	Oct.	Nov.	Dec.	Year
Reykjavík	2.5	2.5	3.0	1.3	1.7	0.1	1.1	1.0	2.3	2.1	2.5	2.2	22.3
Galtarviti	0.9	0.5	0.8	0.6	1.9	1.3	1.5	1.1	0.9	0.5	0.4	0.2	10.6
Grímsey	0.8	0.3	0.3	0.5	0.6	0.4	0.0	0.3	0.6	0.4	0.2	0.2	4.6
Akureyri	2.4	2.4	1.1	1.5	1.2	0.9	1.0	0.3	1.5	0.6	0.7	1.2	14.8
Grímsstadir	1.8	2.0	2.0	2.3	1.7	1.1	1.5	2.0	1.1	1.3	0.4	1.0	18.2
Hallormsstadur	2.0	3.8	2.7	2.3	2.0	1.9	1.8	0.9	1.3	0.9	1.8	2.3	23.7
Dalatangi	0.8	1.5	0.3	0.5	1.2	0.8	0.6	0.4	0.8	0.3	0.5	0.7	8.4
Hólar (Höfn)	2.2	3.1	3.2	1.6	1.6	0.7	0.1	0.4	1.6	1.7	3.6	3.7	23.5
Kirkjubaejarkl.	3.0	2.6	3.8	1.8	0.8	0.2	1.2	0.9	2.0	2.2	3.8	3.8	26.1
Vestmannaeyjar	2.7	1.9	2.9	2.3	1.1	0.1	1.6	1.4	2.0	2.4	2.4	2.4	23.2

the coasts, but lower at inland stations where it is as low as 9 – 12 days in some months. The annual values show a distinct difference between the coasts and interior, with the highest number of overcast days, 247, at Grímsey off the north coast. It is about or over 200 days at the coasts, but less than 200 at inland stations, with 148 (Grímsstadir) the lowest value in the table.

Clear days are few, as expected. In some months, they hardly occur and in the most favourable regions one can expect only 3 – 4 clear days per month on the average. The lowest number of clear days is found at the northwest, north, and east coasts (Galtarviti, Grímsey, Dalatangi). On the other hand, the number is relatively high inland, especially in the northeast and east and in southern Iceland where it is highest (Kirkjubaejark-laustur).

Visibility and fog

Iceland is sparsely populated with little industrial activity, and even domestic heating is to a large extent accomplished in a smokeless way, i.e., with water from hot springs. Consequently, the amount of dust particles in the air is usually small, and in fine weather the visibility is extremely good—mountains 100 – 200 km away may be clearly seen. It happens, however, that southerly or southeasterly winds bring industrial dust from the British Isles or from the continent to Iceland. The weather in most parts will then be hazy with comparatively poor visibility. The main causes of limited visibility are, though, precipitation, especially snowfall, or fog. In winter, visibility less than 10 km, therefore, is more frequent than in summer. Very poor visibility (less than 5 km) is for the same reason (snow) more frequent in winter at Akureyri in the north than at Reykjavik where only a small part of the precipitation falls as snow. Table IX shows the annual frequencies of visibility within certain limits in percent of all observations.

The frequency of fog (horizontal visibility less than 1 km) in Iceland is quite variable. The annual mean number of days with fog is on the whole much higher in northeastern and eastern Iceland (40 – 60 days at the coasts) than in other parts of the country. The reason is without doubt the influence of the cold East Iceland Current which flows in a southerly and southeasterly direction along the east coast. In other parts, the annual number is usually between 5 and 17 days. Days with fog are by far more numerous in summer than in winter, with maximum in July, 8 – 10 days in the regions of most frequent occurrence.

TABLE IX

ANNUAL FREQUENCY OF VISIBILITY WITHIN CERTAIN LIMITS 1961–1970, IN PERCENT OF ALL OBSERVATIONS

	< 1 km	1–5 km	6–10 km	11–20 km	> 20 km
Reykjavík	1	4	10	18	67
Akureyri	2	6	5	7	80

Sunshine and radiation

Duration of bright sunshine has been measured at 6 weather stations in Iceland for a considerable period, but only 2 stations have 30-year normals—Reykjavík with 1,249 h per year and Akureyri with 962 h. These figures are not directly comparable because of differences in topography.

At all sunshine-recording stations, the maximum possible duration of sunshine, S_o, has been found for the 10 years 1958—1967 by plotting all values from clear days and drawing a best-fitting curve so that the majority of the single values lies on or below the curve. In March through September, the fraction of possible monthly mean sunshine, S/S_o, for the same years is mostly in the interval 25—40% at these stations.

Global radiation, G, has been recorded at Reykjavík since July 1957. Ten-year means for 1958—1967 show an annual mean value of $3.1 \cdot 10^9$ J m^{-2} with the highest monthly value in July, $2.1 \cdot 10^2$ J m^{-2} s^{-1}. The month of highest solar altitude, June, showed lower radiation than both May and July because of unfavourable weather conditions in this period.

The fraction of possible radiation, G/G_o, (G_o is global radiation on clear days, found in the same way as S_o), gives an estimate of how much clear sky radiation reaches the ground. Annual mean value of G/G_o at Reykjavík is 57% with a maximum of 63% in August and a minimum of 51% in June. On the average, 43% of the clear sky radiation is lost because of clouds.

Mean global radiation for 1958—1967 has been calculated for 5 stations in Iceland recording duration of sunshine and for 30 stations observing cloud cover (EINARSSON, 1969). This was done by means of equations of regression between G/G_o, on the one hand, and relative duration of sunshine, S/S_o, or mean cloud cover at 0900, 1500, and 2100 GMT, on the other. These equations were based on the values of radiation, cloud cover, and duration of sunshine observed at Reykjavík. According to these computations, annual values of global radiation at the different stations range from $2.5 \cdot 10^9$ to $3.6 \cdot 10^9$ J m^{-2}.

Maximum insolation is found in southern Iceland west of the icecap Mýrdalsjökull, and from there a zone of relatively high radiation reaches north of the glacier Vatnajökull to another maximum in the interior of northeastern Iceland. A minimum area in the inner parts of the district Skagafjördur and the western highland Kjölur is also characteristic. The radiation shows considerable variation and is influenced by local differences in cloud cover and weather conditions.

Potential evapotranspiration

Potential evapotranspiration, E_p, has been estimated with the aid of Penman's equation for a grass-covered surface (EINARSSON, 1972). The calculations were based on the computations of global radiation mentioned in the foregoing section and meteorological data for 28 stations from 1958—1967. Penman's method has been rather widely used in neighbouring countries and it was considered worthwhile to make estimates also for Iceland.

Comparisons at Reykjavík between the evaporation from a class "A" pan and potential evapotranspiration according to Penman have shown that the ratio, class "A"/E_p, is 1.10 on the average for the summer (June—September), being lowest in June but increasing towards autumn.

The calculations of E_p and E_o, the evaporation from an open water surface, showed that the ratio, E_p/E_o, is for Icelandic stations on the order of $0.81-0.87$, the annual average value for all stations being 0.84.

By far, the greatest part of the potential evapotranspiration, or some $75-95\%$, takes place in the summer half of the year, April–September. The monthly maximum values, which in $1958-1967$ occurred in July at the southwest coast but in June elsewhere, reach 100 mm in places. On the other hand, small negative values appear at 5 stations in January and 4 stations in November and December, most of them situated some distance from the coast.

The annual values of E_p lie mainly in the range $360-540$ mm, and minimum and maximum zones are found in the same areas as in the case of global radiation. The area of minimum is in the western highlands near Kjölur, whereas there are areas of maximum evaporation at the southwest coast and in the eastern part of the country north of Vatnajökull.

The potential water balance, i.e., the difference between precipitation and E_p, has been calculated for the year and for the two shorter periods April–September and May–August, using precipitation normals for $1931-1960$ (Fig. 6) and the above values of E_p.

The main characteristics of the precipitation distribution are reflected in the distribution of potential water balance. By far, the largest part of the country has a positive annual balance, and the highest values are found, as expected, in the southeastern part. Only in the interior of northeastern Iceland north of Vatnajökull is the annual balance negative, as low as about -100 mm.

Considering the summer half of the year, April–September, and excluding accumulated snow which melts in spring and also water in the soil, one finds that most of the lowlands in the northern part of Iceland show a slight negative balance. This is not surprising, as the greatest part of the potential evapotranspiration takes place in this period, whereas precipitation is generally less in summer than in winter.

In the still shorter period of May–August, which includes the most important part of the growing season, some areas in the southwest show a negative balance. It must be remembered, however, that correct precipitation measurements would probably alter this somewhat.

Acknowledgements

The author is indebted to Mrs. Adda Bára Sigfúsdóttir and Mr. Flosi Hrafn Sigurdsson of the Icelandic Meteorological Office and to Professor Thorbjörn Karlsson for reading the manuscript and offering valuable suggestions.

References

Anonymous, 1924–1981. *Vedráttan*, monthly and annual climatological summaries. Vedurstofa Íslands, Reykjavík.

Bergthórsson, P., 1969. An estimate of drift ice and temperature in Iceland in 1000 years. *Jökull*, Jöklar-annsóknafélag Íslands, Reykjavík, 19: 94–101.

EINARSSON, M. Á., 1969. *Global Radiation in Iceland*. Vedurstofa Íslands, Reykjavík, 29 pp.

EINARSSON, M. Á., 1972. *Evaporation and Potential Evapotranspiration in Iceland*. Vedurstofa Íslands, Reykjavík, 25 pp.

EINARSSON, M. Á., 1976. *Vedurfar á Íslandi*. Idunn, Reykjavík, 150 pp.

EINARSSON, Th., 1969. Loftslag, sjávarhiti og hafís á forsögulegum tíma. In: M. Á. Einarsson (Editor), *Haf-ísinn*. Almenna bókafélagid, Reykjavík, pp. 389—402.

EYTHÓRSSON, J. and H. SIGTRYGGSSON, 1971. The climate and weather of Iceland. In: *The Zoology of Iceland*, Vol. 1, Part 3. Ejnar Munksgaard, Copenhagen and Reykjavík, 62 pp.

ÓLAFSSON, Ó. E. and P. BERGTHÓRSSON, 1972. Ofvidri á Íslandi. *Vedrid*, Reykjavík, 17: 50—59.

SIGFÚSDÓTTIR, A. B., 1970. Hitafar á Íslandi. *Búnadarbladid Freyr*, Reykjavík, 66: 217—219.

THORODDSEN, Th., 1916—1917. *Árferdi á Íslandi í thúsund ár*. Hid íslenzka fraedafélag, Copenhagen, 432 pp.

Appendix — Climatic tables

TABLE X

CLIMATIC TABLE FOR REYKJAVÍK
Latitude 64°08′N, longitude 21°56′W, elevation 13 m

| Month | Mean sta. press. (mbar) | Temperature (°C) | | extremes | | Mean vap. press. (mbar) | Precipitation (mm) | |
		daily mean	mean daily range	max.	min.		mean	max. in 24 h
Jan.	997.6	−0.4	5.2	10.0	−17.1	5.0	90	36
Feb.	1003.0	−0.1	5.6	10.1	−15.6	5.3	65	40
Mar.	1004.7	1.5	5.8	14.2	−14.3	5.3	65	57
Apr.	1006.7	3.1	5.8	15.2	−12.0	6.2	53	22
May	1012.4	6.9	6.2	20.6	−7.2	7.5	42	19
June	1009.6	9.5	5.9	20.7	−0.2	9.5	41	30
July	1006.8	11.2	5.7	23.4	3.9	10.6	48	31
Aug.	1006.0	10.8	5.8	21.4	−0.4	10.2	66	35
Sept.	1003.6	8.6	5.4	20.1	−2.7	9.0	72	49
Oct.	1000.5	4.9	5.0	15.7	−10.2	7.5	97	37
Nov.	999.8	2.6	4.6	11.5	−10.5	5.6	85	44
Dec.	996.5	0.9	5.1	11.4	−15.3	4.9	81	55
Annual	1003.9	5.0	5.5	23.4	−17.1	7.2	805	57
Rec.(yrs.)	30	30	30	30	30	10[*1]	30	30

| Month | Number of days with | | | Mean cloud-iness (oktas) | Mean sun-shine (h) | Wind | | Solar radiation (J m^{-2} s^{-1}) |
	precip. (⩾0.1 mm)	thunder-storm	fog			most freq. direct.	mean speed (m s^{-1})	
Jan.	20	0.4	0.6	5.6	21	E	6.8	5.8
Feb.	17	0.3	1.1	5.6	57	E	7.1	25.7
Mar.	17	0.1	0.5	5.7	106	E	6.8	79.4
Apr.	18	0.1	0.7	5.8	138	E	6.2	140.0
May	15	0.1	0.9	5.8	185	E	5.5	208.8
June	14	0.0	1.1	5.8	189	SE	5.4	199.6
July	15	0.1	1.1	6.1	178	N, WNW	4.9	212.2
Aug.	17	0.1	0.7	5.9	159	SE	4.8	170.0
Sept.	19	0.1	0.9	6.0	105	E	5.7	87.2
Oct.	20	0.1	0.5	5.8	71	E	6.1	39.2
Nov.	19	0.2	0.8	5.7	32	E	6.8	10.2
Dec.	21	0.2	0.8	5.7	8	E	6.7	1.9
Annual	212	1.7	9.7	5.8	1249	E	6.1	98.3
Rec.(yrs.)	30	30	30	30	30	7[*2]	10[*1]	10[*1]

[*1] Period 1958–1967.
[*2] Period 1965–1971.

TABLE XI

CLIMATIC TABLE FOR AKUREYRI
Latitude 65°41′N, longitude 18°05′W, elevation 4 m

Month	Mean sta. press. (mbar)	Temperature (°C)				Mean vap. press. (mbar)	Precipitation (mm)	
		daily mean	mean daily range	extremes max.	min.		mean	max. in 24 h
Jan.	1001.1	−1.5	6.5	14.0	−20.3	4.6	45	17
Feb.	1005.8	−1.6	6.2	13.2	−19.8	4.8	42	21
Mar.	1008.1	−0.3	6.5	16.0	−22.1	4.8	42	27
Apr.	1009.7	1.7	6.6	16.0	−16.9	5.9	32	16
May	1014.9	6.3	7.3	21.5	−9.0	7.1	15	24
June	1011.5	9.3	6.8	28.6	−3.0	8.9	22	19
July	1008.8	10.9	6.4	24.3	0.5	10.1	35	27
Aug.	1008.0	10.3	6.5	25.0	−2.2	9.6	39	52
Sept.	1005.8	7.8	6.3	22.0	−8.4	8.1	46	92
Oct.	1003.0	3.6	5.8	17.6	−11.2	6.8	57	30
Nov.	1002.8	1.3	5.6	15.5	−15.5	5.3	45	27
Dec.	1000.1	−0.5	6.0	13.4	−20.8	4.6	54	33
Annual	1006.6	3.9	6.4	28.6	−22.1	6.7	474	92
Rec.(yrs.)	30	30	30	30	30	10*1	30	30

Month	Number of days with			Mean cloud-iness (oktas)	Mean sun-shine (h)	Wind	
	precip. (≥0.1 mm)	thunder-storm	fog			most freq. direct.	mean speed (m s⁻¹)
Jan.	13	0.0	0.0	6.0	6	S	3.8
Feb.	13	0.0	0.3	6.0	32	S	3.8
Mar.	11	0.0	0.4	6.0	76	S	3.2
Apr.	11	0.03	0.4	6.2	105	S	2.9
May	7	0.0	1.1	5.7	172	N	3.1
June	8	0.03	1.5	5.9	172	N	3.5
July	11	0.1	1.8	6.2	147	NNW	3.3
Aug.	11	0.0	1.3	6.3	113	N	3.3
Sept.	12	0.0	1.4	6.2	75	N	3.2
Oct.	14	0.0	0.6	6.0	51	SE, S	3.2
Nov.	13	0.0	0.4	6.1	13	SE, S	3.8
Dec.	15	0.03	0.3	6.1	0	SE, S	3.6
Annual	139	0.2	9.5	6.1	962	S	3.4
Rec.(yrs.)	30	30	30	30	30	7*2	10*1

*1 Period 1958–1967.
*2 Period 1965–1971.

TABLE XII

CLIMATIC TABLE FOR HÓLAR Í HORNAFIRDI[1]
Latitude 64°18′N, longitude 15°12′W, elevation 16 m

Month	Mean sta. press. (mbar)	Temperature (°C)				Mean vap. press. (mbar)	Precipitation (mm)	
		daily mean	mean daily range	extremes			mean	max. in 24 h
				max.	min.			
Jan.	999.2	0.3	5.0	11.1	−13.2	4.9	191	134
Feb.	1003.5	0.0	5.0	10.0	−14.6	5.0	115	107
Mar.	1006.1	1.5	5.2	12.0	−13.5	5.1	132	78
Apr	1007.0	3.0	5.5	17.1	−12.2	6.0	108	90
May	1013.0	6.5	5.6	21.1	−7.9	7.2	90	107
June	1010.0	9.3	5.0	26.6	−1.5	8.9	83	84
July	1007.0	10.9	5.4	25.5	2.6	10.2	93	61
Aug.	1006.5	10.4	5.5	20.2	0.6	9.5	116	56
Sept.	1004.1	8.2	5.4	17.9	−3.9	8.6	162	122
Oct.	1001.4	4.9	4.8	17.6	−7.5	7.3	170	100
Nov.	1000.9	2.7	4.9	15.2	−9.6	5.5	187	106
Dec.	998.1	1.2	4.9	10.9	−13.0	4.7	185	92
Annual	1004.7	4.9	5.2	26.6	−14.6	6.9	1632	134
Rec.(yrs.)	30	30	9[2]	30	30	10[3]	30	30

Month	Number of days with			Mean cloud-iness (oktas)	Mean sun-shine (h)	Wind	
	precip. (⩾0.1 mm)	thunder-storm	fog			most freq. direct.	mean speed (m s⁻¹)
Jan.	17	0.4	0.6	5.4	32	N	5.6
Feb.	13	0.2	0.5	5.2	72	N	5.7
Mar.	16	0.1	0.8	5.6	116	N	5.6
Apr.	13	0.03	0.7	5.6	133	E	5.2
May	12	0.0	1.4	5.9	181	E	4.6
June	11	0.0	1.6	6.1	145	SW	4.1
July	16	0.03	1.7	6.6	149	SE	3.2
Aug.	16	0.03	2.1	6.3	140	SE	3.3
Sept.	16	0.03	1.9	5.9	110	N	4.1
Oct.	17	0.2	1.6	5.6	79	N	4.6
Nov.	16	0.1	0.9	5.6	49	N	5.7
Dec.	19	0.2	0.3	5.4	19	N	5.8
Annual	182	1.3	14.1	5.8	1225	N	4.8
Rec.(yrs.)	30	30	30	30	10[3]	7[4]	10[3]

[1] From June 1965: Höfn í Hornafirdi, 64°15′N 15°11′W, elevation 8 m.
[2] Period 1957–1965.
[3] Period 1958–1967.
[4] Period 1965–1971.

Reference Index

ABBOT, W. V., *see* RAO, M. S. V. et al.
ADAMS III, H., *see* MATHER, J. R. et al.
ADEKODUN, J. A., 115, 122
ALAKA, M. A., 224, 253
ALBRECHT, F., 48, 49, 122, 320, 335, 348, 362, 470, 475
ALEXANDER, R. C. and MOBLEY, R. L., 30, 122
ALLISON, L. J., STERANKA, J., HOLUB, R. J., HANSEN, J., GODSHALL, F. A. and PRABHAKARA, C., 351, 362
ALVAREZ, J. A. and THOMPSON, A. H., 283, 362
ANANTHAKRISHNAN, R., 656
ANANTHAKRISHNAN, R. and RANGARAJAN, S., 650, 654
ANDERSSEN, E. C., 326, 362
ANDERSSEN, T., 239, 253
ANDREWS, J. C., *see* NILSSON, C. S. et al.
ANONYMOUS, 693
ASTAPENKO, P. D., 281, 299, 362
ATKINSON, G. D. and SADLER, J. C., 237, 253
AUGSTEIN, E., 114, 122
AUGSTEIN, E., RIEHL, H., OSTAPOFF, F. and WAGNER, V., 218, 219, 220, 253
AUSTIN, T. S., 286, 362

BADGLEY, F. I., *see* FLEAGLE, R. G. et al.
BAINES, G. B., *see* MARAGOS, S. C. et al.
BAKAYEV, V. G., 573, 588
BAKER, T., *see* GORDON, A. L. et al.
BARGMAN, D. J., 654
BARNES, J. C., 319, 362
BARNETT, T. P., 360, 361, 362
BARRETT, E. C., 12, 42, 49, 65, 68, 122
BATES, J. R., 286, 362
BAUMGARTNER, A. and REICHEL, E., 48, 49, 54, 55, 56, 122, 320, 362
BAUMGARTNER, A., MAYER, H. and METZ, W., 60, 122
BECKÉR, E. E., *see* NORQUIST, D. C. et al.
BECKER, R., 36, 122
BENNETT, E. B., *see* WYRTKI, K. and BENNETT, E. B.
BERGERON, T., 308, 362
BERGHAUS, H., 2, 122
BERGTHÓRSSON, P., 679, 693
BERGTHÓRSSON, P., *see* ÓLAFSSON, Ó. E. and BERGTHÓRSSON, P.
BERLAGE, H. P., 349, 351, 352, 362
BERNHARDT, F. and PHILIPPS, H., 60, 61, 62
BEVERIDGE, P. J., *see* MARAGOS, S. C. et al.
BJERKNES, J., 32, 56, 115, 123, 234, 253, 351, 353, 355, 358, 359, 360, 362, 472, 473, 474, 475
BLACK, J. N., 336, 362

BLISS, E. W., *see* WALKER, G. T. and BLISS, E. W.
BOER, G. J. and KYLE, A. C., 356, 362
BOER, G. J., *see* NEWELL, R. E. et al.
BÖHNECKE, G., 253
BOLAND, F. M. and HAMON, B. V., 272, 362
BOLAND, F. M., *see* CRESSWELL, G. E. et al.
BOOTH, A. L., *see* LEESE, J. A. et al.
BOTNIKOV, V. N., 276, 362
BOURKE, R. H., *see* DORMAN, C. E. and BOURKE, R. H.
BREITENGROSS, J. P., 55, 123
BRIDGMAN, H. A., 345, 362
BRIER, G. W., 75, 123
BROCKS, K., 198, 253
BROMANN, E., 93, 123
BROOKFIELD, H. C. and HART, D., 266, 362
BROOKS, C. E. P., 76, 81, 83, 123
BROWN, P. R., 200, 253
BROWNE, M. L., 296, 363
BRUCKNER, I., *see* SCHMETTAU, S. and BRUCKNER, I.
BRUMMER, B., 253
BRYAN, K., 196, 253
BUCHAN, A., 2, 123, 513, 588
BUDD, W. F., 279, 363
BUDYKO, M. I., 48, 60, 61, 69, 123, 243, 244, 245, 246, 247, 248, 253, 335, 336, 338, 363, 470, 475, 578, 579, 580, 581, 588, 607, 654
BUENOS AIRES, DIRECCION DE METEOROLOGIA, GEOFISICA E HIDROLOGIA, 83, 123
BUNKER, A. F., 194, 195, 198, 210, 243, 247, 249, 250, 253, 617, 654
BUNKER, A. F. and CHAFFEE, M., 656
BUNKER, A. F. and WORTHINGTON, L. V., 241, 253
BUREAU OF METEOROLOGY, AUSTRALIA, 266, 363
BURLING, R. W., 277, 363
BURT, W. V., *see* QUINN, W. H. and BURT, W. V.; *see also* QUINN, W. H. et al.

CALLAHAN, J. E., 272, 363
CAMPBELL, W. J., WEEKS, W. F., RAMSEIER, R. O. and GLOERSEN, P., 277, 363
CAMPBELL, W. J., *see* GLOERSEN, P. et al.
CANE, M. A. and SARACHEK, E. S., 361, 363
CAPURRO, L. R. A., 268, 363
CARLETON, A. M., 514, 563, 588
CAUCHARD, G. and INCHAUSPÉ, 363
CAVIEDES, C. N., 359, 363
CENTRAL METEOROLOGICAL OBSERVATORY, 475
CHACKO, O., *see* MANI, A. et al.

CHAFFEE, M., *see* BUNKER, A. F. and CHAFFEE, M.
CHANG, C. P., 288, 363
CHANG, C. P., *see* WALLACE, J. M. and CHANG, C. P.
CHANG, T. C., *see* GLOERSEN, P. et al.
CHARNEY, J., 286, 363
CHARNOCK, H., FRANCIS, J. R. D. and SHEPPARD, P. A., 219, 253
CHASE, J., 204, 253
CHAUSSARD, A. and LAPLACE, L., 654
CHAUSSARD, A., *see* GERMAIN, H. and CHAUSSARD, A.
CHERVIN, R. M., *see* JULIAN, P. R. and CHERVIN, R. M.
CHIEN, C. W., *see* NEIBURGER, M. et al.
CLARKE, L., *see* LAEVASTU, T. et al.
COHEN, T. J. and SWEETSER, E. I., 253
COLEBROOK, J. H., *see* DICKSON, R. et al.
COLEBROOK, J. M., 253
COLLINS, J. A., *see* PRESCOTT, J. A. and COLLINS, J. A.
COLÓN, J. A., 224, 225, 226, 254, 630, 657
CORNEJO-GARRIDO, A. G. and STONE, P. H., 69, 123
COVEY, D. L. and HASTENRATH, S., 358, 363
CRADDOCK, J. M., 227, 230, 254
CRESSWELL, G. E., GOLDING, T. J. and BOLAND, F. M., 276, 363
CROMWELL, T., MONTGOMERY, R. B. and STROUP, E. D., 271, 363
CRUTCHER, H. L. and HOXIT, L. R., 290, 363
CRUTCHER, H. L. and QUAYLE, R. G., 224, 226, 254, 289, 363
CRUTCHER, H. L., JENNE, R. L., TALJAARD, J. J. and VAN LOON, H., 101, 123, 265, 331, 333, 334, 363, 531, 547, 588
CRUTCHER, H. L., *see* JENNE, R. L. et al.; *see also* TALJAARD, J. J. et al.; VAN LOON, H. et al.
CURTIN, D. G., *see* WEBSTER, P. J. and CURTIN, D. G.
CUSHING, D. H., 254

DAMON, P. E. and KUNEN, S. M., 34, 123
DANIEL, H., 2, 123
DANSKE METEOROL. INST., 254
DAS, S. C., 293, 363
DAVIDOVA, N. G., 121, 123
DAVIDOVA, N. G., *see* DUBENTSOV, V. R. and DAVIDOVA, N. G.
DAVIS JR., R. A., 22, 123
DAYTON, G. V., *see* NEWELL, R. E. et al.
DEACON, E. L., 66, 123
DEACON, E. L. and WEBB, E. K., 9, 123
DEACON, G. E. R., 29, 30, 123, 268, 270, 272, 275, 276, 363, 508, 509, 510, 588
DE BILT, KON. NED. METEOROL. INST., 123
DEFANT, A., 239, 241, 254, 268, 272, 363, 508, 588
DEFENSE MAPPING AGENCY HYDROGRAPHIC CENTER (U.S.A.), 270, 363
DELISLE, J. F., 335, 363
DERYCKE, R. J., 277, 363
DESBOIS, M., *see* MOREL, P. and DESBOIS, M.
DESIKAN, V., *see* MANI, A. et al.
DEUTSCHES HYDROGRAPHISCHES INSTITUT, HAMBURG, 3, 4, 31, 86, 94, 100, 114, 116, 118, 120, 124, 189, 191, 608, 655, 656

DEUTSCHE SEEWARTE, 431, 475
D'HAUTESERRE, A., 289, 310, 363, 375
DICKSON, R. and LAMB, H. H., 200, 254
DICKSON, R. and LEE, A., 233, 254
DICKSON, R., LAMB, H. H., MALMBERG, S. A. and COLEBROOK, J. H., 200, 254
DIETRICH, G., 28, 123, 508, 588
DINGLE, W. R. J., *see* ZILLMAN, J. W. and DINGLE, W. R. J.
DIXIT, C. M., *see* RAMAN, C. R. V. and DIXIT, C. M.
DONGUY, J. R., *see* PICKARD, G. L. et al.
DOPPLICK, T. G., 345, 363
DORMAN, C. E. and BOURKE, R. H., 320, 347, 348, 363
DORMAN, C. E., PAULSON, C. A. and QUINN, W. H., 467, 476
DROZDOV, O. A., 49, 123, 320, 364
DUBENTSOV, V. R. and DAVIDOVA, N. G., 573, 588
DUNBAR, M., 196, 254
DUNN, G. E., 225, 254
DUNN, G. E. and MILLER, B. I., 224, 254

EARNEST, C. L., *see* WINSTON, J. S. et al.
EBDON, R. A., *see* WRIGHT, P. B. and EBDON, R. A.
EICKERMANN, W. and FLOHN, H., 75, 123
EINARSSON, M. Á., 674, 681, 682, 685, 688, 692, 694
EINARSSON, TH., 678, 694
ELIOT, J., 656
ELLIOTT, W. P. and REED, R. K., 467, 476
ELLIOTT, W. P., *see* REED, R. K. and ELLIOTT, W. P.
ELLIS, J. S., *see* VONDER HAAR, T. H. and ELLIS, J. S.
ELMS, J. D., *see* QUAYLE, R. G. and ELMS, J. D.
EMERY, W. J., 275, 364
EMMONS, G. and MONTGOMERY, R. B., 619, 655
EMON, J., 629, 638, 655
ENDO, M., *see* HAYASHI, R. and ENDO, M.
ERICKSEN, C. O. and WINSTON, J. S., 310, 364
EYTHÓRSSON, J. and SIGTRYGGSSON, H., 675, 680, 694

FABRICIUS, A. F., 81, 83, 84, 123, 310, 364
FEAN, C. R., 42, 123
FEDDES, R. G., *see* MILLER, D. B. and FEDDES, R. G.
FICKER, H. VON, 114, 129, 197, 218, 219, 256, 630, 655
FINDLATER, J., 617, 655, 656
FLEAGLE, R. G., BADGLEY, F. I. and HSUEH, Y., 655
FLEMING, R. H., *see* SVERDRUP, H. U. et al.
FLETCHER, J. O., 12, 124, 279, 364
FLETCHER, R. D., 610, 655
FLOHN, H., 32, 100, 124, 351, 356, 364, 655
FLOHN, H. and HINKELMANN, K., 359, 364
FLOHN, H., *see* EICKERMANN, W. and FLOHN, H.
FORTAK, H. G., 124
FOX, P. T., *see* JAMES, R. W. and FOX, P. T.
FRANCIS, J. R. D., *see* CHARNOCK, H. et al.
FRANCIS, K. A., *see* IYER, V. D. and FRANCIS, K. A.
FRANK, N. L., 224, 254
FRASER, R., 269, 364
FREEMAN, J. C., 287, 364
FRIEHE, C. A. and SCHMITT, K. F., 338, 364
FROST, R. and STEPHENSON, P. M., 657

700

GABITES, J. F., 289, 290, 345, 364
GALLÉ, P. H., 644, 655, 657
GARSTANG, M., 607, 655
GATES, W. L., 1, 4, 9, 28, 34, 124
GAUNTLETT, D. J., SEAMAN, R. S., KININMONTH, W. R. and LANGFORD, J. C., 282, 364
GERMAIN, H. and CHAUSSARD, A., 655
GIBBS, W. J., 364
GIBBS, W. J., GOTLEY, A. V. and MARTIN, A. R., 281, 310, 364
GILBERT, J., LAMBERT, J. and GOYER, G., 575, 588
GIOVANNELLI, J. and ROBERT, J., 289, 364
GIRARD, J. and RIGNOT, D., 310, 364
GLOERSEN, P., WILHEIT, T. T., CHANG, T. C., NORDBERG, W. and CAMPBELL, W. J., 511, 588
GLOERSEN, P., *see* CAMPBELL, W. J. et al.; *see also* ZWALLY, H. J. and GLOERSEN, P.
GODFREY, J. S., 360, 364
GODFREY, J. S., *see* HAMON, B. V. and GODFREY, J. S.
GODSHALL, F. A., *see* ALLISON, L. J. et al.; *see also* LEESE, J. A. et al.
GOLDBERG, R. D., *see* GORDON, A. L. and GOLDBERG, R. D.
GOLDING, T. J., *see* CRESSWELL, G. E. et al.
GOPAL RAO, S., *see* RAMDAS, L. A. et al.
GORDON, A. H., 612, 627, 638, 655
GORDON, A. H. and TAYLOR, R. C., 286, 364, 612, 655
GORDON, H. A., 206, 254
GORDON, A. L., 29, 124, 268, 270, 272, 275, 276, 277, 364
GORDON, A. L. and GOLDBERG, R. D., 28, 34, 124
GORDON, A. L. and MOLINELLI, E., 268, 272, 364
GORDON, A. L., MOLINELLI, E. and BAKER, T., 269, 275, 364
GORDON, A. L., *see* TAYLOR, H. W. et al.
GORSKOV, C. G., 3, 124
GOTLEY, A. V., *see* GIBBS, W. J. et al.
GOYER, G., *see* GILBERT, J. et al.
GRAY, T. I., *see* WINSTON, J. S. et al.
GRAY, W. M., 288, 289, 290, 364
GRAY, W. M., *see* RUPRECHT, E. and GRAY, W. M.; *see also* WILLIAMS, K. T. and GRAY, W. M.
GREAT BRITAIN, H.M. STATIONARY OFFICE, 432, 476
GREAT BRITAIN, HYDROGRAPHIC OFFICE, 508, 588
GRIFFITHS, J. F., 94, 124
GROEN, P., 62, 124
GRUBER, A., 286, 364
GRUBER, A., *see* WINSTON, J. S. et al.
GRUNDLINGH, M. L., 508, 588
GUILLEN, O., *see* WOOSTER, W. S. and GUILLEN, O.
GUYMER, L. B., 283, 304, 365

HAGGARD, W., 267
HALLEY, E., 2, 124
HAMBURG, DEUTSCHES HYDROGRAPHISCHES INST., 3, 4, 31, 86, 94, 100, 114, 116, 118, 120, 124, 189, 191, 608, 655, 656
HAMON, B. V., 269, 272, 365
HAMON, B. V. and GODFREY, J. S., 272, 365
HAMON, B. V. and TRANTER, D. J., 272, 365

HAMON, B. V., *see* BOLAND, F. M. and HAMON, B. V.
HANN, J., 513, 588
HANSEN, J., *see* ALLISON, L. J. et al.
HANSEN, K. J., *see* KORNFIELD, J. A. et al.
HANTEL, M., 607, 655
HANZAWA, M. and INOUE, T., 472, 476
HANZAWA, M. and TOURNIER, T., 476
HARRIS, T. F. W. and STAVROPOULOS, C. C., 508, 509, 588
HART, D., *see* BROOKFIELD, H. C. and HART, D.
HARWOOD, W. A., 603, 645, 655
HASLER, A. F., *see* KORNFIELD, J. and HASLER, A. F.; *see also* KORNFIELD, J. A. et al.
HASSELMANN, D., *see* WUCKNITZ, J. et al.
HASSELMANN, K., 9, 33, 124
HASTENRATH, S., 66, 124, 358, 365
HASTENRATH, S. and HELLER, L., 358, 365
HASTENRATH, S. and LAMB, P. J., 66, 124, 220, 254, 273, 335, 338, 365
HASTENRATH, S., *see* COVEY, D. L. and HASTENRATH, S.
HAURWITZ, F. D., 345, 365
HAYASHI, R. and ENDO, M., 476
HAYDEN, B. P., *see* RESIO, D. T. and HAYDEN, B. P.
HEATH, R. A., 276, 365
HEDGPETH, J. W., *see* WOOSTER, W. S. and HEDGPETH, J. W.
HEISE, G. and HÖHN, R., 114, 124
HEISE, G., *see* RUDLOFF, W. et al.
HELLER, L., *see* HASTENRATH, S. and HELLER, L.
HELLERMAN, S., 210, 254
HENIN, C., *see* PICKARD, G. L. et al.
HENNING, D., 74, 75, 106, 107, 124
HEPWORTH, M. W., 281, 365
HERDMAN, H. F. P., *see* MACKINTOSH, N. A. and HERDMAN, H. F. P.
HICKEY, B., 361, 365
HILL, H. W., 286, 365
HINKELMANN, K., *see* FLOHN, H. and HINKELMANN, K.
HODGE, D., *see* VISHER, S. S. and HODGE, D.
HOEFER, A., 92, 124
HÖFLICH, O., 16, 28, 30, 31, 34, 46, 97, 114, 124
HÖFLICH, O., *see* RUDLOFF, W. and HÖFLICH, O.; *see also* RUDLOFF, W. et al.
HOFMEYR, W. L., 105, 124
HOGAN, J., 657
HOGBEN, N. and LAMB, F. E., 24, 124
HÖHN, R., 228, 232, 254
HÖHN, R., *see* HEISE, G. and HÖHN, R.
HOLLAND, J. Z. and RASMUSSEN, E. M., 241, 254
HOLLOWAY, J. L., *see* MANABE, S. et al.
HOLMBOE, J., *see* KIDSON, E. and HOLMBOE, J.
HOLOPAINEN, E. O., 68, 124
HOLTON, J. R. and WALLACE, J. M., 286, 365
HOLUB, R. J., *see* ALLISON, L. J. et al.
HOUTMAN, T. J., 276, 277, 365
HOXIT, L. R., *see* CRUTCHER, H. L. and HOXIT, L. R.
HOYT, D. V., *see* SASAMORI, T. et al.; *see also* VAN LOON, H. et al.
HSUEH, Y., *see* FLEAGLE, R. G. et al.

HUBERT, L. F., 114, 124, 308, 365
HUBERT, L. F., KRUEGER, A. F. and WINSTON, J. S., 286, 322, 365
HUTCHINGS, J. W., 289, 365

INDER, S., 263, 365
INOUE, T., *see* HANZAWA, M. and INOUE, T.
ISELIN, C. O., 34, 124
IVANOV, J. A., 276, 365
IYER, V. D. and FRANCIS, K. A., 655

JACOBS, W. C., 49, 124, 238, 239, 240, 243, 254, 470, 476, 560, 588
JAEGER, L., 49, 51, 125
JAGANNATHAN, P., *see* RAMDAS, L. A. et al.
JAKOBSSON, T., 231, 234, 254
JAKOBSSON, T. E., 254
JAMES, R. W. and FOX, P. T., 30, 125
JAPAN METEOROLOGICAL AGENCY, 445, 476
JEFFERIES, C., *see* MILLER, F. R. and JEFFERIES, C.; *see also* RAMAGE, C. S. et al.
JENKINSON, A. F., 208, 209, 254
JENNE, R. L., CRUTCHER, H. L., VAN LOON, H. and TAL-JAARD, J. J., 101, 107, 125, 265, 266, 314, 315, 323, 328, 330, 365, 547, 551, 588
JENNE, R. L., *see* CRUTCHER, H. L. et al.; *see also* TAL-JAARD, J. J. et al.; VAN LOON, H. and JENNE, R. L.; VAN LOON, H. et al.
JENNINGS, F., *see* WOOSTER, W. S. and JENNINGS, F.
JOHNSON, A. I., *see* LAMB, H. H. and JOHNSON, A. I.
JOHNSON, D. R., *see* ZILLMAN, J. W. and JOHNSON, D. R.
JOHNSON, M. W., *see* SVERDRUP, H. U. et al.
JOHNSON, O. S., *see* NEIBURGER, M. et al.
JONES, J., *see* TAFT, B. and JONES, J.
JOZSEF, S. K., 38, 125
JULIAN, P. R. and CHERVIN, R. M., 353, 365

KACHELHOFFER, S. T., *see* SCHWERDTFEGER, W. and KA-CHELHOFFER, S. T.
KAMENKOVICH, V. M., *see* MONIN, A. S. et al.
KANAMITSU, M., *see* KRISHNAMURTI, T. N. et al.
KANGOS, J., 335, 338, 365
KARELSKY, S., 285, 296, 365, 366
KASAHARA, A., 476
KAUFELD, L. and RUDLOFF, W., 16, 125
KELLAS, W. R., *see* STRETEN, N. A. and KELLAS, W. R.
KELLER, J. L., *see* WEBSTER, P. J. and KELLER, J. L.
KELLY, G. A. M., 283, 366
KERR, I. S., 289, 290, 366
KESHAVAMURTHY, R. N., *see* MILLER, F. R. and KESHAV-AMURTHY, R. N.
KESSLER, A., 64, 125
KIDSON, E., 281, 293, 299, 366
KIDSON, E. and HOLMBOE, J., 281, 366
KIDSON, J. W., 285, 350, 351, 353, 356, 366
KIDSON, J. W., *see* NEWELL, R. E. et al.
KILONSKY, B. J., *see* SADLER, J. C. et al.
KININMONTH, W. R., *see* GAUNTLETT, D. J. et al.
KIRCHER, A., 2, 125

KIRK, T. H., 194, 254
KISS, W. J., *see* KRISHNAMURTI, T. N. et al.
KLEIN, W. J., 211, 213, 254
KNAUSS, J. A., 269, 271, 366
KNOCH, K., 49, 125
KNOCH, K. and SCHULZE, A., 12, 125
KOBE IMPERIAL MARINE OBSERVATORY, 476
KONDO, J., 51, 125
KONDRATYEV, K. Y., 61, 125
KÖNIG, M., *see* WUCKNITZ, J. et al.
KONINKLIJK NEDERLANDSCH METEOROLOGISCH INSTI-TUUT, 265, 273, 366, 431, 476, 508, 514, 588
KÖPPEN, W., 12, 125, 159
KORNFIELD, J. and HASLER, A. F., 308, 309, 317, 366
KORNFIELD, J. A., HASLER, A. F., HANSEN, K. J. and SUOMI, V. E., 308, 366
KORT, V. G., 29, 30, 55, 125, 270, 366
KORT, V. G., *see* MONIN, A. S. et al.
KORZUM, V. I., 320, 366
KRAUS, E. B., 110, 125, 199, 238, 254
KRAUS, E. B. and MORRISON, R. E., 254
KRISHNAMURTHY, V., *see* MANI, A. et al.
KRISHNAMURTI, T. N., 353, 356, 366
KRISHNAMURTI, T. N., KANAMITSU, M., KISS, W. J. and LEE, J. D., 353, 356, 366
KRUEGER, A. F., 356, 366
KRUEGER, A. F. and WINSTON, J. S., 354, 366
KRUEGER, A. F., *see* HUBERT, L. F. et al.
KRÜGERMEYER, L., 58, 125
KRUHL, H., 22, 125
KUETTNER, J. P., 114, 125, 220, 254
KUHLBRODT, E., 34, 125
KUHLBRODT, E. and REGER, J., 125
KUNEN, S. M., *see* DAMON, P. E. and KUNEN, S. M.
KUO, H. L., 115, 125
KUO YANG, R. T. W., *see* QUINN, W. H. et al.
KYLE, A. C., *see* BOERM, G. J. and KYLE, A. C.

LABITZKE, K. and VAN LOON, H., 101, 125
LABITZKE, K., *see* VAN LOON, H. et al.
LAEVASTU, T., 470, 476
LAEVASTU, T., CLARKE, L. and WOLFF, P. M., 48, 51, 113, 125
LAJOIE, F. A., *see* LAMOND, M. H. et al.; *see also* PHILL-POT, H. R. et al.
LAMB, F. E., *see* HOGBEN, N. and LAMB, F. E.
LAMB, H. H., 125, 210, 254, 279, 366
LAMB, H. H. and JOHNSON, A. I., 196, 199, 254, 293, 366
LAMB, H. H., *see* DICKSON, R. and LAMB, H. H.; *see also* DICKSON, R. et al.
LAMB, P. J., 21, 30, 78, 125
LAMB, P. J., *see* HASTENRATH, S. and LAMB, P. J.
LAMBERT, J., *see* GILBERT, J. et al.
LAMOND, M. H., PRICE, P. G., NEAL, A. B. and LAJOIE, F. A., 283, 366
LANGFORD, J. C., 281, 285, 366
LANGFORD, J. C., *see* GAUNTLETT, D. J. et al.
LANGMUIR, I., 25, 125
LAPLACE, L., *see* CHAUSSARD, A. and LAPLACE, L.

LASEUR, N. E., *see* RIEHL, H. et al.

LEE, A., *see* DICKSON, R. and LEE, A. A.

LEE, D. M., *see* MAHER, J. V. and LEE, D. M.

LEE, J. D., *see* KRISHNAMURTI, T. N. et al.

LEESE, J. A., BOOTH, A. L. and GODSHALL, F. A., 308, 319, 366

LEGECKIS, R., 32, 125

LEIGH, R. M., 290, 366

LeMONE, M. A., 114, 125

LENINGRAD, GIDROMETEOROL. INST., 125

LETTAU, H., 360, 367

LISBOA, FASCICULO, 71, 73, 126

LOEWE, F., 79, 126, 562, 588

LONDON, J., *see* SASAMORI, T. et al.; *see also* VAN LOON, H. et al.

LONDON, METEOROL. OFFICE, 71, 73, 76, 77, 81, 83, 126

LOURENSZ, R., 288, 289, 291, 367

LUMB, F. E., 194, 244, 255

LUMLEY, J. L. and PANOFSKY, H. A., 9, 126

LYSGAARD, L., 531, 588

MACKINTOSH, N. A., 273, 275, 276, 367, 510, 515, 522, 589

MACKINTOSH, N. A. and HERDMAN, H. F. P., 86, 126, 277, 367, 511, 589

MADDEN, R. A., *see* VAN LOON, H. and MADDEN, R. A.

MAHER, J. V. and LEE, D. M., 266, 367

MAKSIMOV, I. V., 276, 367

MALKUS, J. S. and RIEHL, H., 287, 367

MALKUS, J. S., *see* RIEHL, H. et al.

MALLORY, O. I., 25, 126

MALMBERG, S. A., 196, 200, 255

MALMBERG, S. A., *see* DICKSON, R. et al.

MALONE, T. F., 476

MANABE, S. and WETHERALD, R., 63, 126

MANABE, S., HOLLOWAY, J. L. and STONE, H. M., 286, 367

MANI, A., CHACKO, O., KRISHNAMURTHY, V. and DESIKAN, V., 657

MANNELLO, L. P., *see* WINSTON, J. S. et al.

MARAGOS, S. C., BAINES, G. B. and BEVERIDGE, P. J., 292, 367

MARKGRAF, H., 91, 126

MARTIN, A. R., *see* GIBBS, W. J. et al.

MARTIN, D. W., 4, 126, 283, 367

MARTIN, D. W. and SCHERER, W., 49, 126

MARTIN, D. W. and SIKDAR, D. N., 114, 126

MARTIN, D. W., *see* ZILLMAN, J. W. and MARTIN, D. W.

MATHER, J. R., ADAMS III, H. and YOSHIOKA, G. A., 213, 216, 217, 218, 255

MAURY, M. F., 2, 126, 196, 255

MAWSON, D., 280, 367

MAYER, H., *see* BAUMGARTNER, A. et al.

MAZEIKA, P. A., 30, 32, 126, 255

McCREARY, J., 361, 367

McCREARY, J. P., *see* WHITE, W. B. and McCREARY, J. P.

McDONALD, W. F., 2, 32, 126, 265, 308, 317, 318, 320, 367, 514, 561, 589, 657

McGUIRK, J. P. and REITER, E. R., 310, 367

MECKING, L., *see* MEINARDUS, W. and MECKING, L.

MEINARDUS, W., 116, 126, 238, 255, 265, 275, 303, 320, 367, 513, 558, 575, 589, 655

MEINARDUS, W. and MECKING, L., 4, 126, 281, 299, 367, 512, 513, 589

MELDRUM, C., 3, 126, 512, 589

MESERVE, J. M., 255

METEOROLOGICAL OFFICE, U. K., 201, 227, 231, 255, 260, 265, 266, 273, 367, 508, 514, 520, 529, 530, 589, 607, 644, 650, 655, 657, 659

METZ, W., *see* BAUMGARTNER, A. et al.

MEUSS, J. F., 2, 126

MEYER, H. H. F., 509, 589

MILES, M. K., 218, 255

MILLER, B. I., *see* DUNN, G. E. and MILLER, B. I.

MILLER, D. B. and FEDDES, R. G., 285, 304, 317, 367

MILLER, F. R. and JEFFERIES, C., 657

MILLER, F. R. and KESHAVAMURTHY, R. N., 609, 610, 655

MILLER, F. R., *see* RAMAGE, C. S. et al.

MILLER, J., *see* THOMPSON, O. E. and MILLER, J.

MILTON, D., 224, 255

MITCHELL, J. M., 34, 126

MOBLEY, R. L., *see* ALEXANDER, R. C. and MOBLEY, R. L.

MOLINELLI, E., *see* GORDON, A. L. and MOLINELLI, E.; *see also* GORDON, A. L. et al.; TAYLOR, H. W. et al.

MÖLLER, F., 238, 255, 320, 348, 367, 560, 589

MONIN, A. S., 241, 255

MONIN, A. S., KAMENKOVICH, V. M. and KORT, V. G., 29, 32, 33, 126

MONTGOMERY, R. B., *see* CROMWELL, T. et al., EMMONS, G. and MONTGOMERY, R. B.

MOREL, P. and DESBOIS, M., 330, 367

MORGAN, J., 44, 126

MORIZE, H., 126

MORRISON, R. E., *see* KRAUS, E. B. and MORRISON, R. E.

MOSCOW, PHYSICAL-GEOGRAPHICAL PART., 126

MUFFATTI, A. H. J., 286, 329, 367

MURAKAMI, T., 357, 367, 476

MURPHY, R. C., 358, 367

MURRAY, R., *see* RATCLIFFE, R. A. S. and MURRAY, R.

MUSK, L. F., 358, 367

NAMIAS, J., 233, 255, 310, 368, 370, 472, 473, 476

NAVATO, A. R., *see* WEARE, B. C. et al.

NAVY MINISTRY, MOSCOW, 238, 255, 439, 476

NAZAROV, V. S., 234, 255, 280, 368

NEAL, A. B., 329, 368

NEAL, A. B., *see* LAMOND, M. H. et al.; *see also* PHILLPOT, H. R. et al.

NEALE, A. A., *see* TRENBERTH, K. E. and NEALE, A. A.

NEIBURGER, M., JOHNSON, O. S. and CHIEN, C. W., 287, 368

NEUMANN, G., 95, 126, 508, 589

NEUWIRTH, F., 52, 126

NEW ZEALAND METEOROLOGICAL SERVICE, 266, 368

NEWELL, R. E., KIDSON, J. W., DAYTON, G. V. and BOER, G. J., 105, 126
NEWELL, R. E., KIDSON, J. W., VINCENT, D. G. and BOER, G. J., 220, 255, 266, 345, 356, 368
NEWELL, R. E., *see* WEARE, B. C. et al.
NEWMAN, B. W., *see* TRELOAR, H. M. and NEWMAN, B. W.
NEWNHAM, E. V., 79, 126, 655, 657
NEWTON, C. W., 54, 127, 345, 368
NEWTON, C. W., *see* PALMÉN, E. and NEWTON, C. W.; *see also* VAN LOON, H. et al.
NICHOLLS, N., 353, 368
NICHOLSON, J. R., 657
NIEUWOLT, S., 9, 51, 66, 101, 113, 115, 117, 127
NILSSON, C. S., ANDREWS, J. C. and SCULLY, P. P., 272, 368
NITTA, T., *see* YANAI, M. and NITTA, T.
NORDBERG, W., *see* GLOERSEN, P. et al.
NORQUIST, D. C., BECKER, E. E. and REED, R. J., 127
NOWLIN JR., W. D., WHITWORTH, T. and PITTSBURG, R. D., 272, 368

ODA, L., *see* SADLER, J. C. et al.
OGAWA, A., *see* YAMANOUCHI, Y. and OGAWA, A.
OKADA, T., 476
ÓLAFSSON, Ó. E. and BERGTHÓRSSON, P., 689, 694
OOSTHUIZEN, C. M., *see* VOWINCKEL, E. and OOSTHUIZEN, C. M.
OOYAMA, K., 476
OSAWA, K., *see* TERADA, K. and OSAWA, K.
OSTAPOFF, F., 276, 368
OSTAPOFF, F., TERBEYEV, Y. and WORTHEM, S., 33, 49, 60, 127
OSTAPOFF, F., *see* AUGSTEIN, E. et al.

PALMÉN, E., 56, 127
PALMÉN, E. and NEWTON, C. W., 118, 127, 287, 368
PALMÉN, E. and SÖDERMANN, D., 56, 127
PALMER, C. E., 281, 287, 368
PALMER, C. W., 575, 589
PANOFSKY, H. A., *see* LUMLEY, J. L. and PANOFSKY, H. A.
PARIS, AIR FRANCE, 127
PATZERT, W., *see* WYRTKI, K. et al.
PAULSON, C. A., *see* DORMAN, C. E. et al.
PAWLEY, W. M., *see* QUINN, W. H. et al.
PAYNE, R. E., 60, 127, 196, 255
PEIXOTO, P., 112, 127
PEPPER, J., 28, 81, 83, 85, 127
PERRY, A. H., 200, 227, 234, 255
PERRY, A. H. and WALKER, J. M., 64, 65, 69, 127
PERRY, J. D., 195, 255
PETERSEN, P., 17, 127
PETTERSSEN, S., 608, 655
PFLUGBEIL, C., 16, 127
PFLUGBEIL, C. and STEINBORN, E., 201, 255, 260
PHILIPPS, H., *see* BERNHARDT, F. and PHILIPPS, H.
PHILLPOT, H. R., 266, 368
PHILLPOT, H. R., PRICE, P. G., NEAL, A. B. and LAJOIE, F. A., 267, 283, 368

PICKARD, G. L., DONGUY, J. R., HENIN, C. and ROUGERIE, F., 269, 368
PIKE, A. C., 286, 368
PITTOCK, A. B., 293, 368
PITTSBURG, R. D., *see* NOWLIN JR., W. D. et al.
POND, S., 338, 368
PORTMAN, D. J. and RYZNAR, E., 657
POSNER, G. S., 359, 368
PRABHAKARA, C., *see* ALLISON, L. J. et al.
PRANTNER, G. D., 299, 368, 574, 589
PREDOEHL, M. C., 277, 368
PRESCOTT, J. A. and COLLINS, J. A., 203, 255
PRETORIA, WEATHER BUREAU, 4, 79, 100, 116, 118, 120, 127
PRICE, P. G., *see* LAMOND, M. H. et al.; *see also* PHILLPOT, H. R. et al.; ZILLMAN, J. W. and PRICE, P. G.
PRIESTLEY, C. H. B., 210, 255
PRIVETT, D. W., 48, 127, 335, 338, 368, 657
PROHASKA, F., 35, 49, 91, 93, 127

QUAYLE, R. G., 198, 255
QUAYLE, R. G. and ELMS, J. D., 24, 97, 127
QUAYLE, R. G., *see* CRUTCHER, H. L. and QUAYLE, R. G.
QUINN, W. H., 358, 360, 361, 368
QUINN, W. H. and BURT, W. V., 351, 368
QUINN, W. H., BURT, W. V. and PAWLEY, W. M., 336, 368
QUINN, W. H., ZOPF, D. O., SHORT, K. S. and KUO YANG, R. T. W., 351, 358, 360, 369; *see also* DORMAN, C. E. et al.; WYRTKI, K. et al.

RADOK, U., 329, 369
RAMAGE, C. S., 265, 286, 290, 329, 358, 359, 360, 361, 369, 607, 608, 609, 619, 630, 644, 651, 655
RAMAGE, C. S. and RAMAN, C. R. V., 514, 589, 657
RAMAGE, C. S., MILLER, F. R. and JEFFERIES, C., 514, 589, 657
RAMAKRISHNAN, K. P., *see* RAMANATHAN, K. R. and RAMAKRISHNAN, K. P.
RAMAN, C. R. V., 610, 651, 655
RAMAN, C. R. V. and DIXIT, C. M., 651, 655, 657
RAMAN, C. R. V. and RAMANATHAN, Y., 608, 656; *see also* RAMAGE, C. S. and RAMAN, C. R. V.
RAMANATHAN, K. R. and RAMAKRISHNAN, K. P., 657
RAMANATHAN, K. R. and VENKITESHWARAN, S. P., 657
RAMANATHAN, Y., *see* RAMAN, C. R. V. and RAMANATHAN, Y.
RAMDAS, L. A., JAGANNATHAN, P. and GOPAL RAO, S., 650, 656
RAMSEIER, R. O., *see* CAMPBELL, W. J. et al.
RANGARAJAN, S., *see* ANANTHAKRISHNAN, R. and RANGARAJAN, S.
RAO, M. S. V., ABBOT, W. V. and THEON, J. S., 561, 589
RAO, Y. P., 641, 656
RASMUSSEN, E. M., 56, 127; *see also* HOLLAND, J. Z. and RASMUSSEN, E. M.
RATCLIFFE, R. A. S., 234, 255
RATCLIFFE, R. A. S. and MURRAY, R., 233, 255
RATISBONA, L. R., 49, 87, 88, 89, 90, 127
REED, R. J., 211, 255

REED, R. J., *see* NORQUIST, D. C. et al.
REED, R. K. and ELLIOTT, W. P., 239, 255
REED, R. K., *see* ELLIOTT, W. P. and REED, R. K.
REED, W. W., 81, 84, 127, 266, 369
REGER, J., 100, 114, 127
REGER, J., *see* KUHLBRODT, E. and REGER, J.
REICHEL, E., *see* BAUMGARTNER, A. and REICHEL, E.
REID, J. L., 270, 271, 369, 433, 476; *see also* WOOSTER, W. S. and REID, J. L.
REITER, E. R., 122, 127; *see also* MCGUIRK J. P. and REITER, E. R.
RENNICK, M. A., 115, 127
RESIO, D. T. and HAYDEN, B. P., 255
REX, D. F., 216, 255
RIEHL, H., 9, 49, 128, 220, 224, 256, 287, 369, 612, 652, 656
RIEHL, H. MALKUS, J. S., YEH, T.-C. and LASEUR, N. E., 219, 256; *see also* AUGSTEIN, E. et al.; MALKUS, J. S. and RIEHL, H.
RIGBY, M., 476
RIGNOT, D., *see* GIRARD, J. and RIGNOT, D.
RIO DE JANEIRO, MINISTERIO DA AGRICULTURA, 73, 128
ROBERT, J., *see* GIOVANNELLI, J. and ROBERT, J.
RODEN, G. I., 348, 369, 607, 656
RODEWALD, M., 200, 256
ROGERS, J. C. and VAN LOON, H., 215, 256
ROGERS, J. C., *see* VAN LOON, H. and ROGERS, J. C.
ROLL, H. U., 9, 128, 243, 256, 619, 656
ROMANOV, Y. A., 657
RONDELEUX, M., 644, 656
ROSENTHAL, S. L., 287, 369
ROSSBY, C. G. and WILLETT, H. C., 310, 369
ROUGERIE, F., *see* PICKARD, G. L. et al.
ROWNTREE, P. R., 353, 369
ROYAL AUSTRALIAN AIR FORCE, 265, 369
ROYAL METEOROLOGICAL SOCIETY, AUSTRALIAN BRANCH, 304, 369
ROYAL NEW ZEALAND AIR FORCE, 265, 369
RUBIN, M. J., 4, 128, 293, 369
RUBIN, M. J. and VAN LOON, H., 122, 128
RUDLOFF, W., 100, 101, 128
RUDLOFF, W. and HÖFLICH, O., 100, 128
RUDLOFF, W., HÖFLICH, O. and HEISE, G., 100, 114, 128
RUDLOFF, W., *see* KAUFELD, L. and RUDLOFF, W.
RUPRECHT, E. and GRAY, W. M., 46, 128
RUSSELL, H. C., 281, 296, 369
RUTHERFORD, G. T., 283, 369
RYZNAR, E., *see* PORTMAN, D. J. and RYZNAR, E.

SADLER, J. C., 220, 221, 256, 266, 321, 329, 331, 369, 612, 641, 649, 656, 657
SADLER, J. C., ODA, L. and KILONSKY, B. J., 42, 128
SADLER, J. C., *see* ATKINSON, G. D. and SADLER, J. C.
SARACHEK, E. S., *see* CANE, M. A. and SARACHEK, E. S.
SASAMORI, T., LONDON, J. and HOYT, D. V., 65, 128, 345, 369
SASAMORI, T., *see* VAN LOON, H. et al.
SATER, J. E., 196, 256
SCHÄFER, P. J., *see* WALDEN, H. and SCHÄFER, P. J.

SCHELL, I. I., 196, 256, 272, 360, 369
SCHERER, W., *see* MARTIN, D. W. and SCHERER, W.
SCHERHAG, R., 4, 128
SCHICK, M., 656
SCHMETTAU, S. and BRUCKNER, I., 1, 128
SCHMIDT, F. H., *see* VAN DIJK, W. et al.
SCHMITT, K. F., *see* FRIEHE, C. A. and SCHMITT, K. F.
SCHMITT, W., 4, 128, 299, 369, 513, 589
SCHMITT, W., *see* TALJAARD, J. J. et al.
SCHNAPAUFF, W., 100, 128
SCHOTT, G., 7, 49, 51, 128, 265, 266, 268, 270, 273, 274, 358, 359, 369, 589, 645, 656, 657
SCHOTT, P. G., 238, 256
SCHÜCK, A., 7, 128
SCHULZE, A., *see* KNOCH, K. and SCHULZE, A.
SCHULZE, B. R., 94, 96, 98, 99, 128
SCHUMACHER, A., 7, 128
SCHUMANN, T. E. W., 7, 128
SCHUMANN, T. E. W. and VAN ROOY, M. P., 205, 206, 256
SCHÜTTE, K., 358, 369
SCHUURMANS, C. J. E., *see* VAN DIJK, W. et al.
SCHWEIGGER, E., 358, 370
SCHWERDTFEGER, W., 5, 13, 16, 86, 87, 92, 105, 128, 266, 316, 358, 370, 522, 528, 589
SCHWERDTFEGER, W. and KACHELHOFFER, S. T., 301, 370
SCULLY, P. P., *see* NILSSON, C. S. et al.
SEAMAN, R. S., *see* GAUNTLETT, D. J. et al.
SEELYE, C. J., 265, 347, 348, 370
SEEWETTERAMT HAMBURG, 657
SEKIGUCHI, T., 320, 370
SHAW, N., 513, 589
SHEPPARD, P. A., 208, 209, 256
SHEPPARD, P. A., *see* CHARNOCK, H. et al.
SHORT, K. S., *see* QUINN, W. H. et al.
SHULEIKIN, 86, 128
SIGFÚSDÓTTIR, A. B., 679, 685, 694
SIGTRYGGSSON, H., *see* EYTHÓRSSON, J. and SIGTRYGGSSON, H.
SIGURDSSON, F. H., 684
SIKDAR, D. N., YOUNG, J. A. and SUOMI, V. E., 288, 370
SIKDAR, D. N., *see* MARTIN, D. W. and SIKDAR, D. N.; *see also* YOUNG, J. A. and SIKDAR, D. N.
SIMPSON, G. C., 243, 256, 281, 370
SIMPSON, R. H., 451, 477, 609, 656
SÖDERMANN, D., *see* PALMÉN, E. and SÖDERMANN, D.
SORKINA, A. I., 214, 256
SOUTH AFRICAN WEATHER BUREAU, 267, 370, 514, 563, 566, 568, 571, 589
STAVROPOULOS, C. C., *see* HARRIS, T. F. W. and STAVROPOULOS, C. C.
STEINBACH, W., 32, 128
STEINBORN, E., *see* PFLUGBEIL, C. and STEINBORN, E.
STEPHENSON, P. M., *see* FROST, R. and STEPHENSON, P. M.
STERANKA, J., *see* ALLISON, L. J. et al.
STEWART, R. W., 128
STOCKS, T., 2 128
STOMMEL, H. and WOOSTER, W. S., 617, 656

STONE, H. M., *see* MANABE, S. et al.
STONE, P. H., *see* CORNEJO-GARRIDO, A. G. and STONE, P. H.
STRANZ, D. and TALJAARD, J. J., 117, 128
STRETEN, N. A., 84, 120, 129, 277, 285, 292, 293, 294, 295, 297, 298, 299, 301, 304, 305, 307, 308, 309, 316, 353, 359, 370, 570, 571, 574, 589
STRETEN, N. A. and KELLAS, W. R., 283, 290, 370
STRETEN, N. A. and TROUP, A. J., 299, 301, 302, 304, 305, 306, 308, 342, 370, 514, 563, 570, 571, 572, 573, 574, 575, 589
STRETEN, N. A., *see* TROUP, A. J. and STRETEN, N. A.
STROKINA, L. A., 243, 248, 250, 251, 256
STROKINA, L. A., *see* ZUBENOK, L. I. and STROKINA, L. A.
STROUP, E. D., *see* CROMWELL, T. et al.; *see also* WYRTKI, K. et al.
STRÜBING, K., 27, 129
STUBBS, M. W., *see* WRIGHT, P. B. and STUBBS, M. W.
SULLIVAN, W., 280, 370
SUMNER, E. J., 215, 216, 217, 256
SUOMI, V. E., *see* KORNFIELD, J. A. et al.; *see also* SIKDAR, D. N. et al.; VONDER HAAR, T. and SUOMI, V. E.
SUPAN, A., 49, 129
SVERDRUP, H. U., 508, 509, 581, 589
SVERDRUP, H. U., JOHNSON, M. W. and FLEMING, R. H., 268, 370, 477, 508, 589
SWEETSER, E. I., *see* COHEN, T. J. and SWEETSER, E. I.

TAFT, B. and JONES, J., 271, 370
TALJAARD, J. J., 16, 30, 113, 116, 118, 119, 120, 121, 129, 285, 292, 293, 295, 296, 299, 300, 301, 304, 305, 306, 370, 371, 508, 513, 515, 518, 546, 547, 550, 551, 563, 565, 566, 570, 571, 573, 574, 575, 577, 590, 629, 656
TALJAARD, J. J. and VAN LOON, H., 116, 120, 129, 283, 299, 303, 371, 563, 565, 570, 573, 590
TALJAARD, J. J., SCHMITT, W. and VAN LOON, H., 304, 371, 575, 590
TALJAARD, J. J., JENNE, R. L., VAN LOON, H. and CRUTCHER, H. L., 325, 371, 531, 547, 550, 576, 590
TALJAARD, J. J., VAN LOON, H., CRUTCHER, H. L. and JENNE, R. L., 38, 101, 103, 107, 110, 129, 265, 281, 284, 311, 312, 313, 314, 323, 324, 325, 326, 327, 371, 514, 523, 525, 528, 531, 532, 534–537, 540, 541, 547–549, 590
TALJAARD, J. J., *see* CRUTCHER, H. L. et al.; *see also* JENNE, R. L. et al.; STRANZ, D. and TALJAARD, J. J.; VAN LOON, H. et al.
TAUBER, G. M., 30, 37, 129, 198, 256
TAUNTON, HYDROGRAPHER OF THE NAVY, 71, 73, 76, 78, 81, 83, 86, 94, 129
TAYLOR, H. W., GORDON, A. L. and MOLINELLI, E., 275, 371
TAYLOR, R. C., 321, 347, 348, 371
TAYLOR, R. C., *see* GORDON, A. H. and TAYLOR, R. C.
TEISSERENC DE BORT, L., 513, 590
TERADA, K. and OSAWA, K., 477
TERBEYEV, Y., *see* OSTAPOFF, F. et al.
THEON, J. S., *see* RAO, M. S. V. et al.

THOMPSON, A. H., *see* ALVAREZ, J. A. and THOMPSON, A. H.; *see also* VAN LOON, H. and THOMPSON, A. H.
THOMPSON, B. W., 651, 656, 657
THOMPSON, O. E. and MILLER, J., 223, 256
THORODDSEN, TH., 679, 694
TOKIOKA, T., *see* YANAI, M. and TOKIOKA, T.
TOLSTIKOV, E. I., 270, 336, 371
TOURNIER, T., *see* HANZAWA, M. and TOURNIER, T.
TRANTER, D. J., *see* HAMON, B. V. and TRANTER, D. J.
TRELOAR, H. M. and NEWMAN, B. W., 265, 371
TRENBERTH, K. E., 285, 295, 308, 310, 351, 352, 353, 354, 355, 371
TRENBERTH, K. E. and NEALE, A. A., 283, 371
TRESHNIKOV, A. F., 277, 371
TREWARTHA, G. T., 129
TROUP, A. J., 351, 352, 355, 371
TROUP, A. J. and STRETEN, N. A., 283, 371
TROUP, A. J., *see* STRETEN, N. A. and TROUP, A. J.
TSUCHIYA, M., 269, 271, 371
TUCKER, G. B., 49, 129, 220, 238, 239, 240, 256, 347, 371

UNESCO, 477
U.S. DEPARTMENT OF COMMERCE, 266, 267, 371
U.S. HYDROGRAPHIC OFFICE, 431, 477, 657, 658
U.S. HYDROGRAPHIC OFFICE, *see* U.S. WEATHER BUREAU AND U.S. HYDROGRAPHIC OFFICE
U.S. NAVY, 265, 273, 317, 371, 458, 471, 477, 508, 510, 514, 561, 578, 590
U.S. WEATHER BUREAU, 201, 237, 238, 256, 260, 607, 658, 659
U.S. WEATHER BUREAU AND U.S. HYDROGRAPHIC OFFICE, 256
U.S.S.R. MINISTRY OF DEFENCE, 265, 269, 270, 272, 273, 274, 317, 318, 319, 320, 321, 322, 331, 332, 337, 339, 340, 341, 347, 371
USMANOV, R. F., 656

VAN DIJK, W., 607, 629, 656
VAN DIJK, W., SCHMIDT, F. H. and SCHUURMANS, C. J. E., 214, 256
VAN DUIJNEN MONTIJN, A., 607, 658, 659
VAN HAMME, J. L., 327, 371
VAN LOON, H., 16, 33, 35, 41, 42, 67, 79, 80, 103, 105, 108, 110, 117, 118, 120, 121, 122, 129, 285, 296, 298, 299, 300, 301, 303, 304, 306, 307, 312, 317, 319, 320, 322, 372, 508, 509, 510, 513, 516, 527, 531, 536, 540, 542, 543, 547, 551, 556, 558, 559, 560, 562, 563, 564, 570, 575, 576, 590
VAN LOON, H. and JENNE, R. L., 102, 129, 324, 326, 372
VAN LOON, H. and MADDEN, R. A., 354, 358, 372
VAN LOON, H. and ROGERS, J. C., 214, 256, 354, 372
VAN LOON, H. and THOMPSON, A. H., 4, 129
VAN LOON, H., TALJAARD, J. J., JENNE, R. L. and CRUTCHER, H. L., 101, 105, 129, 265, 372, 547, 552, 554, 558, 591
VAN LOON, H., TALJAARD, J. J., SASAMORI, T., LONDON, J., HOYT, D. V., LABITZKE, K. and NEWTON, C. W., 265, 281, 372, 514, 591, 658

VAN LOON, H., *see* CRUTCHER, H. L. et al.; *see also* JENNE, R. L. et al.; LABITZKE, K. and VAN LOON, H.; ROGERS, J. C. and VAN LOON, H.; RUBIN, M. J. and VAN LOON, H.; TALJAARD, J. J. and VAN LOON, H.; TALJAARD, J. J. et al.; VOWINCKEL, E. and VAN LOON, H.

VAN ROOY, M. P., 514, 591

VAN ROOY, M. P., *see* SCHUMANN, T. E. W. and VAN ROOY, M. P.

VARENIUS, B., 1, 129

VARNADORE, M. S., *see* WINSTON, J. S. et al.

VENKITESHWARAN, S. P., *see* RAMANATHAN, K. R. and VENKITESHWARAN, S. P.

VENKATESWARAN, S. V., 656

VENTER, R. J., 591

VERPLOEGH, G., 198, 256, 656

VIEBROCK, H., 335, 338, 372

VINCENT, D. G., *see* NEWELL, R. E. et al.

VISHER, S. S. and HODGE, D., 288, 372

VONDER HAAR, T. H. and ELLIS, J. S., 345, 372

VONDER HAAR, T. and SUOMI, V. E., 65, 129, 345, 372

VON FICKER, H., 114, 129, 197, 218, 219, 256, 630, 655

VOWINCKEL, E., 32, 43, 86, 129, 265, 372, 512, 514, 522, 540, 562, 591

VOWINCKEL, E. and OOSTHUIZEN, C. M., 46, 47, 130, 562, 591

VOWINCKEL, E. and VAN LOON, H., 512, 514, 515, 563, 591

VUORELA, L. A., 114, 115, 130

WADATI, K., 477

WAGNER, V., 220, 256

WAGNER, V., *see* AUGSTEIN, E. et al.

WALDEN, H., 23, 130

WALDEN, H. and SCHÄFER, P. J., 25, 130

WALKER, G. T., 115, 130, 348, 349, 372

WALKER, G. T. and BLISS, E. W., 349, 372

WALKER, J. M., *see* PERRY, A. H. and WALKER, J. M.

WALLACE, J. M., 75, 106, 130

WALLACE, J. M. and CHANG, C. P., 288, 372

WALLACE, J. M., *see* HOLTON, J. R. and WALLACE, J. M.

WALLACE JR., J. A., 656

WALTER, A., 658

WARREN, B. A., 269, 372

WASHINGTON, U.S. NAVY, 8, 22, 24, 30, 36, 42, 47, 81, 83, 86, 94, 130

WEARE, B. C., 256

WEARE, B. C., NAVATO, A. R. and NEWELL, R. E., 361, 373

WEATHER BUREAU, SOUTH AFRICA, 658

WEBB, E. K., *see* DEACON, E. L. and WEBB, E. K.

WEBSTER, P. J., 357, 373

WEBSTER, P. J. and CURTIN, D. G., 310, 330, 373

WEBSTER, P. J. and KELLER, J. L., 310, 373

WEEKS, W. F., *see* CAMPBELL, W. J. et al.

WEGENER, K., 92, 130

WEICKMANN JR., L., 610, 656

WEINERT, R. A., 329, 373

WELLS, A. J., 658

WETHERALD, R., *see* MANABE, S. and WETHERALD, R.

WEXLER, H., 276, 373

WHITE, W. B. and MCCREARY, J. P., 360, 373

WHITING, G. C., 211, 257

WHITWORTH, T., *see* NOWLIN JR., W. D. et al.

WILHEIT, T. T., *see* GLOERSEN, P. et al.

WILLETT, H. C., *see* ROSSBY, C. G. and WILLETT, H. C.

WILLIAMS, K. T. and GRAY, W. M., 287, 373

WILLIAMS, R., *see* WYRTKI, K. et al.

WINSTON, J. S., GRUBER, A., GRAY, T. I., VARNADORE, M. S., EARNEST, C. L. and MANNELLO, L. P., 345, 346, 373

WINSTON, J. S., *see* ERICKSEN, C. O. and WINSTON, J. S.; *see also* HUBERT, L. F. et al.; KRUEGER, A. F. and WINSTON, J. S.

WÖLCKEN, K., 93, 130

WOLFF, P. M., *see* LAEVASTU, T. et al.

WOOSTER, W. S., 269, 272, 358, 359, 373

WOOSTER, W. S. and GUILLEN, O., 358, 359, 373

WOOSTER, W. S. and HEDGPETH, J. W., 373

WOOSTER, W. S. and JENNINGS, F., 358, 373

WOOSTER, W. S. and REID, J. L., 272, 373

WOOSTER, W. S., *see* STOMMEL, H. and WOOSTER, W. S.

WORLD METEOROLOGICAL ORGANIZATION, 1, 2, 3, 4, 6, 7, 8, 14, 30, 48, 49, 51, 52, 71, 73, 79, 81, 84, 100, 113, 114, 115, 130, 131, 477

WORTHEM, S., *see* OSTAPOFF, F. et al.

WORTHINGTON, L. V., 247, 257

WORTHINGTON, L. V., *see* BUNKER, A. F. and WORTHINGTON, L. V.

WRIGHT, A. D. F., 298, 299, 373

WRIGHT, P. B., 351, 353, 358, 373

WRIGHT, P. B. and EBDON, R. A., 653, 656, 658

WRIGHT, P. B. and STUBBS, M.W., 658

WUCKNITZ, J., 58, 131

WUCKNITZ, J., HASSELMANN, D. and KÖNIG, M., 51, 131

WÜST, G., 38, 48, 49, 51, 53, 54, 131, 238, 239, 257

WYRTKI, K., 269, 271, 272, 273, 276, 335, 338, 358, 359, 360, 373, 374, 638, 656

WYRTKI, K. and BENNETT, E. B., 315, 316, 374

WYRTKI, K., STROUP, E., PATZERT, W., WILLIAMS, R. and QUINN, W., 361, 374

YAMANOUCHI, Y. and OGAWA, A., 477

YANAI, M., 477

YANAI, M. and NITTA, T., 287, 374

YANAI, M. and TOKIOKA, T., 477

YASUNARI, T., 309, 374

YEH, T. C., 651, 656

YEH, T.C., *see* RIEHL, H. et al.

YOSHIOKA, G. A., *see* MATHER, J. R. et al.

YOUNG, J. A. and SIKDAR, D. N., 288, 374

YOUNG, J. A., *see* SIKDAR, D. N. et al.

ZILLMAN, J. W., 18, 64, 131, 277, 283, 319, 335, 336, 338, 343, 347, 357, 374, 510, 578, 591

ZILLMAN, J. W. and DINGLE, W. R. J., 335, 374

ZILLMAN, J. W. and JOHNSON, D. R., 356, 357, 374

ZILLMAN, J. W. and MARTIN, D. W., 304, 316, 374

ZILLMAN, J. W. and PRICE, P. G., 283, 374

ZOPF, D. O., *see* QUINN, W. H. et al.

ZUBENOK, L. I. and STROKINA, L. A., 241, 242, 257

ZWALLY, H. J. and GLOERSEN, P., 511, 591

Geographical Index

Abidjan, 474
Accra, 28
Africa, equatorial coast, 115
—, North, 603
—, southwest coast, 54, 59, 61, 96–98
—, west coast, 33, 35, 41, 46, 53, 57, 64, 94, 95, 106, 107, 238
Agalega, 627, 638, 652, 653, 668
Agulhas Bank, 507
Agulhas Plateau, 507
Akureyri, 673, 682, 683, 684, 687, 688, 689, 690, 691, 692, 696
Alaska, 442, 445, 447
Aleutian Islands, 437, 442, 445, 447, 458
Alofi, 396
Amazon, 27, 55, 56, 67
Amazon Basin, 110, 115, 119
Amazon estuary, 87, 88
Amery Ice Shelf, 506, 541
Amini Divi, 627, 647, 650, 664
Amundsen Bay, 506
Amundsen Sea, 263, 278
Andes, 12, 15, 34, 47, 49, 54, 83, 93, 105, 120, 564
Angola, 25, 39, 95
Antarctic Ice Shelf, 27
Antarctic Ocean, 15, 28, 32, 34, 41, 43, 265, 512
Antarctic Peninsula, 120
Antarctic Trough, 307, 322
Antarctica, 11, 100, 102, 103, 119, 120, 263, 266, 278, 301, 316, 324, 506, 511
Apia, 351, 397
Arabia, 608, 627
Arabian Sea, 606, 608, 609, 616, 619, 627, 628, 630, 640, 644, 649, 650, 651, 653
Arafura Sea, 263
Arctic Sea, 11, 470
Argentina, 42, 45, 46, 47, 53, 61
Ascension Island, 5, 25, 75–77, 78, 152

Baffin Bay, 211
Baffin Island, 213
Banda Sea, 263
Barbados, 224
Barents Sea, 196
Bass Strait, 344
Batavia, *see* Djakarta
Bay of Bengal, 606, 608, 616, 627, 649, 650, 651, 653
Belem, 87

Bellingshausen Sea, 263, 278, 303
Bering Sea, 469, 470
Bering Strait, 470
Berufjördur, 674
Bismarck Sea, 263
Bogotá, 474
Bora Bora, 398
Borgarfjördur, 682
Bouvet Current, 511, 516
Brazil, 26, 35, 46, 55, 115, 118
—, equatorial coast, 49
—, north coast, 47, 87, 111, 114
—, northeast coast, 88, 351
—, southeast coast, 89
Burma, 649

Cabo Corrientes, 32, 39, 64, 67, 92
— Frio, 28, 39, 42, 45, 47, 52, 63, 88, 89, 90, 96
— San Antonio, 33, 36
— San Juan, 58, 93, 94
— de Sao Roque, 33, 47, 87, 88, 89, 115
Campbell Island, 310, 319, 399
Canada, west, 447
Canton Island, 401, 473, 474, 475
Cape Agulhas, 505
— Batterbee, 506
— Farewell, 677
— of Good Hope, 16, 99
— Guardafui, 618
— Hatteras, 211, 217, 235, 241, 259
— Horn, 55
Cape Province, 98–100
Cape Town, 536
Caribbean Sea, 224, 235
Car Nicobar, 650
Caroline Islands, 455, 467, 495, 497, 500
Casey, 513, 533
Cayenne, 234, 260
Ceylon, 650
Chatham Island, 427
Chichijima, 487
Chile, 293, 303
Cocos Island, 652, 654, 670
Congo River, 27, 55, 94, 95
Cook Islands, 263, 415
Coral Sea, 263, 265, 289, 292, 315
Crozet Islands, 505, 507, 509, 564, 583, 585
Crozet Plateau, 507

Cuba, 217

Dakar, 224, 260
Dalatangi, 682, 689, 690, 691
Darwin, 351, 354
Davis Station, 513
Davis Strait, 196, 204, 215, 228, 238
Denmark Strait, 196
Diego Garcia, 612, 627, 647, 652, 653, 667
Diego Suarez, 638
Djakarta, 349, 351
Drake Passage, 55, 93, 117, 120, 279, 303
Drake Strait, 26
Duala, 53
Dutch Harbor, 482

East China Sea, 440, 441, 444, 455, 471
Easter Island, 295, 351, 354, 404
Edinbourgh, 78
Efate Island, 426
Egilsstadir, 687
Emden, 227
Eniwetok, 494
"Ethiopic Sea", 1

Falkland Islands, 5, 81–83, 156
Fanning Island, 501
Fernando de Noronha, 5, 73–75, 87, 151
Fiji Islands, 266, 408, 421, 424
Fiji Sea, 263
Florida, 217
French Polynesia, 289, 310
Frobisher Bay, 258
Funafuti Island, 292, 402
Funchal, 259

Gabon, 94
Galápagos Islands, 263, 503
Galtarviti, 689, 690, 691
Gan, 474, 612, 647, 652
Ganges Delta, 608
George V Land, 278, 280
Georgetown, 5, 76, 77
Georgia, 218
Goose, 227, 258
Gough Island, 5, 80–81, 155, 564
Grand Banks, 212, 238
Grande Comore, 627, 638, 652, 653, 669
Great Australian Bight, 263, 295, 296, 303, 344
Green Mountain, 76, 77
Greenland, 211, 212, 214, 676
Greenland Sea, 196, 200
Grimsey, 673, 687, 690, 691
Grímsstadir, 683, 684, 686, 687, 690, 691
Grytviken, 5, 83, 84, 85, 157
Guadalcanal Island, 403
Guam, 493
Gulf of Aden, 644
—— Alaska, 471

—— Guinea, 45, 47, 49, 53, 64
—— Mexico, 211, 217, 223, 224
—— Oman, 627

Hachijojima, 451, 484
Hallormsstadur, 687, 690
Hao, 290
Harmukotan, 463
Hawaii, 490
Heard Island, 505, 507, 513, 523, 532, 536, 543, 547, 550, 560, 561, 564, 594, 598, 601
Herald Point, 78, 79, 80
Hikuera, 290
Hilo, 490
Himalayas, 605, 630
Höfn, 687, 689, 690
Hokkaido, 464
Hólar, 687, 690
Holár i Hornafirdi, 683, 684, 697
Honiara, 403
Honolulu, 489
Hornbjargsviti, 686, 687
Horta, 259
Hutts Gate, 5, 77, 78
Hvalfjördur, 689
Hveravellir, 674, 689
Iceland, 213, 673–697
India, 649, 650
Irako, 450
Irminger Sea, 200
Ishigakijima, 451
Isla de Pascua, 404

Jamestown, 5, 77, 78
Japan, 433, 441, 442, 447, 455, 467
Java Sea, 263
Jersey, 218
Johnson Island, 492
Juan Fernandez Island, 405

Kalahari Desert, 627, 652
Kamchatka, 444, 457, 463, 465, 467
Kawajalein, 496
Keflavík, 674, 689
Kerguelen Island, 505, 507, 509, 513, 516, 523, 530, 531, 532, 536, 540, 543, 546, 547, 560, 561, 568, 595, 597, 600, 647
Kerguelen Plateau, 507
Kiribati Islands, 263
Kirkjubaejarklaustur, 687, 690, 691
Kjölur, 692, 693
Koror, 498
Kurile Islands, 457, 463, 465
Kvisker, 684, 685

Lambert Glacier, 506, 541
La Plata, 55
Leeward Islands, 398

Little America, 280
Lobito, 39
Lomé, 28
Lord Howe Island, 406
Luanda, 39, 43, 96
Lüderitz, 64, 97

Macquarie Islands, 310, 319, 407, 568
Macquarie Ridge, 272
Madagascar, 606, 638
Mahé, 627, 638, 652, 653, 666
Majuro, 499
Malagasy, 541
Marcus Island, 488
Marion Island, 505, 509, 513, 523, 530, 531, 532, 536, 540, 543, 546, 547, 550, 560, 561, 564, 568, 583, 592, 596, 599
Marokau, 290
Marshall Islands, 496, 499
Mauritius, 629, 638, 652, 654, 670
Mawson, 513, 537, 542, 543, 544
Miami, 259
Midway Island, 486
Minicoy, 612, 627, 647, 650, 651, 652, 665
Mirnyi, 513
Miyakojima, 451
Moçamedes, 39
Molodezhnaya, 513, 533
Montevideo, 39
Moroni, 669
Mount Actaeon, 77
Mount Hodges, 85
Mount Paget, 83
Mozambique Channel, 627
Murotomisaki, 451
Mýrdalsjökull, 684, 692

Naha, 451
Nairobi, 474
Nandi, 408
Nantucket Island, 238
Napuka, 290
Nauru Island, 409
Naze, 451
Nemuro, 464
New Amsterdam Island, 505, 513, 523, 530, 531, 532, 536, 540, 543, 546, 547, 550, 560, 561, 593, 596, 599, 647
New Caledonia, 310, 411
Newfoundland, 212
New Guinea, 265, 266, 475
New Hebrides, 266, 322
New York, 218
New Zealand, 266, 293, 311, 353
Niue Island, 396
Norfolk Island, 410
North Island, 291
North Sea, 228, 238
Norwegian Sea, 234

Noumea, 411
Nuku Alofa, 412

Oahu, 489
Oates Land, 278
Ocean Island, 413
"Ocean Meridional", 1
O Pico, 73
Öræfajökull, 673

Palau Islands, 498
Pampas, 92
Panama Canal, 7
Papeete, 414
Paraná River, 27, 55
Patagonian Shelf, 33, 35, 49, 93-94, 121
Paternoster Point, 96
Perth, 536
Peru, 358
Philippines, 441, 447, 467
Piarco, 259
Pico de Tomé, 71, 73
Pitcairn Island, 416
Plaisance, 671
Ponape, 500
Ponte Noire, 67
Port-aux-Français, 524, 595
Port Nolloth, 97
Port Stanley, 5, 81, 83
Port Victoria, 666
Prince Edward Island, 507
Prydz Bay, 506
Punta Albina, 28, 42, 95
— Rasa, 58, 64

Queen Elizabeth Islands, 211
Queensland, 289

Raoul Island, 417
Rapa, 419
Rarotonga Island, 420
Raufarhöfn, 687, 689
Recife, 90
Resolute Bay, 212
Reykjavik, 258, 673, 683, 684, 685, 687, 688, 689, 690, 691, 692, 695
Rio de Janeiro, 48, 53, 90, 91
Rio Negro, 92
Rio de la Plata, 27, 32, 91-92
Rodriguez, 647
Ross Sea, 120, 263, 278, 307
Rotuma Island, 421
Ryukyu Islands, 432

Sable Island, 234, 259
Sahara, 653
Sakhalin, 457
Salvador, 89
Samoa, 360, 397

San Andres, 224
San Cristobal Island, 422
San Juan, 259
Sandy Point, 80
Santa Cruz Island, 423
São Paulo, 39
Sao Tomé, 5, 71–73, 94, 150, 227, 234, 260
Saudárkrókur, 689
Sea of Japan, 441, 455, 457, 467
Sea of Okhotsk, 441, 455, 458, 469
Seymour Island, 503
Shemya Island, 483
Sidumúli, 687
Singapore, 474
Skagafjördur, 692
Snaefellsnes, 682
Solomon Islands, 265, 266, 360
Solomon Sea, 263
Somalia, 608, 627
South Africa, 42, 45, 506, 564
South China Sea, 440, 441, 442, 444
South Georgia, 5, 83–86
South Island, 329
South Pole, 107, 108
St. Helena, 5, 25, 77–78, 153
Stykkishólmur, 673, 674, 678, 679, 680
Subantarctic Trough, 301
Suva, 424
Swakopmund, 42, 67, 98
Syowa, 513, 533

Tahiti, 295, 351, 354, 414
Takaroa, 425
Tarawa, 502
Tasman Sea, 263, 265, 270, 296, 298, 303, 328, 329, 335
Tasmania, 303
Teigarhorn, 674, 683
Terre Adélie, 278

Thóroddsstadir, 687
Thorshavn, 258
Thyrill, 689
Tibet, 630
Timor Sea, 263, 638
Tokushima, 451
Tonga, 266
Tongatapu Island, 412
Torishima, 485
Tristan da Cunha, 5, 78–80, 121, 154
Truk, 497
Tuamotu Archipelago, 289, 290, 425
Tubuai Islands, 419
Tulear, 629

Uruguay, 47, 48
— River, 27
Uwajima, 451

Valdés Peninsula, 94
Valentia, 258
Vatnajökull, 673, 684, 692
Vestmannaeyjar, 673, 683, 684, 686, 687, 688, 689, 690
Vila, 426
Viti Levu, 408

Waitangi, 427
Wake Island, 491
Walvis Bay, 97
Weddell Sea, 27, 120, 511, 516
West Pakistan, 608
Wilkes, 513
Willis Island, 429

Yakushima, 451
Yap, 495
Yonakunijima, 451
Yucatan Peninsula, 223

Subject Index

Aerology, *see* Upper-air elements and circulation
Air–sea temperature difference, *see* Temperature, air
Albedo, 60, 579
Aleutian Low, 437
Antarctic Convergence, *see* Discontinuities in ocean
Antarctic Divergence, *see* Discontinuities in ocean
Anticyclones, *see* Highs

Beaufort scale, 17, 19
Blocking, 119, 215–216, 298, 564
Bowen ratio, 58–59
Buoys, drifting, 284, 513

Climate anomalies, 231–234, 471–475
— classification, 12–13, 36
— definition, 1
— elements, 14
— fluctuations, 33–34, 83, 199–200, 678–680
— fundamentals, 8–12
— marine features, 11–12
— processes, 9
— ships' holds, 32
— zones, 10–11, 13, 36, 248–252
Clouds, 42–46, 237, 317–319, 459–461, 562–563, 689–691
—, Atlantic cloud band, 115
—, Cacimbo, Angolan coast
—, clusters, 287–288, 460–461
—, satellite observations, 43
—, South Pacific cloud band, 305–306, 308–310
—, variability, 45
Comfort limit, 39, 40
Continental influences, 8, 11, 12, 83, 122, 193, 455, 506, 541, 603, 604
Cross seas, *see* Waves, ocean
Currents, ocean, 25–27, 31, 269–273, 432–433, 508–509
—, Agulhas, 27, 508, 509
—, Alaska, 433
—, Antarctic Circumpolar, 26, 272, 509, 516
—, Benguela, 26, 31
—, Bouvet, 516
—, Brazil, 26, 32
—, California, 432, 433
—, Cape Horn, 272
—, Cromwell, 271
—, East Australian, 271
—, East Wind Drift, 509
—, Equatorial Counter, 271, 606
—, Equatorial North, 26, 432

—, Equatorial South, 26, 271
—, Falkland, 26, 32, 81
—, Guinea, 26
—, Greenland, 675
—, Gulf Stream, 199
—, Humboldt, 271, 272
—, Iceland, 675
—, Irminger, 675
—, Kuroshio, 432, 433
—, North Atlantic Drift, 675
—, Oyashio, 433
—, Peru, *see* Humboldt
—, Peru coastal, 272
—, response to monsoon reversal, 605–607
—, Somali, 606
Cyclogenesis, 120, 301–303, 445, 570–571; *see also* Lows
Cyclones, *see* Lows

Dewpoint, *see* Humidity
Density, air, 36
—, water, 27, 28
Discontinuities in ocean
—, Antarctic Convergence, 29, 84, 275–276, 342, 509, 510
—, Antarctic Divergence, 29, 32, 509, 510
—, Australian Subantarctic Front, 277, 342, 510
—, Subtropical Convergence, 29
Doldrums, *see* Wind

Easterly waves, 287
El Nino, 358–361
Equatorial trough, 610–617, 628, 638, 641, 649, 650, 651, 653; *see also* Intertropical Convergence Zone
Equivalent temperature, *see* Humidity
Evaporation, 51–53, 239–241, 471, 581, 583, 692–693
—, effective, 51
Evapotranspiration, 692–693

Foehn, *see* Wind
Fog, 46, 47, 238, 319, 464–465, 691; *see also* Visibility
Freak waves, *see* Waves, ocean
Front, Antarctic, 306
—, climatic, 304, 306, 576
—, Polar, 121, 304, 575–576
Fronts in ocean, *see* Discontinuities in ocean
Frost, 683–684

Gales, *see* Wind
Geopotential height, *see* Upper air elements and circulation

Global radiation, *see* Radiation
Gusts, *see* Wind

Hadley circulation, 10
Half-yearly oscillation, 16, 80, 307, 542
— of subantarctic trough, 542
Haze, 628
Heat balance, *see* Heat budget
Heat budget, 48–70, 241–247, 332–348, 578–587, 640–641
—, atmosphere, 69, 70, 344–348
—, atmosphere–earth, 68–70
—, diurnal variation, surface, 343–344
—, Indian Ocean, January–July, 640
—, seasonal variation, 649
—, surface, 66–68, 335–344, 586–587, 626, 637, 640
—, surface, local, 89, 91, 92, 94, 95, 96, 98, 100
Heat flux, across Equator, 75
—, condensation heating, atmosphere, 69, 346–348
—, diabatic heating, atmosphere, 348
—, latent heat, 57, 245, 338–339, 346–348, 471, 581
—, loss or gain by ocean, 247
—, sensible heat, 57, 245, 339–340, 581, 584
—, storage in ocean, 69, 341–342
—, total, ocean to atmosphere, 59–60
Heating of atmosphere, *see also* Heat flux
—, condensation heating, 69, 346–348
—, diabatic heating, 348
Highs, 216, 296, 564–570
—, blocking, 215–216, 298, 564
—, frequency, 116, 566–567
—, genesis, 564
—, intensity, 569
—, seasonal movement, 570
—, subtropical, 116, 292–296, 434–437, 539–541, 630
—, tracks, 116–117, 565–566
Historical map series, 4, 267, 514
Historical survey, 1–4, 196–198, 263–265, 281–285, 511–514, 673–674
Humidity, 37–41, 234–237, 458–459, 689
—, dew point, 234, 312
—, equivalent temperature, 39
—, gradient, vertical, 51
—, measurement of, 37–38
—, relative, 40–41, 458
—, upper air, 110–113, 329; *see also* Upper air elements and circulation
—, vapour pressure, 38–39, 234–235, 459, 689
—, vapour pressure, vertical gradient, 51
Hurricanes, *see* Lows

Ice, 277–281, 469–470
—, and annual temperature range, 528–529
—, icebergs, 27, 81, 86, 280–281, 470
—, pack ice, 196, 277–279, 469–470, 510–511, 675
Icebergs, *see* Ice
Icelandic Low, 193, 676
Insolation, *see* Radiation
Intertropical Convergence Zone, 11, 12, 15, 18, 19, 20, 21, 31, 43–45, 114, 220–221, 285, 286, 287; *see also*

Equatorial Trough
Inversion, trade wind, 13, 219–220, 629

Jet streams, *see* Wind

Katabatic wind, 316
Kona storms, 451, 453
Köppen's classification, 12–13, 36
Latent heat, *see* Heat flux
Local winds, Bergwind, 96, 97
—, Cambueiros, 89
—, Cape Southeaster, 99
—, Harmattan, 95
—, Minuano, 90
—, Pampeiros, 90, 92
—, Sudestado, 92
—, Terral, 92
—, Viracao, 90
—, Virazon, 92
Long wave radiation, *see* Radiation
Lows, and frontal zone, 118
—, cyclogenesis, 120, 301–303, 445, 570–571
—, frequency, 118–119, 300, 573–575
—, Kona, 451–453
—, speed of movement, 120, 303, 304, 575
—, subtropical, 447, 451, 453, 608, 609
—, tracks, 119–120, 205, 303–304, 445, 573
—, tropical systems, 213, 223–225, 288–292, 447–451, 611, 612, 629, 630, 641, 644–645

Marsden squares, 197
Mist, 47, 628
Mixed layer, 28, 195
Monsoons, 439, 441–444, 603–605
—, Australian summer, 628
—, Australian winter, 638
—, depressions, 630
—, northeast, 627–628
—, southwest, 630–638
Monsoon depressions, 630

Near-equatorial trough, *see* Equatorial trough, *and* Intertropical Convergence Zone
Normal distribution of wind speed, 19
— of cloudiness, 45

Observations, comparison between merchant and ocean weather ships, 198
—, problems, 6–7, 684, 687
—, selected areas, 8
Ocean fronts, *see* Discontinuities in ocean
Oceanic Polar Front, *see* Discontinuities in ocean

Pack ice, *see* Ice
Polar Front, ocean, *see* Discontinuities in ocean
Precipitation, 47–48, 49–51, 238–239, 319–322, 465–467, 560–563, 605, 684, 687
—, amount, 49, 239, 467, 684–685, 686
—, annual variation, 48, 238–239, 560–561

—, estimates of, 49
—, frequency, 47–48, 321–322, 465, 561, 562, 686
—, intensity, 50, 467
—, measurement of, 49
—, orographic effect, 72, 684
—, snow, 48, 467, 686–687
Pressure, 14–16, 203–206, 284–285, 292–308, 312–314, 434–437, 539–546
—, annual variation, 15–16, 203–205, 312, 543–546
—, diurnal wave, 84, 87
—, extremes, 676
—, half-yearly oscillation, 16, 80, 307, 542
—, variability, 16, 205, 546

Quasi-biennial oscillation, 75

Radiation, balance of atmosphere, 65–66, 345
—, balance of surface, 64–65, 335–338, 581, 583
—, global, 60, 336, 581, 692
—, longwave, 62–63, 335, 579, 580, 582
—, shortwave, net, 60–62, 245, 335, 580
—, solar, 60–62, 244, 337, 579, 580
—, terrestrial, *see* Longwave
Relative humidity, *see* Humidity
Relative topography, *see* Upper-air elements and circulation
Residence time of water vapour, 113
Rollers, *see* Waves, ocean

Salinity, 54–55, 195, 274
Sand storms, 97
Satellite observations, 43, 44, 283, 513
Sea ice, *see* Ice
Sea level pressure, *see* Pressure
Sea surface temperature, *see* Temperature, sea
Semiannual oscillation, *see* Half-yearly oscillation
Sensible heat, *see* Heat flux
Ships' holds and weather, 32
Shortwave radiation, *see* Radiation
Snow, *see* Precipitation
Somali jet, 617
Southern Oscillation, 348–361
— and South Pacific Convergence Zone, 353
South Pacific Convergence Zone, 308–310
— and Southern Oscillation, 353
Storage of heat in ocean, 69, 341–342
Stratosphere, 102, 106
Strong wind, *see* Wind
Subantarctic front, Australasian, *see* Discontinuities in ocean
Subantarctic trough, 15, 307–308, 541–542
Subtropical Convergence, *see* Discontinuities in ocean
Subtropical cyclones, 451, 453, 608–609
Subtropical High, *see* Highs
Sultriness limit, 39
Sunshine duration, 691–692
Surf, *see* Waves, ocean
Swell, *see* Waves, ocean
Synoptic map series, 4, 267, 514

Synoptic climatology, *see* Synoptic regimes
Synoptic regimes, Iceland region, 676–678
—, North Atlantic, 210–225
—, North Indian Ocean, 608–619, 645–654
—, North Pacific, 434–451
—, South Atlantic, 113–122
—, South Indian Ocean, 563–578
—, South Pacific, 284–310

Teleconnections, 214–215, 348–361
Temperature, air, 34–37, 225–228, 310–312, 453–457, 521–531, 680–684; *see also* Upper-air elements and circulation
—, air–sea difference, 36–37, 228–231, 457–458, 526–527, 621, 632
—, annual range, 35, 227, 312, 527–528, 558–559, 680
—, annual variation, 35, 523–526
—, deviation from latitude mean, 528–529
—, diurnal variation, 36, 682
—, extremes, 529–530, 531, 683
—, gradients, 225, 525–526
— measurement of, 34
— variability, 36, 227–228, 531, 620, 682
Temperature, sea, 30–34, 194, 273–274, 457, 458, 514–521
—, annual range, 32, 274, 519–520
—, annual variation, 33, 274, 519–520
—, anomalies, 472, 473
—, climate variation, 33–34
—, deviation from latitude mean, 518–519
—, diurnal variation, 33
—, extremes, 520
—, measurement of, 30
—, variability, 33
Thermocline, 28, 195
Thunderstorms, 48
Trade winds, 19, 21, 97, 114, 218–220, 285–287, 314–315, 438, 628, 629, 630, 638, 639, 653–654
Tropical cyclones, *see* Lows
Tropical synoptic events, 114–116, 220–225, 285–292, 447–451, 610–617, 641–645
Tropopause, *see* Upper-air elements and circulation
Typhoons, *see* Lows, *and* Tropical synoptic events

Upper-air elements and circulation, 101–113, 323–331, 531–539, 547–551, 556–560
—, absolute humidity, 112
—, dew point, 331
—, easterlies, 630
—, geopotential height, 101–104, 323–328, 547–551
—, jet streams, 106, 329, 549, 617
—, specific humidity, 112
—, temperature, 107–110, 331–333, 531–539
—, thickness, 107–110, 203, 204, 577–578
—, thickness anomalies, movement of, 577–578
—, tropopause, 331
—, vapour pressure, 110–111, 112
—, wind, 74, 75, 105–107, 329–331, 452–453, 556–560, 630
—, zonal harmonic waves, 326–328

Upwelling, Somali coast, 617
—, southwest coast of Africa, 31, 97
—, west coast of South America, 273

Vapour pressure, *see* Humidity, *and* Upper-air elements
  and circulation
Visibility, 46–47, 238, 628, 691; *see also* Fog

Walker circulation, 32, 355–358, 474–475
Water balance, 30, 49, 53–56, 69, 693
—, flux of water vapour in, 55–56
—, role of ocean currents in, 55
Water masses, 28–29
Wave, easterly, 285
Waves, ocean, 22–25
—, cross seas, 77
—, freak, 25, 99
—, height, 23, 24, 87, 89, 90, 92, 96, 97, 99, 211
—, Kelvin, 360
—, length, 24
—, period, 23
—, rollers, 77, 78
—, sea, 23
—, steepness, 24
—, surf, 25
—, swell, 23, 77, 78, 96, 99

Waves, zonal harmonic, 326–328
Weather bias, 7, 22
Weather ships, 197
Westerlies, 117–122, 213–218, 296–307, 439, 444–445,
  551–556
Wind, 17–22, 206–210, 314–316, 438–445, 551–556, 687–
  689
—, Beaufort scale, 17, 19
—, comparison between actual and geostrophic, 208
—, direction, 19–22, 554–555
—, doldrums, 18, 628
—, foehn, 85
—, gusts, 689
—, katabatic, 316
—, local, *see* Local winds
—, measurement of, 17
—, speed, 17–19, 555
—, steadiness, 21
—, stress, 210
—, strong wind and gale, 22, 216, 316, 444, 688
—, summer maximum, 551
—, trades, *see* Trade winds
—, westerlies, *see* Westerlies

Zonal index, 121